FORMULAS FOR NATURAL FREQUENCY AND MODE SHAPE

FORMULAS FOR NATURAL FREQUENCY AND MODE SHAPE

Robert D. Blevins Ph.D.

KRIEGER PUBLISHING COMPANY
MALABAR, FLORIDA
1995

Original Edition 1979
Reprint Edition 1984
Reissue 1993, 1995 with minor corrections

Printed and Published by
KRIEGER PUBLISHING COMPANY
KRIEGER DRIVE
MALABAR, FLORIDA 32950

Library of Congress Cataloging-In-Publication Data
Blevins, Robert D.
Formulas for natural frequency and mode shape.

Reprint. Originally published: New York : Van Nostrand Reinhold, c1979.
Includes bibliographical references and index.
1. Structural dynamics - Handbooks, manuals, etc.
2. Vibrations - Handbooks, manuals, etc. 3. Hydraulics - Handbooks, manuals, etc. I. Title.
[TA654.B54 1984] 620.3 84-12583
ISBN 0-89464-894-2

10 9 8 7 6 5 4 3 2

PREFACE

The purpose of this book is to provide a summary of formulas and principles on the vibration of structural and fluid systems. It is intended to be a reference book for engineers, designers, and students who have had some introduction to the theory of vibrations. However, anyone with a grasp of basic physics and an electronic calculator should have little difficulty in applying the formulas presented here.

Vibrations of structures have been known since man first heard wind ruffle the leaves of trees. Quantitative knowledge of vibrations was mostly limited to empirical descriptions of pendulums and stringed instruments until the development of calculus by Sir Isaac Newton and Gottfried Wilhelm Leibnitz in the late 1600's. The first instance in which the normal modes of continuous systems were determined involved the modes of a hanging chain, which were described in terms of Bessel functions by Daniel Bernoulli in 1732. The beauty and intricacy of modal patterns were actually visualized in 1787 when Ernst Chladni developed the method of placing sand on a vibrating plate. Mathematical description of vibrating plates proved more elusive. In 1809 Napoleon Bonaparte presented the Paris Institute of Science with the sum of 3000 francs to be given as a prize for a satisfactory mathematical theory of the vibration of plates. This prize was finally awarded in 1816 to Mademoiselle Sophie Germain, who first derived the correct differential equation but obtained erroneous boundary conditions. The theory of plate vibration was completed in 1850 by Gustav Kirchhoff. During the 1850's, calculus was applied to the vibration analysis of a number of practically important structural systems. This led to Lord Rayleigh's publication of *The Theory of Sound* (1st ed., 1877), which remains in print today. Lord Rayleigh, born John William Strutt, independently developed his own laboratory and devoted himself to science. His fellow countrymen must have thought him a bit odd as he investigated the vibration modes of their church bells.

A. E. H. Love's *A Treatise on the Mathematical Theory of Elasticity* (4th ed., 1926) and Horace Lamb's *The Dynamical Theory of Sound* (2nd ed., 1925) together with Rayleigh's *The Theory of Sound* form the basis of modern vibration analysis. Solutions developed in each of these books are presented here. During the early and middle 1900's, the techniques presented in these books were applied to increasingly complex systems. Sophisticated approximate techniques such as those employed by Stephen P. Timoshenko in *Vibration Problems in Engineering* (1st ed., 1928) also appeared during that period.

The advent of reliable electronic computers in the 1950's and the widespread installation of these computers in the early 1960's led to two new parallel paths for analyzing complex systems. First, the computer made it possible to generate approximate semi-closed form solutions which rely on classical solution techniques but with numerical evaluation of certain terms which cannot be expressed in closed form. Second, the development of large digital computers has made it feasible to simulate systems directly using finite element models. Today it is possible to simulate virtually any well-defined linear system on a large general purpose digital computer and obtain its natural frequencies, mode shapes, and response without resort to a theoretical treatment. Of course, the result is purely numerical, and physical insight into the nature of the solution must still be obtained through classical reasoning.

A vibration analysis generally follows four steps. First, the structure or system of interest is identified, its boundary conditions are estimated, and its interfaces with other systems are plotted. Second, the natural frequencies and mode shapes of the structure are determined by analysis or direct experimental measurement. Third, the time dependent loads on the structure are estimated. Fourth, these loads are applied to an analytical model of the structure to determine its response. The crucial steps in the vibration analysis are the identification of the structure and the determination of its natural frequencies and mode shapes.

The aim of this book is to provide formulas for the natural frequencies and mode shapes of a wide range of structures in easily used form so that the analyst can rapidly obtain his result without either searching the literature or spending the hours ordinarily required to successfully complete a finite element numerical simulation.

This book was written by searching through the cornucopia of solutions available in the literature for those solutions whose practicality and generality make them useful tools for the engineer or designer. In order to yield a compact volume, only those solutions which could be adequately presented in a relatively small space have been included. Chapters 1 through 5 present definitions, symbols, instruction in units, basic principles, and geometric properties of rigid structures. Chapter 6 is devoted to systems with finite number of degrees of freedom: the spring-mass systems. Chapter 7 considers the dynamics of cable systems. Chapters 8 through 12 present results for beams, curved beams, membranes, plates, and shells. Some practical information on the stress analysis of these structures is also included. Chapters 13 and 14 are devoted to vibration in fluid systems and the effect of a surrounding fluid on the vibration of structural systems. Chapter 15 reviews the finite element computer codes presently available for general vibration analysis. Chapter 16 presents data on the properties of materials which are useful as inputs for vibration analysis.

The solutions presented in this book span the technical literature from the second edition of Lord Rayleigh's *The Theory of Sound*, published in 1894, to the journals of 1978. While many of the solutions presented here can be traced to the pre-1930 volumes of Horace Lamb, A. E. H. Love, and Lord Rayleigh, the majority of the results in this book were generated after 1960. Notable among the more recent results are the cable solutions of H. Max Irvine, the multispan beam solutions of Daniel J. Gorman, the plate solutions of Arthur W. Leissa, and the cylindrical shell solutions of C. B. Sharma and D. J. Johns. The formats used in this book were adapted from those developed by Raymond J. Roark, and Chapters 11 and 12 were born in the compilations of Arthur W. Leissa. Those familiar with the work of these two fine analysts will recognize its reflection in this book.

The following individuals reviewed various chapters in this book:

I. T. Almajan
C. D. Babcock
T. K. Caughey
S. S. Chen
S. C. Cheng
H. D. Chiger
L. Halvers
M. T. Jakub
K. P. Kerney
A. C. Lewis
F. Raichlen
T. Sarpkaya
C. B. Sharma
M. W. Wambsganss
H. H. Woo

Their comments, criticisms, and corrections were invaluable. I am indebted to Alberta Gordon and her staff at Van Nostrand Reinhold for their care in the publication of this book. Most of all I would like to thank Boecky Yalof, who diligently edited the manuscript as it moved from handwritten scrawl to printer's copy.

This book is dedicated to the individuals who developed the solutions presented here; it is far more their creation than mine.

R.D.B.
La Jolla, California

CONTENTS

FORMULAS FOR NATURAL FREQUENCY AND MODE SHAPE

1
DEFINITIONS

Added Mass–The mass of fluid entrained by a moving structure as it vibrates in a fluid. The added mass of many slender structures is comparable to the mass of fluid displaced by the structure. The natural frequency of vibrations in the presence of added fluid mass is lower than that which would be observed in a vacuum. See Chapter 14.

Beam–A structure whose cross-sectional properties and deflection vary along only a single axis. A slender beam is a beam whose characteristic cross-sectional dimensions are much less than the span of the beam and the distance between vibration nodes; therefore, the inertia associated with local rotation is overshadowed by the inertia developed in displacement and the deformation due to shearing of the cross section is overshadowed by bending deformations.

Boundary Condition–A constraint applied to a structure independent of time. Boundary conditions can be classified as either geometric or kinetic. Geometric boundary conditions arise from geometric constraints. For example, the displacement of a structure at a joint pinned to a rigid wall is zero. Kinetic boundary conditions arise from force or moments applied to a structure; for example, a pinned joint permits free rotation, so the kinetic boundary condition at a pinned joint is zero moment. (See Pinned Boundary, Clamped Boundary, Free Boundary, and Sliding Boundary.)

Bulk Modulus of Elasticity–The ratio of the tensile or compressive stress, equal in all directions (i.e., hydrostatic pressure), to the relative change in volume. $B = E/[3(1 - 2\nu)]$ for an isotropic elastic material, that is, a material whose properties are the same in all directions. (Definitions of symbols are given in Chapter 2.)

Cable–A massive string. A uniform, massive one-dimensional structure which can bear only tensile loads parallel to its own axis. The bending rigidity of cables is zero. Cables, unlike chains, may stretch in response to tensile loads.

Cable Modulus–The rate of change in the longitudinal stress (axial force over cross-sectional area) in a cable for a small unit longitudinal strain. If the cable is a solid elastic rod, the cable modulus will be equal to the modulus of elasticity of the rod material. If the cable is woven from fibers, the cable modulus will be less than the modulus of elasticity of the component fibers. Typically, the cable modulus of woven steel cables is about 50% of the modulus of elasticity of the steel fibers.

Center of Gravity–The point on which a body can be balanced. The sum over a body of all elements of mass multiplied by the distance from any axis through the center of gravity is zero. The center of gravity is also called the center of mass. (See Chapter 5.)

Centroid—The geometric center of a plane area. The sum over a plane area of all elements of area multiplied by the distance from any axis through the centroid is zero. (See Chapter 5.)

Chain—A uniform, massive one-dimensional structure which can bear only tensile loads parallel to its own axis. The bending rigidity of chains is zero. Chains, unlike cables, do not stretch in response to tensile loads.

Clamped Boundary—A geometric boundary condition such that the structure can neither displace nor rotate along a given boundary.

Concentrated Mass (Point Mass)—A point in space with finite mass but zero moment of inertia for rotation about its center of mass.

Damping—The ability of a structure to absorb vibrational energy. Damping can be generated within the material of the structure (material damping), by the fluid surrounding the structure (fluid damping), or by the impact and scraping at joints (structural damping).

Deformation—The displacement of a structure from its equilibrium position.

Density—The mass per unit volume of a material.

Elastic—A term applied to a material if deformations of the material increase linearly with increasing load without regard to the sign or magnitude of the load. Many real materials of structural importance are elastic for loads below the onset of yielding.

Free Boundary—A boundary along which no restraints are applied to a structure. For example, the tip of a freely vibrating cantilever is a free boundary.

Isotropic—A term applied to a material whose properties are unchanged by rotation of the axis of measurement. Only two elastic constants, the modulus of elasticity (E) and Poisson's ratio (ν), are required to completely specify the elastic behavior of an isotropic material.

Linear—A term applied to a structure or material if all deformations increase in proportion to the load without regard to the sign, magnitude, distribution, or direction of the load. Many structures of practical importance are linear for loads below a maximum linear limit. Nonlinear behavior in a structure is ordinarily due to either a material nonlinearity such as yielding or a geometric nonlinearity such as buckling.

Membrane—A thin, massive, elastic uniform sheet which can support only tensile loads in its own plane. A membrane may be flat like a drum head or curved like a soap bubble. A one-dimensional membrane is a cable. A massless one-dimensional membrane is a string.

Mode Shape (Eigenvector)—A function defined over a structure which describes the relative displacement of any point on the structure as the structure vibrates in a single mode. A mode shape is associated with each natural frequency of a structure. If the deflection of a linear vibrating structure in some direction is denoted by Y(x, t), where x is a point on the structure and t is time, then if the structure vibrates only in the k mode, the deflection can be written as

$$Y(x, t) = \tilde{y}_k(x)\, y_k(t),$$

where $\tilde{y}_k(x)$ is the mode shape, which is a function only of space, and $y_k(t)$ is a function only of time. If the structure vibrates in a number of modes, the total displacement is the sum of the modal displacements:

$$Y(x, t) = \sum_{i=1}^{N} \tilde{y}_i(x)\, y_i(t).$$

Modulus of Elasticity (Young's Modulus)–The rate of change of normal stress for a unit normal strain in a given material. The modulus of elasticity has units of pressure. For most materials, within the limits of linear elasticity, the modulus of elasticity is independent of the sign of the applied stress. Some materials, such as wood, have a directional modulus of elasticity.

Moment of Inertia of a Body–The sum of the products obtained by multiplying each element of mass within a body by the square of its distance from a given axis. (See Product of Inertia of a Body and Chapter 5.)

Moment of Inertia of a Section–The sum of the products obtained by multiplying each element of area within a section by the square of its distance from a given axis. (See Product of Inertia of a Section and Chapter 5.)

Natural Frequency (Eigenvalue)–The frequency at which a linear elastic structure will tend to vibrate once it has been set into motion. A structure can possess many natural frequencies. The lowest of these is called the fundamental natural frequency. Each natural frequency is associated with a mode shape of deformation. Natural frequency can be defined either in terms of cycles per second (hertz) or radians per second. There are 2π radians per cycle.

Neutral Axis–The axis of zero stress in the cross section of a structure. The neutral axis must pass through the centroid of the cross section of homogeneous beams if the axial load is zero so that the beam supports only a bending load.

Node–A point on a structure which does not deflect during vibration in a given mode. Anti-node is a point on a structure where deflection is maximum during vibration in a given mode.

Orthotropic–A term applied to a thin lamina if the material properties of the lamina possess two mutually perpendicular planes of symmetry. Four material constants are required to specify the elastic behavior of an orthotropic lamina. Common examples of orthotropic lamina are sheets of fiber-reinforced plastic or the thin plys of wood that are glued together to form plywood.

Pinned Boundary–A boundary condition such that the structure is free to rotate but not displace along a given boundary.

Plate–A thin two-dimensional elastic structure which is composed of material in the vicinity of a flat two-dimensional sheet. A plate without bending rigidity is a membrane.

Point Mass (Concentrated Mass)–A point in space having mass, but zero moment of inertia for rotation about its center of mass.

Poisson's Ratio–The ratio of the lateral shrinkage (expansion) to the longitudinal expansion (shrinkage) of a bar of a given material which has been placed under a uniform longitudinal tensile (compressive) load. Poisson's ratio is ordinarily near 0.3 and is dimensionless. Some materials, such as wood, have a directional Poisson's ratio. For most materials, within the limits of elasticity, Poisson's ratio is independent of the sign of the applied stress.

Product of Inertia of a Body–The sum of the products obtained by multiplying each element of mass of a body by the distances from two mutually perpendicular axes. (See Chapter 5.)

Product of Inertia of a Section–The sum of the products obtained by multiplying each element of area of a section by the distances from two mutually perpendicular axes. (See Chapter 5.)

Radius of Gyration of a Body–The square root of the quantity formed by dividing the mass moment of inertia of a body by the mass of the body. (See Chapter 5.)

Radius of Gyration of a Section–The square root of the quantity formed by dividing the area moment of inertia of a section by the area of the section. (See Chapter 5.)

Rotary Inertia–The inertia associated with local rotation of a structure. For example, the rotary inertia of a spinning top maintains its rotation.

Seiching–The system of waves in a harbor which is produced as the harbor responds sympathetically to waves in the open sea (also see Sloshing).

Shear Beam–A beam whose deformation in shear substantially exceeds the flexural deformation.

Shear Coefficient–A dimensionless quantity, dependent on the shape of the cross section of a beam, which is introduced into approximate beam theory to account for the fact that shear stress and shear strain are not uniformly distributed over the cross section. The shear coefficient is generally defined as the ratio of the average shear strain over the beam cross section to the shear strain at the centroid. See Section 8.2.

Shear Modulus–The rate change in the shear stress of a material with a unit shear strain. For most materials the shear modulus is independent of the sign of the applied stress, although some materials, such as wood, may have a directional shear modulus. $G = E/[2(1 + \nu)]$ for an isotropic elastic material, that is, a material whose properties are the same in all directions. (Definitions of symbols are given in Chapter 2.)

Shell–A thin elastic structure whose material is confined to the close vicinity of a curved surface, the middle surface of the shell. A curved plate is a shell. A shell without rigidity in bending is a membrane.

Sliding Boundary–A boundary condition such that a structure is free to displace in a given direction along a boundary but rotation is prevented.

Sloshing–The system of surface waves formed in a liquid-filled tank or basin as the liquid is excited.

Speed of Sound--The speed at which very small pressure fluctuations propagate in a infinite fluid or solid.

Spring Constant (Deflection)–The change in load on a linear elastic structure required to produce a unit increment of deflection.

Spring Constant (Torsion)–The change in moment (torque) on a linear elastic structure required to produce a unit increment in rotation.

String–A massless one-dimensional structure which can only bear tension parallel with its own axis. A string is a massless cable.

Viscosity–The ability of a fluid to resist shearing deformation. The viscosity of a linear (Newtonian) fluid is defined as the ratio between the shear stress applied to a fluid and the shearing strain that results. Kinematic viscosity is defined as viscosity divided by fluid density.

2

SYMBOLS

Throughout this book, definitions of symbols are given at the top of each table and in the text. In some cases special symbols have been defined. The symbols listed below have been consistently applied in all cases. These symbols generally follow those used in the literature. One exception is that here I is used to denote all area moments of inertia of sections and J is used to denote all mass moments of inertia of bodies.

A	area ($length^2$)
B	bulk modulus (force/area)
C	center of gravity or centroid, also torsion constant ($length^4$)
E	modulus of elasticity (force/area)
G	shear modulus (force/area)
I	area moment of inertia ($length^4$)
J	mass moment of inertia (mass × $length^2$)
K	shear coefficient (dimensionless)
L	length
M	mass
P	load (force)
S	tension per unit length of edge (force/length) or length
T	tension (force)
X, Y, Z	mutually orthogonal displacements (length)
c	speed of sound (length/time)
f	frequency (hertz)
g	acceleration of gravity (length/$time^2$) or grams
k	deflection spring constant (force/length)
m	mass per unit length (mass/length)
p	load per unit length (force/length) or pressure (Chapter 14)
x, y, z	mutually orthogonal coordinates (length)
$\tilde{x}, \tilde{y}, \tilde{z}$	mode shapes associated with the X, Y, and Z displacements, respectively (dimensionless)
α	angle (radians) or dimensionless constant
γ	mass per unit area (mass/$length^2$) or ratio of specific heats (dimensionless)
ϵ	strain (dimensionless)
θ	rotation (radians)
$\tilde{\theta}$	mode shapes associated with θ rotation (dimensionless)
μ	material density (mass/$length^3$)
ν	Poisson's ratio
π	= 3.1415926

ρ	fluid density (mass/length^3)
σ	stress (force/area) or beam mode shape parameter, Chapter 8 (dimensionless)
ω	frequency (radians/second)
$\mathcal{J}$	Bessel function of first kind
$\mathfrak{M}$	moment (force $\times$ length)
$\mathcal{Y}$	Bessel function of second kind
$\mathcal{k}$	torsion spring constant (moment/angle)

GRAPHIC SYMBOLS

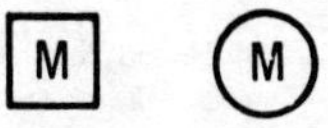

Point Mass

Deflection Spring

Torsion Spring (Table 6-1 only)

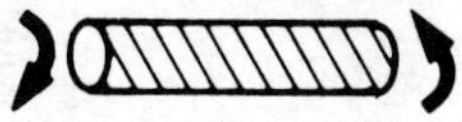
Torsion Spring

Slender Uniform Beam

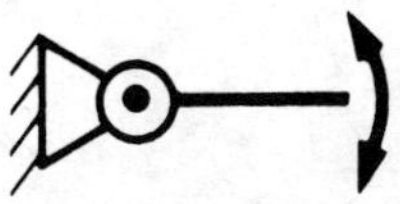
Pivot

Cable

String

Chain

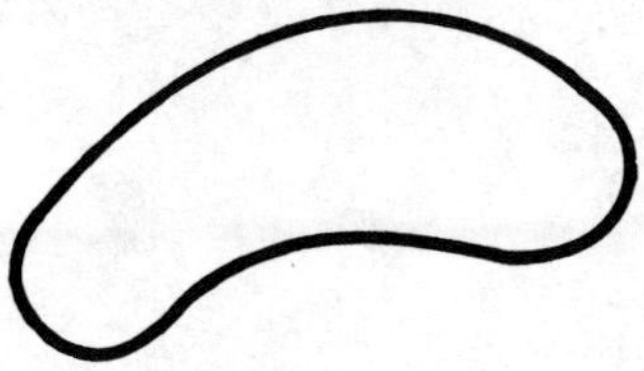
Rigid Body

Infinitely Rigid Wall

Pinned Support

 Clamped Support

 Pinned Support on Frictionless Rollers

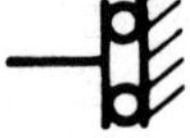 Sliding Support (Deflection but no Rotation)

3

UNITS

The formulas presented in this book will yield the correct result with any consistent set of units. A consistent set of units is one in which Newton's second law, force equals mass times acceleration, is identically satisfied without the introduction of scale or conversion factors. One unit of force applied to one unit of mass must result in one unit of acceleration in a consistent set of units:

$$1 \text{ unit force} = 1 \text{ unit mass} \times 1 \text{ unit acceleration.}$$

For example, if the kilogram is chosen as the unit of force and the unit of acceleration is one centimeter per second per second, then the unit of mass must be such that one kilogram of force will accelerate it at one centimeter per second per second. The unit of mass which satisfies this criterion weighs 1/980.7 kilogram on the surface of the earth and can be expressed as kg/g, where kg is one kilogram of force and g is the acceleration of gravity on the surface of the earth, 980.7 centimeters/second^2. This unit of mass has units of kilograms-second^2/centimeter. Note that it is wrong to attempt to use kilogram (or pound) for both a unit of force and a unit of mass, although it is possible to define a consistent set of units with kilogram (or pound) as a unit of either force or mass.

There is not an engineer or student alive who has not made an error by using an inconsistent set of units. These errors can be avoided by using any of the consistent sets of units presented in Table 3-1. While lack of an intuitive physical feel for quantities such as a newton, dyne, or slug may be reason to convert the final result of a calculation to a more intuitive, but inconsistent, system, it is important to remember that correct dynamic analysis demands consistent units.

Often the engineer who has worked with "English" units has considerable difficulty in converting to the metric system. It may be helpful to remember that a smallish apple weighs about one newton. If we make this apple into apple jelly and spread it evenly over a table one meter square, the resultant pressure is one pascal.

The following abbreviations are used in this book:

Space and Time		*Force*		*Frequency*		*Mass*	
degree	deg	newton	N	hertz	Hz	gram	g
radian	rad	kilonewton	kN			kilogram	kg
meter	m	pascal	Pa			pound	lb
centimeter	cm	decibel	db				
inch	in.	dyne	dyn				
foot	ft						
second	sec						

Table 3-1. Consistent Sets of Units.

	Force	Mass	Length	Time	Pressure	Density	g[a]
1	newton[b]	kilogram	meter	second	pascal[c]	$\frac{\text{kilogram}}{\text{meter}^3}$	$\frac{\text{9.807 meters}}{\text{second}^2}$
2	kilogram	$\frac{\text{kilogram}}{g}$	meter	second	$\frac{\text{kilogram}}{\text{meter}^2}$	$\frac{\text{kilogram}}{g\text{-meter}^3}$	$\frac{\text{9.807 meters}}{\text{second}^2}$
3	kilogram	$\frac{\text{kilogram}}{g}$	centimeter	second	$\frac{\text{kilogram}}{\text{centimeter}^2}$	$\frac{\text{kilogram}}{g\text{-centimeter}^3}$	$\frac{\text{980.7 centimeters}}{\text{second}^2}$
4	dyne[d]	gram	centimeter	second	$\frac{\text{dyne}}{\text{centimeter}^2}$	$\frac{\text{gram}}{\text{centimeter}^3}$	$\frac{\text{980.7 centimeters}}{\text{second}^2}$
5	gram	$\frac{\text{gram}}{g}$	centimeter	second	$\frac{\text{gram}}{\text{centimeter}^2}$	$\frac{\text{gram}}{g\text{-centimeter}^3}$	$\frac{\text{980.7 centimeters}}{\text{second}^2}$
6	pound	slug[e]	foot[f]	second	$\frac{\text{pound}}{\text{foot}^2}$	$\frac{\text{slug}}{\text{foot}^3}$	$\frac{\text{32.17 feet}}{\text{second}^2}$
7	pound	$\frac{\text{pound}}{g}$	inch[g]	second	$\frac{\text{pound}}{\text{inch}^2}$	$\frac{\text{pound}}{g\text{-inch}^3}$	$\frac{\text{386.1 inches}}{\text{second}^2}$

[a]g = acceleration due to gravity at the surface of the earth.
[b]Newton = force required to accelerate 1 kilogram of mass at 1 meter/second^2 = 0.2248 pound. One kilogram of mass weighs 9.807 newtons on the surface of the earth.
[c]Pascal = newton/meter^2 = 10^{-5} bar = 10 dynes/centimeter^2 = 1.4503×10^{-4} pounds/inch^2.
[d]Dyne = force required to accelerate 1 gram of mass at 1 centimeter/second^2 = 10^{-5} newton = 2.248×10^{-6} pounds.
[e]Slug = pound/g. One slug of mass weighs 32.17 pounds on the surface of the earth.
[f]Foot = 12 inches = 0.3048 meter.
[g]Inch = 2.54 centimeters.

4

PRINCIPLES AND ANALYTICAL METHODS

Any structure with mass and elasticity will possess one or more natural frequencies of vibration. The natural frequencies are the result of cyclic exchanges of kinetic and potential energy within the structure. The kinetic energy is associated with velocity of structural mass, while the potential energy is associated with storage of energy in the elastic deformations of a resilient structure. Just as a ball bouncing off a hard-wood floor exchanges the potential energy at the apex of its flight for kinetic energy as it plummets, so an elastic structure exchanges the potential energy of elastic deformation for velocity of vibration as it vibrates back and forth. The rate of energy exchange between the potential and kinetic forms of energy is the natural frequency.

If a structure is linear, that is, its deformation is proportional to the load it bears regardless of the magnitude, distribution, or direction of load, and has constant mass, then it can be shown that the natural frequency of the structure is independent of the amplitude of vibration. For example, consider the spring-body system shown in Fig. 4-1. If the rigid body is displaced from its equilibrium position by an amount Y_0, then the change in the potential energy of the spring is:

$$\Delta PE = \int_0^{Y_0} F\, dY,$$

$$= \int_0^{Y_0} kY\, dY = \tfrac{1}{2} kY_0^2. \qquad (4\text{-}1)$$

k is the spring constant of the spring, the rate of change of the spring force, F, with deformation, Y. If we assume the vibration is harmonic in time, t, at frequency f,

$$Y = Y_0 \sin (2\pi f)t, \qquad (4\text{-}2)$$

then the velocity of the body at any time is:

$$\dot{Y} = Y_0\ (2\pi f) \cos (2\pi f)t, \qquad (4\text{-}3)$$

where $\dot{Y}$ denotes the derivative of Y with respect to time. The velocity of the body is maximum at $t = n/(2f)$, $n = 1, 2, \ldots$, when the deformation of the system is zero, $Y = 0$. The maximum kinetic energy of the rigid body is:

$$\Delta KE = \tfrac{1}{2}\ M\dot{Y}_{max}^2 = \tfrac{1}{2}\ MY_0^2\ (2\pi f)^2. \qquad (4\text{-}4)$$

Note that both the potential energy (Eq. 4-1) and kinetic energy (Eq. 4-3) are proportional to amplitude squared and that the kinetic energy is a function of frequency, while the potential energy is independent of frequency.

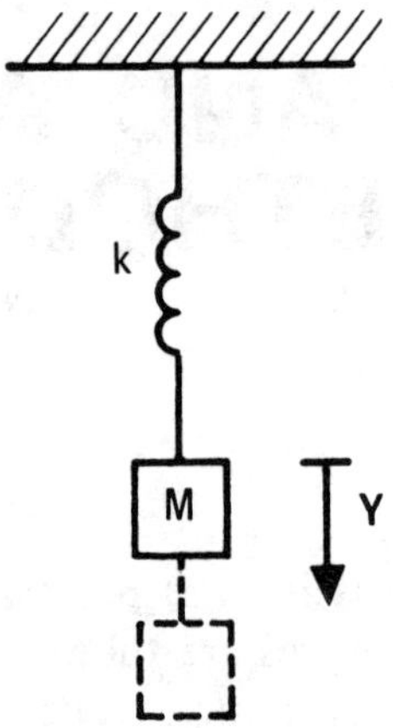

Fig. 4-1. Spring-body system. k is the spring constant. M is the mass.

If we assume that the spring is massless and the body is perfectly rigid, then all the potential energy in the system is associated with spring deformation and all the kinetic energy in the system is associated with velocity of the mass. Since the system is free of external forces, the total energy of the system is constant. Thus, the maximum potential energy of the system, which occurs when $Y = \pm Y_0$, must equal the maximum kinetic energy, which occurs when $Y = 0$:

$$\Delta KE = \Delta PE. \tag{4-5}$$

Substituting Eqs. 4-1 and 4-4 into this equation and solving for f gives:

$$f = \frac{1}{2\pi}\left(\frac{k}{M}\right)^{1/2}, \tag{4-6}$$

where the natural frequency, f, has units of cycles per unit time (hertz). The natural frequency of the spring-body system increases with the stiffness of the spring and decreases with increasing mass of the body. The natural frequency of all linear structures increases with increasing stiffness and decreases with increasing mass.

The natural frequency of the spring-body system can be obtained in a more rigorous fashion by writing and solving the equation of motion exactly. Since the force applied to the body by the spring and gravity must equal the mass times the acceleration of the body, the equation of motion of the body is:

$$-ky - Mg = M\ddot{Y}. \tag{4-7}$$

ky is the force of the spring on the body and mg is the force of gravity on the body, g being the acceleration of gravity. These forces are negative since they act downward. $M\ddot{Y}$ is the product of the mass and the acceleration of the body. Equation 4-7 may be rewritten as:

$$M\ddot{Y} + kY = -Mg. \tag{4-8}$$

This equation has the following solution:

$$Y = A \sin \omega t + B \cos \omega t - \frac{Mg}{k}, \tag{4-9}$$

where the natural frequency is

$$\omega = 2\pi f = \left(\frac{k}{M}\right)^{1/2} \tag{4-10}$$

and A and B are constants with the units of length. The solution is comprised of components which oscillate with circular frequency ω (radians/second) and a component which is independent of time. The oscillatory component represents harmonic motion at the natural frequency, while the static component, $-Mg/k$, is simply the static stretching of the spring under the force of gravity. The natural frequency of the spring-body system is independent of mean deformation due to gravity.

There is always some damping in real structures which will make free vibrations decay with time, and there is some amplitude beyond which the structure no longer behaves linearly. For real structures the concept of natural frequency must be tempered by some knowledge of the differences between the ideal mathematical model and the actual structure. The discrepancies between a linear model and a real structure are often due to neglected (1) linear and (2) nonlinear effects. Some linear effects which are often neglected are the effect of shearing deformation in slender structures and the effect of surrounding fluid. Nonlinear effects which are often neglected are plasticity due to yielding and the amplitude dependence of damping.

The vast majority of solutions presented in this book are exact solutions to linear equations of motion. The equations of motion were derived from basic principles based on certain ideal behavior, such as beams that only bend and do not shear and strings that never yield. The diagrams and explanations are presented to clarify which idealizations have been used.

If a model does not closely approximate a real structure, then the natural frequency of the structure can often be estimated by using bracketing assumptions. For example, if the edge of a plate is riveted at broad intervals to a heavy beam, then the boundary condition on the plate is probably intermediate between a clamped and a pinned edge. It is useful to use both approximations in the analysis to bracket the natural frequency. Similarly, if the material properties of a structure are not well known, then the natural frequency can be bracketed by high and low estimates of the material properties.

The emergence of dynamic finite element computer programs in recent years has made possible frequency analysis of very complex structures. These programs are replacing approximate methods of analysis of complex structures. Finite element programs have not replaced closed form solutions because a frequency can be calculated from a closed form result in a matter of minutes, while computer programs generally require hours to set up and run successively. Moreover, finite element programs have created a new niche for closed form solutions, that of providing limiting case checks on complex computer models.

5

GEOMETRIC PROPERTIES OF PLANE AREAS AND SOLID BODIES

5.1. PLANE AREAS

Figure 5.1 shows an area lying in the x–y plane. A is the total area enclosed by the boundary, and dA is a small element of that area. The x and y axes are mutually perpendicular with origin at point 0. The z axis is perpendicular to the x-y plane and passes through the point 0. The mutually perpendicular axes r and s and r′ and s′ are rotated counterclockwise with respect to the x and y axes by the angles θ and θ', respectively. The r′ and s′ axes are principal axes, that is, axes about which the product of inertia is zero. C marks the centroid of the area, which is located at coordinates x_C and y_C with respect to the x–y axis. C is also the origin of the x′–y′ axes, which are translated with respect to the x-y axes.

The location of the centroid with respect to the x-y axes is given by:

$$x_C = \frac{\int_A x \, dA}{A}$$

$$y_C = \frac{\int_A y \, dA}{A}$$

The neutral axis is the axis along which the stress is zero. This axis must pass through the centroid of the cross section of homogeneous structures if the axial load on the structure is zero, so that the structure supports only a bending load. Thus, the neutral axis of beams which flex but do not extend must pass through the centroid of the cross section.

The area moments of inertia (I_x, I_y), the area product of inertia (I_{xy}) about the x-y axes, and the polar area moment of inertia about the z axis (I_{zz}) are defined by the following formulas:

$$I_x = \int_A y^2 \, dA,$$

$$I_y = \int_A x^2 \, dA,$$

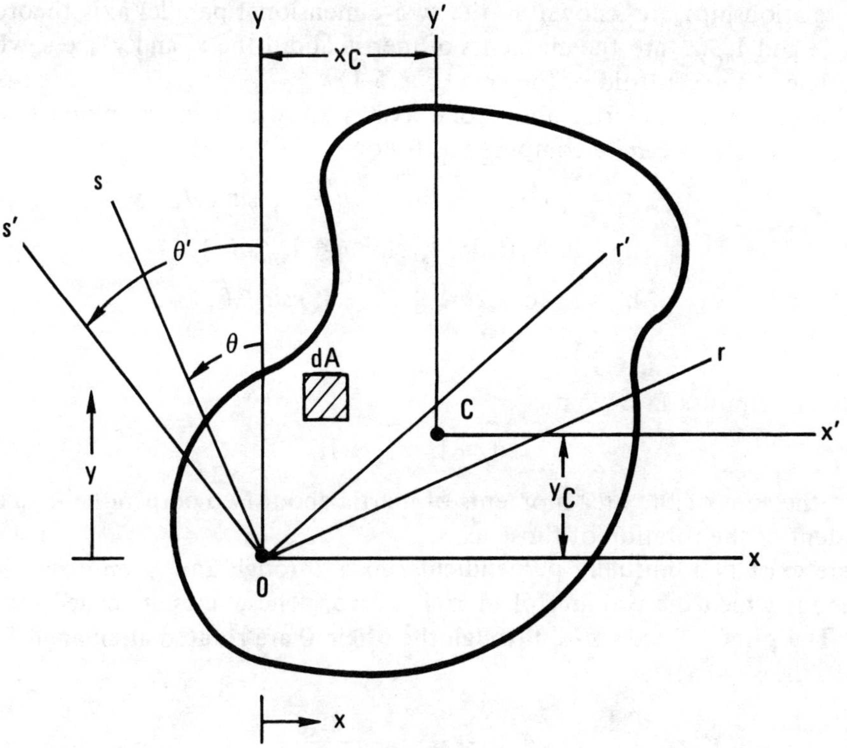

Fig. 5-1. A plane section.

$$I_{xy} = \int_A xy \, dA,$$

$$I_z = I_x + I_y = \int_A (x^2 + y^2) \, dA.$$

The radii of gyration of the area are defined by:

$$r_x = \left(\frac{I_x}{A}\right)^{1/2},$$

$$r_y = \left(\frac{I_y}{A}\right)^{1/2}.$$

If the moment of inertia about an axis which passes through the centroid is known, then the moment of inertia about a parallel axis which is translated can easily be computed as follows:

$$I_x = I_{x_C} + y_C^2 A,$$

$$I_y = I_{y_C} + x_C^2 A,$$

$$I_{xy} = I_{x_C y_C} + x_C y_C A,$$

$$I_z = I_{z_C} + x_C^2 A + y_C^2 A.$$

These relationships are known as the two-dimensional parallel axis theorem. I_{x_C}, I_{y_C}, I_{z_C}, and $I_{x_C y_C}$ are the moments of inertia about the x′ and y′ axes, which have their origin at the centroid of the area (Fig. 5-1).

If the moment of inertia about one axis is known, then the moment of inertia about a rotated axis can be computed as follows:

$$I_r = I_x \cos^2 \theta + I_y \sin^2 \theta - I_{xy} \sin 2\theta,$$

$$I_s = I_y \cos^2 \theta + I_x \sin^2 \theta + I_{xy} \sin 2\theta,$$

$$I_{rs} = I_{xy} \cos 2\theta - \tfrac{1}{2} (I_y - I_x) \sin 2\theta,$$

$$I_z = I_z .$$

The above formulas show that

$$I_r + I_s = I_x + I_y ,$$

or that the sum of the area moments of inertia about two perpendicular axes are independent of the rotation of those axes.

There exist two mutually perpendicular axes through any given origin in a plane area about which the product of inertia is zero. These axes are called the principal axes. The principal axes r′–s′ through the origin 0 are rotated at an angle θ' with respect to the x–y axes:

$$\theta' = \frac{1}{2} \text{arc tan} \frac{2I_{xy}}{I_y - I_x} .$$

The area moments of inertia about the principal axes are:

$$I_{r'} = \tfrac{1}{2} (I_x + I_y) \pm \tfrac{1}{2} [(I_y - I_x)^2 + 4I_{xy}^2]^{1/2},$$

$$I_{s'} = \tfrac{1}{2} (I_x + I_y) \mp \tfrac{1}{2} [(I_y - I_x)^2 + 4I_{xy}^2]^{1/2},$$

$$I_{r's'} = 0.$$

The principal moments of inertia are the maximum and minimum moments of inertia which can be measured about any axis which passes through a given point. An axis of symmetry is always a principal axis. The product of inertia is zero for any two axes, one of which is a principal axis. If the principal moments of inertia are equal, then the moments of inertia about any rotated axis through the origin of the principal axes are equal to the principal moments of inertia. For example, if r′ – s′ are principal axes (Fig. 5-1) and $I_{r'} = I_{s'}$, then $I_{r'} = I_r = I_s = I_{s'}$ regardless of θ or θ'.

The moment of inertia of complex sections can often be computed by subdividing the sections into simpler component sections. The moment of inertia of a complex section is the sum of the moments of inertia of the component sections less the moments of inertia of voids. For example, the moment of inertia of an I-beam cross section can be computed by summing the moments of inertia of three rectangles in space. The moments of inertia of common structural cross sections are usually given in manufacturers' data books (Ref. 5-1) or construction manuals (Refs. 5-2 through 5-4). Table 5-1 gives the moments of inertia for a number of practically important plane sections.

Table 5-1. Properties of Plane Sections.

Section	Area (A) and Distance from Axes to Centroid (C), x_C, y_C.	Area Moments of Inertia About Principal Axes	Area Products of Inertia
1. Right Triangle 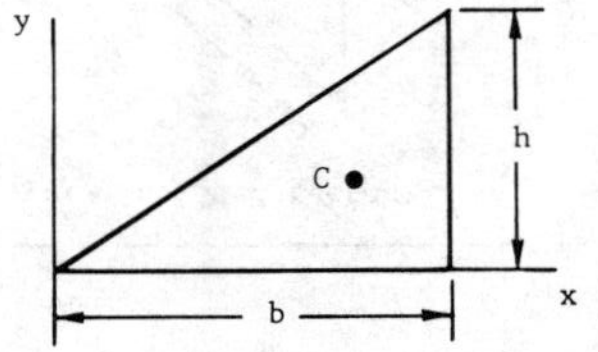	$x_C = \frac{2}{3}b$ $y_C = \frac{1}{3}h$ $A = \frac{1}{2}bh$	$I_{x_C} = \frac{bh^3}{36}$ $I_{y_C} = \frac{b^3h}{36}$ $I_x = \frac{bh^3}{12}$ $I_y = \frac{b^3h}{4}$	$I_{x_Cy_C} = \frac{A}{36}hb = \frac{h^2b^2}{72}$ $I_{xy} = \frac{A}{4}hb = \frac{h^2b^2}{8}$
2. Triangle 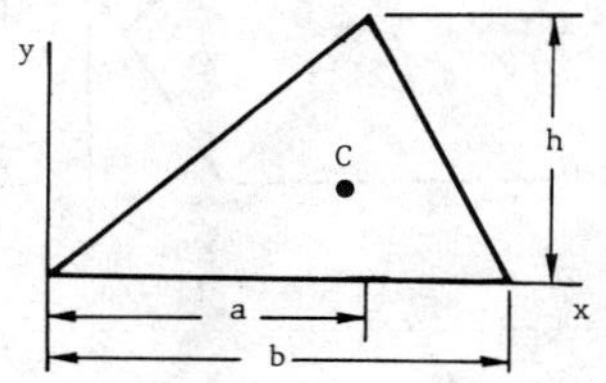	$x_C = \frac{1}{3}(a + b)$ $y_C = \frac{1}{3}h$ $A = \frac{1}{2}bh$	$I_{x_C} = \frac{bh^3}{36}$ $I_{y_C} = \frac{bh}{36}(b^2 - ab + a^2)$ $I_x = \frac{bh^3}{12}$ $I_y = \frac{bh}{12}(b^2 + ab + a^2)$	$I_{x_Cy_C} = \frac{Ah}{36}(2a - b) = \frac{bh^2}{72}(2a - b)$ $I_{xy} = \frac{Ah}{12}(2a + b) = \frac{bh^2}{24}(2a + b)$
3. Square 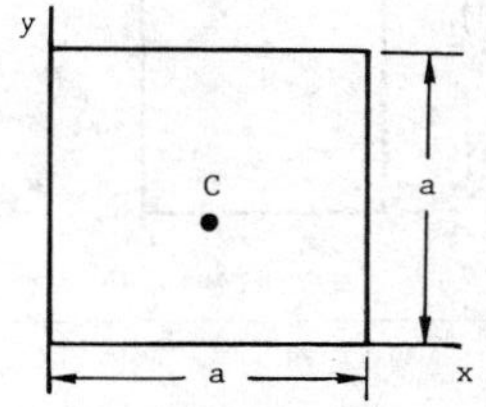	$x_C = \frac{1}{2}a$ $y_C = \frac{1}{2}a$ $A = a^2$	$I_{x_C} = I_{y_C} = \frac{a^4}{12}$ $I_x = I_y = \frac{a^4}{3}$	$I_{x_Cy_C} = 0$ $I_{xy} = \frac{A}{4}a^2 = \frac{a^4}{4}$

Table 5-1. Properties of Plane Sections. *(Continued)*

Section	Area (A) and Distance from Axes to Centroid (C), x_C, y_C.	Area Moments of Inertia About Principal Axes	Area Products of Inertia
4. Rectangle 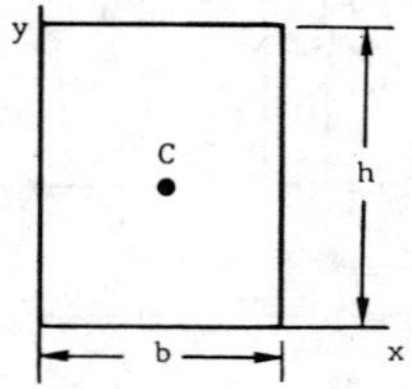	$x_C = \frac{1}{2}b$ $y_C = \frac{1}{2}h$ $A = bh$	$I_{x_C} = \frac{bh^3}{12}$ $I_{y_C} = \frac{b^3h}{12}$ $I_x = \frac{bh^3}{3}$ $I_y = \frac{b^3h}{3}$	$I_{x_Cy_C} = 0$ $I_{xy} = \frac{A}{4}bh = \frac{b^2h^2}{4}$
5. Parallelogram 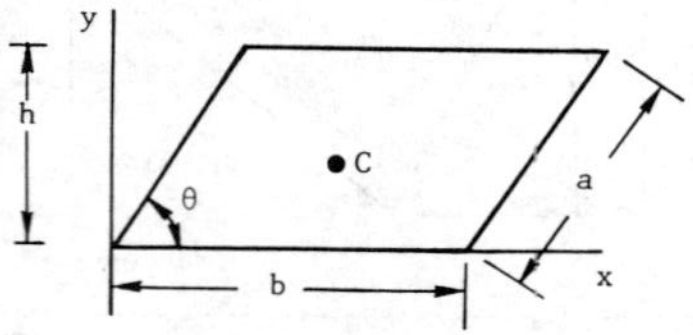	$x_C = \frac{1}{2}(b + a\cos\theta)$ $y_C = \frac{1}{2}(a\sin\theta)$ $= \frac{h}{2}$ $A = ab\sin\theta$ $= bh$	$I_{x_C} = \frac{a^3b}{12}\sin^3\theta$ $I_{y_C} = \frac{ab}{12}\sin\theta\,(b^2 + a^2\cos^2\theta)$ $I_x = \frac{a^3b}{3}\sin^3\theta$ $I_y = \frac{ab}{3}\sin\theta\,(b + a\cos\theta)^2$ $- \frac{a^2b^2}{6}\sin\theta\cos\theta$	$I_{x_Cy_C} = \frac{a^3b}{12}\sin^2\theta\cos\theta$
6. Inclined Rectangle 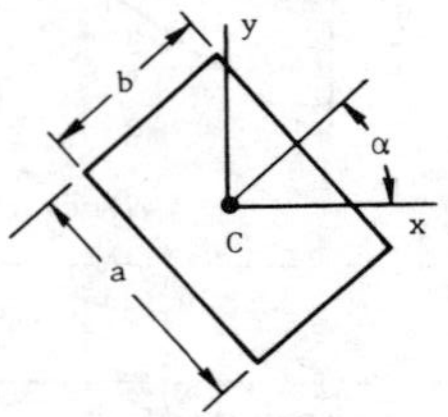	$x_C = 0$ $y_C = 0$ $A = ab$	$I_x = \frac{ab}{12}(a^2\cos^2\alpha + b^2\sin^2\alpha)$ $I_y = \frac{ab}{12}(a^2\sin^2\alpha + b^2\cos^2\alpha)$	$I_{xy} = \frac{ab}{24}(a^2 - b^2)\sin 2\alpha$

7. Diamond 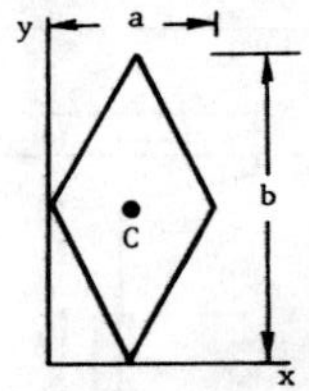	$x_C = \frac{a}{2}$ $y_C = \frac{b}{2}$ $A = \frac{ab}{2}$	$I_{x_C} = \frac{ab^3}{48}$ $I_{y_C} = \frac{a^3 b}{48}$	
8. Hollow Rectangle 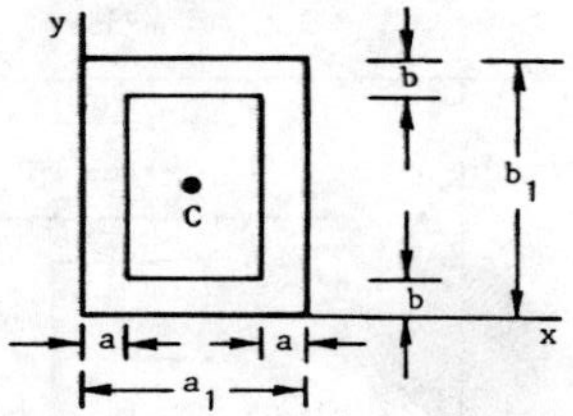	$x_C = \frac{a_1}{2}$ $y_C = \frac{b_1}{2}$ $A = 2a_1 b + 2ab_1 - 4ab$	$I_{x_C} = \frac{a_1 b_1^3 - (a_1 - 2a)(b_1 - 2b)^3}{12}$ $I_{y_C} = \frac{a_1^3 b_1 - (a_1 - 2a)^3 (b_1 - 2b)}{12}$	
9. Thin Hollow Rectangle 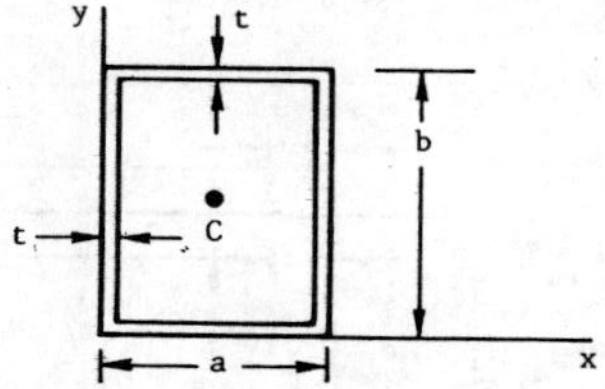	$x_C = \frac{a}{2}$ $y_C = \frac{b}{2}$ $A = 2at + 2bt$	$I_{x_C} = \frac{tb^3}{6} + \frac{tab^2}{2}$ $I_{y_C} = \frac{ta^3}{6} + \frac{ta^2 b}{2}$	
10. Channel 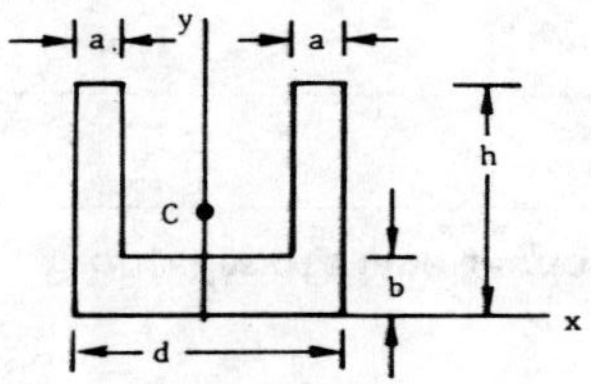	$x_C = 0$ $y_C = \frac{2ah^2 + db^2 - 2ab^2}{2A}$ $A = 2ah + db - 2ab$	$I_x = \frac{2}{3}ah^3 + \frac{1}{3}(d - 2a)b^3$ $I_{y_C} = I_y = \frac{d^3 h - (h - b)(d - 2a)^3}{12}$	

Table 5-1. Properties of Plane Sections. ***(Continued)***

Section	Area (A) and Distance from Axes to Centroid (C), x_C, y_C.	Area Moments of Inertia About Principal Axes	Area Products of Inertia
11. Tapered Channel 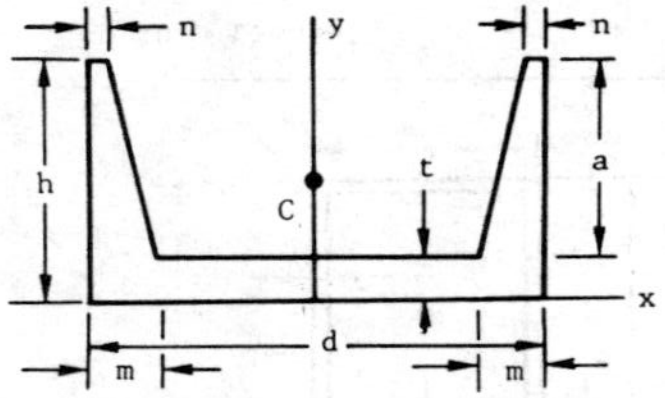	$x_C = 0$ $y_C = \frac{6h^2n + 3ct^2 + 2a(m-n)(h+2t)}{6A}$ $(c = d - 2n)$ $A = dt + a(m + n)$	$I_x = \frac{2nh^3 + et^3 + \frac{m-n}{2a}(h^4 - t^4)}{3}$ $I_y = \frac{hd^3 - \frac{a}{8(m-n)}(c^4 - e^4)}{12}$ $(c = d - 2n,\ e = d - 2m)$	
12. I Section 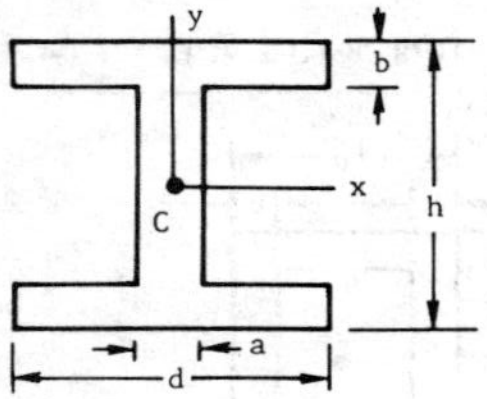	$x_C = 0$ $y_C = 0$ $A = 2db + ah - 2ab$	$I_x = \frac{dh^3 - (d-a)(h-2b)^3}{12}$ $I_y = \frac{2bd^3 + (h-2b)a^3}{12}$	
13. Tapered I Section 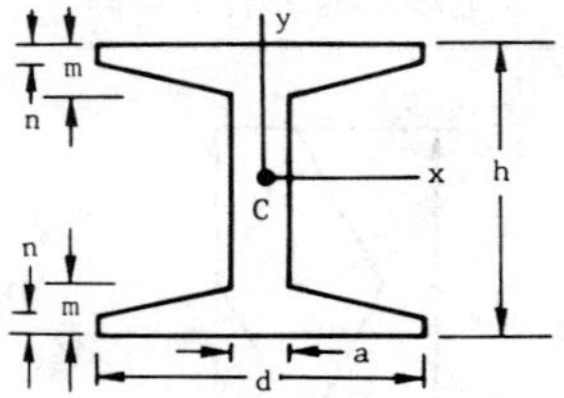	$x_C = 0$ $y_C = 0$ $A = ha + (d - a)(m + n)$	$I_x = \frac{dh^3 - \frac{(d-a)}{8(m-n)}(c^4 - e^4)}{12}$ $I_y = \frac{2nd^3 + ea^3 + \frac{m-n}{2(d-a)}(d^4 - a^4)}{12}$ $(c = h - 2n,\ e = h - 2m)$	

14. T Section 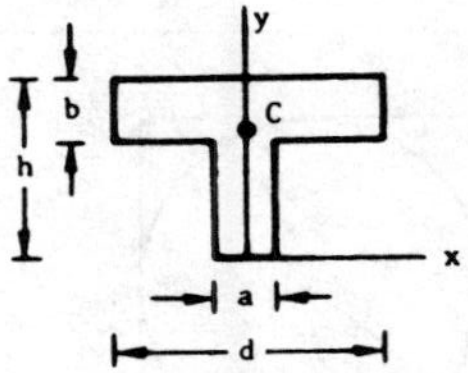	$x_C = 0$ $y_C = h - \frac{ah^2 + db^2 - ab^2}{2A}$ $A = db + ah - ab$	$I_{x_C} = \frac{1}{3}[d(h - y_C)^3$ $- (d - a)(h - b - y_C)^3 + ay_C^3]$ $I_{y_C} = I_y = \frac{d^3b + (h - b)a^3}{12}$ $I_x = \frac{1}{3}dh^3 - \frac{1}{3}(d - a)(h - b)^3$	
15. L Section 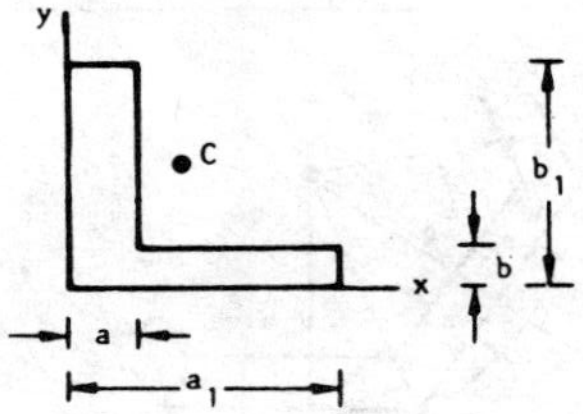	$y_C = \frac{ab_1^2 + a_1b^2 - ab^2}{2A}$ $x_C = \frac{a_1^2b + a^2b_1 - a^2b}{2A}$ $A = a_1b + ab_1 - ab$	$I_{x_C} = \frac{1}{3}[a_1y_C^3 - (a_1 - a)(y_C - b)^3$ $+ a(b_1 - y_C)^3]$ $I_{y_C} = \frac{1}{3}[b_1x_C^3 - (b_1 - b)(x_C - a)^3$ $+ b(a_1 - x_C)^3]$ $I_x = \frac{1}{3}ab_1^3 + \frac{1}{3}(a_1 - a)b^3$ $I_y = \frac{1}{3}ba_1^3 + \frac{1}{3}(b_1 - b)a^3$	
16. Z Section 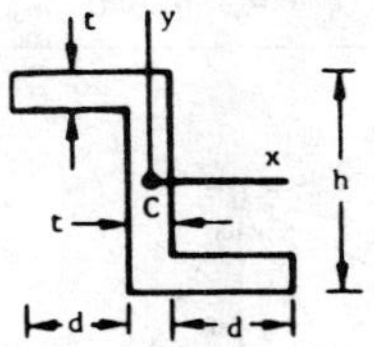	$x_C = 0$ $y_C = 0$ $A = t(h + 2d)$	$I_x = \frac{(d + t)h^3 - d(h - 2t)^3}{12}$ $I_y = \frac{h(2d + t)^3 - 2d^3(h - t)}{12}$ $- \frac{6d(d + t)^2(h - t)}{12}$	
17. Trapezoid 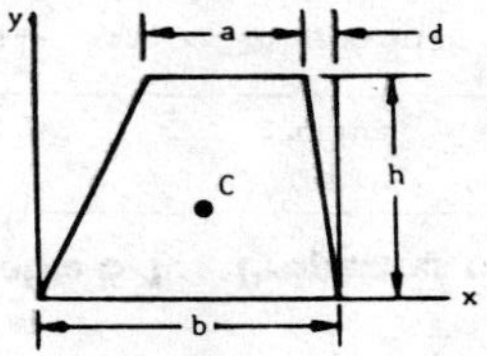	$x_C = \frac{2b^2 + 2ba - bd - 2ad - a^2}{3(a + b)}$ $y_C = \frac{1}{3}h\left(\frac{2a + b}{a + b}\right)$ $A = \frac{1}{2}h(a + b)$	$I_{x_C} = \frac{h^3(a^2 + 4ab + b^2)}{36(a + b)}$ $I_x = \frac{h^3}{12}(3a + b)$ $I_{y_C} = \frac{h}{36(a + b)}[b^4 + a^4 + 2ab(a^2 + b^2)$ $- d(b^3 + 3b^2a - 3ba^2 - a^3)$ $+ d^2(b^2 + 4ab + a^2)]$	$I_{x_Cy_C} = \frac{h^2}{72(a + b)}[a(3b^2 - 3ab - a^2)$ $+ b^3 - d(2b^2 + 8ab + 2a^2)]$

Table 5-1. Properties of Plane Sections. (*Continued*)

Section	Area (A) and Distance from Axes to Centroid (C), x_C, y_C.	Area Moments of Inertia About Principal Axes	Area Products of Inertia
18. Regular Polygon with n Sides 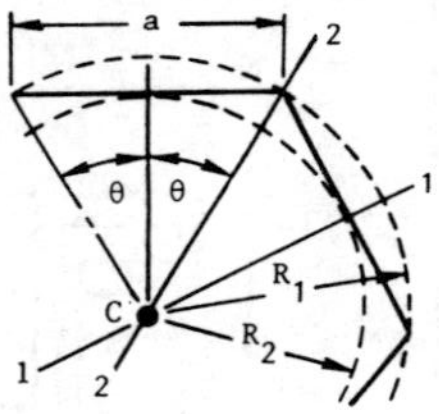	$\left(R_1 = \frac{a}{2 \sin \theta}\right.$ $\left.R_2 = \frac{a}{2 \tan \theta}\right)$ $A = \frac{1}{4}a^2 n \cot \theta$	$I_1 = \frac{A(6R_1^2 - a^2)}{24}$ $I_2 = \frac{A(12R_2^2 + a^2)}{48}$	
19. Hollow Regular Polygon with n Sides 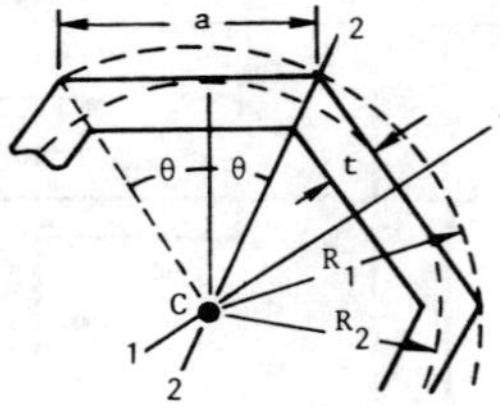	$\left(R_1 = \frac{a}{2 \sin \theta}\right.$ $\left.R_2 = \frac{a}{2 \tan \theta}\right)$ $A = nat\left(1 - \frac{t \tan \theta}{a}\right)$	$I_1 = \frac{na^3t}{8}\left(\frac{1}{3} + \frac{1}{\tan^2 \theta}\right) \cdot \left[1 - 3\frac{t \tan \theta}{a} + 4\left(\frac{t \tan \theta}{a}\right)^2 - 2\left(\frac{t \tan \theta}{a}\right)^3\right]$ $I_2 = I_1$	
20. Circle 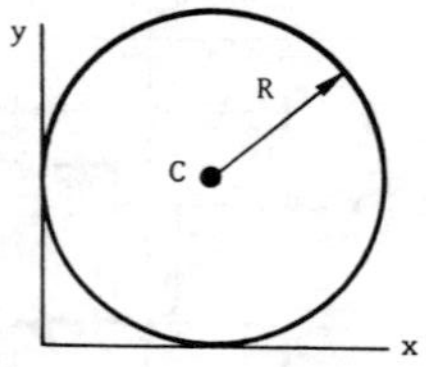	$x_C = R$ $y_C = R$ $A = \pi R^2$	$I_{x_C} = I_{y_C} = \frac{1}{4}\pi R^4$ $I_x = I_y = \frac{5}{4}\pi R^4$	$I_{x_C y_C} = 0$ $I_{xy} = \pi R^4$

21. Semicircle 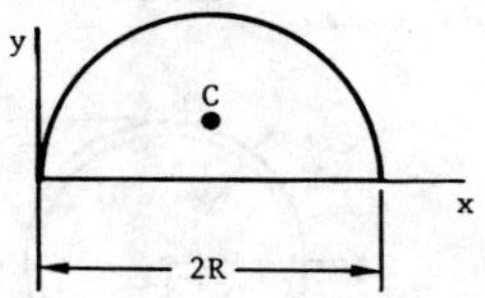	$x_C = R$ $y_C = \frac{4R}{3\pi}$ $A = \frac{1}{2}\pi R^2$	$I_{x_C} = \frac{R^4(9\pi^2 - 64)}{72\pi}$ $I_{y_C} = \frac{1}{8}\pi R^4$ $I_x = \frac{1}{8}\pi R^4$ $I_y = \frac{5}{8}\pi R^4$	$I_{x_C y_C} = 0$ $I_{xy} = \frac{2}{3}R^4$
22. Circular Sector 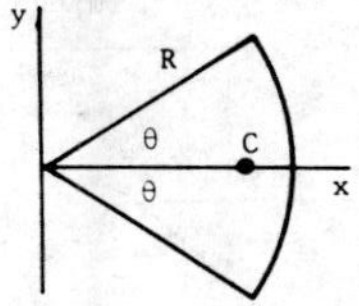	$x_C = \frac{2R}{3}\frac{\sin\theta}{\theta}$ $y_C = 0$ $A = R^2\theta$	$I_x = \frac{1}{4}R^4(\theta - \sin\theta\cos\theta)$ $I_y = \frac{1}{4}R^4(\theta + \sin\theta\cos\theta)$	$I_{x_C y_C} = 0$ $I_{xy} = 0$
23. Circular Segment 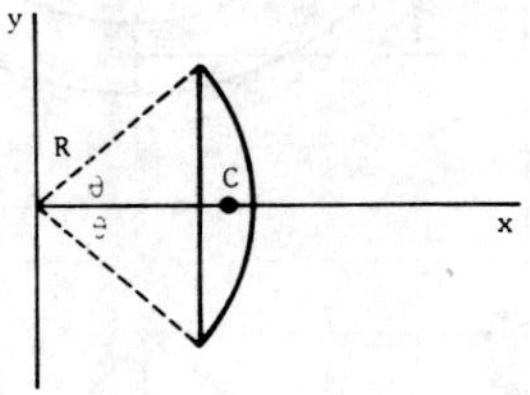	$x_C = \frac{2R}{3}\left(\frac{\sin^3\theta}{\theta - \sin\theta\cos\theta}\right)$ $y_C = 0$ $A = R^2\left(\theta - \frac{1}{2}\sin 2\theta\right)$	$I_x = \frac{AR^2}{4}\left[1 - \frac{2\sin^3\theta\cos\theta}{3(\theta - \sin\theta\cos\theta)}\right]$ $I_y = \frac{AR^2}{4}\left[1 + \frac{2\sin^3\theta\cos\theta}{\theta - \sin\theta\cos\theta}\right]$	$I_{x_C y_C} = 0$ $I_{xy} = 0$
24. Annulus 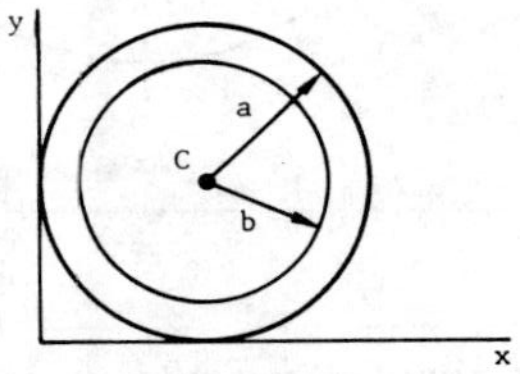	$x_C = a$ $y_C = a$ $A = \pi(a^2 - b^2)$	$I_{x_C} = I_{y_C} = \frac{\pi}{4}(a^4 - b^4)$ $I_x = I_y = \frac{5}{4}\pi a^4 - \pi a^2 b^2 - \frac{\pi}{4}b^4$	$I_{x_C y_C} = 0$ $I_{xy} = Aa^2 = \pi a^2(a^2 - b^2)$

Table 5-1. Properties of Plane Sections. *(Continued)*

Section	Area (A) and Distance from Axes to Centroid (C), x_C, y_C.	Area Moments of Inertia About Principal Axes	Area Products of Inertia
25. Semiannulus 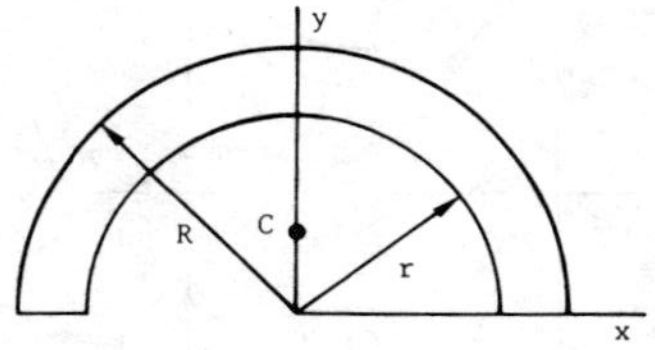	$x_C = 0$ $y_C = \frac{4R}{3\pi}\left(\frac{r}{R} + \frac{R}{R + r}\right)$ $A = \frac{\pi}{2}(R^2 - r^2)$	$I_x = \frac{\pi}{8}(R^4 - r^4)$ $I_y = I_x$	
26. Sector of Annulus 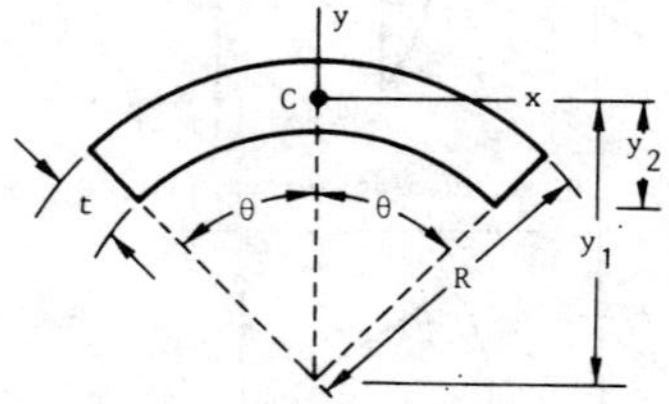	$x_C = 0$ $A = \theta t(2R - t)$ $y_C = 0$ $y_1 = \frac{2R \sin\theta}{3\theta}\left(1 - \frac{t}{R} + \frac{1}{2 - t/R}\right)$ $y_2 = R\left[\frac{2 \sin\theta}{3\theta(2 - t/R)} + \left(1 - \frac{t}{R}\right)\frac{2 \sin\theta - 3\theta \cos\theta}{3\theta}\right]$	$I_x = R^3 t\left[\left(1 - \frac{3t}{2R} + \frac{t^2}{R^2} - \frac{t^3}{4R^3}\right)\right.$ $\cdot\left(\theta + \sin\theta\cos\theta - \frac{2 \sin^2\theta}{\theta}\right)$ $\left. + \frac{t^2 \sin^2\theta}{3R^2\theta(2 - t/R)}\left(1 - \frac{t}{R} + \frac{t^2}{6R^2}\right)\right]$ $I_y = R^3 t\left(1 - \frac{3t}{2R} + \frac{t^2}{R^2} - \frac{t^3}{4R^3}\right)$ $\cdot(\theta - \sin\theta\cos\theta)$	
27. Thin Annulus 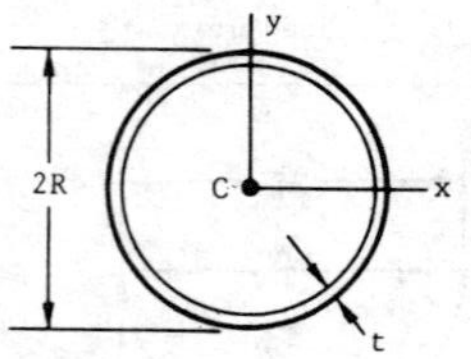	$x_C = 0$ $y_C = 0$ $A = 2\pi Rt$	$I_x = \pi R^3 t$ $I_y = \pi R^3 t$	$I_{x_C y_C} = 0$ $I_{xy} = 2\pi R^3 t$

28. Sector of Thin Annulus 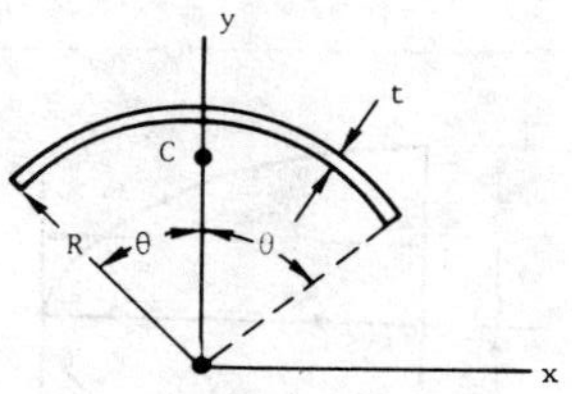	$x_C = 0$ $y_C = R\frac{\sin\theta}{\theta}$ $A = 2\theta Rt$	$I_{x_C} = R^3t\left(\theta + \sin\theta\cos\theta - \frac{2\sin^2\theta}{\theta}\right)$ $I_{y_C} = R^3t(\theta - \sin\theta\cos\theta)$	
29. Ellipse 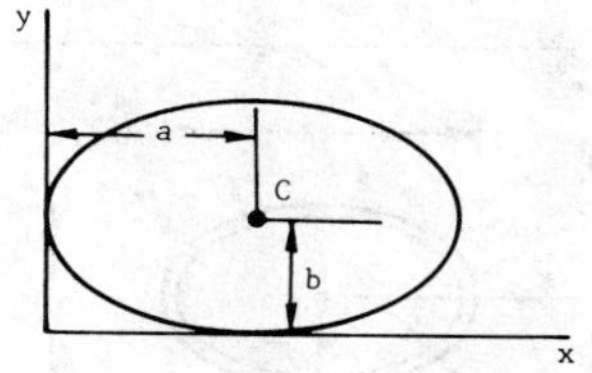	$x_C = a$ $y_C = b$ $A = \pi ab$	$I_{x_C} = \frac{\pi}{4}ab^3$ $I_{y_C} = \frac{\pi}{4}a^3b$ $I_x = \frac{5}{4}\pi ab^3$ $I_y = \frac{5}{4}\pi a^3b$	$I_{x_Cy_C} = 0$ $I_{xy} = Aab = \pi a^2b^2$
30. Semi-ellipse 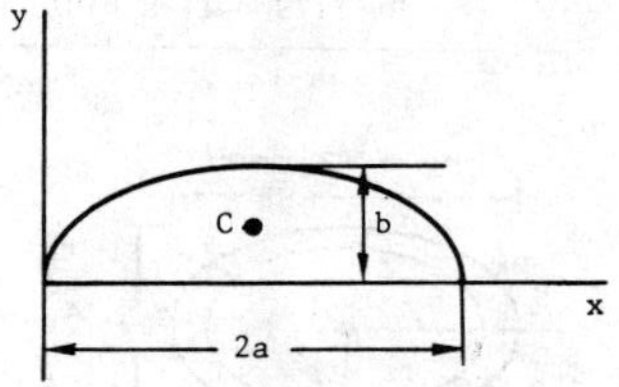	$x_C = a$ $y_C = \frac{4b}{3\pi}$ $A = \frac{1}{2}\pi ab$	$I_{x_C} = \frac{ab^3}{72\pi}(9\pi^2 - 64)$ $I_{y_C} = \frac{\pi}{8}a^3b$ $I_x = \frac{\pi}{8}ab^3$ $I_y = \frac{5}{8}\pi a^3b$	$I_{x_Cy_C} = 0$ $I_{xy} = \frac{2}{3}a^2b^2$
31. Quarter Ellipse 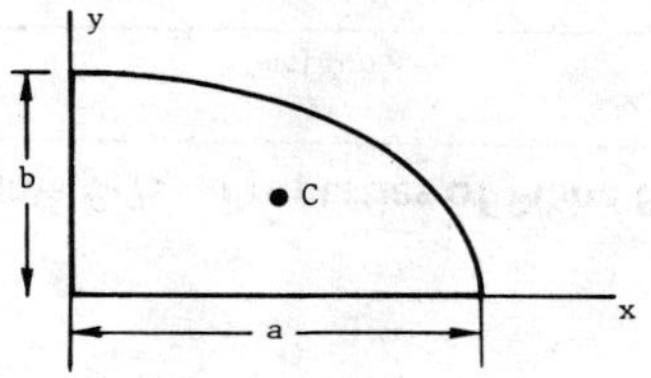	$x_C = \frac{4}{3\pi}a$ $y_C = \frac{4}{3\pi}b$ $A = \frac{1}{4}\pi ab$	$I_{x_C} = ab^3\left(\frac{\pi}{16} - \frac{4}{9\pi}\right)$ $I_{y_C} = a^3b\left(\frac{\pi}{16} - \frac{4}{9\pi}\right)$ $I_x = \frac{\pi ab^3}{16}$ $I_y = \frac{\pi a^3b}{16}$	

Table 5-1. Properties of Plane Sections. (*Continued*)

Section	Area (A) and Distance from Axes to Centroid (C), x_C, y_C.	Area Moments of Inertia About Principal Axes	Area Products of Inertia
32. Hollow Ellipse 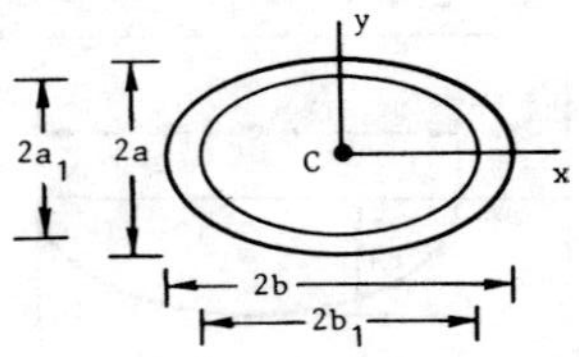	$x_C = 0$ $y_C = 0$ $A = \pi(ab - a_1b_1)$	$I_x = \frac{\pi}{4}(ba^3 - b_1a_1^3)$ $I_y = \frac{\pi}{4}(ab^3 - a_1b_1^3)$	
33. Thin Hollow Ellipse 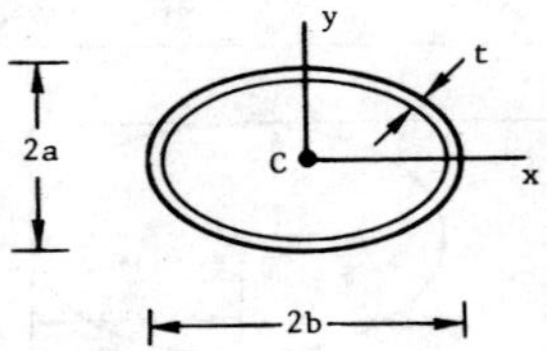	$x_C = 0$ $y_C = 0$ $A = \pi t(a + b)$	$I_x = \frac{\pi}{4}a^2(a + 3b)t$ $I_y = \frac{\pi}{4}b^2(b + 3a)t$	
34. Parabola 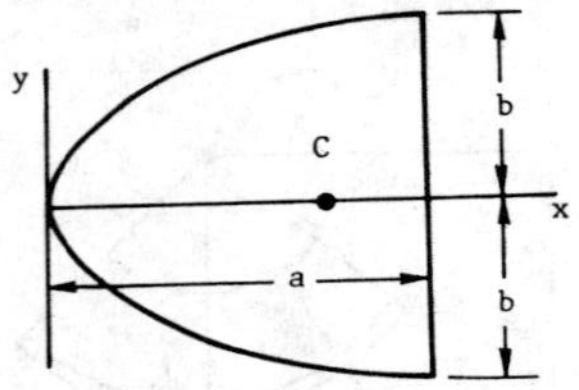	$x_C = \frac{3}{5}a$ $y_C = 0$ $A = \frac{4}{3}ab$	$I_{x_C} = I_x = \frac{4}{15}ab^3$ $I_{y_C} = \frac{16}{175}a^3b$ $I_y = \frac{4}{7}a^3b$	$I_{x_C y_C} = 0$ $I_{xy} = 0$

35. Semi-parabola 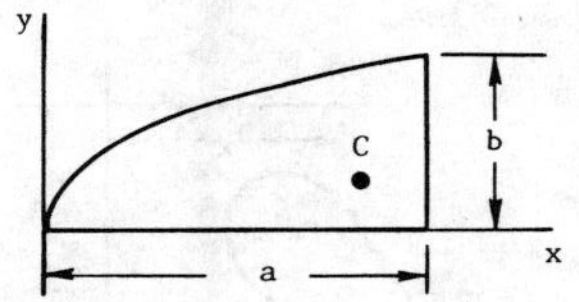	$x_C = \frac{3}{5}a$ $y_C = \frac{3}{8}b$ $A = \frac{2}{3}ab$	$I_x = \frac{2}{15}ab^3$ $I_y = \frac{2}{7}a^3b$	$I_{xy} = \frac{A}{4}ab = \frac{1}{6}a^2b^2$
36. n^{th} Degree Parabola 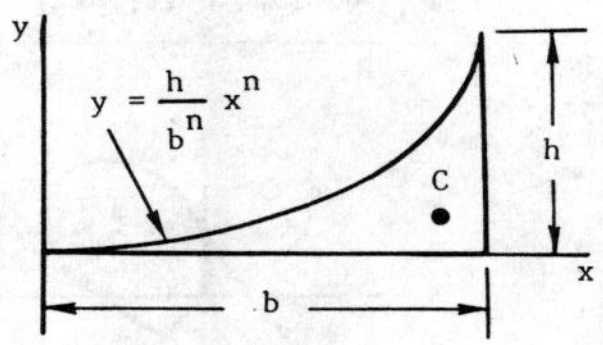	$x_C = \frac{n+1}{n+2}b$ $y_C = \frac{h}{2}\left(\frac{n+1}{2n+1}\right)$ $A = \frac{bh}{n+1}$	$I_x = \frac{bh^3}{3(3n+1)}$ $I_y = \frac{hb^3}{n+3}$	
37. n^{th} Degree Parabola 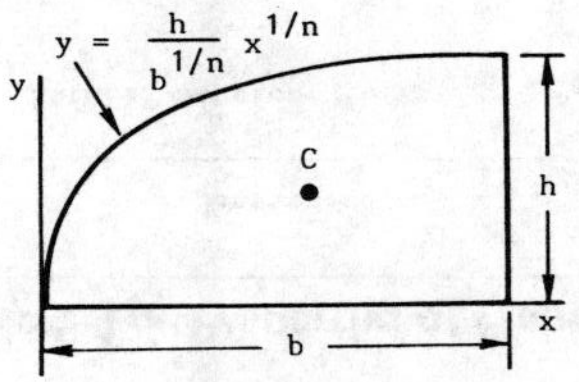	$x_C = \frac{n+1}{2n+1}b$ $y_C = \frac{n+1}{2(n+2)}h$ $A = \frac{n}{n+1}bh$	$I_x = \frac{n}{3(n+3)}bh^3$ $I_y = \frac{n}{3n+1}hb^3$	

Table 5-1. Properties of Plane Sections. (*Continued*)

Section	Area (A) and Distance from Axes to Centroid (C), x_C, y_C.	Area Moments of Inertia About Principal Axes	Area Products of Inertia
38. Rotated Section 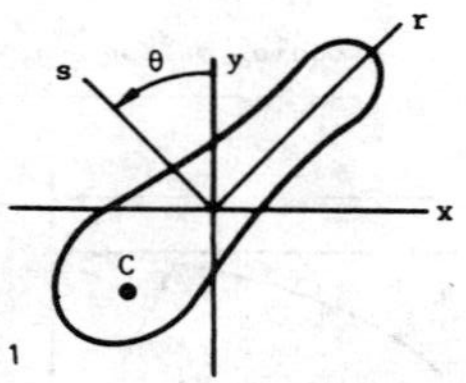	– –	$I_x = I_r \cos^2\theta + I_s \sin^2\theta$ $+ I_{rs} \sin 2\theta$ $I_y = I_r \sin^2\theta + I_s \cos^2\theta$ $- I_{rs} \sin 2\theta$	$I_{xy} = I_{rs} \cos 2\theta$ $- \frac{1}{2}(I_s - I_r) \sin 2\theta$ $\left(I_{xy} = 0 \text{ for } \theta = \frac{1}{2} \text{arc tan} \frac{2I_{rs}}{I_s - I_r}\right)$
39. Translated Section 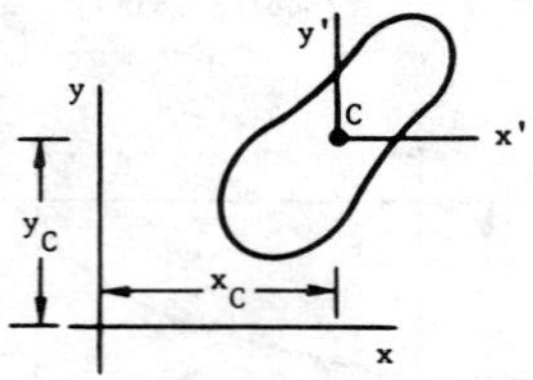	– – A	$I_x = I_{x_C} + y_C^2 A$ $I_y = I_{y_C} + x_C^2 A$ I_{x_C}, I_{y_C} and $I_{x_C y_C}$ are moments of inertia associated with x' and y' axes which are parallel to x and y axes, respectively.	$I_{xy} = I_{x_C y_C} + x_C y_C A$

5.2. SOLID BODIES

Figure 5-2 shows a solid body of volume V and density μ. dV is a small element of volume. The body is rigid, but the density may vary over the volume. The x–y–z axes form an orthogonal right-handed coordinate system with origin 0. C marks the center of gravity of the body and the origin of the coordinate system x′–y′–z′. The x–y–z coordinate system is translated with respect to the x′–y′–z′ system.

The mass of the body is:

$$M = \int_V \mu \, dV.$$

The location of the center of gravity with respect to the x–y–z coordinates is:

$$x_C = \frac{\int_V \mu x \, dV}{M},$$

$$y_C = \frac{\int_V \mu y \, dV}{M},$$

$$z_C = \frac{\int_V \mu z \, dV}{M}.$$

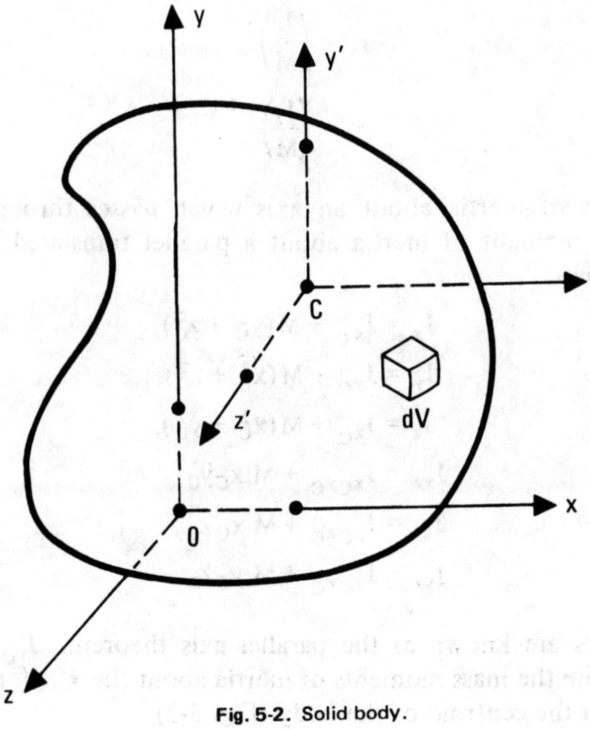

Fig. 5-2. Solid body.

The mass moments of inertia and mass products of inertia about the x, y, and z axes are defined by the following formulas:

$$J_x = \int_V \mu (y^2 + z^2)\, dV,$$

$$J_y = \int_V \mu (x^2 + z^2)\, dV,$$

$$J_z = \int_V \mu (x^2 + y^2)\, dV,$$

$$J_{xy} = \int_V \mu\, xy\, dV,$$

$$J_{xz} = \int_V \mu\, xz\, dV,$$

$$J_{yz} = \int_V \mu\, yz\, dV.$$

The radii of gyration of the body are:

$$r_x = \left(\frac{J_x}{M}\right)^{1/2},$$

$$r_y = \left(\frac{J_y}{M}\right)^{1/2},$$

$$r_z = \left(\frac{J_z}{M}\right)^{1/2}.$$

If the moment of inertia about an axis which passes through the centroid is known, then the moment of inertia about a parallel translated axis can easily be computed as follows:

$$J_x = J_{x_C} + M(y_C^2 + z_C^2),$$
$$J_y = J_{y_C} + M(x_C^2 + z_C^2),$$
$$J_z = J_{z_C} + M(x_C^2 + y_C^2),$$
$$J_{xy} = J_{x_C y_C} + M\, x_C y_C,$$
$$J_{xz} = J_{x_C z_C} + M\, x_C z_C,$$
$$J_{yz} = J_{y_C z_C} + M\, y_C z_C.$$

These relationships are known as the parallel axis theorem. J_{x_C}, J_{y_C}, J_{z_C}, $J_{x_C y_C}$, $J_{x_C z_C}$, and $J_{y_C z_C}$ are the mass moments of inertia about the x′, y′, and z′ axes, which have their origin at the centroid of the body (Fig. 5-2).

The mass moment of inertia about a rotated axis may be computed from the mass moments of inertia about the base coordinate system and the unit vector in the direction of the rotated axis. If the unit vector in the direction of the rotated axis is $\hat{r}$, shown in Fig. 5-3, then $\hat{r}$ can be written as:

$$\hat{r} = l\hat{i} + m\hat{j} + n\hat{k},$$

where $\hat{i}$, $\hat{j}$, and $\hat{k}$ are unit vectors in the x, y, and z coordinate directions, respectively, and l, m, and n are scalers such that:

$$l^2 + m^2 + n^2 = 1.$$

The moment of inertia about the $\hat{r}$ axis is (Ref. 5-5):

$$J_r = l^2 J_x + m^2 J_y + n^2 J_z - 2lm\, J_{xy} - 2ln\, J_{xz} - 2mn\, J_{yz}.$$

If a unit vector $\hat{s}$ is defined such that $\hat{s}$ is perpendicular to $\hat{r}$,

$$\hat{s} = l'\hat{i} + m'\hat{j} + n'\hat{k},$$

then it follows that

$$ll' + mm' + nn' = 0$$

and

$$l'^2 + m'^2 + n'^2 = 1.$$

The mass product of inertia with respect to the r–s axes is (Ref. 5-5):

$$J_{rs} = -ll'\, J_x - mm'\, J_y - nn'\, J_z + (lm' + ml')\, J_{xy} + (ln' + nl')\, J_{xz} + (mn' + nm')\, J_{yz}.$$

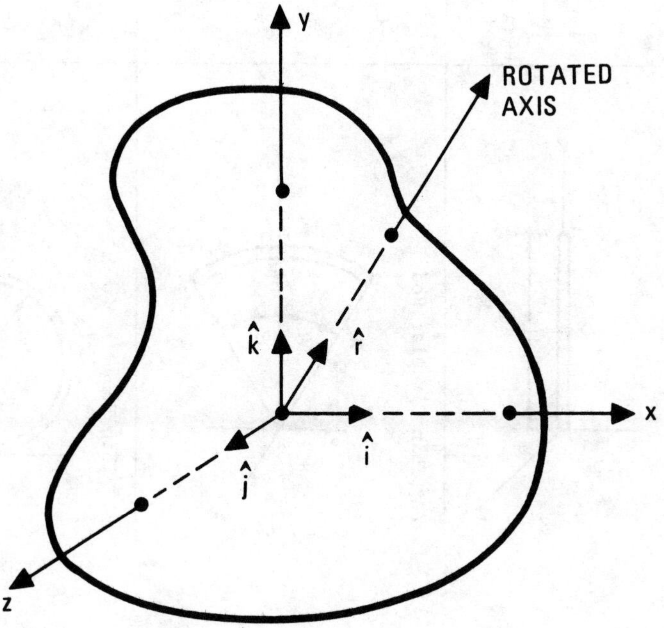

Fig. 5-3. Rotated axis and unit vectors.

Table 5-2. Properties of Homogeneous Bodies.

Notation: C locates centroid (center of mass and gravity); μ = density (mass/volume)

	Mass (M) & Location of Center of Mass (C)	Mass Moment of Inertia	Mass Products of Inertia
1. Thin Rod 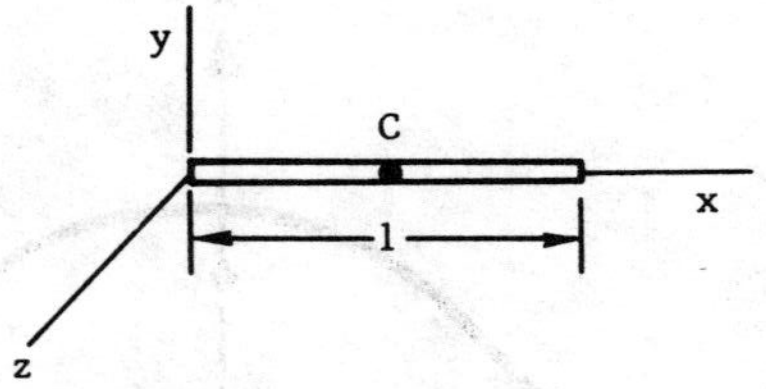	$x_C = \frac{L}{2}$ $y_C = 0$ $M = \mu A l$ (A = area of cross section)	$J_x = J_{x_C} = 0$ $J_{y_C} = J_{z_C} = \frac{M}{12} L^2$ $J_y = J_z = \frac{M}{3} L^2$	$J_{x_C y_C}$, etc. = 0 J_{xy}, etc. = 0
2. Thin Circular Rod 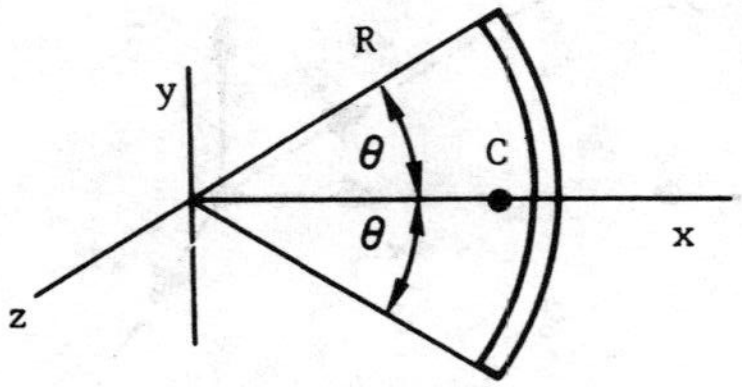	$x_C = \frac{R \sin\theta}{\theta}$ $y_C = 0$ $z_C = 0$ $M = 2\mu R\theta A$ (A = area of cross section)	$J_x = J_{x_C} = \frac{MR^2(\theta - \sin\theta\cos\theta)}{2\theta}$ $J_y = \frac{MR^2(\theta + \sin\theta\cos\theta)}{2\theta}$ $J_z = MR^2$	$J_{x_C y_C}$, etc. = 0 J_{xy}, etc. = 0
3. Thin Hoop 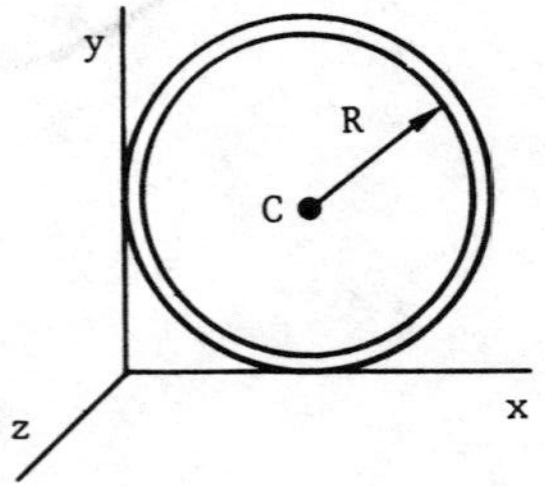	$x_C = R$ $y_C = R$ $z_C = 0$ $M = 2\pi\mu RA$ (A = area of cross section)	$J_{x_C} = J_{y_C} = \frac{MR^2}{2}$ $J_{z_C} = MR^2$ $J_x = J_y = \frac{3}{2} MR^2$ $J_z = 3\,MR^2$	$J_{x_C y_C}$, etc. = 0 $J_{xy} = MR^2$ $J_{xz} = J_{yz} = 0$

4. Torus 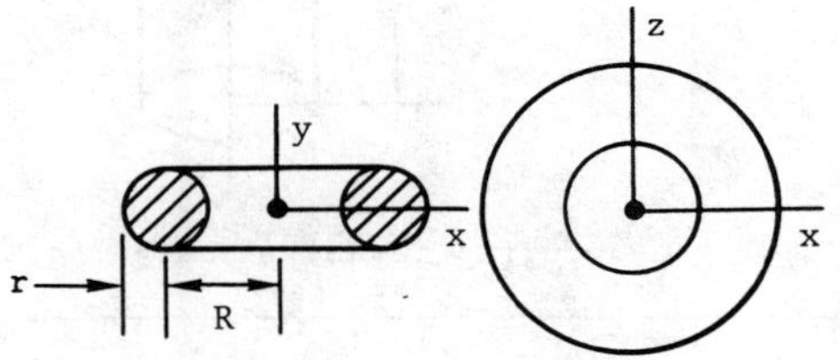	$x_C = 0$ $y_C = 0$ $z_C = 0$ $M = 2\mu\pi^2 Rr^2$	$J_x = J_z = \frac{M}{2}(R^2 + \frac{5}{4} r^2)$ $J_y = \frac{M}{4}(4 R^2 + 3 r^2)$	J_{xy}, etc. = 0
5. Cube 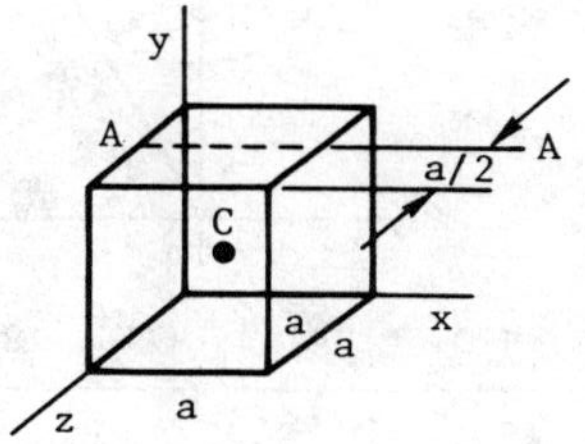	$x_C = \frac{1}{2} a$ $y_C = \frac{1}{2} a$ $z_C = \frac{1}{2} a$ $M = \mu a^3$	$J_{x_C} = J_{y_C} = J_{z_C} = \frac{Ma^2}{6}$ $J_x = J_y = J_z = \frac{2}{3} Ma^2$ $J_{AA} = \frac{5}{12} Ma^2$	$J_{x_C y_C}$, etc. = 0 J_{xy}, etc. = $\frac{Ma^2}{4}$
6. Rectangular Prism 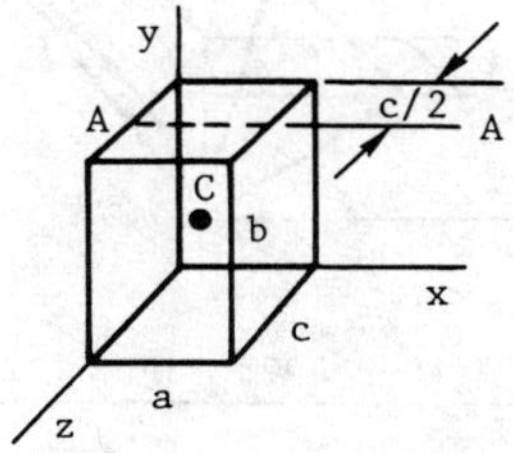	$x_C = \frac{a}{2}$ $y_C = \frac{b}{2}$ $z_C = \frac{c}{2}$ $M = \mu abc$	$J_{x_C} = \frac{M}{12}(b^2 + c^2)$ $J_x = \frac{M}{3}(b^2 + c^2)$ $J_{AA} = \frac{M}{12}(4 b^2 + c^2)$	$J_{x_C y_C}$, etc. = 0 $J_{xy} = \frac{Mab}{4}$ $J_{xz} = \frac{Mac}{4}$ $J_{yz} = \frac{Mbc}{4}$

Table 5-2. Properties of Homogeneous Bodies. *(Continued)*

	Mass (M) & Location of Center of Mass (C)	Mass Moment of Inertia	Mass Products of Inertia
7. Right Rectangular Pyramid 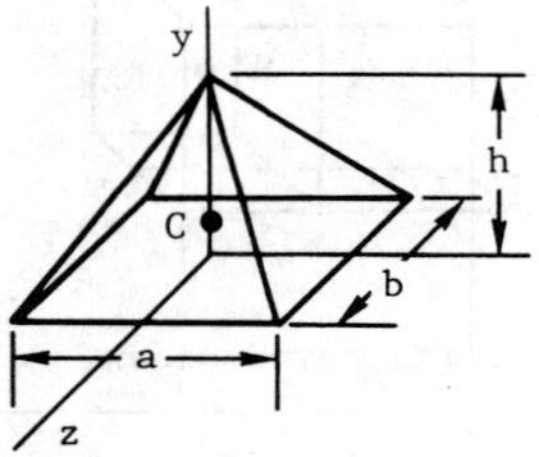	$x_C = 0$ $y_C = \frac{h}{4}$ $z_C = 0$ $M = \frac{1}{3}\ \mu abh$	$J_{x_C} = \frac{M}{80}\ (4\ b^2 + 3\ h^2)$ $J_{y_C} = J_y = \frac{M}{20}\ (a^2 + b^2)$ $J_x = \frac{M}{20}\ (b^2 + 2\ h^2)$ $J_z = \frac{M}{20}\ (a^2 + 2\ h^2)$	$J_{x_C y_C}$, etc. = 0 J_{xy}, etc. = 0
8. Right Circular Cone 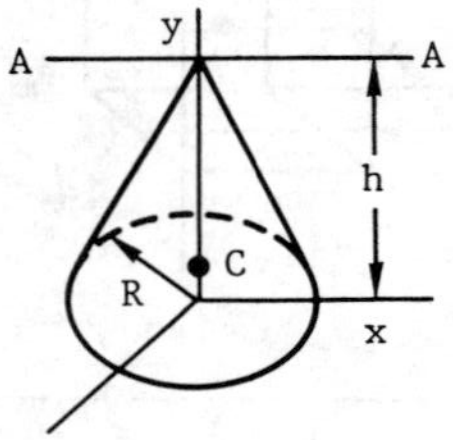	$x_C = 0$ $y_C = \frac{h}{4}$ $z_C = 0$ $M = \frac{\pi}{3}\ \mu R^2 h$	$J_{x_C} = J_{z_C} = \frac{3}{80}\ M\ (4\ R^2 + h^2)$ $J_{y_C} = J_y = \frac{3}{10}\ MR^2$ $J_x = J_z = \frac{1}{20}\ M\ (3\ R^2 + 2\ h^2)$ $J_{AA} = \frac{3}{20}\ M\ (R^2 + 4\ h^2)$	$J_{x_C y_C}$, etc. = 0 J_{xy}, etc. = 0
9. Right Circular Cylinder 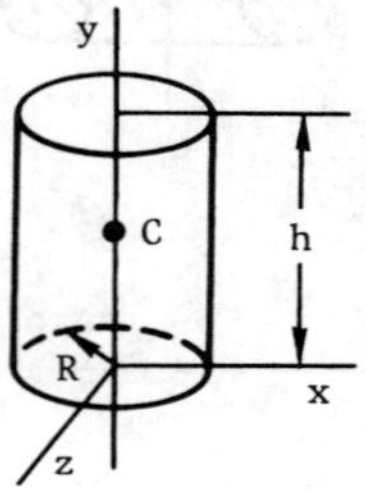	$x_C = 0$ $y_C = \frac{h}{2}$ $z_C = 0$ $M = \mu\pi R^2 h$	$J_{x_C} = J_{z_C} = \frac{M}{12}\ (3\ R^2 + h^2)$ $J_{y_C} = J_y = \frac{MR^2}{2}$ $J_x = J_z = \frac{M}{12}\ (3\ R^2 + 4\ h^2)$	$J_{x_C y_C}$, etc. = 0 J_{xy}, etc. = 0

10. Hollow Right Circular Cylinder 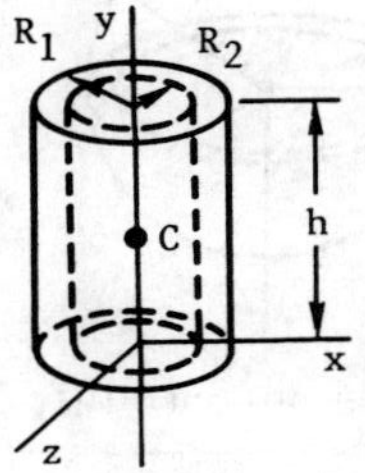	$x_C = 0$ $y_C = \frac{h}{2}$ $z_C = 0$ $M = \mu\pi h\ (R_1^2 - R_2^2)$	$J_{x_C} = J_{z_C} = \frac{M}{12}\ (3\ R_1^2 + 3\ R_2^2 + h^2)$ $J_{y_C} = J_y = \frac{M}{2}\ (R_1^2 + R_2^2)$ $J_x = J_z = \frac{M}{12}\ (3\ R_1^2 + 3\ R_2^2 + 4\ h^2)$	$J_{x_C y_C}$, etc. = 0 J_{xy}, etc. = 0
11. Thin Hollow Right Circular Cylinder 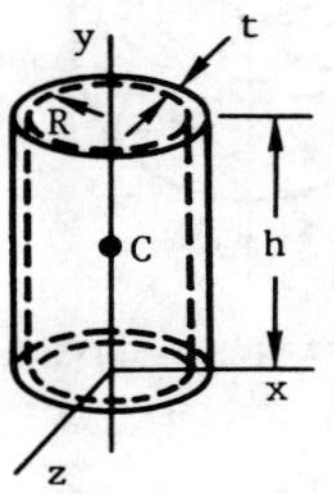	$x_C = 0$ $y_C = \frac{h}{2}$ $z_C = 0$ $M = 2\ \mu\pi Rt$	$J_{x_C} = J_{z_C} = \frac{M}{12}\ (6\ R^2 + h^2)$ $J_{y_C} = J_y = MR^2$ $J_x = J_z = \frac{M}{6}\ (3\ R^2 + 2\ h^2)$	$J_{x_C y_C}$, etc. = 0 J_{xy}, etc. = 0
12. Sphere 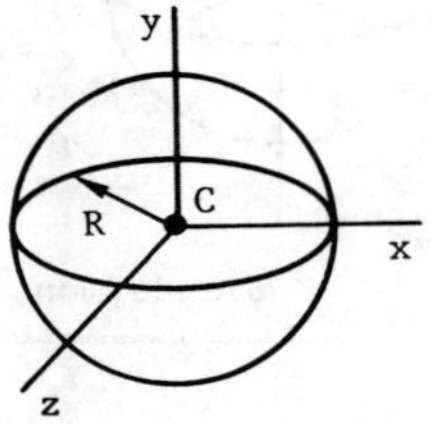	$x_C = 0$ $y_C = 0$ $z_C = 0$ $M = \frac{4}{3}\ \mu\pi R^3$	$J_x = J_y = J_z = \frac{2}{5}\ MR^2$	$J_{x_C y_C}$, etc. = 0

Table 5-2. Properties of Homogeneous Bodies. *(Continued)*

	Mass (M) & Location of Center of Mass (C)	Mass Moment of Inertia	Mass Products of Inertia
13. Hemisphere 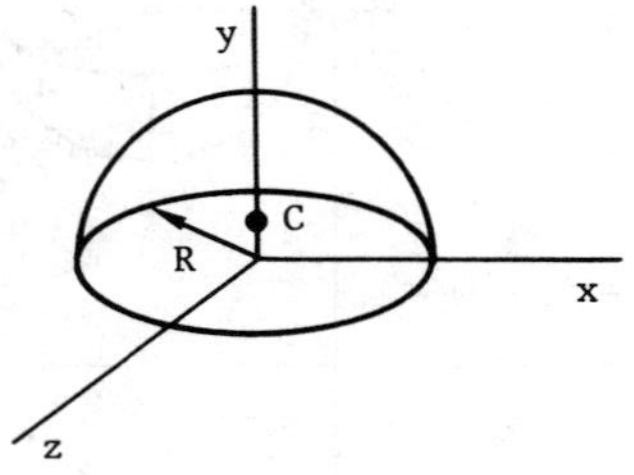	$x_C = 0$ $y_C = \frac{3}{8} R$ $z_C = 0$ $M = \frac{2}{3} \mu\pi R^3$	$J_x = J_y = J_z = \frac{2}{5} MR^2$	$J_{x_C y_C}$, etc. = 0 J_{xy}, etc. = 0
14. Hollow Sphere 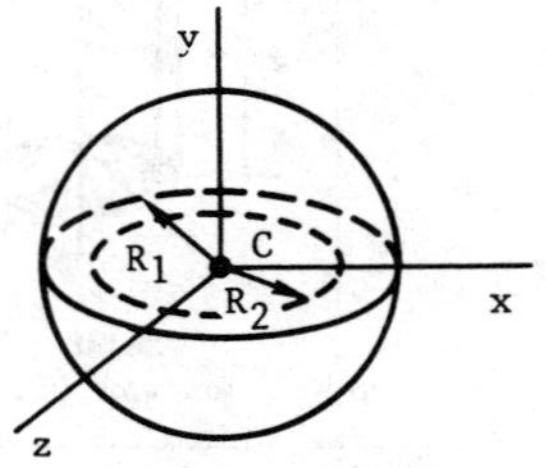	$x_C = 0$ $y_C = 0$ $z_C = 0$ $M = \frac{4}{3} \mu\pi (R_1^3 - R_2^3)$	$J_x = J_y = J_z = \frac{2}{5} M \left(\frac{R_1^5 - R_2^5}{R_1^3 - R_2^3}\right)$	J_{xy}, etc. = 0
15. Thin Hollow Sphere 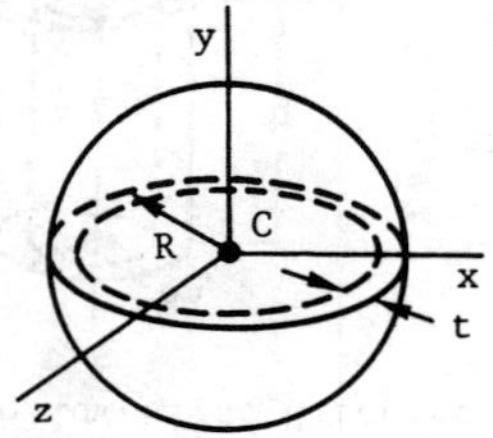	$x_C = 0$ $y_C = 0$ $z_C = 0$ $M = 4 \mu\pi R^2 t$	$J_x = J_y = J_z = \frac{2}{3} MRt$	J_{xy}, etc. = 0

<table>
<tr><td>16. Ellipsoid
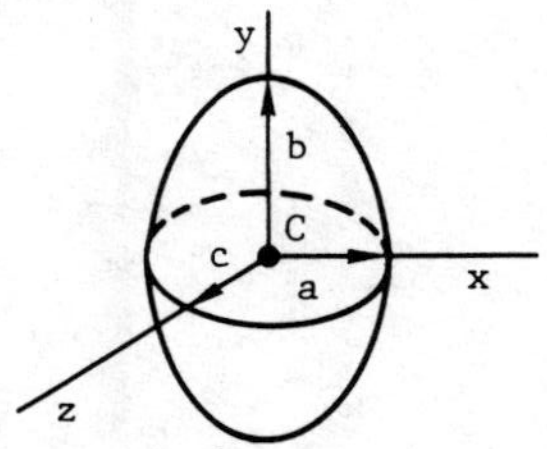</td><td>$x_C = 0$
$y_C = 0$
$z_C = 0$
$M = \frac{4}{3}\mu\pi abc$</td><td>$J_x = \frac{M}{5}(b^2 + c^2)$
$J_y = \frac{M}{5}(a^2 + c^2)$
$J_z = \frac{M}{5}(a^2 + b^2)$</td><td>$J_{xy}$, etc. = 0</td></tr>
<tr><td>17. Paraboloid of Revolution
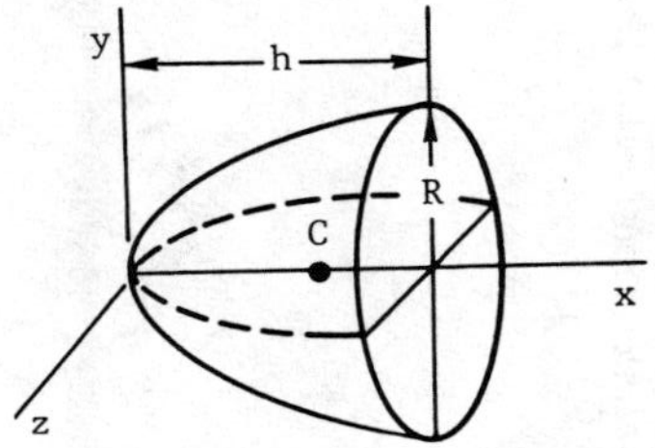</td><td>$x_C = \frac{2}{3}h$
$y_C = 0$
$z_C = 0$
$M = \frac{\pi}{2}\mu R^2 h$</td><td>$J_x = J_{x_C} = \frac{1}{3}MR^2$
$J_{y_C} = J_{z_C} = \frac{M}{18}(3R^2 + h^2)$
$J_y = J_z = \frac{M}{6}(R^2 + 3h^2)$</td><td>$J_{x_C y_C}$, etc. = 0
J_{xy}, etc. = 0</td></tr>
<tr><td>18. Elliptic Paraboloid
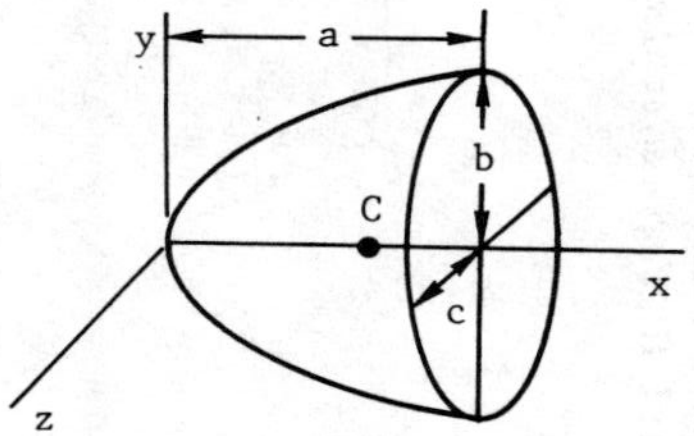</td><td>$x_C = \frac{2}{3}a$
$y_C = 0$
$z_C = 0$
$M = \frac{\pi}{2}\mu abc$</td><td>$J_{x_C} = J_x = \frac{M}{6}(b^2 + c^2)$
$J_{y_C} = \frac{M}{18}(3c^2 + a^2)$
$J_{z_C} = \frac{M}{18}(3b^2 + a^2)$
$J_y = \frac{M}{6}(c^2 + 3a^2)$
$J_z = \frac{M}{6}(b^2 + 3a^2)$</td><td>$J_{x_C y_C}$, etc. = 0
J_{xy}, etc. = 0</td></tr>
</table>

There exists a set of axes through any given origin about which the products of inertia of a body are zero. These axes are called the principal axes of the body. The orientation of the principal axes can generally be found by setting the products of inertia to zero and solving for the vector coefficients l, m, n, l′, m′, n′. However, any axis of symmetry must be a principal axis. If one of the principal axes is known, then the orientation of the remaining axes can be found by the techniques discussed for plane sections. The principal moments of inertia are the maximum and minimum moments of inertia which can be measured about any axis that passes through a given point.

The mass moments of inertia of complex bodies can often be computed by subdividing the body into homogeneous rectangular prisms, cylinders, and spheres and then summing the mass moments of inertia of the component bodies and subtracting off the moments of inertia of voids. The mass moments of inertia of various homogeneous (i.e., μ = constant) bodies are given in Table 5-2.

The mass moments of inertia of many solid bodies of uniform thickness and density, such as plates and disks, can easily be adapted from the area moments of inertia given in Table 5-1 for plane sections. For example, the mass moment of inertia of a disk of radius R and thickness t about an axis through its center of gravity and perpendicular to the face of the disk is:

$$J_z = \mu t \left(I_{x_C} + I_{y_C}\right) = \frac{\pi}{2} \mu t R^4,$$

where I_{x_C} and I_{y_C} are given in frame 20 of Table 5-1.

REFERENCES

5-1. *Alcoa Structural Handbook*, Aluminum Company of America, Pittsburgh, Pa.

5-2. *Manual of Steel Construction*, American Institute of Steel Construction, Inc., New York, N.Y.

5-3. *Timber Construction Manual*, American Institute of Timber Construction, John Wiley & Sons, N.Y., 1974.

5-4. *Aluminum Construction Manual*, Aluminum Association, New York, N.Y.

5-5. Shames, I. H., *Engineering Mechanics, Dynamics*, Prentice-Hall, Englewood Cliffs, N.J., 1966, pp. 549–551.

6

SPRING AND PENDULUM SYSTEMS

6.1. ASSUMPTIONS

Spring and pendulum systems are the simplest dynamic systems. Unlike continuous structures such as cables or beams, these systems possess a finite number of natural frequencies and their dynamic behavior is described by second order ordinary differential equations. The practical importance of spring and pendulum systems lies in their ability to model and illustrate the behavior of more complex structures, rather than their own application.

The general assumptions used in the analysis of spring and pendulum systems in this chapter are:

1. Point masses have zero moment of inertia for rotation about the center of mass of the point mass.
2. All masses are rigid bodies.
3. Springs, gears, strings, and pivots are massless unless otherwise noted.
4. All springs are linear.
5. Deflections of pendulums are sufficiently small so that the restoring torque of gravity is proportional to the deflection of the pendulum.
6. The tension preload in string-mass systems is much greater than dynamic variation in tension during vibration.
7. The deformation of pulley cables is negligible.

6.2. SPRING CONSTANTS

Spring constants are defined as the force required to produce a unit deformation in a linear structure. Spring constants for the deflection and torsion of structural springs are given in Table 6-1. In all cases the rate of spring loading is assumed to be well below the fundamental natural frequency of the spring. Thus, the spring constant can be accurately predicted from static analysis. All the spring beams in Table 6-1 are assumed to be uniform and homogeneous, and have a length-to-depth ratio of at least 10 to 1 so that the approximations used in the slender beam stress analysis hold true.

Any linear structure subject to a static load can be represented by an equivalent spring. Reference 6-1 presents a large number of formulas for deflection of loaded structures which can be interpreted in terms of spring constants. Reference 6-2 presents numerous data for mechanical springs.

Table 6-1. Spring Stiffnesses.

Notation: I = area moment of inertia; $\mathfrak{M}$ = moment; θ = angle of rotation;
E = modulus of elasticity; G = shear modulus; see Table 3-1 for consistent sets of units

Geometry	Spring Constant (force/unit deflection at point of load)
1. Axial Spring 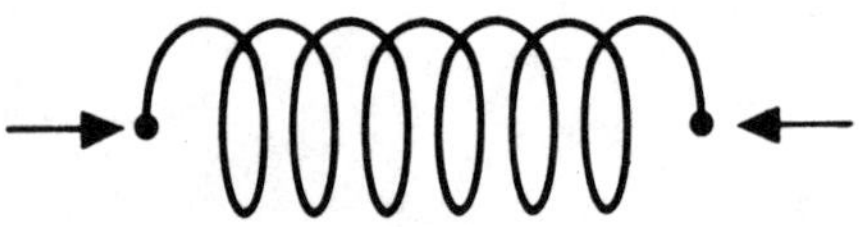	k
2. Two Axial Springs in Series 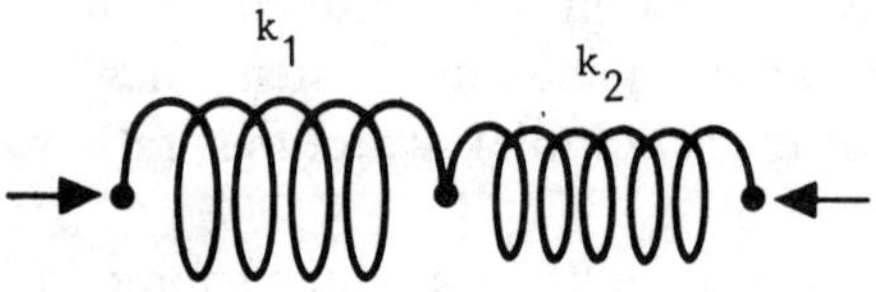	$\frac{k_1 k_2}{k_1 + k_2}$
3. N Axial Springs in Series 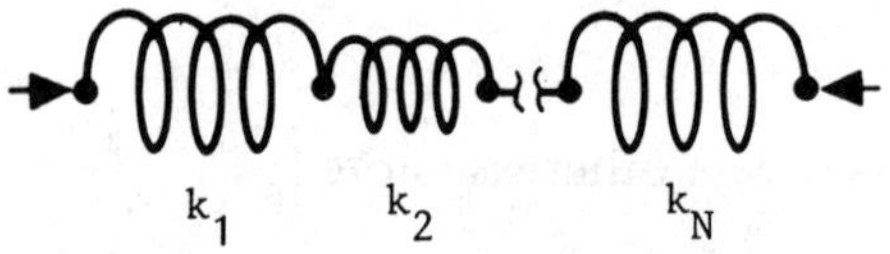	$\frac{1}{\frac{1}{k_1} + \frac{1}{k_2} + \cdots + \frac{1}{k_N}}$
4. Two Axial Springs in Parallel, Tilting Bar 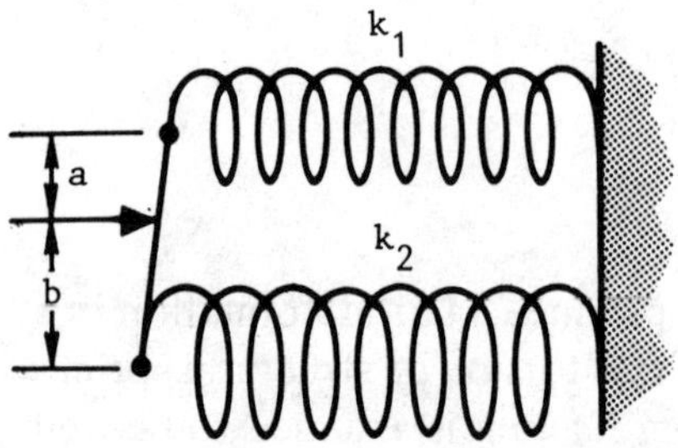	$\frac{k_1 k_2 (a + b)^2}{k_1 a^2 + k_2 b^2}$
5. Two Axial Springs in Parallel 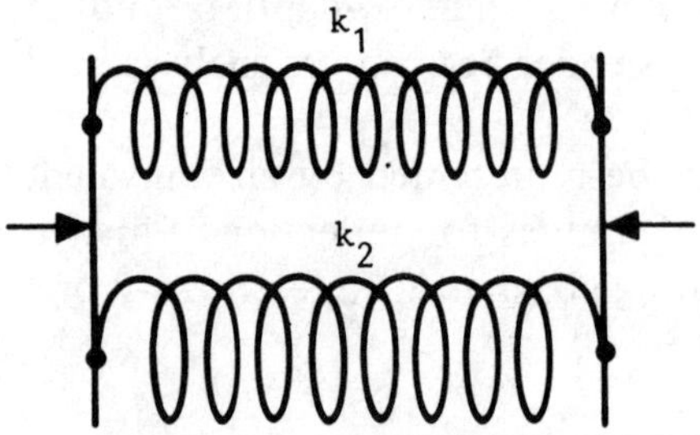	$k_1 + k_2$ Bars do not tilt.

Table 6-1. Spring Stiffness. (*Continued*)

Geometry	Spring Constant (force/unit deflection at point of load)
6. Two Parallel Axial Springs, One Series Axial Spring 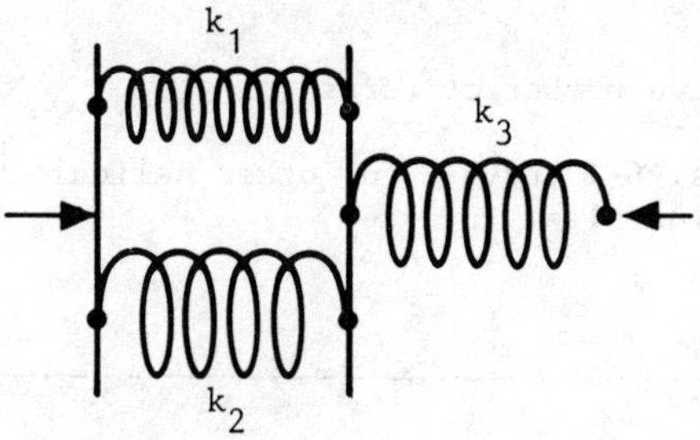	$\frac{k_1k_3 + k_2k_3}{k_1 + k_2 + k_3}$ Bars do not tilt.
7. N Axial Springs in Parallel 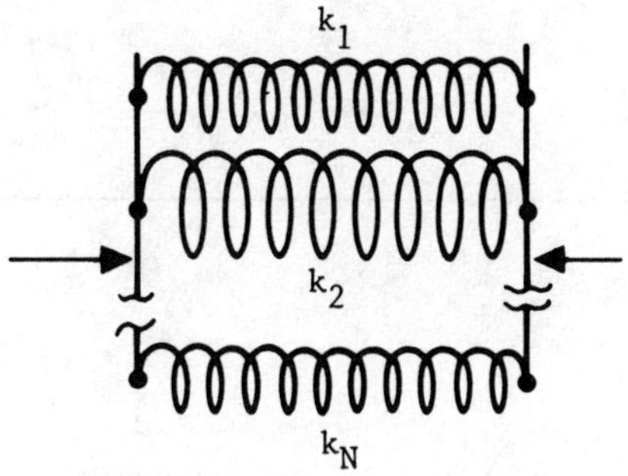	$k_1 + k_2 + \cdots k_N$ Bars do not tilt.
8. Stretching Bar 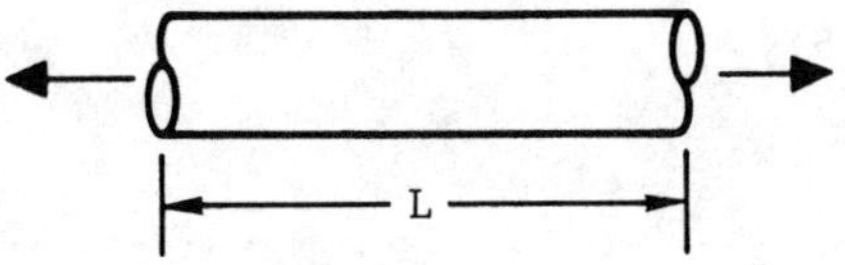	$\frac{EA}{L}$ A = cross-sectional area E = elastic modulus
9. Tapered Bar 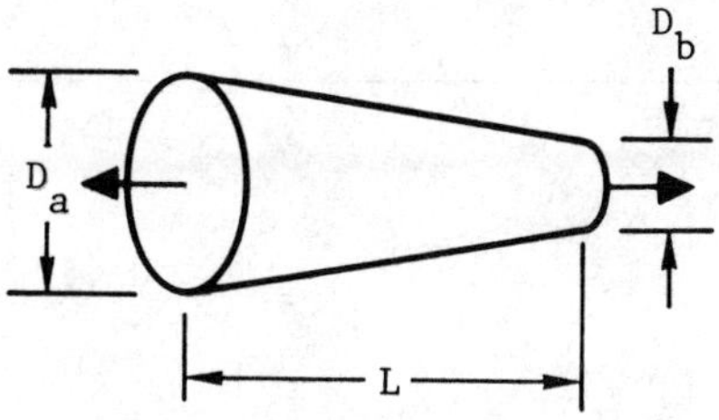	$\frac{\pi E D_a D_b}{4L}$

Table 6-1. Spring Stiffness. *(Continued)*

Geometry	Spring Constant (force/unit deflection at point of load)
10. Axial Helical Spring 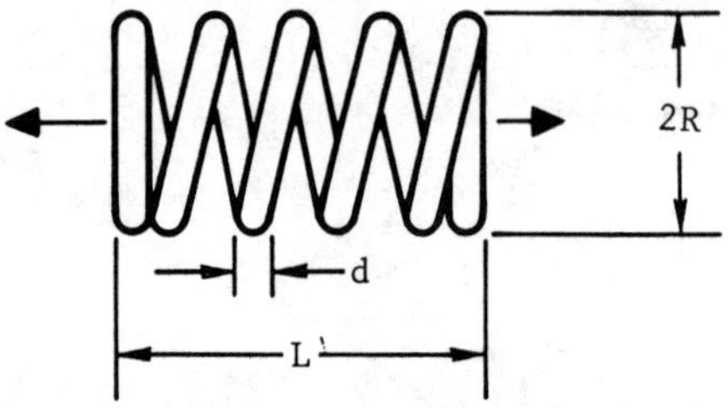	$\frac{Gd^4}{64nR^3}$ for $L >> R$ n = active number of turns See Refs. 6-1 and 6-2 for other helical spring formulas.
11. Cantilever Beam, End Load 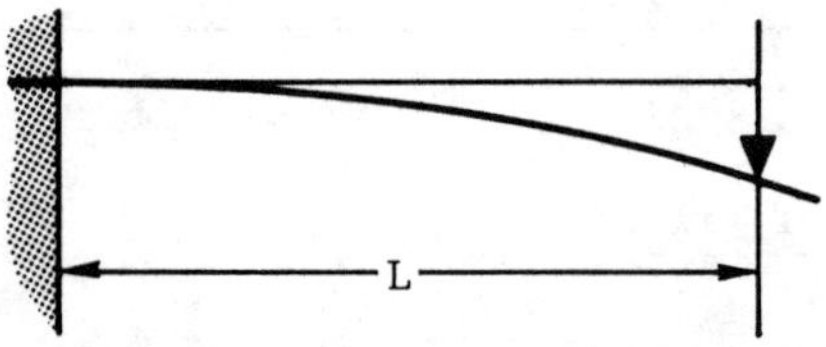	$\frac{3EI}{L^3}$
12. Pinned-Pinned Beam, Center Load 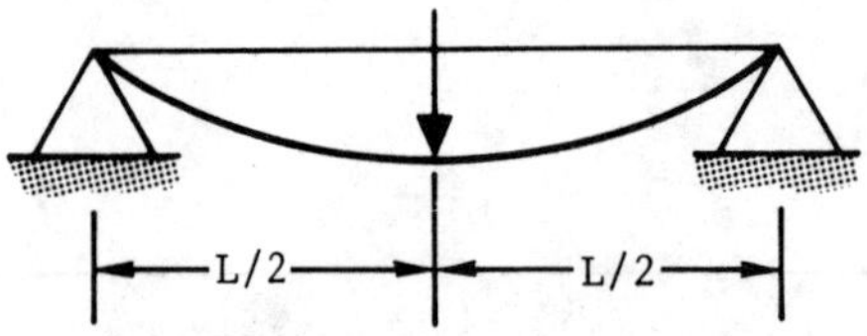	$\frac{48EI}{L^3}$
13. Pinned-Pinned Beam, Off-Center Load 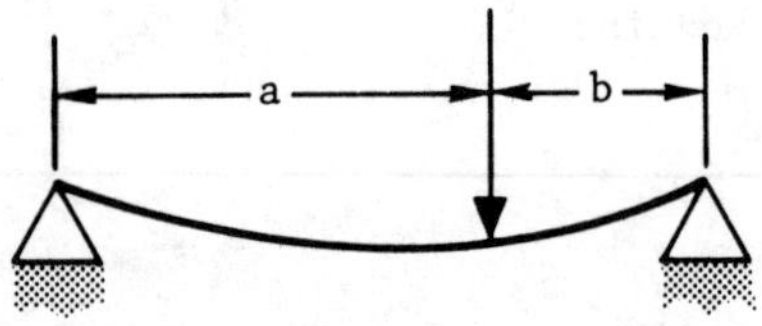	$\frac{3EI\ (a + b)}{a^2b^2}$
14. Clamped-Clamped Beam, Center Load 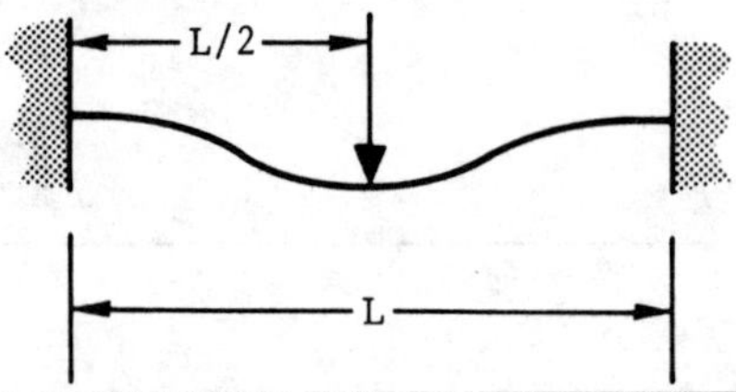	$\frac{192EI}{L^3}$

Table 6-1. Spring Stiffness. *(Continued)*

Geometry	Spring Constant (force/unit deflection at point of load)
15. Clamped-Clamped Beam, Off-Center Load 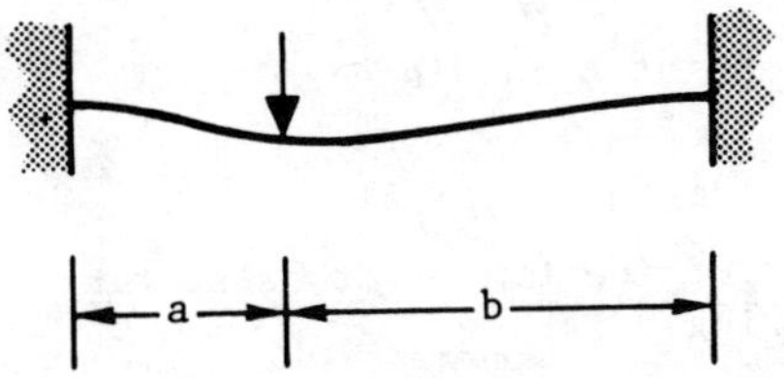 	$\frac{3EI\ (a+b)^3}{a^3 b^3}$
16. Clamped-Pinned Beam, Center Load 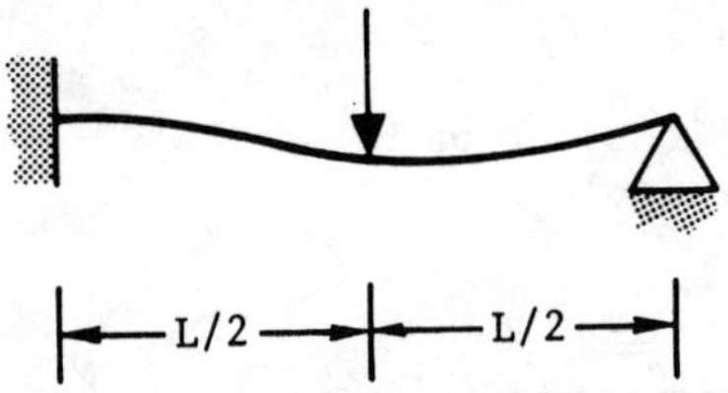 	$\frac{768EI}{7L^3}$
17. Two-Beam Support 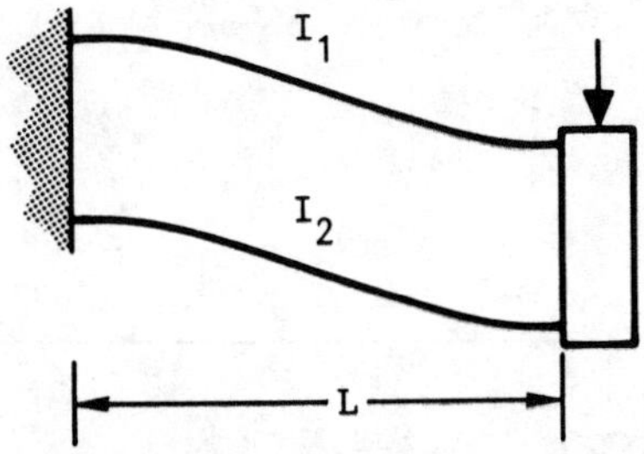 	$\frac{12E\ (I_1 + I_2)}{L^3}$ Beams are same length and they are offset. The ends of the beam define two planes that are always parallel.
18. N Beam Support, $N \geq 2$ 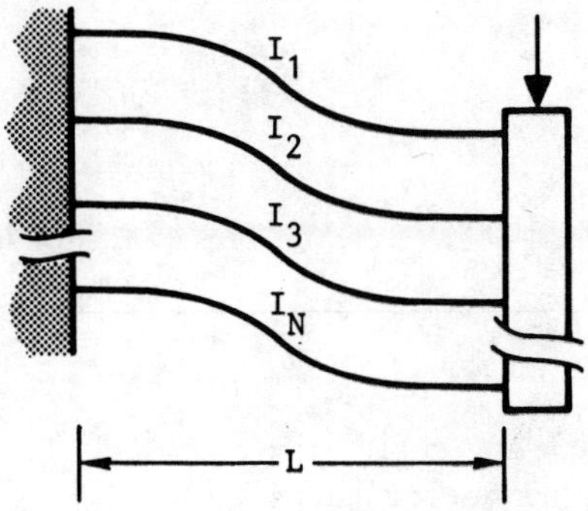 	$\frac{12EI_S}{L^3}$ $I_S = I_1 + I_2 + \cdots + I_N$ The beams are of equal length and they are offset. The ends of the beams define two planes which are always parallel.

Table 6-1. Spring Stiffness. *(Continued)*

Geometry	Spring Constant (moment/angular deflection in radians)
19. Spiral Torsion Spring 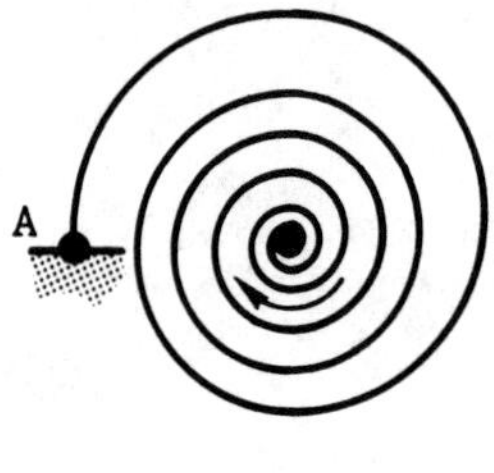	$\frac{EI}{L}$ if point A is clamped $0.8\ \frac{EI}{L}$ if point A is pinned L = total length of spiral I = moment of inertia of cross section of spring Ref. 6-2
20. Torsion of a Uniform Shaft 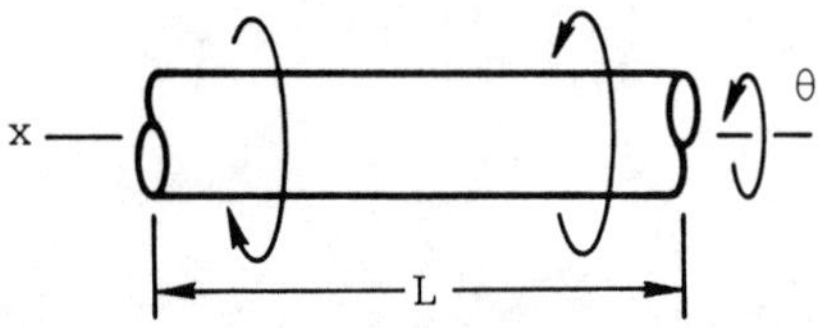	$\frac{GC}{L}$ C = torsion constant of cross section (Table 8-18)
21. Torsion of a Tapered Circular Shaft 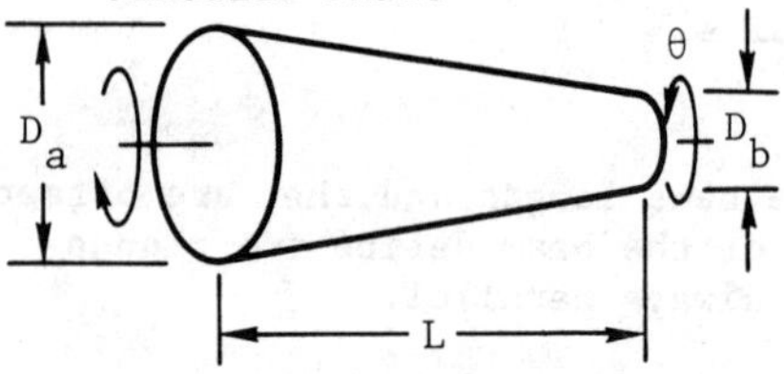	$\frac{3\pi}{32}\ \frac{D_b^4\ G}{L[D_b/D_a + (D_b/D_a)^2 + (D_b/D_a)^3]}$
22. Bending Helical Spring 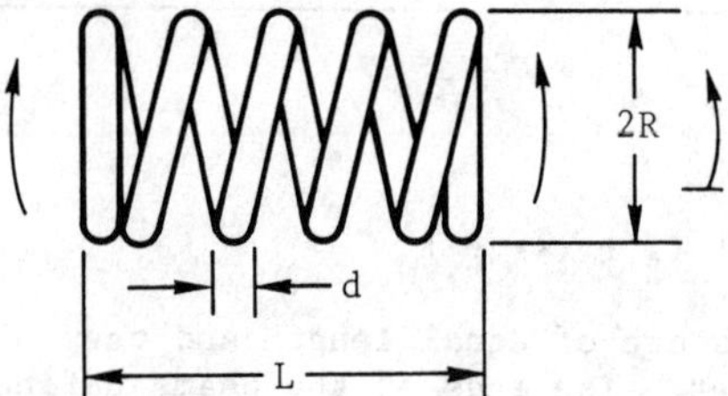	$\frac{E\ d^4}{64nR}\left(\frac{1}{1 + E/2G}\right)$ for L >> R n = active number of turns
23. Torsion of a Helical Spring 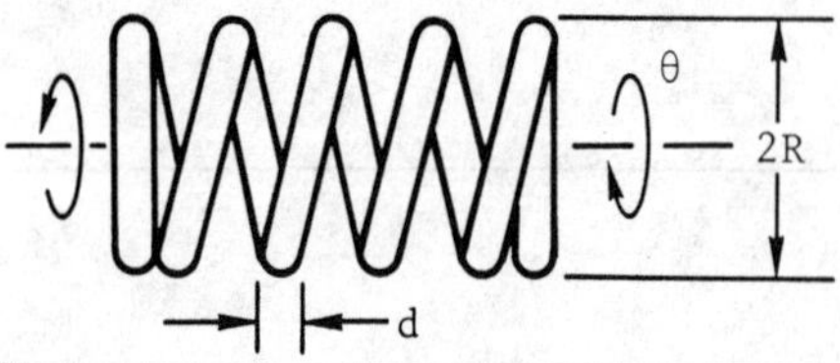	$\frac{E\ d^4}{64nR}$ n = active number of turns

Table 6-1. Spring Stiffness. *(Continued)*

Geometry	Spring Constant (moment/angular deflection in radians)
24. Torsion of Two Springs in Series 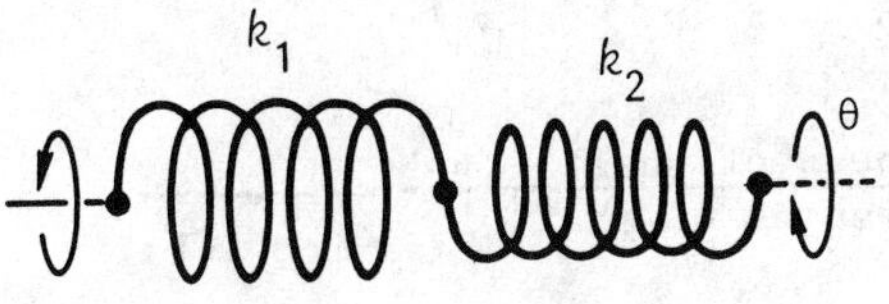	$\dfrac{k_1 k_2}{k_1 + k_2}$
25. Torsion of N Springs in Series 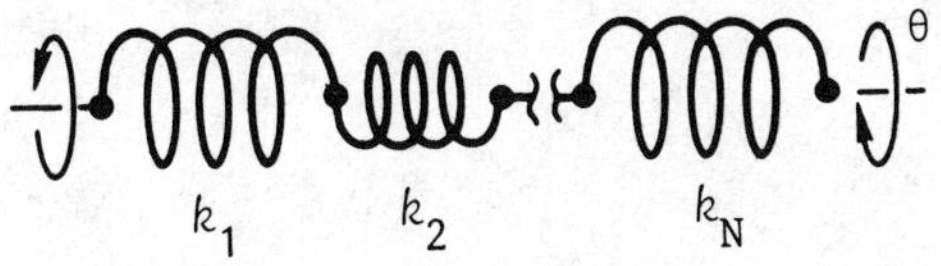	$\dfrac{1}{\frac{1}{k_1} + \frac{1}{k_2} + \cdots \frac{1}{k_N}}$
26. Torsion of Two Nested Springs (in parallel) 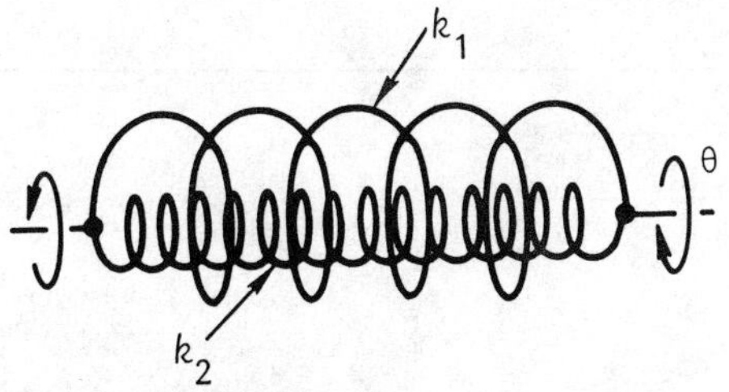	$k_1 + k_2$
27. Torsion of N Nested Springs (in parallel) 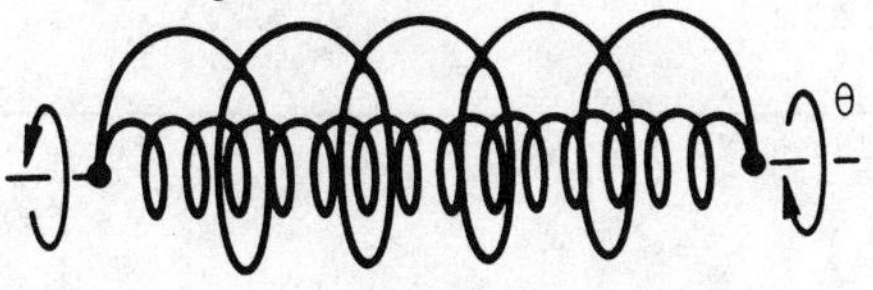	$k_1 + k_2 + \cdots k_N$
28. Torsion of Two Geared Shafts 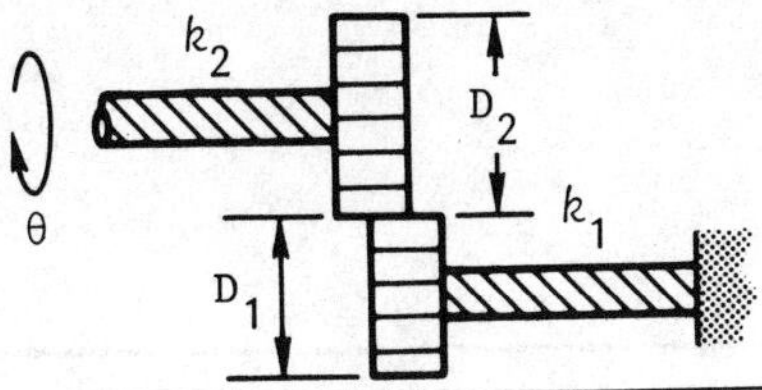	$\dfrac{k_1 k_2}{k_1 + n^2 k_2}$ $n = \dfrac{D_1}{D_2} = \dfrac{\text{speed of shaft 2}}{\text{speed of shaft 1}}$

Table 6-1. Spring Stiffness. (*Continued*)

Geometry	Spring Constant (moment/angular deflection in radians)
29. Torsion of N Geared Shafts 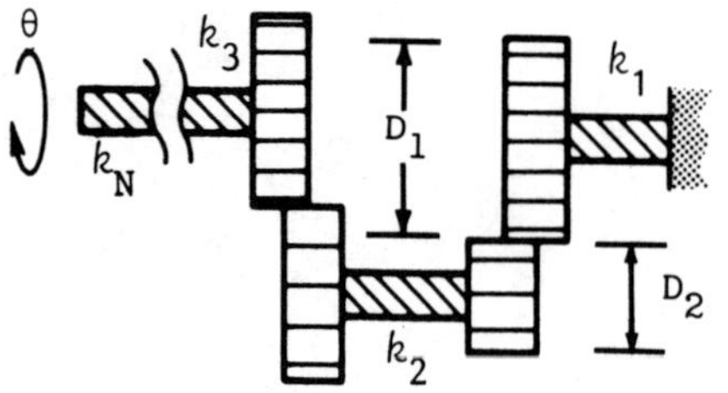	$\left(\frac{n_1^2}{k_1}+\frac{n_2^2}{k_2}+\frac{n_3^2}{k_3}+\ldots+\frac{n_N^2}{k_N}\right)^{-1}$ $n_i = \frac{\text{speed of shaft } i + 1}{\text{speed of shaft } i} = \frac{D_i}{D_{i+1}}$
30. Torsion of Pinned-Pinned Beam, Center 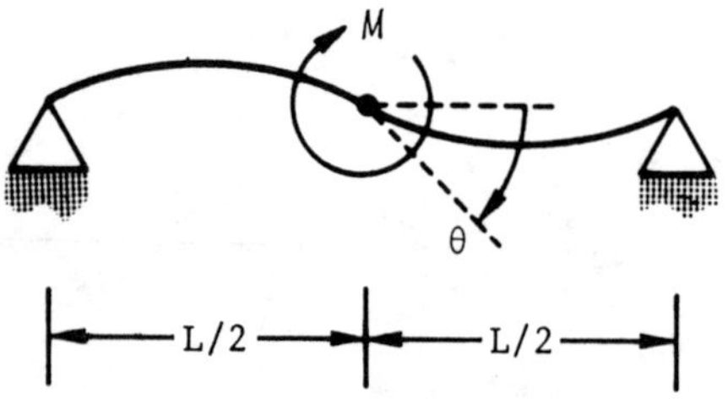	$\frac{12EI}{L}$
31. Torsion of Pinned-Pinned Beam, Off Center 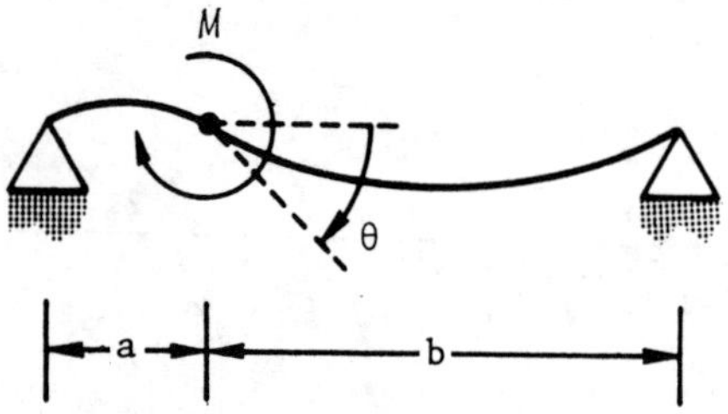	$\frac{EI}{\frac{a^2}{a+b}+\frac{b}{3}-\frac{2a}{3}}$
32. Torsion of Clamped-Clamped Beam, Center 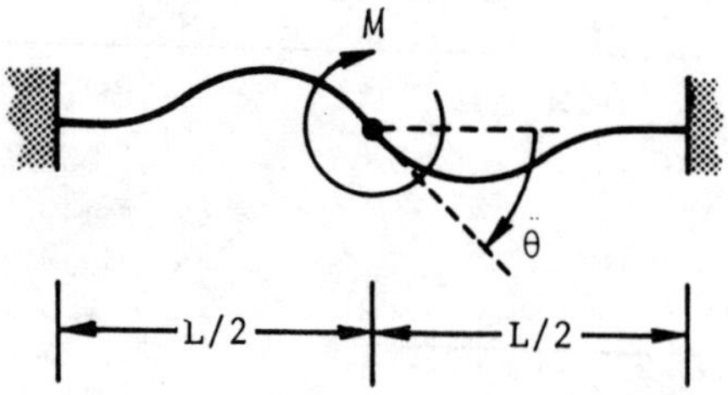	$\frac{16EI}{L}$

Table 6-1. Spring Stiffness. *(Continued)*

Geometry	Spring Constant (moment/angular deflection in radians)
33. Torsion of Clamped-Clamped Beam, Off Center 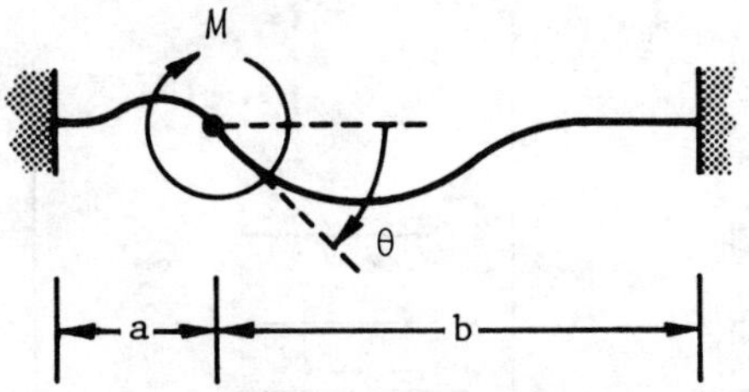	$\dfrac{EI}{a - \dfrac{4a^2}{a + b} + \dfrac{6a^3}{(a + b)^2} - \dfrac{3a^4}{(a + b)^3}}$
34. Torsion of Clamped-Pinned Beam, End 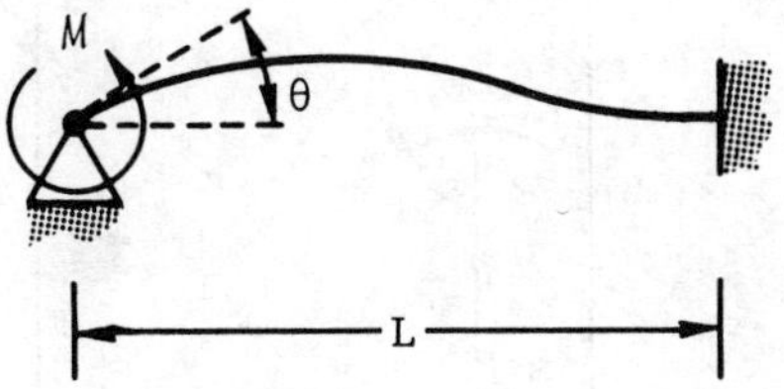	$\dfrac{4EI}{L}$
35. Torsion of Clamped-Pinned Beam, Off Center 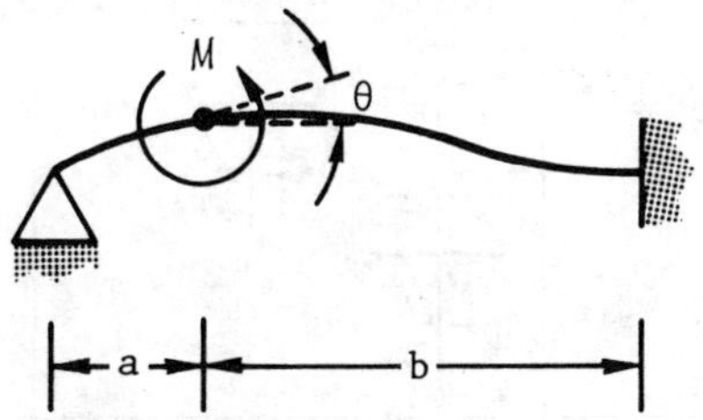	$\dfrac{4EI\ (a + b)^3}{4b\ (a + b)^3 - 3\ (2a + b)^2\ b^2}$

6.3. BEHAVIOR OF SPRING-MASS SYSTEMS

Each natural frequency of a linear structure can be associated with a mode shape that characterizes the form of free vibrations of the structure. The mode shape gives the relative displacement of points on the structure as the structure vibrates in a given mode. The total displacement is the sum of the modal displacements. For example, the free vibrations of the system of two equal masses and two equal springs shown in frame 2 of Table 6-2 can be expressed as the sum of harmonic vibrations at the natural frequencies f_1 and f_2:

$$x_1 = A_1 \sin (2\pi f_1 t + \phi_1) + A_2 \sin (2\pi f_2 t + \phi_2),$$

$$x_2 = B_1 \sin (2\pi f_1 t + \phi_1) + B_2 \sin (2\pi f_2 t + \phi_2).$$

Table 6-2. Mass, Spring Systems.

Notation: f = frequency (Hz); E = modulus of elasticity; I = area moment of inertia about the cross section (Table 5-1); k = deflection spring constant (force/deflection); T = tension in wire; x, y = displacement; $\tilde{x}$, $\tilde{y}$ = mode shapes associated with x and y. See Table 3-1 for consistent sets of units. Springs, wires, and beams are massless unless otherwise noted.

Geometry	Natural Frequency (hertz), f_i	Mode Shape and Remarks
1. Mass, Spring 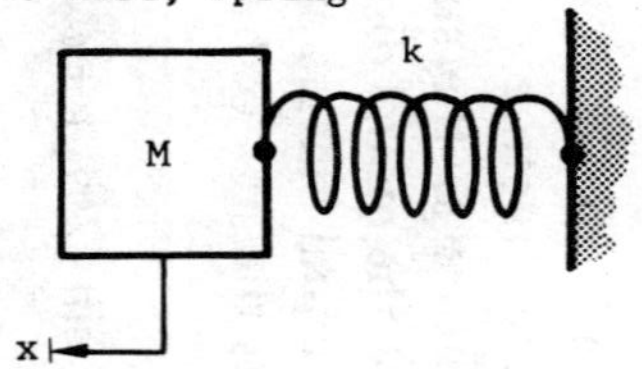	$\frac{1}{2\pi}\left(\frac{k}{M}\right)^{1/2}$	--
2. Two Equal Masses, Two Equal Springs 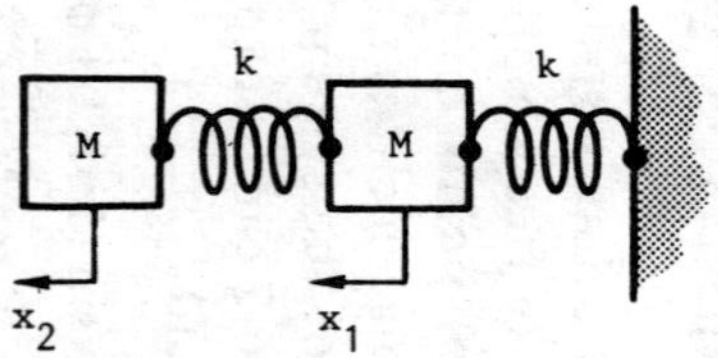	$\frac{(3-5^{1/2})^{1/2}}{2^{3/2}\pi}\left(\frac{k}{M}\right)^{1/2}, \quad \frac{(3+5^{1/2})^{1/2}}{2^{3/2}\pi}\left(\frac{k}{M}\right)^{1/2}$	$\begin{bmatrix}\tilde{x}_1\\ \tilde{x}_2\end{bmatrix} = \begin{bmatrix}1\\ \frac{1+5^{1/2}}{2}\end{bmatrix}, \begin{bmatrix}1\\ \frac{1-5^{1/2}}{2}\end{bmatrix}$
3. Two Unequal Masses, Two Unequal Springs 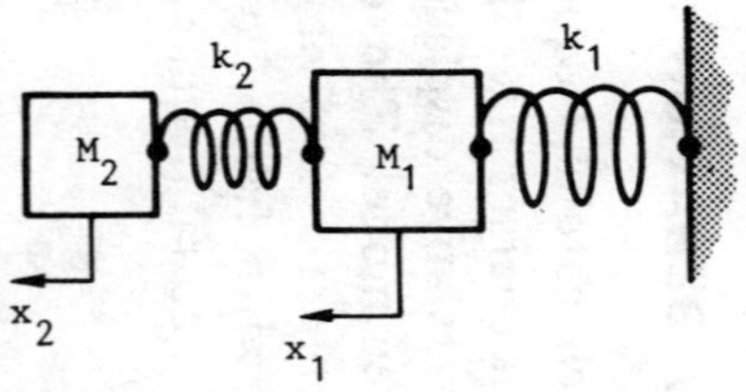	$\frac{1}{2^{3/2}\pi}\left\{\frac{k_1}{M_1}+\frac{k_2}{M_1}+\frac{k_2}{M_2}\mp\left[\left(\frac{k_1}{M_1}+\frac{k_2}{M_1}+\frac{k_2}{M_2}\right)^2 - \frac{4k_1k_2}{M_1M_2}\right]^{1/2}\right\}^{1/2}$	$\begin{bmatrix}\tilde{x}_1\\ \tilde{x}_2\end{bmatrix}_i = \begin{bmatrix}1\\ 1+\frac{k_1}{k_2}-\frac{M_1}{k_2}(2\pi f_i)^2\end{bmatrix}$ $i = 1,2$

4. Three Equal Masses, Three Equal Springs 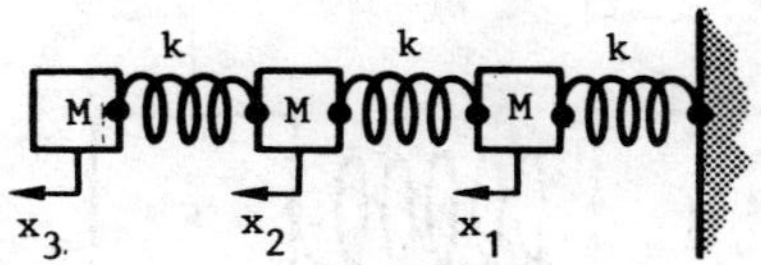	$0.07082\left(\frac{k}{M}\right)^{1/2}$, $0.1985\left(\frac{k}{M}\right)^{1/2}$, $0.2868\left(\frac{k}{M}\right)^{1/2}$	$\begin{bmatrix}\tilde{x}_1\\ \tilde{x}_2\\ \tilde{x}_3\end{bmatrix} = \begin{bmatrix}1\\ 1.802\\ 2.247\end{bmatrix}, \begin{bmatrix}1\\ 0.445\\ -0.802\end{bmatrix}, \begin{bmatrix}1\\ -1.247\\ 0.555\end{bmatrix}$
5. N Equal Masses, N Equal Springs 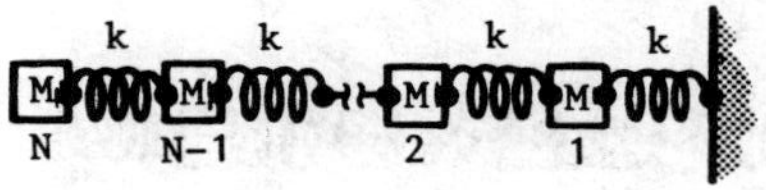	$\frac{\alpha_{N,i}}{2\pi}\left(\frac{k}{M}\right)^{1/2}$; $i = 1,2,\cdots,N$ $\alpha_{N,i} = 2 \sin\left[\frac{(2i-1)}{(2N+1)}\frac{\pi}{2}\right]$	Ref. 6-4
6. Mass, Two Equal Springs 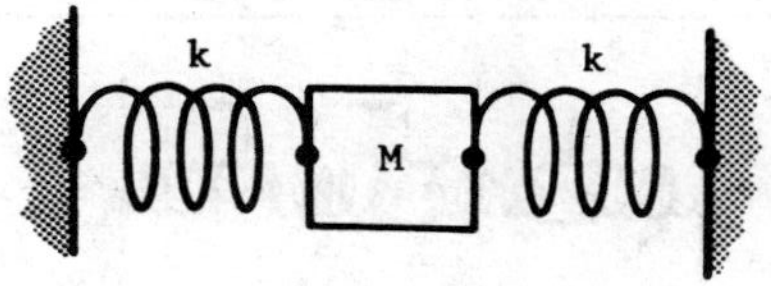	$\frac{1}{2\pi}\left(\frac{2k}{M}\right)^{1/2}$	--
7. Mass, Two Unequal Springs 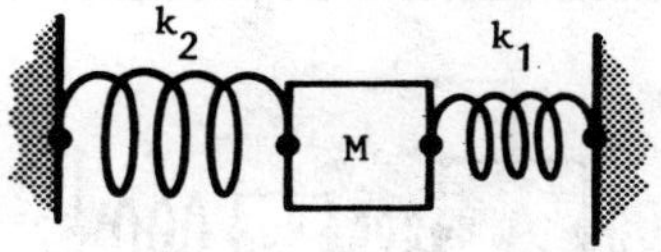	$\frac{1}{2\pi}\left(\frac{k_1 + k_2}{M}\right)^{1/2}$	--
8. Two Equal Masses, Three Equal Springs 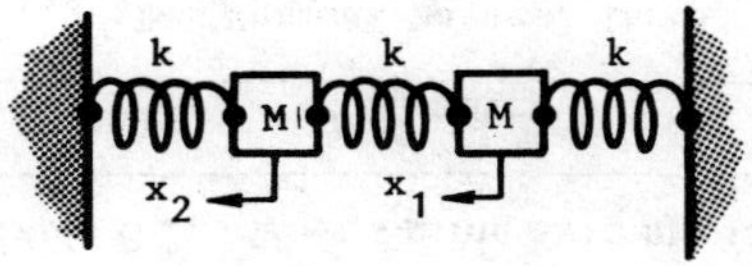	$\frac{1}{2\pi}\left(\frac{k}{M}\right)^{1/2}$, $\frac{1}{2\pi}\left(\frac{3k}{M}\right)^{1/2}$	$\begin{bmatrix}\tilde{x}_1\\ \tilde{x}_2\end{bmatrix} = \begin{bmatrix}1\\ 1\end{bmatrix}, \begin{bmatrix}1\\ -1\end{bmatrix}$

Table 6-2. Mass, Spring Systems. *(Continued)*

Geometry	Natural Frequency (hertz), f_i	Mode Shape and Remarks
9. Two Unequal Masses, Three Unequal Springs 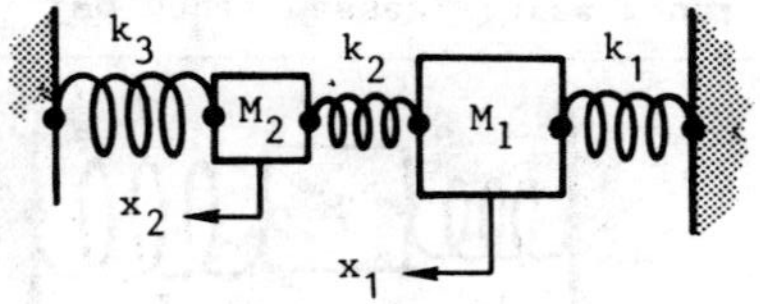	$\frac{1}{2^{3/2}\pi}\left\{\frac{k_1+k_2}{M_1}+\frac{k_2+k_3}{M_2}\pm\left[\left(\frac{k_1+k_2}{M_1}+\frac{k_2+k_3}{M_2}\right)^2 - \frac{4}{M_1M_2}\left(k_1k_2+k_2k_3+k_1k_3\right)\right]^{1/2}\right\}^{1/2}$	$\begin{bmatrix}\tilde{x}_1\\ \tilde{x}_2\end{bmatrix}_i = \begin{bmatrix}1\\ 1+\frac{k_1}{k_2}-\frac{M_1}{k_2}(2\pi f_i)^2\end{bmatrix}$ $i=1,2$
10. Three Equal Masses, Four Equal Springs 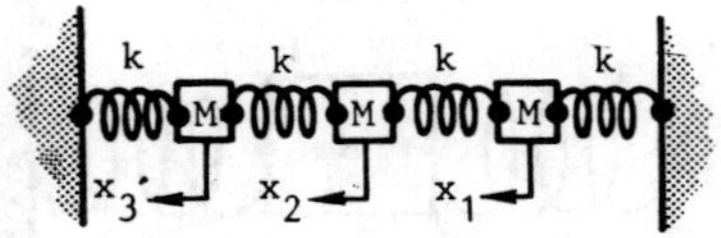	$\frac{(2-2^{1/2})^{1/2}}{2\pi}\left(\frac{k}{M}\right)^{1/2}$, $\frac{2^{1/2}}{2\pi}\left(\frac{k}{M}\right)^{1/2}$, $\frac{(2+2^{1/2})^{1/2}}{2\pi}\left(\frac{k}{M}\right)^{1/2}$	$\begin{bmatrix}\tilde{x}_1\\ \tilde{x}_2\\ \tilde{x}_3\end{bmatrix} = \begin{bmatrix}1\\ 2^{1/2}\\ 1\end{bmatrix}, \begin{bmatrix}1\\ 0\\ -1\end{bmatrix}, \begin{bmatrix}1\\ -2^{1/2}\\ 1\end{bmatrix}$
11. N Equal Masses, N + 1 Equal Springs 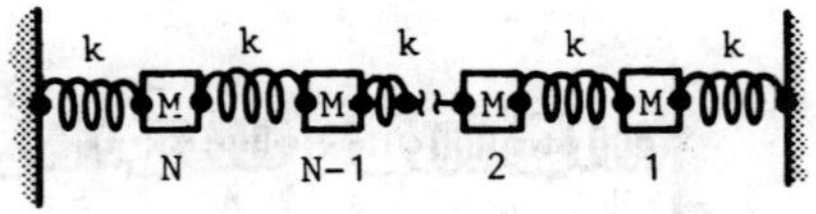	$\frac{\alpha_{N,i}}{2\pi}\left(\frac{k}{M}\right)^{1/2}$; $i = 1,2,3,\cdots,N$ $\alpha_{N,i} = 2\sin\left[\frac{i}{(N+1)}\frac{\pi}{2}\right]$	Ref. 6-4.
12. Two Equal Masses, One Spring 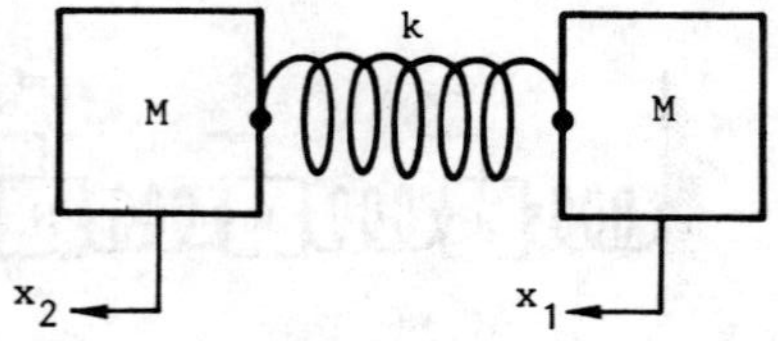	$0,\quad \frac{1}{2\pi}\left(\frac{2k}{M}\right)^{1/2}$	$\begin{bmatrix}\tilde{x}_1\\ \tilde{x}_2\end{bmatrix} = \begin{bmatrix}1\\ 1\end{bmatrix}, \begin{bmatrix}1\\ -1\end{bmatrix}$

13. Two Unequal Masses, One Spring 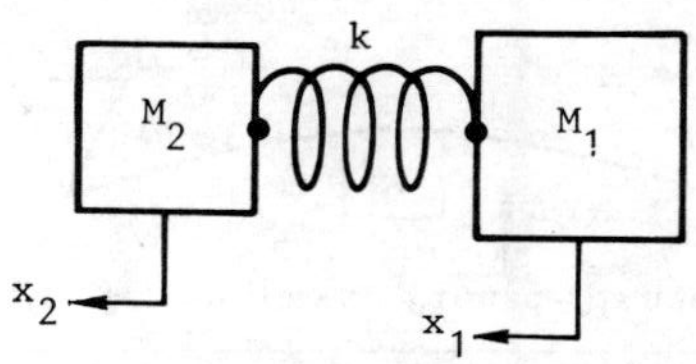	$0, \quad \frac{1}{2\pi}\left[\frac{k\ (M_1 + M_2)}{M_1 M_2}\right]^{1/2}$	$\begin{bmatrix}\tilde{x}_1 \\ \tilde{x}_2\end{bmatrix} = \begin{bmatrix}1 \\ 1\end{bmatrix}, \begin{bmatrix}1 \\ \frac{-M_1}{M_2}\end{bmatrix}$
14. Three Equal Masses, Two Equal Springs 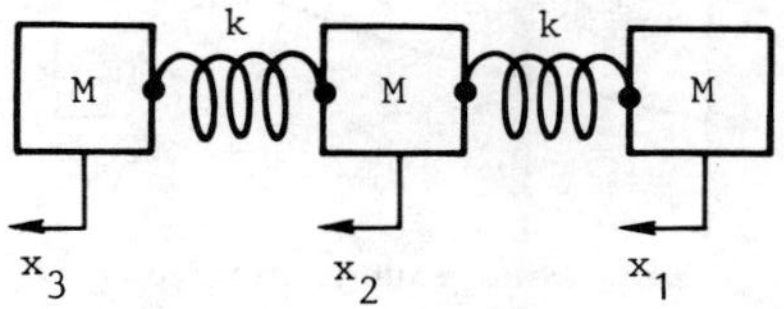	$0, \quad \frac{1}{2\pi}\left(\frac{k}{M}\right)^{1/2}, \quad \frac{1}{2\pi}\left(\frac{3k}{M}\right)^{1/2}$	$\begin{bmatrix}\tilde{x}_1 \\ \tilde{x}_2 \\ \tilde{x}_3\end{bmatrix} = \begin{bmatrix}1 \\ 1 \\ 1\end{bmatrix}, \begin{bmatrix}1 \\ 0 \\ -1\end{bmatrix}, \begin{bmatrix}1 \\ -2 \\ 1\end{bmatrix}$
15. Three Unequal Masses, Two Unequal Springs 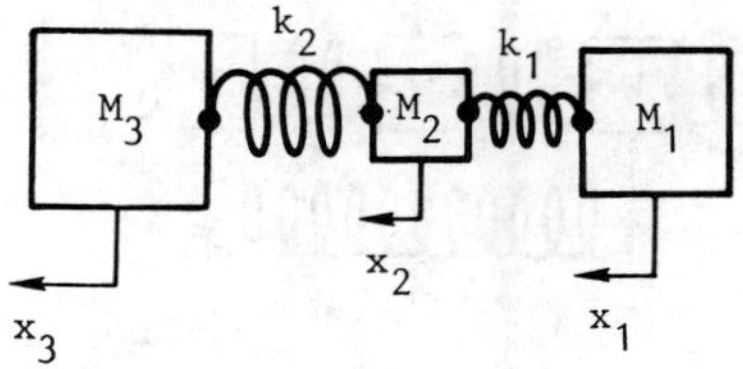	$0, \quad \frac{1}{2^{3/2}\pi}\left\{\frac{k_1 + k_2}{M_2} + \frac{k_1}{M_1} + \frac{k_2}{M_3} \mp \left[\left(\frac{k_1 + k_2}{M_2} + \frac{k_1}{M_1} + \frac{k_2}{M_3}\right)^2 - 4k_1k_2\left(\frac{1}{M_1M_2} + \frac{1}{M_2M_3} + \frac{1}{M_1M_3}\right)\right]^{1/2}\right\}^{1/2}$	$\begin{bmatrix}\tilde{x}_1 \\ \tilde{x}_2 \\ \tilde{x}_3\end{bmatrix}_i = \begin{bmatrix}1 \\ 1 - \frac{M_1}{k_1}(2\pi f_i)^2 \\ \left[1 - \frac{M_1}{k_1}(2\pi f_i)^2\right] \Big/ \left[1 - \frac{M_3}{k_2}(2\pi f_i)^2\right]\end{bmatrix}$ i=1,2,3
16. Four Equal Masses, Three Equal Springs 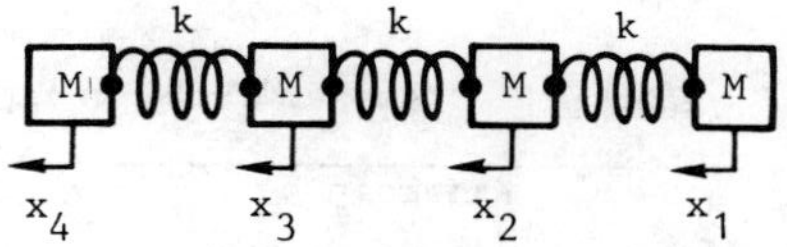	$0, \quad \frac{(2 - 2^{1/2})^{1/2}}{2\pi}\left(\frac{k}{M}\right)^{1/2}, \quad \frac{2^{1/2}}{2\pi}\left(\frac{k}{M}\right)^{1/2},$ $\frac{(2 + 2^{1/2})^{1/2}}{2\pi}\left(\frac{k}{M}\right)^{1/2}$	$\begin{bmatrix}\tilde{x}_1 \\ \tilde{x}_2 \\ \tilde{x}_3 \\ \tilde{x}_4\end{bmatrix} = \begin{bmatrix}1 \\ 1 \\ 1 \\ 1\end{bmatrix}, \begin{bmatrix}1 \\ -1 + 2^{1/2} \\ 1 - 2^{1/2} \\ -1\end{bmatrix}, \begin{bmatrix}1 \\ -1 \\ -1 \\ 1\end{bmatrix}, \begin{bmatrix}1 \\ -1 - 2^{1/2} \\ 1 + 2^{1/2} \\ -1\end{bmatrix}$

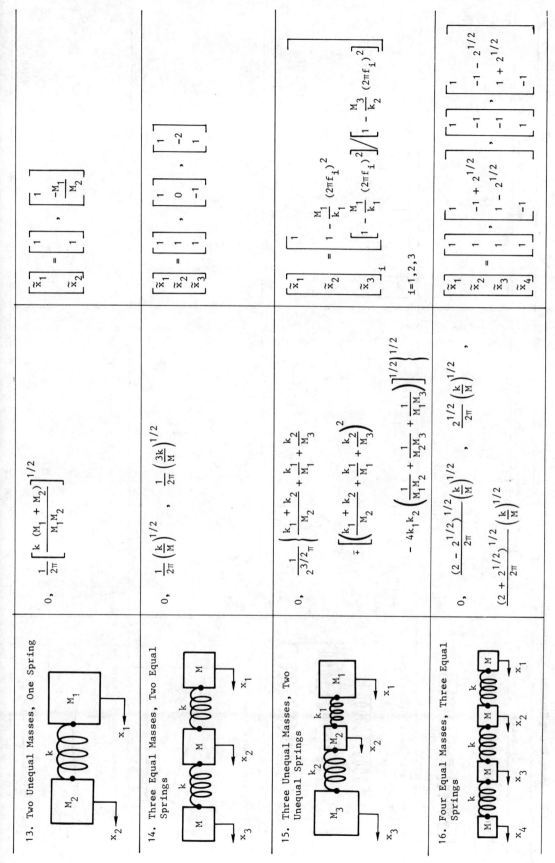

Table 6-2. Mass, Spring Systems. *(Continued)*

Geometry	Natural Frequency (hertz), f_i	Mode Shape and Remarks
17. N Equal Masses, N - 1 Equal Springs 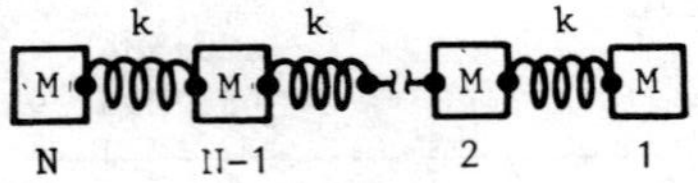	$\frac{\alpha_{N,i}}{2\pi}\left(\frac{k}{M}\right)^{1/2}$ $i = 1,2,3,\cdots,N$ $\alpha_{N,i} = 2\sin\left[\frac{(i-1)}{N}\frac{\pi}{2}\right]$	Ref. 6-4
18. Three Equal Masses, Six Equal Springs 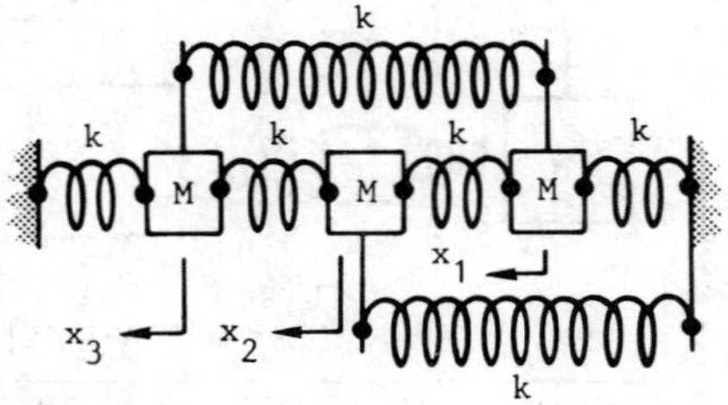	$\frac{1}{2\pi}\left(\frac{k}{M}\right)^{1/2}$, $\frac{1}{\pi}\left(\frac{k}{M}\right)^{1/2}$, $\frac{1}{\pi}\left(\frac{k}{M}\right)^{1/2}$	$\begin{bmatrix}\tilde{x}_1\\ \tilde{x}_2\\ \tilde{x}_3\end{bmatrix} = \begin{bmatrix}1\\1\\1\end{bmatrix}, \begin{bmatrix}1\\0\\-1\end{bmatrix}, \begin{bmatrix}1\\-2\\1\end{bmatrix}$
19. End Mass, Cantilever Beam 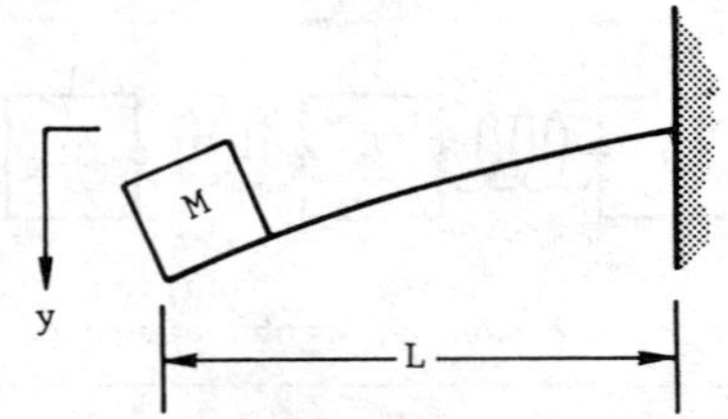	$\frac{1}{2\pi}\left(\frac{3EI}{ML^3}\right)^{1/2}$	--
20. Center Mass, Pinned-Pinned Beam 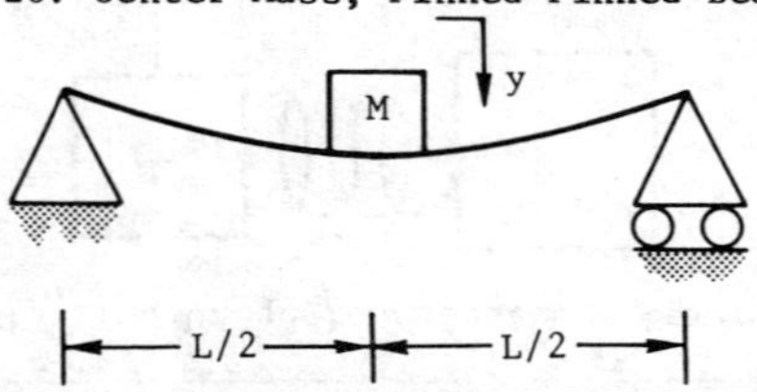	$\frac{2}{\pi}\left(\frac{3EI}{ML^3}\right)^{1/2}$	--

21. Off-Center Mass, Pinned-Pinned Beam 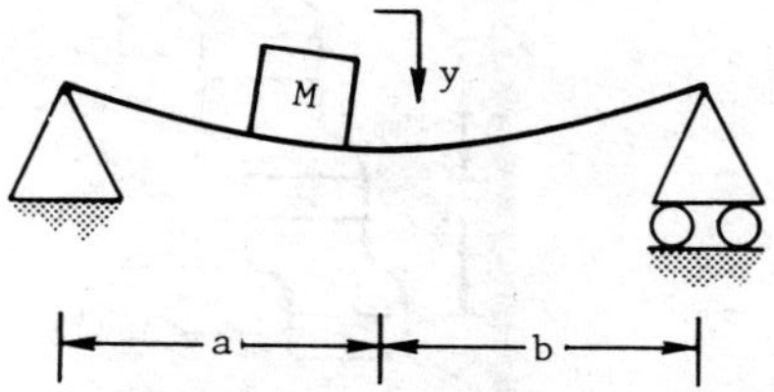	$\frac{1}{2\pi}\left[\frac{3EI\ (a+b)}{Ma^2b^2}\right]^{1/2}$	--
22. Center Mass, Clamped-Clamped Beam 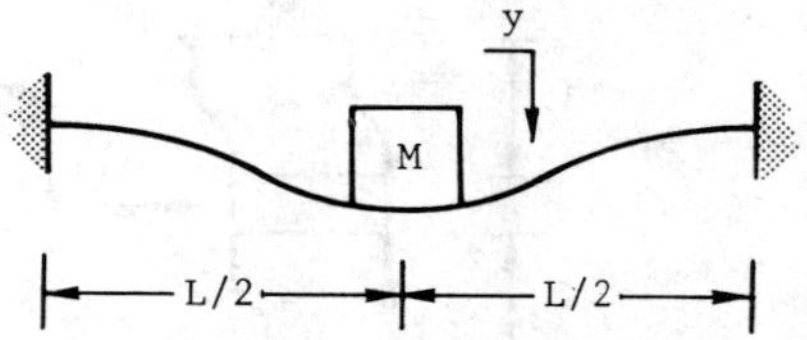	$\frac{4}{\pi}\left(\frac{3EI}{ML^3}\right)^{1/2}$	--
23. Off-Center Mass, Clamped-Clamped Beam 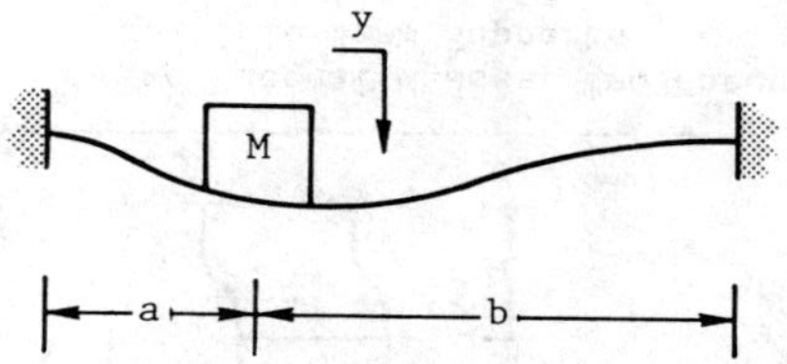	$\frac{1}{2\pi}\left[\frac{3EI}{M}\ \frac{(a+b)^3}{a^3b^3}\right]^{1/2}$	--
24. Mass, Multiple Beam Support 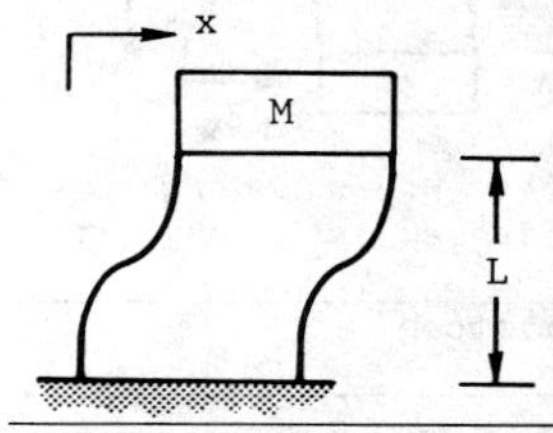	$\frac{1}{\pi}\left(\frac{3EI_s}{ML^3}\right)^{1/2}$	$I_s = \Sigma I_i$ See footnote at end of table.

Table 6-2. Mass, Spring Systems. *(Continued)*

Geometry	Natural Frequency (hertz), f_i	Mode Shape and Remarks
25. Two Equal Masses, Multiple Beam Supports 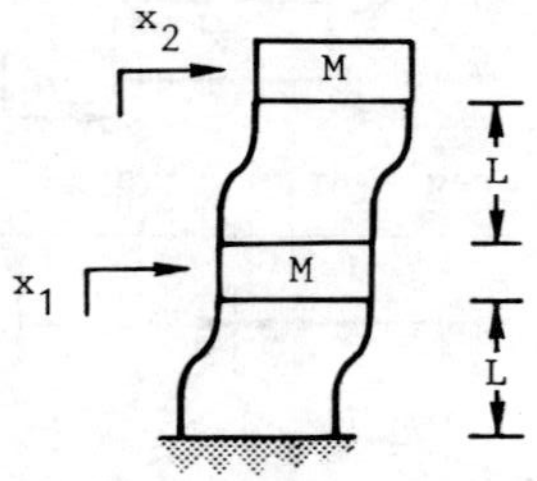	$\frac{(3 - 5^{1/2})^{1/2}}{2^{1/2}\pi}\left(\frac{3EI_s}{ML^3}\right)^{1/2}$, $\frac{(3 + 5^{1/2})^{1/2}}{2^{1/2}\pi}\left(\frac{3EI_s}{ML^3}\right)^{1/2}$	$\begin{bmatrix}\tilde{x}_1 \\ \tilde{x}_2\end{bmatrix}_i = \begin{bmatrix}1 \\ \frac{1 + 5^{1/2}}{2}\end{bmatrix}, \begin{bmatrix}1 \\ \frac{1 - 5^{1/2}}{2}\end{bmatrix}$ $I_s = \Sigma I_i$ See footnote at end of table.
26. Two Unequal Masses, Two Unequal Multiple Beam Supports 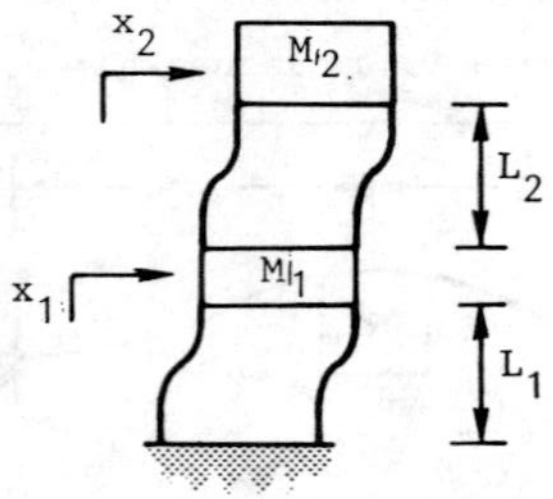	See frame 3 with: $k_1 = \frac{12EI_s}{L_1^3}$ $k_2 = \frac{12EI_s}{L_2^3}$	See frame 3 with: $k_1 = \frac{12EI_s}{L_1^3}$ $I_s = \Sigma I_i$ $k_2 = \frac{12EI_s}{L_2^3}$ See footnote at end of table.
27. Three Equal Masses, Three Equal Multiple Beam Supports 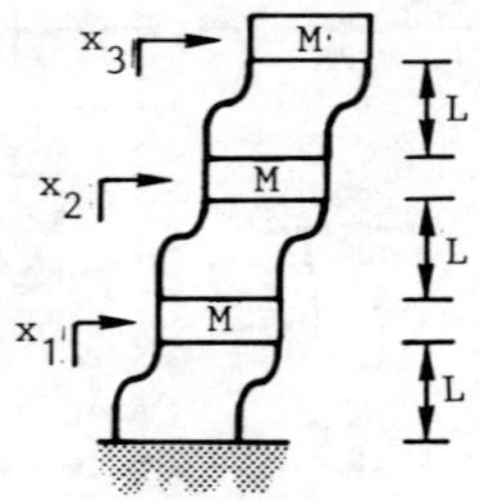	$0.07082\left(\frac{12EI_s}{ML^3}\right)^{1/2}$, $0.1985\left(\frac{12EI_s}{ML^3}\right)^{1/2}$, $0.2868\left(\frac{12EI_s}{ML^3}\right)^{1/2}$	$\begin{bmatrix}\tilde{x}_1 \\ \tilde{x}_2 \\ \tilde{x}_3\end{bmatrix}_1 = \begin{bmatrix}1.000 \\ 1.802 \\ 2.247\end{bmatrix}, \begin{bmatrix}1.000 \\ 0.445 \\ -0.802\end{bmatrix}, \begin{bmatrix}1.000 \\ -1.247 \\ 0.555\end{bmatrix}$ $I_s = \Sigma I_i$ See footnote at end of table.

28. N Equal Masses, N Equal Multiple Beam Supports 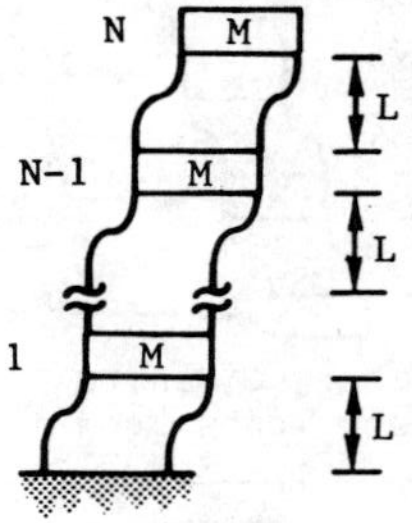	See frame 5 with: $k = \frac{12EI_s}{L^3}$	$I_s = \Sigma I_i$ See footnote at end of table.
29. Two Equal Masses, Multiple Beam Support 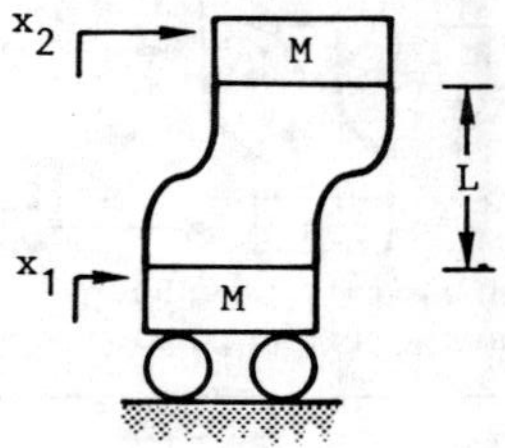	$0, \quad \frac{1}{2\pi}\left(\frac{24EI_s}{ML^3}\right)^{1/2}$	$\begin{bmatrix} \tilde{x}_1 \\ \tilde{x}_2 \end{bmatrix} = \begin{bmatrix} 1 \\ 1 \end{bmatrix}, \begin{bmatrix} 1 \\ -1 \end{bmatrix}$ $I_s = \Sigma I_i$ See footnote at end of table.
30. Two Unequal Masses, Multiple Beam Support 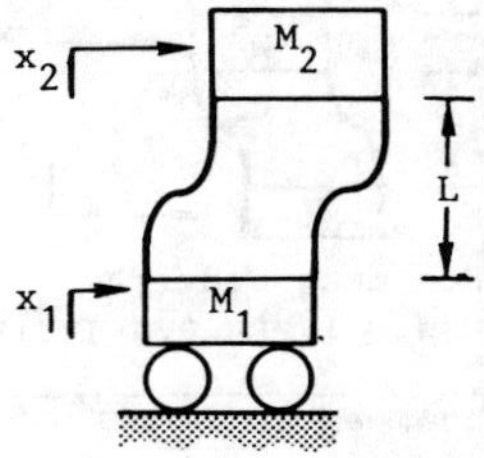	$0, \quad \frac{1}{\pi}\left[\frac{3EI_s\,(M_1 + M_2)}{L^3 M_1 M_2}\right]^{1/2}$	$\begin{bmatrix} \tilde{x}_1 \\ \tilde{x}_2 \end{bmatrix}_i = \begin{bmatrix} 1 \\ 1 \end{bmatrix}, \begin{bmatrix} 1 \\ \frac{-M_1}{M_2} \end{bmatrix}$ $I_s = \Sigma I_i$ See footnote at end of table.

Table 6-2. Mass, Spring Systems. *(Continued)*

Geometry	Natural Frequency (hertz), f_i	Mode Shape and Remarks
31. Three Equal Masses, Two Equal Multiple Beam Supports 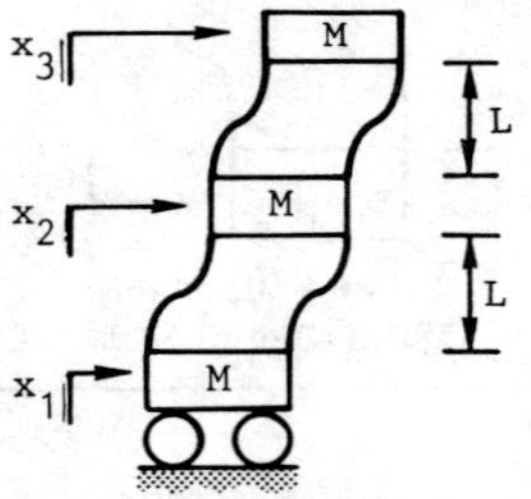	$0, \ \frac{1}{\pi}\left(\frac{3EI_s}{ML^3}\right)^{1/2}, \ \frac{3}{\pi}\left(\frac{EI_s}{ML^3}\right)^{1/2}$	$\begin{bmatrix}\tilde{x}_1\\ \tilde{x}_2\\ \tilde{x}_3\end{bmatrix}_i = \begin{bmatrix}1\\1\\1\end{bmatrix}, \begin{bmatrix}1\\0\\-1\end{bmatrix}, \begin{bmatrix}1\\-2\\1\end{bmatrix}$ $I_s = \Sigma I_i$ See footnote at end of table.
32. Three Unequal Masses, Two Unequal Multiple Beam Supports 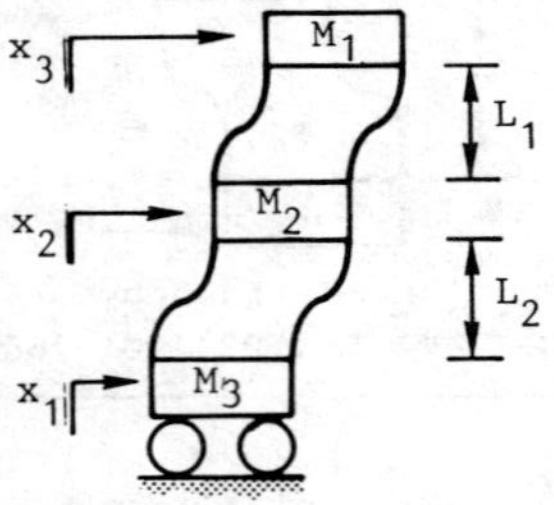	See frame 15 with: $k_1 = \frac{12EI_s}{L_1^3}$ $k_2 = \frac{12EI_s}{L_2^3}$	See frame 15 with: $k_1 = \frac{12EI_s}{L_1^3}$ $\quad I_s = \Sigma I_i$ $k_2 = \frac{12EI_s}{L_2^3}$ See footnote at end of table.
33. Four Equal Masses, Three Equal Multiple Beam Supports 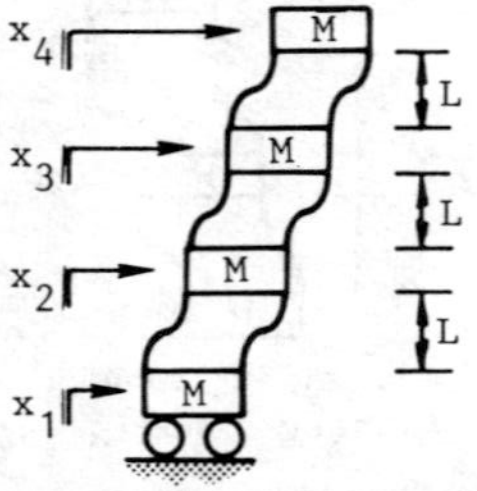	See frame 16 with: $k = \frac{12EI_s}{L^3}$	See frame 16 with: $k = \frac{12EI_s}{L^3}$ $I_s = \Sigma I_i$ See footnote at end of table.

34. N Equal Masses, N - 1 Equal Multiple Beam Supports 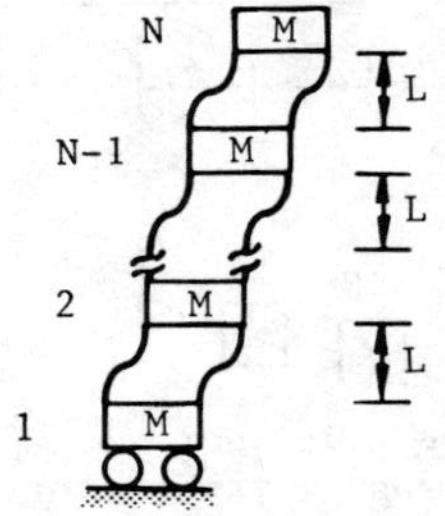	See frame 17 with: $k = \frac{12EI_s}{L^3}$	$I_s = \Sigma I_i$ See footnote at end of table.
35. Center Mass, Tensioned Wire 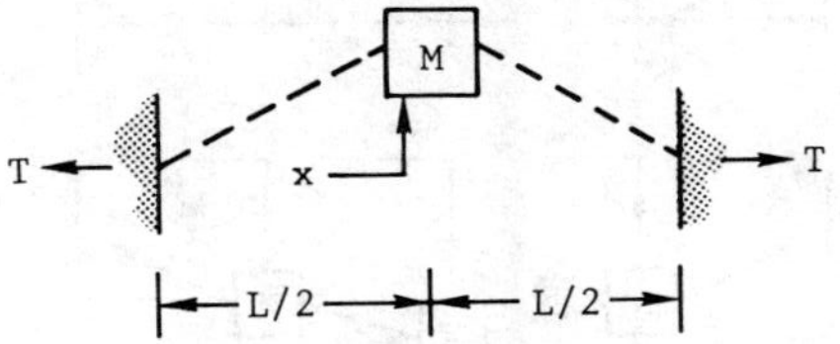	$\frac{1}{\pi}\left(\frac{T}{ML}\right)^{1/2}$	--
36. Off-Center Mass, Tensioned Wire 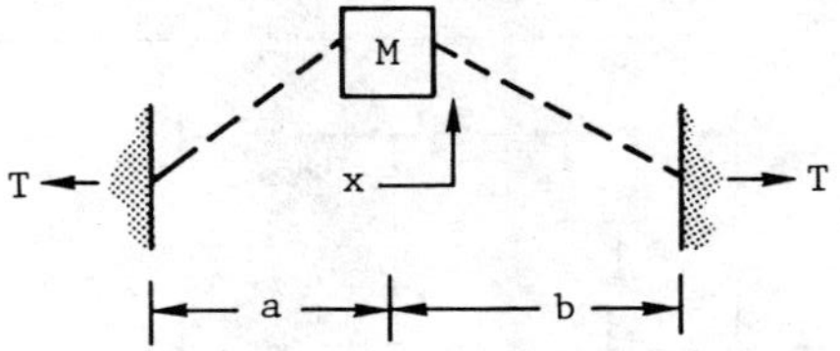	$\frac{1}{2\pi}\left[\frac{T\ (a+b)}{Mab}\right]^{1/2}$	--
37. Two Equal Masses, Equal Lengths of Tensioned Wire 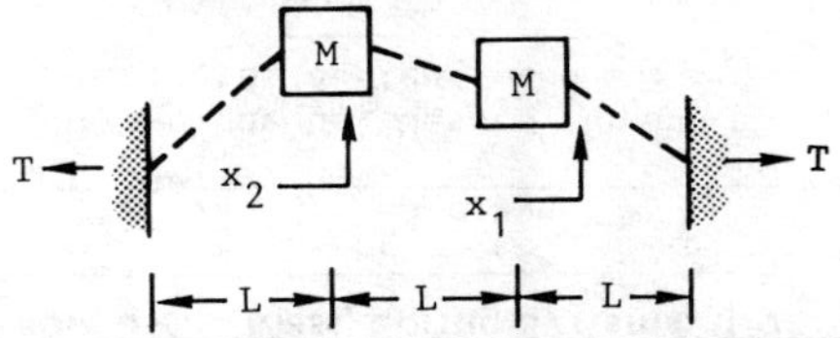	$\frac{1}{2\pi}\left(\frac{T}{ML}\right)^{1/2}$, $\frac{1}{2\pi}\left(\frac{3T}{ML}\right)^{1/2}$	$\begin{bmatrix}\tilde{x}_1 \\ \tilde{x}_2\end{bmatrix}_i = \begin{bmatrix}1 \\ 1\end{bmatrix} , \begin{bmatrix}1 \\ -1\end{bmatrix}$

Table 6-2. Mass, Spring Systems. (*Continued*)

Geometry	Natural Frequency (hertz), f_i	Mode Shape and Remarks
38. Two Unequal Masses, Unequal Lengths of Tensioned Wire 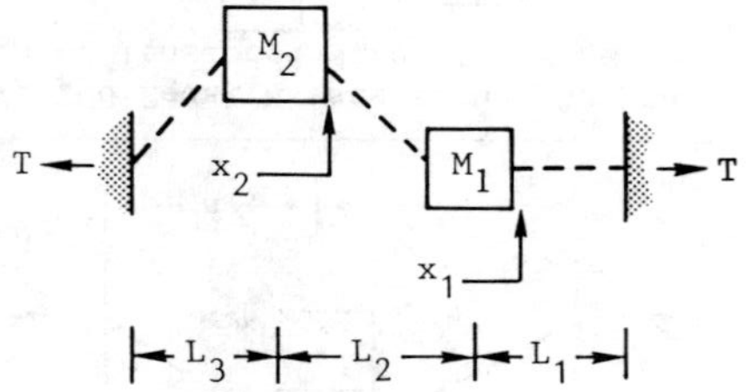	See frame 9 with: $k_1 = \frac{T}{L_1}$, $k_2 = \frac{T}{L_2}$, $k_3 = \frac{T}{L_3}$	See frame 9 with: $k_1 = \frac{T}{L_1}$, $k_2 = \frac{T}{L_2}$, $k_3 = \frac{T}{L_3}$
39. Three Equal Masses, Equal Lengths of Tensioned Wire 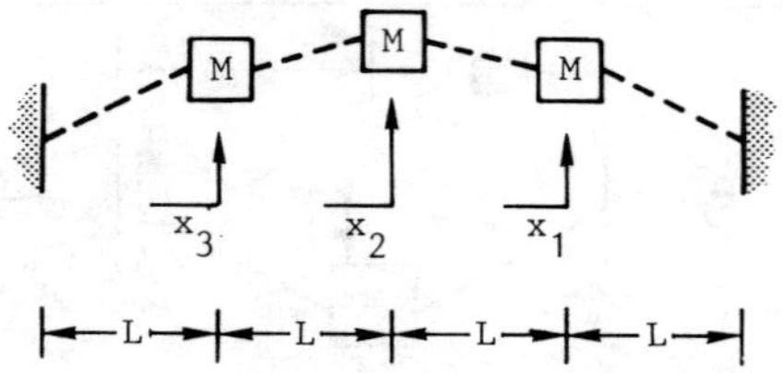	$\frac{(2 - 2^{1/2})^{1/2}}{2\pi}\left(\frac{T}{ML}\right)^{1/2}$, $\frac{2^{1/2}}{2\pi}\left(\frac{T}{ML}\right)^{1/2}$, $\frac{(2 + 2^{1/2})^{1/2}}{2\pi}\left(\frac{T}{ML}\right)^{1/2}$	$\begin{bmatrix}\tilde{x}_1\\ \tilde{x}_2\\ \tilde{x}_3\end{bmatrix} = \begin{bmatrix}1\\ 2^{1/2}\\ 1\end{bmatrix}, \begin{bmatrix}1\\ 0\\ -1\end{bmatrix}, \begin{bmatrix}1\\ -2^{1/2}\\ 1\end{bmatrix}$
40. N Equal Masses, Equal Lengths of Tensioned Wire 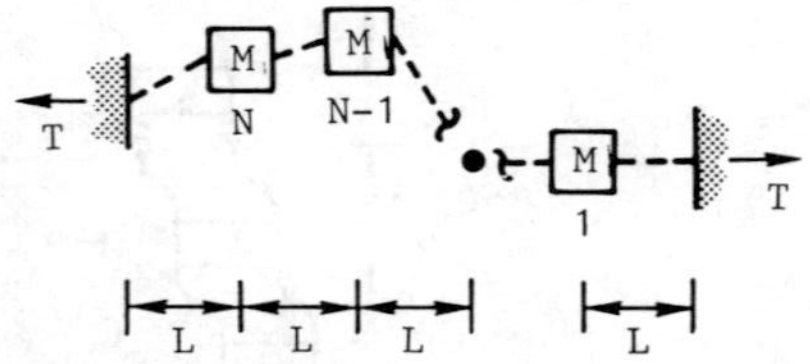	See frame 11 with: $k = \frac{T}{L}$	--

Note: In cases 24-34, I_s is the sum of the area moment of inertia of all beams between each floor. For example, if there are two columns between each floor as shown in the diagrams, $I_s = I_1 + I_2$. In cases 26, 30, and 32, I_s may vary between stories.

x_1 and x_2 are the displacements of the masses. ϕ_1 and ϕ_2 are phase angles which are determined by the means used to set the structure in motion. The constants A_1, A_2, B_1, and B_2 specify the amplitude of vibration. During free vibrations, the ratios A_1/B_1 and A_2/B_2 are fixed by the mode shapes. For this system Table 6-2, frame 2 gives:

$$\frac{B_1}{A_1} = \tilde{x}_{21} = \frac{1 + 5^{1/2}}{2} = 1.618,$$

$$\frac{B_2}{A_2} = \tilde{x}_{22} = \frac{1 - 5^{1/2}}{2} = -0.618.$$

Thus, if either A_1, A_2 or B_1, B_2, or B_1, A_2 or B_2, A_1 can be found, then the remaining two constants are uniquely determined by the mode shape. Note that these mode shapes are orthogonal:

$$MA_1A_2 + MB_1B_2 = M(1 - 1) = 0.$$

The orthogonality condition for a general n coordinate system is:

$$\sum_{i=1}^{n} M_i\, A_i^k A_i^j = 0 \qquad j \neq k,$$

where M_i is the i mass, A_i^k is the modal displacement of the i mass in the k mode, and A_i^j is the modal displacement of the i mass in the j mode ($j \neq k$). This orthogonality condition applies for all systems in Tables 6-2, 6-3, and 6-4. The modes of many continuous systems are orthogonal as well.

Tables 6-2, 6-3, and 6-4 present natural frequencies and mode shapes of point mass-spring systems, rigid body-torsion systems, and pendulum systems. These systems possess one natural frequency for each mass in the system and there is a unique mode shape for each natural frequency. The majority of the formulas presented in these tables were developed by writing the linear equations of motion for each system and then solving these equations exactly. An example of this analysis is presented in the following paragraphs. Approximate techniques are discussed in Refs. 6-3 through 6-12. References 6-9 and 6-12 are particularly rich in example problems. References 6-13 through 6-17 provide additional background material.

6.4. EXAMPLE OF EXACT ANALYSIS

Consider the system of two equal point masses of mass M and two equal springs with spring constant k shown in frame 2 of Table 6-2. The mass multiplied by the acceleration of each point mass is equal to the forces on the mass:

$$M\ddot{x}_1 = -kx_1 + k(x_2 - x_1),$$
$$M\ddot{x}_2 = -k(x_2 - x_1). \tag{6-1}$$

These equations can be written as:

$$M\ddot{x}_1 + 2kx_1 - kx_2 = 0,$$
$$M\ddot{x}_2 + kx_2 - kx_1 = 0. \tag{6-2}$$

In matrix format these equations are:

$$[M]\{\ddot{\underline{x}}\} + [K]\{\underline{x}\} = 0, \tag{6-3}$$

where

$$[M] = \begin{bmatrix} M & 0 \\ 0 & M \end{bmatrix}, \qquad [K] = \begin{bmatrix} 2k & -k \\ -k & k \end{bmatrix}, \qquad \{\underline{x}\} = \begin{Bmatrix} x_1 \\ x_2 \end{Bmatrix}.$$

The dot (·) over a variable denotes derivative with respect to time. If the free vibrations are assumed to be harmonic in time,

$$\{\underline{x}\} = \{\tilde{\underline{x}}\} \sin \omega t, \tag{6-4}$$

where $\{\tilde{\underline{x}}\}$ is a vector that is independent of time which specifies the mode shape of the vibrations. If Eq. 6-4 is substituted into Eq. 6-3, the matrix equation of motion becomes:

$$[-\omega^2 [M] + [K]]\ \{\tilde{\underline{x}}\} = 0. \tag{6-5}$$

This equation has nontrivial solution only if the matrix to the left of the vector $\{\tilde{\underline{x}}\}$ has a zero determinant:

$$|-\omega^2 [M] + [K]| = 0,$$

$$\begin{vmatrix} -\omega^2 M + 2k & -k \\ -k & -\omega^2 M + k \end{vmatrix} = 0, \tag{6-6}$$

$$\omega^4 - \frac{3k}{M}\omega^2 + \frac{k^2}{M^2} = 0.$$

Thus, the natural frequencies of vibration, in hertz, are:

$$f = \frac{\omega}{2\pi} = \frac{1}{2\pi}\left(\frac{k}{M}\right)^{1/2}\left(\frac{3 \mp 5^{1/2}}{2}\right)^{1/2} \tag{6-7}$$

The mode shapes associated with the natural frequencies are determined by alternately substituting each natural frequency into Eq. 6-5 and solving for the associated eigenvector $\{\tilde{\underline{x}}\}$ which specifies the mode shape. The result is:

$$\{\tilde{\underline{x}}\} = \begin{Bmatrix} \tilde{x}_1 \\ \tilde{x}_2 \end{Bmatrix}_i = \begin{pmatrix} 1 \\ \dfrac{1 + 5^{1/2}}{2} \end{pmatrix}, \begin{pmatrix} 1 \\ \dfrac{1 - 5^{1/2}}{2} \end{pmatrix}, \quad i = 1, 2$$

where the first vector is associated with the lowest natural frequency and the second vector is associated with the second natural frequency.

For vibration at the lowest natural frequency,

$$\tilde{x}_{21} = \frac{x_2}{x_1} = \frac{1 + 5^{1/2}}{2} > 0$$

and the two masses move in the same direction simultaneously. For vibration at the second natural frequency,

$$\tilde{x}_{22} = \frac{x_2}{x_1} = \frac{1 - 5^{1/2}}{2} < 0$$

and the two masses move in opposite directions.

The total response of free vibrations of the system is the sum of the modal vibrations:

$$x_1 = A \sin(2\pi f_1 t + \phi_1) + B \sin(2\pi f_2 t + \phi_2),$$

$$x_2 = \frac{A(1 + 5^{1/2})}{2} \sin(2\pi f_1 t + \phi_1) + \frac{B(1 - 5^{1/2})}{2} \sin(2\pi f_2 t + \phi_2),$$

where A, B, ϕ_1, and ϕ_2 are constants which are determined by the means used to set the system in motion and f_1 and f_2 are determined by Eq. 6-7.

6.5. CABLE-MASS EXAMPLE

Consider the cable-pulley-mass system shown in Fig. 6-1. The steel cable is tied at the roof of a construction site. The cable runs from the roof over a rigidly held pul-

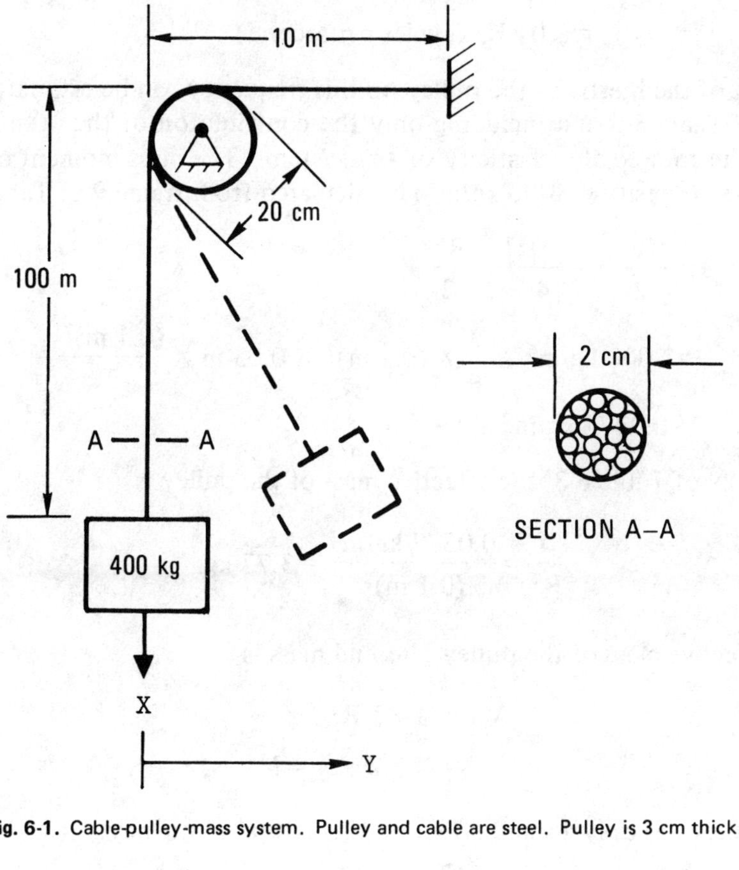

Fig. 6-1. Cable-pulley-mass system. Pulley and cable are steel. Pulley is 3 cm thick.

ley to a mass load 100 meters below. The problem is to find the natural frequencies of this system. In this example the units of frame 1 of Table 3-1 will be used.

The system has two modes of vibration: (1) an extensional mode corresponding to Y = 0 and X oscillating (Fig. 6-1) as the cable stretches along its own axis, and (2) a pendulum mode corresponding to Y oscillating as the mass swings back and forth. The extensional mode will be analyzed first.

For a first approximation to the extensional mode, the effects of the pulley and mass of the cable are ignored and the system reduces to the simple spring-mass system of frame 1 in Table 6-2. The effective spring constant of the 100-meter length of cable is estimated from frame 8 of Table 6-1 using a steel modulus of 2×10^{11} pascals and assuming the entire cross section of the cable is composed of steel. Thus,

$$k = \frac{EA}{L} = \frac{2 \times 10^{11}\ \text{Pa}}{110\ \text{m}} \times \frac{\pi}{4}(2 \times 10^{-2}\ \text{m})^2,$$

$$= 571{,}200\ \frac{\text{N}}{\text{m}}.$$

The natural frequency is calculated from frame 1 of Table 6-2:

$$f = \frac{1}{2\pi}\left(\frac{k}{M}\right)^{1/2} = \frac{1}{2\pi}\left(\frac{571{,}200\ \text{N/m}}{400\ \text{kg}}\right)^{1/2},$$

$$= 6.01\ \text{Hz (cycles per second)}.$$

The effect of the inertia of the pulley on this frequency can be estimated by using frame 29 of Table 6-3 and including only the contribution of the 10-meter length of cable at the roof to the elasticity of the system. The mass moment of inertia of the steel pulley (density = 8000 kg/m^3) is calculated from frame 9 of Table 5-2:

$$J = \frac{MR^2}{2} = \frac{\mu\pi D^2 t}{4} \times \frac{R^2}{2},$$

$$= 8000\ \text{kg/m}^3 \times \frac{\pi}{4} \times (0.2\ \text{m})^2 \times 0.03\ \text{m} \times \frac{(0.1\ \text{m})^2}{2},$$

$$= 0.0377\ \text{kg/m}^2.$$

Using frame 29 of Table 6-3, the effective mass of the pulley is

$$\frac{J}{R^2} = \frac{0.0377\ \text{kg/m}^2}{(0.1\ \text{m})^2} = 3.77\ \text{kg}.$$

The total effective mass of the pulley plus end mass is

$$M_e = M + J/R^2,$$

$$= 403.77\ \text{kg} \approx M.$$

Thus, the pulley has only a small effect on the natural frequency.

Table 6-3. Rigid Body, Torsion Spring Systems.

Notation: C = center of mass; f = frequency (Hz); J = mass moment of inertia of body about pivot; k = deflection spring constant (force/displacement); k = torsion spring constant (moment/angular rotation); x = displacement; θ = angular rotation (radians); $\tilde{x}$, $\tilde{\theta}$ = mode shapes associated with x and θ. See Table 3-1 for consistent sets of units. Springs, levers, and gears are massless unless otherwise noted.

Geometry	Natural Frequency (hertz), f_i	Mode Shape and Remarks
1. Body, Spring 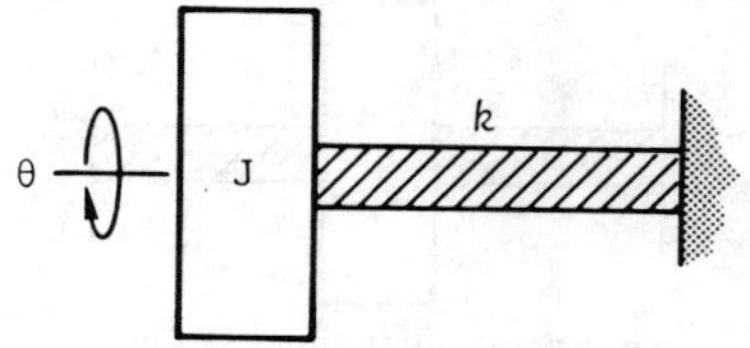	$\frac{1}{2\pi}\left(\frac{k}{J}\right)^{1/2}$	--
2. Two Equal Bodies, Two Equal Springs 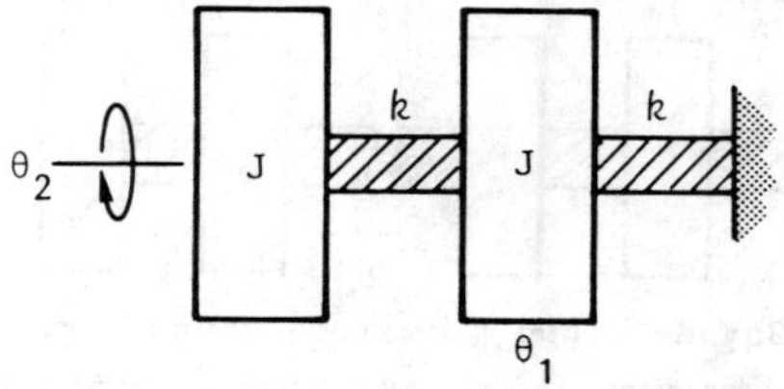	$\frac{(3 - 5^{1/2})^{1/2}}{2^{3/2}\pi}\left(\frac{k}{J}\right)^{1/2}$, $\frac{(3 + 5^{1/2})^{1/2}}{2^{3/2}\pi}\left(\frac{k}{J}\right)^{1/2}$	$\begin{bmatrix}\tilde{\theta}_1\\ \tilde{\theta}_2\end{bmatrix} = \begin{bmatrix}1\\ \frac{1 + 5^{1/2}}{2}\end{bmatrix}$, $\begin{bmatrix}1\\ \frac{1 - 5^{1/2}}{2}\end{bmatrix}$
3. Two Unequal Bodies, Two Unequal Springs 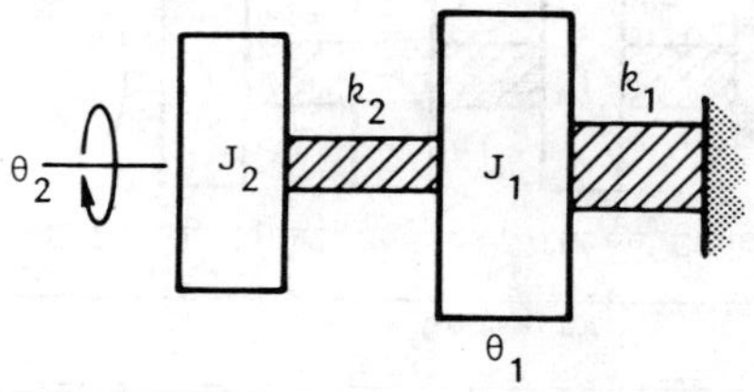	$\frac{1}{2^{3/2}\pi}\left\{\frac{k_1}{J_1} + \frac{k_2}{J_1} + \frac{k_2}{J_2} \mp \left[\left(\frac{k_1}{J_1} + \frac{k_2}{J_1} + \frac{k_2}{J_2}\right)^2 - \frac{4k_1k_2}{J_1J_2}\right]^{1/2}\right\}^{1/2}$	$\begin{bmatrix}\tilde{\theta}_1\\ \tilde{\theta}_2\end{bmatrix}_i = \begin{bmatrix}1\\ 1 + \frac{k_1}{k_2} - \frac{J_1}{k_2}(2\pi f_i)^2\end{bmatrix}$ $i = 1,2$

Table 6-3. Rigid Body, Torsion Spring Systems. *(Continued)*

Geometry	Natural Frequency (hertz), f_i	Mode Shape and Remarks
4. Three Equal Bodies, Three Equal Springs 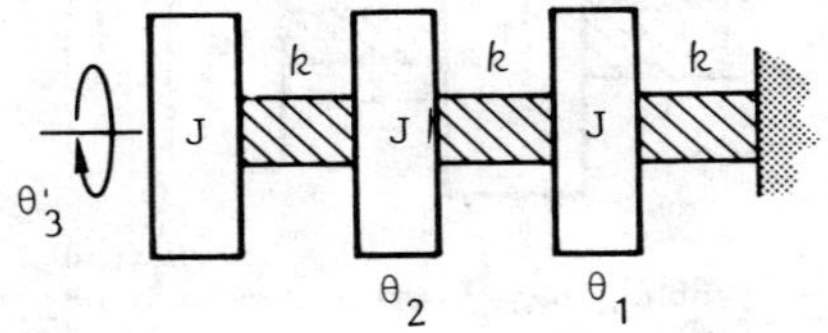	$0.07083\left(\frac{k}{J}\right)^{1/2}$, $0.1985\left(\frac{k}{J}\right)^{1/2}$, $0.2868\left(\frac{k}{J}\right)^{1/2}$	$\begin{bmatrix}\tilde{\theta}_1\\ \tilde{\theta}_2\\ \tilde{\theta}_3\end{bmatrix} = \begin{bmatrix}1\\ 1.802\\ 2.247\end{bmatrix}, \begin{bmatrix}1\\ 0.445\\ -0.802\end{bmatrix}, \begin{bmatrix}1\\ -1.247\\ 0.555\end{bmatrix}$
5. N Equal Bodies, N Equal Springs 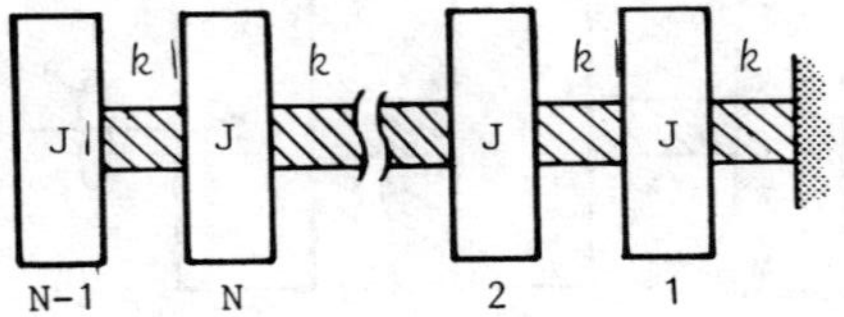	$\frac{\alpha_{N,i}}{2\pi}\left(\frac{k}{J}\right)^{1/2}$; $i = 1,2,3,\cdots,N$ $\alpha_{N,i} = 2\sin\left[\frac{(2i-1)}{(2N+1)}\frac{\pi}{2}\right]$	Ref. 6-4
6. Body, Two Equal Springs 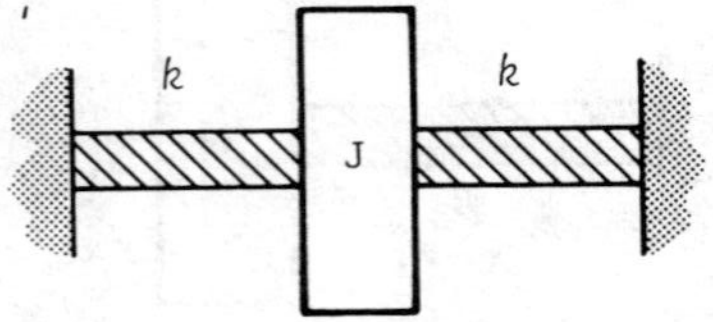	$\frac{1}{2\pi}\left(\frac{2k}{J}\right)^{1/2}$	--
7. Body, Two Unequal Springs 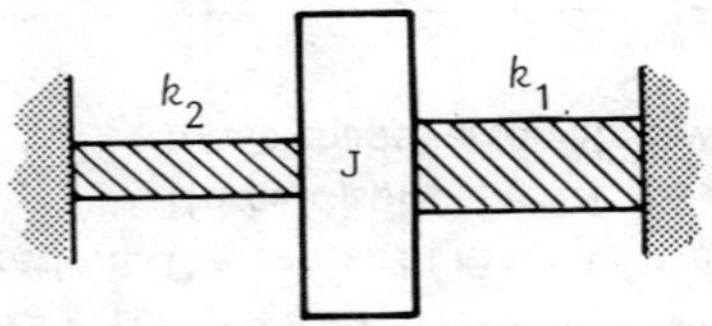	$\frac{1}{2\pi}\left(\frac{k_1 + k_2}{J}\right)^{1/2}$	--

8. Two Equal Bodies, Three Equal Springs 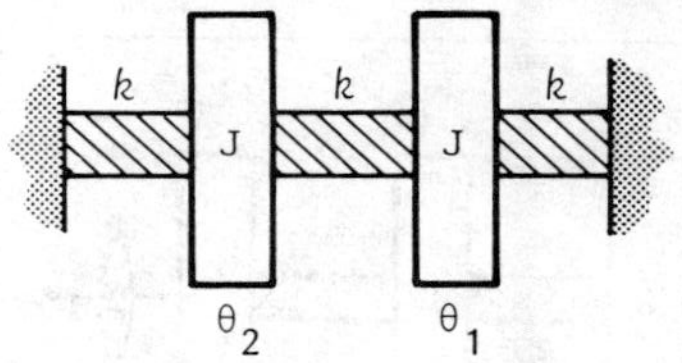	$\frac{1}{2\pi}\left(\frac{k}{J}\right)^{1/2}$, $\frac{1}{2\pi}\left(\frac{3k}{J}\right)^{1/2}$	$\begin{bmatrix}\tilde{\theta}_1\\ \tilde{\theta}_2\end{bmatrix} = \begin{bmatrix}1\\ 1\end{bmatrix}$, $\begin{bmatrix}1\\ -1\end{bmatrix}$
9. Two Unequal Bodies, Three Unequal Springs 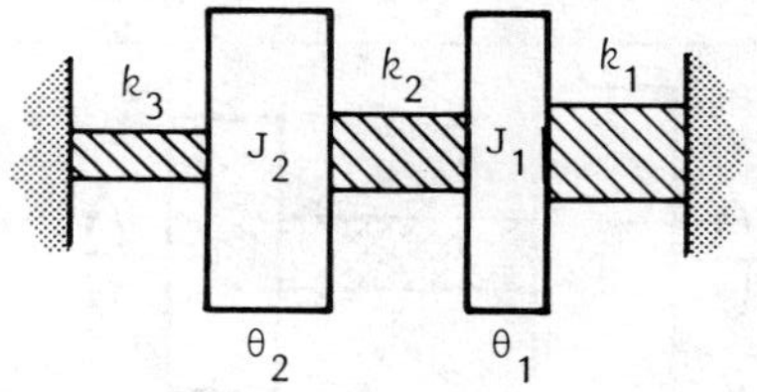	$\frac{1}{2^{3/2}\pi}\left\{\frac{k_1 + k_2}{J_1} + \frac{k_2 + k_3}{J_2} \mp \left[\left(\frac{k_1 + k_2}{J_1} + \frac{k_2 + k_3}{J_2}\right)^2 - \frac{4}{J_1 J_2}(k_1 k_2 + k_2 k_3 + k_1 k_3)\right]^{1/2}\right\}^{1/2}$	$\begin{bmatrix}\tilde{\theta}_1\\ \tilde{\theta}_2\end{bmatrix}_i = \begin{bmatrix}1\\ 1 + \frac{k_1}{k_2} - \frac{J_1}{k_2}(2\pi f_i)^2\end{bmatrix}$ $i = 1,2$
10. Three Equal Bodies, Four Equal Springs 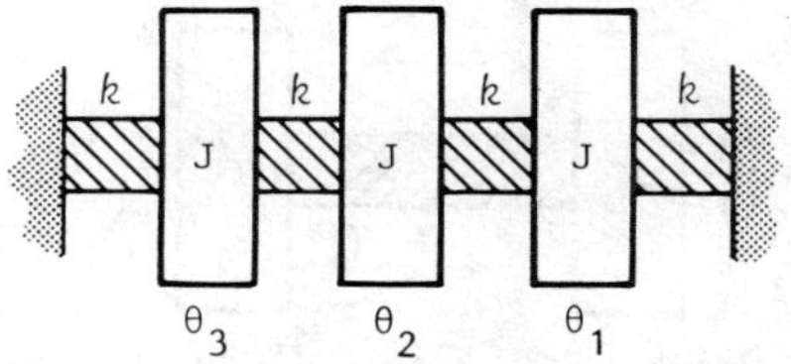	$\frac{(2 - 2^{1/2})^{1/2}}{2\pi}\left(\frac{k}{J}\right)^{1/2}$, $\frac{2^{1/2}}{2\pi}\left(\frac{k}{J}\right)^{1/2}$, $\frac{(2 + 2^{1/2})^{1/2}}{2\pi}\left(\frac{k}{J}\right)^{1/2}$	$\begin{bmatrix}\tilde{\theta}_1\\ \tilde{\theta}_2\\ \tilde{\theta}_3\end{bmatrix} = \begin{bmatrix}1\\ 2^{1/2}\\ 1\end{bmatrix}$, $\begin{bmatrix}1\\ 0\\ -1\end{bmatrix}$, $\begin{bmatrix}1\\ -2^{1/2}\\ 1\end{bmatrix}$
11. N Equal Bodies, N + 1 Equal Springs 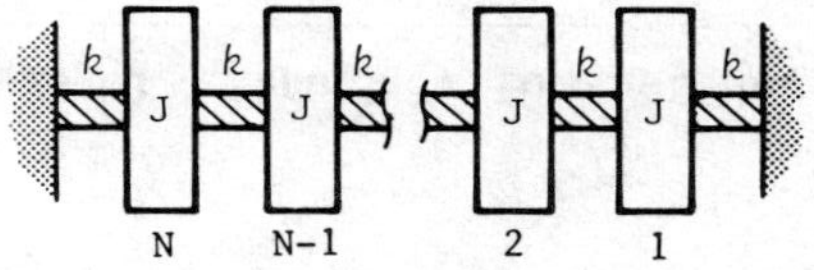	$\frac{\alpha_{N,i}}{2\pi}\left(\frac{k}{J}\right)^{1/2}$; $i = 1,2,3,\cdots,N$ $\alpha_{N,i} = 2 \sin\left[\frac{i}{(N + 1)}\frac{\pi}{2}\right]$	Ref. 6-4

Table 6-3. Rigid Body, Torsion Spring Systems. (*Continued*)

Geometry	Natural Frequency (hertz), f_i	Mode Shape and Remarks
12. Two Equal Bodies, One Spring 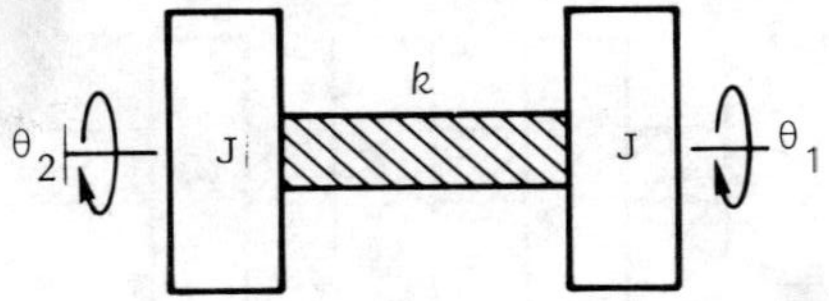	$0, \quad \frac{1}{2\pi}\left(\frac{2k}{J}\right)^{1/2}$	$\begin{bmatrix}\tilde{\theta}_1\\ \tilde{\theta}_2\end{bmatrix} = \begin{bmatrix}1\\ 1\end{bmatrix}, \begin{bmatrix}1\\ -1\end{bmatrix}$
13. Two Unequal Bodies, One Spring 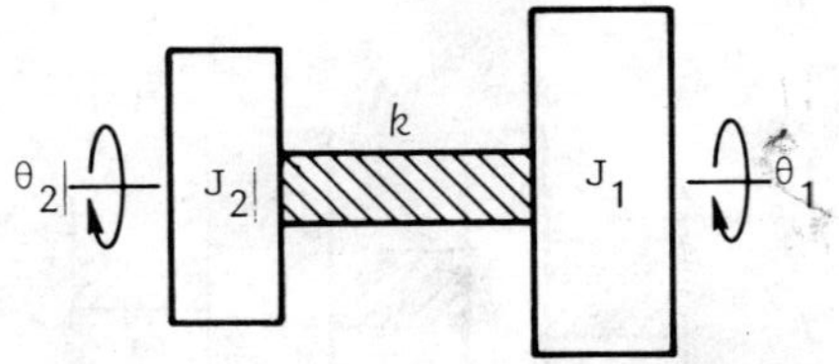	$0, \quad \frac{1}{2\pi}\left[\frac{k\,(J_1 + J_2)}{J_1 J_2}\right]^{1/2}$	$\begin{bmatrix}\tilde{\theta}_1\\ \tilde{\theta}_2\end{bmatrix} = \begin{bmatrix}1\\ 1\end{bmatrix}, \begin{bmatrix}1\\ -\frac{J_1}{J_2}\end{bmatrix}$
14. Three Equal Bodies, Two Equal Springs 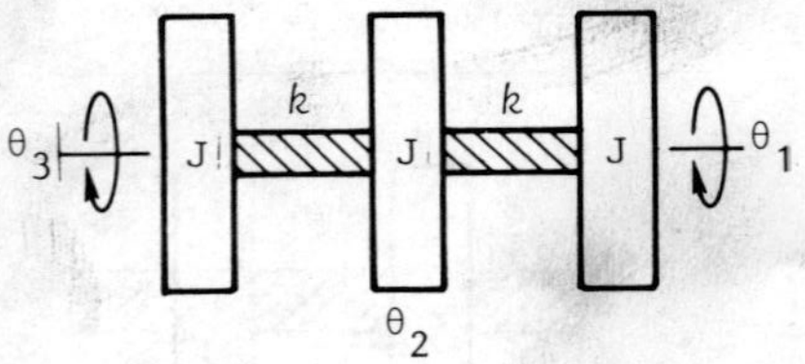	$0, \quad \frac{1}{2\pi}\left(\frac{k}{J}\right)^{1/2}, \quad \frac{1}{2\pi}\left(\frac{3k}{J}\right)^{1/2}$	$\begin{bmatrix}\tilde{\theta}_1\\ \tilde{\theta}_2\\ \tilde{\theta}_3\end{bmatrix} = \begin{bmatrix}1\\ 1\\ 1\end{bmatrix}, \begin{bmatrix}1\\ 0\\ -1\end{bmatrix}, \begin{bmatrix}1\\ -2\\ 1\end{bmatrix}$

15. Three Unequal Bodies, Two Unequal Springs 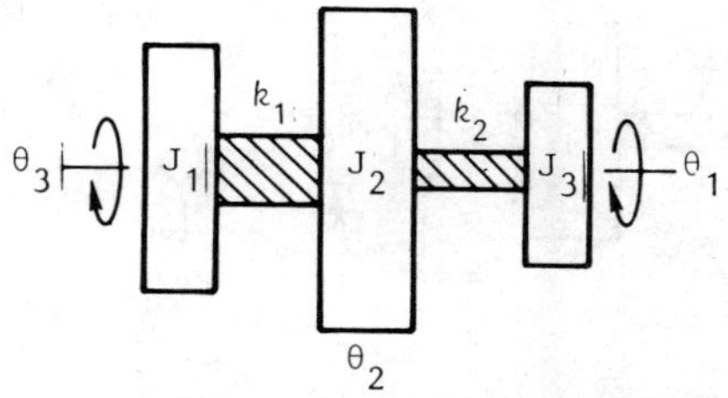	$0,\quad \frac{1}{2^{3/2}\pi}\left\{\frac{k_1+k_2}{J_2}+\frac{k_1}{J_1}+\frac{k_2}{J_3}\mp\left[\left(\frac{k_1+k_2}{J_2}+\frac{k_1}{J_1}+\frac{k_2}{J_3}\right)^2-4k_1k_2\left(\frac{1}{J_1J_2}+\frac{1}{J_2J_3}+\frac{1}{J_1J_3}\right)\right]^{1/2}\right\}^{1/2}$	$\begin{bmatrix}\tilde\theta_1\\ \tilde\theta_2\\ \tilde\theta_3\end{bmatrix}_i=\begin{bmatrix}1\\ 1-\frac{J_1}{k_1}(2\pi f_i)^2\\ \left[1-\frac{J_1}{k_1}(2\pi f_i)^2\right]\Big/\left[1-\frac{J_3}{J_2}(2\pi f_i)^2\right]\end{bmatrix}$ $i = 1,2,3$
16. Four Equal Bodies, Three Equal Springs 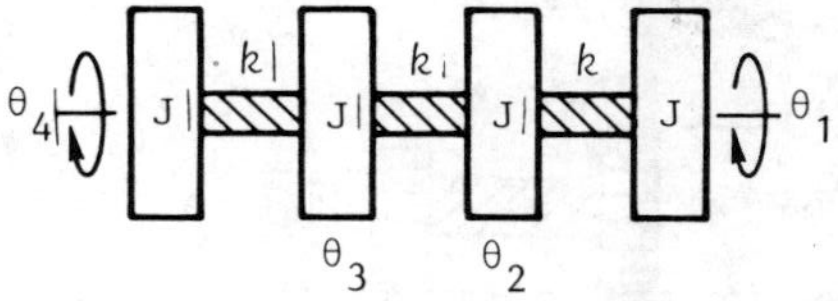	$0,\quad \frac{(2-2^{1/2})^{1/2}}{2\pi}\left(\frac{k}{J}\right)^{1/2},\quad \frac{2^{1/2}}{2\pi}\left(\frac{k}{J}\right)^{1/2},$ $\frac{(2+2^{1/2})^{1/2}}{2\pi}\left(\frac{k}{J}\right)^{1/2}$	$\begin{bmatrix}\tilde\theta_1\\ \tilde\theta_2\\ \tilde\theta_3\\ \tilde\theta_4\end{bmatrix}=\begin{bmatrix}1\\ 1\\ 1\\ 1\end{bmatrix},\begin{bmatrix}1\\ -1+2^{1/2}\\ 1-2^{1/2}\\ -1\end{bmatrix},\begin{bmatrix}1\\ -1\\ -1\\ -1\end{bmatrix},\begin{bmatrix}1\\ -1-2^{1/2}\\ 1+2^{1/2}\\ -1\end{bmatrix}$
17. N Equal Bodies, N − 1 Equal Springs 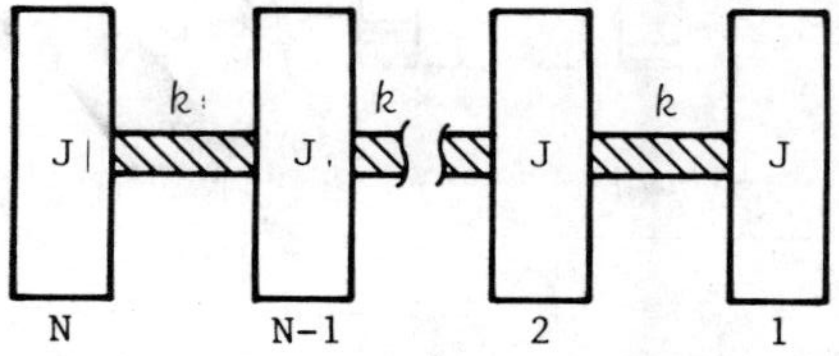	$\frac{\alpha_{N,i}}{2\pi}\left(\frac{k}{J}\right)^{1/2}\quad ;\quad i = 1,2,3,\cdots,N$ $\alpha_{N,i} = 2\sin\left[\frac{(i-1)}{N}\frac{\pi}{2}\right]$	Ref. 6-4
18. One-Body, Three-Spring, Geared System 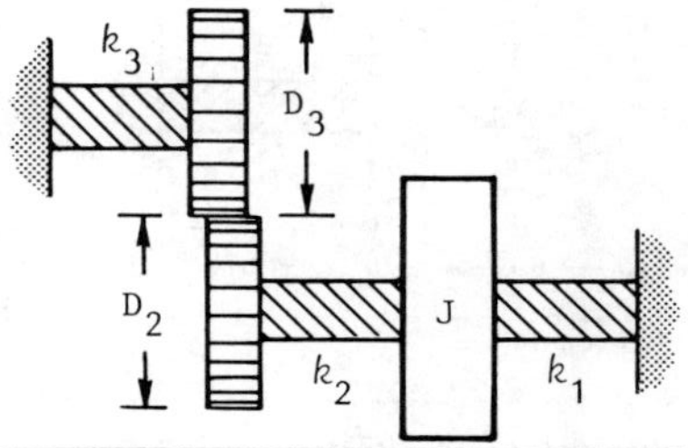	$\frac{1}{2\pi}\left(\frac{k_1+k_{eq}}{J}\right)^{1/2}$	$k_{eq}=\frac{k_2k_3}{n^2k_2+k_3}$ $n=\frac{D_3}{D_2}=\frac{\text{speed of shaft 2}}{\text{speed of shaft 3}}$ Gears are massless.

Table 6-3. Rigid Body, Torsion Spring Systems. (*Continued*)

Geometry	Natural Frequency (hertz), f_i	Mode Shape and Remarks
19. Two Massive Gears, Two Springs 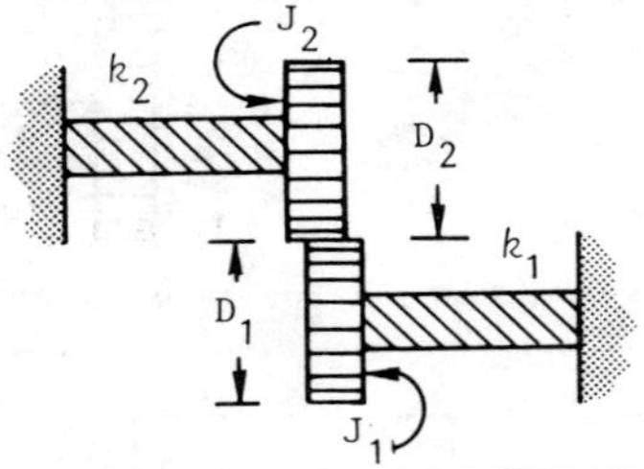	$\frac{1}{2\pi}\left(\frac{k_1 + n^2 k_2}{J_1 + n^2 J_2}\right)^{1/2}$	$n = \frac{D_1}{D_2}$
20. Four Massive Gears, Three Springs 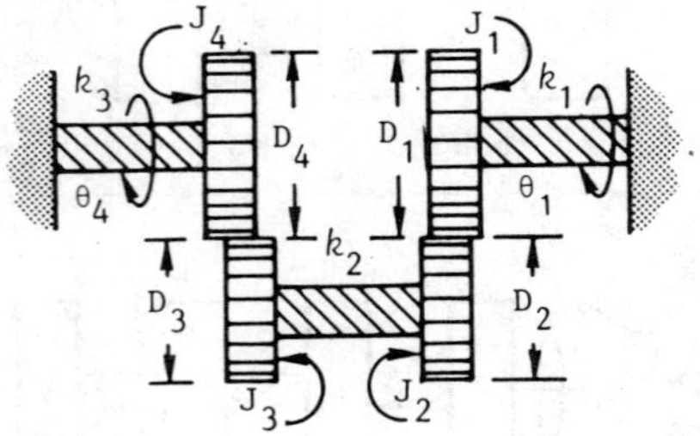	$\frac{1}{2^{3/2}\pi}\left\{\frac{k'_3}{J'_4} + \frac{k'_1}{J'_1} \mp \left[\left(\frac{k'_3}{J'_4} + \frac{k'_1}{J'_1}\right)^2 - \frac{4}{J'_1 J'_4}\left(k'_1 k'_3 - n_1^2 n_4^2 k_2^2\right)\right]^{1/2}\right\}^{1/2}$ $J'_1 = J_1 + n_1^2 J_2$, $k'_3 = k_3 + n_4^2 k_2$ $J'_4 = J_4 + n_4^2 J_3$, $k'_1 = k_1 + n_1^2 k_2$	$\begin{bmatrix}\tilde{\theta}_1 \\ \tilde{\theta}_4\end{bmatrix}_i = \begin{bmatrix}1 \\ \frac{k'_1}{n_1 n_4 k_2} - \frac{J'_1}{n_1 n_4 k_2}(2\pi f_i)^2\end{bmatrix}$ i=1,2 $n_1 = \frac{D_1}{D_2}$, $n_4 = \frac{D_4}{D_3}$
21. Two-Body, Three-Spring, Geared System 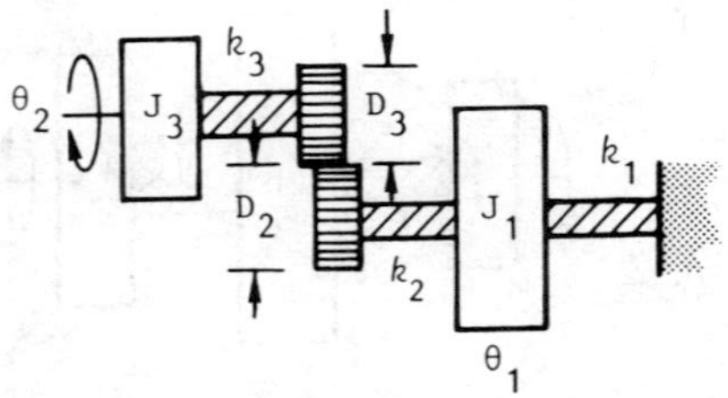	$\frac{1}{2^{3/2}\pi}\left\{\frac{k_1}{J_1} + \frac{n^2 k_{eq}}{J_1} + \frac{k_{eq}}{J_3} \mp \left[\left(\frac{k_1}{J_1} + \frac{n^2 k_{eq}}{J_1} + \frac{k_{eq}}{J_3}\right)^2 - \frac{4k_1 k_{eq}}{J_1 J_3}\right]^{1/2}\right\}^{1/2}$ $k_{eq} = \frac{k_2 k_3}{k_2 + n^2 k_3}$, $n = \frac{D_2}{D_3}$	$\begin{bmatrix}\tilde{\theta}_1 \\ \tilde{\theta}_3\end{bmatrix}_i = \begin{bmatrix}1 \\ -n^2 - \frac{k_1}{k_{eq}} + \frac{J_1}{k_{eq}}(2\pi f_i)^2\end{bmatrix}$ i=1,2 Gears are massless.

22. Two-Body, Two-Spring Geared System 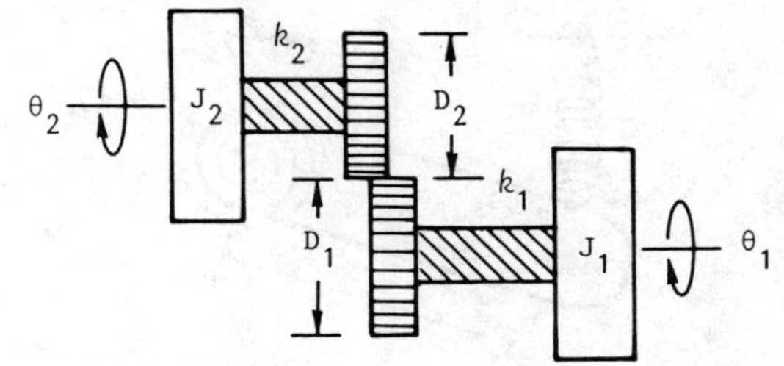	$0, \quad \frac{1}{2\pi}\left[\frac{\left(J_1 + n^2 J_2\right) k_1 k_2}{\left(k_1 + n^2 k_2\right) J_1 J_2}\right]^{1/2}$ $n = \frac{D_1}{D_2} = \frac{\text{speed of shaft 2}}{\text{speed of shaft 1}}$	$\begin{bmatrix}\tilde{\theta}_1 \\ \tilde{\theta}_2\end{bmatrix} = \begin{bmatrix}1 \\ -n\end{bmatrix}, \begin{bmatrix}1 \\ \frac{J_1}{J_2}\end{bmatrix}$ Gears are massless.
23. Two-Body, Two-Massive-Gear, Two-Spring System 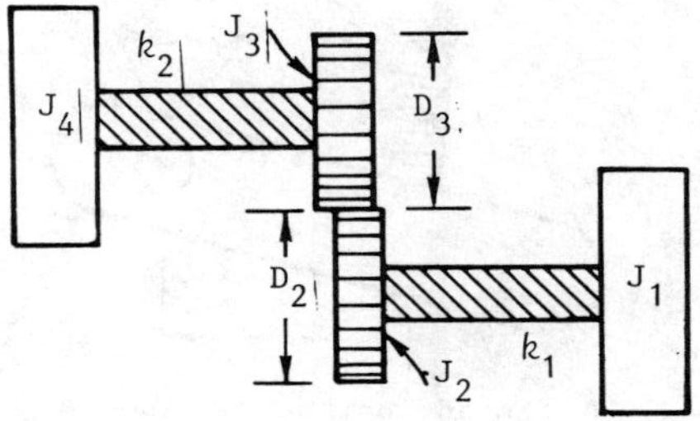	$0, \quad \frac{1}{2^{3/2}\pi}\left\{w_a^2 + w_b^2 \pm \left[\left(w_a^2 - w_b^2\right)^2 + 4w_c^2 w_d^2\right]\right\}^{1/2}$	$w_a^2 = \frac{k_1}{J_1} + \frac{k_1}{J_2 + n^2 J_3}, \quad n = \frac{D_2}{D_3}$ $w_b^2 = \frac{k_2}{J_4} + \frac{k_2 n^2}{J_2 + n^2 J_3},$ $w_c^2 = \frac{k_1}{J_2 + n^2 J_3}, \quad w_d^2 = \frac{k_2 n^2}{J_2 + n^2 J_3}$ Ref. 6-4
24. Mass, Extension Spring, Rigid Lever 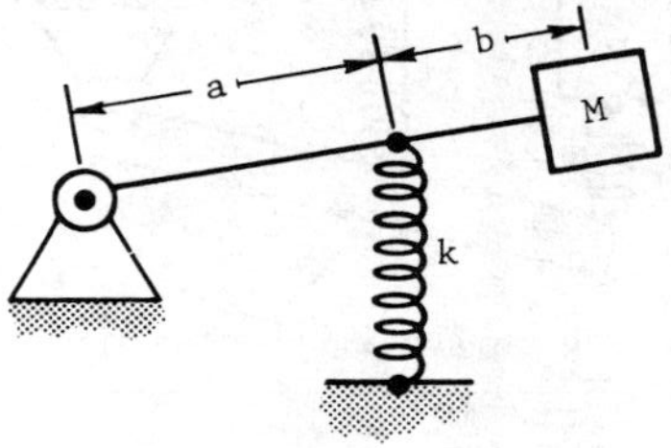	$\frac{1}{2\pi}\left[\frac{ka^2}{M\ (a + b)^2}\right]^{1/2}$	--

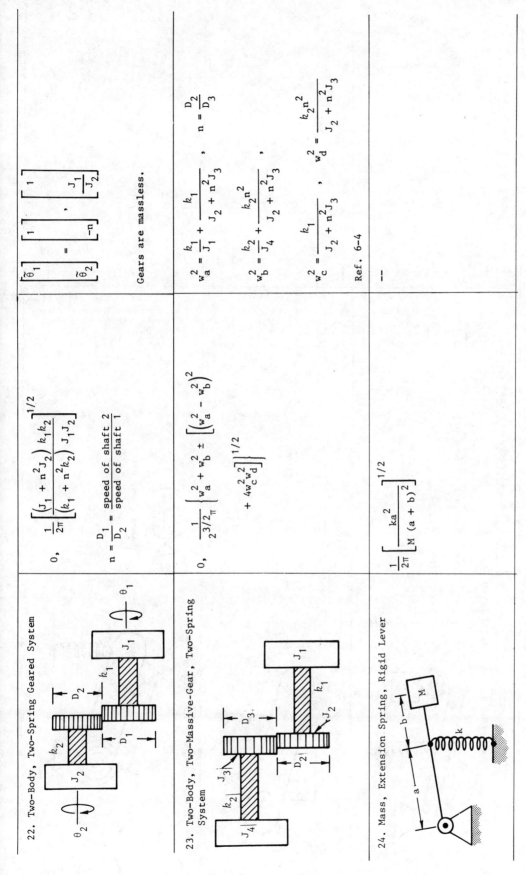

Table 6-3. Rigid Body, Torsion Spring Systems. (*Continued*)

Geometry	Natural Frequency (hertz), f_i	Mode Shape and Remarks
25. Mass, Two Extension Springs, Rigid Lever 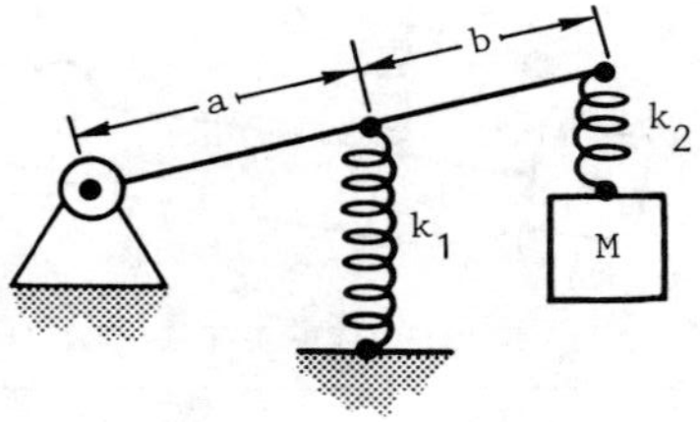	$\frac{1}{2\pi}\left\{\frac{k_1 k_2 a}{M\,[k_1 a + k_2\,(a+b)]}\right\}^{1/2}$	--
26. Body, Extension Spring 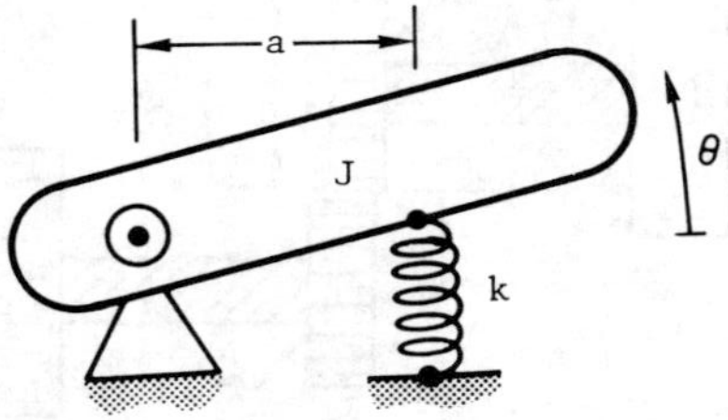	$\frac{1}{2\pi}\left(\frac{ka^2}{J}\right)^{1/2}$ J taken about pivot.	--
27. Body, Extension and Torsion Spring 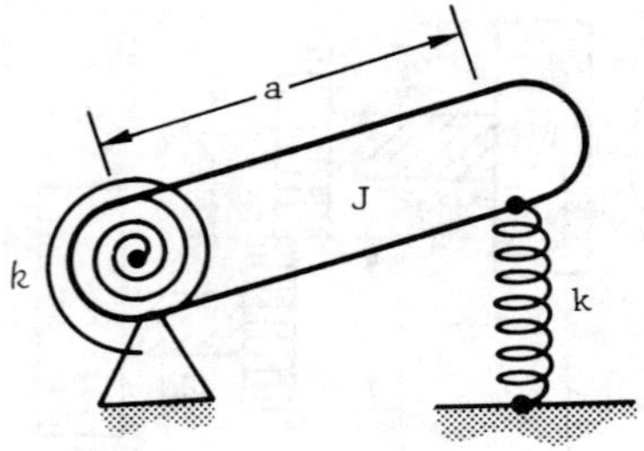	$\frac{1}{2\pi}\left(\frac{ka^2 + k}{J}\right)^{1/2}$ J taken about pivot.	--

28. Body, Two Extension Springs 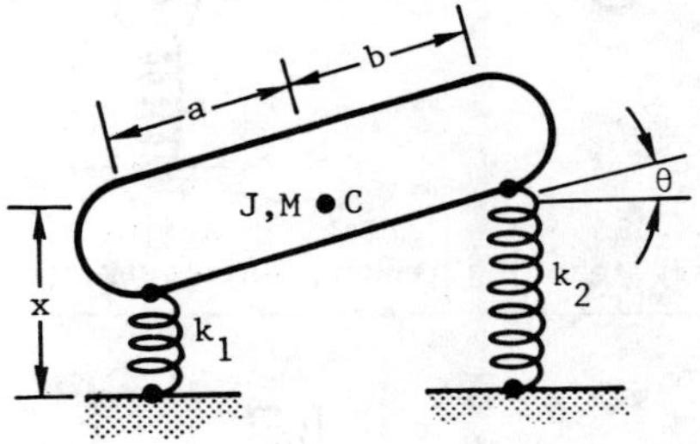	$\frac{1}{2^{3/2}\pi}\left\{\frac{k_1 + k_2}{M} + \frac{k_1a^2 + k_2b^2}{J} \pm \left[\left(\frac{k_1 + k_2}{M} + \frac{k_1a^2 + k_2b^2}{J}\right)^2 - \frac{4k_1k_2\ (a + b)^2}{JM}\right]^{1/2}\right\}^{1/2}$ J taken about center of mass.	$\begin{bmatrix}\tilde{x} \\ \tilde{\theta}\end{bmatrix}_i = \begin{bmatrix}1 \\ \frac{k_1 + k_2 - M(2\pi f_i)^2}{k_1a - k_2b}\end{bmatrix}$ i=1,2
29. Mass, Spring, Pulley 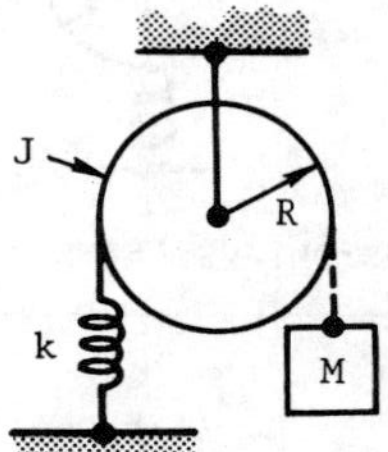	$\frac{1}{2\pi}\left(\frac{k}{M + J/R^2}\right)^{1/2}$ J taken about center of pulley.	--
30. Roller, Spring 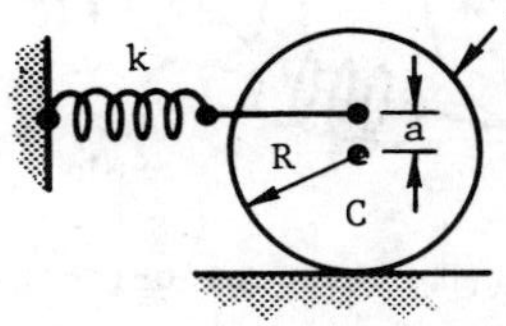	$\frac{1}{2\pi}\left[\frac{k\ (R + a)^2}{MR^2 + J}\right]^{1/2}$ J taken about center of curvature.	--

Table 6-3. Rigid Body, Torsion Spring Systems. *(Continued)*

Geometry	Natural Frequency (hertz), f_i	Mode Shape and Remarks
31. Two Anchored Pulleys, Two Springs 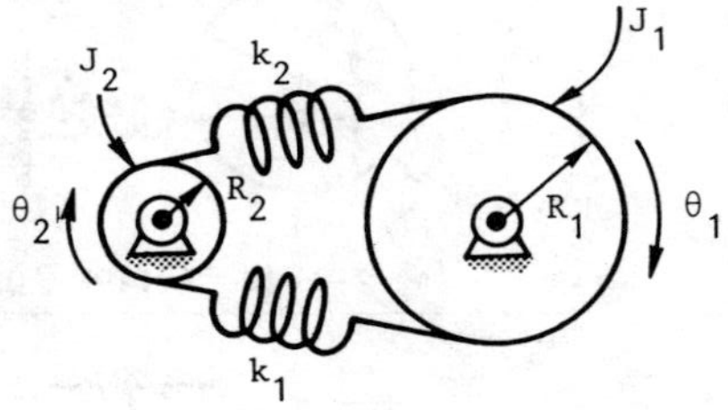	$0, \quad \frac{1}{2\pi}\left[\frac{(k_1 + k_2)\ (R_1^2 J_2 + R_2^2 J_1)}{J_1 J_2}\right]^{1/2}$	$\begin{bmatrix}\tilde{\theta}_1 \\ \tilde{\theta}_2\end{bmatrix} = \begin{bmatrix}1 \\ \frac{R_1}{R_2}\end{bmatrix}, \begin{bmatrix}1 \\ \frac{-R_2 J_1}{R_1 J_2}\end{bmatrix}$
32. Mass, Spring-Supported Pulley 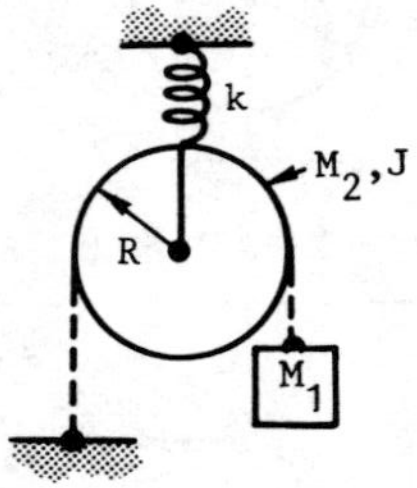	$\frac{1}{2\pi}\left(\frac{k}{M_2 + 4M_1 + J/R^2}\right)^{1/2}$	--
33. Mass, Two Unequal Springs, Massive Pulley 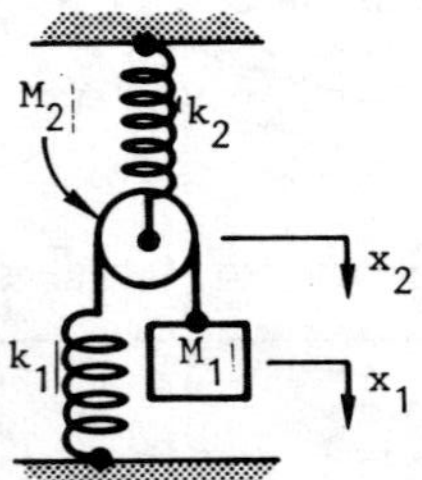	$\frac{1}{2^{3/2}\pi}\left\{\frac{k_1}{M_1} + \frac{k_2}{M_2} + \frac{4k_1}{M_2} \mp \left[\left(\frac{k_1}{M_1} + \frac{k_2}{M_2} + \frac{4k_1}{M_2}\right)^2 - \frac{4k_1 k_2}{M_1 M_2}\right]^{1/2}\right\}^{1/2}$	$\begin{bmatrix}\tilde{x}_1 \\ \tilde{x}_2\end{bmatrix}_i = \begin{bmatrix}1 \\ \frac{1}{2} - \frac{M_1}{2k_1}(2\pi f_i)^2\end{bmatrix}$ $i=1,2$ Pulley has no rotational inertia, i.e., $J = 0$.

34. Mass, Suspended Massive Pulley 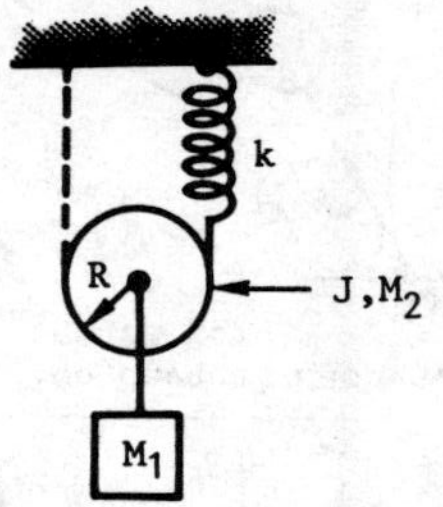	$\frac{1}{\pi}\left(\frac{k}{M_1 + M_2 + J/R^2}\right)^{1/2}$ J taken about center of pulley.	--
35. Mass, Two Unequal Springs, Massive Pulley 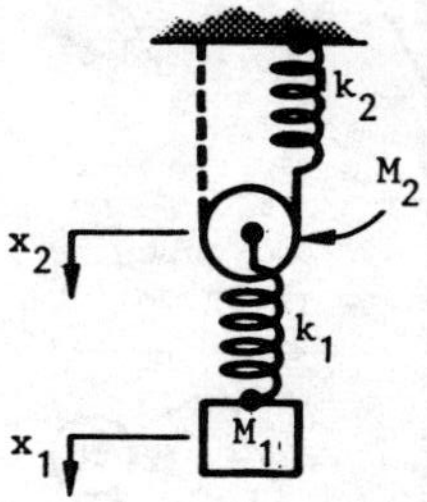	$\frac{1}{2^{3/2}\pi}\left\{\frac{k_1}{M_1} + \frac{k_1}{M_2} + \frac{4k_2}{M_2} \mp \left[\left(\frac{k_1}{M_1} + \frac{k_1}{M_2} + \frac{4k_2}{M_2}\right)^2 - \frac{16k_1k_2}{M_1M_2}\right]^{1/2}\right\}^{1/2}$	$\begin{bmatrix}\tilde{x}_1 \\ \tilde{x}_2\end{bmatrix}_i = \begin{bmatrix}1 \\ 1 - \frac{M_1}{k_1}(2\pi f_i)^2\end{bmatrix}$ i=1,2 Pulley has no rotational inertia, i.e., J = 0.
36. Body, Torsion and Displacement Springs 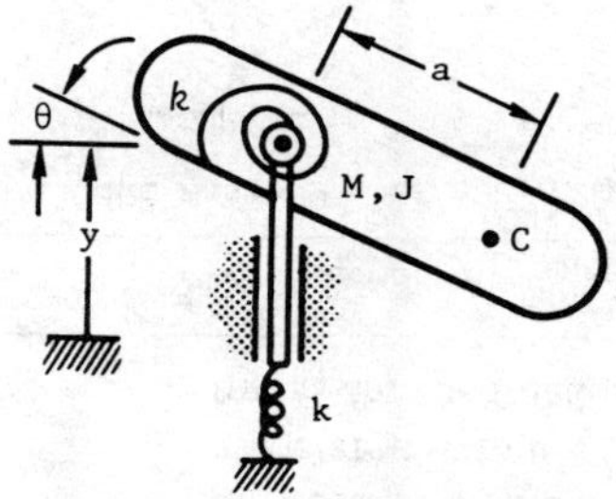	$\frac{\left\{f_y^2 + f_\theta^2 \mp \left[(f_y^2 + f_\theta^2)^2 - 4f_y^2 f_\theta^2(1 - Ma^2/J)\right]^{1/2}\right\}^{1/2}}{2^{1/2}(1 - Ma^2/J)^{1/2}}$ f_y and f_θ are the uncoupled frequencies (hertz): $f_y = \frac{1}{2\pi}\left(\frac{k}{M}\right)^{1/2}$, $f_\theta = \frac{1}{2\pi}\left(\frac{k}{J}\right)^{1/2}$	$\begin{bmatrix}\tilde{y} \\ \tilde{\theta}\end{bmatrix}_i = \begin{bmatrix}1 \\ \frac{1}{a}\left[1 - \frac{k}{M(2\pi f_i)^2}\right]\end{bmatrix}$ i=1,2 J taken about pivot. When a=0 the system is uncoupled and $\tilde{y}$ and $\tilde{\theta}$ are independent.

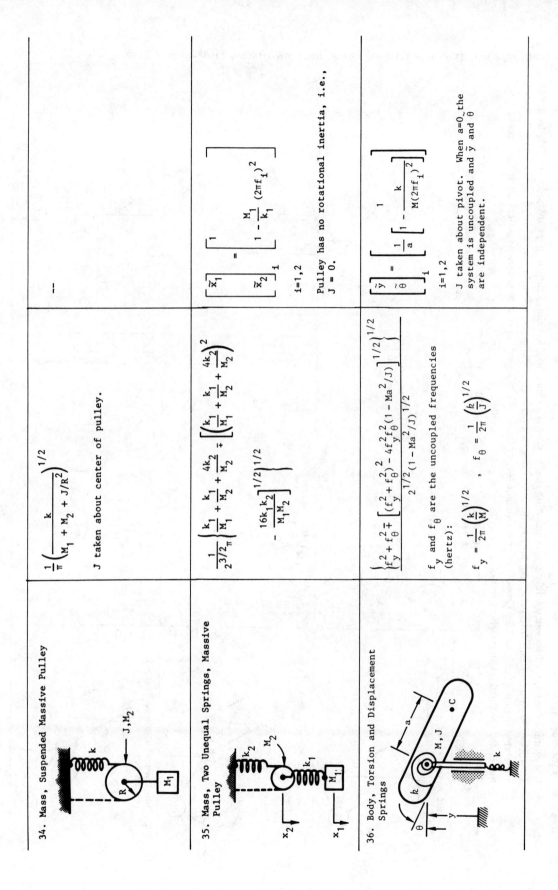

Table 6-4. Pendulum Systems.

Notation: C = center of mass; f = frequency (Hz); g = acceleration due to gravity (Table 3-1); k = deflection spring constant (force/deflection); $\mathscr{k}$ = torsion spring constant (moment/angular rotation); x = displacement; θ, ϕ = angular deflection (radians); $\tilde{\theta}$, $\tilde{\phi}$, $\tilde{x}$ = mode shapes associated with θ, ϕ, and x. See Table 3-1 for consistent sets of units. All connecting rods are massless unless otherwise noted; formulas are for small angular deflections only.

Geometry	Natural Frequency (hertz), f_i	Mode Shape and Remarks
1. One Mass 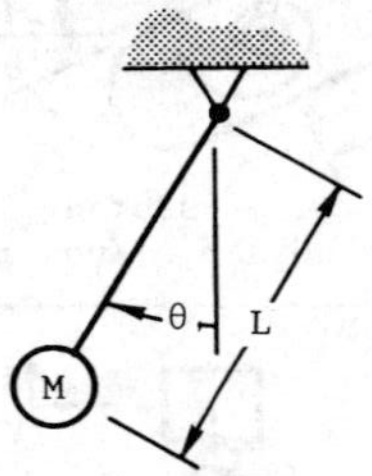	$\frac{1}{2\pi}\left(\frac{g}{L}\right)^{1/2}$	--
2. Two Equal Masses, Equal-Length Rods 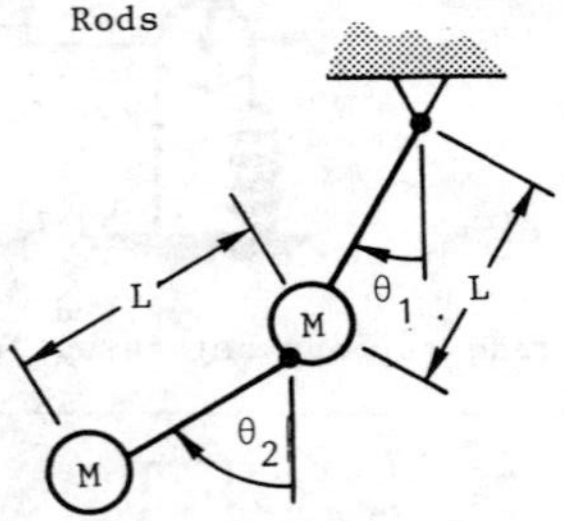	$\frac{1}{2\pi}\left(2 \mp 2^{1/2}\right)^{1/2}\left(\frac{g}{L}\right)^{1/2}$	$\begin{bmatrix}\tilde{\theta}_1\\ \tilde{\theta}_2\end{bmatrix} = \begin{bmatrix}1\\ 2^{1/2}\end{bmatrix}, \begin{bmatrix}1\\ -2^{1/2}\end{bmatrix}$
3. Two Unequal Masses, Unequal-Length Rods 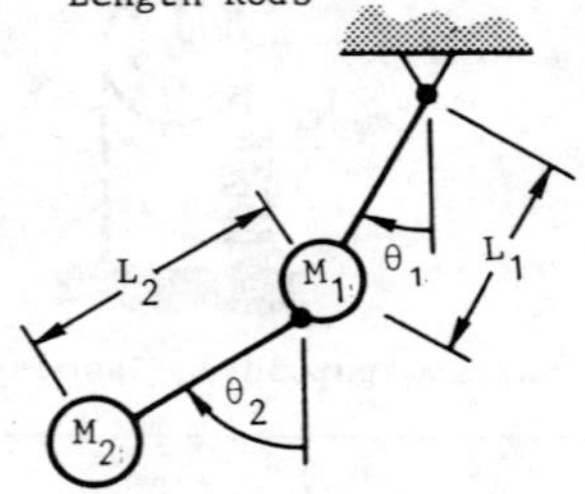	$\frac{g^{1/2}}{2^{3/2}\pi}\left\{\left(1+\frac{M_2}{M_1}\right)\left(\frac{1}{L_1}+\frac{1}{L_2}\right) \mp \left[\left(1+\frac{M_2}{M_1}\right)^2 \left(\frac{1}{L_1}+\frac{1}{L_2}\right)^2 - \frac{4}{L_1L_2}\left(1+\frac{M_2}{M_1}\right)\right]^{1/2}\right\}^{1/2}$	$\begin{bmatrix}\tilde{\theta}_1\\ \tilde{\theta}_2\end{bmatrix}_i = \begin{bmatrix}1\\ \dfrac{L_1\,(2\pi f_i)^2}{g - L_2\,(2\pi f_i)^2}\end{bmatrix}$ i = 1,2

4. Three Equal Masses, Equal-Length Rods 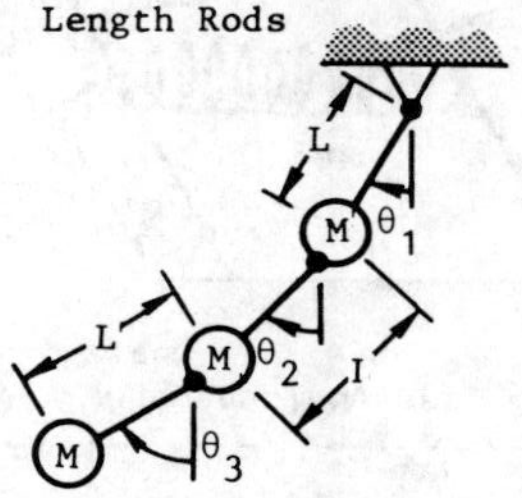	$0.1026\left(\frac{g}{L}\right)^{1/2}$, $0.2411\left(\frac{g}{L}\right)^{1/2}$, $0.3992\left(\frac{g}{L}\right)^{1/2}$	$\begin{bmatrix}\tilde{\theta}_1\\ \tilde{\theta}_2\\ \tilde{\theta}_3\end{bmatrix} = \begin{bmatrix}0.2549\\ 0.5842\\ 1.0\end{bmatrix}, \begin{bmatrix}-0.9567\\ -1.2943\\ 1.0\end{bmatrix}, \begin{bmatrix}8.2019\\ -5.2900\\ 1.0\end{bmatrix}$
5. Chain 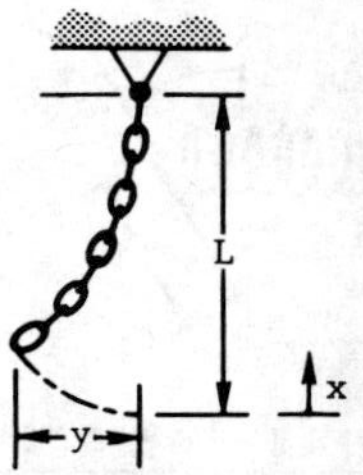	$\frac{\lambda_i}{2\pi}\left(\frac{g}{L}\right)^{1/2}$; $i=1,2,3,\cdots$ $\lambda_i = 1.2026, 2.7602, 4.3266, 5.8955, \cdots, [J_o(2\lambda_i) = 0]$	$\tilde{y}_i(x) = J_o\left[2\lambda_i\left(\frac{x}{L}\right)^{1/2}\right]$ Ref. 6-18 Note: J_o = Bessel function of first kind and zero order
6. Chain with End Mass 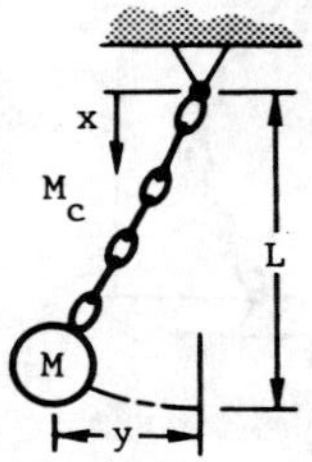	$\frac{1}{2\pi}\left(\frac{g}{L}\frac{M + M_c/2}{M + M_c/3}\right)^{1/2}$ M_c = mass of chain First mode only	$\tilde{y}(x) \approx \frac{x}{L}$

Table 6-4. Pendulum Systems. *(Continued)*

<table>
<tr><th>Geometry</th><th>Natural Frequency (hertz), f_i</th><th>Mode Shape and Remarks</th></tr>
<tr><td>7. Chain with Constraint
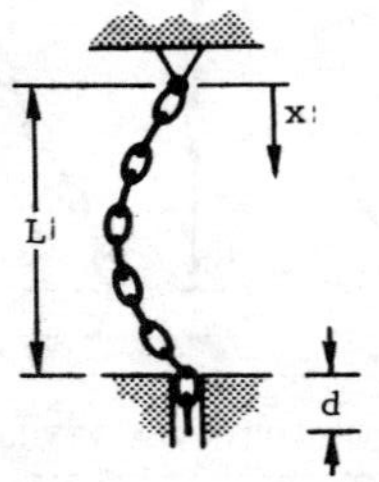</td><td>$\frac{\lambda_i}{2\pi}\left(\frac{g}{L}\right)^{1/2}$; $i = 1,2,3$

d + L = total length of chain</td><td><table><tr><th></th><th colspan="5">d/L</th></tr><tr><th></th><th>0</th><th>0.01</th><th>0.1</th><th>1.0</th><th>10.0</th></tr><tr><td>λ_1</td><td>1.203</td><td>1.648</td><td>2.108</td><td>3.787</td><td>10.177</td></tr><tr><td>λ_2</td><td>2.760</td><td>3.410</td><td>4.268</td><td>7.582</td><td>20.354</td></tr><tr><td>λ_3</td><td>4.327</td><td>5.160</td><td>6.418</td><td>11.375</td><td>30.531</td></tr></table>Ref. 6-19</td></tr>
<tr><td>8. Two Equal Masses, Coupled System
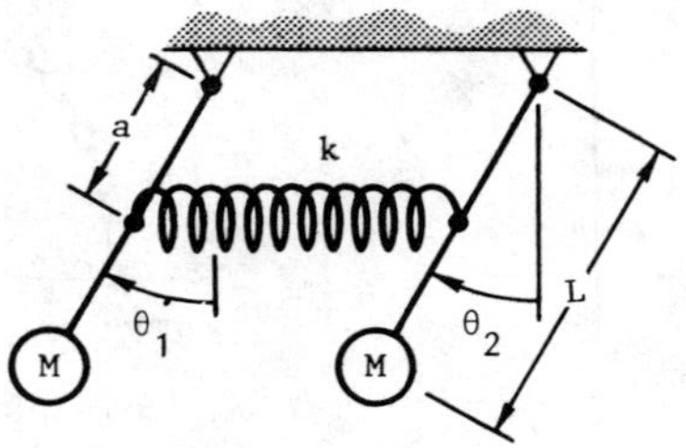</td><td>$\frac{1}{2\pi}\left(\frac{g}{L}\right)^{1/2}$, $\frac{1}{2\pi}\left(\frac{g}{L} + \frac{2ka^2}{ML^2}\right)^{1/2}$</td><td>$\begin{bmatrix}\tilde{\theta}_1\\ \tilde{\theta}_2\end{bmatrix} = \begin{bmatrix}1\\ 1\end{bmatrix} , \begin{bmatrix}1\\ -1\end{bmatrix}$</td></tr>
<tr><td>9. Two Unequal Masses, Coupled System
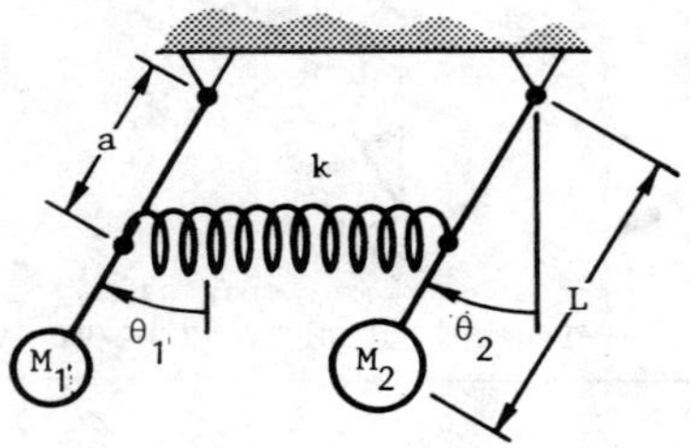</td><td>$\frac{1}{2^{3/2}\pi}\left(\frac{2g}{L} + \frac{a^2k}{L^2}\left(\frac{1}{M_1} + \frac{1}{M_2}\right) \mp \left\{\left[\frac{2g}{L} + \frac{a^2k}{L^2}\left(\frac{1}{M_1} + \frac{1}{M_2}\right)\right]^2 - 4\left[\frac{g^2}{L^2} + \frac{ga^2k}{L^3}\left(\frac{1}{M_1} + \frac{1}{M_2}\right)\right]\right\}^{1/2}\right)^{1/2}$</td><td>$\begin{bmatrix}\tilde{\theta}_1\\ \tilde{\theta}_2\end{bmatrix}_i = \begin{bmatrix}1\\ 1 + \frac{M_1 gL}{a^2k} - \frac{M_1 L^2}{a^2k}(2\pi f_i)^2\end{bmatrix}$

$i = 1,2$</td></tr>
</table>

10. Three Equal Masses, Coupled System 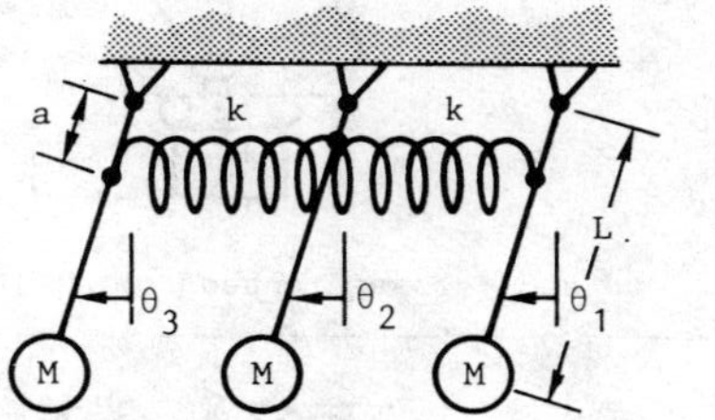	$\frac{1}{2\pi}\left(\frac{g}{L}\right)^{1/2}$, $\frac{1}{2\pi}\left(\frac{g}{L}+\frac{ka^2}{ML^2}\right)^{1/2}$, $\frac{1}{2\pi}\left(\frac{g}{L}+\frac{3ka^2}{ML^2}\right)^{1/2}$	$\begin{bmatrix}\tilde{\theta}_1\\ \tilde{\theta}_2\\ \tilde{\theta}_3\end{bmatrix} = \begin{bmatrix}1\\1\\1\end{bmatrix}, \begin{bmatrix}1\\0\\-1\end{bmatrix}, \begin{bmatrix}1\\-2\\1\end{bmatrix}$
11. Mass with Deflection Spring 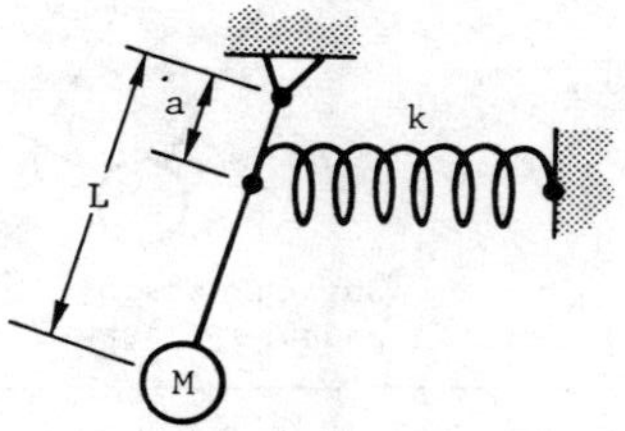	$\frac{1}{2\pi}\left(\frac{g}{L}+\frac{ka^2}{ML^2}\right)^{1/2}$	--
12. Mass with Torsion Spring 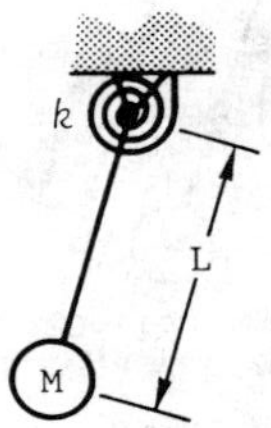	$\frac{1}{2\pi}\left(\frac{g}{L}+\frac{k}{ML^2}\right)^{1/2}$	--

Table 6-4. Pendulum Systems. *(Continued)*

Geometry	Natural Frequency (hertz), f_i	Mode Shape and Remarks
13. Mass, Inverted System, Deflection Spring 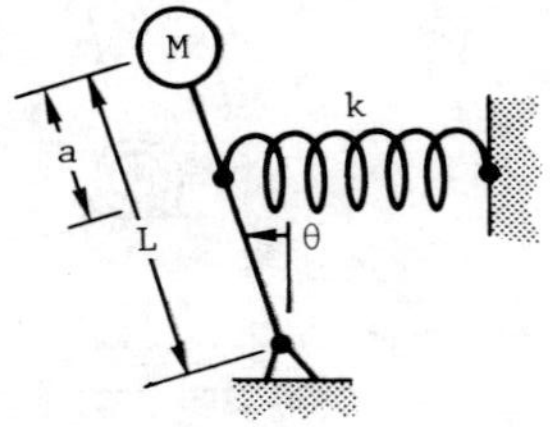	$\frac{1}{2\pi}\left(\frac{ka^2}{ML^2} - \frac{g}{L}\right)^{1/2}$	--
14. Mass, Inverted System, Torsion Spring 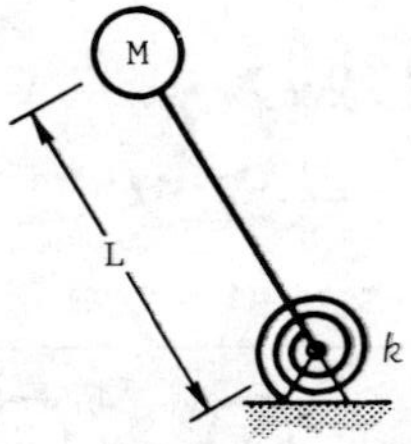	$\frac{1}{2\pi}\left(\frac{k}{ML^2} - \frac{g}{L}\right)^{1/2}$	--
15. Two Unequal Masses, One on Rollers 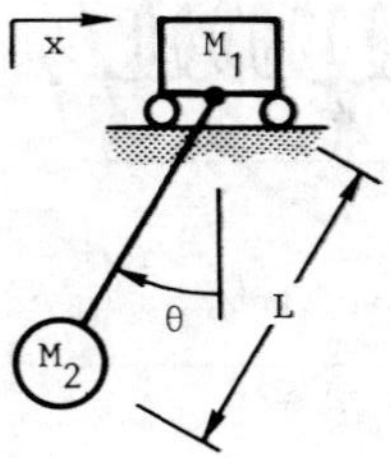	$0, \quad \frac{1}{2\pi}\left[\frac{g}{L}\left(1 + \frac{M_2}{M_1}\right)\right]^{1/2}$	$\begin{bmatrix}\tilde{x}\\ \tilde{\theta}\end{bmatrix} = \begin{bmatrix}1\\ 0\end{bmatrix}, \begin{bmatrix}1\\ -\left(1 + \frac{M_1}{M_2}\right)\frac{1}{L}\end{bmatrix}$

16. Two Unequal Masses, One on Rollers, Spring 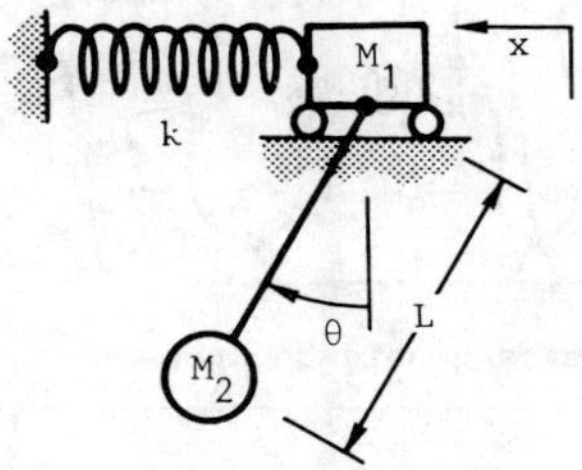	$\frac{1}{2^{3/2}\pi}\left(\left(1+\frac{M_2}{M_1}\right)\frac{g}{L}+\frac{k}{M_1}\mp\left\{\left[\left(1+\frac{M_2}{M_1}\right)\frac{g}{L}+\frac{k}{M_1}\right]^2-\frac{4kg}{M_1L}\right\}^{1/2}\right)^{1/2}$	$\begin{bmatrix}\tilde{x}\\ \tilde{\theta}\end{bmatrix}_i=\begin{bmatrix}1\\ \frac{k}{M_2L\,(2\pi f_i)^2}-\left(1+\frac{M_1}{M_2}\right)\frac{1}{L}\end{bmatrix}$ i = 1,2
17. Two Equal Masses, Torsion Bar 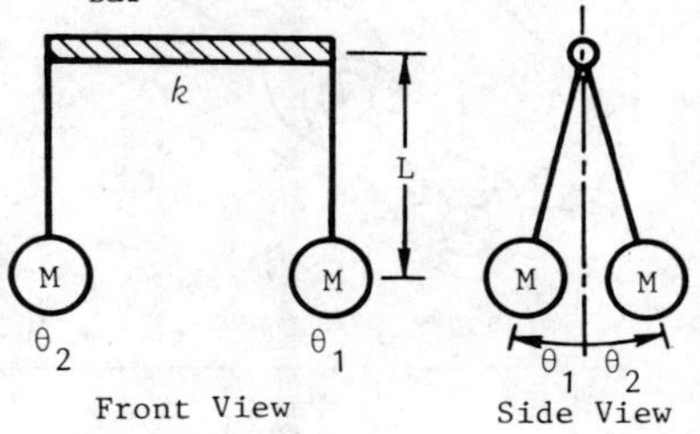	$\frac{1}{2\pi}\left(\frac{g}{L}\right)^{1/2}$, $\frac{1}{2\pi}\left(\frac{g}{L}+\frac{2k}{ML^2}\right)^{1/2}$	$\begin{bmatrix}\tilde{\theta}_1\\ \tilde{\theta}_2\end{bmatrix}=\begin{bmatrix}1\\ 1\end{bmatrix},\begin{bmatrix}1\\ -1\end{bmatrix}$
18. Mass, Inclined Axis of Rotation 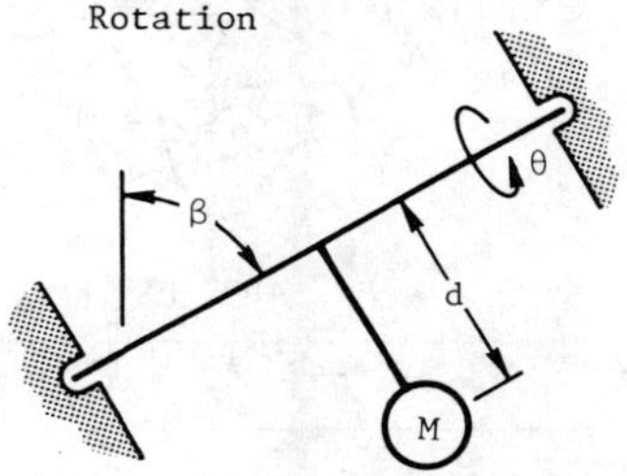	$\frac{1}{2\pi}\left(\frac{g\sin\beta}{d}\right)^{1/2}$	--

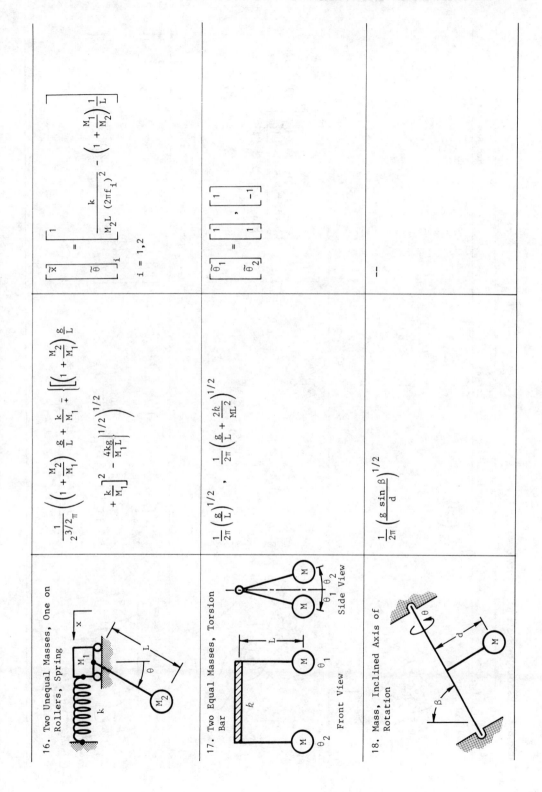

Table 6-4. Pendulum Systems. *(Continued)*

Geometry	Natural Frequency (hertz), f_i	Mode Shape and Remarks
19. Rigid Body 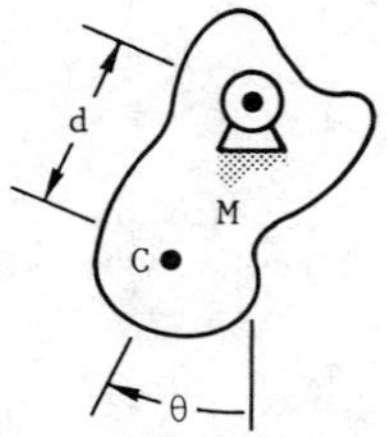	$\frac{1}{2\pi}\left(\frac{Mgd}{J}\right)^{1/2}$	--
20. Two Pinned Rigid Bodies 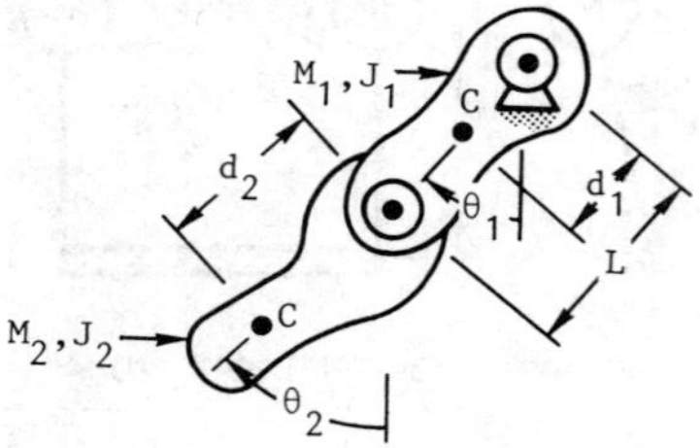	$\frac{g^{1/2}}{2^{3/2}\pi a^{1/2}}\left[b \mp (b^2 - 4ac)^{1/2}\right]^{1/2}$ $a = J_1'J_2' - M_2^2 d_2^2 L^2$ $b = J_1'M_2d_2 + J_2'(M_1d_1 + M_2L)$ $c = M_2d_2(M_1d_1 + M_2L)$	$\begin{bmatrix}\tilde{\theta}_1 \\ \tilde{\theta}_2\end{bmatrix}_i = \begin{bmatrix} 1 \\ \frac{M_1gd_1 + M_2gL}{M_2d_2L(2\pi f_i)^2} - \frac{J_1'}{M_2d_2L}\end{bmatrix}$ $i = 1,2$ $J_1' = J_1 + M_1d_1^2 + M_2L^2$ $J_2' = J_2 + M_2d_2^2$ J_1, J_2 about centers of gravity, C.
21. Two-Body Coupled System 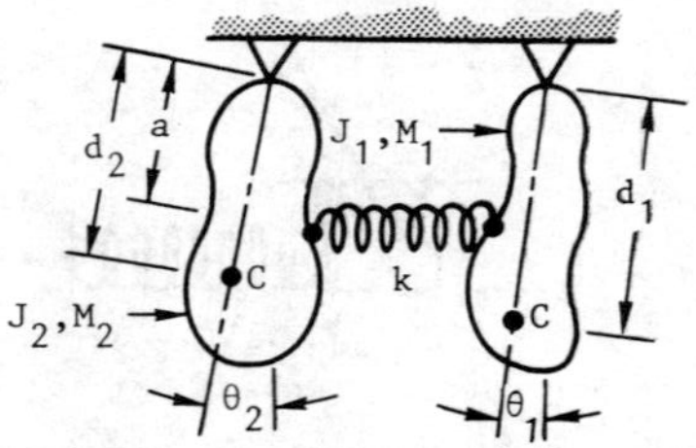	$\frac{1}{2^{3/2}\pi}\left[b \mp (b^2 - 4c)^{1/2}\right]^{1/2}$ $b = \frac{ka^2}{J_1} + \frac{ka^2}{J_2} + \frac{gM_1d_1}{J_1} + \frac{gM_2d_2}{J_2}$ $c = \frac{g^2M_1M_2d_1d_2}{J_1J_2} + \frac{ka^2gM_1d_1}{J_1J_2} + \frac{ka^2gM_2d_2}{J_1J_2}$	$\begin{bmatrix}\tilde{\theta}_1 \\ \tilde{\theta}_2\end{bmatrix}_i = \begin{bmatrix} 1 \\ 1 + \frac{gM_1d_1}{ka^2} - \frac{J_1}{ka^2}(2\pi f_i)^2\end{bmatrix}$ $i = 1,2$

22. Two Bodies, Torsion Bar 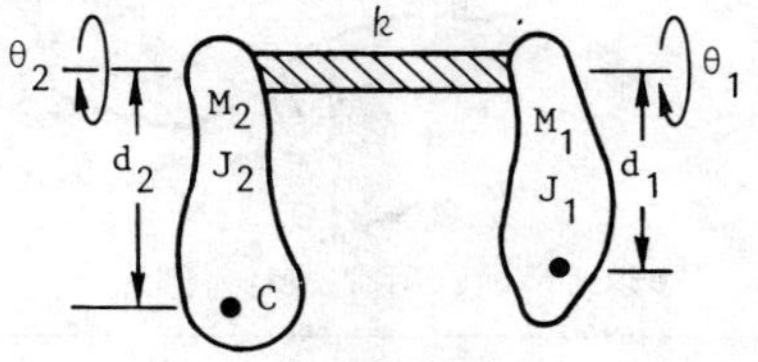 Front View	$\frac{1}{2^{3/2}\pi}\left[b \mp (b^2 - 4c)^{1/2}\right]^{1/2}$ $b = \frac{k}{J_1} + \frac{k}{J_2} + \frac{gd_2M_2}{J_2} + \frac{gd_1M_1}{J_1}$ $c = \frac{g^2 d_1 d_2 M_1 M_2}{J_1 J_2} + \frac{kgd_1M_1}{J_1J_2} + \frac{kgd_2M_2}{J_1J_2}$	$\begin{bmatrix}\tilde{\theta}_1 \\ \tilde{\theta}_2\end{bmatrix}_i = \begin{bmatrix} 1 \\ 1 + \frac{gd_1M_1}{k} - \frac{J_1}{k}(2\pi f_i)^2\end{bmatrix}$ $i = 1,2$
23. Rigid Body, Inverted 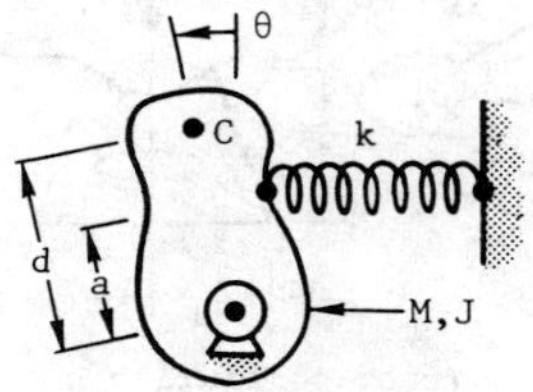	$\frac{1}{2\pi}\left(\frac{ka^2}{J} - \frac{Mgd}{J}\right)^{1/2}$	--
24. String-Suspended Body 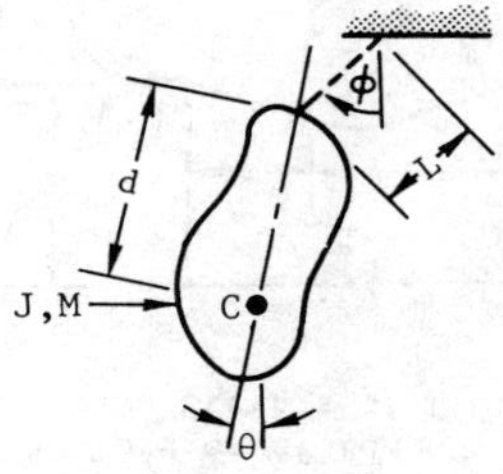	$\frac{g^{1/2}}{2^{3/2}\pi J^{1/2}}\left(Md + (J + Md^2)/L \mp \left\{\left[Md + (J + Md^2)/L\right]^2 - 4JMd/L\right\}^{1/2}\right)^{1/2}$	$\begin{bmatrix}\tilde{\phi} \\ \tilde{\theta}\end{bmatrix}_i = \begin{bmatrix} 1 \\ \frac{g}{d\,(2\pi f_i)^2} - \frac{L}{d}\end{bmatrix}$ $i = 1,2$ J about center of gravity, C.

Table 6-4. Pendulum Systems. (*Continued*)

Geometry	Natural Frequency (hertz), f_i	Mode Shape and Remarks
25. Torsion of Body Suspended by Two Strings of Equal Length 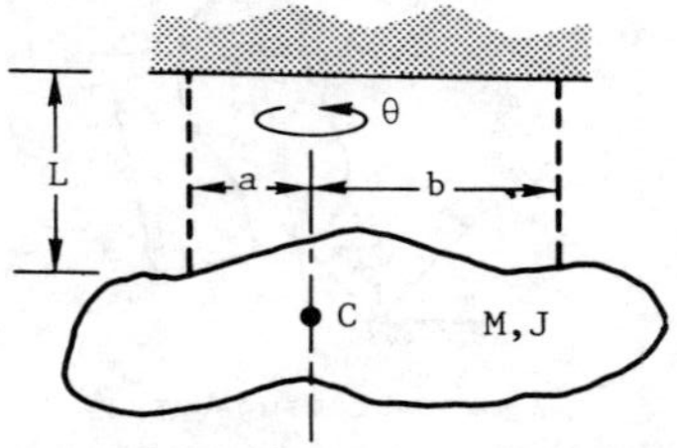	$\frac{1}{2\pi}\left(\frac{Mgab}{JL}\right)^{1/2}$	-- J about center of gravity, C. Rotation is about center of gravity. Strings are vertical.
26. Torsion of Body Suspended by N Strings of Equal Length 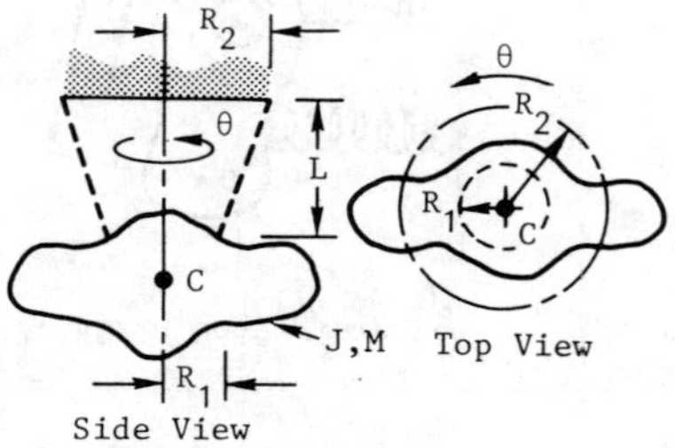	$\frac{1}{2\pi}\left(\frac{MgR_1R_2}{JL}\right)^{1/2}$	Result is independent of number of suspension strings. Rotation is about center of gravity, C. Strings are splayed between circles of radius R_1 at body and R_2 at attachment at angular intervals such that each string bears the same tension. J about center of gravity, C.
27. Rigid Body, Inclined Axis 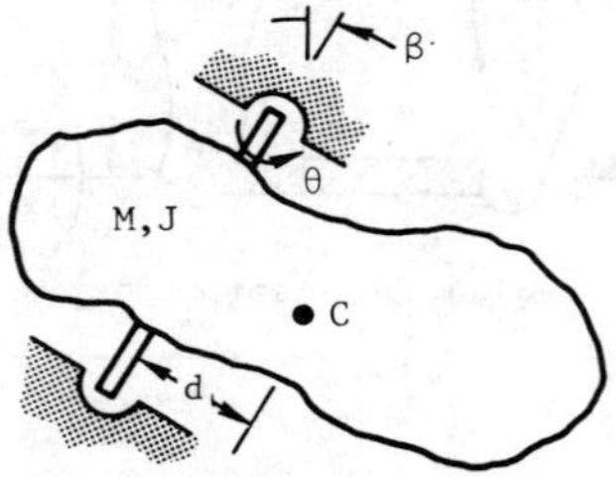	$\frac{1}{2\pi}\left(\frac{Mgd\ \sin\beta}{J}\right)^{1/2}$	--

28. Ball or Cylinder in Trough 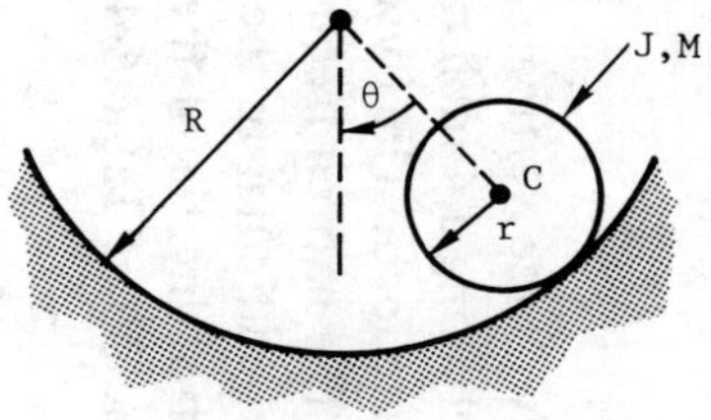	$\frac{1}{2\pi}\left[\frac{Mg}{(R - r)\ (M + J/r^2)}\right]^{1/2}$	J about center of ball. Ball does not slip.
29. Body with Flat Bottom Balanced on a Semicircle 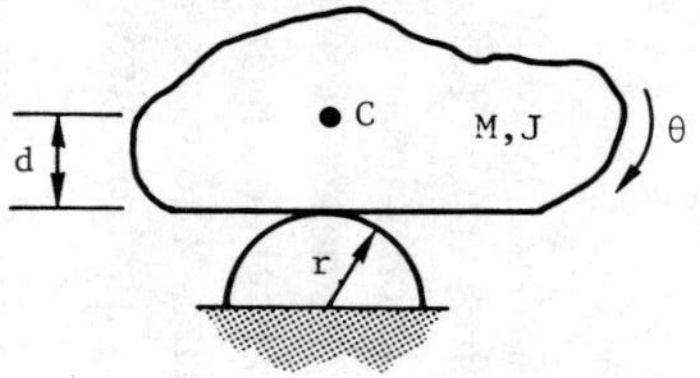	$\frac{1}{2\pi}\left[\frac{(r - d)\ Mg}{J + Md^2}\right]^{1/2}$	J about center of gravity, C. Body does not slip.
30. Body with Rounded Bottom 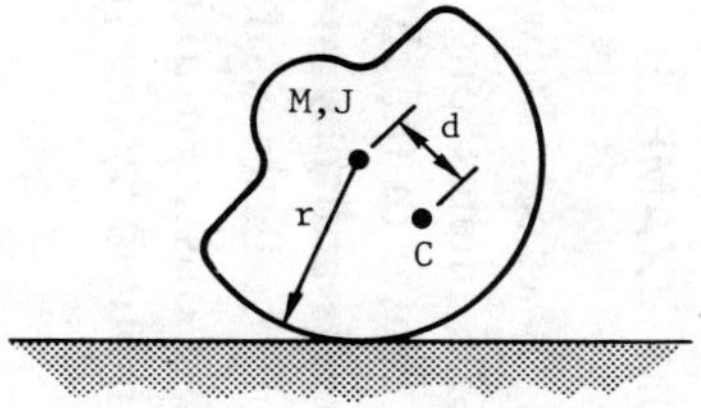	$\frac{1}{2\pi}\left[\frac{Mgd}{J + M\ (r - d)^2}\right]^{1/2}$	J about center of gravity, C. Body does not slip.

So far in this analysis, the mass of the cable has been neglected. A 110-meter length of steel cable, 2 centimeters in diameter, will have a mass of:

$$M_C = \frac{\mu \pi D_C^2}{4} \times L,$$

$$= 8000 \text{ kg/m}^2 \times \frac{\pi}{4} \times (0.02 \text{ m})^2 \times 110 \text{ m},$$

$$= 276.5 \text{ kg}.$$

Since the mass of the cable is comparable to the mass of the solid object at the end of the cable, it is reasonable to believe the cable mass will significantly affect the natural frequency of the system. If the pulley is neglected, the natural frequency of a spring-mass can be estimated from frame 1 of Table 8-8:

$$f = \frac{1}{2\pi}\left(\frac{k}{M + 0.33\ M_C}\right)^{1/2},$$

$$= \frac{1}{2\pi}\left(\frac{571{,}200 \text{ N/m}^2}{400 \text{ kg} + 0.33 \times 276 \text{ kg}}\right)^{1/2},$$

$$= 5.42 \text{ Hz}.$$

This figure is the best estimate of the extensional natural frequency of the system in Fig. 6-1.

The pendulum mode can be analyzed by neglecting the stretching of the cable and using frame 6 of Table 6-4. The fundamental natural frequency of the pendulum mode is found to be:

$$f_1 = \frac{1}{2\pi}\left(\frac{9.807 \text{ m/sec}^2}{100 \text{ m}} \ \frac{400 \text{ kg} + 276 \text{ kg}/2}{400 \text{ kg} + 276 \text{ kg}/3}\right)^{1/2},$$

$$= 0.0521 \text{ Hz} = 1/19.2 \text{ sec}.$$

6.6. PENDULUM EXAMPLE

A circular steel flywheel, shown in Fig. 6-2, is suspended by three vertical wires which are each 9.5 inches long. The wires attach to the wheel along a 4-inch-radius circle about the center of the flywheel. Each wire bears the same load. The flywheel weighs 3.50 pounds and has an overall diameter of 8 inches. The natural frequency of small torsional oscillations of the flywheel about an axis perpendicular to the face of the flywheel is observed to be 1.4 Hz. The mass moment of inertia of the flywheel can be computed by inverting the equation given in frame 26 of Table 6-4:

$$J = \frac{Mg\ R^2}{(2\pi f)^2\ L}.$$

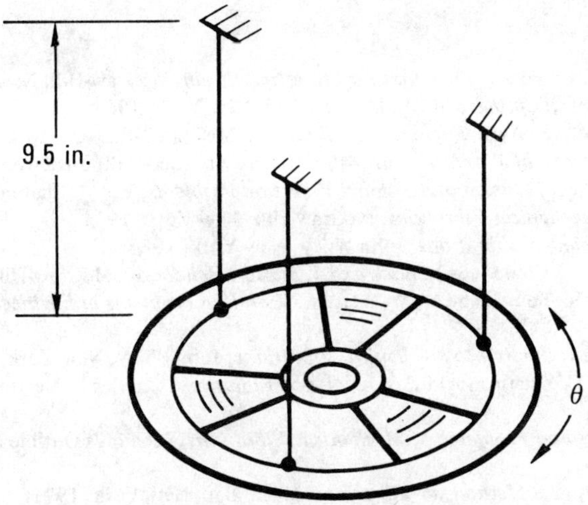

Fig. 6-2. Circular flywheel suspended by three wires.

Using the units of frame 7 of Table 3-1, the required parameters are:

$$f = 1.4 \text{ Hz},$$
$$L = 9.5 \text{ in.},$$
$$g = 386.1 \text{ in./sec}^2,$$
$$M = 3.5 \text{ lb}/g = 9.058 \times 10^{-3} \text{ lb-sec}^2\text{/in.},$$
$$R = 4.0 \text{ in.}$$

Thus,

$$J = 0.07618 \text{ lb-in.-sec}^2.$$

J can be put in more conventional units such that pound is a measure of mass by multiplying this equation by g:

$$J = 29.4 \text{ lb-in.}^2.$$

The polar mass moment of inertia of a disk of thickness t about an axis through its center of gravity and perpendicular to the face of the disk can be adapted from the formulas given in frame 20 of Table 5-1:

$$J = \mu t(I_{x_C} + I_{y_C}) = \frac{\pi}{2}\mu t R^4.$$

If the disk is made of steel ($\mu = 0.289$ pound/inch^3) and is 4.0 inches in radius, then the required thickness to yield $J = 29.4$ pound-inch^2 is:

$$t = 0.253 \text{ in.}$$

REFERENCES

6-1. Roark, R. J., and W. C. Young, *Formulas for Stress and Strain*, McGraw-Hill, New York, 1975.
6-2. Wahl, A. M., *Mechanical Springs*, 2d ed., McGraw-Hill, New York, 1963.
6-3. Jacobsen, L. S., and R. S. Ayre, *Engineering Vibrations*, McGraw-Hill, New York, 1958.
6-4. Thomson, W. T., *Theory of Vibration with Applications*, Prentice-Hall, Englewood Cliffs, N.J., 1972.
6-5. Wilson, W. K., *Practical Solution of Torsional Vibration Problems*, Vol. 1, Chapman & Hall, London, 1956.
6-6. Den Hartog, J. P., *Mechanical Vibrations*, McGraw-Hill, New York, 1934.
6-7. Church, A. H., *Mechanical Vibrations*, John Wiley, New York, 1963.
6-8. Harris, C. M., and C. E. Crede (eds.) *Shock and Vibration Handbook*, McGraw-Hill, New York, 1961.
6-9. Timoshenko, S., D. H. Young, and W. Weaver, Jr., *Vibration Problems in Engineering*, 4th ed., John Wiley, New York, 1974.
6-10. Steidel, R. F., *An Introduction to Mechanical Vibrations*, John Wiley, New York, 1971.
6-11. Volterra, E., and E. C. Zachmanoglou, *Dynamics of Vibrations*, Charles E. Merrill Books, Columbus, Ohio, 1965.
6-12. Seto, W. W., *Theory and Problems of Mechanical Vibrations*, Schaum's Outline Series, McGraw-Hill, New York, 1964.
6-13. Meirovitch, L., *Analytical Methods in Vibrations*, Macmillan, New York, 1971.
6-14. Housner, G. W., and D. E. Hudson, *Applied Mechanics, Dynamics*, D. Van Nostrand, Princeton, N.J., 1950.
6-15. Myklestad, N. O., *Vibration Analysis*, McGraw-Hill, New York, 1944.
6-16. Bishop, R. E. D., and D. C. Johnson, *The Mechanics of Vibration*, Cambridge University Press, Cambridge, England, 1960.
6-17. Hurty, W. C., and M. F. Rubinstein, *Dynamics of Structures*, Prentice-Hall, Englewood Cliffs, N.J., 1964.
6-18. Lamb, H., *The Dynamical Theory of Sound*, Dover Press, New York, 1960 (first published in 1925), pp. 84–86.
6-19. Huang, T., and D. W. Doreing, "Frequencies of a Hanging Chain," *J. Am. Acoust. Soc.* **45**, 1046–1049 (1969).

7

CABLES AND CABLE TRUSSES

7.1. CABLES

General Case. Since the bending rigidity of slender cables is near zero, they can support only tensile loads. Cables can consist of a single strand, as in a guitar string; they can be woven from many strands, as in a rope or structural cable; or they can consist of a series of extensible links, as in an extensible chain. The natural frequencies of cables are influenced by both the mass and elasticity of the cable and the tendency of the cable to sag between supports.

The general assumptions used in the analysis of cables in this section are:

1. The cables are uniform and elastic.
2. Cables support only tension loads. The mean tension in the cable is much greater than the fluctuating component of tension during vibration.
3. Cables are either straight or sag into shallow parabolas with the maximum depth of the sag being no greater than $\frac{1}{8}$ of the cable span.
4. The vibration amplitude is small compared with the sag of the cable and the distance between vibration nodes.
5. The cables are supported at both ends by rigid tie-downs.

Most cables of practical importance are not straight; they sag to support their own weight and external loads. The sag of a cable has considerable influence on the natural frequencies and mode shapes of the cable. The difference between the fundamental natural frequency of a straight cable and one with a small sag can be as large as 300%. Thus, considerable attention will be paid in this chapter to the influence of sag on the natural frequencies and mode shapes of cables.

Mean Sag. A uniform elastic cable with mean tension T_0 which bears a uniform load of p force per unit length along its span will sag into a catenary as shown in Fig. 7-1. If the catenary is shallow, so that the maximum depth of the catenary is no greater than about $\frac{1}{8}$ of the cable span, L, then the mean tension in the cable is nearly constant along the span, and the catenary can be well approximated by a shallow parabola (Ref. 7-1),

$$\overline{Y} = \frac{pL^2}{2T_0}\left[\frac{x}{L} - \left(\frac{x}{L}\right)^2\right], \tag{7-1}$$

which is shown in Fig. 7-1. The maximum sag of the cable at midspan (x = L/2) is

$$d = \frac{pL^2}{8T_0}. \tag{7-2}$$

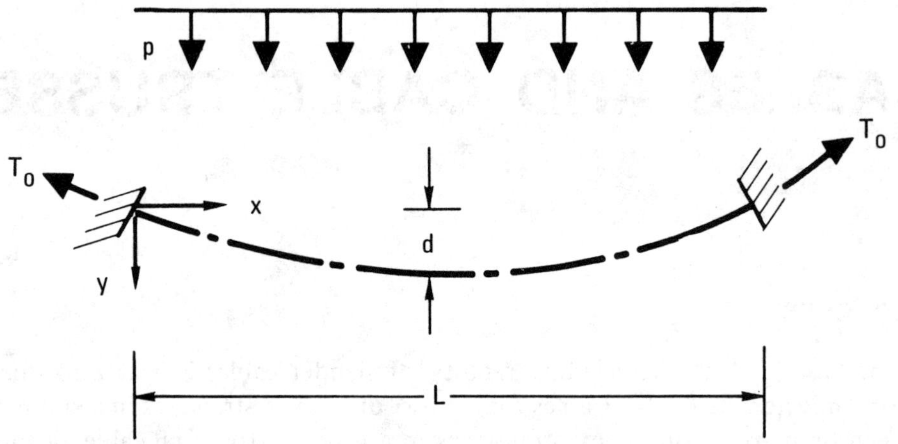

Fig. 7-1. Cable hanging in a shallow parabola.

The load per unit length on the cable, p, is ordinarily the sum of the weight per unit length of the cable plus the weight per unit length of any structures, such as bridge decks, that the cable supports. The rigidity of supported structures is neglected in the analysis.

An additional equation is required to complete the static solution. If the length of the straight, unstressed cable is L_c, then the following equation relates the mid-span mean sag d to L_c and the cable span L, if the cable does not stretch significantly under load (Ref. 7-4):

$$\frac{d}{L} = \left(\frac{5}{24}\right)^{1/2} \left[1 - \left(1 - \frac{18}{5}\frac{L_c - L}{L}\right)^{1/2}\right]^{1/2}.$$

If L_c and L are known, this equation can be used to compute d and the mean tension can then be found from Eq. 7-2. Other static solutions are given in Refs. 7-1 through 7-4.

Natural Frequencies and Mode Shapes. If the cable is given a small arbitrary displacement from static equilibrium, the resulting vibrations will have three components (Fig. 7-2): (1) horizontal motion in the plane of the catenary, X(x, t); (2) vertical motion in the plane of the catenary, Y(x, t); and (3) transverse motion out of the plane of the catenary, Z(x, t). The spatial form of the free vibrations associated with each of these components is given by the mode shapes $\tilde{x}(x)$, $\tilde{y}(x)$, and $\tilde{z}(x)$, respectively. The total displacement of the cable during free vibration is the sum of the mean sag and the dynamic displacements in each of the modes:

$$X(x, t) = \sum_i A_i \tilde{x}_i(x) \cos(\omega_i t + \phi_i),$$

$$Y(x, t) = \overline{Y}(x) + \sum_i B_i \tilde{y}_i(x) \cos(\omega_i t + \psi_i),$$

$$Z(x, t) = \sum_i C_i \tilde{z}_i(x) \cos(\omega_i t + \zeta_i),$$

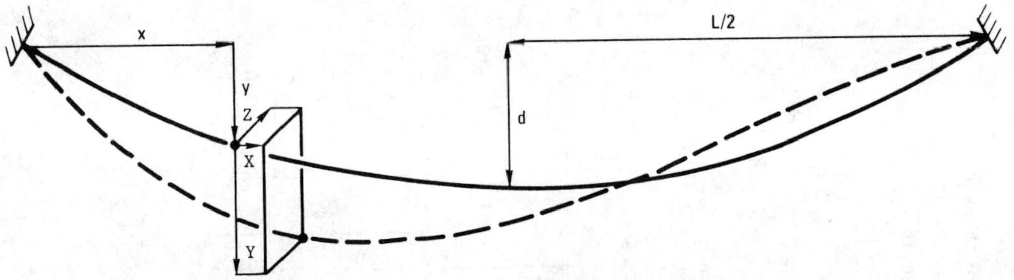

Fig. 7-2. Definition diagram for cable dynamic deflections X, Y, and Z.

where ω_i is the circular natural frequency of the cable in the i mode. A_i, B_i, and C_i are constants with the units of length and ϕ_i, ψ_i, and ζ_i are phase angles. $\overline{Y}$ is the mean sag (Eq. 7-1).

The natural frequencies and mode shapes of both straight cables and cables which have sagged into shallow catenaries are given in Table 7-1. Note that the natural frequencies of straight cables are independent of the direction of transverse (Y or Z) vibration, but the natural frequencies of sagging cables are a function of the direction of vibration relative to the plane of the catenary. By comparing frames 3 and 4 of Table 7-1 with frame 1, it can be seen that the sag has considerable influence on the natural frequencies and mode shapes. The cable sag couples the longitudinal displacement (X) and the in-plane displacement (Y) and can result in substantial shifts in the natural frequencies of the symmetric modes in the plane of the catenary. The form of typical symmetric and antisymmetric in-plane modes of sagging cables is shown in Fig. 7-3. The symmetric in-plane modes are generally nonsinusoidal and are a function of the parameter α.

The parameter α is a measure of the sag of the cable:

$$\alpha^2 = \left(\frac{8d}{L}\right)^2 \frac{EA}{T_0} \frac{L}{L_e}, \tag{7-3}$$

where E is the cable modulus (see following paragraph), A is the area of the cross section, d is the midspan sag (Fig. 7-1, Eq. 7-2), and T_0 is the mean cable tension. L_e is the virtual cable length,

$$L_e = \int_0^L \left(\frac{ds}{dx}\right)^3 dx, \tag{7-4}$$

where ds is an element along the cable. For a parabolic catenary,

$$L_e \approx L\left[1 + 8\left(\frac{d}{L}\right)^2\right].$$

L_e is not much different from L since $(d/L)^2 \ll 1$. α^2 goes to zero as the midspan deflection goes to zero.

The cable modulus, E, is defined as the proportionality constant in the relation-

Table 7-1. Cables.

Notation: L = span of cable; T_0 = mean tension in cable; X, Y, Z = displacements of cable (Fig. 7-2); $\overline{Y}$ = sag of cable (Fig. 7-1, Eq. 7-1); d = midspan mean deflection (Eq. 7-2); m = mass of cable and supported structures per unit length span; p = load per unit length of span; $\tilde{x}, \tilde{y}, \tilde{z}$, = mode shapes associated with X, Y, and Z displacements; x = distance from cable tie-downs along the span; see Fig. 7-2; see Table 3-1 for consistent sets of units.

Description of Mean Deflection ($\overline{Y}$) and Mode Shape	Natural Frequency (hertz), f_i	Mode Shape
1. Straight Cable ($\overline{Y} = 0$)	$\frac{i}{2L}\left(\frac{T_0}{m}\right)^{1/2}$; $i = 1, 2, 3, \ldots$	$\tilde{x} = 0$ $\tilde{y} = \sin(i\pi x/L)$ $\tilde{z} = \sin(i\pi x/L)$
2. Shallow Parabolic Sag, $\overline{Y}$, Given by Eq. 7-1; Out-of-Plane Mode	$\frac{i}{2L}\left(\frac{T_0}{m}\right)^{1/2}$; $i = 1, 2, 3, \ldots$	$\begin{bmatrix}\tilde{x}\\ \tilde{y}\\ \tilde{z}\end{bmatrix}_i = \begin{bmatrix}0\\ 0\\ \sin(i\pi x/L)\end{bmatrix}$; $i = 1, 2, 3, \ldots$, (Ref. 7-4)
3. Shallow Parabolic Sag, $\overline{Y}$, Given by Eq. 7-1; In-Plane Antisymmetric Mode	$\frac{i}{L}\left(\frac{T_0}{m}\right)^{1/2}$; $i = 1, 2, 3, \ldots$	$\begin{bmatrix}\tilde{x}\\ \tilde{y}\\ \tilde{z}\end{bmatrix}_i = \begin{bmatrix}-\frac{1}{2}\left(\frac{pL}{T_0}\right)\left[\left(1-\frac{2x}{L}\right)\sin\left(\frac{2i\pi x}{L}\right)+\frac{1}{i\pi}\left(1-\cos\frac{2i\pi x}{L}\right)\right]\\ \sin(2i\pi x/L)\\ 0\end{bmatrix}$; $i = 1, 2, 3, \ldots$ (Ref. 7-4)
4. Shallow Parabolic Sag, $\overline{Y}$, Given by Eq. 7-1; In-Plane Symmetric Mode	$\frac{\lambda_i}{2L}\left(\frac{T_0}{m}\right)^{1/2}$; $i = 1, 2, 3, \ldots$ λ_1 from Fig. 7-5. Other λ_i from Eqs. 7-5 through 7-7.	$\begin{bmatrix}\tilde{x}\\ \tilde{y}\\ \tilde{z}\end{bmatrix}_i = \begin{bmatrix}\frac{pL}{T_0}\left(\frac{(\pi\lambda_i)^2}{\alpha^2}\frac{L_x}{L_e}-\frac{1}{2}\left(1-\frac{2x}{L}\right)\left[1-\tan\left(\frac{\pi\lambda_i}{2}\right)\sin\left(\frac{\pi\lambda_i x}{L}\right)-\cos\left(\frac{\pi\lambda_i x}{L}\right)\right]-\frac{1}{\pi\lambda_i}\left\{\left(\frac{\pi\lambda_i x}{L}\right)-\tan\frac{\pi\lambda_i}{2}\cdot\left[1-\cos\left(\frac{\pi\lambda_i x}{L}\right)\right]-\sin\left(\frac{\pi\lambda_i x}{L}\right)\right\}\right)\\ 1-\tan\left(\frac{\pi\lambda_i}{2}\right)\sin\left(\frac{\pi\lambda_i x}{L}\right)-\cos\left(\frac{\pi\lambda_i x}{L}\right)\\ 0\end{bmatrix}$; $i = 1, 2, 3, \ldots$ $L_e = L\left[1+8\left(\frac{d}{L}\right)^2\right]$ $L_x = L\left\{\left(\frac{x}{L}\right)+24\left(\frac{d}{L}\right)^2\left[\left(\frac{x}{L}\right)-2\left(\frac{x}{L}\right)^2+\frac{4}{3}\left(\frac{x}{L}\right)^3\right]\right\}$ (Ref. 7-4)

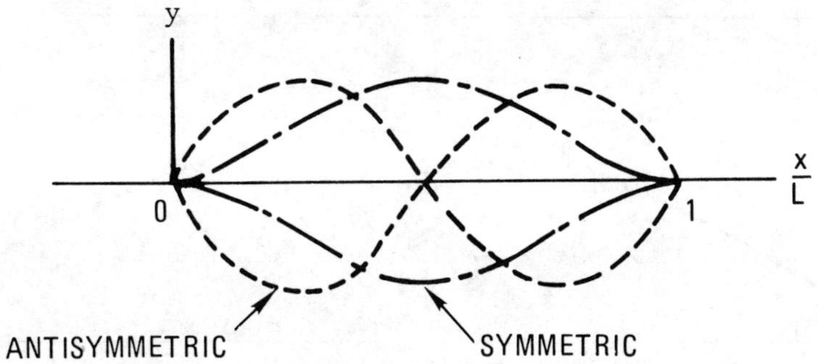

Fig. 7-3. First symmetric and antisymmetric in-plane cable modes.

ship between the force applied to the cable and the longitudinal extension of the cable:

$$F = EA \frac{\Delta L}{L},$$

where F is the force applied to the ends of a straight segment of cable of length L which extends elastically by the length ΔL under the action of the force. A is the cross-sectional area of the cable; however, only the product EA enters this formula and other formulas in this chapter. If the cable is composed of a solid wire, the cable modulus is identical to the modulus of elasticity of the material in the wire. If the cable is constructed from individual links, a chain, for example, the cable modulus can be obtained from stress analysis of the individual links. If the cable is woven from fibers, the cable modulus will be a function of the modulus of elasticity of the fibers, the orientation of the fibers in the cable, and whether the ends of the cable are constrained against rotation (Ref. 7-5). Often the cable moduli of helically woven wire cables are 30% to 50% of the modulus of elasticity of the component wires.

The parameter α (Eq. 7-3) governs the influence of the mean sag on the natural frequencies and mode shapes of the cable. If $\alpha^2 \approx 0$, then the sag in the cable does not influence the natural frequencies and mode shapes of the cable and the results for sagging cables given in frames 2, 3, and 4 of Table 7-1 approach the straight cable case of frame 1 of Table 7-1. If $\alpha^2 > 0$, but $\alpha^2 < 4\pi^2$, the natural frequency of the first symmetric in-plane mode is less than the natural frequency of the first antisymmetric in-plane mode; if $\alpha^2 = 4\pi^2$, the first symmetric and first antisymmetric in-plane modes have the same natural frequencies and the first symmetric in-plane mode is tangent to the mean cable catenary at the tie-downs, as shown in Fig. 7-4; if $\alpha^2 > 4\pi^2$, the natural frequency of the first symmetric in-plane mode is greater than the natural frequency of the first antisymmetric in-plane mode and the first symmetric mode has two internal modes (Fig. 7-4). This has been observed experimentally (Refs. 7-4, 7-8).

The nondimensional frequency parameter λ in frame 4 of Table 7-1, which governs the natural frequency of in-plane symmetric modes, is found from the positive

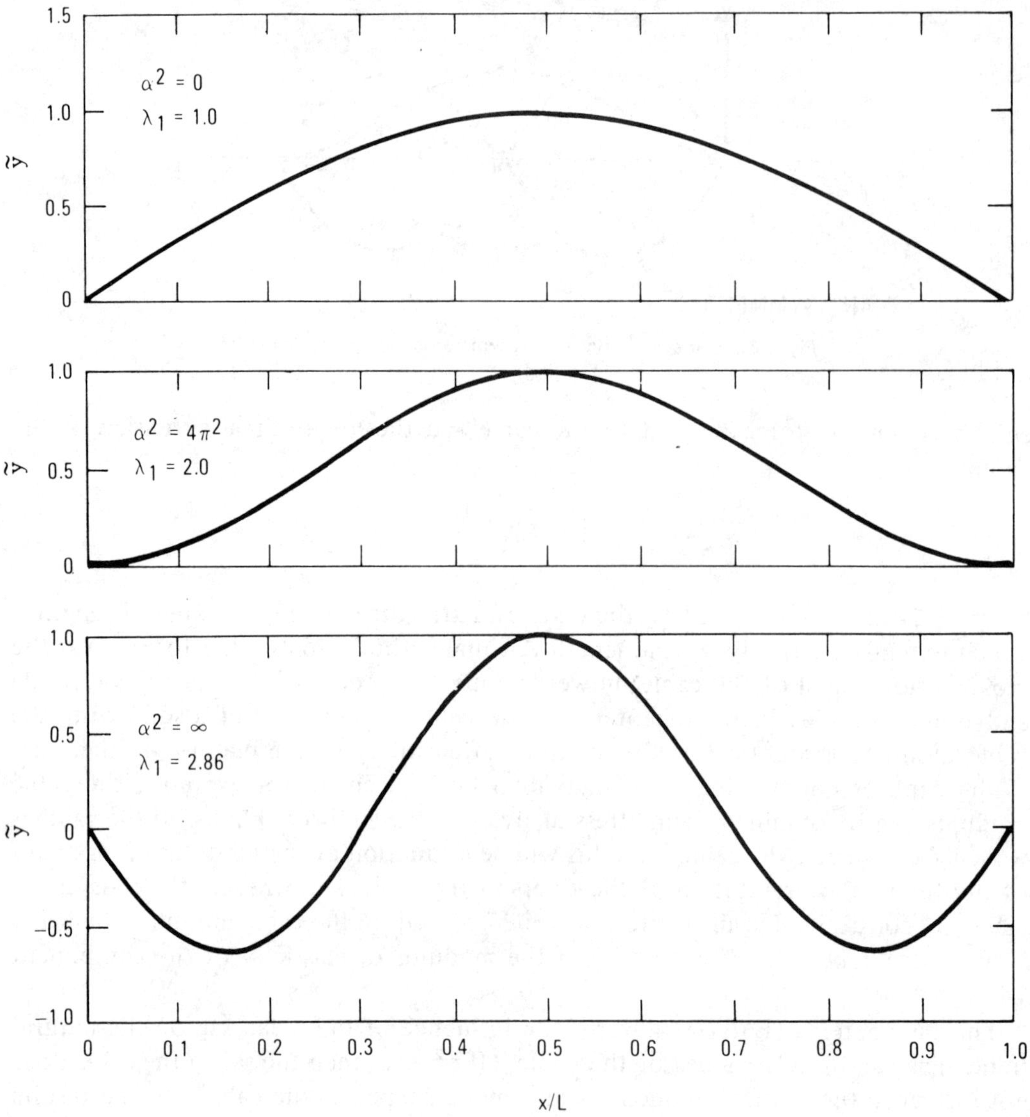

Fig. 7-4. Mode shape of first symmetric in-plane cable mode showing influence of the parameter α (Eq. 7-3).

nonzero roots of the following transcendental equation:

$$\tan\left(\frac{\pi\lambda}{2}\right) = \frac{\pi\lambda}{2} - \frac{4}{\alpha^2}\left(\frac{\pi\lambda}{2}\right)^3. \tag{7-5}$$

If α^2 is much greater than 100,

$$\begin{aligned} \lambda_1 &= 2.861, \\ \lambda_2 &= 4.9181, \\ \lambda_3 &= 6.9418, \\ \lambda_i &= 2i + 1; \qquad i = 4, 5, 6, \ldots \end{aligned} \tag{7-6}$$

If α^2 is much less than 1.0,

$$\lambda_i = 2i - 1; \qquad i = 1, 2, 3, \ldots \tag{7-7}$$

Figure 7-5 gives the first positive nonzero root of Eq. 7-5 for $1 < \alpha^2 < 200$. Higher roots of Eq. 7-5 can be bracketed using Eqs. 7-6 and 7-7.

When the cable vibrates in a symmetric in-plane mode, a first-order fluctuation is induced in cable tension. The amplitude of the fluctuating component of cable tension, T_D, is given by:

$$\frac{T_D}{T_0} = \frac{(\pi\lambda_i)^2}{8d} B_i \tag{7-8}$$

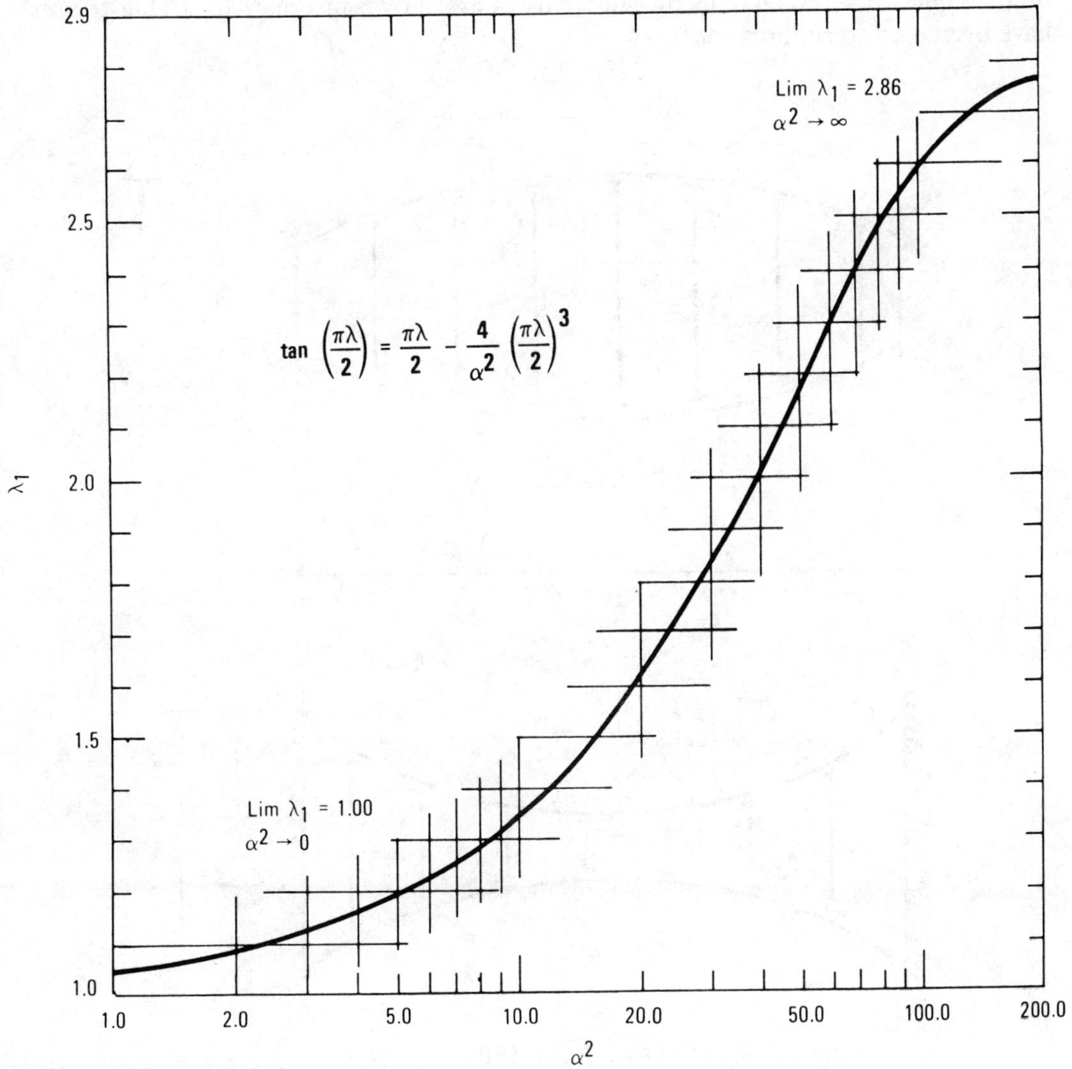

Fig. 7-5. First positive real root to Eq. 7-5 and Eq. 7-12. α^2 is given by Eq. 7-3 for single cables and by Eq. 7-10 for cable trusses.

for vibrations in the i symmetric in-plane mode. B_i is the amplitude of the dynamic in-plane deflection in the i mode $[y_i = B_i \tilde{y}_i(x) \sin(\omega_i t + \phi_i)]$ and $\tilde{y}_i(x)$ is the mode shape of the i mode given in frame 4 of Table 7-1. d is the midspan sag (Eq. 7-2). The fluctuation in tension is zero to first order for free vibration in antisymmetric in-plane modes.

7.2. CABLE TRUSSES

General Case. A cable truss consists of two uniform pretensioned cables anchored at both ends and separated by a series of light-weight vertical spacers. Since the pre-tension in the cables is usually high, the geometry of the truss is largely determined by the span of the cables and the height of the spacers rather than the sag of the cable truss. Examples of cable trusses are shown in Fig. 7-6. Cable trusses have been used to support roofs.

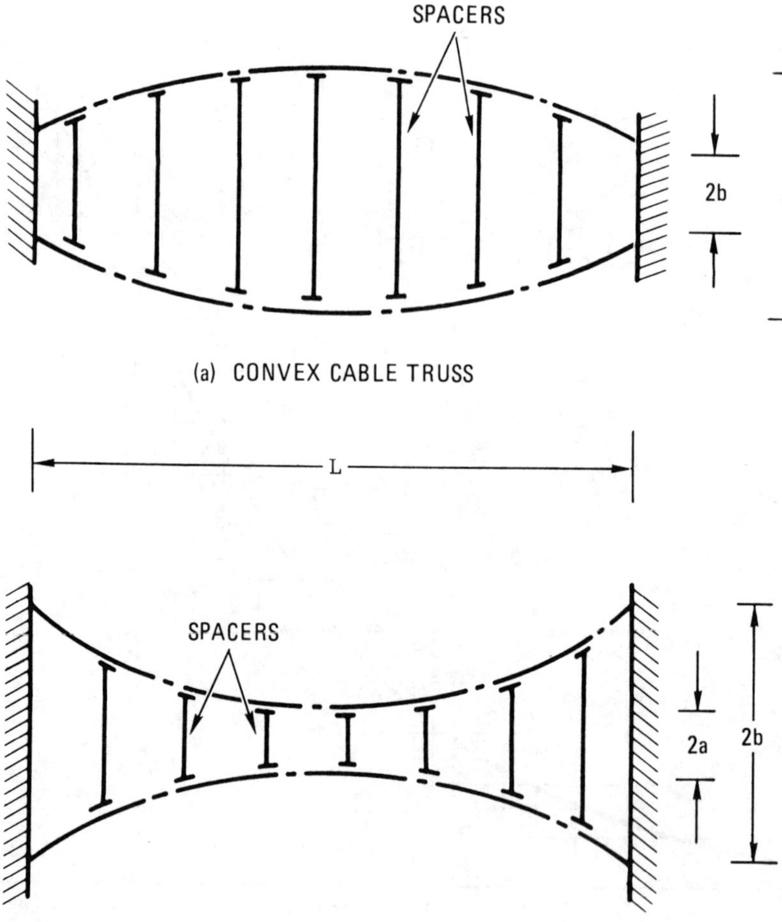

Fig. 7-6. Cable trusses.

The general assumptions used in the derivation of the solutions presented in this section are:

1. The cable trusses are either bi-convex or bi-concave and the cables are described by slender parabolas,

$$\bar{Y} = \pm b\left\{1 + 4\left(\frac{a-b}{b}\right)\left[\frac{x}{L} - \left(\frac{x}{L}\right)^2\right]\right\}$$

   such that

$$\frac{|a-b|}{L} \leqslant \frac{1}{8},$$

   where 2a is the spacing between the cables at midspan, 2b is the spacing between cables at the tie-downs, L is the span of the cable truss, and x is the distance along the centerline of the cable truss (Fig. 7-6).
2. The deflections of the cables due to mean load are much less than $2|a - b|$.
3. The cables are uniform and identical.
4. Dynamic deflections are small compared with the mean deflection of the cable truss and the distance between vibration nodes.
5. The variation in tension in the cables during vibration is small compared with the pre-tension.
6. The cable tie-downs are rigid.

As in the analysis of cables, the natural frequencies and mode shapes of the cable trusses are dependent on the mean position of the cables. With single cables the mean position of the cables (sag) is determined by the mean load on the cable. With cable trusses the mean position of the cables is determined by the geometry of the cable truss, since it is assumed that the sag is much less than the spacing imposed by the tie-downs and vertical spacers.

Mean Sag. If the top cable in a cable truss bears a uniform vertical load of p per unit of span, then vertical deflection of the cable truss in the direction of p is (Ref. 7-6):

$$\delta\bar{Y}(x) = \frac{pL^2}{2T_0}\left(1 - \frac{\alpha^2}{\alpha^2 + 12}\right)\left[\frac{x}{L} - \left(\frac{x}{L}\right)^2\right], \tag{7-9}$$

where the parameter α is

$$\alpha^2 = \frac{8^2(a-b)^2}{L^2}\,\frac{EA}{T_0}\,\frac{L}{L_e}. \tag{7-10}$$

T_0 is the mean cable tension, E is the cable modulus (discussed in Section 7.1), A is the cross-sectional area of the cable, and L_e is the virtual cable length (Eq. 7-4), which is approximately equal to the span of the cables, L. Note that α^2 is a function of the geometry of the cable truss but not the mean sag of the truss. The change in the horizontal component of cable tension in the top cable due to the downward-acting load p is:

$$\Delta T = \frac{pL^2}{16(b-a)} \frac{\alpha^2}{\alpha^2 + 12}.$$

The change in tension of the bottom cable is $-\Delta T$. One of the cables will go slack if the change in tension equals the mean tension, $\Delta T = T$. This occurs for a mean load per unit length of:

$$p = 16T_0 \frac{(b-a)}{L^2} \frac{\alpha^2 + 12}{\alpha^2}.$$

If a circular roof is supported by cable trusses on the radii, then the mean load on the cable trusses will be in the form of a triangular block maximum at the tie-downs and zero in the middle, as shown in Fig. 7-7. The vertical deflection of the cable truss under this triangular block load with a maximum load of p_0 per unit span at the tie-downs is (Ref. 7-6)

$$\delta\overline{Y}(r) = \frac{p_0 L^2}{2T_0} \left\{ \frac{1}{3} \left[\frac{1}{8} - \left(\frac{r}{L}\right)^3 \right] - \frac{3}{64} \frac{\alpha^2}{(\alpha^2 + 12)} \left[1 - 4\left(\frac{r}{L}\right)^2 \right] \right\} \qquad (7\text{-}11)$$

for $r/L > 0$, where α^2 is given by Eq. 7-10. For $\alpha^2 > 96$, the midspan of the cable truss will rise under the triangular block load. The change in the horizontal component of tension in the top cable in the cable truss due to the triangular block load is:

$$\Delta T = \frac{3}{128} \frac{p_0 L^2}{(b-a)} \frac{\alpha^2}{\alpha^2 + 12}.$$

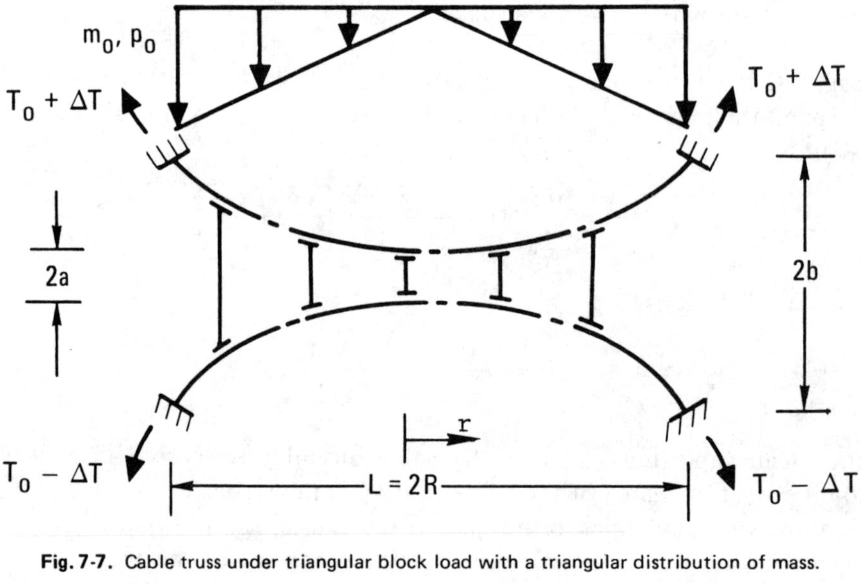

Fig. 7-7. Cable truss under triangular block load with a triangular distribution of mass.

The change in tension in the bottom cable is $-\Delta T$. One of the cables will go slack if the change in tension equals the mean tension, $\Delta T = T$. This occurs for a maximum load per unit length of:

$$p_0 = \frac{128}{3} \frac{T_0(b-a)}{L^2} \frac{\alpha^2 + 12}{\alpha^2}.$$

Additional static solutions are given in Refs. 7-6 and 7-7.

Mode Shapes and Natural Frequencies. The mode shapes and natural frequencies of cable trusses for vibration in the plane of the cable truss are given in Table 7-2. The natural frequencies and mode shapes of these cable trusses are a function of the geometry of the cable truss rather than the mean deflection of the truss, since it was assumed that the mean deflection is much less than the spacing imposed by the tie-downs and vertical spacers. As the spacers become very close, α (Eq. 7-10) approaches zero and the results for the cable truss (frames 2, 3, and 4 of Table 7-2) approach those of a straight cable (frame 1 of Table 7-2).

The natural frequency of cable trusses in symmetric modes can be found from frame 2 of Table 7-1, where the constant λ is found from the positive non-zero roots of the following transcendental equation:

$$\tan\left(\frac{\pi\lambda}{2}\right) = \frac{\pi\lambda}{2} - \frac{4}{\alpha^2}\left(\frac{\pi\lambda}{2}\right)^3. \qquad (7\text{-}12)$$

α is given by Eq. 7-10. This equation is identical to the corresponding equation for single cables (Eq. 7-5) except for the redefinition of α. Solutions to Eq. 7-12 can be found in Eqs. 7-6 and 7-7 and Fig. 7-5.

The parameter α (Eq. 7-10) governs the effect of the shape of the cable truss on the cable truss dynamics. If $\alpha^2 > 0$ but $\alpha^2 < 4\pi^2$, the natural frequency of the first symmetric mode is less than the natural frequency of the first antisymmetric mode; if $\alpha^2 > 4\pi^2$, the natural frequency of the first symmetric mode is greater than the natural frequency of the first antisymmetric mode.

When the cable truss vibrates in a symmetric in-plane mode, a first-order fluctuation is induced in the cable tension. The dynamic fluctuation in tension, T_D, in the top cable of the cable truss as the cable vibrates in the i symmetric mode is given by:

$$\frac{T_D}{T_0} = \frac{(\pi\lambda_i)^2}{8(b-a)} B_i.$$

B_i is the amplitude of the vertical dynamic deflection in the i mode $[y_i = B_i\tilde{y}_i(x)\sin(\omega_i t + \phi_i)]$ and $\tilde{y}_i(x)$ is the mode shape of the i symmetric mode given in frame 3 of Table 7-2. y_i is positive in the direction of the applied load p. The fluctuation in tension in the bottom cable of the cable truss is $-T_D$. The fluctuation in tension is zero to first order for vibration in antisymmetric modes.

The natural frequency of cable trusses with triangular distribution of mass (Fig. 7-7) is given in frame 4 of Table 7-2 for symmetric in-plane modes. This geometry approximates the load and mass distribution of a circular roof supported

Table 7-2. Cable Trusses—In-Plane Modes.

Notation: $\mathcal{J}_k$ = Bessel function of first kind and k order; L = span of cable truss; R = one-half span of cable truss; T_0 = tension preload in cables; m = mass of truss and supported structures per unit span; r = distance from midspan; $\tilde{y}$ = mode shape for displacement in plane of truss; x = distance from tie-down; α = (see Eq. 7-10); see Table 3-1 for consistent sets of units.

Description	Natural Frequency (hertz), f_i	Mode Shape
1. Closely Spaced Cables ($\alpha = 0$) Without Sag Uniform Mass Distribution	$\frac{i}{2L}\left(\frac{2T_o}{m}\right)^{1/2}$; $i = 1,2,3,\ldots$	$\tilde{y}_i = \sin(i\pi x/L)$ $\tilde{x} = \tilde{z} = 0$ (Ref. 7-6)
2. Antisymmetric Modes ($\alpha \neq 0$) Uniform Mass Distribution	$\frac{i}{L}\left(\frac{2T_o}{m}\right)^{1/2}$; $i = 1,2,3,\ldots$	$\tilde{y}_i = \sin(2i\pi x/L)$; $i = 1,2,3,\ldots$ $\tilde{x} = \tilde{z} = 0$ (Ref. 7-6)
3. Symmetric Modes ($\alpha \neq 0$) Uniform Mass Distribution	$\frac{\lambda_i}{2L}\left(\frac{2T_o}{m}\right)^{1/2}$; $i = 1,2,3,\ldots$ λ_1 from Fig. 7-5. Other λ_i given by Eq. 7-12 or Eqs. 7-5 and 7-6.	$\tilde{y}_i = 1 - \tan\left(\frac{\pi\lambda_i}{2}\right)\sin\left(\frac{\lambda_i \pi x}{L}\right) - \cos\left(\frac{\lambda_i \pi x}{L}\right)$; $i = 1,2,3,\ldots$ $\tilde{x} = \tilde{z} = 0$ (Ref. 7-6)
4. Symmetric Modes ($\alpha \neq 0$) Triangular Distribution of Mass (Fig. 7-7)	$\frac{3}{4\pi}\frac{\lambda_i}{R}\left(\frac{2T_o}{m_o}\right)^{1/2}$; $i = 1,2,3,\ldots$ λ_i given by Eq. 7-13.	$\tilde{y}_i = 1 - \frac{r^{1/2}J_{-1/3}(\lambda_i r^{3/2}/R^{3/2})}{R^{1/2}J_{-1/3}(\lambda_i)}$ $i = 1,2,3,\ldots$ $\tilde{x} = \tilde{z} = 0$ (Ref. 7-6)

by cable trusses on the radii. The constant λ in frame 4 of Table 7-2 is given by the positive non-zero roots to the following equation:

$$\frac{\mathcal{J}_{2/3}(\lambda)}{\mathcal{J}_{-1/3}(\lambda)} = \frac{3}{4}\lambda - \frac{27\lambda^3}{8\alpha^2}, \qquad (7\text{-}13)$$

where $\mathcal{J}_{2/3}$ is the Bessel function of the first kind and two-thirds order and $\mathcal{J}_{-1/3}$ is the Bessel function of the first kind and minus one-third order. In Eq. 7-13, α^2 is defined by

$$\alpha^2 = \frac{36(a-b)^2}{R^2}\frac{EA}{T_0}\frac{R}{R_e}, \qquad (7\text{-}14)$$

where R, a, and b are as shown in Fig. 7-7 and R_e is the virtual roof radius (Eq. 7-4) and is approximately equal to R. E is the modulus of elasticity of the cable and A is the cross section of the cable. T_0 is the mean cable tension. If α^2 is much less than 1, then the positive first root of Eq. 7-13 is $\lambda_1 = 1.87$. If α^2 is much greater than 100, then the first positive root of Eq. 7-13 is $\lambda_1 = 4.72$. All other first positive roots for Eq. 7-13 for various α must lie between these two values.

7.3. CABLE EXAMPLE

Consider a cable spanning 915 m (3000 ft) which sags 76 m (250 ft) at midspan and supports 4400 N/m (300 lb/ft). The tension in this cable can easily be calculated from Eq. 7-2 to be:

$$T_0 = 6.06 \times 10^6 \text{ N}.$$

If the modulus of the cable is 1.8×10^{11} N/m^2 (26×10^6 psi) and the cross-sectional area of the cable is 0.161 m^2 (250 in.^{2}), then α^2 is easily calculated from Eq. 7-3 to be:

$$\alpha^2 = 2 \times 10^3 >> 4\pi^2.$$

The natural frequency of the first transverse mode is calculated from frame 2 of Table 7-1 using the units of frame 1 of Table 3-1:

$$f_1\,(\text{transverse}) = 0.063 \text{ Hz} = 1/15.8 \text{ sec}.$$

m = 449 kg/m was used in the calculation. The first vertical antisymmetric mode has a natural frequency of (frame 3 of Table 7-1):

$$f_1\,(\text{vertical, antisymmetric}) = 0.13 \text{ Hz} = 1/7.9 \text{ sec}.$$

The frequency of the first symmetric vertical mode can be found from frame 4 of Table 7-1 where λ_i, given by Eq. 7-6 or Fig. 7-5, is 2.86. Thus,

$$f_1\,(\text{vertical, symmetric}) = 0.18 \text{ Hz} = 1/5.5 \text{ sec}.$$

The natural frequency of the first vertical symmetric mode is greater than the natural frequency of the first vertical antisymmetric mode since $\alpha^2 > 4\pi^2$.

7.4. CABLE TRUSS EXAMPLE

Assume that the roof of a rectangular sports arena is to be supported by a series of bi-concave cable trusses with the following specifications using the units of frame 1 of Table 3-1: $L = 61$ m, $a = 0.61$ m, $b = 3.05$ m, $E = 1.04 \times 10^{11}$ N/m^2, $A = 3.22 \times 10^{-3}$ m^2, $T_0 = 1.11 \times 10^6$ N. The cable trusses support a load of $p = 5830$ N/m. α^2 is calculated from Eq. 7-10 to be:

$$\alpha^2 = 30.9.$$

Since $\alpha^2 < 4\pi^2$, the first symmetric mode has the lowest natural frequency. The natural frequency of the first antisymmetric vertical mode is given by frame 1 of Table 7-2:

$$f_1\,(\text{vertical, antisymmetric}) = 1.0 \text{ Hz}.$$

A mass distribution of $m = 0.594$ kg/m was used in this calculation. The natural frequency of the first vertical symmetric mode is found from frame 2 of Table 7-2 to be:

$$f_1\,(\text{vertical, symmetric}) = 0.925 \text{ Hz},$$

where $\lambda_1 = 1.85$ from Fig. 7-5.

REFERENCES

7-1. Irvine, H. M., "Statics of Suspended Cables," *ASCE, J. Eng. Mech. Div.* **101,** 187–205 (1975).

7-2. Irvine, H. M., "Post-Elastic Response of Suspended Cables," *ASCE, J. Eng. Mech. Div.* **101,** 725–738 (1975).

7-3. Wilson, A. J., and R. J. Wheen, "Inclined Cables Under Load Design Expressions," *ASCE, J. Struc. Div.* **103,** 1061–1078 (1977).

7-4. Irvine, H. M., and T. K. Caughey, "The Linear Theory of Free Vibrations of a Suspended Cable," *Proc. Roy. Soc. London Series A*, **341,** 299–315 (1974).

7-5. Costello, G. A., and J. W. Phillips, "Effective Modulus of Twisted Wire Cables," *ASCE, J. Eng. Mech. Div.* **102,** 171-181 (1976).

7-6. Irvine, H. M., "Statics and Dynamics of Cable Trusses," *ASCE, J. Eng. Mech. Div.* **101,** EM4, 429–446 (1975).

7-7. Irvine, H. M., and P. C. Jennings, "Lateral Stability of Cable Truss," *ASCE, J. Eng. Mech. Div.* **101,** EM4, 403-416 (1975).

7-8. Ramberg, S. E., and O. M. Griffin, "Free Vibration of Taut and Slack Marine Cables," *ASCE, J. Struc. Div.* **103,** 2079-2092 (1977). Also see discussion of this paper by A. S. Richardson Jr., M. L. Gambhir, and B. V. Bachelor, *ASCE, J. Struc. Div.* **104,** 1038-1041 (1978).

8

STRAIGHT BEAMS

8.1. TRANSVERSE VIBRATION OF FLEXURE BEAMS

8.1.1. General Case

General Assumptions. A straight elastic beam possesses both the mass and stiffness to resist bending. During transverse vibration, the beam flexes perpendicular to its own axis to alternately store potential energy in the elastic bending of the beam and then release it into the kinetic energy of transverse motion. Pipes, pilings, and entire buildings can experience these transverse flexural vibrations.

The general assumptions used in the analysis of beams in Section 8.1 are:

1. The beams are uniform along the span.
2. The beams are composed of a linear, homogeneous, isotropic elastic material.
3. The beams are slender. The dimensions of the beam cross section are much less than the length of the beam or the distance between vibration nodes. Rotary inertia and shear deformation are not considered.
4. Only deformations normal to the undeformed beam axis are considered. Plane sections remain plane.
5. No axial loads are applied to the beam.
6. The shear center of the beam cross section coincides with the center of mass (i.e., the plane of vibration is also a plane of symmetry of the beam), so that rotation and translation of the beam are uncoupled.

In Sections 8.1.1, 8.1.2, 8.1.3, 8.1.7, and 8.1.8, all of these assumptions are in effect. In the remaining sections of Chapter 8, one or more of the assumptions are relaxed. For example, Sections 8.1.4 and 8.1.5 consider the effect of axial load on the beam (assumption 5), Section 8.1.6 considers nonuniform beams (assumption 1), and Section 8.2 incorporates the effect of shear deformation (assumption 3).

Stress and Strain. As the beam deflects transverse to its own axis, the elements on the convex side of the beam lengthen and the elements on the concave side shorten, as can be seen in Fig. 8-1(a). There exists an axis through the beam cross section where elements are neither lengthened nor shortened. This axis is the neutral axis of the cross section. The neutral axis must pass through the centroid of the cross section unless loads are applied axially to the beam. Plane surfaces through the beam remain plane as the beam deflects. The stresses and strains in the beam are proportional to the distance from the neutral axis. The maximum stresses and strains arise at the greatest distance in the cross section from the neutral axis.

If a beam's longitudinal axis coincides with the x axis and its y axis measures

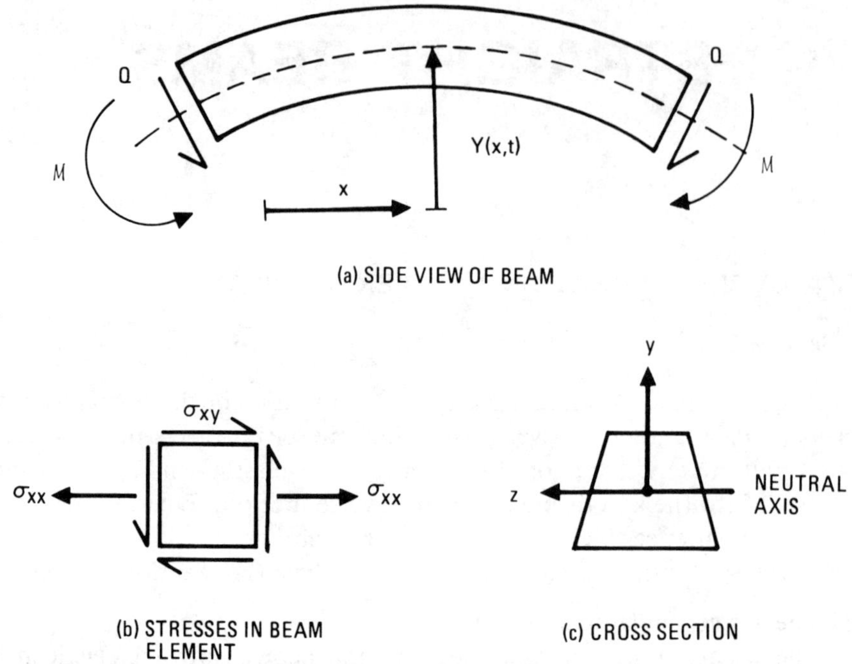

Fig. 8-1. Deformation, stresses, and coordinates for a flexure beam.

distance from the neutral axis, as shown in Fig. 8-1, then the normal longitudinal strain (ϵ_x) and the lateral strains (ϵ_z and ϵ_y) induced by the transverse deflection Y(x, t) of the beam in the x-y plane are:

$$\epsilon_x = -y\frac{\partial^2 Y}{\partial x^2},$$

$$\epsilon_z = \nu y\frac{\partial^2 Y}{\partial x^2},$$

$$\epsilon_y = \nu y\frac{\partial^2 Y}{\partial x^2},$$

$$\epsilon_{xy} = \epsilon_{xz} = \epsilon_{yz} = 0. \tag{8-1}$$

The lateral expansion of the beam on the concave side and lateral contraction of the beam on the convex side which correspond to ϵ_z are due to the Poisson's ratio effect. ν is Poisson's ratio. The transverse shear strain over the cross section ϵ_{xy} has been neglected in the flexure beam theory.

These strains correspond to the following stresses in a beam composed of a homogeneous, isotropic material:

$$\sigma_{xx} = E\epsilon_x = -Ey\frac{\partial^2 Y}{\partial x^2},$$

$$\sigma_{yy} = \sigma_{zz} = 0,$$

$$\sigma_{xy} = \sigma_{xz} = \sigma_{yz} = 0. \tag{8-2}$$

The first subscript refers to the normal to the face on which the stress acts. The second subscript refers to the direction of the stress. E is the elastic modulus. σ_{xx} is the normal stress applied parallel to the axis. This stress is associated with the following bending moment in the beam:

$$\mathfrak{M} = \int_A y\sigma_{xx}\, dA = -EI\frac{\partial^2 Y}{\partial x^2}. \tag{8-3}$$

I is the area moment of inertia of the beam cross section about the neutral axis:

$$I = \int_A y^2\, dA,$$

where the integration is made over the area (A) of the cross section. (See Table 5-1.) Equations 8-2 and 8-3 can be combined to express the stress in an element of the beam as a function of the moment borne by the beam:

$$\sigma_{xx} = \frac{\mathfrak{M} y}{I}, \tag{8-4}$$

where y is the distance of the element from the neutral axis.

Equilibrium of the beam cross section requires that the transverse shear force supported by the beam be equal to the derivative of the bending moment in the beam with respect to the longitudinal axis:

$$Q_x = \frac{\partial \mathfrak{M}}{\partial x}.$$

The transverse shear force supported by the beam is the resultant of the transverse shear stresses and can be related to the deformation of the beam through Eq. 8-3:

$$Q_x = \int_A \sigma_{xy}\, dA = -EI\frac{\partial^3 Y}{\partial x^3}. \tag{8-5}$$

Although the shear deformation ϵ_{xy} and the shear stress σ_{xy} have been neglected, equilibrium considerations dictate that the integral of the shear stress over the cross section cannot be zero. The shear stress (σ_{xy}) will generally have a parabolic distribution over simple closed cross sections with the maximum at the centroid and zero at the edges of the cross section. An approximate expression for the maximum shear stress at the centroid of simple closed cross sections such as rectangular, circular, and elliptical is:

$$\sigma_{xy}\Big|_{\text{centroid}} = \frac{Q_x}{KA} = -\frac{EI}{KA}\frac{\partial^3 Y}{\partial x^3},$$

where Q_x is the transverse shearing force supported by the cross section, A is the area of the cross section, and K is a dimensionless constant, approximately equal to 0.8, which is discussed in Section 8.2 and presented in Table 8-14. The average shear stress over the cross section is:

$$\sigma_{xy}\Big|_{\text{average}} = \frac{Q_x}{A} = -\frac{EI}{A}\frac{\partial^3 Y}{\partial x^3}.$$

Boundary Conditions. The mathematical expressions for various boundary conditions which can be applied to the beam are:

$$\text{Free Boundary:} \qquad \frac{\partial^2 Y}{\partial x^2} = \frac{\partial^3 Y}{\partial x^3} = 0,$$

$$\text{Clamped Boundary:} \qquad Y = \frac{\partial Y}{\partial x} = 0,$$

$$\text{Pinned Boundary:} \qquad Y = \frac{\partial^2 Y}{\partial x^2} = 0,$$

$$\text{Sliding Boundary:} \qquad \frac{\partial Y}{\partial x} = \frac{\partial^3 Y}{\partial x^3} = 0. \tag{8-6}$$

Y(x, t) is the transverse deformation of the beam and x is the spanwise coordinate. These boundary conditions are applied for all times at the spanwise location of the boundary.

The free boundary condition corresponds to no geometric restraint or load on the beam. Thus, the moment (Eq. 8-3) and shear load (Eq. 8-5) in the beam are zero at a free boundary. The clamped boundary condition prevents the beam from rotation and displacement. The pinned boundary condition prevents displacement but allows rotation. The moment (Eq. 8-3) is zero at the pinned boundary. The sliding boundary condition allows displacement but not rotation. The shear load (Eq. 8-5) is zero at a sliding boundary.

Natural Frequencies, Mode Shapes, and Deformation. The mode shapes and natural frequencies of a slender beam are a function of an integer index (i) which may be associated with the number of flexural half-waves in the mode shape. For each i there is a natural frequency and mode shape. If the beam vibrates freely, then the total transverse deformation is the sum of the modal deformations:

$$Y(x, t) = \sum_{i=1}^{N} A_i \tilde{y}_i(x) \sin(2\pi f_i t + \phi_i). \tag{8-7}$$

$\tilde{y}_i(x)$ is the mode shape associated with the i vibration mode and f_i is the natural frequency of that mode in hertz. A_i is a constant with the units of length and ϕ_i is a phase angle. A_i and ϕ_i are determined by the means used to set the beam in motion. The natural frequency in hertz can generally be expressed in the form:

$$f_i = \frac{\lambda_i^2}{2\pi L^2} \left(\frac{EI}{m}\right)^{1/2}; \qquad i = 1, 2, 3 \ldots,$$

where λ_i is a dimensionless parameter which is a function of the boundary conditions applied to the beam, L is the length of the beam, and m is the mass per unit length of the beam. Once Y(x, t) has been found from Eq. 8-7, Eqs. 8-1 through 8-5 can be used to determine the stresses, strains, and resultant forces in the beam.

The longitudinal deformation of the beam is of order Y^2 for small transverse deformations of slender straight beams, and so the longitudinal deformation is neglected in the linear theory.

Sandwich Beams. A laminated sandwich beam is shown in Fig. 8-2. The beam consists of uniform layers of material which are glued together. Each layer is assumed to be homogeneous. The beam cross section is symmetric about the z axis (neutral axis). The analysis of sandwich beams is considerably more difficult than the analysis of homogeneous beams because large shear strains can develop between the layers (Ref. 8-37). However, if the sandwich beam is slender, that is, the depth of the beam is small compared with the length of the beam and the distance between vibration nodes, then it is reasonable to assume that plane sections of the beam remain plane during vibration. The natural frequencies of these slender sandwich beams can be computed using the formulas developed for homogeneous beams by defining the sandwich beam equivalent stiffness,

$$EI = \sum_k 2E_k b \int_{d_k}^{d_{k+1}} y^2 dy,$$

$$= \frac{2b}{3} \sum_{k=0,1,2\ldots} E_k (d_{k+1}^3 - d_k^3), \tag{8-8}$$

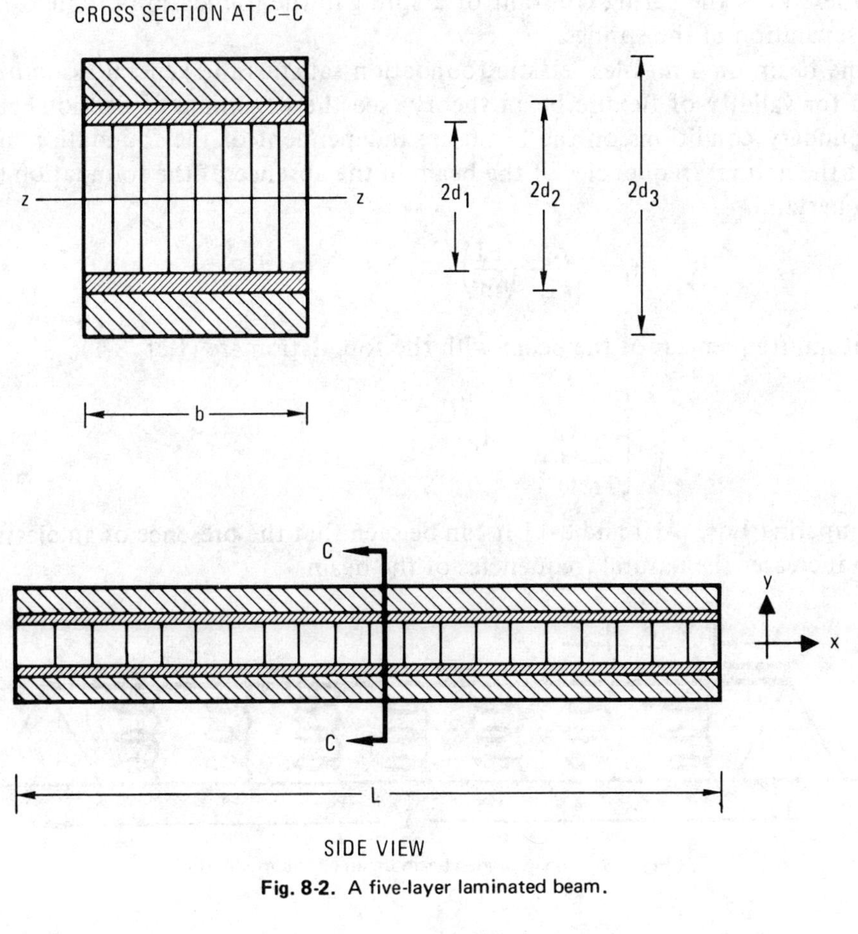

Fig. 8-2. A five-layer laminated beam.

and the equivalent mass per unit length,

$$m = \sum_k m_k = \sum_k 2b\mu_k \int_{d_k}^{d_{k+1}} dy,$$

$$= 2b \sum_{k=0,1,2\ldots} \mu_k (d_{k+1} - d_k). \tag{8-9}$$

The distance from the neutral axis of the beam to the seam between the k and the k + 1 layers in the beam is d_k ($d_o = 0$). E_k and μ_k are the modulus of elasticity and the density of the k layer, and b is the width of the beam. m_k is the mass per unit length of each layer (Ref. 8-6). The natural frequencies of sandwich beams can be found from the formulas for the natural frequencies of homogeneous beams by substituting equivalent stiffness (EI) and equivalent mass per unit length (m) for the stiffness and mass per unit length of the homogeneous beams.

Beams on Elastic Foundations. A beam on an elastic foundation is shown in Fig. 8-3. The elastic foundation could consist of a series of springs, as in Fig. 8-3, or simply an elastic pad. The foundation modulus, E_f, is defined as the ratio of the load per unit spanwise length applied to the foundation to the deformation this load produces. E_f has units of force per unit area. $E_f = k_f/\Delta$ for the beam shown in Fig. 8-3, where k_f is the spring constant of a spring in the foundation (Table 6-1) and Δ is the separation of the springs.

If the beam on a massless elastic foundation satisfies the general assumptions required for validity of flexure beam theory (see the beginning of Section 8.1.1) and the boundary conditions on the beam are independent of the foundation modulus, then if the natural frequencies of the beam in the absence of the foundation ($E_f = 0$) are, in hertz,

$$f_i = \frac{\lambda_i^2}{2\pi L^2}\left(\frac{EI}{m}\right)^{1/2}; \qquad i = 1, 2, 3 \ldots, \tag{8-10}$$

the natural frequencies of the beam with the foundation are (Ref. 8-1)

$$f_i = \left(f_i\Big|^2_{\substack{\text{without}\\ \text{foundation}\\ (E_f=0)}} + \frac{E_f}{4\pi^2 m}\right)^{1/2}; \qquad i = 1, 2, 3 \ldots. \tag{8-11}$$

By comparing Eqs. 8-10 and 8-11 it can be seen that the presence of an elastic foundation increases the natural frequencies of the beam.

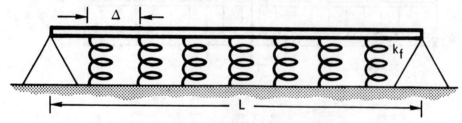

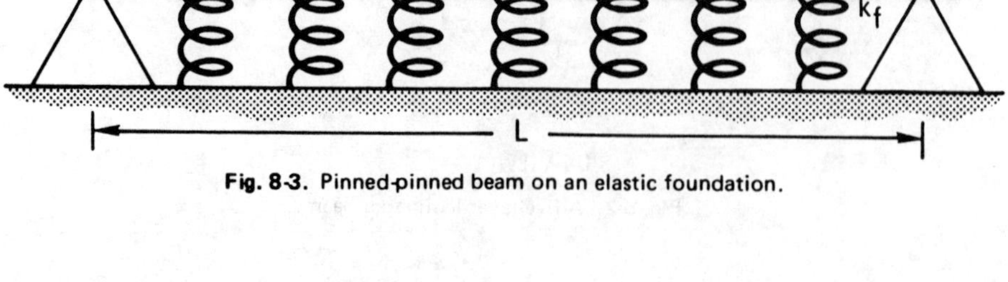

Fig. 8-3. Pinned-pinned beam on an elastic foundation.

For example, the natural frequencies of the pinned-pinned beam on the spring foundation shown in Fig. 8-3 ($E_f = k_f/\Delta$) are:

$$f_i = \left(\frac{i^4 \pi^4}{4\pi^2 L^4} \frac{EI}{m} + \frac{k_f}{4\pi^2 \Delta m}\right)^{1/2}; \qquad i = 1, 2, 3, \ldots.$$

The mode shapes of beams on elastic foundations are unchanged by the presence of the foundation (Ref. 8-1).

Simplification for Tubes. The formula describing the natural frequency of transverse vibration of slender beams can be simplified for tubular beams. If the mass of fluid inside a tube and the added mass of entrained fluid outside a tube can be neglected in comparison with the mass of the tube, then frame 24 of Table 5-1 gives:

$$\frac{I}{m} = \frac{I}{\mu A} = \frac{D_o^2 + D_i^2}{16\mu}.$$

μ is the mass density of the tube material, A is the cross section of the tube, and D_o and D_i are the outer and inner diameters of the tube, respectively. Using this result, the formula for the natural frequency of tubular beams becomes:

$$f_i = \frac{\lambda_i^2}{2\pi L^2}\left(\frac{EI}{m}\right)^{1/2},$$

$$= \frac{\lambda_i^2}{8\pi L^2}\left[\frac{E(D_o^2 + D_i^2)}{\mu}\right]^{1/2}. \tag{8-12}$$

One consequence of this formula is that the natural frequencies of thin-walled tubes are nearly independent of the wall thickness. In addition, since the ratio of elastic modulus to density, E/μ, is approximately constant for common engineering metals such as aluminum and steel, the natural frequencies of many tubes are nearly independent of the metal used in their construction.

8.1.2. Single-Span Beams

Natural Frequencies and Mode Shapes. The natural frequencies and mode shapes of transverse vibration of uniform single-span beams are given in Table 8-1. The mode shapes corresponding to the first three modes of the beams of Table 8-1 are shown in Fig. 8-4. The mode shapes and derivatives of the first five modes of the clamped-free, clamped-pinned, clamped-clamped, and clamped-sliding beam are listed in Table 8-2. Note that mode shapes and derivatives of the free-free, free-pinned, and free-sliding beams can easily be adapted from Table 8-2 by using the transformations given in the following subsection. The beam mode shapes of frames 1, 2, 3, 4, 6, 7, and 8 of Table 8-1 are very sensitive to the values of λ_i and σ_i in the higher modes. Chang and Craig have noted (Ref. 8-2) that changes in σ_i as small as 10^{-6} can result in a significant change in the computed mode shapes. Thus, it is advisable to employ a high degree of numerical accuracy in λ_i and σ_i when computing the mode shapes of higher modes of beams with these boundary conditions.

Table 8-1. Single-Span Beams.

Notation: x = distance along span of beam; m = mass per unit length of beam; E = modulus of elasticity; I = area moment of inertia of beam about neutral axis (Table 5-1); L = span of beam; see Table 3-1 for consistent sets of units

Natural Frequency (hertz); $f_i = \dfrac{\lambda_i^2}{2\pi L^2}\left(\dfrac{EI}{m}\right)^{1/2}$; i=1,2,3...

Description[a]	λ_i; i=1,2,3...	Mode Shape, $\tilde{y}_i\left(\frac{x}{L}\right)$	σ_i; i=1,2,3...
1. Free-Free 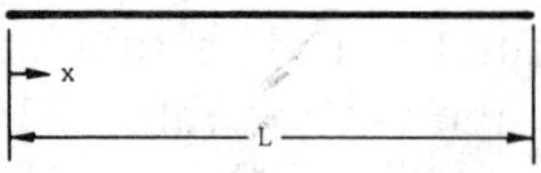	4.73004074 7.85320462 10.9956078 14.1371655 17.2787597 $(2i+1)\frac{\pi}{2}$; i>5	$\cosh\frac{\lambda_i x}{L} + \cos\frac{\lambda_i x}{L} - \sigma_i\left(\sinh\frac{\lambda_i x}{L} + \sin\frac{\lambda_i x}{L}\right)$	0.982502215 1.000777312 0.999966450 1.000001450 0.999999937 ≈1.0 for i>5 See Ref. 8-2
2. Free-Sliding 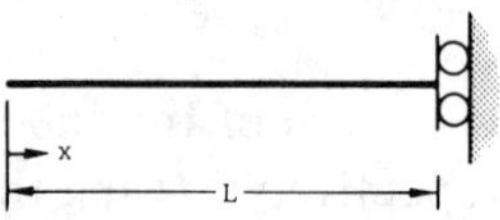	2.36502037 5.49780392 8.63937983 11.78097245 14.92256510 $(4i-1)\frac{\pi}{4}$; i>5	$\cosh\frac{\lambda_i x}{L} + \cos\frac{\lambda_i x}{L} - \sigma_i\left(\sinh\frac{\lambda_i x}{L} + \sin\frac{\lambda_i x}{L}\right)$	0.982502207 0.999966450 0.999999933 0.999999993 0.999999993 1.0; i>5
3. Clamped-Free 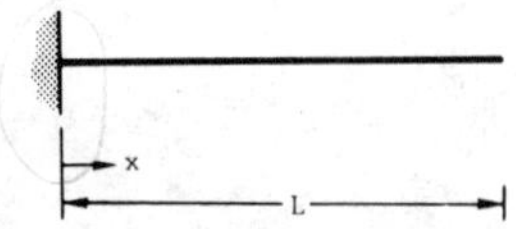	1.87510407 4.69409113 7.85475744 10.99554073 14.13716839 $(2i-1)\frac{\pi}{2}$; i>5	$\cosh\frac{\lambda_i x}{L} - \cos\frac{\lambda_i x}{L} - \sigma_i\left(\sinh\frac{\lambda_i x}{L} - \sin\frac{\lambda_i x}{L}\right)$	0.734095514 1.018467319 0.999224497 1.000033553 0.999998550 ≈1.0; i>5 See Ref. 8-2
4. Free-Pinned 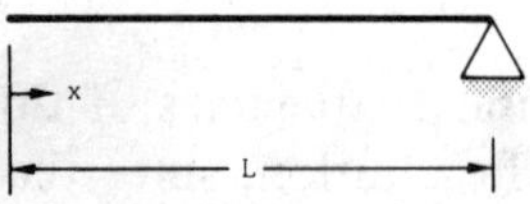	3.92660231 7.06858275 10.21017612 13.35176878 16.49336143 $(4i+1)\frac{\pi}{4}$; i>5	$\cosh\frac{\lambda_i x}{L} + \cos\frac{\lambda_i x}{L} - \sigma_i\left(\sinh\frac{\lambda_i x}{L} + \sin\frac{\lambda_i x}{L}\right)$	1.000777304 1.000001445 1.000000000 1.000000000 1.000000000 1.0; i>5
5. Pinned-Pinned 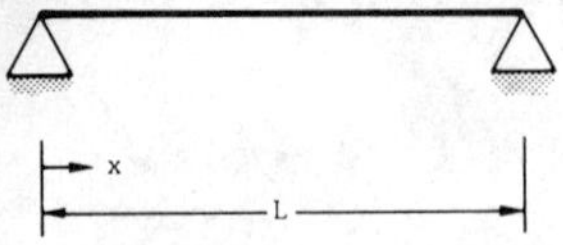	$i\pi$	$\sin\frac{i\pi x}{L}$	--
6. Clamped-Pinned 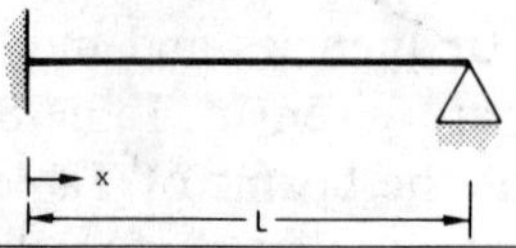	3.92660231 7.06858275 10.21017612 13.35176878 16.49336143 $(4i+1)\frac{\pi}{4}$; i>5	$\cosh\frac{\lambda_i x}{L} - \cos\frac{\lambda_i x}{L} - \sigma_i\left(\sinh\frac{\lambda_i x}{L} - \sin\frac{\lambda_i x}{L}\right)$	1.000777304 1.000001445 1.000000000 1.000000000 1.000000000 1.0; i>5
7. Clamped-Clamped 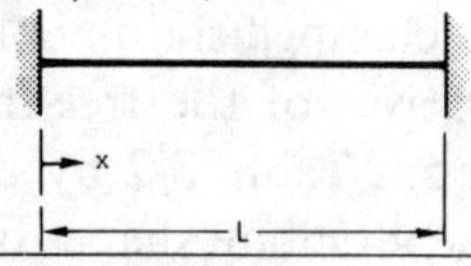	4.73004074 7.85320462 10.9956079 14.1371655 17.2787597 $(2i+1)\frac{\pi}{2}$; i>5	$\cosh\frac{\lambda_i x}{L} - \cos\frac{\lambda_i x}{L} - \sigma_i\left(\sinh\frac{\lambda_i x}{L} - \sin\frac{\lambda_i x}{L}\right)$	0.982502215 1.000777312 0.999966450 1.000001450 0.999999937 1.0; i>5 See Ref. 8-2
8. Clamped-Sliding 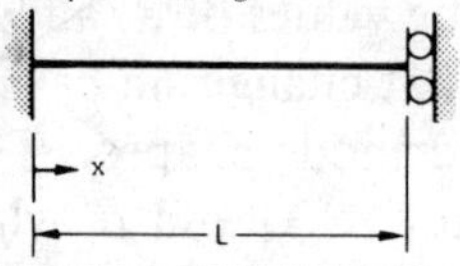	2.36502037 5.49780392 8.63937983 11.78097245 14.92256510 $(4i-1)\frac{\pi}{4}$; i>5	$\cosh\frac{\lambda_i x}{L} - \cos\frac{\lambda_i x}{L} - \sigma_i\left(\sinh\frac{\lambda_i x}{L} - \sin\frac{\lambda_i x}{L}\right)$	0.982502207 0.999966450 0.999999933 0.999999993 0.999999993 1.0; i>5

Table 8-1. Single-Span Beams. *(Continued)*

Natural Frequency (hertz); $f_i = \frac{\lambda_i^2}{2\pi L^2}\left(\frac{EI}{m}\right)^{1/2}$; i=1,2,3...

Description[(a)]	λ_i; i=1,2,3...	Mode Shape, $\tilde{y}_i\left(\frac{x}{L}\right)$	σ_i; i=1,2,3...
9. Sliding-Pinned 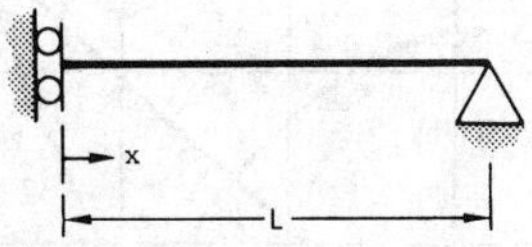	$(2i - 1)\frac{\pi}{2}$	$\cos\frac{(2i - 1)\pi x}{2L}$	--
10. Sliding-Sliding 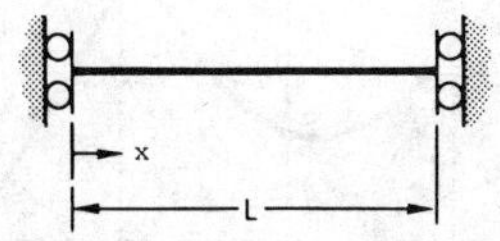	$i\pi$	$\cos\frac{i\pi x}{L}$	--

[(a)] The boundary conditions are defined mathematically in Eq. 8-6.

The dimensionless natural frequency parameters λ_i and σ_i can be numerically computed from the following formulas:

Boundary Conditions	Transcendental Equation for λ	Formula for σ_i
1. Free-free	$\cos\lambda\cosh\lambda = 1$	$\frac{\cosh\lambda_i - \cos\lambda_i}{\sinh\lambda_i - \sin\lambda_i}$
2. Free-sliding	$\tan\lambda + \tanh\lambda = 0$	$\frac{\sinh\lambda_i - \sin\lambda_i}{\cosh\lambda_i + \cos\lambda_i}$
3. Clamped-free	$\cos\lambda\cosh\lambda + 1 = 0$	$\frac{\sinh\lambda_i - \sin\lambda_i}{\cosh\lambda_i + \cos\lambda_i}$
4. Free-pinned	$\tan\lambda = \tanh\lambda$	$\frac{\cosh\lambda_i - \cos\lambda_i}{\sinh\lambda_i - \sin\lambda_i}$
5. Clamped-pinned	Same as free-pinned	
6. Clamped-clamped	Same as free-free	
7. Clamped-sliding	Same as free-sliding	

The boundary conditions are defined mathematically by Eq. 8-6. Since the longitudinal displacement of straight beams during transverse vibration is of second order ($\sim Y^2$), the presence or absence of longitudinal constraints does not affect the natural frequencies in Table 8-1.

The rigid body modes corresponding to the $\lambda = 0$ solution for the free-free, free-sliding, free-pinned, sliding-pinned, and sliding-sliding beams have been omitted from Table 8-1.

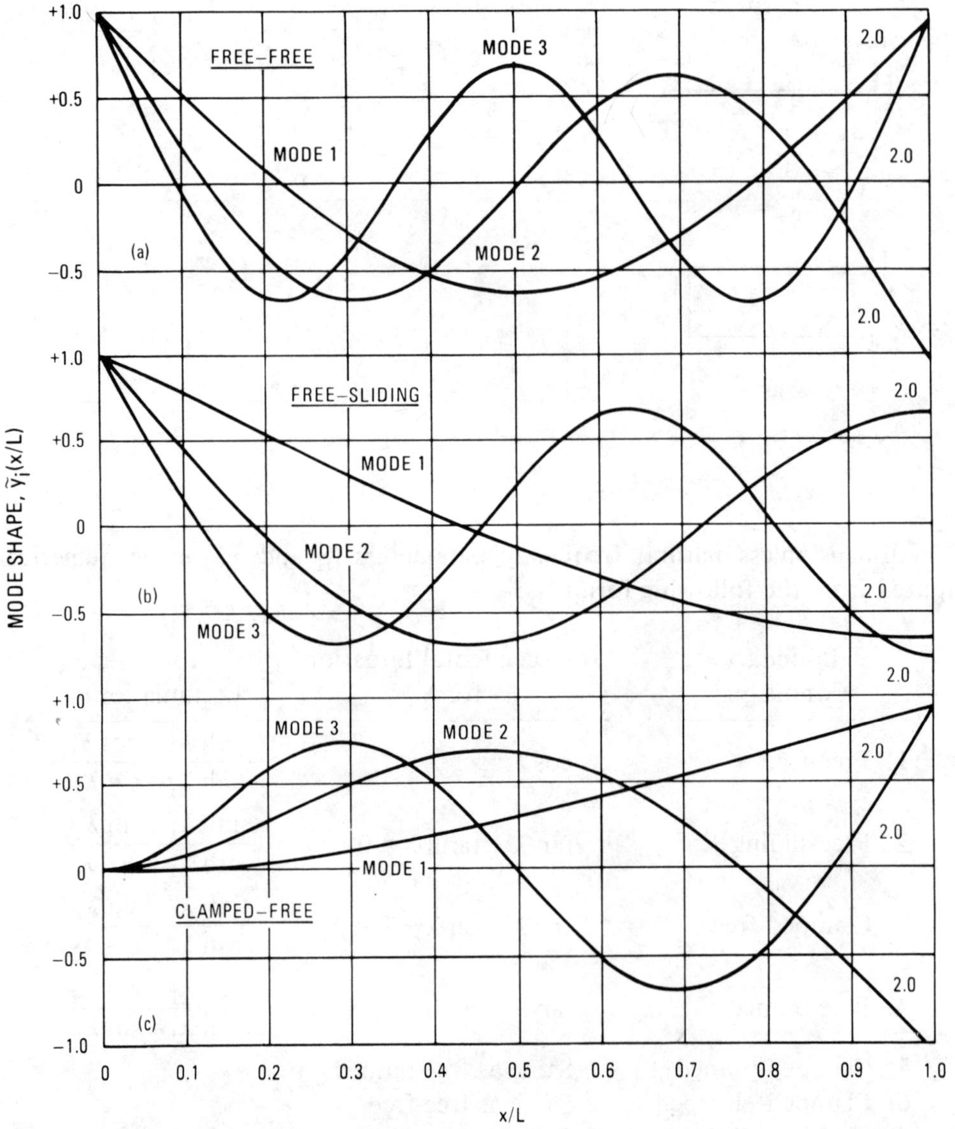

Fig. 8-4. Normalized mode shapes of first three modes of straight slender beams with the various boundary conditions corresponding to Table 8-1. These mode shapes have been divided by the normalization constants given in the graphs.

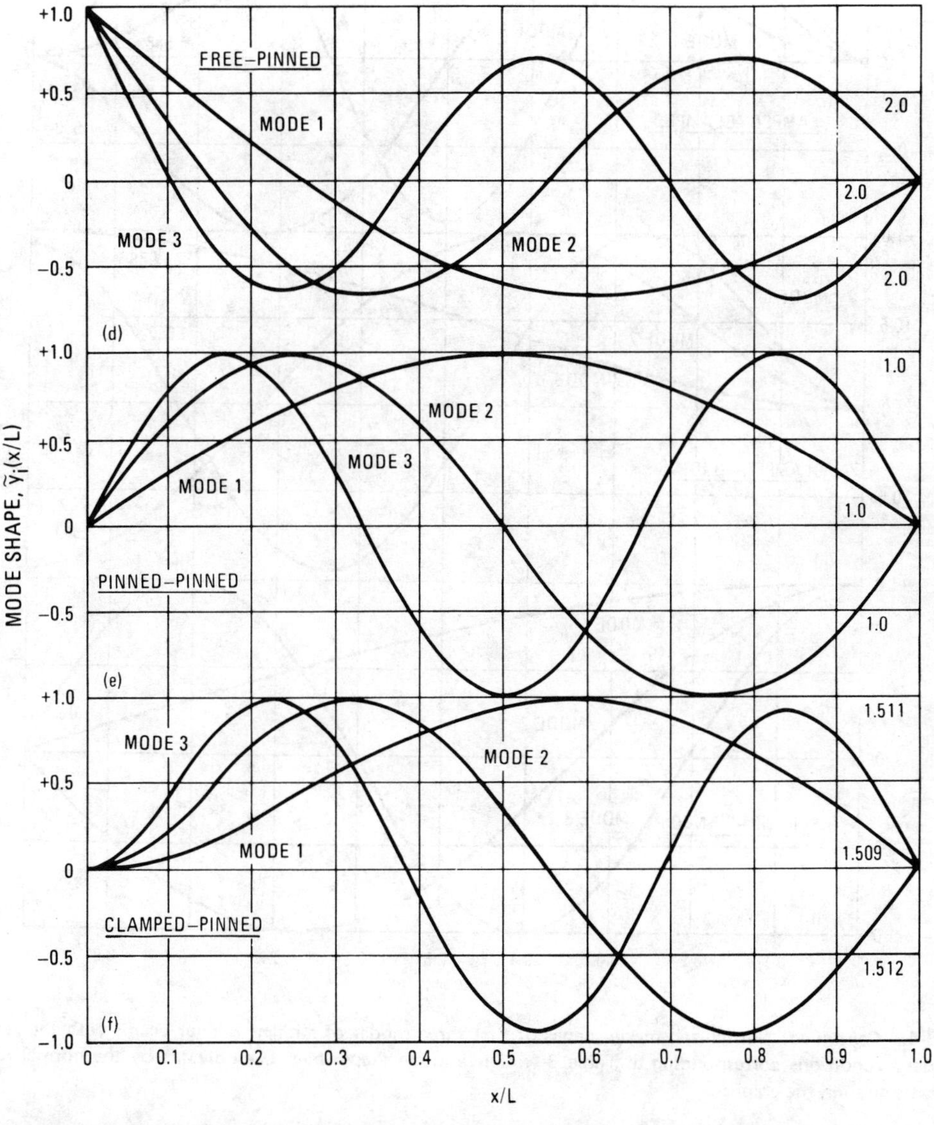

Fig. 8-4. (*Continued*) Normalized mode shapes of first three modes of straight slender beams with the various boundary conditions corresponding to Table 8-1. These mode shapes have been divided by the normalization constants given in the graphs.

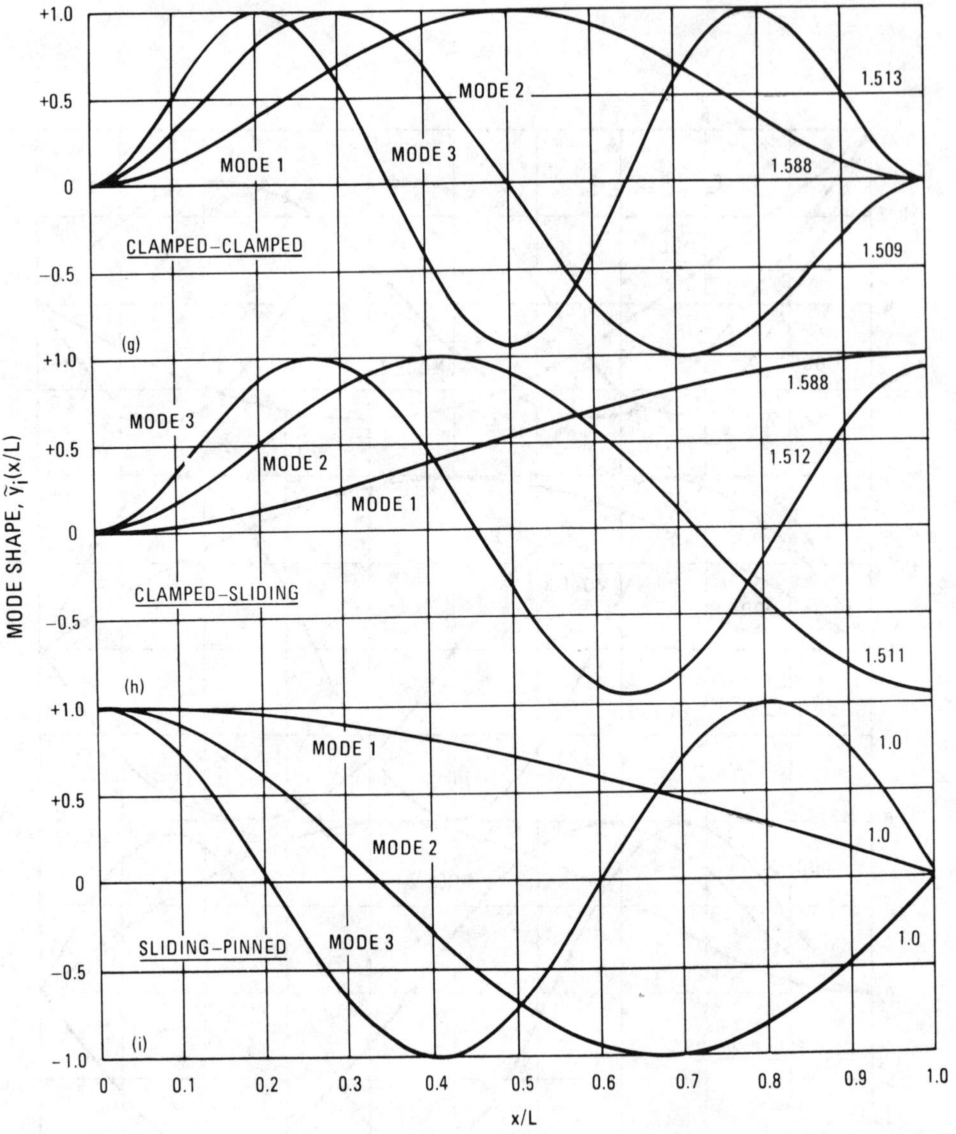

Fig. 8-4. *(Continued)* Normalized mode shapes of first three modes of straight slender beams with the various boundary conditions corresponding to Table 8-1. These mode shapes have been divided by the normalization constants given in the graphs.

Properties of the Mode Shapes. The mode shapes of single-span beams have the following properties:

1. The mode shapes, $\tilde{y}_i$, are functions of the spanwise coordinate $\lambda_i x/L$. Thus,

$$\tilde{y}_i' = \frac{d\tilde{y}_i}{d\left(\frac{\lambda_i x}{L}\right)} = \frac{L}{\lambda_i}\frac{d\tilde{y}_i}{dx}$$

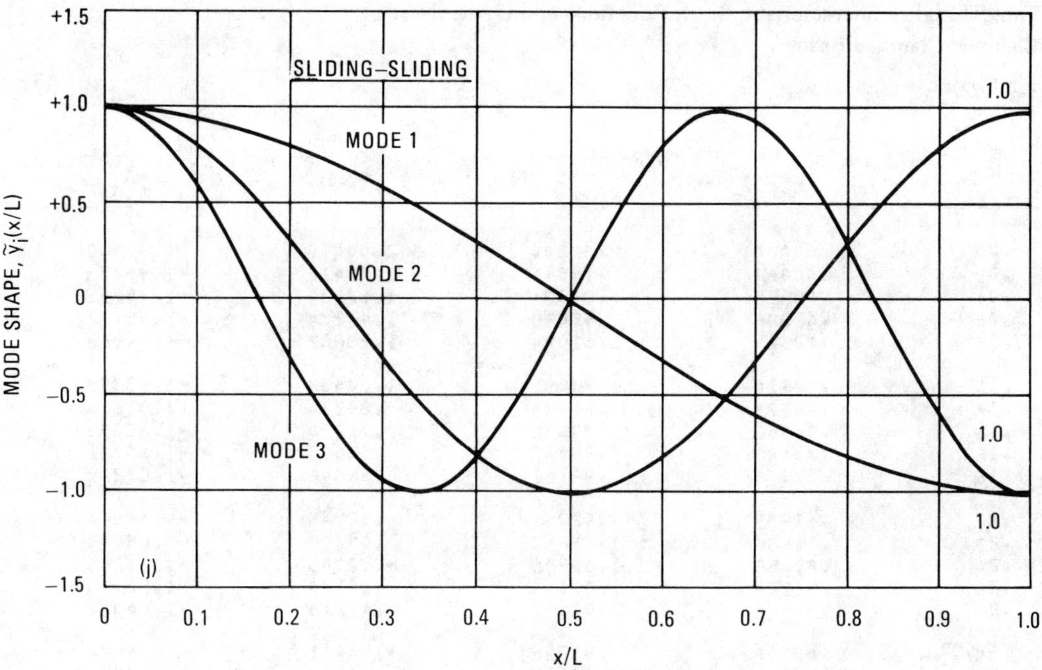

Fig. 8-4. *(Continued)* Normalized mode shapes of first three modes of straight slender beams with the various boundary conditions corresponding to Table 8-1. These mode shapes have been divided by the normalization constants given in the graphs.

2. For any of the beams listed in Table 8-1,

$$\frac{d^n\tilde{y}_i}{d\left(\frac{\lambda_i x}{L}\right)^n} = \frac{d^{n+4}\tilde{y}_i}{d\left(\frac{\lambda_i x}{L}\right)^{n+4}}; \qquad n = 0, 1, 2, 3, \ldots, \quad i = 1, 2, 3, \ldots$$

3. The beams are orthogonal over the span

$$\int_0^L \tilde{y}_i(x)\,\tilde{y}_j(x)\,dx = \int_0^L \tilde{y}_i''(x)\,\tilde{y}_j''(x)\,dx$$

$$= \begin{cases} 0, & i \neq j \\ \dfrac{L}{2}, & i = j \quad \text{pinned-pinned, sliding-pinned, sliding-sliding} \\ L, & i = j \quad \text{all other beams in Table 8-1} \end{cases}$$

$\tilde{y}_i$ and $\tilde{y}_j$ are the mode shapes associated with any two modes of any one of the beams given in Table 8-1.

Table 8-2(a). Characteristic Beam Functions and Derivatives.
Clamped-Clamped Beam

First Mode

$\frac{x}{L}$	$\tilde{y}_1$	$\tilde{y}_1' = \frac{L}{\lambda_1}\frac{d\tilde{y}_1}{dx}$	$\tilde{y}_1'' = \frac{L^2}{\lambda_1^2}\frac{d^2\tilde{y}_1}{dx^2}$	$\tilde{y}_1''' = \frac{L^3}{\lambda_1^3}\frac{d^3\tilde{y}_1}{dx^3}$
.00	.00000	.00000	2.00000	-1.96500
.02	.00867	.18041	1.81412	-1.96473
.04	.03358	.34324	1.62832	-1.96285
.06	.07306	.48850	1.44284	-1.95792
.08	.12545	.61624	1.25802	-1.94862
.10	.18910	.72655	1.07433	-1.93383
.12	.26237	.81956	.89234	-1.91254
.14	.34363	.89546	.71270	-1.88393
.16	.43126	.95450	.53615	-1.84732
.18	.52370	.99702	.36346	-1.80219
.20	.61939	1.02342	.19545	-1.74814
.22	.71684	1.03418	.03299	-1.68494
.24	.81458	1.02986	-.12305	-1.61250
.26	.91124	1.01113	-.27180	-1.53085
.28	1.00546	.97870	-.41240	-1.44017
.30	1.09600	.93338	-.54401	-1.34074
.32	1.18168	.87608	-.66581	-1.23296
.34	1.26141	.80774	-.77704	-1.11735
.36	1.33419	.72942	-.87699	-.99452
.38	1.39913	.64219	-.96500	-.86516
.40	1.45545	.54723	-1.04050	-.73007
.42	1.50246	.44574	-1.10297	-.59008
.44	1.53962	.33897	-1.15201	-.44611
.46	1.56647	.22821	-1.18728	-.29911
.48	1.58271	.11478	-1.20854	-.15007
.50	1.58815	-.00000	-1.21564	.00000
.52	1.58271	-.11478	-1.20854	.15007
.54	1.56647	-.22821	-1.18728	.29911
.56	1.53962	-.33897	-1.15201	.44611
.58	1.50246	-.44574	-1.10297	.59008
.60	1.45545	-.54723	-1.04050	.73007
.62	1.39913	-.64219	-.96500	.86516
.64	1.33419	-.72942	-.87699	.99452
.66	1.26141	-.80774	-.77704	1.11735
.68	1.18168	-.87608	-.66581	1.23296
.70	1.09600	-.93338	-.54401	1.34074
.72	1.00546	-.97870	-.41240	1.44017
.74	.91124	-1.01113	-.27180	1.53085
.76	.81458	-1.02986	-.12305	1.61250
.78	.71684	-1.03418	.03299	1.68494
.80	.61939	-1.02342	.19545	1.74814
.82	.52370	-.99702	.36346	1.80219
.84	.43126	-.95450	.53615	1.84732
.86	.34363	-.89546	.71270	1.88393
.88	.26237	-.81956	.89234	1.91254
.90	.18910	-.72655	1.07433	1.93383
.92	.12545	-.61624	1.25802	1.94862
.94	.07306	-.48850	1.44284	1.95792
.96	.03358	-.34324	1.62832	1.96285
.98	.00867	-.18041	1.81412	1.96473
1.00	.00000	.00000	2.00000	1.96500

Table 8-2(a). Clamped-Clamped Beam. (*Continued*)

Second Mode

$\frac{x}{L}$	$\tilde{y}_2$	$\tilde{y}_2' = \frac{L}{\lambda_2}\frac{d\tilde{y}_2}{dx}$	$\tilde{y}_2'' = \frac{L^2}{\lambda_2^2}\frac{d^2\tilde{y}_2}{dx^2}$	$\tilde{y}_2''' = \frac{L^3}{\lambda_2^3}\frac{d^3\tilde{y}_2}{dx^3}$
.00	.00000	.00000	2.00000	-2.00155
.02	.02338	.28944	1.68568	-2.00031
.04	.08834	.52955	1.37202	-1.99203
.06	.18715	.72055	1.06060	-1.97079
.08	.31214	.86296	.75386	-1.93187
.10	.45574	.95776	.45486	-1.87177
.12	.61058	1.00643	.16712	-1.78813
.14	.76958	1.01105	-.10555	-1.67975
.16	.92601	.97427	-.35923	-1.54651
.18	1.07363	.89940	-.59009	-1.38932
.20	1.20675	.79029	-.79450	-1.21002
.22	1.32032	.65138	-.96917	-1.01128
.24	1.41006	.48755	-1.11133	-.79652
.26	1.47245	.30410	-1.21875	-.56977
.28	1.50485	.10661	-1.28992	-.33555
.30	1.50550	-.09916	-1.32402	-.09872
.32	1.47357	-.30736	-1.32106	.13566
.34	1.40913	-.51224	-1.28181	.36247
.36	1.31313	-.70820	-1.20786	.57665
.38	1.18740	-.88997	-1.10158	.77340
.40	1.03456	-1.05270	-.96606	.94823
.42	.85794	-1.19210	-.80507	1.09714
.44	.66151	-1.30449	-.62295	1.21670
.46	.44974	-1.38693	-.42455	1.30414
.48	.22752	-1.43727	-.21508	1.35742
.50	.00000	-1.45420	-.00000	1.37533
.52	-.22752	-1.43727	.21508	1.35742
.54	-.44974	-1.38693	.42455	1.30414
.56	-.66151	-1.30449	.62295	1.21670
.58	-.85794	-1.19210	.80507	1.09714
.60	-1.03456	-1.05270	.96606	.94823
.62	-1.18740	-.88997	1.10158	.77340
.64	-1.31313	-.70820	1.20786	.57665
.66	-1.40913	-.51224	1.28181	.36247
.68	-1.47357	-.30736	1.32106	.13566
.70	-1.50550	-.09916	1.32402	-.09872
.72	-1.50485	.10661	1.28992	-.33555
.74	-1.47245	.30410	1.21875	-.56977
.76	-1.41006	.48755	1.11133	-.79652
.78	-1.32032	.65138	.96917	-1.01128
.80	-1.20675	.79029	.79450	-1.21002
.82	-1.07363	.89940	.59009	-1.38932
.84	-.92601	.97427	.35923	-1.54651
.86	-.76958	1.01105	.10555	-1.67975
.88	-.61058	1.00643	-.16712	-1.78813
.90	-.45574	.95776	-.45486	-1.87177
.92	-.31214	.86296	-.75386	-1.93187
.94	-.18715	.72055	-1.06060	-1.97079
.96	-.08834	.52955	-1.37202	-1.99203
.98	-.02338	.28944	-1.68568	-2.00031
1.00	.00000	.00000	-2.00000	-2.00155

Table 8-2(a). Clamped-Clamped Beam. *(Continued)*

Third Mode

$\frac{x}{L}$	$\tilde{y}_3$	$\tilde{y}_3' = \frac{L}{\lambda_3}\frac{d\tilde{y}_3}{dx}$	$\tilde{y}_3'' = \frac{L^2}{\lambda_3^2}\frac{d^2\tilde{y}_3}{dx^2}$	$\tilde{y}_3''' = \frac{L^3}{\lambda_3^3}\frac{d^3\tilde{y}_3}{dx^3}$
.00	.00000	.00000	2.00000	-1.99993
.02	.04482	.39147	1.56038	-1.99658
.04	.16510	.68646	1.12323	-1.97469
.06	.33975	.88609	.69428	-1.91998
.08	.54803	.99303	.28189	-1.82280
.10	.77005	1.01202	-.10393	-1.67794
.12	.98720	.95005	-.45253	-1.48447
.14	1.18265	.81648	-.75348	-1.24534
.16	1.34190	.62284	-.99738	-.96697
.18	1.45317	.38256	-1.17658	-.65867
.20	1.50783	.11049	-1.28573	-.33199
.22	1.50059	-.17760	-1.32221	-.00003
.24	1.42971	-.46574	-1.28637	.32333
.26	1.29690	-.73832	-1.18164	.62424
.28	1.10719	-.98086	-1.01443	.88956
.30	.86863	-1.18057	-.79386	1.10762
.32	.59186	-1.32695	-.53144	1.26880
.34	.28949	-1.41222	-.24051	1.36606
.36	-.02444	-1.43171	.06439	1.39528
.38	-.33527	-1.38399	.36811	1.35553
.40	-.62836	-1.27099	.65569	1.24912
.42	-.88987	-1.09783	.91301	1.08148
.44	-1.10739	-.87257	1.12747	.86096
.46	-1.27060	-.60585	1.28859	.59841
.48	-1.37174	-.31031	1.38852	.30668
.50	-1.40600	.00000	1.42238	-.00000
.52	-1.37174	.31031	1.38852	-.30668
.54	-1.27060	.60585	1.28859	-.59841
.56	-1.10739	.87257	1.12747	-.86096
.58	-.88987	1.09783	.91301	-1.08148
.60	-.62836	1.27099	.65569	-1.24912
.62	-.33527	1.38399	.36811	-1.35553
.64	-.02444	1.43171	.06439	-1.39528
.66	.28949	1.41222	-.24051	-1.36606
.68	.59186	1.32695	-.53144	-1.26880
.70	.86863	1.18057	-.79386	-1.10762
.72	1.10719	.98086	-1.01443	-.88956
.74	1.29690	.73832	-1.18164	-.62424
.76	1.42971	.46574	-1.28637	-.32333
.78	1.50059	.17760	-1.32221	.00003
.80	1.50783	-.11049	-1.28573	.33199
.82	1.45317	-.38256	-1.17658	.65867
.84	1.34190	-.62284	-.99738	.96697
.86	1.18265	-.81648	-.75348	1.24534
.88	.98720	-.95005	-.45253	1.48447
.90	.77005	-1.01202	-.10393	1.67794
.92	.54803	-.99303	.28189	1.82280
.94	.33975	-.88609	.69428	1.91998
.96	.16510	-.68646	1.12323	1.97469
.98	.04482	-.39147	1.56038	1.99658
1.00	.00000	.00000	2.00000	1.99993

Table 8-2(a). Clamped-Clamped Beam. (*Continued*)

Fourth Mode

$\frac{x}{L}$	$\tilde{y}_4$	$\tilde{y}_4' = \frac{L}{\lambda_4}\frac{d\tilde{y}_4}{dx}$	$\tilde{y}_4'' = \frac{L^2}{\lambda_4^2}\frac{d^2\tilde{y}_4}{dx^2}$	$\tilde{y}_4''' = \frac{L^3}{\lambda_4^3}\frac{d^2\tilde{y}_4}{dx^3}$
.00	.00000	.00000	2.00000	-2.00000
.02	.07241	.48557	1.43502	-1.99300
.04	.25958	.81207	.87658	-1.94824
.06	.51697	.98325	.33937	-1.83960
.08	.80177	1.00789	-.15633	-1.65333
.10	1.07449	.90088	-.58802	-1.38736
.12	1.30078	.68345	-.93412	-1.05012
.14	1.45308	.38242	-1.17673	-.65879
.16	1.51208	.02893	-1.30380	-.23724
.18	1.46765	-.34350	-1.31068	.18649
.20	1.31923	-.70122	-1.20092	.58286
.22	1.07549	-1.01271	-.98634	.92349
.24	.75348	-1.25091	-.68630	1.18364
.26	.37700	-1.39515	-.32640	1.34442
.28	-.02537	-1.43265	.06348	1.39439
.30	-.42268	-1.35944	.45136	1.33056
.32	-.78413	-1.18058	.80569	1.15876
.34	-1.08158	-.90972	1.09776	.89319
.36	-1.29186	-.56793	1.30395	.55537
.38	-1.39857	-.18205	1.40755	.17245
.40	-1.39351	.21752	1.40010	-.22494
.42	-1.27726	.59923	1.28198	-.60506
.44	-1.05919	.93288	1.06244	-.93759
.46	-.75676	1.19208	.75879	-1.19604
.48	-.39407	1.35629	.39504	-1.35983
.50	.00000	1.41251	-.00000	-1.41592
.52	.39407	1.35629	-.39504	-1.35983
.54	.75676	1.19208	-.75879	-1.19604
.56	1.05919	.93288	-1.06244	-.93759
.58	1.27726	.59923	-1.28198	-.60506
.60	1.39351	.21752	-1.40010	-.22494
.62	1.39857	-.18205	-1.40755	.17245
.64	1.29186	-.56793	-1.30395	.55537
.66	1.08158	-.90972	-1.09776	.89319
.68	.78413	-1.18058	-.80569	1.15876
.70	.42268	-1.35944	-.45136	1.33056
.72	.02537	-1.43265	-.06348	1.39439
.74	-.37700	-1.39515	.32640	1.34442
.76	-.75348	-1.25091	.68630	1.18364
.78	-1.07549	-1.01271	.98634	.92349
.80	-1.31923	-.70122	1.20092	.58286
.82	-1.46765	-.34350	1.31068	.18649
.84	-1.51208	.02893	1.30380	-.23724
.86	-1.45308	.38242	1.17673	-.65879
.88	-1.30078	.68345	.93412	-1.05012
.90	-1.07449	.90088	.58802	-1.38736
.92	-.80177	1.00789	.15633	-1.65333
.94	-.51697	.98325	-.33937	-1.83960
.96	-.25958	.81207	-.87658	-1.94824
.98	-.07241	.48557	-1.43502	-1.99300
1.00	.00000	-.00000	-2.00000	-2.00000

Table 8-2(a). Clamped-Clamped Beam. (*Continued*)

Fifth Mode

$\frac{x}{L}$	$\tilde{y}_5$	$\tilde{y}_5' = \frac{L}{\lambda_5}\frac{d\tilde{y}_5}{dx}$	$\tilde{y}_5'' = \frac{L^2}{\lambda_5^2}\frac{d^2\tilde{y}_5}{dx^2}$	$\tilde{y}_5''' = \frac{L^3}{\lambda_5^3}\frac{d^3\tilde{y}_5}{dx^3}$
.00	.00000	.00000	2.00000	-2.00000
.02	.10567	.57181	1.30996	-1.98743
.04	.36791	.90694	.63409	-1.90894
.06	.70631	1.01517	.00291	-1.72440
.08	1.04591	.91867	-.54391	-1.42067
.10	1.32178	.65359	-.96646	-1.00892
.12	1.48381	.26880	-1.23231	-.52030
.14	1.50043	-.17781	-1.32241	-.00021
.16	1.36090	-.62465	-1.23490	.49865
.18	1.07551	-1.01269	-.98632	.92350
.20	.67360	-1.29164	-.61047	1.22851
.22	.19959	-1.42540	-.15491	1.38072
.24	-.29269	-1.39597	.32432	1.36434
.26	-.74658	-1.20525	.76897	1.18287
.28	-1.10952	-.87470	1.12537	.85886
.30	-1.33938	-.44262	1.35061	.43141
.32	-1.40954	.04046	1.41749	-.04838
.34	-1.31208	.51781	1.31772	-.52340
.36	-1.05881	.93326	1.06282	-.93721
.38	-.67987	1.23790	.68273	-1.24067
.40	-.22020	1.39584	.22226	-1.39777
.42	.26575	1.38850	-.26425	-1.38982
.44	.72046	1.21684	-.71933	-1.21771
.46	1.09011	.90119	-1.08923	-.90172
.48	1.33098	.47892	-1.33023	-.47917
.50	1.41457	-.00000	-1.41386	.00000
.52	1.33098	-.47892	-1.33023	.47917
.54	1.09011	-.90119	-1.08923	.90172
.56	.72046	-1.21684	-.71933	1.21771
.58	.26575	-1.38850	-.26425	1.38982
.60	-.22020	-1.39584	.22226	1.39777
.62	-.67987	-1.23790	.68273	1.24067
.64	-1.05881	-.93326	1.06282	.93721
.66	-1.31208	-.51781	1.31772	.52340
.68	-1.40954	-.04046	1.41749	.04838
.70	-1.33938	.44262	1.35061	-.43141
.72	-1.10952	.87470	1.12537	-.85886
.74	-.74658	1.20525	.76897	-1.18287
.76	-.29269	1.39597	.32432	-1.36434
.78	.19959	1.42540	-.15491	-1.38072
.80	.67360	1.29164	-.61048	-1.22851
.82	1.07551	1.01269	-.98632	-.92350
.84	1.36090	.62465	-1.23490	-.49865
.86	1.50043	.17781	-1.32241	.00021
.88	1.48381	-.26880	-1.23231	.52030
.90	1.32178	-.65359	-.96646	1.00892
.92	1.04591	-.91867	-.54391	1.42067
.94	.70631	-1.01517	.00291	1.72440
.96	.36791	-.90694	.63409	1.90894
.98	.10567	-.57181	1.30996	1.98743
1.00	.00000	.00000	2.00000	2.00000

Table 8-2(b). Characteristic Beam Functions and Derivatives.
Clamped-Free Beam

First Mode

$\frac{x}{L}$	$\tilde{y}_1$	$\tilde{y}_1' = \frac{L}{\lambda_1}\frac{d\tilde{y}_1}{dx}$	$\tilde{y}_1'' = \frac{L^2}{\lambda_1^2}\frac{d^2\tilde{y}_1}{dx^2}$	$\tilde{y}_1''' = \frac{L^3}{\lambda_1^3}\frac{d^3\tilde{y}_1}{dx^3}$
.00	.00000	.00000	2.00000	-1.46819
.02	.00139	.07397	1.94494	-1.46817
.04	.00552	.14588	1.88988	-1.46805
.06	.01231	.21572	1.83483	-1.46773
.08	.02168	.28350	1.77980	-1.46710
.10	.03355	.34921	1.72480	-1.46607
.12	.04784	.41287	1.66985	-1.46455
.14	.06449	.47446	1.61496	-1.46245
.16	.08340	.53400	1.56016	-1.45969
.18	.10451	.59148	1.50549	-1.45617
.20	.12774	.64692	1.45096	-1.45182
.22	.15301	.70031	1.39660	-1.44656
.24	.18024	.75167	1.34247	-1.44032
.26	.20936	.80100	1.28859	-1.43302
.28	.24030	.84832	1.23500	-1.42459
.30	.27297	.89364	1.18175	-1.41497
.32	.30730	.93696	1.12889	-1.40410
.34	.34322	.97832	1.07646	-1.39191
.36	.38065	1.01771	1.02451	-1.37834
.38	.41953	1.05516	.97309	-1.36334
.40	.45977	1.09070	.92227	-1.34685
.42	.50131	1.12435	.87209	-1.32884
.44	.54408	1.15612	.82262	-1.30924
.46	.58800	1.18606	.77391	-1.28801
.48	.63301	1.21418	.72604	-1.26512
.50	.67905	1.24052	.67905	-1.24052
.52	.72604	1.26512	.63301	-1.21418
.54	.77391	1.28801	.58800	-1.18606
.56	.82262	1.30924	.54408	-1.15612
.58	.87209	1.32884	.50131	-1.12435
.60	.92227	1.34685	.45977	-1.09070
.62	.97309	1.36334	.41953	-1.05516
.64	1.02451	1.37834	.38065	-1.01771
.66	1.07646	1.39191	.34322	-.97832
.68	1.12889	1.40410	.30730	-.93696
.70	1.18175	1.41497	.27297	-.89364
.72	1.23500	1.42459	.24030	-.84832
.74	1.28859	1.43302	.20936	-.80100
.76	1.34247	1.44032	.18024	-.75167
.78	1.39660	1.44656	.15301	-.70031
.80	1.45096	1.45182	.12774	-.64692
.82	1.50549	1.45617	.10451	-.59148
.84	1.56016	1.45969	.08340	-.53400
.86	1.61496	1.46245	.06449	-.47446
.88	1.66985	1.46455	.04784	-.41287
.90	1.72480	1.46607	.03355	-.34921
.92	1.77980	1.46710	.02168	-.28350
.94	1.83483	1.46773	.01231	-.21572
.96	1.88988	1.46805	.00552	-.14588
.98	1.94494	1.46817	.00139	-.07397
1.00	2.00000	1.46819	-.00000	.00000

Table 8-2(b). Clamped-Free Beam. *(Continued)*

Second Mode

$\frac{x}{L}$	$\tilde{y}_2$	$\tilde{y}_2' = \frac{L}{\lambda_2}\frac{d\tilde{y}_2}{dx}$	$\tilde{y}_2'' = \frac{L^2}{\lambda_2^2}\frac{d^2\tilde{y}_2}{dx^2}$	$\tilde{y}_2''' = \frac{L^3}{\lambda_2^3}\frac{d^3\tilde{y}_2}{dx^3}$
.00	.00000	.00000	2.00000	-2.03693
.02	.00853	.17879	1.80878	-2.03667
.04	.03301	.33962	1.61764	-2.03483
.06	.07174	.48253	1.42680	-2.03002
.08	.12305	.60755	1.23661	-2.02097
.10	.18526	.71475	1.04750	-2.00658
.12	.25670	.80428	.86004	-1.98590
.14	.33573	.87631	.67485	-1.95814
.16	.42070	.93108	.49261	-1.92267
.18	.51002	.96892	.31409	-1.87901
.20	.60211	.99020	.14007	-1.82682
.22	.69544	.99539	-.02864	-1.76592
.24	.78852	.98501	-.19123	-1.69625
.26	.87992	.95970	-.34687	-1.61792
.28	.96827	.92013	-.49475	-1.53113
.30	1.05227	.86707	-.63410	-1.43625
.32	1.13068	.80136	-.76419	-1.33373
.34	1.20236	.72389	-.88431	-1.22416
.36	1.26626	.63565	-.99384	-1.10821
.38	1.32141	.53764	-1.09222	-.98667
.40	1.36694	.43093	-1.17895	-.86040
.42	1.40209	.31665	-1.25365	-.73034
.44	1.42619	.19593	-1.31600	-.59749
.46	1.43871	.06995	-1.36578	-.46291
.48	1.43920	-.06012	-1.40289	-.32772
.50	1.42733	-.19307	-1.42733	-.19307
.52	1.40289	-.32772	-1.43920	-.06012
.54	1.36578	-.46291	-1.43871	.06995
.56	1.31600	-.59749	-1.42619	.19593
.58	1.25365	-.73034	-1.40209	.31665
.60	1.17895	-.86040	-1.36694	.43093
.62	1.09222	-.98667	-1.32141	.53764
.64	.99384	-1.10821	-1.26626	.63565
.66	.88431	-1.22416	-1.20236	.72389
.68	.76419	-1.33373	-1.13068	.80136
.70	.63410	-1.43625	-1.05227	.86707
.72	.49475	-1.53113	-.96827	.92013
.74	.34687	-1.61792	-.87992	.95970
.76	.19123	-1.69625	-.78852	.98501
.78	.02864	-1.76592	-.69544	.99539
.80	-.14007	-1.82682	-.60211	.99020
.82	-.31409	-1.87901	-.51002	.96892
.84	-.49261	-1.92267	-.42070	.93108
.86	-.67485	-1.95814	-.33573	.87631
.88	-.86004	-1.98590	-.25670	.80428
.90	-1.04750	-2.00658	-.18526	.71475
.92	-1.23661	-2.02097	-.12305	.60755
.94	-1.42680	-2.03002	-.07174	.48253
.96	-1.61764	-2.03483	-.03301	.33962
.98	-1.80878	-2.03667	-.00853	.17879
1.00	-2.00000	-2.03693	.00000	.00000

Table 8-2(b). Clamped-Free Beam. (*Continued*)

Third Mode

$\frac{x}{L}$	$\tilde{y}_3$	$\tilde{y}_3' = \frac{L}{\lambda_3}\frac{d\tilde{y}_3}{dx}$	$\tilde{y}_3'' = \frac{L^2}{\lambda_3^2}\frac{d^2\tilde{y}_3}{dx^2}$	$\tilde{y}_3''' = \frac{L^3}{\lambda_3^3}\frac{d^3\tilde{y}_3}{dx^3}$
.00	.00000	.00000	2.00000	-1.99845
.02	.02339	.28953	1.68610	-1.99721
.04	.08839	.52979	1.37287	-1.98892
.06	.18727	.72099	1.06188	-1.96766
.08	.31237	.86367	.75558	-1.92871
.10	.45614	.95879	.45702	-1.86854
.12	.61120	1.00785	.16973	-1.78480
.14	.77049	1.01291	-.10245	-1.67629
.16	.92728	.97665	-.35563	-1.54286
.18	1.07535	.90237	-.58594	-1.38540
.20	1.20901	.79394	-.78975	-1.20575
.22	1.32324	.65580	-.96375	-1.00656
.24	1.41376	.49285	-1.10515	-.79124
.26	1.47707	.31040	-1.21172	-.56380
.28	1.51055	.11405	-1.28189	-.32872
.30	1.51248	-.09041	-1.31485	-.09085
.32	1.48203	-.29711	-1.31055	.14479
.34	1.41932	-.50026	-1.26974	.37310
.36	1.32534	-.69422	-1.19397	.58908
.38	1.20196	-.87368	-1.08556	.78797
.40	1.05185	-1.03374	-.94753	.96533
.42	.87841	-1.17003	-.78359	1.11722
.44	.68568	-1.27881	-.59802	1.24030
.46	.47822	-1.35704	-.39555	1.33189
.48	.26103	-1.40247	-.18131	1.39005
.50	.03938	-1.41367	.03938	1.41367
.52	-.18131	-1.39005	.26103	1.40247
.54	-.39555	-1.33189	.47822	1.35704
.56	-.59802	-1.24030	.68568	1.27881
.58	-.78359	-1.11722	.87841	1.17003
.60	-.94753	-.96533	1.05185	1.03374
.62	-1.08556	-.78797	1.20196	.87368
.64	-1.19397	-.58908	1.32534	.69422
.66	-1.26974	-.37310	1.41932	.50026
.68	-1.31055	-.14479	1.48203	.29711
.70	-1.31485	.09085	1.51248	.09041
.72	-1.28189	.32872	1.51055	-.11405
.74	-1.21172	.56380	1.47707	-.31040
.76	-1.10515	.79124	1.41376	-.49285
.78	-.96375	1.00656	1.32324	-.65580
.80	-.78975	1.20575	1.20901	-.79394
.82	-.58594	1.38540	1.07535	-.90237
.84	-.35563	1.54286	.92728	-.97665
.86	-.10245	1.67629	.77049	-1.01291
.88	.16973	1.78480	.61120	-1.00785
.90	.45702	1.86854	.45614	-.95879
.92	.75558	1.92871	.31237	-.86367
.94	1.06188	1.96766	.18727	-.72099
.96	1.37287	1.98892	.08839	-.52979
.98	1.68610	1.99721	.02339	-.28953
1.00	2.00000	1.99845	.00000	.00000

Table 8-2(b). Clamped-Free Beam. *(Continued)*

Fourth Mode

$\frac{x}{L}$	$\tilde{y}_4$	$\tilde{y}_4' = \frac{L}{\lambda_4}\frac{d\tilde{y}_4}{dx}$	$\tilde{y}_4'' = \frac{L^2}{\lambda_4^2}\frac{d^2\tilde{y}_4}{dx^2}$	$\tilde{y}_4''' = \frac{L^3}{\lambda_4^3}\frac{d^3\tilde{y}_4}{dx^3}$
.00	.00000	.00000	2.00000	-2.00007
.02	.04482	.39147	1.56035	-1.99672
.04	.16510	.68645	1.12317	-1.97482
.06	.33974	.88606	.69420	-1.92012
.08	.54801	.99298	.28179	-1.82294
.10	.77002	1.01194	-.10407	-1.67809
.12	.98714	.94994	-.45270	-1.48463
.14	1.18256	.81633	-.75368	-1.24552
.16	1.34177	.62264	-.99762	-.96717
.18	1.45299	.38230	-1.17687	-.65891
.20	1.50758	.11017	-1.28608	-.33228
.22	1.50027	-.17801	-1.32262	-.00038
.24	1.42928	-.46625	-1.28688	.32290
.26	1.29634	-.73895	-1.18226	.62370
.28	1.10648	-.98164	-1.01518	.88888
.30	.86774	-1.18153	-.79478	1.10676
.32	.59073	-1.32813	-.53258	1.26772
.34	.28808	-1.41368	-.24191	1.36469
.36	-.02621	-1.43351	.06264	1.39357
.38	-.33748	-1.38622	.36594	1.35339
.40	-.63112	-1.27376	.65299	1.24643
.42	-.89330	-1.10126	.90964	1.07812
.44	-1.11166	-.87683	1.12327	.85675
.46	-1.27592	-.61115	1.28336	.59315
.48	-1.37836	-.31690	1.38199	.30011
.50	-1.41424	-.00819	1.41424	-.00819
.52	-1.38199	.30011	1.37836	-.31690
.54	-1.28336	.59315	1.27592	-.61115
.56	-1.12327	.85675	1.11166	-.87683
.58	-.90964	1.07812	.89330	-1.10126
.60	-.65299	1.24643	.63112	-1.27376
.62	-.36594	1.35339	.33748	-1.38622
.64	-.06264	1.39357	.02621	-1.43351
.66	.24191	1.36469	-.28808	-1.41368
.68	.53258	1.26772	-.59073	-1.32813
.70	.79478	1.10676	-.86774	-1.18153
.72	1.01518	.88888	-1.10648	-.98164
.74	1.18226	.62370	-1.29634	-.73895
.76	1.28688	.32290	-1.42928	-.46625
.78	1.32262	-.00038	-1.50027	-.17801
.80	1.28608	-.33228	-1.50758	.11017
.82	1.17687	-.65891	-1.45299	.38230
.84	.99762	-.96717	-1.34177	.62264
.86	.75368	-1.24552	-1.18256	.81633
.88	.45270	-1.48463	-.98714	.94994
.90	.10407	-1.67809	-.77002	1.01194
.92	-.28179	-1.82294	-.54801	.99298
.94	-.69420	-1.92012	-.33974	.88606
.96	-1.12317	-1.97482	-.16510	.68645
.98	-1.56035	-1.99672	-.04482	.39147
1.00	-2.00000	-2.00007	.00000	.00000

Table 8-2(b). Clamped-Free Beam. (*Continued*)

Fifth Mode

$\frac{x}{L}$	$\tilde{y}_5$	$\tilde{y}_5' = \frac{L}{\lambda_5}\frac{d\tilde{y}_5}{dx}$	$\tilde{y}_5'' = \frac{L^2}{\lambda_5^2}\frac{d^2\tilde{y}_5}{dx^2}$	$\tilde{y}_5''' = \frac{L^3}{\lambda_5^3}\frac{d^3\tilde{y}_5}{dx^3}$
.00	.00000	.00000	2.00000	-2.00000
.02	.07241	.48557	1.43502	-1.99300
.04	.25958	.81207	.87659	-1.94824
.06	.51697	.98325	.33938	-1.83959
.08	.80177	1.00789	-.15633	-1.65332
.10	1.07449	.90089	-.58801	-1.38736
.12	1.30078	.68345	-.93411	-1.05011
.14	1.45309	.38243	-1.17672	-.65878
.16	1.51209	.02895	-1.30378	-.23723
.18	1.46767	-.34348	-1.31066	.18651
.20	1.31925	-.70119	-1.20090	.58289
.22	1.07553	-1.01267	-.98631	.92352
.24	.75353	-1.25086	-.68626	1.18368
.26	.37706	-1.39509	-.32634	1.34448
.28	-.02529	-1.43257	.06355	1.39446
.30	-.42257	-1.35934	.45146	1.33066
.32	-.78399	-1.18045	.80582	1.15889
.34	-1.08141	-.90954	1.09793	.89337
.36	-1.29162	-.56770	1.30418	.55561
.38	-1.39826	-.18174	1.40786	.17276
.40	-1.39309	.21794	1.40051	-.22452
.42	-1.27670	.59978	1.28253	-.60451
.44	-1.05846	.93361	1.06317	-.93686
.46	-.75579	1.19304	.75976	-1.19507
.48	-.39278	1.35757	.39632	-1.35855
.50	.00170	1.41421	.00170	-1.41421
.52	.39632	1.35855	-.39278	-1.35757
.54	.75976	1.19507	-.75579	-1.19304
.56	1.06317	.93686	-1.05846	-.93361
.58	1.28253	.60451	-1.27670	-.59978
.60	1.40051	.22452	-1.39309	-.21794
.62	1.40786	-.17276	-1.39826	.18174
.64	1.30418	-.55561	-1.29162	.56770
.66	1.09793	-.89337	-1.08141	.90954
.68	.80582	-1.15889	-.78399	1.18045
.70	.45146	-1.33066	-.42257	1.35934
.72	.06355	-1.39446	-.02529	1.43257
.74	-.32634	-1.34448	.37706	1.39509
.76	-.68626	-1.18368	.75353	1.25086
.78	-.98631	-.92352	1.07553	1.01267
.80	-1.20090	-.58289	1.31925	.70119
.82	-1.31066	-.18651	1.46767	.34348
.84	-1.30378	.23723	1.51209	-.02895
.86	-1.17672	.65878	1.45309	-.38243
.88	-.93411	1.05011	1.30078	-.68345
.90	-.58801	1.38736	1.07449	-.90089
.92	-.15633	1.65332	.80177	-1.00789
.94	.33938	1.83959	.51697	-.98325
.96	.87659	1.94824	.25958	-.81207
.98	1.43502	1.99300	.07241	-.48557
1.00	2.00000	2.00000	-.00000	.00000

Table 8-2(c). Characteristic Beam Functions and Derivatives.
Clamped-Pinned Beam

First Mode

$\frac{x}{L}$	$\tilde{y}_1$	$\tilde{y}_1' = \frac{L}{\lambda_1}\frac{d\tilde{y}_1}{dx}$	$\tilde{y}_1'' = \frac{L^2}{\lambda_1^2}\frac{d^2\tilde{y}_1}{dx^2}$	$\tilde{y}_1''' = \frac{L^3}{\lambda_1^3}\frac{d^3\tilde{y}_1}{dx^3}$
.00	.00000	.00000	2.00000	-2.00155
.02	.00601	.15089	1.84282	-2.00140
.04	.02338	.28944	1.68568	-2.00031
.06	.05114	.41566	1.52869	-1.99745
.08	.08834	.52955	1.37202	-1.99203
.10	.13400	.63116	1.21589	-1.98336
.12	.18715	.72055	1.06060	-1.97079
.14	.24685	.79778	.90647	-1.95379
.16	.31214	.86296	.75386	-1.93187
.18	.38208	.91623	.60318	-1.90464
.20	.45574	.95776	.45486	-1.87177
.22	.53221	.98775	.30935	-1.83299
.24	.61058	1.00643	.16712	-1.78813
.26	.68999	1.01409	.02866	-1.73706
.28	.76958	1.01105	-.10555	-1.67975
.30	.84852	.99764	-.23500	-1.61621
.32	.92601	.97427	-.35923	-1.54651
.34	1.00129	.94137	-.47775	-1.47082
.36	1.07363	.89940	-.59009	-1.38932
.38	1.14233	.84886	-.69582	-1.30229
.40	1.20675	.79029	-.79450	-1.21002
.42	1.26626	.72427	-.88574	-1.11288
.44	1.32032	.65138	-.96917	-1.01128
.46	1.36841	.57226	-1.04447	-.90566
.48	1.41006	.48755	-1.11133	-.79652
.50	1.44486	.39794	-1.16950	-.68437
.52	1.47245	.30410	-1.21875	-.56977
.54	1.49253	.20675	-1.25894	-.45330
.56	1.50485	.10661	-1.28992	-.33555
.58	1.50922	.00439	-1.31162	-.21715
.60	1.50550	-.09916	-1.32402	-.09872
.62	1.49363	-.20332	-1.32714	.01910
.64	1.47357	-.30736	-1.32106	.13566
.66	1.44537	-.41057	-1.30589	.25033
.68	1.40913	-.51224	-1.28181	.36247
.70	1.36498	-.61167	-1.24904	.47145
.72	1.31313	-.70820	-1.20786	.57665
.74	1.25384	-.80117	-1.15858	.67750
.76	1.18740	-.88997	-1.10158	.77340
.78	1.11418	-.97400	-1.03725	.86382
.80	1.03456	-1.05270	-.96606	.94823
.82	.94899	-1.12556	-.88849	1.02616
.84	.85794	-1.19210	-.80507	1.09714
.86	.76193	-1.25187	-.71636	1.16078
.88	.66151	-1.30449	-.62295	1.21670
.90	.55724	-1.34961	-.52547	1.26458
.92	.44974	-1.38693	-.42455	1.30414
.94	.33962	-1.41622	-.32086	1.33515
.96	.22752	-1.43727	-.21508	1.35742
.98	.11409	-1.44996	-.10789	1.37084
1.00	.00000	-1.45420	-.00000	1.37533

Table 8-2(c). Clamped-Pinned Beam. (*Continued*)

Second Mode

$\frac{x}{L}$	$\tilde{y}_2$	$\tilde{y}_2' = \frac{L}{\lambda_2}\frac{d\tilde{y}_2}{dx}$	$\tilde{y}_2'' = \frac{L^2}{\lambda_2^2}\frac{d^2\tilde{y}_2}{dx^2}$	$\tilde{y}_2''' = \frac{L^3}{\lambda_2^3}\frac{d^3\tilde{y}_2}{dx^3}$
.00	.00000	.00000	2.00000	-2.00000
.02	.01904	.26276	1.71729	-1.99909
.04	.07241	.48557	1.43502	-1.99300
.06	.15446	.66857	1.15424	-1.97727
.08	.25958	.81207	.87658	-1.94824
.10	.38223	.91666	.60415	-1.90305
.12	.51697	.98325	.33937	-1.83960
.14	.65851	1.01310	.08494	-1.75656
.16	.80177	1.00789	-.15633	-1.65333
.18	.94193	.96966	-.38158	-1.53001
.20	1.07449	.90088	-.58802	-1.38736
.22	1.19534	.80441	-.77300	-1.22676
.24	1.30078	.68345	-.93412	-1.05012
.26	1.38759	.54152	-1.06927	-.85985
.28	1.45308	.38242	-1.17673	-.65879
.30	1.49510	.21017	-1.25518	-.45011
.32	1.51208	.02893	-1.30380	-.23724
.34	1.50305	-.15704	-1.32224	-.02381
.36	1.46765	-.34350	-1.31068	.18649
.38	1.40611	-.52625	-1.26983	.38993
.40	1.31923	-.70122	-1.20092	.58286
.42	1.20839	-.86456	-1.10569	.76180
.44	1.07549	-1.01271	-.98634	.92349
.46	.92292	-1.14243	-.84553	1.06496
.48	.75348	-1.25091	-.68630	1.18364
.50	.57035	-1.33577	-.51204	1.27736
.52	.37700	-1.39515	-.32640	1.34442
.54	.17715	-1.42770	-.13323	1.38365
.56	-.02537	-1.43265	.06348	1.39439
.58	-.22661	-1.40978	.25968	1.37654
.60	-.42268	-1.35944	.45136	1.33056
.62	-.60973	-1.28256	.63460	1.25745
.64	-.78413	-1.18058	.80569	1.15876
.66	-.94244	-1.05549	.96112	1.03650
.68	-1.08158	-.90972	1.09776	.89319
.70	-1.19882	-.74612	1.21281	.73173
.72	-1.29186	-.56793	1.30395	.55537
.74	-1.35888	-.37866	1.36931	.36769
.76	-1.39857	-.18205	1.40755	.17245
.78	-1.41019	.01800	1.41789	-.02643
.80	-1.39351	.21752	1.40010	-.22494
.82	-1.34890	.41256	1.35450	-.41911
.84	-1.27726	.59923	1.28198	-.60506
.86	-1.18004	.77383	1.18398	-.77904
.88	-1.05919	.93288	1.06244	-.93759
.90	-.91715	1.07323	.91977	-1.07752
.92	-.75676	1.19208	.75879	-1.19604
.94	-.58123	1.28706	.58271	-1.29078
.96	-.39407	1.35629	.39504	-1.35983
.98	-.19902	1.39839	.19951	-1.40183
1.00	.00000	1.41251	-.00000	-1.41592

Table 8-2(c). Clamped-Pinned Beam. (*Continued*)

Third Mode

$\frac{x}{L}$	$\tilde{y}_3$	$\tilde{y}_3' = \frac{L}{\lambda_3}\frac{d\tilde{y}_3}{dx}$	$\tilde{y}_3'' = \frac{L^2}{\lambda_3^2}\frac{d^2\tilde{y}_3}{dx^2}$	$\tilde{y}_3''' = \frac{L^3}{\lambda_3^3}\frac{d^3\tilde{y}_3}{dx^3}$
.00	.00000	.00000	2.00000	-2.00000
.02	.03886	.36671	1.59173	-1.99731
.04	.14410	.65019	1.18531	-1.97961
.06	.29879	.85122	.78508	-1.93509
.08	.48626	.97168	.39742	-1.85535
.10	.69037	1.01491	.03009	-1.73537
.12	.89584	.98593	-.30845	-1.57331
.14	1.08857	.89148	-.60968	-1.37037
.16	1.25604	.74002	-.86560	-1.13046
.18	1.38759	.54152	-1.06927	-.85985
.20	1.47476	.30725	-1.21523	-.56678
.22	1.51147	.04939	-1.29988	-.26098
.24	1.49419	-.21934	-1.32168	.04683
.26	1.42202	-.48616	-1.28137	.34551
.28	1.29662	-.73864	-1.18195	.62397
.30	1.12212	-.96520	-1.02863	.87171
.32	.90489	-1.15556	-.82867	1.07934
.34	.65324	-1.30107	-.59110	1.23893
.36	.37703	-1.39512	-.32637	1.34445
.38	.08727	-1.43330	-.04597	1.39199
.40	-.20439	-1.41364	.23807	1.37996
.42	-.48616	-1.33665	.51362	1.30919
.44	-.74658	-1.20525	.76897	1.18287
.46	-.97504	-1.02471	.99329	1.00646
.48	-1.16223	-.80235	1.17711	.78747
.50	-1.30050	-.54726	1.31263	.53513
.52	-1.38422	-.26994	1.39411	.26005
.54	-1.41001	.01818	1.41807	-.02625
.56	-1.37687	.30521	1.38344	-.31179
.58	-1.28624	.57929	1.29160	-.58465
.60	-1.14194	.82907	1.14631	-.83344
.62	-.95000	1.04421	.95356	-1.04778
.64	-.71844	1.21582	.72134	-1.21873
.66	-.45691	1.33678	.45927	-1.33915
.68	-.17628	1.40210	.17821	-1.40403
.70	.11174	1.40907	-.11017	-1.41064
.72	.39519	1.35742	-.39391	-1.35870
.74	.66227	1.24931	-.66123	-1.25036
.76	.90188	1.08924	-.90103	-1.09010
.78	1.10404	.88387	-1.10335	-.88458
.80	1.26035	.64175	-1.25979	-.64233
.82	1.36432	.37294	-1.36386	-.37341
.84	1.41160	.08860	-1.41124	-.08899
.86	1.40025	-.19943	-1.39996	.19910
.88	1.33072	-.47918	-1.33049	.47891
.90	1.20590	-.73904	-1.20573	.73881
.92	1.03098	-.96820	-1.03085	.96800
.94	.81323	-1.15713	-.81313	1.15695
.96	.56168	-1.29798	-.56162	1.29782
.98	.28680	-1.38491	-.28677	1.38476
1.00	.00000	-1.41429	-.00000	1.41414

Table 8-2(c). Clamped-Pinned Beam. (*Continued*)

Fourth Mode

$\frac{x}{L}$	$\tilde{y}_4$	$\tilde{y}_4' = \frac{L}{\lambda_4}\frac{d\tilde{y}_4}{dx}$	$\tilde{y}_4'' = \frac{L^2}{\lambda_4^2}\frac{d^2\tilde{y}_4}{dx^2}$	$\tilde{y}_4''' = \frac{L^3}{\lambda_4^3}\frac{d^3\tilde{y}_4}{dx^3}$
.00	.00000	.00000	2.00000	-2.00000
.02	.06496	.46278	1.46633	-1.99408
.04	.23451	.78357	.93791	-1.95600
.06	.47105	.96521	.42662	-1.86287
.08	.73820	1.01441	-.05091	-1.70171
.10	1.00204	.94270	-.47581	-1.46893
.12	1.23237	.76665	-.82947	-1.16955
.14	1.40407	.50751	-1.09559	-.81599
.16	1.49825	.19041	-1.26206	-.42659
.18	1.50306	-.15704	-1.32223	-.02380
.20	1.41422	-.50624	-1.27577	.36779
.22	1.23501	-.82944	-1.12901	.72343
.24	.97582	-1.10140	-.89465	1.02023
.26	.65324	-1.30107	-.59110	1.23893
.28	.28879	-1.41295	-.24121	1.36537
.30	-.09274	-1.42807	.12917	1.39164
.32	-.46510	-1.34455	.49299	1.31666
.34	-.80250	-1.16772	.82386	1.14636
.36	-1.08149	-.90963	1.09785	.89328
.38	-1.28266	-.58823	1.29518	.57571
.40	-1.39201	.22602	1.40159	.21644
.42	-1.40200	.15152	1.40934	-.15886
.44	-1.31209	.51780	1.31771	-.52342
.46	-1.12877	.84697	1.13308	-.85127
.48	-.86513	1.11580	.86843	-1.11909
.50	-.53994	1.30530	.54246	-1.30782
.52	-.17628	1.40210	.17821	-1.40403
.54	.20000	1.39937	-.19853	-1.40084
.56	.56222	1.29734	-.56109	-1.29847
.58	.88466	1.10326	-.88379	-1.10413
.60	1.14445	.83092	-1.14379	-.83159
.62	1.32317	.49963	-1.32266	-.50014
.64	1.40813	.13289	-1.40774	-.13328
.66	1.39330	-.24329	-1.39301	.24300
.68	1.27973	-.60226	-1.27950	.60203
.70	1.07546	-.91855	-1.07529	.91837
.72	.79497	-1.16974	-.79484	1.16960
.74	.45814	-1.33802	-.45804	1.33792
.76	.08884	-1.41146	-.08876	1.41138
.78	-.28675	-1.38486	.28681	1.38480
.80	-.64202	-1.26010	.64206	1.26005
.82	-.95177	-1.04601	.95180	1.04598
.84	-1.19405	-.75779	1.19407	.75776
.86	-1.35169	-.41585	1.35171	.41583
.88	-1.41351	-.04443	1.41352	.04441
.90	-1.37513	.33014	1.37514	-.33015
.92	-1.23928	.68130	1.23929	-.68131
.94	-1.01558	.98417	1.01559	-.98417
.96	-.71989	1.21727	.71990	-1.21728
.98	-.37317	1.36409	.37317	-1.36409
1.00	.00000	1.41421	.00000	-1.41422

Table 8-2(c). Clamped-Pinned Beam. *(Continued)*

Fifth Mode

$\frac{x}{L}$	$\tilde{y}_5$	$\tilde{y}_5' = \frac{L}{\lambda_5}\frac{d\tilde{y}_5}{dx}$	$\tilde{y}_5'' = \frac{L^2}{\lambda_5^2}\frac{d^2\tilde{y}_5}{dx^2}$	$\tilde{y}_5''' = \frac{L^3}{\lambda_5^3}\frac{d^3\tilde{y}_5}{dx^3}$
.00	.00000	.00000	2.00000	-2.00000
.02	.09685	.55098	1.34119	-1.98902
.04	.33974	.88607	.69424	-1.92005
.06	.65851	1.01311	.08494	-1.75655
.08	.98717	.95000	-.45262	-1.48455
.10	1.26755	.72628	-.88320	-1.11064
.12	1.45308	.38243	-1.17672	-.65879
.14	1.51200	-.03273	-1.31329	-.16597
.16	1.42950	-.46599	-1.28662	.32312
.18	1.20840	-.86454	-1.10567	.76182
.20	.86819	-1.18105	-.79432	1.10719
.22	.44239	-1.37825	-.38928	1.32514
.24	-.02533	-1.43261	.06352	1.39442
.26	-.48616	-1.33665	.51362	1.30919
.28	-.89158	-1.09954	.91132	1.07980
.30	-1.19872	-.74602	1.21291	.73183
.32	-1.37505	-.31360	1.38526	.30340
.34	-1.40200	.15152	1.40934	-.15886
.36	-1.27698	.59950	1.28226	-.60478
.38	-1.01369	.98227	1.01748	-.98607
.40	-.64068	1.25871	.64340	-1.26144
.42	-.19828	1.39912	.20025	-1.40109
.44	.26570	1.38846	-.26429	-1.38987
.46	.70119	1.22792	-.70018	-1.22894
.48	1.06118	.93487	-1.06045	-.93560
.50	1.30683	.54093	-1.30630	-.54146
.52	1.41161	.08861	-1.41123	-.08899
.54	1.36423	-.37331	-1.36395	.37304
.56	1.16977	-.79500	-1.16957	.79481
.58	.84919	-1.13099	-.84905	1.13085
.60	.43707	-1.34505	-.43697	1.34495
.62	-.02218	-1.41408	.02225	1.41400
.64	-.47902	-1.33063	.47907	1.33058
.66	-.88421	-1.10371	.88425	1.10368
.68	-1.19405	-.75779	1.19407	.75776
.70	-1.37513	-.33015	1.37515	.33013
.72	-1.40793	.13308	1.40794	-.13310
.74	-1.28891	.58196	1.28892	-.58197
.76	-1.03091	.96809	1.03092	-.96810
.78	-.66175	1.24983	.66176	-1.24984
.80	-.22123	1.39680	.22123	-1.39680
.82	.24315	1.39315	-.24314	-1.39316
.84	.68130	1.23928	-.68130	-1.23929
.86	1.04600	.95178	-1.04600	-.95178
.88	1.29790	.56165	-1.29790	-.56165
.90	1.40985	.11096	-1.40985	-.11096
.92	1.36978	-.35170	-1.36978	.35170
.94	1.18201	-.77644	-1.18201	.77644
.96	.86678	-1.11745	-.86678	1.11745
.98	.45809	-1.33797	-.45809	1.33797
1.00	.00000	-1.41421	.00000	1.41421

Table 8-2(d). Characteristic Beam Functions and Derivatives. Clamped-Sliding Beam

First Mode

$\frac{x}{L}$	$\tilde{y}_1$	$\tilde{y}_1' = \frac{L}{\lambda_1}\frac{d\tilde{y}_1}{dx}$	$\tilde{y}_1'' = \frac{L^2}{\lambda_1^2}\frac{d^2\tilde{y}_1}{dx^2}$	$\tilde{y}_1''' = \frac{L^3}{\lambda_1^3}\frac{d^3\tilde{y}_1}{dx^3}$
.00	.00000	.00000	2.00000	-1.96500
.02	.00220	.09240	1.90705	-1.96497
.04	.00867	.18041	1.81412	-1.96473
.06	.01920	.26402	1.72120	-1.96409
.08	.03358	.34324	1.62832	-1.96285
.10	.05160	.41806	1.53552	-1.96085
.12	.07306	.48850	1.44284	-1.95792
.14	.09774	.55456	1.35032	-1.95389
.16	.12545	.61624	1.25802	-1.94862
.18	.15597	.67357	1.16600	-1.94198
.20	.18910	.72655	1.07433	-1.93383
.22	.22464	.77521	.98308	-1.92405
.24	.26237	.81956	.89234	-1.91254
.26	.30210	.85964	.80218	-1.89920
.28	.34363	.89546	.71270	-1.88393
.30	.38675	.92707	.62399	-1.86666
.32	.43126	.95450	.53615	-1.84732
.34	.47698	.97781	.44927	-1.82585
.36	.52370	.99702	.36346	-1.80219
.38	.57123	1.01221	.27882	-1.77629
.40	.61939	1.02342	.19545	-1.74814
.42	.66799	1.03072	.11348	-1.71769
.44	.71684	1.03418	.03299	-1.68494
.46	.76576	1.03387	-.04588	-1.64988
.48	.81458	1.02986	-.12305	-1.61250
.50	.86313	1.02225	-.19839	-1.57282
.52	.91124	1.01113	-.27180	-1.53085
.54	.95873	.99657	-.34318	-1.48663
.56	1.00546	.97870	-.41240	-1.44017
.58	1.05127	.95760	-.47938	-1.39152
.60	1.09600	.93338	-.54401	-1.34074
.62	1.13952	.90617	-.60618	-1.28786
.64	1.18168	.87608	-.66581	-1.23296
.66	1.22235	.84323	-.72279	-1.17610
.68	1.26141	.80774	-.77704	-1.11735
.70	1.29873	.76976	-.82847	-1.05679
.72	1.33419	.72942	-.87699	-.99452
.74	1.36769	.68685	-.92252	-.93061
.76	1.39913	.64219	-.96500	-.86516
.78	1.42841	.59561	-1.00434	-.79828
.80	1.45545	.54723	-1.04050	-.73007
.82	1.48016	.49722	-1.07339	-.66063
.84	1.50246	.44574	-1.10297	-.59008
.86	1.52230	.39294	-1.12920	-.51854
.88	1.53962	.33897	-1.15201	-.44611
.90	1.55436	.28401	-1.17139	-.37293
.92	1.56647	.22821	-1.18728	-.29911
.94	1.57593	.17175	-1.19968	-.22478
.96	1.58271	.11478	-1.20854	-.15007
.98	1.58679	.05747	-1.21387	-.07510
1.00	1.58815	.00000	-1.21564	.00000

Table 8-2(d). Clamped-Sliding Beam. (*Continued*)

Second Mode

$\frac{x}{L}$	$\tilde{y}_2$	$\tilde{y}_2' = \frac{L}{\lambda_2}\frac{d\tilde{y}_2}{dx}$	$\tilde{y}_2'' = \frac{L^2}{\lambda_2^2}\frac{d^2\tilde{y}_2}{dx^2}$	$\tilde{y}_2''' = \frac{L^3}{\lambda_2^3}\frac{d^3\tilde{y}_2}{dx^3}$
.00	.00000	.00000	2.00000	-1.99993
.02	.01165	.20782	1.78011	-1.99950
.04	.04482	.39147	1.56038	-1.99658
.06	.09685	.55099	1.34121	-1.98895
.08	.16510	.68646	1.12323	-1.97469
.10	.24694	.79807	.90725	-1.95215
.12	.33975	.88609	.69428	-1.91998
.14	.44095	.95090	.48542	-1.87713
.16	.54803	.99303	.28189	-1.82280
.18	.65852	1.01314	.08500	-1.75648
.20	.77005	1.01202	-.10393	-1.67794
.22	.88034	.99062	-.28355	-1.58719
.24	.98720	.95005	-.45253	-1.48447
.26	1.08861	.89154	-.60958	-1.37029
.28	1.18265	.81648	-.75348	-1.24534
.30	1.26761	.72637	-.88309	-1.11054
.32	1.34190	.62284	-.99738	-.96697
.34	1.40415	.50763	-1.09546	-.81588
.36	1.45317	.38256	-1.17658	-.65867
.38	1.48799	.24953	-1.24014	-.49683
.40	1.50783	.11049	-1.28573	-.33199
.42	1.51214	-.03255	-1.31310	-.16581
.44	1.50059	-.17760	-1.32221	-.00003
.46	1.47308	-.32266	-1.31319	.16360
.48	1.42971	-.46574	-1.28637	.32333
.50	1.37080	-.60491	-1.24229	.47744
.52	1.29690	-.73832	-1.18164	.62424
.54	1.20872	-.86419	-1.10534	.76212
.56	1.10719	-.98086	-1.01443	.88956
.58	.99341	-1.08679	-.91014	1.00516
.60	.86863	-1.18057	-.79386	1.10762
.62	.73427	-1.26098	-.66709	1.19583
.64	.59186	-1.32695	-.53144	1.26880
.66	.44302	-1.37759	-.38865	1.32575
.68	.28949	-1.41222	-.24051	1.36606
.70	.13306	-1.43035	-.08886	1.38930
.72	-.02444	-1.43171	.06439	1.39528
.74	-.18117	-1.41621	.21734	1.38396
.76	-.33527	-1.38399	.36811	1.35553
.78	-.48493	-1.33540	.51483	1.31039
.80	-.62836	-1.27099	.65569	1.24912
.82	-.76388	-1.19149	.78896	1.17250
.84	-.88987	-1.09783	.91301	1.08148
.86	-1.00482	-.99110	1.02630	.97721
.88	-1.10739	-.87257	1.12747	.86096
.90	-1.19633	-.74365	1.21525	.73418
.92	-1.27060	-.60585	1.28859	.59841
.94	-1.32930	-.46083	1.34658	.45533
.96	-1.37174	-.31031	1.38852	.30668
.98	-1.39741	-.15609	1.41389	.15428
1.00	-1.40600	.00000	1.42238	.00000

Table 8-2(d). Clamped-Sliding Beam. *(Continued)*

Third Mode

$\frac{x}{L}$	$\tilde{y}_3$	$\tilde{y}_3' = \frac{L}{\lambda_3}\frac{d\tilde{y}_3}{dx}$	$\tilde{y}_3'' = \frac{L^2}{\lambda_3^2}\frac{d^2\tilde{y}_3}{dx^2}$	$\tilde{y}_3''' = \frac{L^3}{\lambda_3^3}\frac{d^3\tilde{y}_3}{dx^3}$
.00	.00000	.00000	2.00000	-2.00000
.02	.02814	.31572	1.65450	-1.99835
.04	.10567	.57181	1.30996	-1.98743
.06	.22232	.76860	.96867	-1.95958
.08	.36791	.90694	.63409	-1.90894
.10	.53246	.98836	.31054	-1.83135
.12	.70631	1.01517	.00291	-1.72440
.14	.88031	.99058	-.28363	-1.58726
.16	1.04591	.91867	-.54391	-1.42067
.18	1.19534	.80441	-.77300	-1.22675
.20	1.32178	.65359	-.96646	-1.00892
.22	1.41947	.47270	-1.12053	-.77164
.24	1.48381	.26880	-1.23231	-.52030
.26	1.51147	.04939	-1.29988	-.26098
.28	1.50043	-.17781	-1.32241	-.00021
.30	1.45002	-.40503	-1.30025	.25526
.32	1.36090	-.62465	-1.23490	.49865
.34	1.23501	-.82944	-1.12901	.72343
.36	1.07551	-1.01269	-.98632	.92350
.38	.88664	-1.16844	-.81160	1.09341
.40	.67360	-1.29164	-.61047	1.22851
.42	.44239	-1.37825	-.38928	1.32514
.44	.19959	-1.42540	-.15491	1.38072
.46	-.04782	-1.43144	.08542	1.39385
.48	-.29269	-1.39597	.32432	1.36434
.50	-.52789	-1.31986	.55450	1.29326
.52	-.74658	-1.20525	.76897	1.18287
.54	-.94236	-1.05541	.96120	1.03658
.56	-1.10952	-.87470	1.12537	.85886
.58	-1.24316	-.66841	1.25650	.65509
.60	-1.33938	-.44262	1.35061	.43141
.62	-1.39538	-.20398	1.40483	.19455
.64	-1.40954	.04046	1.41749	-.04838
.66	-1.38148	.28345	1.38818	-.29011
.68	-1.31208	.51781	1.31772	-.52341
.70	-1.20344	.73657	1.20819	-.74127
.72	-1.05881	.93326	1.06282	-.93721
.74	-.88253	1.10204	.88592	-1.10535
.76	-.67987	1.23790	.68273	-1.24067
.78	-.45688	1.33681	.45930	-1.33912
.80	-.22020	1.39584	.22226	-1.39777
.82	.02309	1.41324	-.02134	-1.41484
.84	.26575	1.38850	-.26425	-1.38982
.86	.50053	1.32238	-.49924	-1.32346
.88	.72046	1.21684	-.71933	-1.21771
.90	.91895	1.07503	-.91796	-1.07572
.92	1.09011	.90119	-1.08923	-.90172
.94	1.22883	.70049	-1.22803	-.70088
.96	1.33098	.47892	-1.33023	-.47917
.98	1.39351	.24308	-1.39280	-.24321
1.00	1.41457	-.00000	-1.41386	.00000

Table 8-2(d). Clamped-Sliding Beam. *(Continued)*

Fourth Mode

$\frac{x}{L}$	$\tilde{y}_4$	$\tilde{y}_4' = \frac{L}{\lambda_4}\frac{d\tilde{y}_4}{dx}$	$\tilde{y}_4'' = \frac{L^2}{\lambda_4^2}\frac{d^2\tilde{y}_4}{dx^2}$	$\tilde{y}_4''' = \frac{L^3}{\lambda_4^3}\frac{d^3\tilde{y}_4}{dx^3}$
.00	.00000	.00000	2.00000	-2.00000
.02	.05116	.41573	1.52901	-1.99590
.04	.18721	.72077	1.06124	-1.96923
.06	.38223	.91666	.60415	-1.90305
.08	.61089	1.00714	.16843	-1.78646
.10	.84906	.99870	-.23333	-1.61443
.12	1.07449	.90089	-.58802	-1.38736
.14	1.26755	.72628	-.88320	-1.11064
.16	1.41191	.49020	-1.10824	-.79388
.18	1.49510	.21018	-1.25518	-.45010
.20	1.50899	-.09478	-1.31943	-.09478
.22	1.45002	-.40503	-1.30025	.25526
.24	1.31924	-.70120	-1.20091	.58288
.26	1.12212	-.96520	-1.02863	.87171
.28	.86819	-1.18105	-.79432	1.10719
.30	.57038	-1.33574	-.51202	1.27738
.32	.24429	-1.41986	-.19818	1.37375
.34	-.09274	-1.42807	.12917	1.39164
.36	-.42263	-1.35939	.45141	1.33061
.38	-.72755	-1.21719	.75029	1.19445
.40	-.99102	-1.00898	1.00898	.99102
.42	-1.19872	-.74602	1.21291	.73183
.44	-1.33939	-.44262	1.35060	.43141
.46	-1.40542	-.11539	1.41428	.10653
.48	-1.39330	.21773	1.40030	-.22473
.50	-1.30380	.53843	1.30933	-.54396
.52	-1.14194	.82907	1.14631	-.83344
.54	-.91673	1.07365	.92018	-1.07710
.56	-.64068	1.25871	.64340	-1.26144
.58	-.32906	1.37406	.33122	-1.37622
.60	.00085	1.41336	.00085	-1.41506
.62	.33081	1.37447	-.32947	-1.37581
.64	.64257	1.25954	-.64151	-1.26060
.66	.91888	1.07496	-.91804	-1.07580
.68	1.14445	.83092	-1.14379	-.83159
.70	1.30683	.54093	-1.30630	-.54146
.72	1.39701	.22102	-1.39659	-.22144
.74	1.41002	-.11112	-1.40969	.11079
.76	1.34513	-.43714	-1.34487	.43689
.78	1.20592	-.73903	-1.20571	.73882
.80	1.00008	-1.00008	-.99992	.99992
.82	.73899	-1.20588	-.73886	1.20575
.84	.43707	-1.34505	-.43696	1.34495
.86	.11100	-1.40989	-.11092	1.40982
.88	-.22120	-1.39683	.22127	1.39677
.90	-.54117	-1.30659	.54122	1.30654
.92	-.83123	-1.14414	.83128	1.14411
.94	-1.07536	-.91847	1.07540	.91845
.96	-1.26006	-.64205	1.26009	.64203
.98	-1.37512	-.33015	1.37515	.33014
1.00	-1.41420	.00000	1.41423	.00000

Table 8-2(d). Clamped-Sliding Beam. (*Continued*)

Fifth Mode

$\frac{x}{L}$	$\tilde{y}_5$	$\tilde{y}_5' = \frac{L}{\lambda_5}\frac{d\tilde{y}_5}{dx}$	$\tilde{y}_5'' = \frac{L^2}{\lambda_5^2}\frac{d^2\tilde{y}_5}{dx^2}$	$\tilde{y}_5''' = \frac{L^3}{\lambda_5^3}\frac{d^3\tilde{y}_5}{dx^3}$
.00	.00000	.00000	2.00000	-2.00000
.02	.08021	.50787	1.40372	-1.99180
.04	.28552	.83865	.81551	-1.93968
.06	.56365	.99721	.25328	-1.81414
.08	.86472	.99484	-.25859	-1.60097
.10	1.14332	.85051	-.69359	-1.30024
.12	1.36090	.59093	-1.02722	-.92462
.14	1.48788	.24938	-1.24030	-.49696
.16	1.50536	-.13627	-1.32167	-.04743
.18	1.40612	-.52624	-1.26982	.38994
.20	1.19469	-.88182	-1.09356	.78069
.22	.88664	-1.16844	-.81161	1.09341
.24	.50688	-1.35844	-.45121	1.30277
.26	.08727	-1.43330	-.04597	1.39199
.28	-.33638	-1.38511	.36702	1.35446
.30	-.72755	-1.21719	.75029	1.19445
.32	-1.05238	-.94367	1.06925	.92680
.34	-1.28266	-.58823	1.29518	.57571
.36	-1.39842	-.18189	1.40771	.17260
.38	-1.38971	.23970	1.39660	-.24659
.40	-1.25752	.63948	1.26263	-.64460
.42	-1.01369	.98227	1.01748	-.98607
.44	-.67990	1.23788	.68271	-1.24069
.46	-.28574	1.38379	.28783	-1.38587
.48	.13386	1.40716	-.13231	-1.40871
.50	.54177	1.30599	-.54062	-1.30714
.52	.90188	1.08924	-.90103	-1.09010
.54	1.18233	.77612	-1.18169	-.77675
.56	1.35830	.39432	-1.35783	-.39479
.58	1.41421	-.02239	-1.41386	.02204
.60	1.34513	-.43715	-1.34487	.43689
.62	1.15713	-.81328	-1.15694	.81308
.64	.86685	-1.11752	-.86671	1.11738
.66	.49994	-1.32297	-.49984	1.32286
.68	.08884	-1.41146	-.08876	1.41138
.70	-.33011	-1.37517	.33017	1.37511
.72	-.71987	-1.21729	.71991	1.21725
.74	-1.04598	-.95180	1.04601	.95177
.76	-1.27961	-.60215	1.27963	.60213
.78	-1.40010	-.19927	1.40011	.19926
.80	-1.39680	.22123	1.39681	-.22124
.82	-1.27000	.62216	1.27001	-.62217
.84	-1.03091	.96809	1.03092	-.96810
.86	-.70068	1.22843	.70069	-1.22843
.88	-.30850	1.38015	.30850	-1.38016
.90	.11096	1.40985	-.11096	-1.40986
.92	.52061	1.31490	-.52061	-1.31490
.94	.88423	1.10369	-.88423	-1.10370
.96	1.16967	.79491	-1.16967	-.79491
.98	1.35170	.41584	-1.35169	-.41584
1.00	1.41421	.00000	-1.41421	.00000

4. The periodic nature of beam mode shapes gives rise to the following identities which can be used to extend Table 8-2 to free-free, free-pinned, and free-sliding beams:

   a. Free-free/clamped-clamped:

$$\tilde{y}_i \text{ (free-free)} = \tilde{y}_i'' \text{ (clamped-clamped)}$$
$$\tilde{y}_i' \text{ (free-free)} = \tilde{y}_i''' \text{ (clamped-clamped)}$$
$$\tilde{y}_i'' \text{ (free-free)} = \tilde{y}_i \text{ (clamped-clamped)}$$
$$\tilde{y}_i''' \text{ (free-free)} = \tilde{y}' \text{ (clamped-clamped)}$$

   b. Free-pinned/clamped-pinned:

$$\tilde{y}_i \text{ (free-pinned)} = \tilde{y}_i'' \text{ (clamped-pinned)}$$
$$\tilde{y}_i' \text{ (free-pinned)} = \tilde{y}_i''' \text{ (clamped-pinned)}$$
$$\tilde{y}_i'' \text{ (free-pinned)} = \tilde{y}_i \text{ (clamped-pinned)}$$
$$\tilde{y}_i''' \text{ (free-pinned)} = \tilde{y}_i' \text{ (clamped-pinned)}$$

   c. Free-sliding/clamped-sliding:

$$\tilde{y}_i \text{ (free-sliding)} = \tilde{y}_i'' \text{ (clamped-sliding)}$$
$$\tilde{y}_i' \text{ (free-sliding)} = \tilde{y}_i''' \text{ (clamped-sliding)}$$
$$\tilde{y}_i'' \text{ (free-sliding)} = \tilde{y}_i \text{ (clamped-sliding)}$$
$$\tilde{y}_i''' \text{ (free-sliding)} = \tilde{y}_i' \text{ (clamped-sliding)}$$

5. Properties of the integrals of the mode shapes are given in Appendix C.

8.1.3. Multispan Beams

General Formulation. A multispan beam is a beam with one or more intermediate supports. The natural frequencies of uniform multispan beams can be expressed in the same form as the natural frequencies of single-span beams:

$$f_i = \frac{\lambda_i^2}{2\pi L^2}\left(\frac{EI}{m}\right)^{1/2}; \qquad i = 1, 2, 3, \ldots. \tag{8-13}$$

f_i is the natural frequency of the i mode in hertz. E is the modulus of elasticity, I is the area moment of inertia of the cross section (Table 5-1) about the neutral axis, L is a characteristic span length, m is the mass per unit length of the span, and λ_i is a dimensionless parameter which is a function of the mode number (i), boundary conditions, and the number of spans. Consistent sets of units are given in Table 3-1. The following subsections give λ_i for various multispan beams.

Two-Span Beams. The parameter λ_i is given in Figs. 8-5(a) through 8-5(f) for two-span beams with various boundary conditions. In these cases the characteristic span, L, has been chosen to be the overall length of the beam. Note that the parameter λ_i for the two-span beams approaches the λ_i of single-span beams (Table 8-1) as a/L approaches 0 or 1. For example, Fig. 8-5(b) shows that λ_i approaches π as a/L approaches 1.0.

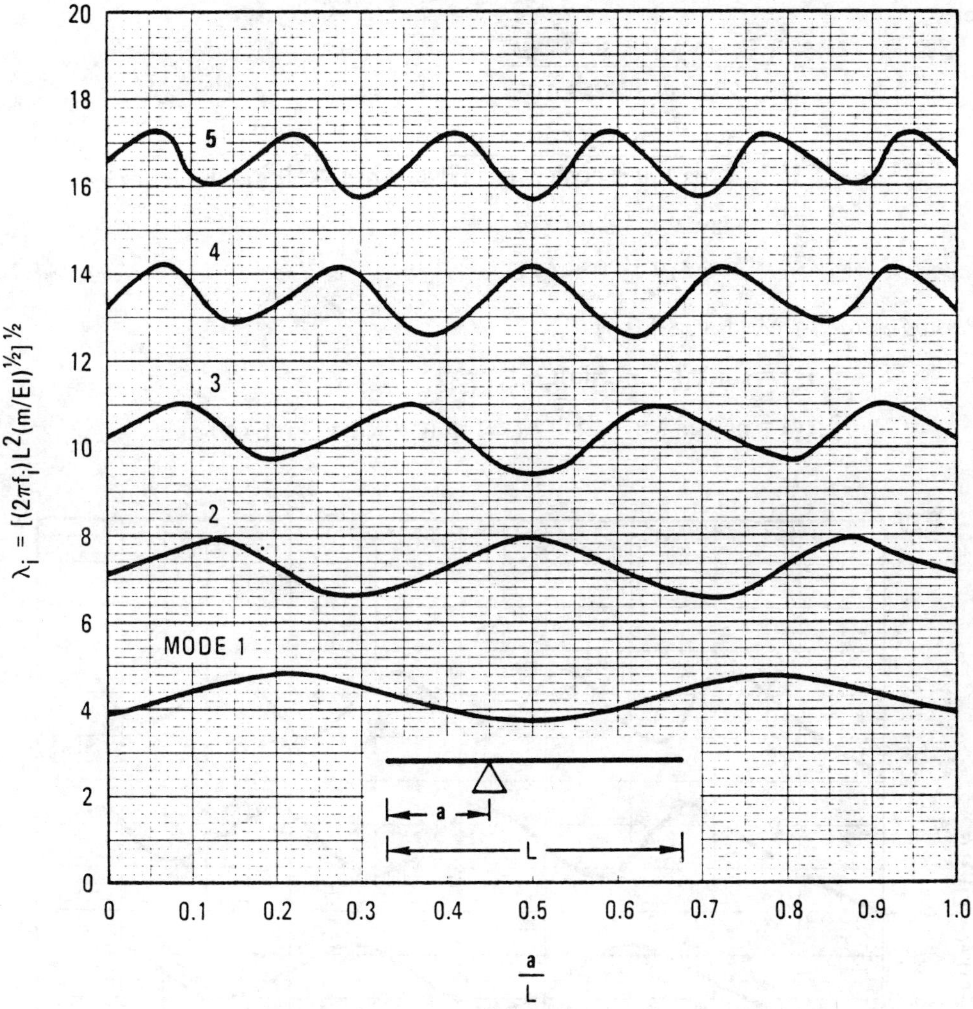

Fig. 8-5(a). Natural frequency parameters of a free-pinned-free two-span beam (Eqs. 8-13, 8-14). (Ref. 8-3)

The mode shapes ($\tilde{y}_i$) of the two-span beams corresponding to Fig. 8-5 are (Ref. 8-3):

$$\tilde{y}_i(\xi) = \begin{cases} \tilde{y}_\alpha(\xi), & 0 \leqslant \xi < \dfrac{a}{L} \\ \tilde{y}_\beta(\xi), & \dfrac{a}{L} \leqslant \xi \leqslant 1 \end{cases}; \qquad i = 1, 2, 3, 4, 5,$$

where

$$\xi = \frac{x}{L}, \qquad \mu = \frac{a}{L}, \qquad \eta = 1 - \frac{a}{L}.$$

Free-pinned-free [Fig. 8-5(a)]

$$\tilde{y}_\alpha(\xi) = \sin \lambda_i \xi + \sinh \lambda_i \xi + \eta_1 \; (\cos \lambda_i \xi + \cosh \lambda_i \xi),$$

$$\tilde{y}_\beta(\xi) = \eta_2 \, [\sin \lambda_i \xi + \sinh \lambda_i \xi + \eta_3 \; (\cos \lambda_i \xi + \cosh \lambda_i \xi)], \qquad (8\text{-}14)$$

where

$$\eta_1 = \frac{-\sin\lambda_i\mu - \sinh\lambda_i\mu}{\cos\lambda_i\mu + \cosh\lambda_i\mu},$$

$$\eta_2 = \frac{-(1+\cos\lambda_i\mu\cosh\lambda_i\mu)(\cos\lambda_i\eta + \cosh\lambda_i\eta)}{(\cos\lambda_i\mu + \cosh\lambda_i\mu)(1+\cos\lambda_i\eta\cosh\lambda_i\eta)},$$

$$\eta_3 = \frac{-\sin\lambda_i\eta - \sinh\lambda_i\eta}{\cos\lambda_i\eta + \cosh\lambda_i\eta}.$$

Pinned-pinned-free [Fig. 8-5(b)]

$$\tilde{y}_\alpha(\xi) = \sin\lambda_i\xi + \eta_1\sinh\lambda_i\xi,$$

$$\tilde{y}_\beta(\xi) = \eta_2\left[\sin\lambda_i\xi + \sinh\lambda_i\xi + \eta_3(\cos\lambda_i\xi + \cosh\lambda_i\xi)\right], \tag{8-15}$$

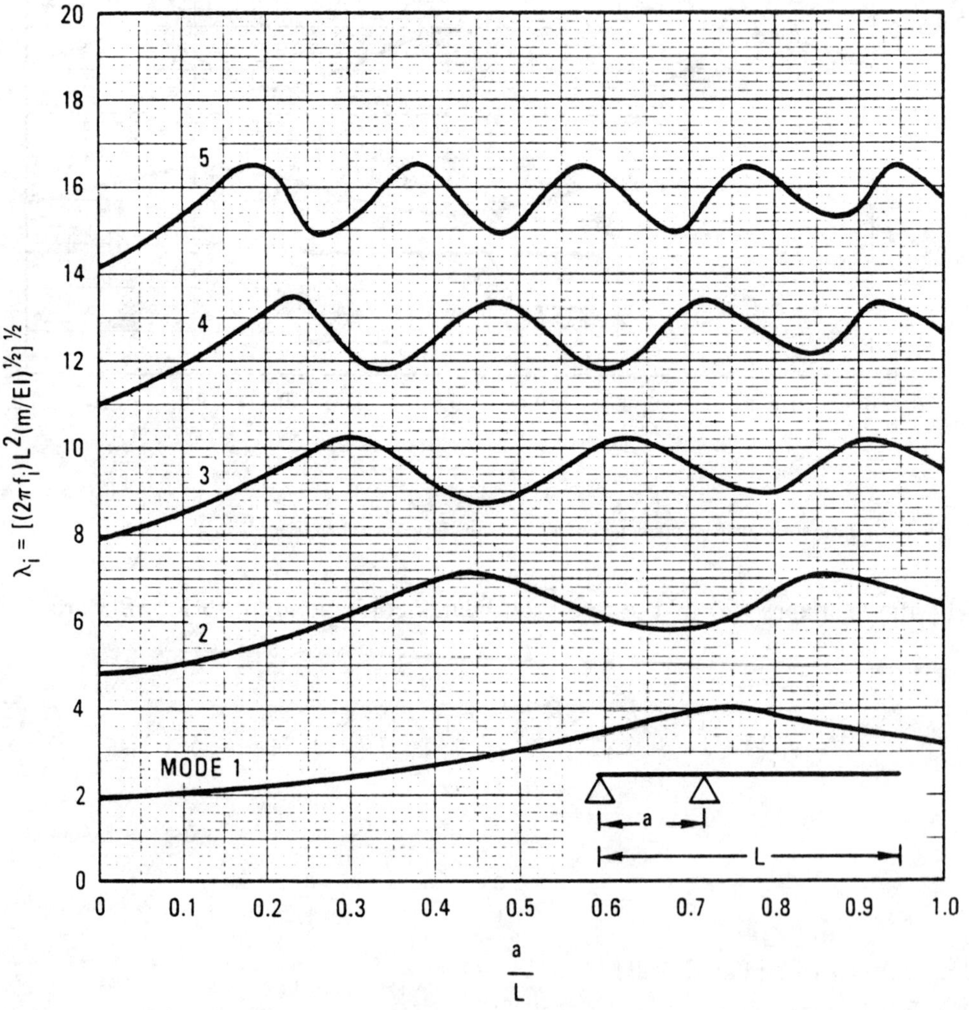

Fig. 8-5(b). Natural frequency parameters of a pinned-pinned-free two-span beam (Eqs. 8-13, 8-15). (Ref. 8-3)

where

$$\eta_1 = \frac{-\sin \lambda_i \mu}{\sinh \lambda_i \mu},$$

$$\eta_2 = \frac{(\sin \lambda_i \mu \cosh \lambda_i \mu - \cos \lambda_i \mu \sinh \lambda_i \mu)(\cos \lambda_i \eta + \cosh \lambda_i \eta)}{2 \sinh \lambda_i \mu (1 + \cos \lambda_i \eta \cosh \lambda_i \eta)},$$

$$\eta_3 = \frac{-\sin \lambda_i \eta - \sinh \lambda_i \eta}{\cos \lambda_i \eta + \cosh \lambda_i \eta}.$$

Pinned-pinned-pinned [Fig. 8-5(c)]

$$\tilde{y}_\alpha(\xi) = \sin \lambda_i \xi + \eta_1 \sinh \lambda_i \xi,$$

$$\tilde{y}_\beta(\xi) = \eta_2 (\sin \lambda_i \xi + \eta_3 \sinh \lambda_i \xi), \qquad (8\text{-}16)$$

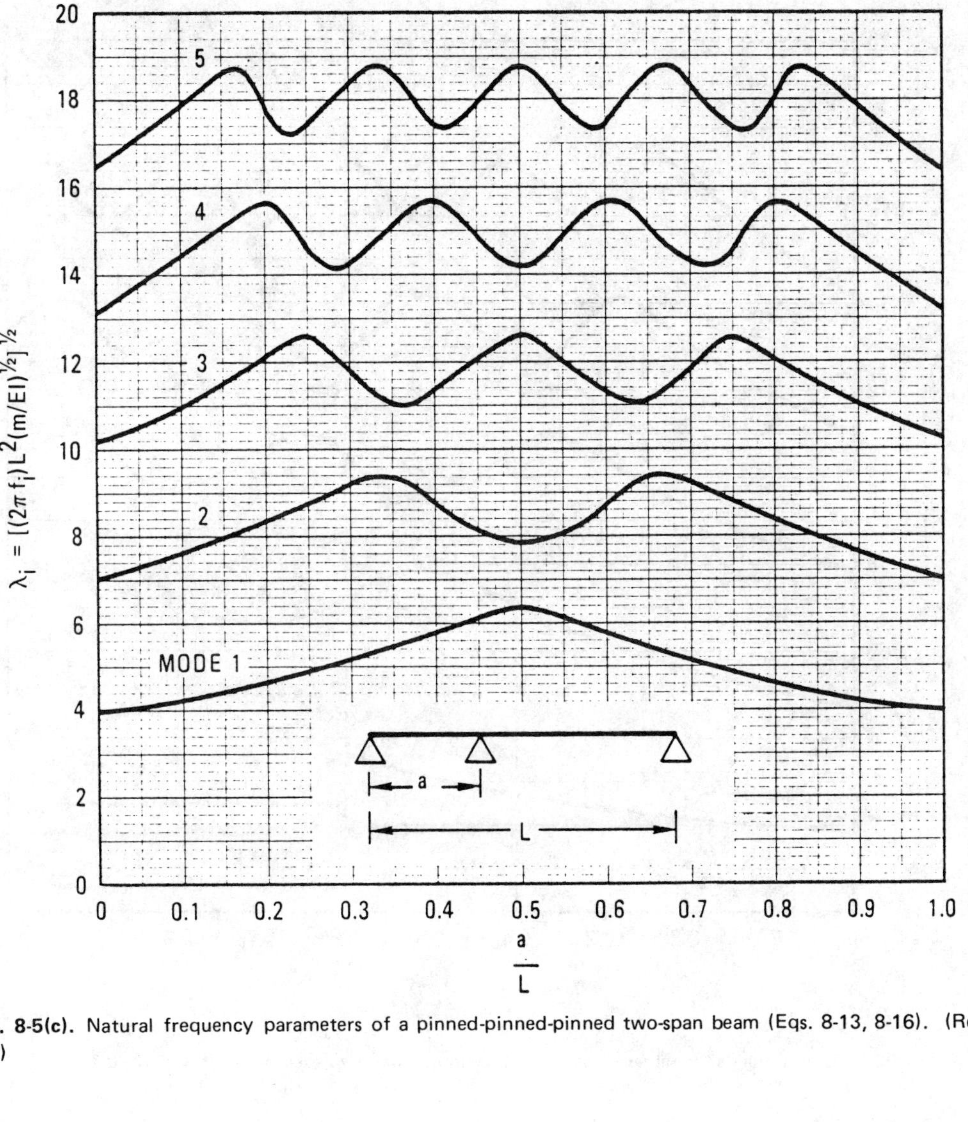

Fig. 8-5(c). Natural frequency parameters of a pinned-pinned-pinned two-span beam (Eqs. 8-13, 8-16). (Ref. 8-3)

where

$$\eta_1 = \frac{-\sin \lambda_i \mu}{\sinh \lambda_i \mu},$$

$$\eta_2 = \frac{\sin \lambda_i \mu}{\sin \lambda_i \eta},$$

$$\eta_3 = \frac{-\sin \lambda_i \eta}{\sinh \lambda_i \eta}.$$

Clamped-pinned-free [Fig. 8-5(d)]

$$\tilde{y}_\alpha(\xi) = \sin \lambda_i \xi - \sinh \lambda_i \xi + \eta_1 (\cos \lambda_i \xi - \cosh \lambda_i \xi),$$

$$\tilde{y}_\beta(\xi) = \eta_2 \left[(\sin \lambda_i \xi + \sinh \lambda_i \xi + \eta_3 (\cos \lambda_i \xi + \cosh \lambda_i \xi)\right], \qquad (8\text{-}17)$$

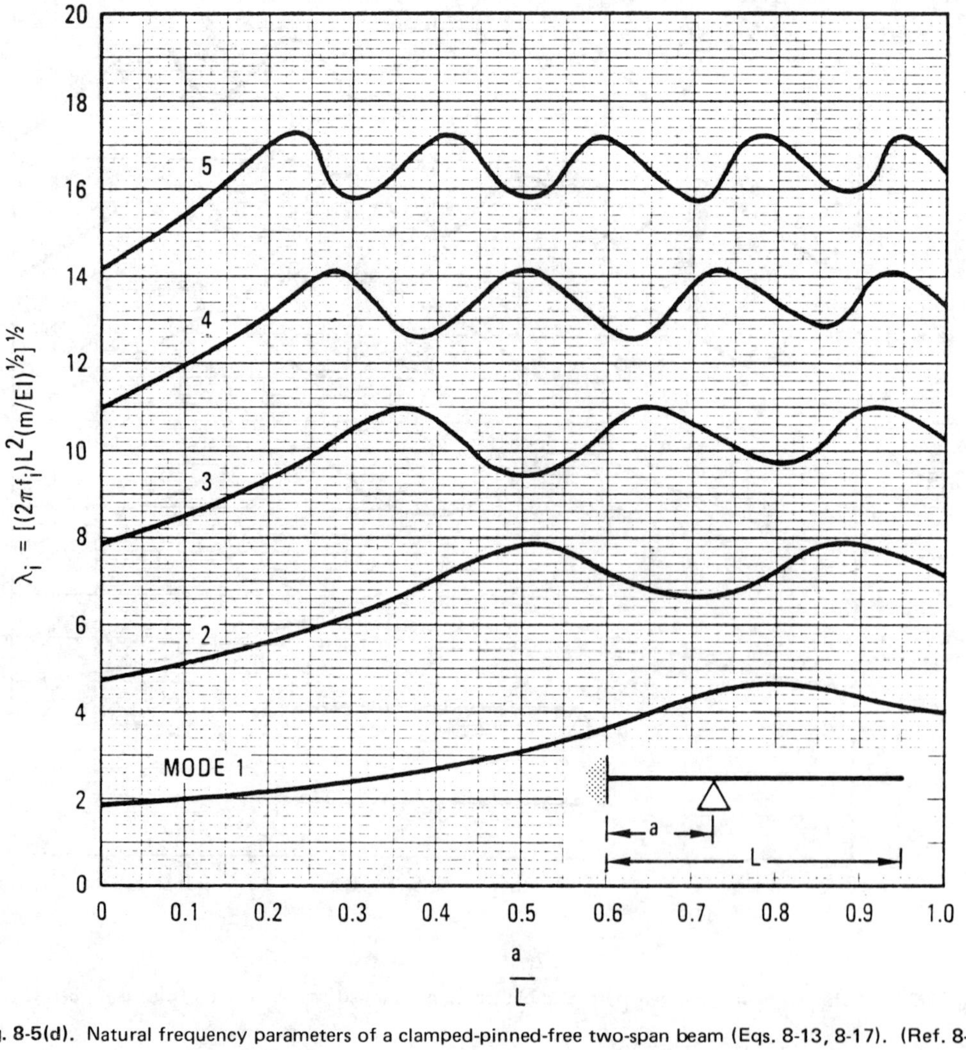

Fig. 8-5(d). Natural frequency parameters of a clamped-pinned-free two-span beam (Eqs. 8-13, 8-17). (Ref. 8-3)

where

$$\eta_1 = \frac{\sinh \lambda_i \mu - \sin \lambda_i \mu}{\cos \lambda_i \mu - \cosh \lambda_i \mu},$$

$$\eta_2 = \frac{(1 - \cos \lambda_i \mu \cosh \lambda_i \mu)(\cos \lambda_i \eta + \cosh \lambda_i \eta)}{(\cosh \lambda_i \mu - \cos \lambda_i \mu)(1 + \cos \lambda_i \eta \cosh \lambda_i \eta)},$$

$$\eta_3 = \frac{-\sin \lambda_i \eta - \sinh \lambda_i \eta}{\cos \lambda_i \eta + \cosh \lambda_i \eta}.$$

Clamped-pinned-pinned [Fig. 8-5(e)]

$$\tilde{y}_\alpha(\xi) = \sin \lambda_i \xi - \sinh \lambda_i \xi + \eta_1 (\cos \lambda_i \xi - \cosh \lambda_i \xi),$$

$$\tilde{y}_\beta(\xi) = \eta_2 (\sin \lambda_i \xi + \eta_3 \sinh \lambda_i \xi), \qquad (8\text{-}18)$$

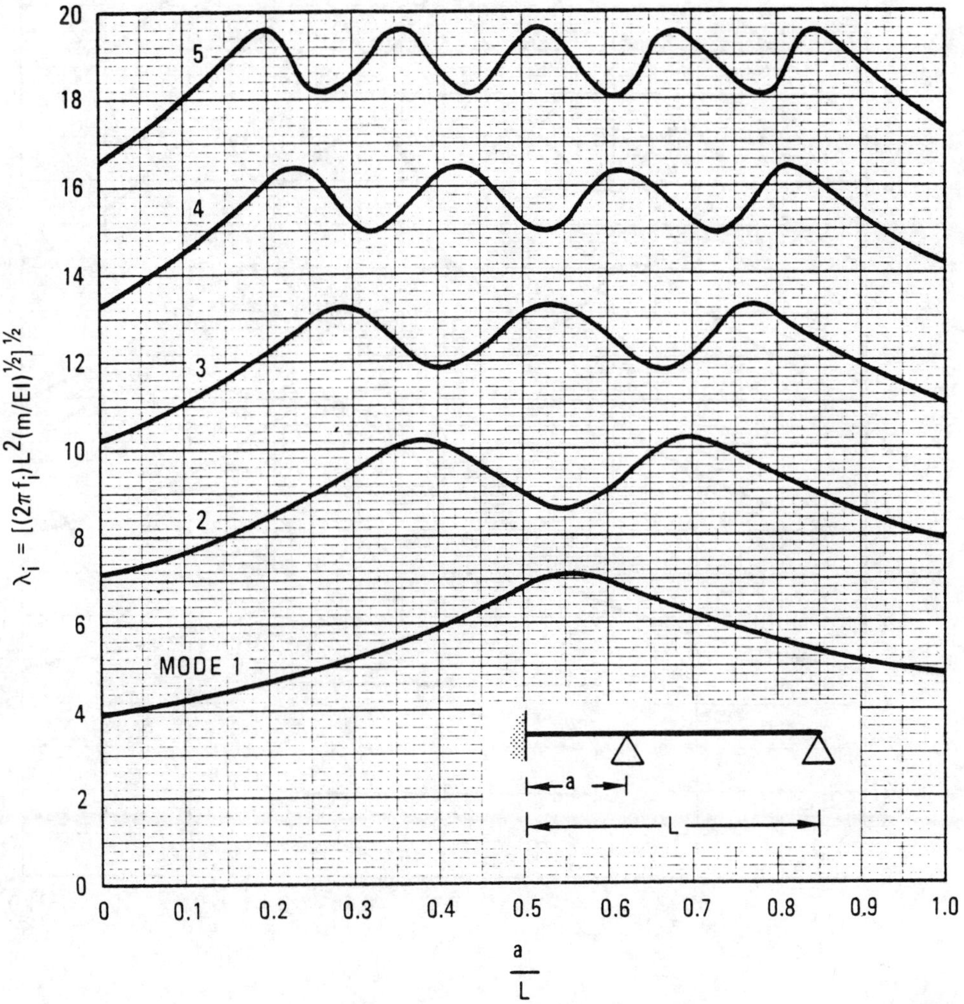

Fig. 8-5(e). Natural frequency parameters of a clamped-pinned-pinned two-span beam (Eqs. 8-13, 8-18). (Ref. 8-3)

where

$$\eta_1 = \frac{\sinh \lambda_i \mu - \sin \lambda_i \mu}{\cos \lambda_i \mu - \cosh \lambda_i \mu},$$

$$\eta_2 = \frac{-2(1 - \cos \lambda_i \mu \cosh \lambda_i \mu)}{(\cos \lambda_i \mu - \cosh \lambda_i \mu)(\cos \lambda_i \eta \sinh \lambda_i \eta - \sin \lambda_i \eta \cosh \lambda_i \eta)},$$

$$\eta_3 = \frac{-\sin \lambda_i \eta}{\sinh \lambda_i \eta}.$$

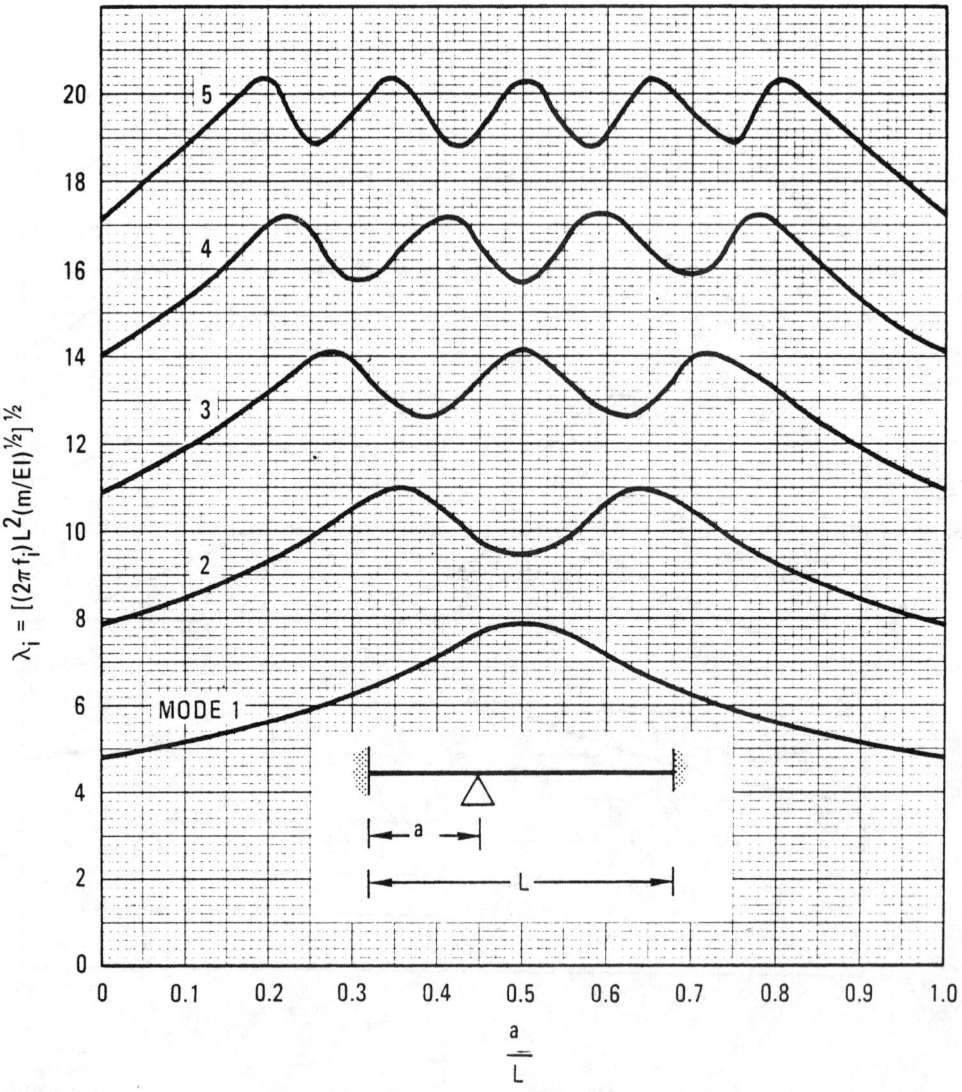

Fig. 8-5(f). Natural frequency parameters of a clamped-pinned-clamped two-span beam (Eqs. 8-13, 8-19). (Ref. 8-3)

Clamped-pinned-clamped [Fig. 8-5(f)]

$$\tilde{y}_\alpha(\xi) = \sin \lambda_i\xi - \sinh \lambda_i\xi + \eta_1 (\cos \lambda_i\xi - \cosh \lambda_i\xi),$$

$$\tilde{y}_\beta(\xi) = \eta_2 [\sin \lambda_i\xi - \sinh \lambda_i\xi + \eta_3 (\cos \lambda_i\xi - \cosh \lambda_i\xi)], \qquad (8\text{-}19)$$

where

$$\eta_1 = \frac{\sinh \lambda_i\mu - \sin \lambda_i\mu}{\cos \lambda_i\mu - \cosh \lambda_i\mu},$$

$$\eta_2 = \frac{(1 - \cos \lambda_i\mu \cosh \lambda_i\mu)(\cosh \lambda_i\eta - \cos \lambda_i\eta)}{(\cos \lambda_i\mu - \cosh \lambda_i\mu)(1 - \cos \lambda_i\eta \cosh \lambda_i\mu)},$$

$$\eta_3 = \frac{\sinh \lambda_i\eta - \sin \lambda_i\eta}{\cos \lambda_i\eta - \cosh \lambda_i\eta}.$$

Caution should be taken when applying these formulas to obtain the mode shapes of the higher modes (i = 2, 3, 4, . . .). Experience with single-span beams indicates that the formulas become increasingly sensitive to small errors in λ_i with increasing mode number.

Figure 8-6 shows the effect of support spacing of a two-span beam with extreme ends clamped on the mode shape of the first mode.

Multispan Beams of Equal Span. The natural frequencies of multispan beams with equal spans are given in Tables 8-3(a) through 8-3(f) for beams with various boundary conditions applied to the extreme ends of the beams. Intermediate supports are assumed to be pinned. Note that here the characteristic span, L, is taken to be the distance between supports. As the number of spans becomes large, λ_1 approaches π

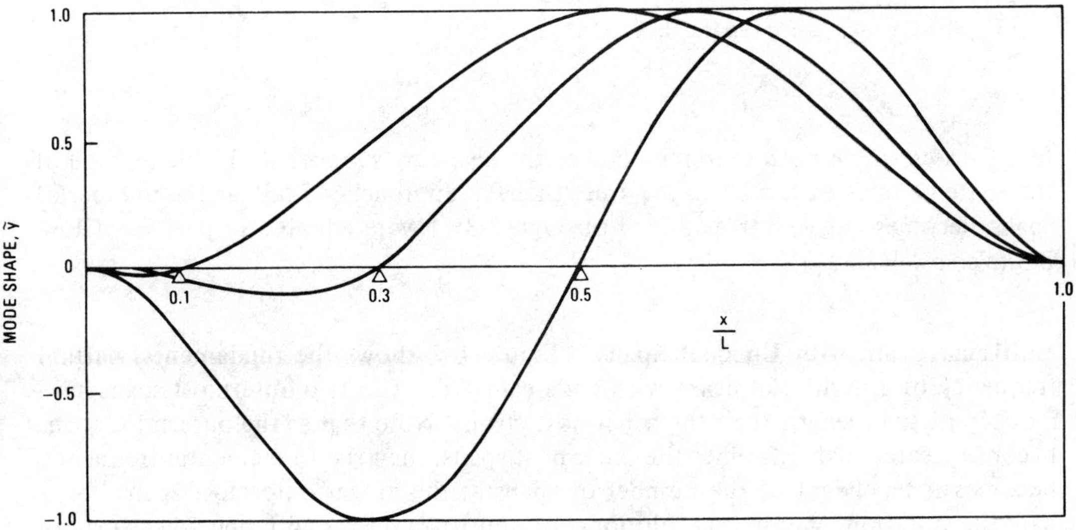

Fig. 8-6. Mode shape of first mode of a beam with extreme ends clamped and an intermediate pinned support for a/L = 0.1, 0.3, and 0.5. (Ref. 8-4)

Table 8-3a. Free-Free Multispan Beam with Pinned Intermediate Supports.

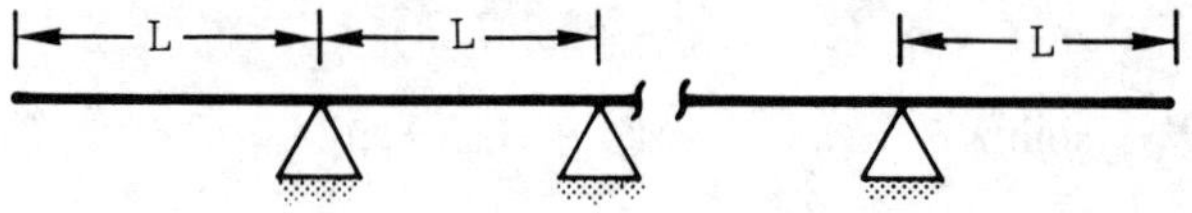

$$f_i = \frac{\lambda_i^2}{2\pi L^2}\left(\frac{EI}{m}\right)^{1/2}$$

$\lambda_i = \lambda_i$ (Number of Spans)

Number of Spans(a)	Mode Number (i)					
	1	2	3	4	5	6
1	4.730	7.853	11.00	14.14	17.28	20.42
2	1.875	3.927	4.694	7.069	7.855	10.21
3	1.412	1.648	3.580	4.273	4.707	6.707
4	1.506	1.571	3.413	3.928	4.438	4.713
5	1.530	1.548	3.324	3.710	4.144	4.528
6	1.537	1.542	3.270	3.568	3.927	4.285
7	1.538	1.540	3.237	3.471	3.770	4.084
8	1.539	1.539	3.215	3.404	3.653	3.926
9	1.539	1.539	3.200	3.353	3.564	3.803
10	1.539	1.539	3.189	3.316	3.496	3.705
11	1.539	1.539	3.180	3.287	3.443	3.626
12	1.539	1.539	3.174	3.265	3.400	3.563
13	1.539	1.539	3.169	3.248	3.365	3.510
14	1.539	1.539	3.166	3.234	3.337	3.466
15	1.539	1.539	3.162	3.222	3.313	3.430

(a) Span = a beam segment of axial length L.

for all cases where both extreme ends of the beam are supported. If one or both of the extreme ends of the beam are free, then λ_1 approaches 1.539 as the number of spans becomes large. Tables 8-3(a) through 8-3(f) were adapted in part from Ref. 8-5 (also see Ref. 8-6).

Multispan Beam with Unequal Spans. Figure 8-7 shows the fundamental natural frequency of a multispan beam with ends clamped. The two outermost spans have a different span length than the innermost spans. Note that as the outermost spans become greater than 1.2 times the innermost spans, then the fundamental frequency becomes independent of the number of spans as the longer outermost spans dominate the vibration. Additional solutions for multispan beams with unequal spans can be found in Ref. 8-6.

Table 8-3b. Free-Pinned Multispan Beam with Pinned Intermediate Supports.

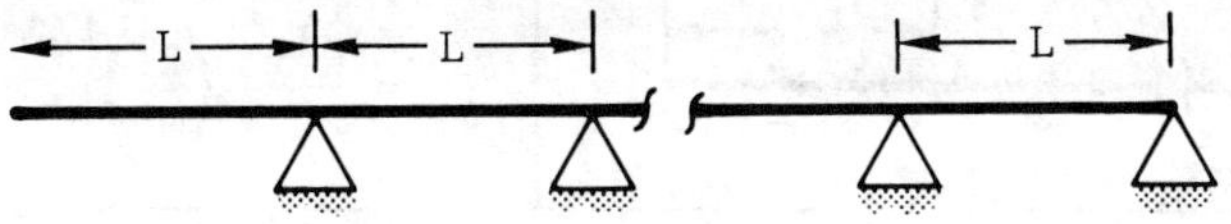

$$f_i = \frac{\lambda_i^2}{2\pi L^2}\left(\frac{EI}{m}\right)^{1/2}$$

Number of Spans[a]	$\lambda_i = \lambda_i$ (Number of Spans)					
	Mode Number (i)					
	1	2	3	4	5	6
1	3.927	7.069	10.21	13.35	16.49	19.63
2	1.505	3.412	4.431	6.541	7.574	9.678
3	1.536	3.270	3.927	4.580	6.410	7.070
4	1.539	3.215	3.653	4.200	4.640	6.358
5	1.539	3.188	3.496	3.926	4.358	4.668
6	1.539	3.173	3.400	3.742	4.112	4.455
7	1.539	3.166	3.337	3.611	3.927	4.241
8	1.539	3.160	3.294	3.519	3.786	4.066
9	1.539	3.156	3.263	3.449	3.679	3.927
10	1.539	3.153	3.241	3.397	3.595	3.814
11	1.539	3.150	3.225	3.357	3.528	3.723
12	1.539	3.150	3.212	3.325	3.475	3.6475
13	1.539	3.148	3.202	3.300	3.432	3.586
14	1.539	3.147	3.194	3.279	3.396	3.534
15	1.539	3.147	3.187	3.263	3.367	3.490

[a] Span = a beam segment of axial length L.

8.1.4. Beams Supporting a Uniform Axial Load

A tensile load which is applied axially to a beam (Fig. 8-8) increases the natural frequencies of the beam. A compressive axial load decreases the natural frequencies of the beam. The effect of axial load on the natural frequencies of the single-span beams given in Table 8-1 can be found by considering the natural frequency parameter λ_i to be a function of a nondimensional load parameter and the beam boundary conditions:

$$\lambda_i = \lambda_i\left(\frac{PL^2}{EI}, \text{boundary conditions}\right).$$

P is the axial force. P is positive if the load is tensile. P is negative if the load is compressive.

Table 8-3c. Clamped-Free Multispan Beam with Pinned Intermediate Supports.

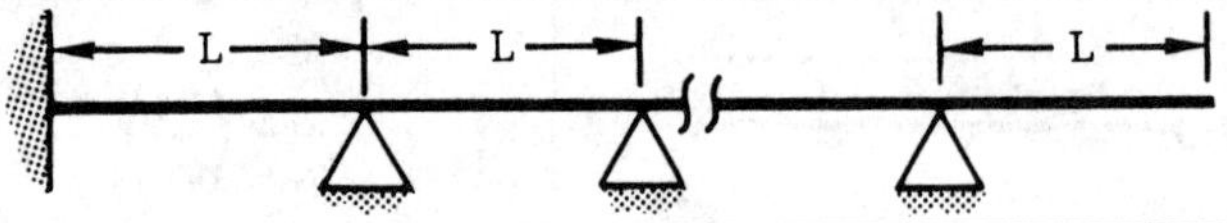

$$f_i = \frac{\lambda_i^2}{2\pi L^2}\left(\frac{EI}{m}\right)^{1/2}$$

Number of Spans(a)	$\lambda_i = \lambda_i$ (Number of Spans)					
	Mode Number (i)					
	1	2	3	4	5	6
1	1.875	4.694	7.855	11.00	14.14	17.28
2	1.570	3.923	4.707	7.058	7.842	10.19
3	1.541	3.570	4.283	4.720	6.707	7.430
4	1.539	3.403	3.928	4.450	4.723	6.545
5	1.539	3.316	3.706	4.148	4.538	4.724
6	1.539	3.265	3.563	3.927	4.292	4.592
7	1.539	3.233	3.466	3.767	4.086	4.389
8	1.539	3.213	3.399	3.649	3.926	4.204
9	1.539	3.198	3.349	3.560	3.802	4.051
10	1.539	3.187	3.312	3.492	3.703	3.927
11	1.539	3.179	3.285	3.439	3.624	3.624
12	1.539	3.173	3.263	3.397	3.559	3.739
13	1.539	3.168	3.245	3.362	3.507	3.669
14	1.539	3.165	3.232	3.334	3.463	3.609
15	1.539	3.162	3.221	3.311	3.427	3.559

(a) Span = a beam segment of axial length L.

Exact solutions for the effect of an axial load on the natural frequency parameter are available for pinned-pinned, sliding-sliding, and sliding-pinned single-span uniform beams. For a pinned-pinned or sliding-sliding single-span beam (Ref. 8-8):

$$f_i = \frac{i^2\pi^2}{2\pi L^2}\left(1 + \frac{PL^2}{EIi^2\pi^2}\right)^{1/2}\left(\frac{EI}{m}\right)^{1/2}; \qquad i = 1, 2, 3, \ldots. \tag{8-20}$$

For a sliding-pinned single-span beam (Ref. 8-8):

$$f_i = \frac{i^2\pi^2}{8\pi L^2}\left(1 + \frac{4PL^2}{EIi^2\pi^2}\right)^{1/2}\left(\frac{EI}{m}\right)^{1/2}; \qquad i = 1, 3, 5, \ldots. \tag{8-21}$$

For other uniform beams, an approximate formula for the effect of axial load on the natural frequency f_i is:

$$\frac{f_i|_{P\neq 0}}{f_i|_{P=0}} = \left(1 + \frac{P}{|P_b|}\frac{\lambda_1^2}{\lambda_i^2}\right)^{1/2}; \qquad i = 1, 2, 3, \ldots, \tag{8-22}$$

Table 8-3d. Pinned-Pinned Multispan Beam with Pinned Intermediate Supports.

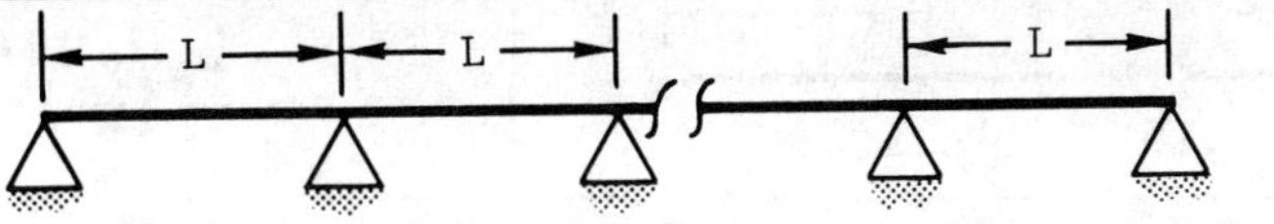

$$f_i = \frac{\lambda_i^2}{2\pi L^2}\left(\frac{EI}{m}\right)^{1/2}$$

Number of Spans[a]	$\lambda_i = \lambda_i$ (Number of Spans)					
	Mode Number (i)					
	1	2	3	4	5	6
1	3.142	6.283	9.425	12.57	15.71	18.85
2	3.142	3.927	6.283	7.068	9.424	10.21
3	3.142	3.557	4.297	4.713	6.707	7.430
4	3.142	3.393	3.928	4.463	6.283	6.545
5	3.142	3.310	3.700	4.152	4.550	6.284
6	3.142	3.260	3.557	3.927	4.293	4.602
7	3.142	3.230	3.460	3.764	4.089	4.394
8	3.142	3.210	3.394	3.645	3.926	4.208
9	3.142	3.196	3.344	3.557	3.800	4.053
10	3.142	3.186	3.309	3.488	3.700	3.927
11	3.142	3.178	3.282	3.436	3.621	3.823
12	3.142	3.173	3.261	3.393	3.557	3.738
13	3.142	3.168	3.244	3.359	3.504	3.666
14	3.141	3.164	3.230	3.332	3.460	3.607
15	3.141	3.161	3.219	3.309	3.424	3.557

[a] Span = a beam segment of axial length L.

where λ_1 and λ_i on the right-hand side of Eq. 8-22 refer to the nondimensional frequency parameters in the absence of axial load (Tables 8-1, 8-3) and P_b is the axial load required to buckle the beam. P_b is given in Table 8-4 for beams with various boundary conditions.

Close inspection of the numerical results presented in Refs. 8-8, 8-9, and 8-10 and Eqs. 8-20 and 8-21 shows that Eq. 8-22 generally predicts the effect of axial load on the natural frequencies to within 1% of the exact natural frequencies for single-span beams subject to axial load within $\pm P_b$. Equation 8-22 is equivalent to the exact solutions for the pinned-pinned, sliding-sliding, and sliding-pinned beams (Eqs. 8-20, 8-21). Equation 8-22 is presented graphically in Fig. 8-9.

If a sufficiently large compressive axial load is applied to the beam such that

$$P = P_b = \begin{cases} -\pi^2 \dfrac{EI}{L^2} & \text{pinned-pinned or sliding-sliding beam} \\ -\pi^2 \dfrac{EI}{4L^2} & \text{sliding-pinned beam} \end{cases},$$

Table 8-3e. Clamped-Pinned Multispan Beam with Pinned Intermediate Supports.

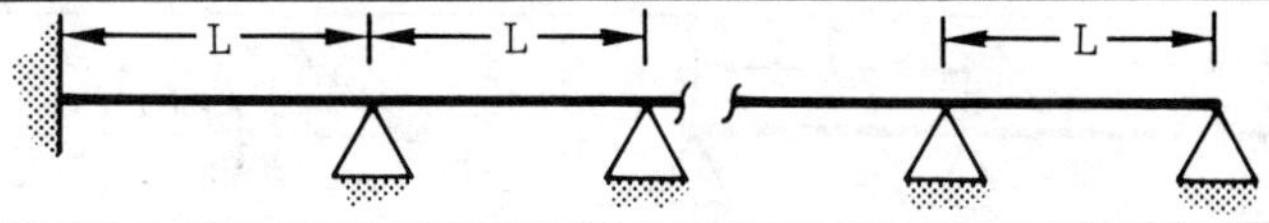

$$f_i = \frac{\lambda_i^2}{2\pi L^2}\left(\frac{EI}{m}\right)^{1/2}$$

$\lambda_i = \lambda_i$ (Number of Spans)

Number of Spans(a)	Mode Number (i)					
	1	2	3	4	5	6
1	3.927	7.069	10.21	13.35	16.49	19.63
2	3.393	4.463	6.545	7.591	9.687	10.73
3	3.261	3.927	4.600	6.410	7.070	7.727
4	3.210	3.645	4.207	4.655	6.357	6.795
5	3.186	3.488	3.926	4.366	4.682	6.332
6	3.173	3.393	3.738	4.115	4.463	4.697
7	3.164	3.331	3.607	3.927	4.247	4.527
8	3.159	3.290	3.514	3.784	4.069	4.341
9	3.156	3.260	3.444	3.675	3.927	4.178
10	3.153	3.239	3.393	3.592	3.813	4.041
11	3.151	3.222	3.354	3.525	3.721	3.927
12	3.149	3.210	3.322	3.472	3.645	3.832
13	3.148	3.200	3.297	3.428	3.583	3.751
14	3.147	3.192	3.277	3.393	3.531	3.684
15	3.147	3.186	3.261	3.364	3.489	3.627

(a) Span = a beam segment of axial length L.

then the fundamental (i = 1) natural frequency goes to zero as the beam buckles. If a very large tensile load is applied to a beam with pinned or clamped ends such that

$$P \gg |P_b|,$$

where P_b is the buckling load (Table 8-4), then natural frequencies approach those of a straight tensioned cable (frame 1 of Table 7-1) because the forces associated with the tension in the beam become much greater than the beam stiffness.

The mode shapes of the pinned-pinned, pinned-sliding, and sliding-sliding beam are unaffected by axial load. The mode shapes of the seven remaining single-span beams listed in Table 8-1 are a function of axial load. However, the change in mode shape produced by a large change in axial load is generally very small, as is shown in Fig. 8-10 for the first mode of a clamped-free beam.

8.1.5. Beams Supporting a Linearly Varying Axial Load

General Case. A tensile load which varies linearly over the span of a beam can be generated by a uniform axial traction which is applied to each spanwise segment of

Table 8-3f. Clamped-Clamped Multispan Beam with Pinned Intermediate Supports.

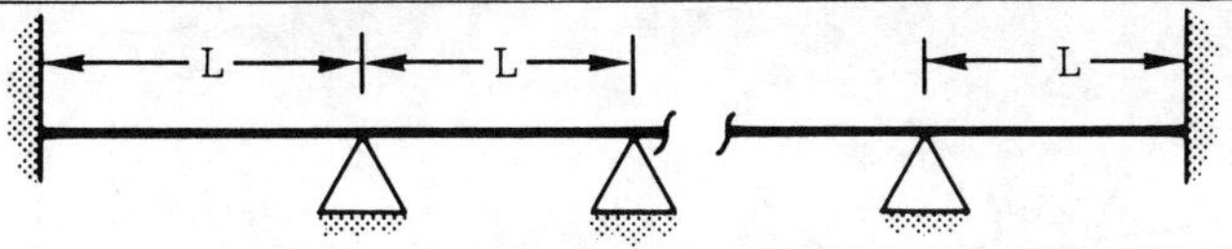

$$f_i = \frac{\lambda_i^2}{2\pi L^2}\left(\frac{EI}{m}\right)^{1/2}$$

Number of Spans[(a)]	$\lambda_i = \lambda_i$ (Number of Spans)					
	Mode Number (i)					
	1	2	3	4	5	6
1	4.730	7.853	11.00	14.14	17.28	20.42
2	3.927	4.730	7.068	7.853	10.21	11.00
3	3.557	4.297	4.730	6.707	7.430	7.853
4	3.393	3.928	4.463	4.730	6.545	7.068
5	3.310	3.700	4.152	4.550	4.730	6.460
6	3.260	3.557	3.927	4.298	4.602	4.730
7	3.230	3.460	3.764	4.089	4.394	4.634
8	3.210	3.394	3.645	3.926	4.208	4.464
9	3.196	3.344	3.557	3.800	4.053	4.298
10	3.186	3.309	3.488	3.700	3.927	4.153
11	3.178	3.282	3.435	3.621	3.823	4.030
12	3.173	3.261	3.393	3.557	3.738	3.927
13	3.168	3.244	3.359	3.504	3.666	3.839
14	3.164	3.230	3.332	3.460	3.607	3.764
15	3.161	3.219	3.309	3.424	3.557	3.701

[(a)] Span = a beam segment of axial length L.

the beam. The tractions can be the result of gravity loads, as on a long drill pipe, or rotational loads, as on a whirling propeller. If one end of the beam is free of axial load, then the traction loads, w, per unit length of the span will generate a reaction wL at the other end of the beam [(Fig. 8-11(a)]. If both ends of the beam can support axial loads, then the axial force in the beam will vary linearly between the two end reactions [Fig. 8-11(b)]. Both of these cases will be considered in the following sections.

The natural frequencies of uniform beams with axial loads which vary linearly along the span are given by:

$$f_i = \frac{\lambda_i^2}{2\pi L^2}\left(\frac{EI}{m}\right)^{1/2}; \qquad i = 1, 2, 3, \ldots. \tag{8-23}$$

f_i is the natural frequency of the i mode in hertz. E is the modulus of elasticity, I is the area moment of inertia of the beam cross section (Table 5-1), m is the mass per unit length of the beam, and L is the overall beam span. Consistent sets of units are given in Table 3-1. The dimensionless parameter λ_i is a function of the boundary

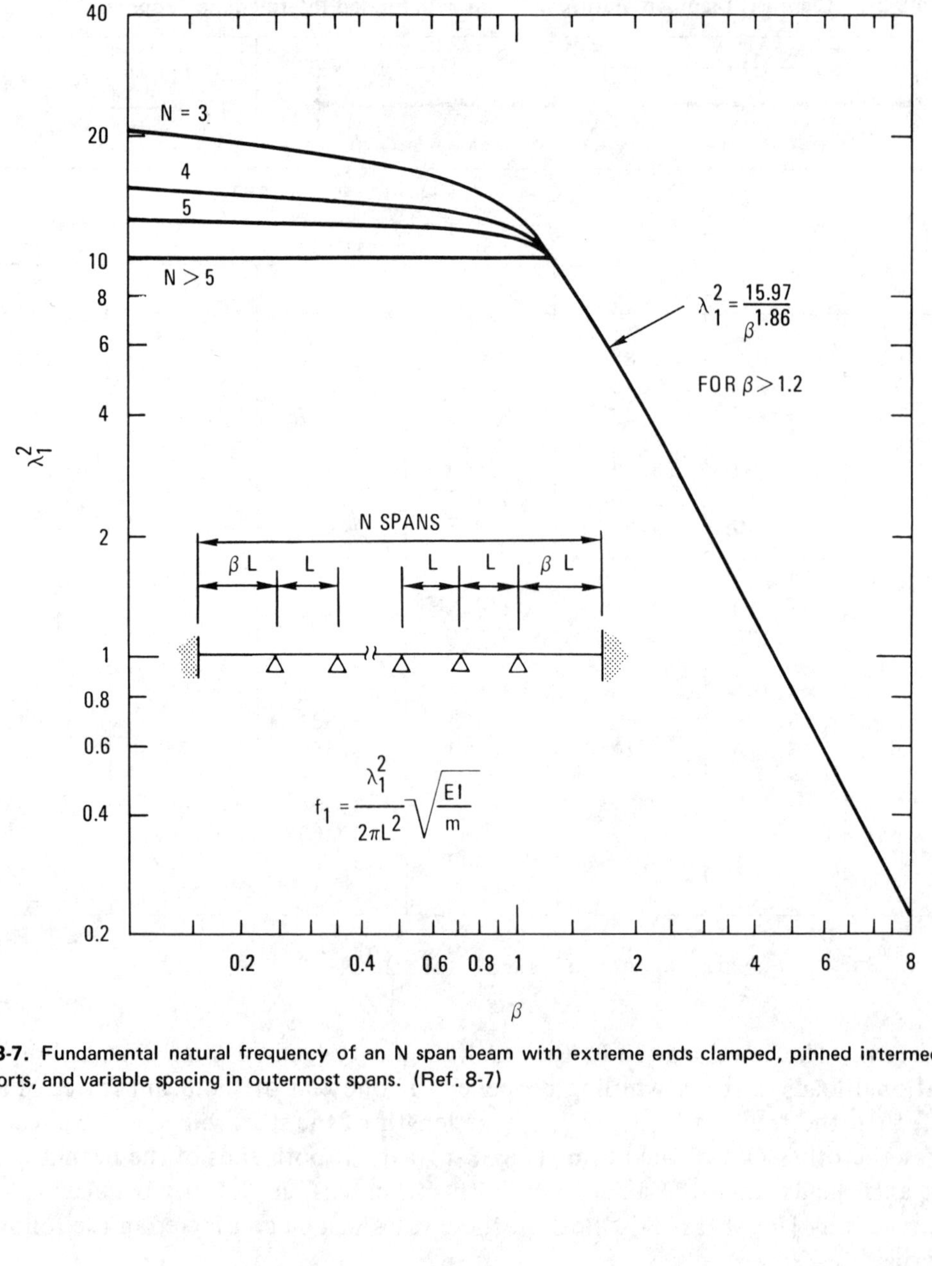

Fig. 8-7. Fundamental natural frequency of an N span beam with extreme ends clamped, pinned intermediate supports, and variable spacing in outermost spans. (Ref. 8-7)

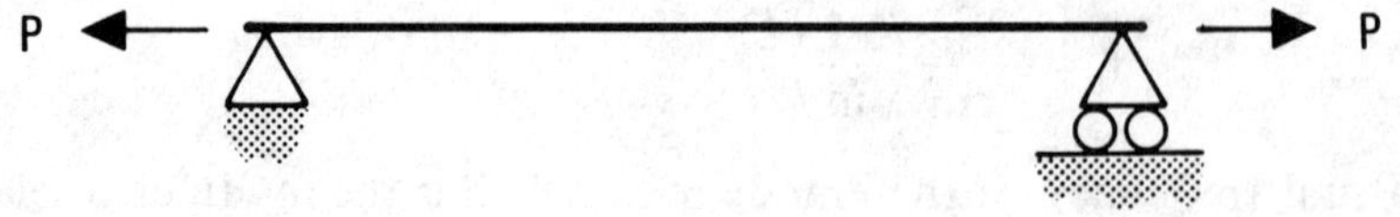

Fig. 8-8. Tensile axial load on a pinned-pinned beam.

Table 8-4. Buckling Load and Mode for Uniform Beams Under Axial Load with Various Boundary Conditions.[a]

Notation: E = modulus of elasticity; I = area moment of inertia; L = span of beam; x = distance along beam

Boundary Conditions	Buckling Load, $\lvert P_b \rvert$	Buckling Mode Shape
1. Free-Free	$\frac{\pi^2 EI}{L^2}$	$\sin \frac{\pi x}{L}$
2. Free-Sliding	$\frac{\pi^2 EI}{4L^2}$	$\sin \frac{\pi x}{2L}$
3. Clamped-Free	$\frac{\pi^2 EI}{4L^2}$	$1 - \cos \frac{\pi x}{2L}$
4. Free-Pinned	$\frac{\pi^2 EI}{L^2}$	$\sin \frac{\pi x}{L}$
5. Pinned-Pinned	$\frac{\pi^2 EI}{L^2}$	$\sin \frac{\pi x}{L}$
6. Clamped-Pinned	$\frac{2.05\pi^2 EI}{L^2}$	--
7. Clamped-Clamped	$\frac{4\pi^2 EI}{L^2}$	$1 - \cos \frac{2\pi x}{L}$
8. Clamped-Sliding	$\frac{\pi^2 EI}{L^2}$	$1 - \cos \frac{\pi x}{L}$
9. Sliding-Pinned	$\frac{\pi^2 EI}{4L^2}$	$\cos \frac{\pi x}{2L}$
10. Sliding-Sliding	$\frac{\pi^2 EI}{L^2}$	$\cos \frac{\pi x}{L}$

[a] Ref. 8-8.

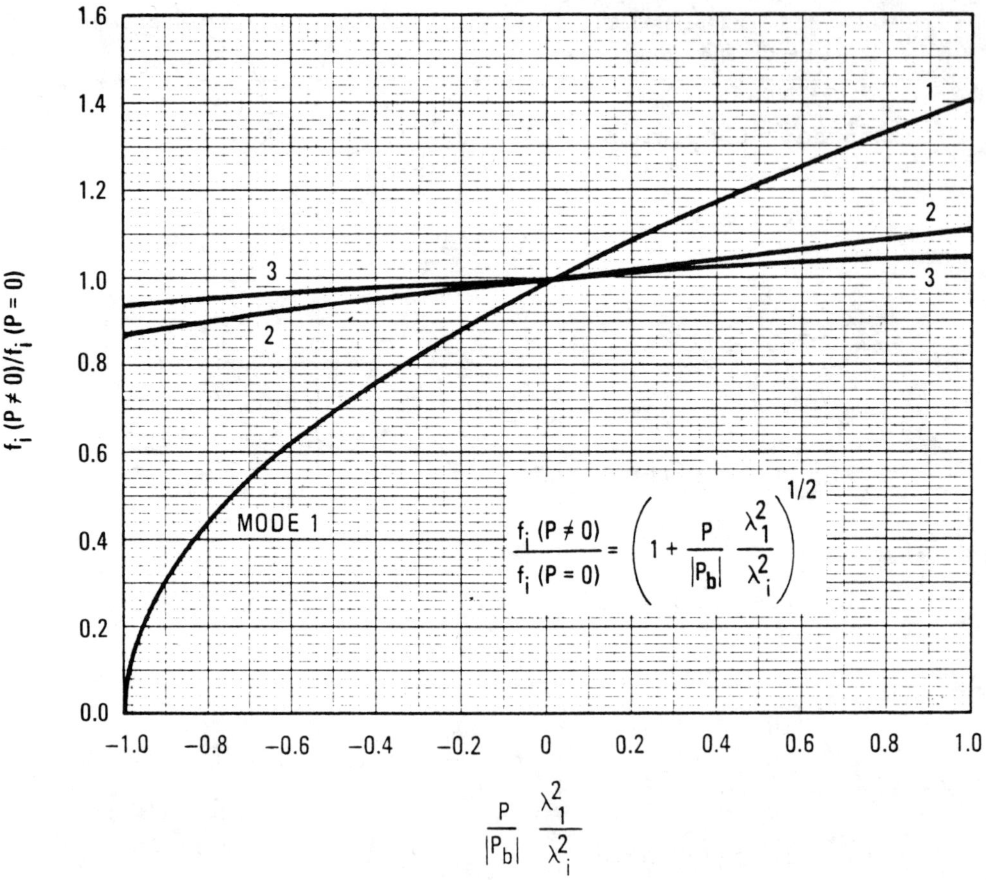

Fig. 8-9. Change in natural frequency (f_i) with axial load (P) as predicted by Eq. 8-22. P_b is the beam buckling load and is given in Table 8-4.

conditions and the axial load. The following sections are devoted to finding λ_i for two cases of linearly varying axial load.

Zero Axial Load on One End of Beam. If the axial load is zero at one end of the beam, as in Fig. 8-11(a), then the constants λ_i (Eq. 8-23) are functions of the beam boundary conditions and the axial traction parameter α:

$$\lambda_i = \lambda_i\ (\alpha, \text{boundary conditions}),$$

where

$$\alpha = \frac{wL^3}{EI}. \tag{8-24}$$

w is the axial traction per unit length of the beam. w is constant over the span of the beam and has units of force per unit length. w is positive if the tractions put the beam in tension and negative if the tractions put the beam in compression. For

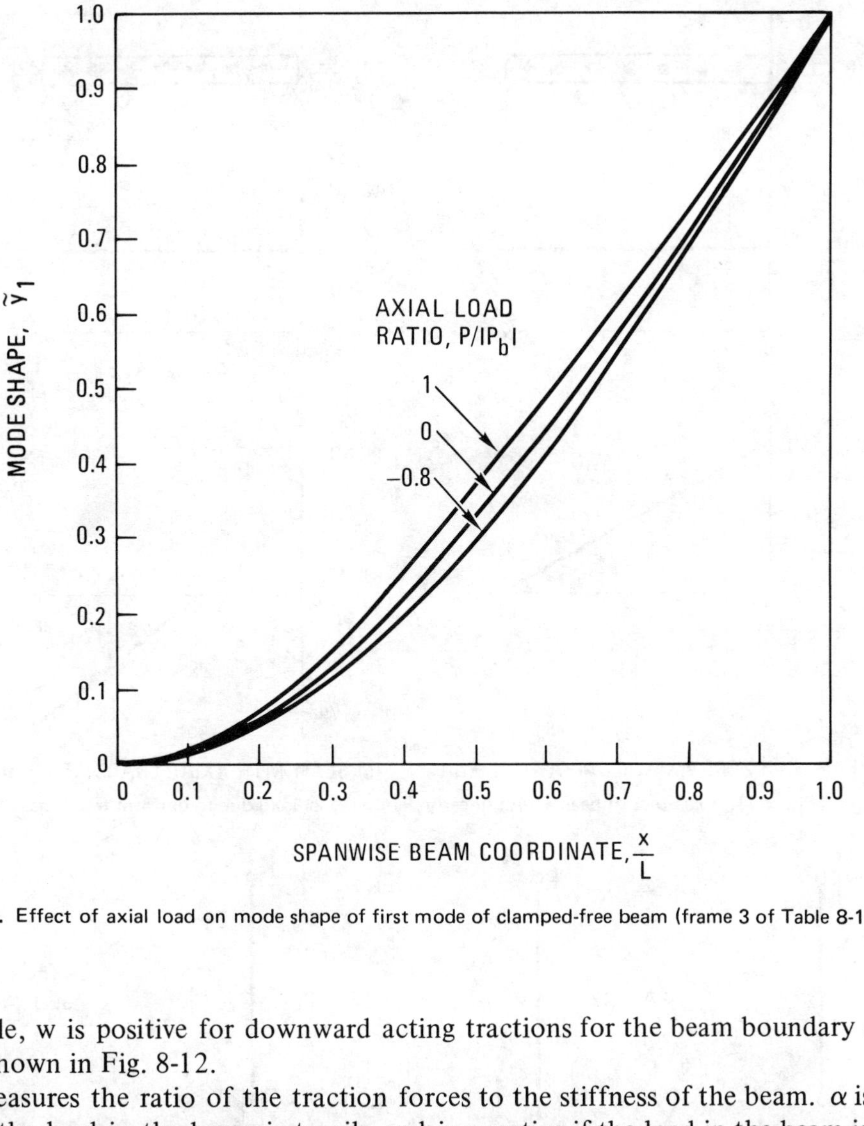

Fig. 8-10. Effect of axial load on mode shape of first mode of clamped-free beam (frame 3 of Table 8-1). (Ref. 8-8)

example, w is positive for downward acting tractions for the beam boundary conditions shown in Fig. 8-12.

α measures the ratio of the traction forces to the stiffness of the beam. α is positive if the load in the beam is tensile and is negative if the load in the beam is compressive. The natural frequencies will increase with increasing tensile load and decrease with increasing compressive load. A sufficiently large compressive load will cause any of the natural frequencies to go to zero ($\lambda_i = 0$) as the beam buckles.

Table 8-5 gives the critical axial traction required to buckle the beams shown in Fig. 8-12. Tables 8-6 and 8-7 and Fig. 8-13 give the natural frequency parameters λ_i (Eq. 8-23) for the first three modes of the beams of Fig. 8-12 for a range of α (Eq. 8-24). Note that $\lambda_i = 0$ can correspond to either onset of buckling or a rigid body rotation mode (case 12). Tables 8-6 and 8-7 and Fig. 8-13 were generated by Huang and Dareing (Ref. 8-11).

As α becomes large, on the order of 1000, the tensile force in the beam becomes much greater than the force associated with bending rigidity and the beams can be

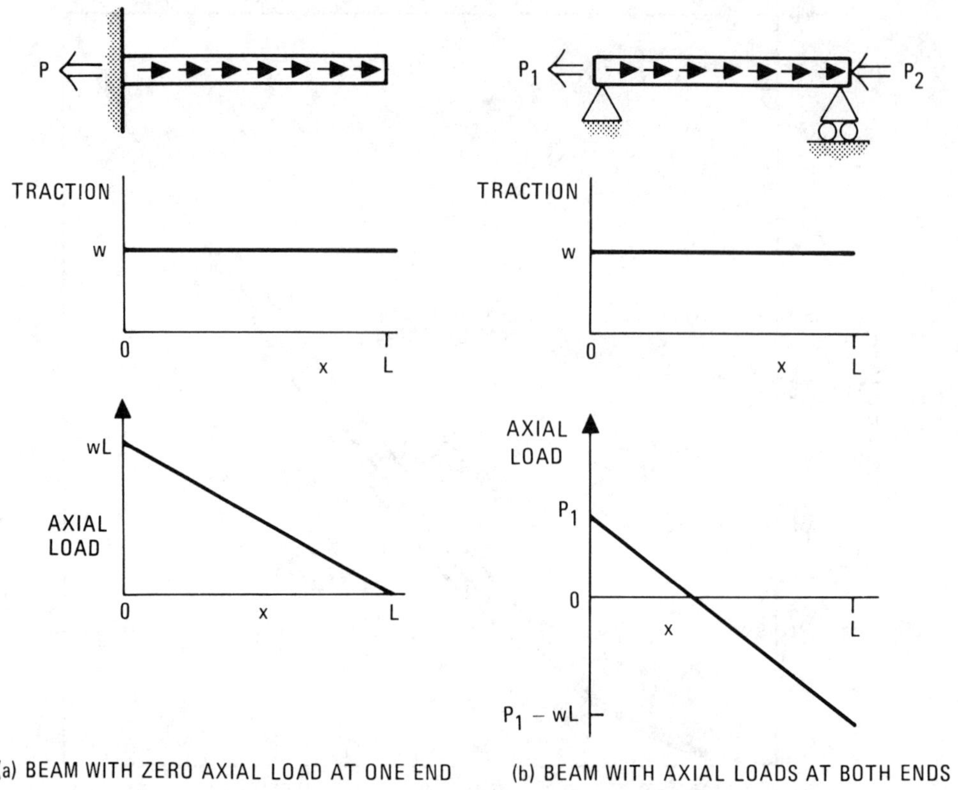

Fig. 8-11. Examples of beams with linearly varying axial load due to uniform tractions.

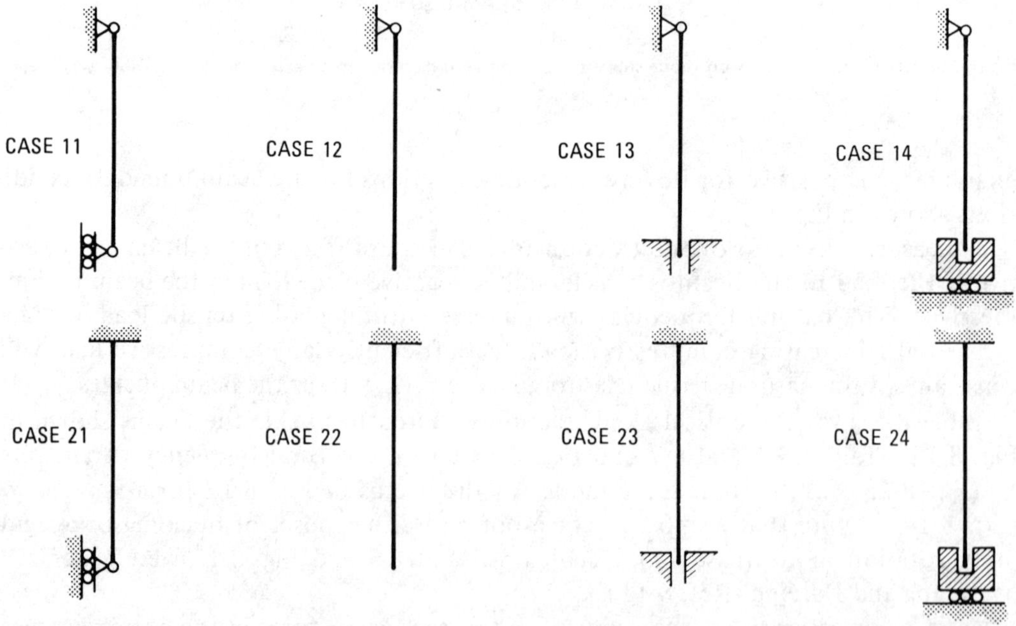

Fig. 8-12. Beam boundary conditions. A downward axial traction is applied to each spanwise element of the beam. The axial load at the lower end of each beam is zero. The boundary conditions associated with cases 13, 14, 23, and 24 do not permit rotation of the lower end of the beam.

Table 8-5. Values of α (Eq. 8-24) for Onset of Buckling ($\lambda_i = 0$).[a]

Case	Mode 1	Mode 2	Mode 3	Mode 4
11	-18.569	-86.431	-196.29	
12	0	-25.638	-95.95	-210.08
13	-30.009	-112.09	-234.99	
14	-3.4766	-44.138	-129.26	
21	-52.501	-129.05	-276.24	
22	-7.8373	-55.977	-148.51	
23	-74.629	-157.03	-325.51	
24	-18.956	-81.887	-189.22	

[a] Ref. 8-11.

modeled as hanging chains. For example, consider a long slender beam which is supported only at the top end. A linearly varying axial tension is produced in the beam as the beam supports its own weight. The traction is:

$$w = mg,$$

where m is the mass per unit length of the beam and g is the acceleration of gravity. If this equation is substituted into Eq. 8-24 and Eqs. 8-24 and 8-23 are combined to eliminate the term EI, the result is:

$$f_i = \frac{1}{2\pi} \frac{\lambda_i^2}{\alpha^{1/2}} \left(\frac{g}{L}\right)^{1/2}; \qquad i = 1, 2, 3, \ldots.$$

As the tension in the beam becomes large (for example, case 12 of Table 8-7 for $\alpha = 1000$), the dimensionless term $\lambda_i^2/\alpha^{1/2}$ approaches a limit:

$$\lim_{\alpha \to \infty} \frac{\lambda_i^2}{\alpha^{1/2}} = \begin{cases} 1.20 \text{ mode } 1 \\ 2.86 \text{ mode } 2. \\ 4.95 \text{ mode } 3 \end{cases}$$

Comparing this result with the natural frequencies given in frame 5 of Table 6-4, it can be seen that as α becomes large the natural frequencies of long hanging beams are accurately predicted by the natural frequencies of hanging chains.

Axial Load at Both Ends of the Beam. If the beam is subject to a uniform spanwise traction and can support axial loads at both its ends, as in Fig. 8-11(b), then the

Table 8-6. Natural Frequency Parameter λ_i (Eq. 8-23) as a Function of α (Eq. 8-24) for $\alpha \leqslant 0$.[a]

α	0	-5	-10	-15	-18		
Case 11 λ_1	3.1416	2.9168	2.6120	2.1089	1.3368		
Case 11 λ_2	6.2832	6.1810	6.0728	5.9581	5.8857		
Case 11 λ_3	9.4248	9.3577	9.2889	9.2185	9.1753		
α	0	-5	-10	-15	-20	-25	-30
Case 13 λ_1	3.9266	3.7660	3.5762	3.3426	3.0347	2.5652	0.5371
Case 13 λ_2	7.0686	6.9801	6.8877	6.7911	6.6899	6.5836	6.4717
Case 13 λ_3	10.2102	10.1489	10.0805	10.0227	9.9575	9.8909	9.8229
α	0	-1	-2	-3	-3.4		
Case 14 λ_1	1.5708	1.4442	1.2700	0.9580	0.6068		
Case 14 λ_2	4.7124	4.6844	4.6558	4.6266	4.6148		
Case 14 λ_3	7.8540	7.8378	7.8214	7.8050	7.7983		
α	0	-10	-20	-30	-40	-50	
Case 21 λ_1	3.9266	3.7342	3.5017	3.2039	2.7753	1.8629	
Case 21 λ_2	7.0686	6.9339	6.7895	6.6339	6.4648	6.2794	
Case 21 λ_3	10.2102	10.1088	10.0039	9.8951	9.7821	9.6645	
α	0	-2	-4	-6	-7.5		
Case 22 λ_1	1.8751	1.7424	1.5694	1.3059	0.8550		
Case 22 λ_2	4.6941	4.6517	4.6081	4.5631	4.5285		
Case 22 λ_3	7.8548	7.8289	7.8027	7.7762	7.7562		
α	0	-20	-40	-60	-74.5		
Case 23 λ_1	4.7300	4.4011	3.9537	3.2127	0.9902		
Case 23 λ_2	7.8532	7.6018	7.3177	6.9900	6.7157		
Case 23 λ_3	10.9956	10.8040	10.6000	10.3820	10.2139		
α	0	-5	-10	-15	-18.5		
Case 24 λ_1	2.3650	2.1919	1.9630	1.6013	0.9335		
Case 24 λ_2	5.4978	5.4160	5.3300	5.2390	5.1721		
Case 24 λ_3	8.6394	8.5812	8.5216	8.4605	8.4169		

(a) Ref. 8-11.

dimensionless frequency parameters λ_i are a function of the beam boundary conditions and two axial load parameters:

$$\lambda_i = \lambda_i\left(\text{boundary conditions}, \frac{wL^3}{EI}, \frac{P_1}{wL}\right).$$

P_1 is the larger of the two axial forces applied at the beam ends. P_1 is positive for forces which act outward from the beam [(Fig. 8-11(b)]. $P_1/(wL)$ is a measure of

Table 8-7. Natural Frequency Parameter λ_i (Eq. 8-23) as a Function of α (Eq. 8-24) for $\alpha > 0$.[a]

α	0	200	400	600	800	1000
Case 11 λ_1	3.1416	5.4652	6.3076	6.8804	7.3261	7.6957
Case 11 λ_2	6.2832	8.4625	9.5456	10.3113	10.9171	11.4243
Case 11 λ_3	9.4248	11.3031	12.4565	13.3145	14.0089	14.5980
Case 12 λ_1	0	4.1293	4.9077	5.4300	5.8342	6.1684
Case 12 λ_2	3.9266	6.6210	7.6994	8.4435	9.0269	9.5131
Case 12 λ_3	7.0686	9.2779	10.4467	11.2868	11.9574	12.5222
Case 13 λ_1	3.9266	6.0973	6.9535	7.5387	7.9947	8.3732
Case 13 λ_2	7.0686	9.1389	10.2306	11.0098	11.6285	12.1474
Case 13 λ_3	10.2102	11.9999	13.1454	14.0087	14.7115	15.3097
Case 14 λ_1	1.5708	4.2112	4.9728	5.4867	5.8856	6.2160
Case 14 λ_2	4.7124	7.0946	8.1293	8.8488	9.4147	9.8873
Case 14 λ_3	7.8540	9.9110	11.0588	11.8913	12.5578	13.1197
Case 21 λ_1	3.9266	5.6429	6.4353	6.9875	7.4212	7.7828
Case 21 λ_2	7.0686	8.7817	9.7720	10.4969	11.0790	11.5706
Case 21 λ_3	10.2102	11.7443	12.7812	13.5811	14.2404	14.8056
Case 22 λ_1	1.8751	4.2155	4.9762	5.4905	5.8898	6.2205
Case 22 λ_2	4.6941	6.8154	7.8375	8.5586	9.1288	9.6062
Case 22 λ_3	7.8548	9.6222	10.6904	11.4853	12.1295	12.6768
Case 23 λ_1	4.7300	6.3202	7.1121	7.6704	8.1109	8.4790
Case 23 λ_2	7.8532	9.4945	10.4851	11.2186	11.8105	12.3115
Case 23 λ_3	10.9956	12.4680	13.4947	14.2966	14.9617	15.3141
Case 24 λ_1	2.3650	4.3116	5.0489	5.5526	5.9453	6.2714
Case 24 λ_2	5.4978	7.3379	8.3004	8.9896	9.5382	9.9992
Case 24 λ_3	8.6394	10.2929	11.3331	12.1155	12.7524	13.2945

[a] Ref. 8-11.

the ratio of the axial load induced at the ends of the beam to the axial load induced by the tractions along the beam.

If $|P_1|/(wL) \gg 1$, then the contribution of the axial loads at the ends of the beam dominates the axial force in the beam and the beam can be analyzed using the methods described in Section 8.1.4. If $P_1/(wL) \approx 1$ or $|P_1|/(wL) \ll 1$, then the axial load on one end of the beam is nearly zero and the natural frequencies of the beam can be analyzed using the techniques described in the previous paragraphs. The dimensionless frequency parameter λ_i (Eq. 8-23) for the fundamental mode of pinned-pinned beams and clamped-clamped beams is given in Figs. 8-14 and 8-15 for an intermediate range of P_1 (wL). Data on higher modes of these geometries can be found in tabular form in Ref. 8-12.

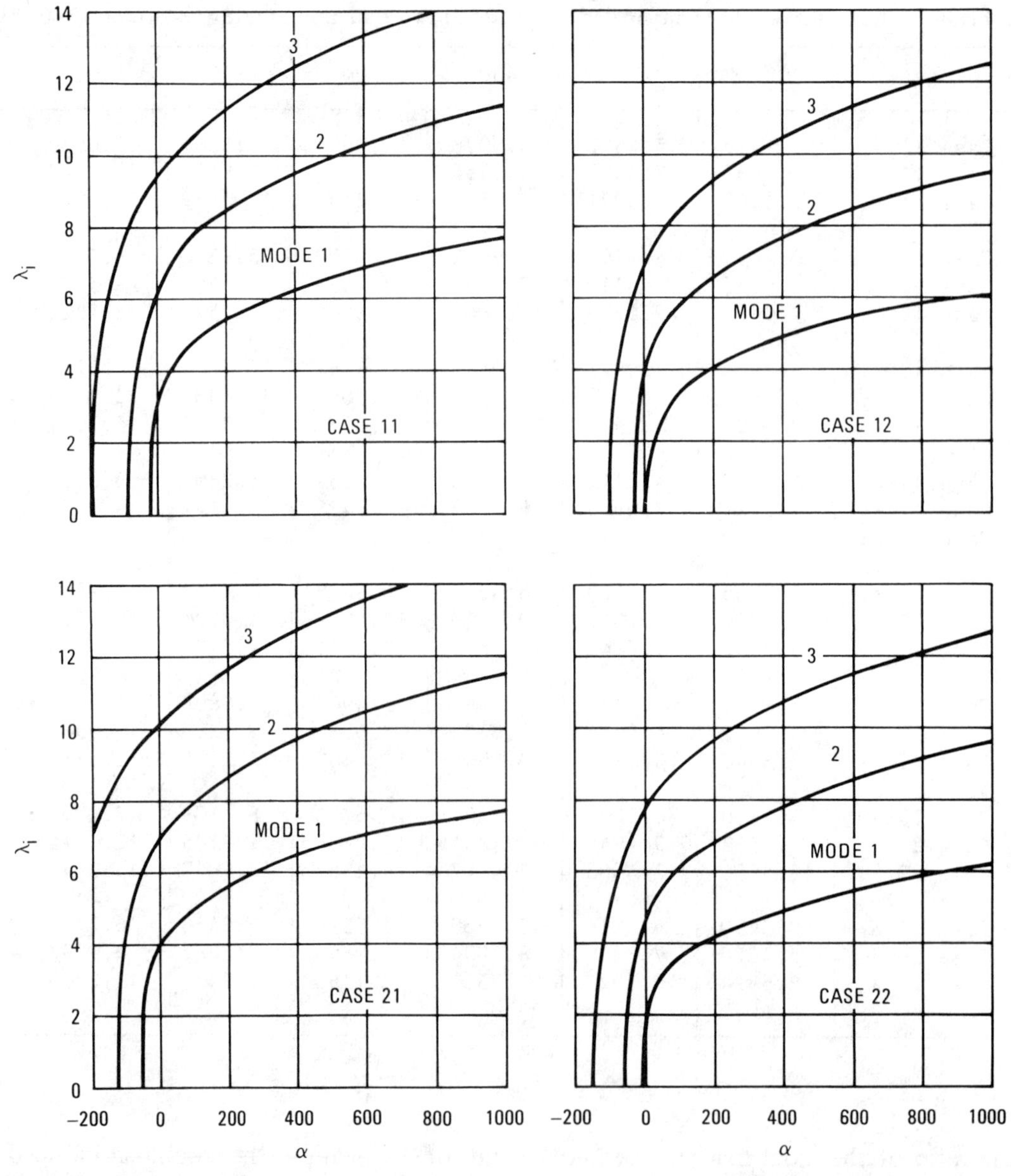

Fig. 8-13. Natural frequency parameter, λ_i (Eq. 8-23), as a function of α (Eq. 8-24) for four of the beams shown in Fig. 8-12. (Ref. 8-11)

8.1.6. Beams with Concentrated Masses

The fundamental natural frequency and mode shape of slender uniform beams with concentrated masses are given in Table 8-8. These approximate formulas are generally within about 1% of the exact solutions. Most of the formulas in Table 8-8 were derived using the energy (Rayleigh) technique which is described in Ref. 8-17 (pp. 31–40). The formulas in Table 8-8 neglect the effect of rotary inertia of the concentrated masses. Analysis of higher modes generally requires consideration of

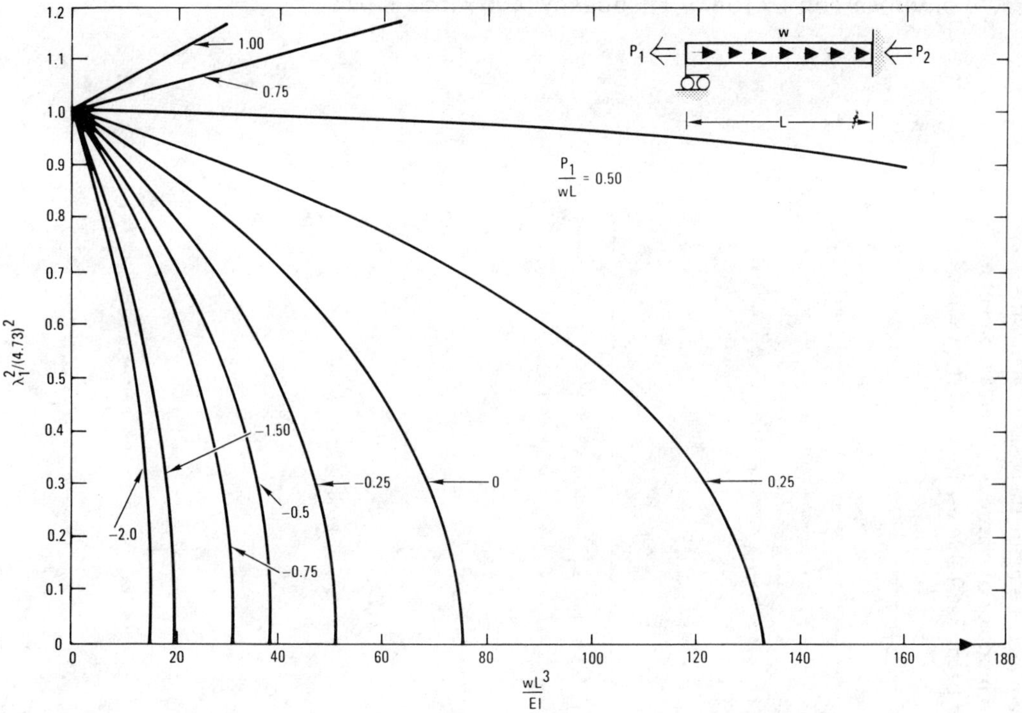

Fig. 8-14. Effect of axial loads on the fundamental frequency of a clamped-clamped beam (Eq. 8-23). (Ref. 8-12)

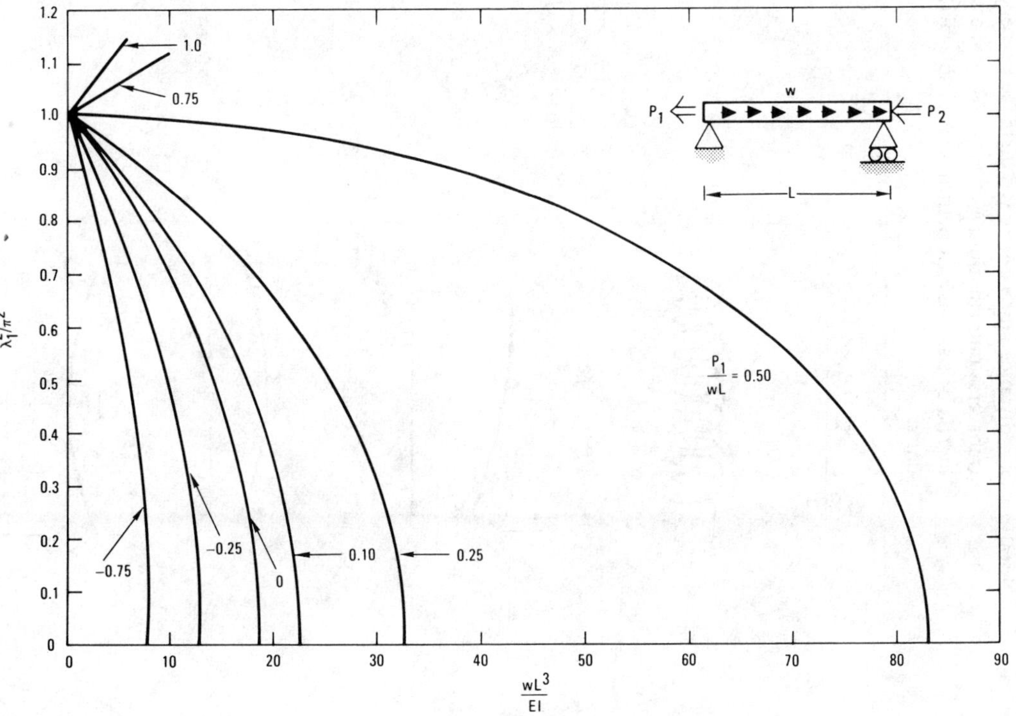

Fig. 8-15. Effect of linearly varying axial loads on fundamental frequency of a pinned-pinned beam (Eq. 8-23). (Ref. 8-12)

Table 8-8. Slender Beams with Concentrated Masses.

Notation: x = distance along beam; y = distance perpendicular to beam axis; $\tilde{y}$ = mode shape associated with transverse deformation; M = mass; M_b = mass of beam; E = modulus of elasticity; I = area moment of inertia of beam about neutral axis (Table 5-1); L = span of beam; see Table 3-1 for consistent sets of units.

Description	Fundamental Natural Frequency, f_1 (hertz)	Mode Shape, $\tilde{y}$ (x)
1. Mass, Helical Spring 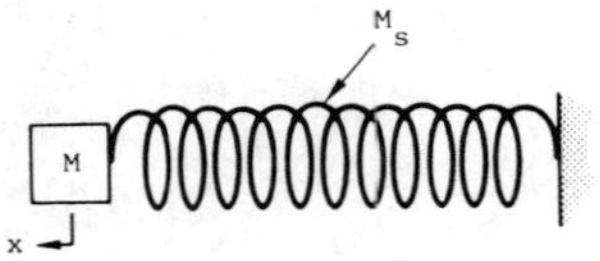	$\frac{1}{2\pi}\left(\frac{k}{M + 0.33\,M_s}\right)^{1/2}$	k = spring constant (Table 6-1) M_s = mass of spring See Frame 4 of Table 8-17 for exact solution.
2. Mass, Cantilever 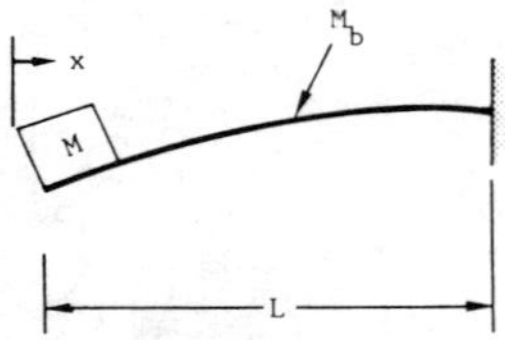	$\frac{1}{2\pi}\left[\frac{3EI}{L^3(M + 0.24\,M_b)}\right]^{1/2}$	$\left(\frac{x}{L}\right)^3 - 3\left(\frac{x}{L}\right) + 2$ See Refs. 8-13, 8-14, 8-15 for higher modes.
3. End Masses, Free-Free Beam 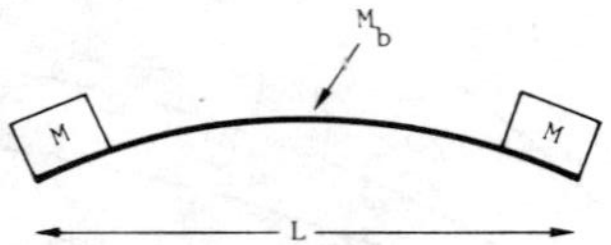	$\frac{\pi}{2}\left\{\frac{EI}{L^3 M_b}\left[1 + \frac{5.45}{1 - 77.4\,(M/M_b)^2}\right]\right\}^{1/2}$	-- See Ref. 8-14 for higher modes.
4. Center Mass, Pinned-Pinned Beam 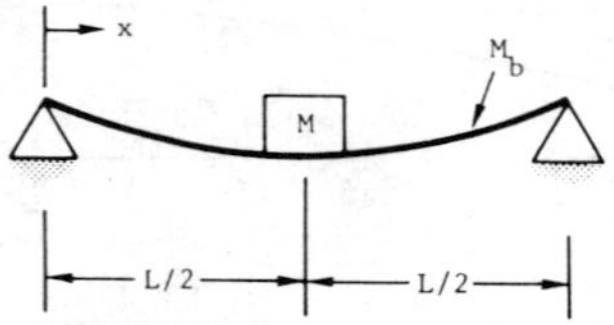	$\frac{2}{\pi}\left[\frac{3EI}{L^3\,(M + 0.49\,M_b)}\right]^{1/2}$	$3\,\frac{x}{L} - 4\left(\frac{x}{L}\right)^3$ See Ref. 8-15 for second mode.

5. Off-Center Mass, Pinned-Pinned Beam 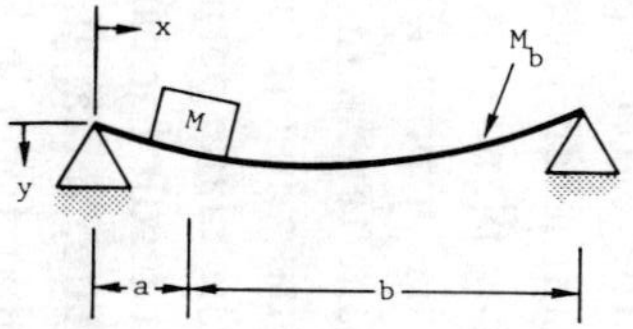	$\frac{1}{2\pi}\left\{\frac{3EI\ (a+b)}{a^2b^2\ [M + (\alpha+\beta)\ M_b]}\right\}^{1/2}$ $\alpha = \frac{a}{a+b}\left[\frac{(2b+a)^2}{12b^2} + \frac{a^2}{28b^2} - \frac{a\ (2b+a)}{10b^2}\right]$ $\beta = \frac{b}{a+b}\left[\frac{(2a+b)^2}{12a^2} + \frac{b^2}{28a^2} - \frac{b\ (2a+b)}{10a^2}\right]$	$\left[2\left(1-\frac{x}{L}\right) - \frac{b^2}{L^2} - \left(1-\frac{x}{L}\right)^2\right]\left(\frac{x}{L}\right)$; $0 \le x \le a$ $\left[\frac{2b}{L} - \frac{b^2}{L^2} - \left(1-\frac{x}{L}\right)^2\right]\left(1-\frac{x}{L}\right)$; $a < x \le L$ $L = a + b$
6. Center Mass, Clamped-Clamped Beam 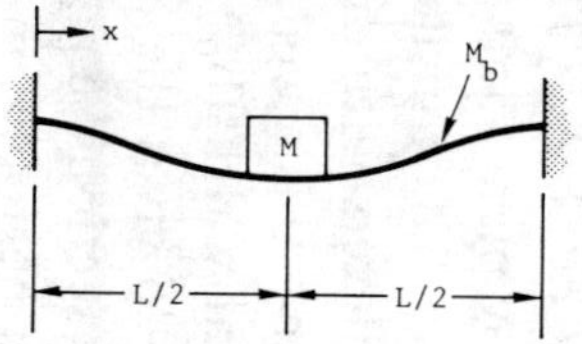	$\frac{4}{\pi}\left[\frac{3EI}{L^3\ (M + 0.37\ M_b)}\right]^{1/2}$	$3\left(\frac{x}{L}\right)^2 - 4\left(\frac{x}{L}\right)^3$; $0 \le \frac{x}{L} \le \frac{1}{2}$ See Ref. 8-16 for second mode.
7. Off-Center Mass, Clamped-Clamped Beam 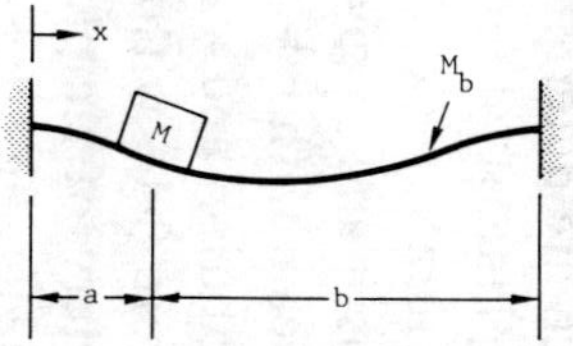	$\frac{4}{\pi}\left(\frac{3EI}{L^3\ [M + (\alpha+\beta)\ M_b]}\right)^{1/2}$ $\alpha = \frac{a}{a+b}\left[\frac{(3a+b)^2}{28b^2} + \frac{9\ (a+b)^2}{20b^2} - \frac{(a+b)(3a+b)}{4b^2}\right]$ $\beta = \frac{b}{a+b}\left[\frac{(3b+a)^2}{28a^2} + \frac{9\ (a+b)^2}{20a^2} - \frac{(a+b)(3b+a)}{4a^2}\right]$	$\left(\frac{x}{L}\right)^2\left(\frac{3a}{L}\frac{x}{L} + \frac{b}{L}\frac{x}{L} - \frac{3a}{L}\right)$; $0 \le x \le a$ $\left(1-\frac{x}{L}\right)^2\left[\frac{3b+a}{L}\left(1-\frac{x}{L}\right) - \frac{3b}{L}\right]$; $a < x \le b$

the rotary inertia of the concentrated masses (Refs. 8-14, 8-15). The effect of an axial load on a cantilever beam with a concentrated load at its tip is described in Ref. 8-8.

8.1.7. Tapered Beams

Formulation. A slender tapered beam can be described by the cross section at the maximum width, the taper in the plane of vibration, the taper in the plane perpendicular to the vibration, and the boundary conditions. The cross section is assumed to be a simple closed section which varies uniformly with the taper. The natural frequencies of slender tapered beams can be expressed as:

$$f_i = \frac{\lambda_i^2}{2\pi L^2}\left(\frac{EI_0}{\mu A_0}\right)^{1/2}; \qquad i = 1, 2, 3, \ldots. \tag{8-25}$$

f_i is the natural frequency of the i mode in hertz. I_0 and A_0 are the area moment of inertia and the area of the cross section at the widest point along the span (Table 5-1). For example, if the beam shown in Fig. 8-16 has a rectangular cross section, then frame 4 of Table 5-1 gives:

$$I_0 = \tfrac{1}{12} b_0 h_0^3,$$

$$A_0 = b_0 h_0,$$

where h_0 is the height of the beam at the support (in the plane of vibration) and b_0 is the width of the beam at the support. L is the overall length of the beam, E is the modulus of elasticity, μ is the material density, and λ_i is a dimensionless constant which is a function of the beam boundary conditions and the taper:

$$\lambda_i = \lambda_i \text{ (boundary conditions, taper)}.$$

Consistent sets of units are given in Table 3-1.

Nonlinearly Tapered Cantilevers. The natural frequencies of the cantilever (clamped-free beam) shown in Fig. 8-16 are determined by Eq. 8-25. The cantilever has two planes of symmetry: (1) the plane of vibration (x–y plane) and (2) the plane perpendicular to the direction of vibration (x–z plane). The cantilever has a power law taper in both planes. Various n and m give cantilevers with different tapers in the two planes. For example, a beam with $n = 1$ and $m = 0$ is a wedge with a linear taper to a point in the plane of vibration and a constant width (b_0) in the plane perpendicular to the vibration. Figure 8-16 with Eq. 8-25 gives the exact result for cantilever beams with a wide variety of simple cross sections, e.g., square, rectangular, diamond, circular, and elliptical. Figure 8-16 can be found in tabular form in Ref. 8-18.

Nonlinearly Tapered Free-Free Beams. The natural frequencies of nonlinearly tapered beams with free-free boundaries can be found from Eq. 8-25 and Fig. 8-17. These beams have three mutually perpendicular planes of symmetry which intersect

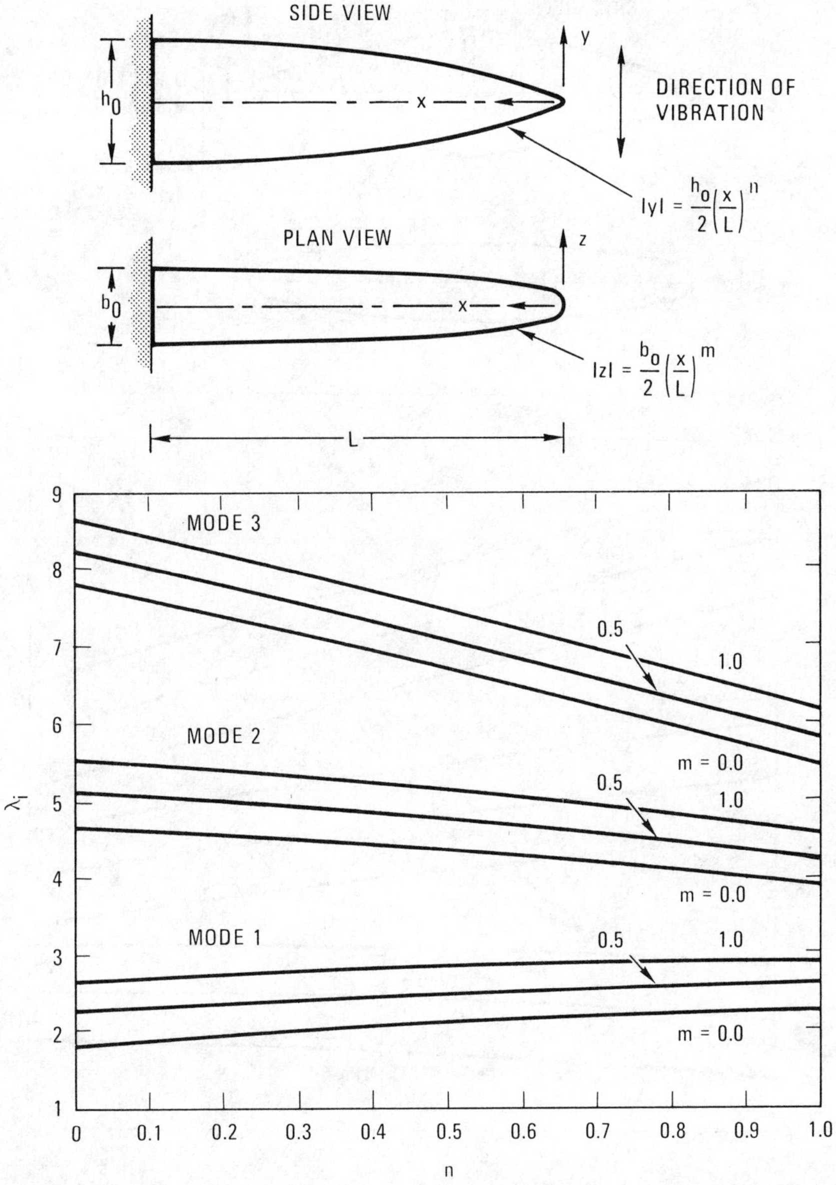

Fig. 8-16. Clamped-free beam with nonlinear taper. The natural frequency is given by Eq. 8-25. Note that m = 0 or n = 0 corresponds to constant thickness and m = 1 or n = 1 corresponds to linear taper to a point. (Ref. 8-18)

at point C: (1) the plane of vibration (x–y plane), (2) the plane perpendicular to the plane of vibration (x–z plane), and (3) the plane which bisects the midspan of the beam. L is the overall length of the beam, and A_0 and I_0 refer to the properties of the beam at midspan (x = L/2). Figure 8-17 gives the exact result for free-free beams with a wide variety of simple cross sections, e.g., square, rectangular, diamond, circular, and elliptical. Figure 8-17 can be found in tabular form in Ref. 8-18.

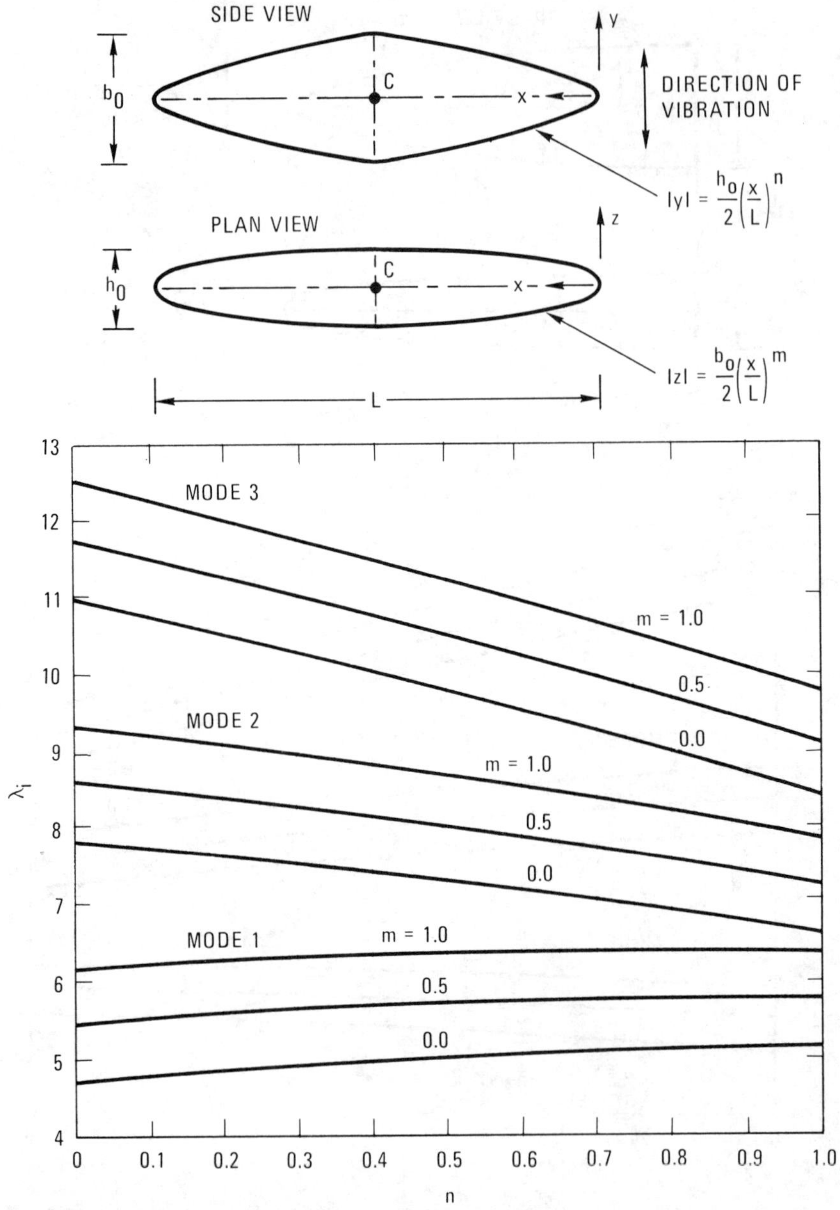

Fig. 8-17. Free-free beam with nonlinear taper. The natural frequencies are given by Eq. 8-25. Note that m = 0 or n = 0 corresponds to constant thickness and m = 1 or n = 1 corresponds to linear taper to each point (Ref. 8-18).

Truncated Linearly Tapered Beams. The natural frequencies of truncated cantilevers with a linear taper in two planes are given by Eq. 8-25 and Fig. 8-18. The ratios h_0/h_1 and b_0/b_1 specify the taper of the cross section. I_0 and A_0 refer to the properties of the cross section at the clamped support. Although Fig. 8-18 was developed for cantilevers with rectangular cross sections, it probably provides a good approximation for most cantilevers with other simple closed cross sections.

Figure 8-18 may be found in tabular form in Ref. 8-19. The mode shapes of these beams are given in Ref. 8-20.

The natural frequencies of slender linearly tapered beams with clamped-pinned, pinned-pinned, and clamped-clamped boundary conditions are given in Figs. 8-19, 8-20, and 8-21 and by Eq. 8-25. A_0 and I_0 in Eq. 8-25 refer to the properties of the cross section at the widest point. The beams of Figs. 8-19, 8-20, and 8-21 are tapered only in the plane of vibration. However, studies of beams with these boundary conditions (Refs. 8-21, 8-22, 8-6) have shown that the natural frequencies are nearly independent of the taper, if any, in the plane perpendicular to the plane of vibration. Thus, Figs. 8-19, 8-20, and 8-21 probably provide a good approximation for beams which have linear tapers perpendicular to the plane of vibration as well as in the plane of vibration. The natural frequencies of linearly tapered beams with various other boundary conditions can be found in Refs. 8-6 and 8-21 through 8-24.

$$f_i = \frac{\lambda_i^2}{2\pi L^2}\left(\frac{EI_0}{\mu A_0}\right)^{1/2} ;$$

$$i = 1, 2, 3 \ldots$$

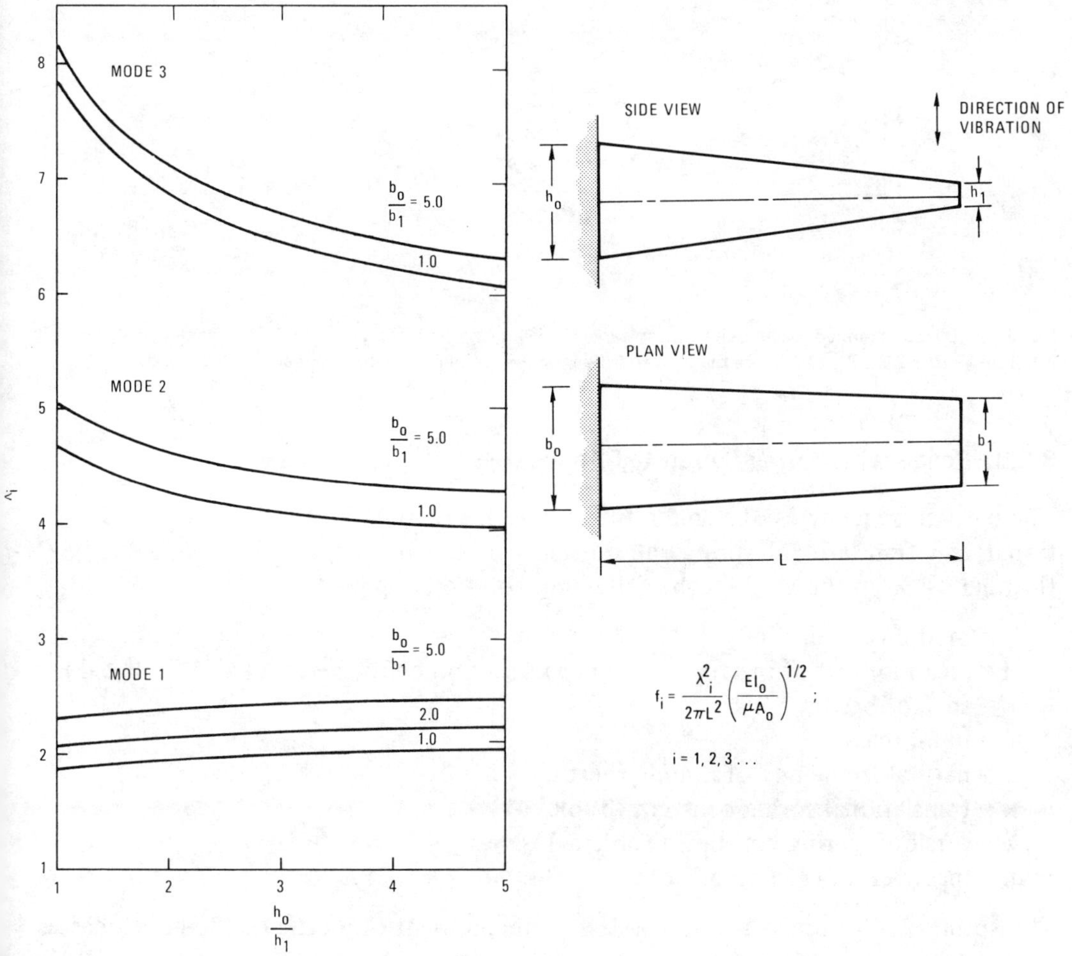

Fig. 8-18. Clamped-free beam with double linear taper and truncated end. The natural frequencies are given by Eq. 8-25 (Ref. 8-19). A_0 and I_0 are the area and the moment of inertia of the beam cross section at the wider end.

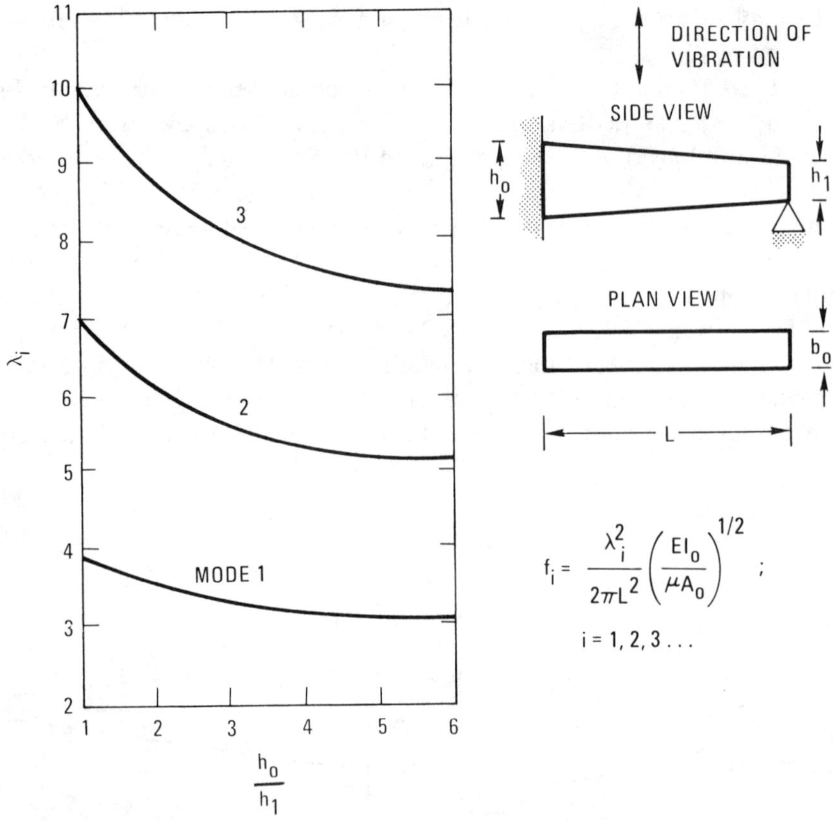

Fig. 8-19. Clamped-pinned beam with linear taper in plane of vibration. Natural frequencies are given by Eq. 8-25 (Ref. 8-21). A_0 and I_0 are the area and the moment of inertia of the beam at the wider end.

8.1.8. Beams with Spring-Supported Boundaries

The natural frequencies of slender uniform beams with various combinations of rotational and translational springs at the ends of the beams are given in Tables 8-9 through 8-13. In these tables the following notation applies:

E = modulus of elasticity,
I = area moment of inertia of beam cross section about neutral axis (Table 5-1),
L = span of beam,
M = point mass,
f_i = natural frequency of i mode (hertz),
k = translational spring constant (Table 6-1),
ℓ = torsional spring constant (Table 6-1),
m = mass per unit length of beam.

The springs are assumed to be massless. The moment of inertia of the point masses is zero for rotation about their centers. Consistent sets of units are given in Table 3-1.

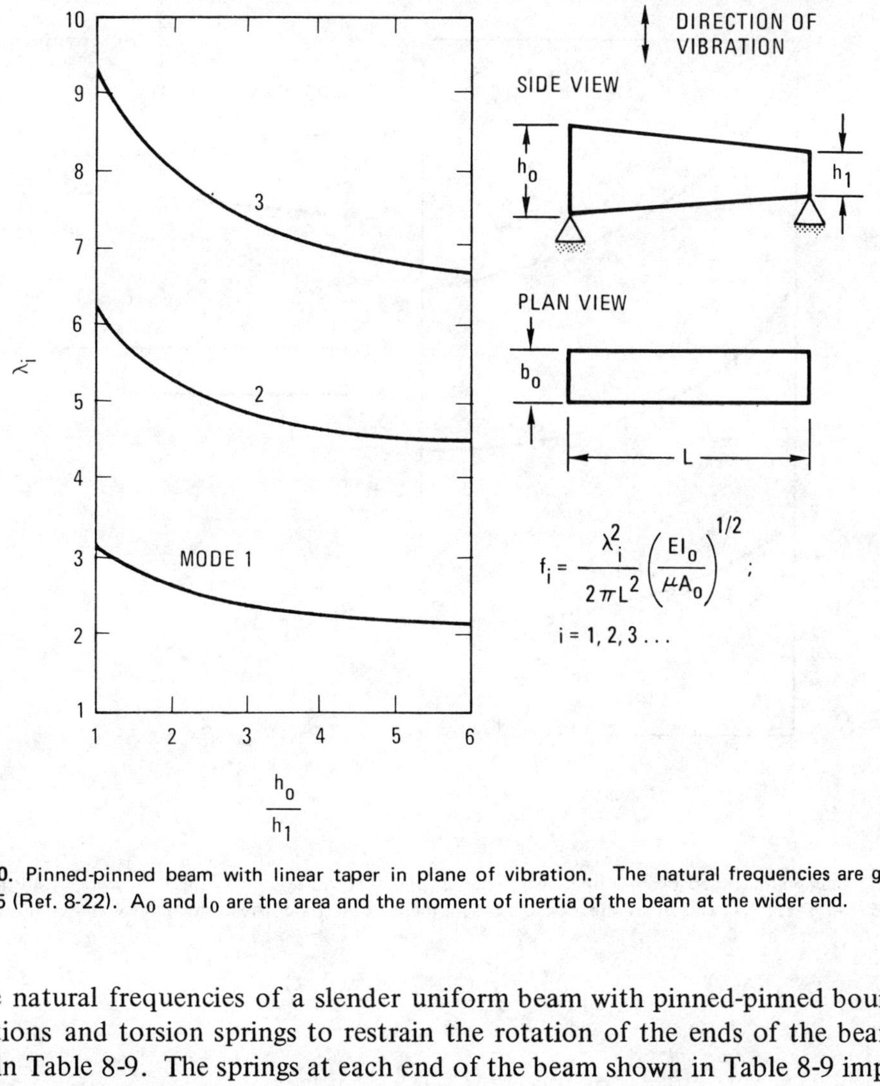

Fig. 8-20. Pinned-pinned beam with linear taper in plane of vibration. The natural frequencies are given by Eq. 8-25 (Ref. 8-22). A_0 and I_0 are the area and the moment of inertia of the beam at the wider end.

The natural frequencies of a slender uniform beam with pinned-pinned boundary conditions and torsion springs to restrain the rotation of the ends of the beam are given in Table 8-9. The springs at each end of the beam shown in Table 8-9 impose a moment on the ends of the beam. The moment applied to the ends of the beam is proportional to the local rotation of the beam:

$$\mathfrak{M}_1 = k_1 \frac{\partial Y(0, t)}{\partial x},$$

$$\mathfrak{M}_2 = k_2 \frac{\partial Y(L, t)}{\partial x},$$

where k_1 and k_2 are the spring constants (Table 6-1), Y(x, t) is the transverse displacement of the beam, and x is the spanwise coordinate measured along the span of the beam from left to right. k_1 and k_2 have units of force x length. The boundary conditions on the beam of Table 8-9 are:

$$Y(0, t) = Y(L, t) = 0,$$

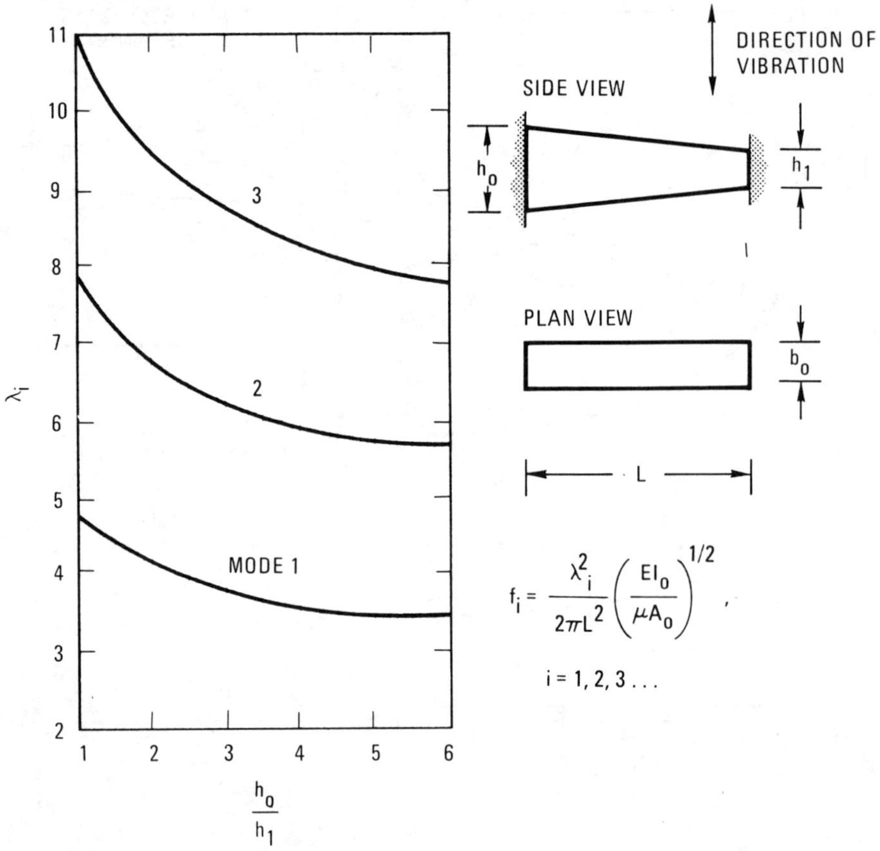

Fig. 8-21. Clamped-clamped beam with linear taper in plane of vibration. Natural frequencies are given by Eq. 8-25 (Ref. 8-22). A_0 and I_0 are the area and the moment of inertia of the beam at the wider end.

$$EI \frac{\partial^2 Y(0,t)}{\partial x^2} = k_1 \frac{\partial Y(0,t)}{\partial x},$$

$$EI \frac{\partial^2 Y(L,t)}{\partial x^2} = k_2 \frac{\partial Y(L,t)}{\partial x}.$$

Note that $k_1 = k_2 = 0$ and $k_1 = k_2 = \infty$ conform to the classical pinned-pinned and clamped-clamped boundary conditions.

The natural frequencies of a slender uniform beam with one end free and the other end pinned and restrained against rotation by a torsion spring are given in Table 8-10. The boundary conditions applied to this beam at the spring-supported end are:

$$Y(0,t) = 0,$$

$$EI \frac{\partial^2 Y(0,t)}{\partial x^2} = k \frac{\partial Y(0,t)}{\partial x},$$

where $x = 0$ is the pinned end of the beam. The mode shape of vibration of this beam is given in Ref. 8-26.

Table 8-9. Natural Frequencies of a Pinned-Pinned Beam with Unequal Torsion Springs at the Pinned Joints. [a]

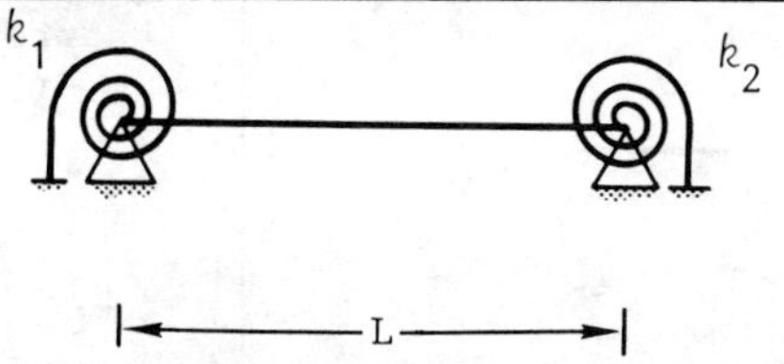

Natural Frequency (hertz),

$$f_i = \frac{\lambda_i^2}{2\pi L^2}\left(\frac{EI}{m}\right)^{1/2}$$

$\frac{k_1 L}{EI}$	$\frac{k_2 L}{EI}$	$\lambda_i = \lambda_i\left(\frac{k_1 L}{EI}, \frac{k_2 L}{EI}\right)$				
		i = 1	i = 2	i = 3	i = 4	i = 5
0.0	0	3.142	6.283	9.425	12.566	15.708
0.01	0	3.143	6.284	9.425	12.566	15.708
0.01	0.01	3.144	6.284	9.425	12.567	15.708
0.01	0.1	3.158	6.291	9.430	12.570	15.711
0.01	1.0	3.274	6.356	9.475	12.604	15.739
0.01	10	3.666	6.688	9.752	12.840	15.942
0.01	100	3.890	7.003	10.119	13.235	16.354
0.01	∞	3.927	7.069	10.210	13.352	16.493
0.1	0	3.157	6.291	9.430	12.570	15.711
0.1	0.01	3.158	6.291	9.430	12.570	15.711
0.1	0.1	3.172	6.298	9.435	12.574	15.714
0.1	1.0	3.285	6.363	9.480	12.608	15.741
0.1	10	3.678	6.694	9.756	12.843	15.945
0.1	100	3.902	7.010	10.123	13.239	16.356
0.1	∞	3.939	7.675	10.215	13.355	16.496
1.0	0	3.273	6.356	9.474	12.604	15.738
1.0	0.01	3.274	6.356	9.475	12.604	15.739
1.0	0.1	3.288	6.363	9.480	12.608	15.741
1.0	1.0	3.398	6.427	9.524	12.642	15.769
1.0	10	3.780	6.754	9.799	12.876	15.972
1.0	100	4.004	7.068	10.165	13.271	16.383
1.0	∞	4.041	7.133	10.256	13.387	16.522
10	0	3.664	6.687	9.751	12.839	15.942
10	0.01	3.666	6.688	9.752	12.840	15.942
10	0.1	3.678	6.694	9.756	12.843	15.945
10	1.0	3.780	6.754	9.799	12.876	15.972
10	10	4.155	7.068	10.065	13.105	16.171
10	100	4.390	7.383	10.429	13.497	16.579
10	∞	4.430	7.449	10.521	13.614	16.719
100	0	3.889	7.003	10.116	13.235	16.353
100	0.01	3.890	7.003	10.119	13.235	16.354
100	0.1	3.902	7.010	10.123	13.239	16.356
100	1.0	4.004	7.068	10.165	13.271	16.383
100	10	4.390	7.383	10.429	13.497	16.579
100	100	4.641	7.710	10.801	13.894	16.990
100	∞	4.685	7.781	10.897	14.014	17.133
∞	∞	4.712	7.854	10.995	14.137	17.278

[a] Ref. 8-25

Table 8-10. Natural Frequencies of a Pinned-Free Beam with a Torsion Spring at the Pinned Joint.[a]

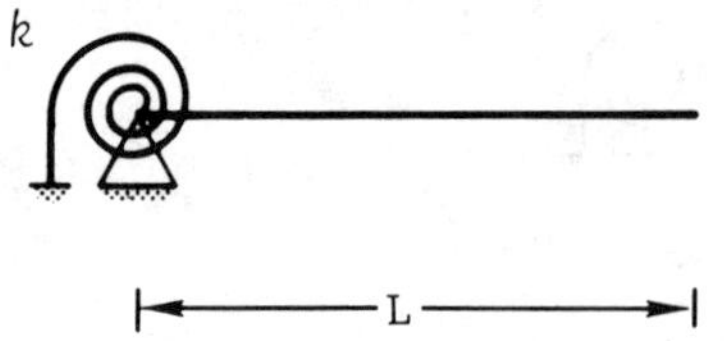

Natural Frequency (hertz),

$$f_i = \frac{\lambda_i^2}{2\pi L^2}\left(\frac{EI}{m}\right)^{1/2}$$

$\frac{kL}{EI}$	$\lambda_i = \lambda_i\ (kL/EI)$					
	i = 1	i = 2	i = 3	i = 4	i = 5	i = 6
0	0	3.927	7.069	10.21	13.35	16.49
0.01	0.4159	3.928	7.069	10.21	13.35	16.49
0.1	0.7357	3.938	7.076	10.22	13.36	16.50
1	1.248	4.031	7.134	10.26	13.39	16.52
10	1.723	4.400	7.451	10.52	13.61	16.72
100	1.857	4.650	7.783	10.90	14.01	17.13
∞	1.875	4.694	7.855	11.00	14.14	17.28

[a] Ref. 8-26.

Table 8-11 gives the fundamental natural frequency of a beam with pinned-free boundaries, a torsion spring at the pinned joint, and a translation spring attached to the free end. The boundary conditions of this beam are:

$$Y(0, t) = 0, \qquad EI\frac{\partial^2 Y(0, t)}{\partial x^2} = k\frac{\partial Y(0, t)}{\partial x},$$

$$\frac{\partial^2 Y(L, t)}{\partial x^2} = 0, \qquad EI\frac{\partial^3 Y(L, t)}{\partial x^3} = kY(L, t).$$

Table 8-12 gives the fundamental natural frequency of a beam with pinned-free boundaries, a torsion spring at the pinned end, and a concentrated mass attached to the free end. The boundary conditions of this beam are:

$$Y(0, t,) = 0, \qquad EI\frac{\partial^2 Y(0, t)}{\partial x^2} = k\frac{\partial Y(0, t)}{\partial x},$$

$$\frac{\partial^2 Y(L, t)}{\partial x^2} = 0, \qquad EI\frac{\partial^3 Y(L, t)}{\partial x^3} = M\frac{\partial^2 Y(L, t)}{\partial t^2}.$$

If the rotational inertia of the mass (M) shown in Table 8-12 had been included in the analysis, then the natural frequency of the beam would be lowered somewhat

Table 8-11. Fundamental Natural Frequency of a Beam with Torsion and Translational Spring Boundaries.[a]

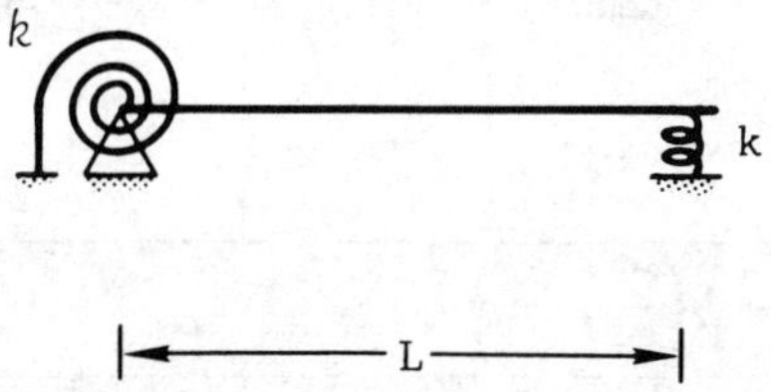

Fundamental Natural Frequency (hertz),

$$f_1 = \frac{\lambda_1^2}{2\pi L^2}\left(\frac{EI}{m}\right)^{1/2}$$

$$\lambda_1 = \lambda_1\left(\frac{kL^3}{EI}, \frac{kL}{EI}\right)$$

kL/EI	kL^3/EI							
	0	0.01	0.1	1	10	100	1000	∞
0	0	0.4162	0.7397	1.3098	2.2313	2.9886	3.1261	3.1416
0.01	0.4159	0.4948	0.7577	1.3134	2.2326	2.9901	3.1277	3.1432
0.1	0.7358	0.7541	0.8782	1.3437	2.2434	3.0030	3.1415	3.1572
1	1.2479	1.2520	1.2870	1.5358	2.3265	3.1084	3.2566	3.2733
10	1.7227	1.7245	1.7406	1.8793	2.5388	3.4412	3.6423	3.6646
100	1.8568	1.8583	1.8720	1.9939	2.6262	3.6133	3.8614	3.8892
1000	1.8732	1.8748	1.8882	2.0084	2.6376	3.6377	3.8940	3.9227
∞	1.8751	1.8766	1.8900	2.0100	2.6389	3.6405	3.8978	3.9266

(a) Ref. 8-27, copyright Academic Press (London) Ltd., reproduced with permission.

(Ref. 8-28). The natural frequencies of a tapered beam with the same boundary conditions as Table 8-12 are given in Ref. 8-30.

The fundamental natural frequency of a pinned-pinned beam with a central point mass and equal torsion spring at each end is given in Table 8-13. The boundary conditions on this beam are:

$$Y(0, t) = Y(L, t) = 0, \qquad \frac{\partial^2 Y(0, t)}{\partial x^2} = k\frac{\partial Y(0, t)}{\partial x},$$

$$\frac{\partial^2 Y(L, t)}{\partial x^2} = k\frac{\partial Y(L, t)}{\partial x}, \qquad EI\frac{\partial^3 Y(L/2, t)}{\partial x^3} = \frac{M}{2}\frac{\partial^2 Y(L/2, t)}{\partial t^2}.$$

The natural frequency of the second mode of this beam can be found in Ref. 8-29.

The natural frequencies of a cantilever beam with a rotational constraint along its span can be found in Ref. 8-42. The natural frequencies of other beams with spring-supported boundaries are given in Ref. 8-6.

Table 8-12. Fundamental Natural Frequency of a Beam with Torsion Spring and Point Mass.[a]

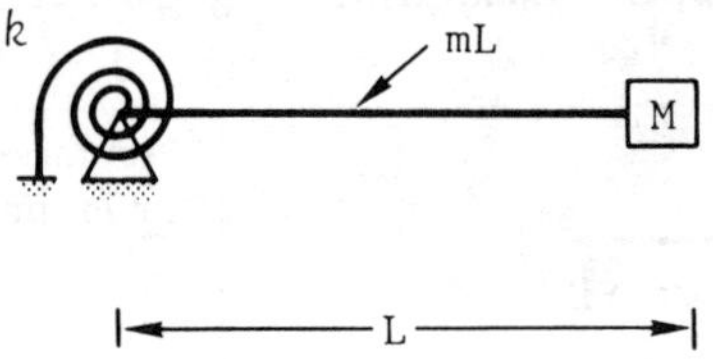

Fundamental Natural Frequency (hertz),

$$f_1 = \frac{\lambda_1^2}{2\pi L^2}\left(\frac{EI}{m}\right)^{1/2}$$

$\lambda_1 = \lambda_1\left(\frac{M}{mL}, \frac{kL}{EI}\right)$

$\frac{kL}{EI}$	$\frac{M}{mL} = 0$	0.01	0.1	1.0	10	100
0	0	0	0	0	0	0
0.01	0.4159	0.4129	0.3895	0.2941	0.1762	0.09983
0.1	0.7378	0.7303	0.6887	0.5194	0.3111	0.1762
1.0	1.2479	1.2381	1.1642	0.8705	0.5194	0.2941
10.0	1.7227	1.7071	1.5912	1.1642	0.6887	0.3895
100.0	1.8568	1.8388	1.7071	1.2381	0.7303	0.4129
∞	1.8751	1.861	1.723	1.247	0.735	0.416

(a) Ref. 8-28.

Table 8-13. Fundamental Natural Frequency of a Beam with Equal Torsion Springs at Each Pinned End and a Central Point Mass.[a]

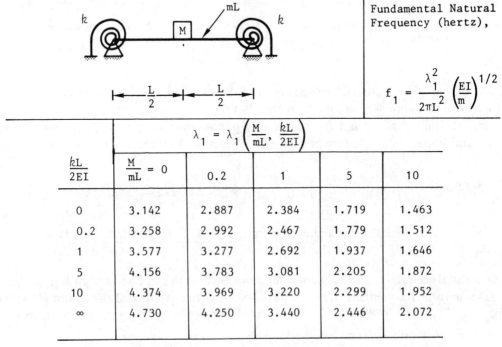

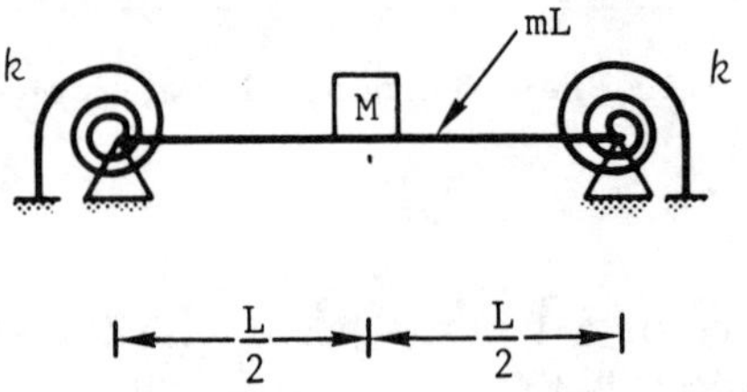

Fundamental Natural Frequency (hertz),

$$f_1 = \frac{\lambda_1^2}{2\pi L^2}\left(\frac{EI}{m}\right)^{1/2}$$

$\lambda_1 = \lambda_1\left(\frac{M}{mL}, \frac{kL}{2EI}\right)$

$\frac{kL}{2EI}$	$\frac{M}{mL} = 0$	0.2	1	5	10
0	3.142	2.887	2.384	1.719	1.463
0.2	3.258	2.992	2.467	1.779	1.512
1	3.577	3.277	2.692	1.937	1.646
5	4.156	3.783	3.081	2.205	1.872
10	4.374	3.969	3.220	2.299	1.952
∞	4.730	4.250	3.440	2.446	2.072

(a) Ref. 8-29.

8.2. TRANSVERSE VIBRATION OF SHEAR BEAMS

General Case. Beams can deform perpendicular to the beam axis either by flexure or by shearing, as is shown in Fig. 8-22. Flexural deformation ordinarily dominates the deformation of slender beams. Shear deformation can be important in short beams or in the higher modes of slender beams. A familiar form of shear deformation of a beam is the shaking of a rectangular piece of gelatin dessert. The shear beams considered in this section are composed of homogeneous isotropic materials and all flexural deformations are neglected.

The shear deformations in a homogeneous beam are the result of shearing of elements in the cross section under the transverse shear stress, σ_{xy} (y is the coordinate perpendicular to the beam axis, x is the coordinate parallel to the beam axis). The shear stress, σ_{xy}, is not uniform across the cross section during shear deformations. The shear stress generally has a parabolic distribution over simple closed cross sections with the maximum shear stress at the centroid of the cross section and zero shear stress at the edges of the cross section. K is the ratio of the average shear strain over the cross section to the maximum shear strain at the centroid. Cowper's values of K (Ref. 8-31) are listed in Table 8-14. The average value of shear stress over the cross section is

$$\sigma_{xy}\Big|_{\text{average}} = KG\,\frac{\partial Y}{\partial x}, \tag{8-26}$$

the maximum shear stress at the centroid is

$$\sigma_{xy}\Big|_{\text{centroid}} = G\,\frac{\partial Y}{\partial x}, \tag{8-27}$$

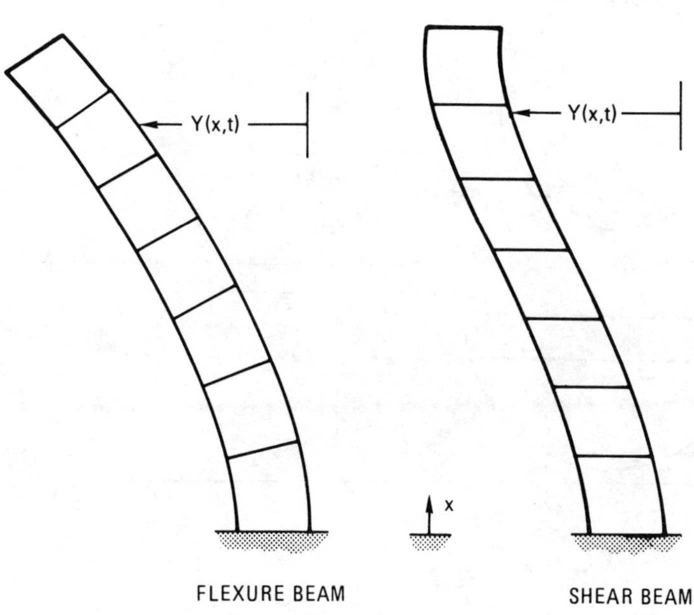

Fig. 8-22. First mode of a clamped-free flexure beam (frame 3 of Table 8-1) and a fixed-free shear beam (frame 2 of Table 8-15).

Table 8-14. Shear Coefficients.[a]

Notation: ν = Poisson's ratio,– – –axis perpendicular to applied shear load

Section	Shear Coefficient, K
1. Circle 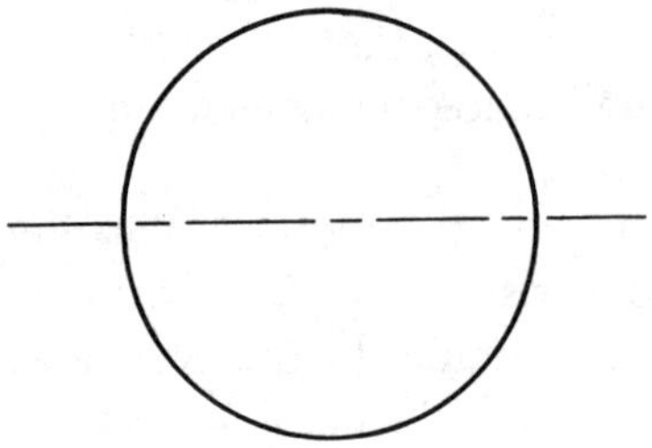	$\frac{6(1+\nu)}{7+6\nu}$
2. Hollow Circle 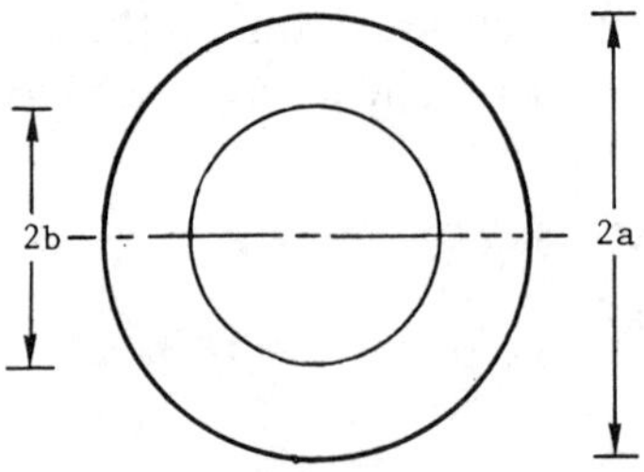	$\frac{6(1+\nu)(1+m^2)^2}{(7+6\nu)(1+m^2)^2+(20+12\nu)\,m^2}$ where $m = b/a$
3. Thin Hollow Circle 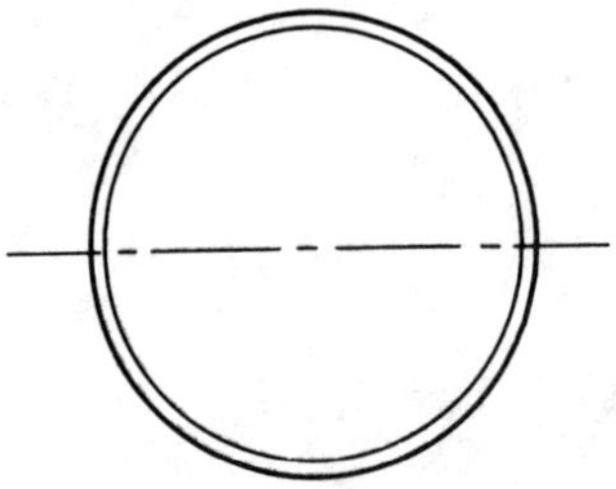	$\frac{2(1+\nu)}{4+3\nu}$
4. Semicircle 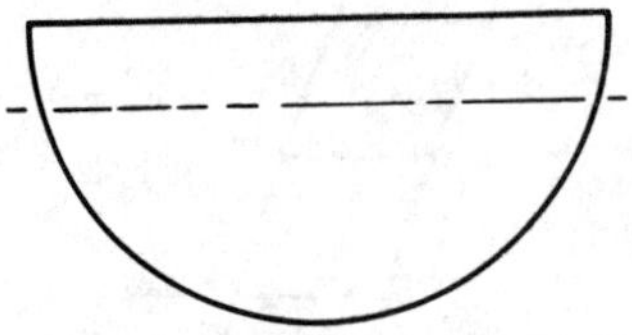	$\frac{1+\nu}{1.305+1.273\nu}$

Table 8-14. Shear Coefficients. (*Continued*)

Section	Shear Coefficient, K
5. Ellipse 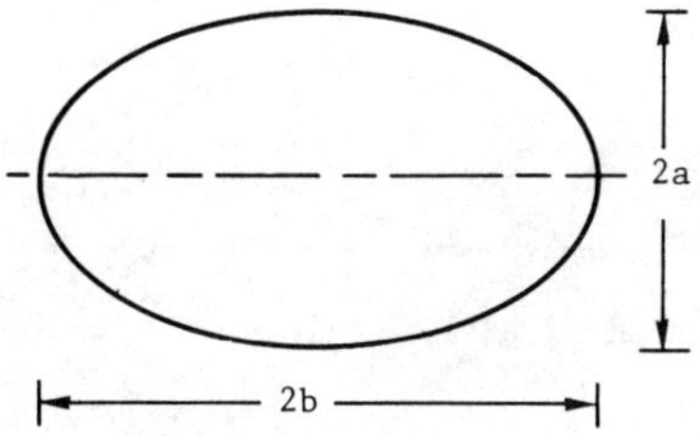	$\dfrac{12(1+\nu)\,a^2(3a^2+b^2)}{(40+37\nu)\,a^4+(16+10\nu)\,a^2b^2+\nu b^4}$
6. Rectangle 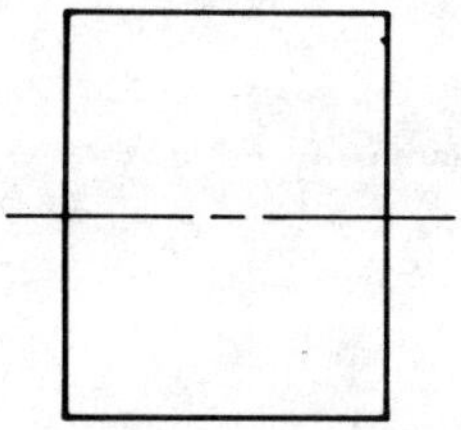	$\dfrac{10(1+\nu)}{12+11\nu}$
7. Thin-Walled Square 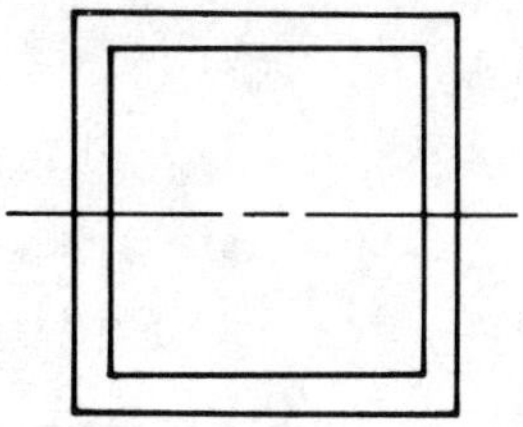	$\dfrac{20(1+\nu)}{48+39\nu}$
8. Thin-Walled Box 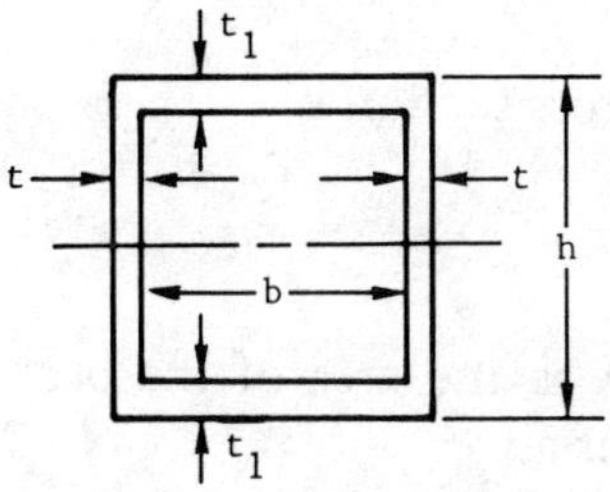	$\dfrac{10(1+\nu)(1+3m^2)}{F(m,n)}$ where $F(m,n) = 12 + 72m + 150m^2 + 90m^3$ $+ \nu(11 + 66m + 135m^2 + 90m^3)$ $+ 10n^2[(3+\nu)m + 3m^2]$, $m = bt_1/ht$, $n = b/h$

Table 8-14. Shear Coefficients. (*Continued*)

Section	Shear Coefficient, K
9. Thin-Walled I-Section 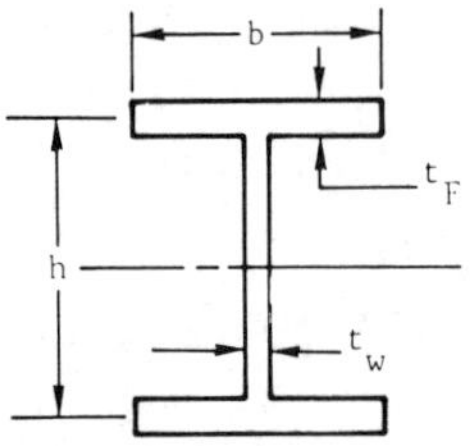	$\dfrac{10(1+\nu)(1+3m)^2}{F(m,n)}$, where $F(m,n) = 12 + 72m + 150m^2 + 90m^3$ $+ \nu(11 + 66m + 135m^2 + 90m^3)$ $+ 30n^2(m + m^2) + 5\nu n^2(8m + 9m^2)$, $m = 2bt_F/(ht_w)$, $n = b/h$
10. Spar and Web 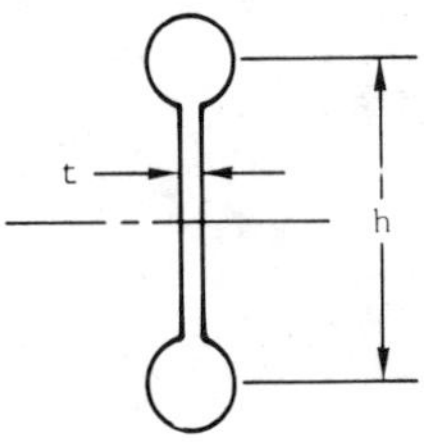	$\dfrac{10(1+\nu)(1+3m)^2}{(12 + 72m + 150m^2 + 90m^3) + \nu(11 + 66m + 135m^2 + 90m^3)}$, where $m = 2A_S/(ht)$, A_S = area of one spar
11. Thin-Walled T-Section 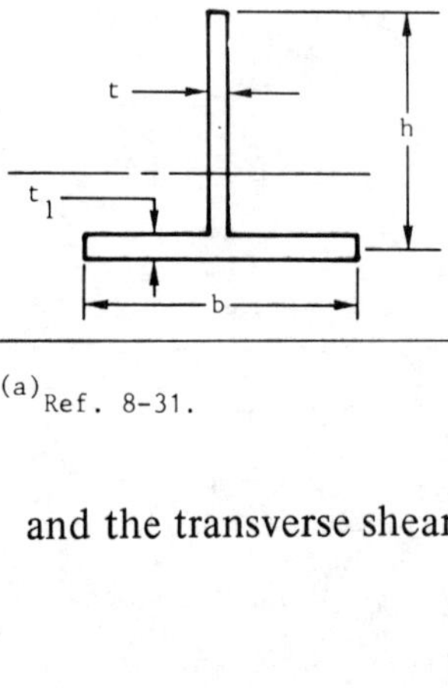	$\dfrac{10(1+\nu)(1+4m)^2}{F(m,n)}$, where $F(m,n) = 12 + 96m + 276m^2 + 192m^3$ $+ \nu(11 + 88m + 248m^2 + 216m^3)$ $+ 30n^2(m + m^2) + 10\nu n^2(4m + 5m^2 + m^3)$ $m = bt_1/(ht)$, $n = b/h$

[a] Ref. 8-31.

and the transverse shear force, Q, supported by the cross section is

$$Q = KAG \frac{\partial Y}{\partial x}. \tag{8-28}$$

Y(x, t) is the transverse deflection of the beam, A is the area of the beam cross section, and G is the shear modulus of the beam material:

$$G = \frac{E}{2(1+\nu)},$$

where E is the elastic modulus and ν is Poisson's ratio for a shear beam composed of a homogeneous isotropic material. The remaining stresses, σ_{xx}, σ_{yy}, etc., are generally neglected in the approximate theory. The shear strains are:

$$\epsilon_{xy} = \frac{\sigma_{xy}}{G}, \tag{8-29}$$

where σ_{xy} is given by Eq. 8-26 or 8-27.

Uniform Shear Beams. Table 8-15 gives the natural frequencies of uniform shear beams, that is, beams whose flexural deformation has been neglected. The natural frequencies of these beams are proportional to $1/L$ rather than $1/L^2$ as for flexural beams (Table 8-1), and the natural frequencies increase linearly with mode number. The boundary conditions, applied at the spanwise point of the boundary of the beams in Table 8-15, are:

Fixed: $Y = 0$ (zero displacement),

Free: $\frac{\partial Y}{\partial x} = 0$ (zero shear stress, Eq. 8-27),

Spring: $kY = KAG \frac{\partial Y}{\partial x}$,

Mass: $-M \frac{\partial^2 Y}{\partial t^2} = KAG \frac{\partial Y}{\partial x}$.

Only one boundary condition on the beam displacement perpendicular to the beam axis $Y(x, t)$ is required at a shear beam boundary, since the equation of motion for shear beams contains only second-order derivatives. (The equation of motion for flexural beams is fourth order. Two boundary conditions are required at each flexural beam boundary.) A is the area of cross section of the beam. k is the translation spring constant, G is the shear modulus, and M is the mass of the point mass. K is the shear coefficient (Table 8-14).

Application to Flexure Beam Theory. The transverse deformation of real beams is the sum of flexure and shear deformations. Shear deformations are usually neglected for analysis of slender beams. However, the shear deformations can become important in the higher modes of beams and in special structures. Figure 8-25 shows the importance of shear deformations on the natural frequencies of a pinned-pinned beam. Figure 8-25 and comparison of Tables 8-1 and 8-15 suggest that the ratio of flexure deformation to shear deformation of uniform beams with simple closed cross sections is approximately:

$$\frac{\text{shear deformation}}{\text{flexure deformation}} = \frac{ih}{L},$$

where i is the mode number, h is a cross-sectional diameter (or typical dimension of the cross section), and L is the beam span. If this ratio approaches 1, then shear deformation will probably have a substantial effect on the natural frequency of the beam and the natural frequency will be substantially lower than that predicted by flexure theory. Generally, numerical solutions are required to incorporate both shear and flexure deformation in the prediction of natural frequencies of beams. However, in some cases the natural frequency of a beam with comparable shear and flexure deformations can be estimated using the Southwell-Dunkerley approximation:

$$\frac{1}{f^2} = \frac{1}{f_F^2} + \frac{1}{f_S^2}, \tag{8-30}$$

Table 8-15. Uniform Shear Beams.

Notation: k = spring constant (Table 6-1); x = spanwise coordinate; y = coordinate in direction of vibration; $\tilde{y}$ = mode shape for deformation in y direction; A = area of cross section; G = shear modulus $\{G = E/[2(1+\nu)]$, where E = modulus of elasticity, ν = Poisson's ratio$\}$; K = shear coefficient (Table 8-14); L = span of beam; M = mass; μ = mass density of beam material; see Table 3-1 for consistent sets of units.

Natural Frequency (hertz), $f_i = \frac{\lambda_i}{2\pi L}\left(\frac{KG}{\mu}\right)^{1/2}$; $i=1,2,3\cdots$

Description	λ_i or Transcendental Equation for λ	Mode Shape, $\tilde{y}_i\left(\frac{x}{L}\right)$
1. Free-Free 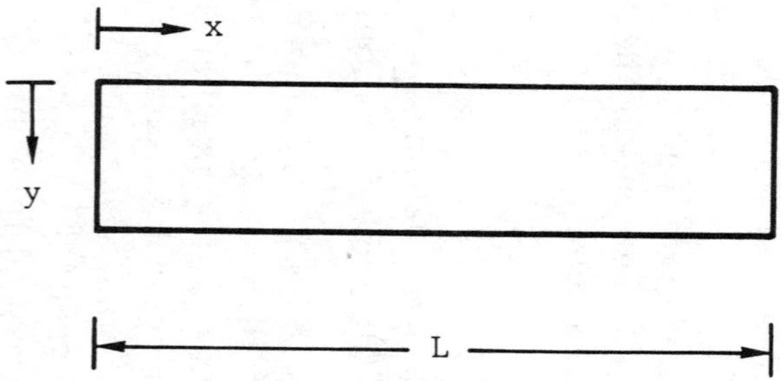	$i\pi$; $i=1,2,3\cdots$	$\cos\frac{i\pi x}{L}$; $i=1,2,3\cdots$
2. Fixed-Free 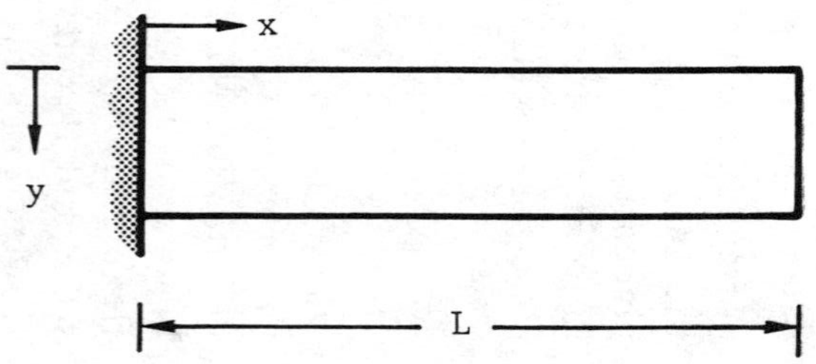	$\frac{(2i-1)\pi}{2}$; $i=1,2,3\cdots$	$\sin\frac{\pi(2i-1)x}{2L}$; $i=1,2,3\cdots$

3. Fixed-Fixed 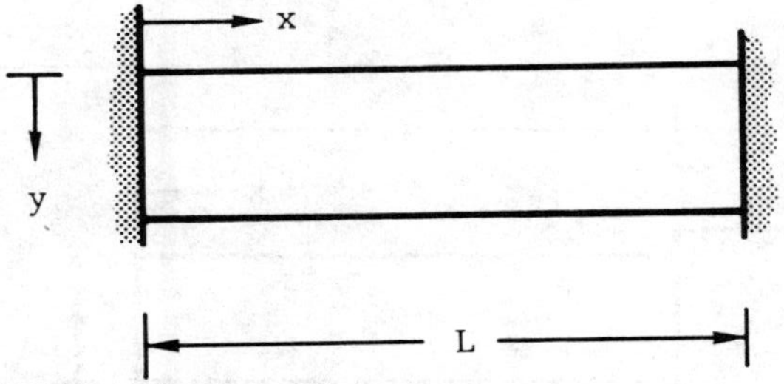	$i\pi$; $i=1,2,3\cdots$	$\sin \frac{i\pi x}{L}$; $i=1,2,3\cdots$
4. Spring-Free 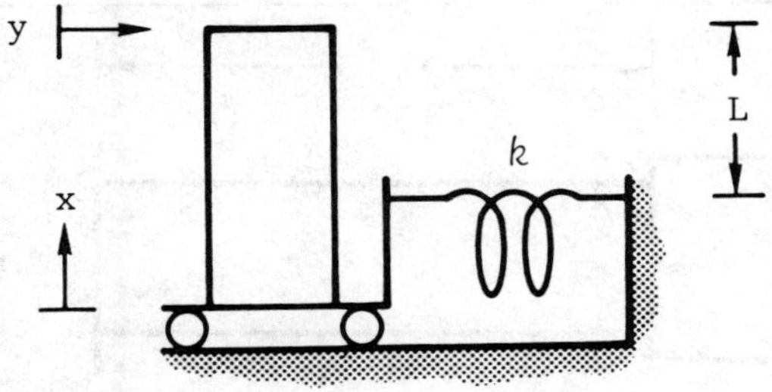	$\tan \lambda = \left(\frac{kL}{KAG}\right)\frac{1}{\lambda}$ See Fig. 8-23 for λ_1	$\cot \lambda_i \cos \frac{\lambda_i x}{L}$ $+ \sin \frac{\lambda_i x}{L}$; $i=1,2,3\cdots$
5. Fixed-Spring 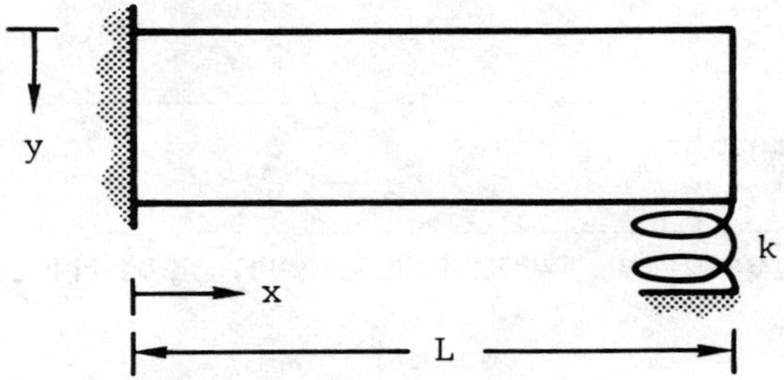	$\cot \lambda = -\left(\frac{kL}{KAG}\right)\frac{1}{\lambda}$ See Fig. 8-24 for λ_1	$\sin \frac{\lambda_i x}{L}$; $i=1,2,3\cdots$

Table 8-15. Uniform Shear Beams. *(Continued)*

Natural Frequency (hertz), $f_i = \frac{\lambda_i}{2\pi L}\left(\frac{KG}{\mu}\right)^{1/2}$; $i=1,2,3\cdots$

Description	λ_i or Transcendental Equation for λ	Mode Shape, $\tilde{y}_i\left(\frac{x}{L}\right)$
6. Fixed-Mass 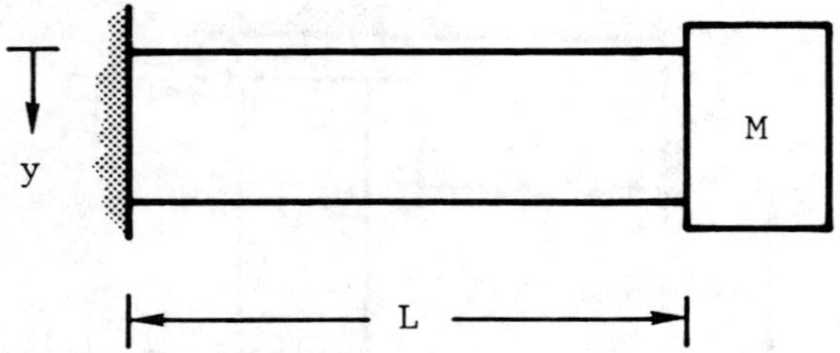	$\cot\lambda = \left(\frac{M}{\mu AL}\right)\lambda$ See Fig. 8-23 for λ_1	$\sin\frac{\lambda_i x}{L}$; $i=1,2,3\cdots$
7. Free-Mass 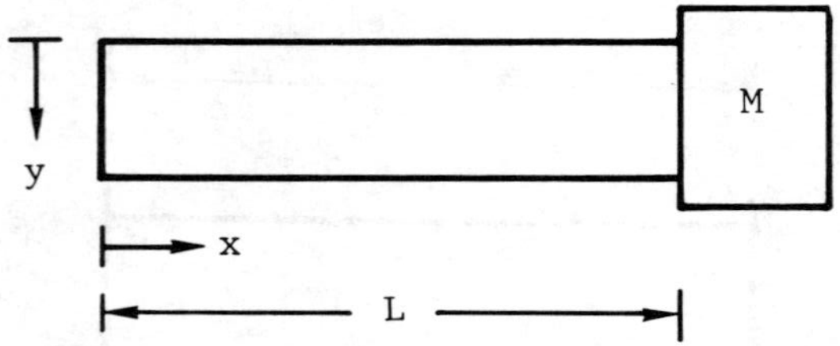	$\tan\lambda = -\left(\frac{M}{\mu AL}\right)\lambda$ See Fig. 8-24 for λ_i	$\cos\frac{\lambda_i x}{L}$; $i=1,2,3\cdots$

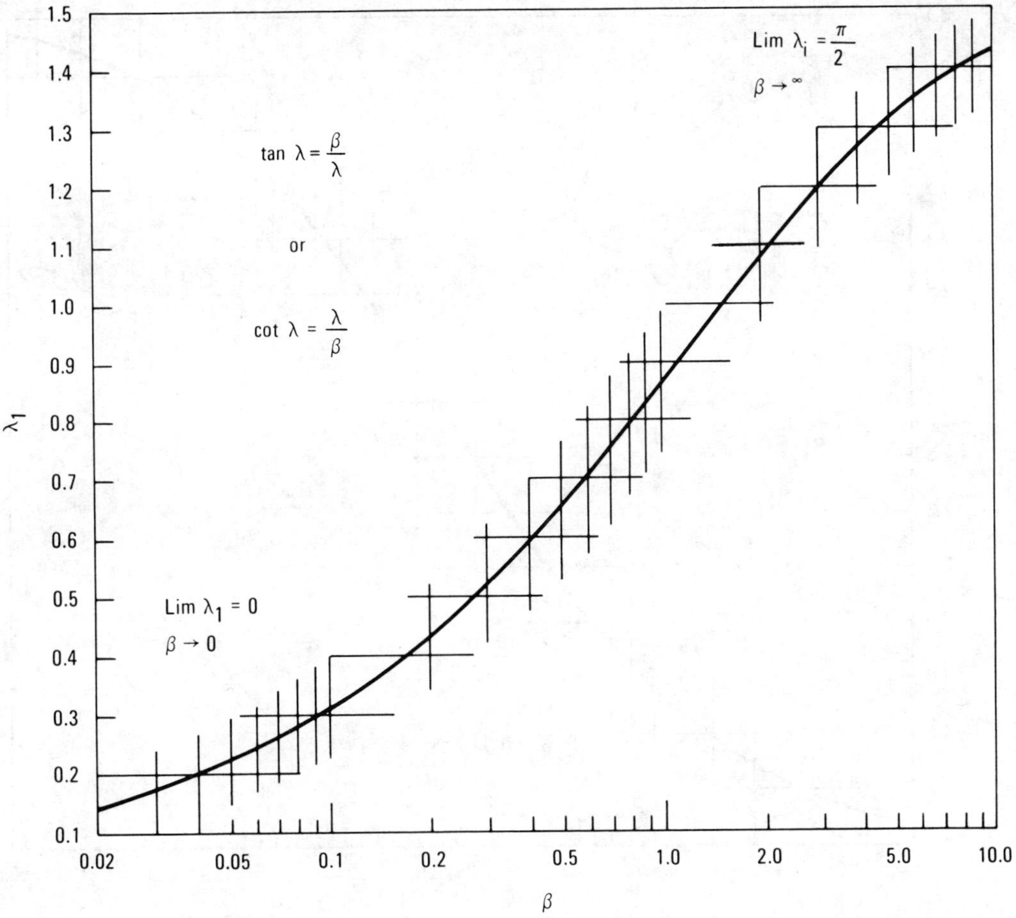

Fig. 8-23. First solution to transcendental equation of frames 4 and 6 of Tables 8-15, 8-16, and 8-19.

where f is the fundamental natural frequency of the beam, f_F is the fundamental natural frequency predicted by flexure theory, and f_S is the fundamental natural frequency as predicted by shear beam theory (Ref. 8-32).

Application to Multistory Buildings. Shear deformation often dominates the deformation of multistory buildings. The fundamental natural frequencies in hertz of many multistory buildings can be approximated by (Refs. 8-33, 8-34):

$$f_1 = c\frac{D^{1/2}}{H}, \tag{8-31}$$

where

$$c = \begin{cases} 20 \text{ ft}^{1/2}/\text{sec} \\ 11 \text{ m}^{1/2}/\text{sec} \end{cases}.$$

D is the building width in the direction of vibration, and H is the building height. A second, simpler formula for the fundamental natural frequency of multistory build-

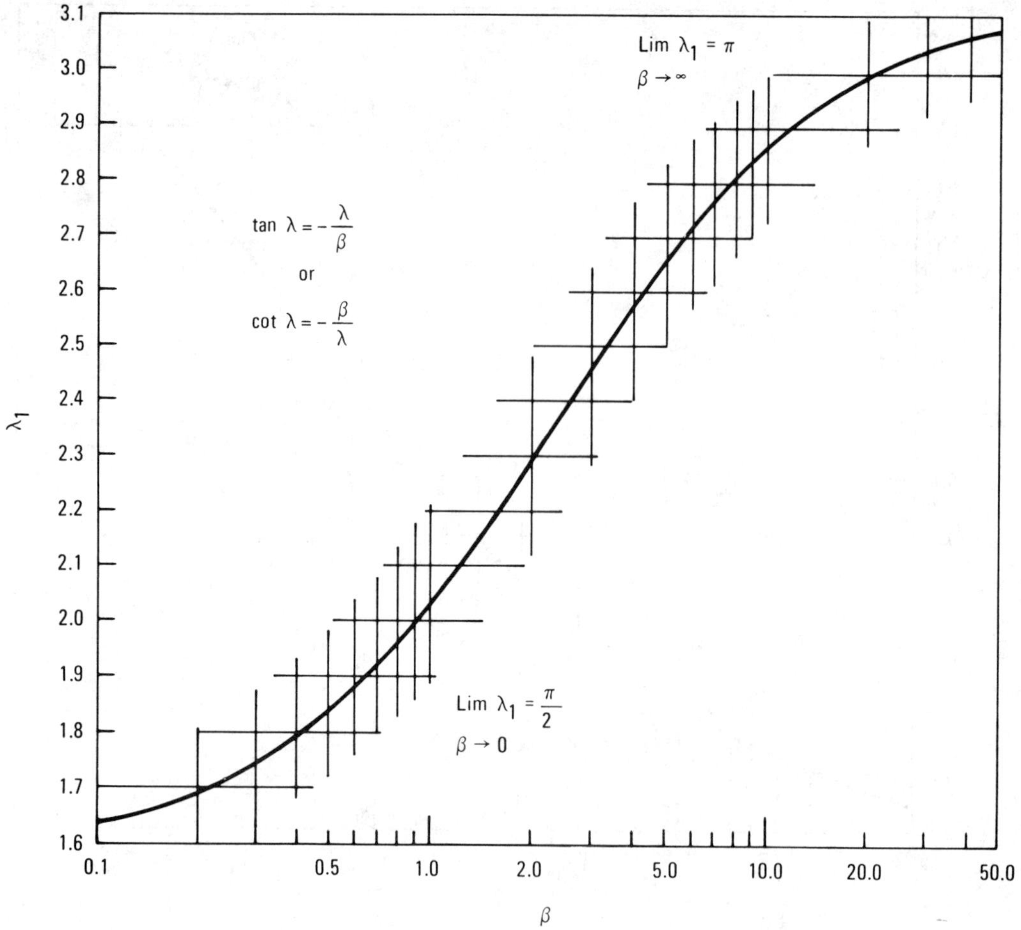

Fig. 8-24. First solution to transcendental equation of frames 5 and 7 of Tables 8-15, 8-16, and 8-19.

ings is $f_1 = 10/N$, where N is the number of stories. Of course, variation in the construction of a building will affect these results, but the empirical formula suggests that a shear beam model ($f \propto 1/H$) rather than a flexure beam model ($f \propto 1/H^2$) is appropriate for buildings. Spring-mass models of buildings are given in Table 6-2.

Shear Deformation and Rotary Inertia. Ordinarily, a discussion of shear deformation includes the effect of rotary inertia. Rotary inertia is the inertia associated with local rotation of the beam cross section during flexural deformation. Rotary inertia is neglected in flexure and shear beam theory. Timoshenko's results suggest that the effect of rotary inertia on natural frequency is generally less than that of shear deformation (Ref. 8-17, pp. 432–435). Both effects tend to lower the natural frequencies of beams from that predicted by flexure beam theory.

It is not generally possible to obtain closed form solutions for beams when shear deformation, flexural deformation, and rotary inertia are considered. However, Goodman and Sutherland (Ref. 8-35), using Timoshenko's formulation, have found

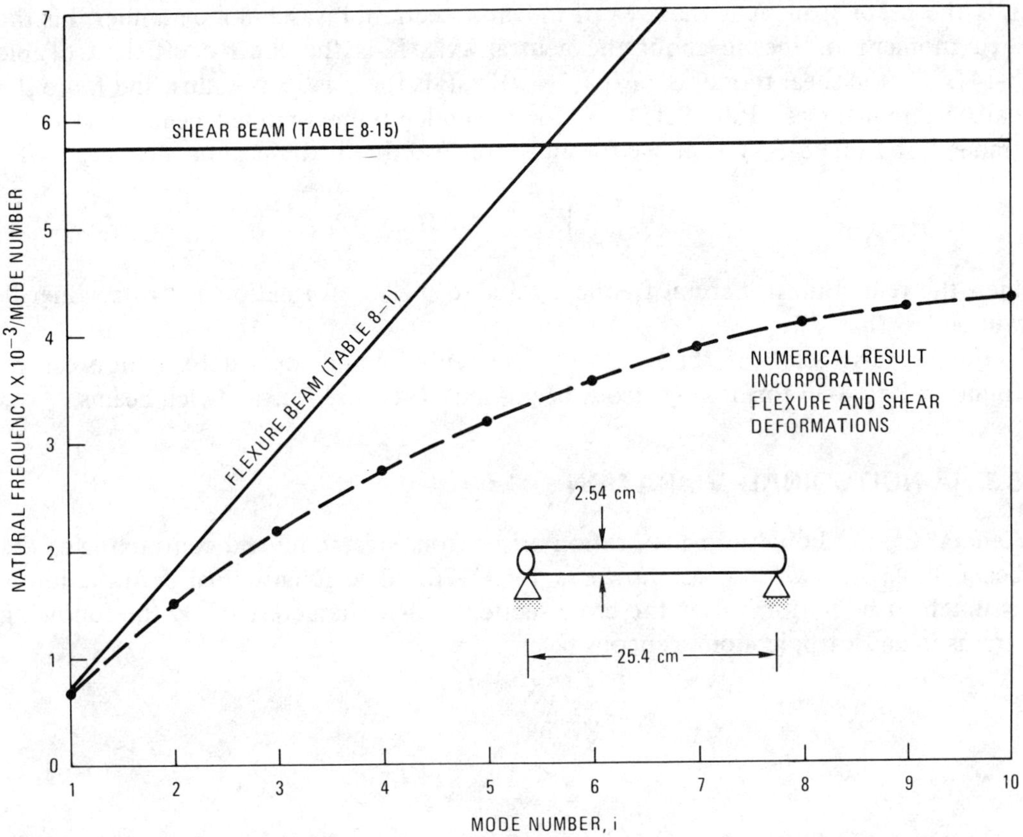

Fig. 8-25. Comparison of numerical result for pinned-pinned steel beam with flexure and shear beam theories. Rotary inertia is neglected.

that the natural frequencies of a pinned-pinned beam, including the effects of rotary inertia and shear deformation, are:

$$\frac{f_i\Big|_{\substack{\text{rotary inertia}\\ \text{+ shear deformation}}}}{f_i\Big|_{\substack{\text{flexure beam}\\ \text{theory (Table 8-1)}}}} = \frac{L}{i}\left(\frac{A}{I}\right)^{1/2} [B - (B^2 - D)^{1/2}]^{1/2}$$

$$= 1 - \frac{i^2\pi^2}{2L^2}\left(\frac{I}{A}\right)\left(1 + \frac{E}{KG}\right), \quad \text{for} \quad \frac{i}{L}\left(\frac{I}{A}\right)^{1/2} << 1,$$

$$i = 1, 2, 3, \ldots \qquad (8\text{-}32)$$

where

$$B = \frac{1}{2\pi^2} + \frac{D}{2}\left[\pi^2 + \left(\frac{L}{i}\right)^2 \left(\frac{A}{I}\right)\right],$$

$$D = \frac{KG}{\pi^4 E}.$$

L is the beam span, A is the area of the cross section, i is the mode number, I is the area moment of inertia about the neutral axis, K is the shear coefficient (Table 8-14), G is the shear modulus $\{E/[2(1+\nu)]\}$, E is the elastic modulus, and f_i are the natural frequencies (Ref. 8-35). f_i for a slender pinned-pinned beam is given in frame 4 of Table 8-1. The above formula indicates that if the beam is slender,

$$\left(\frac{I}{A}\right)^{1/2}\frac{i}{L}<0.05,$$

then the reduction in natural frequency due to shear deformation and rotary inertia will be less than 5%.

Further discussion of the effect of rotary inertia and shear deformation can be found in Ref. 8-36 for homogeneous beams and Ref. 8-37 for sandwich beams.

8.3. LONGITUDINAL VIBRATION

General Case. Longitudinal vibration arises from stretching and contracting of the beam along its own axis as shown in Fig. 8-26. The longitudinal deformation is assumed to be uniform over the cross section and is characterized by the following strains in an isotropic, homogeneous beam:

$$\epsilon_x = \frac{\partial X}{\partial x},$$

$$\epsilon_y = -\nu\frac{\partial X}{\partial x},$$

$$\epsilon_z = -\nu\frac{\partial X}{\partial x},$$

$$\epsilon_{xy} = \epsilon_{xz} = \epsilon_{yz} = 0. \tag{8-33}$$

ϵ_x, ϵ_y, and ϵ_z are the normal strains; ϵ_{xy}, ϵ_{xz}, and ϵ_{yz} are the shear strains. X(x, t) is the deformation of the beam along its own axis and ν is Poisson's ratio. The lateral contraction of the beam corresponding to ϵ_y and ϵ_z is due to the Poisson's ratio

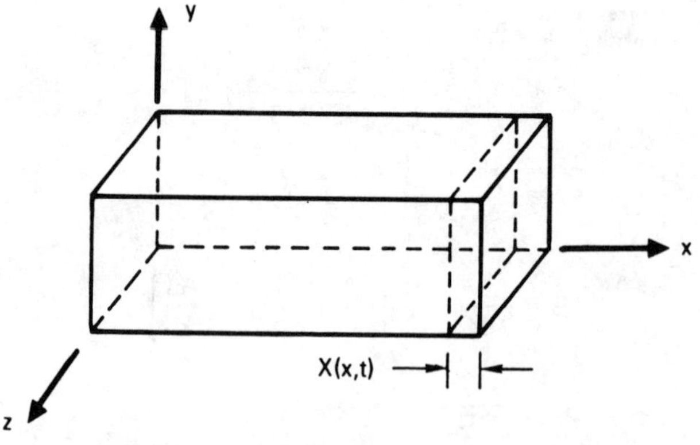

Fig. 8-26. Longitudinal deformation of a beam.

Table 8-16. Longitudinal Vibration of Uniform Beams.

Notation: k = spring constant (Table 6-1); x = spanwise length; $\tilde{x}$ = mode shape for longitudinal deformation; A = area of cross section; E = modulus of elasticity; L = span of beam; M = mass; μ = mass density of beam material; see Table 3-1 for consistent sets of units.

Natural Frequency (hertz), $f_i = \frac{\lambda_i}{2\pi L}\left(\frac{E}{\mu}\right)^{1/2}$; $i=1,2,3\cdots$

Description	λ_i or Transcendental Equation for λ	Mode Shape, $\tilde{x}_i\left(\frac{x}{L}\right)$
1. Free-Free 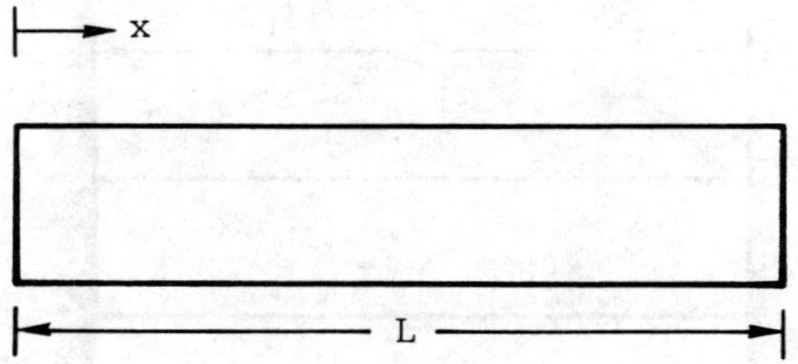	$i\pi$, $i=1,2,3\cdots$	$\cos\frac{i\pi x}{L}$; $i=1,2,3\cdots$
2. Fixed-Free 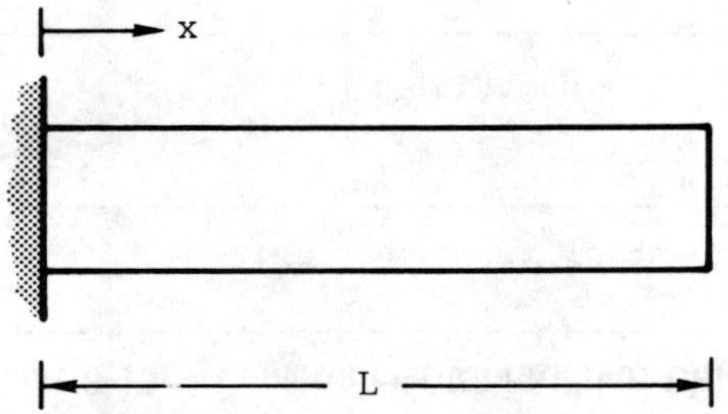	$\frac{(2i-1)\pi}{2}$; $i=1,2,3\cdots$	$\sin\frac{\pi(2i-1)x}{2L}$; $i=1,2,3\cdots$

Table 8-16. Longitudinal Vibration of Uniform Beams. *(Continued)*

Natural Frequency (hertz), $f_i = \frac{\lambda_i}{2\pi L}\left(\frac{E}{\mu}\right)^{1/2}$; i=1,2,3···

Description	λ_i or Transcendental Equation for λ	Mode Shape, $\tilde{x}_i\left(\frac{x}{L}\right)$
3. Fixed-Fixed 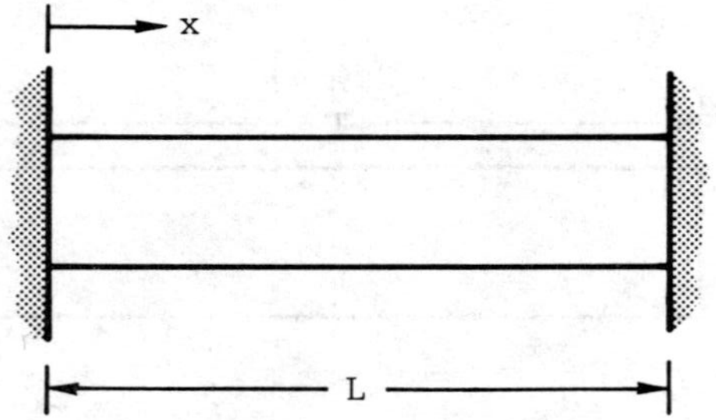	$i\pi$; i=1,2,3···	$\sin\frac{i\pi x}{L}$; i=1,2,3···
4. Spring-Free 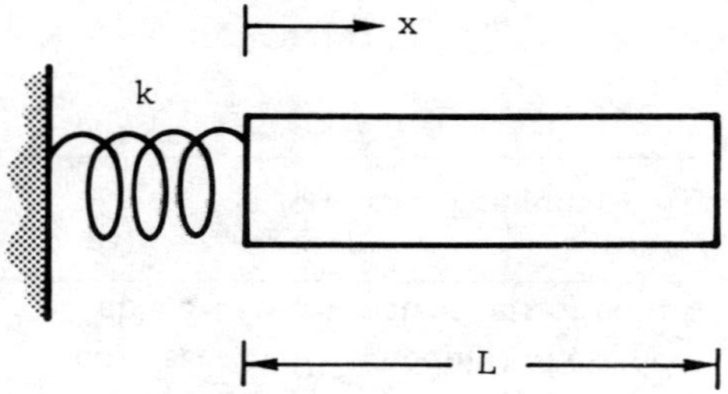	$\tan\lambda = \left(\frac{kL}{AE}\right)\frac{1}{\lambda}$ See Fig. 8-23 for λ_1	$\cot\lambda_i \cos\frac{\lambda_i x}{L}$ $+ \sin\frac{\lambda_i x}{L}$; i=1,2,3···

5. Fixed-Spring 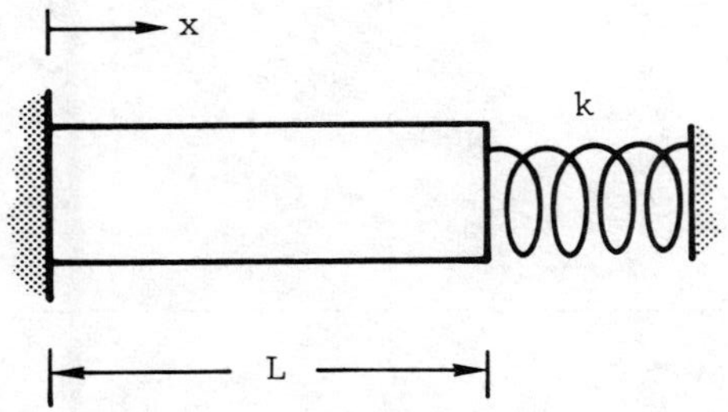	$\cot \lambda = -\left(\frac{kL}{AE}\right)\frac{1}{\lambda}$ See Fig. 8-24 for λ_i (spring is massless)	$\sin \frac{\lambda_i x}{L}$; $i=1,2,3\cdots$
6. Fixed-Mass 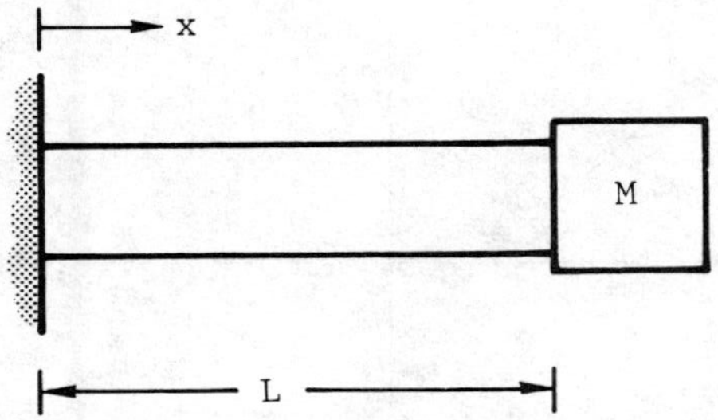	$\cot \lambda = \left(\frac{M}{\mu AL}\right)\lambda$ See Fig. 8-23 for λ_1	$\sin \frac{\lambda_i x}{L}$; $i=1,2,3\cdots$
7. Free-Mass 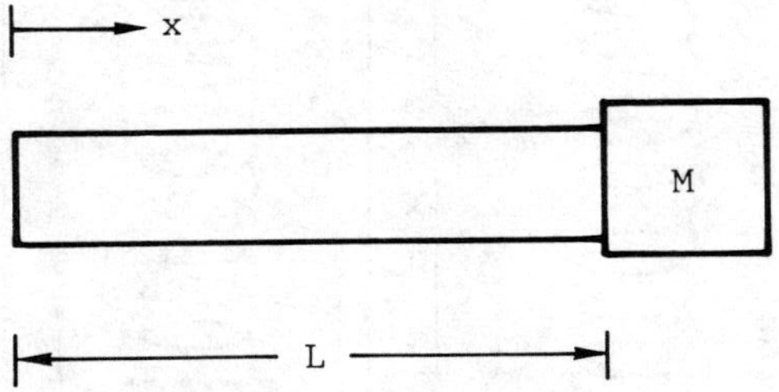	$\tan \lambda = -\left(\frac{M}{\mu AL}\right)\lambda$ See Fig. 8-24 for λ_1	$\cos \frac{\lambda_i x}{L}$; $i=1,2,3\cdots$

Table 8-17. Longitudinal Vibration of Massive Helical Springs.

Notation: k = spring constant (Table 6-1); x = spanwise length; $\tilde{x}_i$ = mode shape for longitudinal deformation; m = mass per unit span of spring; L = span of spring; M = mass; see Table 3-1 for consistent sets of units

Natural Frequency (hertz), $f_i = \frac{\lambda_i}{2\pi}\left(\frac{k}{mL}\right)^{1/2}$; $i=1,2,3\cdots$

Description	λ_i or Transcendental Equation for λ	Mode Shape, $\tilde{x}_i\left(\frac{x}{L}\right)$
1. Free 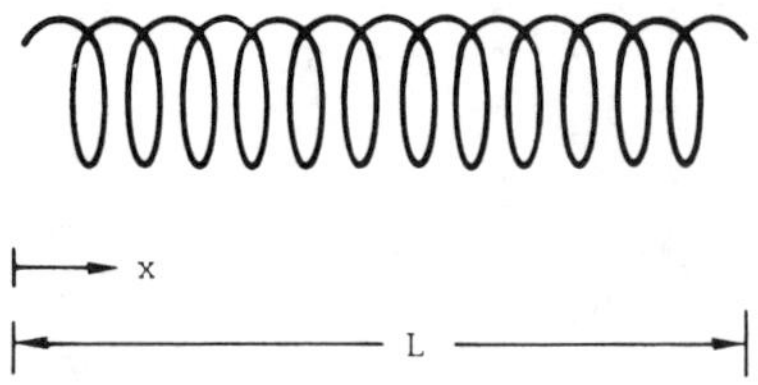	$i\pi;\ i=1,2,3\cdots$	$\cos\frac{i\pi x}{L};\ i=1,2,3\cdots$
2. Fixed-Free 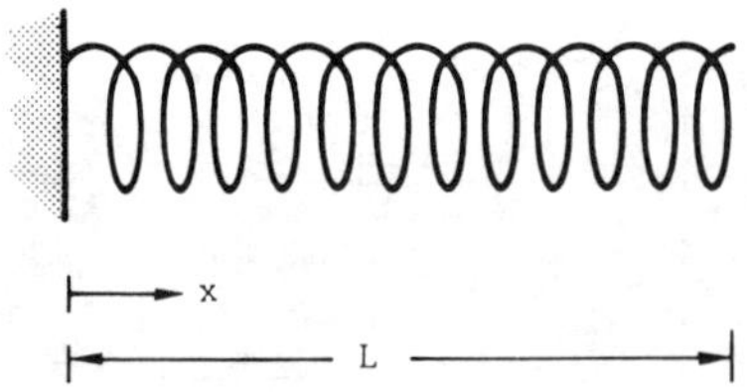	$\frac{(2i-1)\pi}{2};\ i=1,2,3\cdots$	$\sin\frac{(2i-1)\pi x}{2L};\ i=1,2,3\cdots$
3. Fixed-Fixed 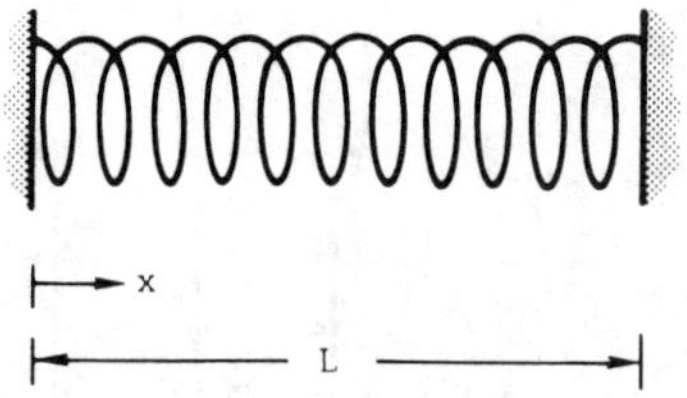	$i\pi;\ i=1,2,3\cdots$	$\sin\frac{i\pi x}{L};\ i=1,2,3\cdots$
4. Fixed-Mass 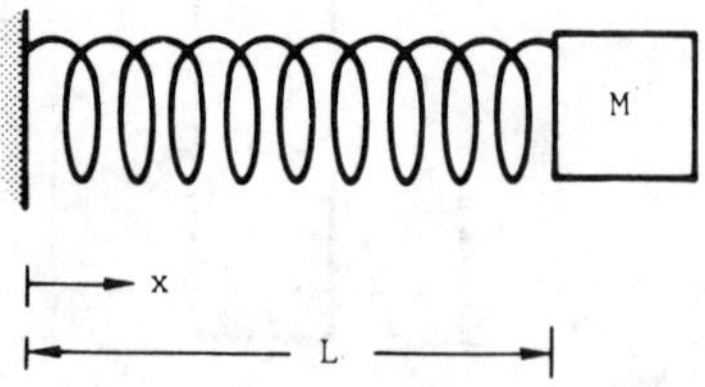	$\cot\lambda = \left(\frac{M}{mL}\right)\lambda$ See Fig. 8-23 for λ_1 See frame 1 of Table 8-8 for approximate solution	$\sin\frac{\lambda_i x}{L};\ i=1,2,3\cdots$
5. Free-Mass 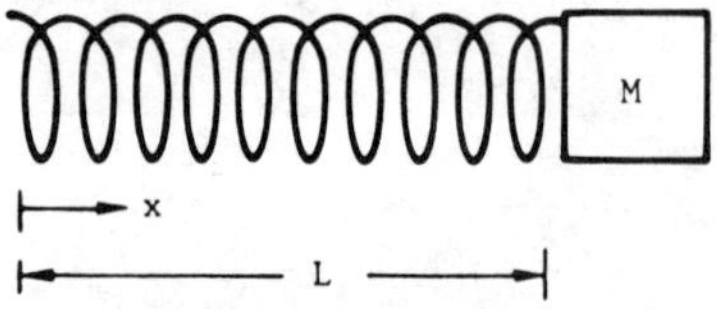	$\tan\lambda = -\left(\frac{M}{mL}\right)\lambda$ See Fig. 8-24 for λ_1	$\cos\frac{\lambda_i x}{L};\ i=1,2,3\cdots$

effect. These strains correspond to the following stresses in an isotropic, homogeneous beam:

$$\sigma_{xx} = E\frac{\partial X}{\partial x},$$

$$\sigma_{yy} = 0,$$

$$\sigma_{zz} = 0,$$

$$\sigma_{xy} = \sigma_{xz} = \sigma_{yz} = 0. \tag{8-34}$$

σ_{xx} is the normal stress along the beam axis. All other stresses are zero.

Common boundary conditions which can be applied to these beams are:

$$\text{Free:} \quad \frac{\partial X}{\partial x} = 0,$$

$$\text{Fixed:} \quad X = 0,$$

$$\text{Spring:} \quad kX = AE\frac{\partial X}{\partial x},$$

$$\text{Mass:} \quad -M\frac{\partial^2 X}{\partial t^2} = AE\frac{\partial X}{\partial x}. \tag{8-35}$$

A free boundary is a boundary with zero longitudinal stress (Eq. 8-34). A fixed boundary is a boundary which does not permit displacement. $X(x, t)$ is the deformation of the beam parallel to the x axis. A is the area of cross section of the beam, E is the elastic modulus, M is mass, and k is the spring longitudinal constant (Table 6-1). These boundary conditions are applied for all time at the spanwise location of the boundary. In practice it is often difficult to achieve the theoretical ideal of the fixed boundary for metallic beams, because the local longitudinal stiffness of the clamp at the boundary is apt to be of the same order as the longitudinal stiffness of the beam.

Tables. The natural frequencies and mode shapes for longitudinal vibration of slender, uniform, homogeneous isotropic beams are given in Table 8-16. The longitudinal natural frequencies of helical springs are given in Table 8-17. The longitudinal natural frequencies of discretized (spring-mass) systems can be found in Table 6-2.

8.4. TORSIONAL VIBRATION OF BEAMS AND SHAFTS

General Case. Torsional vibrations are the result of local twisting of a beam or shaft about its own axis. Table 6-3 considered torsional vibrations of various systems of inertia masses which were interconnected by massless torsional springs. This section emphasizes the torsional vibration of massive beams and shafts.

Exact closed form results for torsional vibrations can generally be obtained only for the case of shafts or tubes with circular cross sections. For these cross sections the shear strain and shear stress in the cross section of a homogeneous isotropic circular shaft are:

$$\epsilon_{x\theta} = r\,\frac{\partial\theta}{x}, \qquad \sigma_{x\theta} = Gr\,\frac{\partial\theta}{\partial x}, \qquad R_i < r < R_o, \tag{8-36}$$

where r is the radial distance from the center of the cross section. θ is the twist angle and x is the spanwise distance along the shaft. G is the shear modulus $\{E/[2(1+\nu)]$, where E is the elastic modulus and ν is Poisson's ratio$\}$. The maximum stresses and strains are developed at the outer radius (R_o) of the cross section. All other stresses and strains are zero. The torque borne by the cross section is:

$$\mathfrak{M} = GI_P\,\frac{\partial\theta}{\partial x}, \tag{8-37}$$

where I_P is the polar area moment of inertia about the axis of the shaft.

Shafts with noncircular cross section generally will warp as the shaft deforms in torsion (Refs. 8-38, 8-39). However, the effect of warping on the deformation is very small for shafts with simple closed cross sections if the cross-sectional dimensions are small compared with the length of the shaft. The results of Gere (Ref. 8-38) imply that the effect of warping of the cross section on the natural frequency of the beam may be neglected even for thin-walled open cross sections if

$$\frac{1}{i}\,\frac{t}{D}\,\frac{L}{D} > 10,$$

where i is the mode number (i = 1, 2, 3, . . .), t is a typical minimum wall thickness of the cross section, D is a diameter of the cross section, and L is the length of the shaft.

The natural frequency of torsion is a function of the torsional constant C. C is defined as the moment required to produce a torsional rotation of 1 radian on a unit spanwise length of shaft divided by the shear modulus:

$$C = \frac{\mathfrak{M}}{G\,\dfrac{\partial\theta}{\partial x}}. \tag{8-38}$$

C is equal to the polar area moment of inertia of the cross section for shafts with circular cross sections (Eq. 8-37) and is less than the polar area moment of inertia for shafts with noncircular cross sections. Values of C for various cross sections are given in Table 8-18. Additional values for C can be found in Ref. 8-40.

It is assumed here that the center of mass of the section coincides with the center of rotation. If the center of mass does not lie on the axis of rotation, then the torsional and flexural natural frequencies of the shaft will be coupled (Ref. 8-17, pp. 471–475).

Table 8-18. Torsional Constants of Various Cross Sections.

Cross Section	Torsional Constant, C
1. Circle 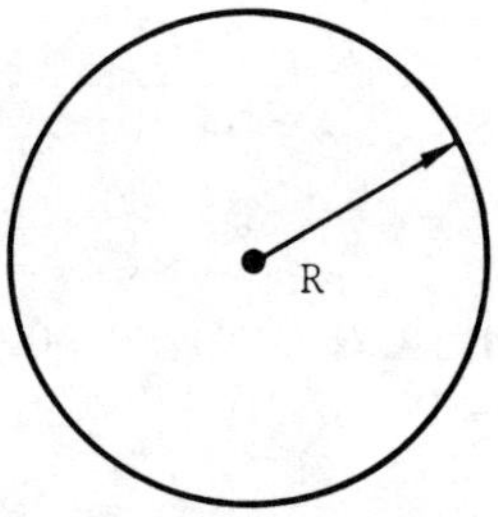	$\frac{\pi R^4}{2}$
2. Hollow Circle 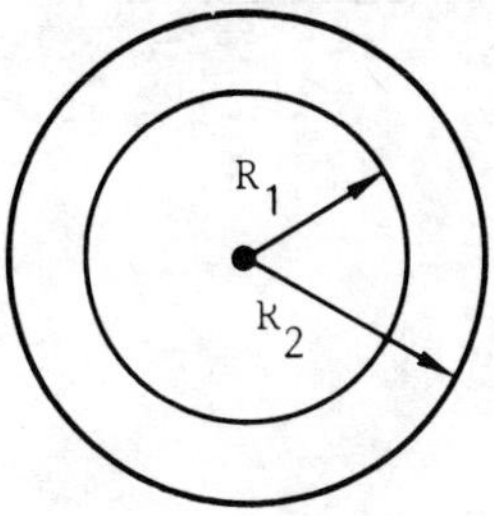	$\frac{\pi}{2}(R_2^4 - R_1^4)$
3. Ellipse 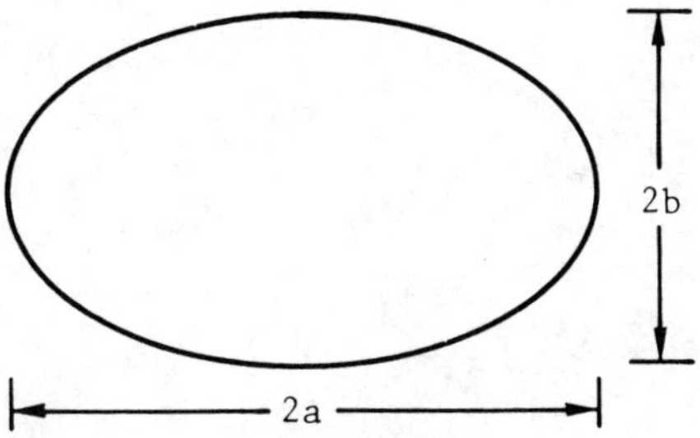	$\frac{\pi a^3 b^3}{a^2 + b^2}$
4. Square 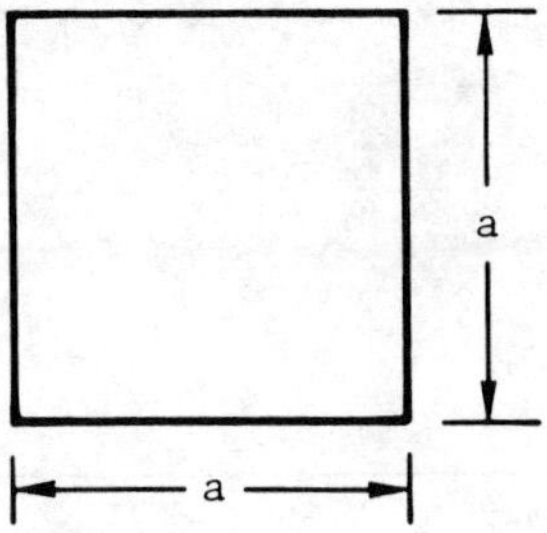	$0.1406a^4$

Table 8-18. Torsional Constants of Various Cross Sections. *(Continued)*

Cross Section	Torsional Constant, C
5. Rectangle 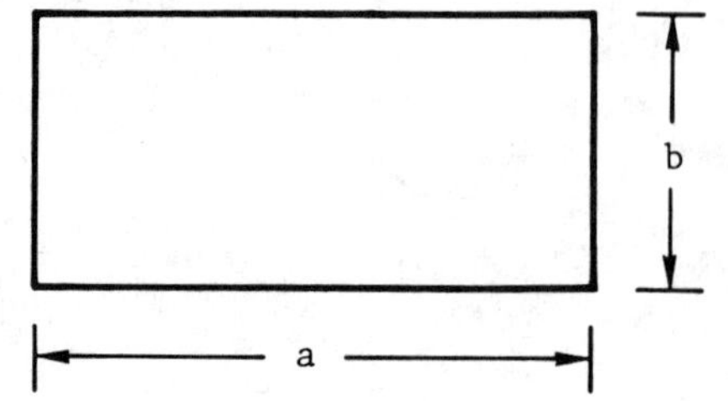 	$\frac{ca^3b^3}{a^2 + b^2}$ $\frac{a}{b}$: 1, 2, 4, 8, ∞ c: 0.281, 0.286, 0.299, 0.312, $\frac{1}{3}$
6. Hollow Rectangle 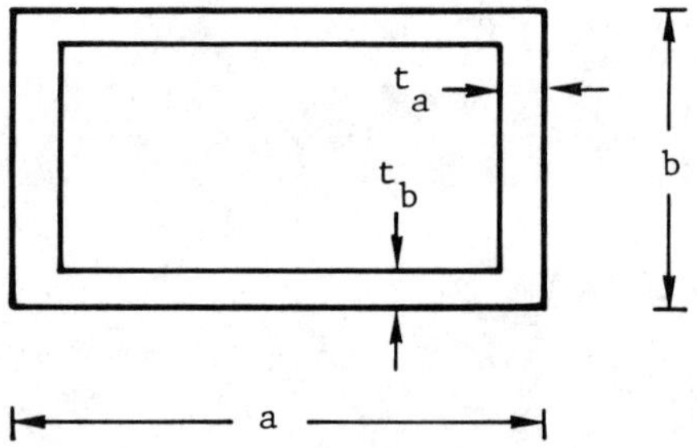 	$\frac{2t_a t_b (a - t_a)^2 (b - t_b)^2}{at_a + bt_b - t_a^2 - t_b^2}$
7. Equilateral Triangle 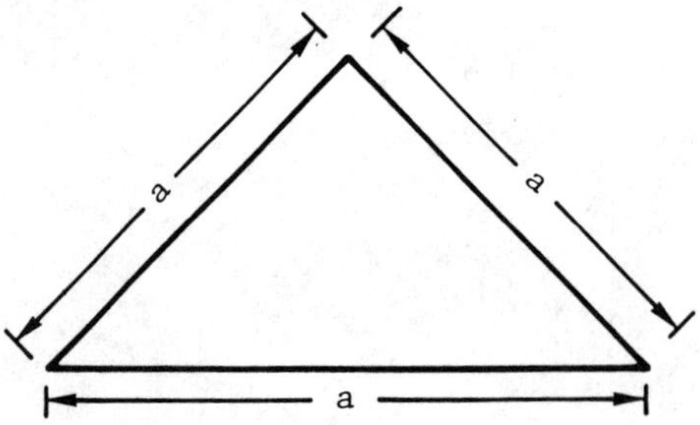 	$\frac{\sqrt{3}}{80} a^4$
8. Hexagon 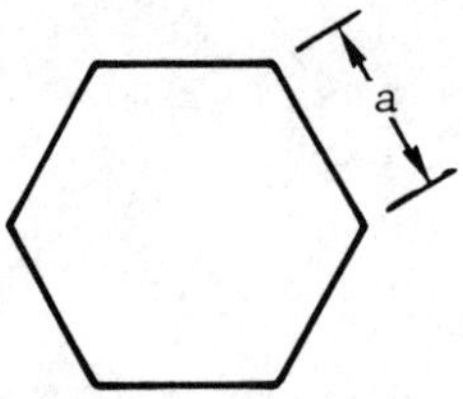 	$1.03a^4$

Table 8-18. Torsional Constants of Various Cross Sections. *(Continued)*

Cross Section	Torsional Constant, C
9. Thin-Walled Open Section, Nonuniform Wall 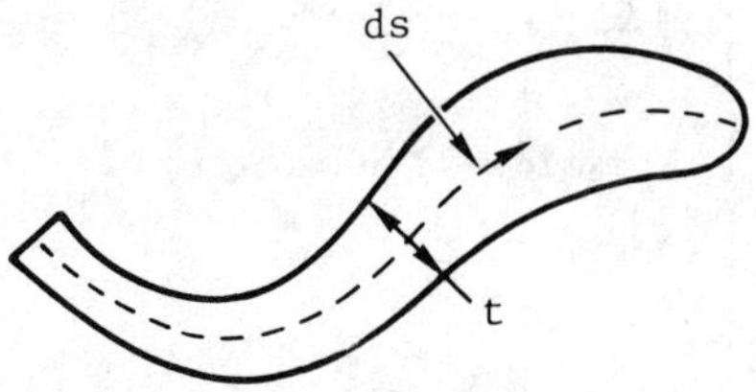	$\frac{1}{3}\int t^3 ds$
10. Thin-Walled Open Section, Composed of Rectangles 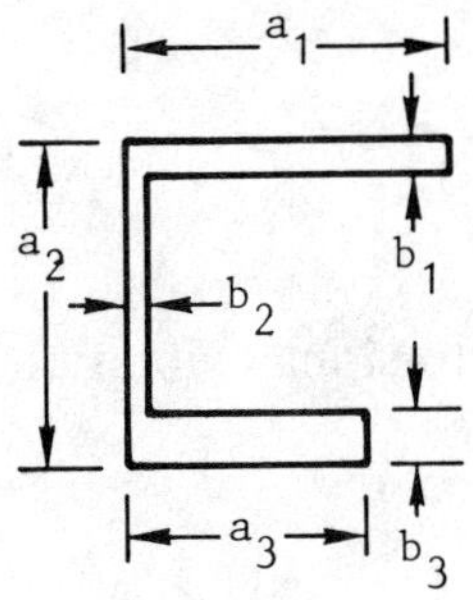	$\sum_i c_i \frac{a_i^3 b_i^3}{a_i^2 + b_i^2}$ c_i is given in frame 5 of this table
11. Thin-Walled Open Section of Uniform Thickness 	$\frac{1}{3} St^3$ S = length of midwall perimeter
12. Thin-Walled Closed Section of Uniform Wall 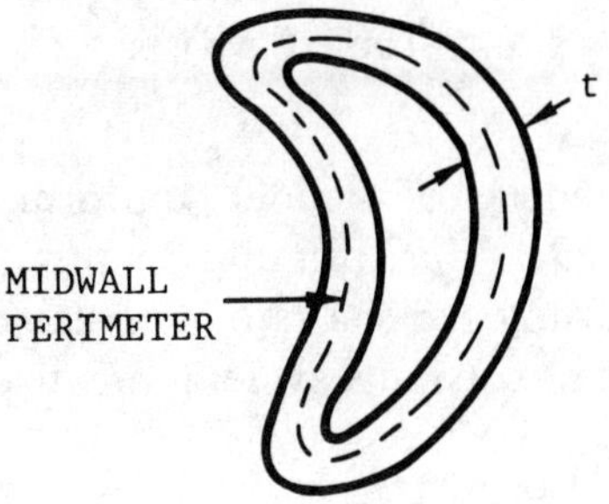	$\frac{4A^2 t}{S}$ A = area enclosed by midwall perimeter S = length of midwall perimeter

Table 8-18. Torsional Constants of Various Cross Sections. (*Continued*)

Cross Section	Torsional Constant, C
13. Thin-Walled Closed Section with Variable Wall 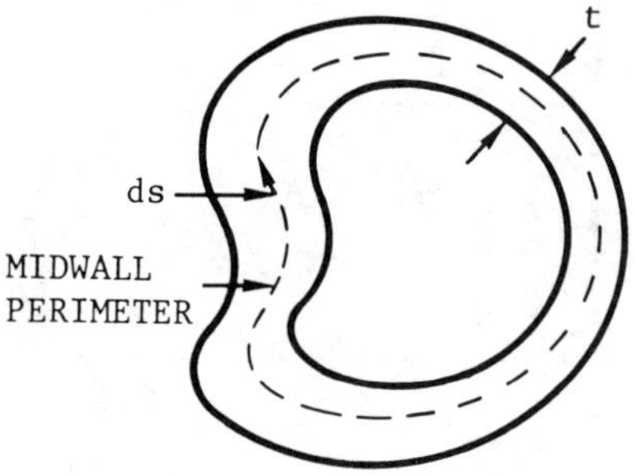	$\dfrac{4A^2}{\int \frac{ds}{t}}$ S = midwall perimeter

The common boundary conditions associated with the torsional vibration of shafts are:

$$\text{Free:} \qquad \frac{\partial\theta}{\partial x} = 0,$$

$$\text{Fixed:} \qquad \theta = 0,$$

$$\text{Spring:} \qquad k\theta = GC\frac{\partial\theta}{\partial x},$$

$$\text{Inertial Mass:} \quad -J\frac{\partial^2\theta}{\partial t^2} = GC\frac{\partial\theta}{\partial x}.$$

A free boundary is a boundary without stress (Eq. 8-36). A fixed boundary does not permit rotation. $\theta(x, t)$ is the deformation of the beam in torsion. G is the shear modulus, J is the polar mass moment of inertia (Table 5-2), k is the torsional spring constant (Table 6-1), and C is the torsional constant (Table 8-18).

Tables. Formulas for the torsional natural frequencies of slender, uniform, isotropic, homogeneous shafts are given in Table 8-19. The formulas in this table neglect warping of the cross section and are exact only for shafts of circular cross section. Formulas for discretized (spring-inertial mass) torsional systems are given in Table 6-3.

Table 8-19. Torsional Vibration of Uniform Shafts.

Notation: k = torsion spring constant (Table 6-1); x = spanwise length; C = torsional constant of cross section (Table 8-18); G = shear modulus $\{E/[2(1+\nu)]$, where E = modulus of elasticity, ν = Poisson's ratio$\}$; I_p = polar area moment of inertia of cross section about axis of torsion (Table 5-1); J = mass moment of inertia about axis of torsion (Table 5-2); θ = angle of twist; $\tilde{\theta}$ = mode shape associated with θ; μ = mass density of shaft material; see Table 3-1 for consistent sets of units.

Natural Frequencies (hertz), $f_i = \frac{\lambda_i}{2\pi L}\left(\frac{CG}{\mu I_p}\right)^{1/2}$; i=1,2,3...

Description	λ_i or Transcendental Equation for λ	Mode Shape, $\tilde{\theta}\left(\frac{x}{L}\right)$
1. Free-Free 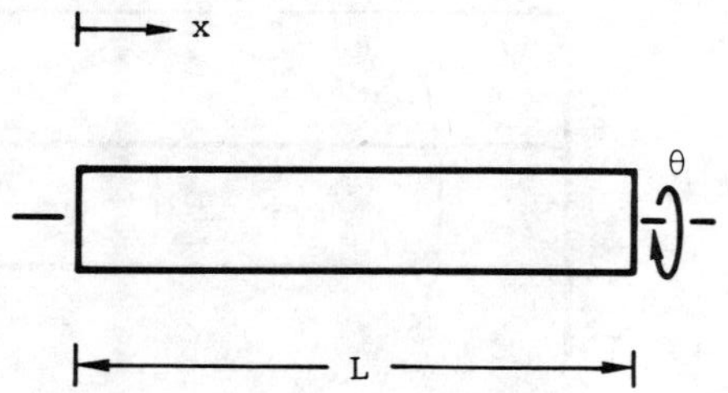	$i\pi$; i=1,2,3···	$\cos \frac{i\pi x}{L}$; i=1,2,3···
2. Fixed-Free 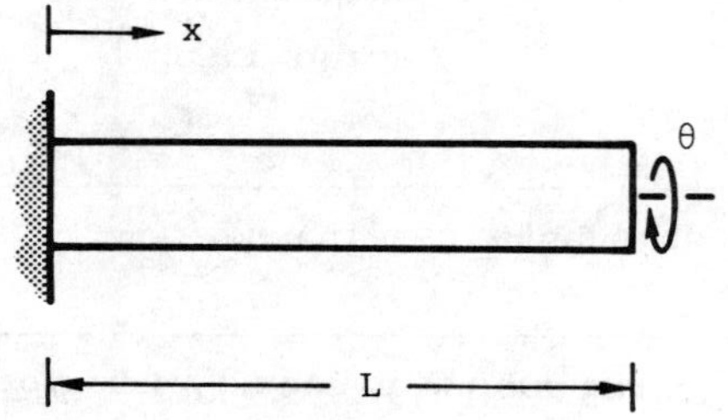	$\frac{(2i-1)\pi}{2}$; i=1,2,3···	$\sin \frac{\pi(2i-1)x}{2L}$; i=1,2,3···

Table 8-19. Torsional Vibration of Uniform Shafts. *(Continued)*

Natural Frequencies (hertz), $f_i = \frac{\lambda_i}{2\pi L}\left(\frac{CG}{\mu I_p}\right)^{1/2}$; i=1,2,3...

Description	λ_i or Transcendental Equation for λ	Mode Shape, $\tilde{\theta}\left(\frac{x}{L}\right)$
3. Fixed-Fixed 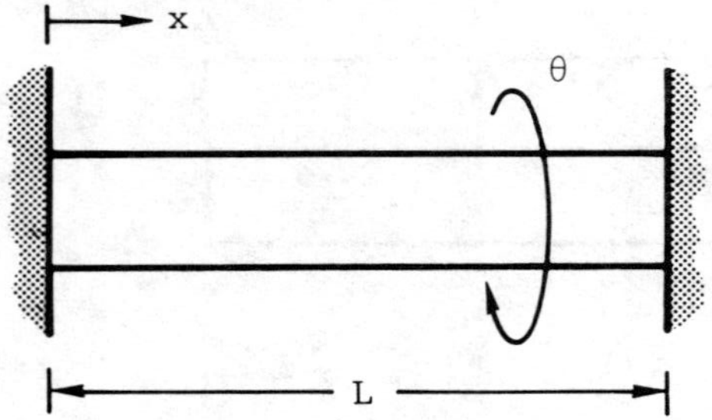	$i\pi$; i=1,2,3···	$\sin\frac{i\pi x}{L}$; i=1,2,3···
4. Spring-Free 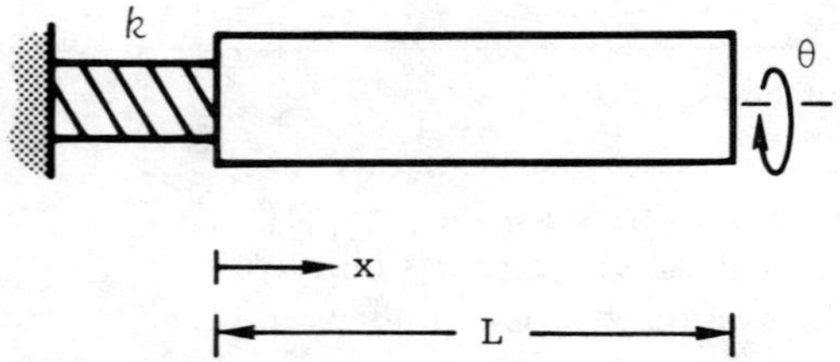	$\tan\lambda = \left(\frac{kL}{GC}\right)\frac{1}{\lambda}$ See Fig. 8-23 for λ_1	$\cot\lambda_i \cos\frac{\lambda_i x}{L}$ $+ \sin\frac{\lambda_i x}{L}$; i=1,2,3···

5. Fixed-Spring 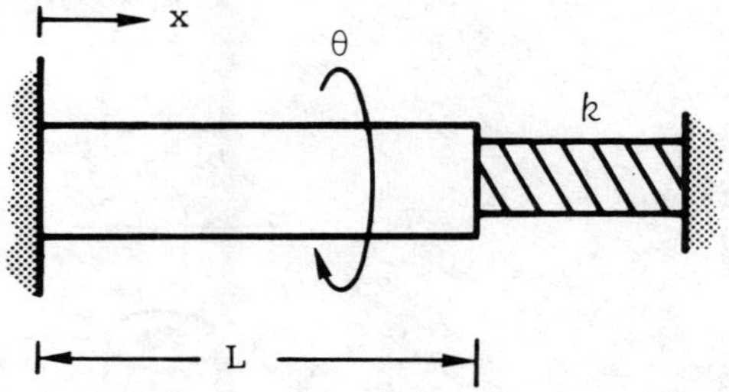	$\cot \lambda = -\left(\frac{kL}{GC}\right)\frac{1}{\lambda}$ See Fig. 8-24 for λ_1	$\sin \frac{\lambda_i x}{L}$; $i=1,2,3\cdots$
6. Fixed-Inertial Mass 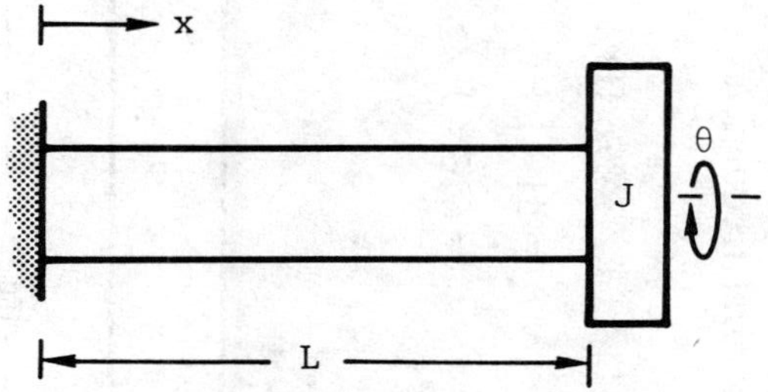	$\cot \lambda = \left(\frac{J}{\mu LC}\right)\lambda$ See Fig. 8-23 for λ_1	$\sin \frac{\lambda_i x}{L}$; $i=1,2,3\cdots$
7. Free-Inertial Mass 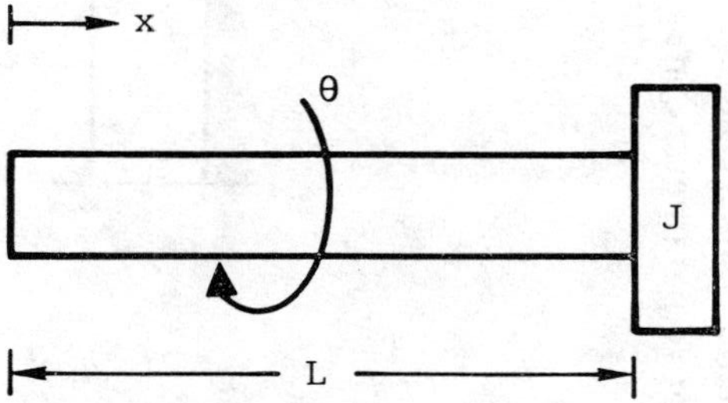	$\tan \lambda = -\left(\frac{J}{\mu LC}\right)\lambda$ See Fig. 8-24 for λ_1	$\cos \frac{\lambda_i x}{L}$; $i=1,2,3\cdots$

8.5. EXAMPLES

8.5.1. Tube Example

Properties. Consider the cantilever steel tube shown in Fig. 8-27. The material and geometric properties of this tube in the system of units of frame 3 of Table 3-1 are:

$$E = 2.04 \times 10^6 \text{ kg/cm}^2,$$

$$\nu = 0.3,$$

$$G = \frac{E}{2(1+\nu)} = 7.86 \times 10^5 \text{ kg/cm}^2,$$

$$K = 0.548 \text{ (frame 2 of Table 8-14)},$$

$$\mu = \frac{8.02 \times 10^{-3} \text{ kg/cm}^3}{980.7 \text{ cm/sec}^2} = 8.17 \times 10^{-6} \text{ kg-sec}^2/\text{cm}^4,$$

$$A = 5.50 \text{ cm}^2 \text{ (frame 24 of Table 5-1)},$$

$$m = 4.49 \times 10^{-5} \text{ kg-sec}^2/\text{cm}^2,$$

$$I = 8.59 \text{ cm}^4 \text{ (frame 24 of Table 5-1)},$$

$$C = I_P = 17.18 \text{ cm}^4,$$

$$L = 100 \text{ cm}.$$

The natural frequencies of flexure, shear, and longitudinal vibrations will be calculated for this beam.

Transverse Vibration. The natural frequencies of flexure vibration are given by frame 3 of Table 8-1. The natural frequencies of the first three modes, using the above parameters, are:

$$f_1 = 34.96 \text{ Hz},$$

$$f_2 = 219.1 \text{ Hz},$$

$$f_3 = 613.4 \text{ Hz}.$$

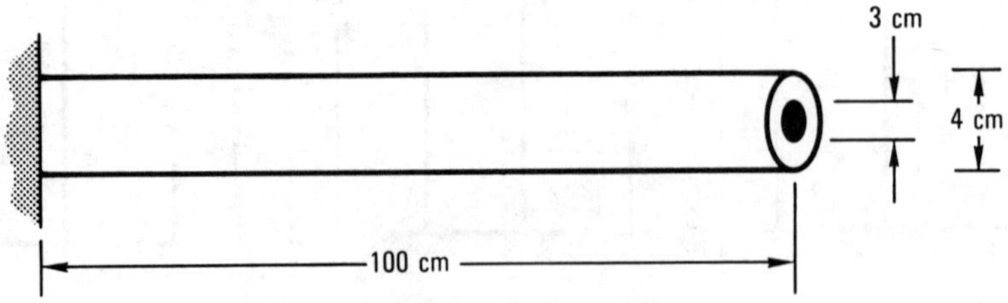

Fig. 8-27. Cantilever steel tube.

The natural frequencies computed from the shear beam model of frame 2 of Table 8-15 are:

$$f_1 = 574.0 \text{ Hz},$$

$$f_2 = 1722 \text{ Hz},$$

$$f_3 = 2870 \text{ Hz}.$$

For this slender beam the flexure beam model of Table 8-1 can be expected to provide a much better estimate of the natural frequencies of the first three modes than the shear beam model of Table 8-15.

Longitudinal Vibration. The natural frequencies of longitudinal vibration are computed from frame 2 of Table 8-16. The first three frequencies are:

$$f_1 = 1249 \text{ Hz},$$

$$f_2 = 3748 \text{ Hz},$$

$$f_3 = 6246 \text{ Hz}.$$

However, the boundary conditions of frame 2 of Table 8-16 imply that the end of the beam is attached to a wall which is much stiffer than the steel of the tube, so the wall will not expand or contract as the longitudinal waves travelling along the tube reach the wall. Since very few materials have a higher elastic modulus than steel, it is doubtful that a fixed-free boundary is appropriate for the tube. If the presence of the wall is neglected, then the longitudinal natural frequencies are found from frame 1 of Table 8-16 to be:

$$f_1 = 2498 \text{ Hz},$$

$$f_2 = 4997 \text{ Hz},$$

$$f_3 = 7495 \text{ Hz}.$$

The actual natural frequencies probably fall between these two sets of solutions.

Torsional Vibration. The torsional natural frequencies are computed from frame 2 of Table 8-19. The first three natural frequencies are:

$$f_1 = 775.4 \text{ Hz},$$

$$f_2 = 2327 \text{ Hz},$$

$$f_3 = 3877 \text{ Hz}.$$

Note that the natural frequencies of longitudinal and torsional vibration of the beam are considerably higher than the natural frequencies of transverse flexure vibration.

Axial Load. It is proposed to use another length of this same pipe as a flag pole. The maximum height to which a piece of this pipe can stand freely without buckling can be computed from Eq. 8-24 and case 22, mode 1 of Table 8-5 using an axial traction of w = mg, the weight per unit length of the pipe, where g is the acceleration of gravity (980.7 cm/sec^2):

$$w = -mg,$$

$$\alpha_{CRIT} = -7.84 = \frac{-mgL^3}{EI},$$

$$L_{CRIT} = \left(\frac{EI\,\alpha_{CRIT}}{-mg}\right)^{1/3} = 1461 \text{ cm}.$$

It is decided to limit the height of the flag pole to 700 cm. The natural frequency of the flag pole can be computed from Eq. 8-23 and Table 8-6 (case 22):

$$L = 700 \text{ cm},$$

$$\alpha = \frac{-mgL^3}{EI} = -0.862,$$

$$\lambda_1 = 1.82, \qquad f_1 = 32.93 \text{ Hz},$$

$$\lambda_2 = 4.68, \qquad f_2 = 218.2 \text{ Hz},$$

$$\lambda_3 = 7.84, \qquad f_3 = 611.4 \text{ Hz}.$$

The linearly varying axial load has only a small effect on the natural frequencies.

8.5.2. Beam-Mass Example

The aluminum beam shown in Fig. 8-28 has a 1-centimeter square cross section and supports a 2-kilogram weight at midspan. The parameters which describe this beam, in the units of frame 3 of Table 3-1, are:

$$E = 7.03 \times 10^5 \text{ kg/cm}^2,$$

$$\mu = \frac{2.77 \times 10^{-3} \text{ kg/cm}^3}{980.7 \text{ cm/sec}^2} = 2.83 \times 10^{-6} \text{ kg-sec}^2/\text{cm}^4,$$

$$m = 2.83 \times 10^{-6} \text{ kg-sec}^2/\text{cm}^2,$$

$$I = 0.0833 \text{ cm}^4 \text{ (frame 3 of Table 5-1)},$$

$$L = 50 \text{ cm},$$

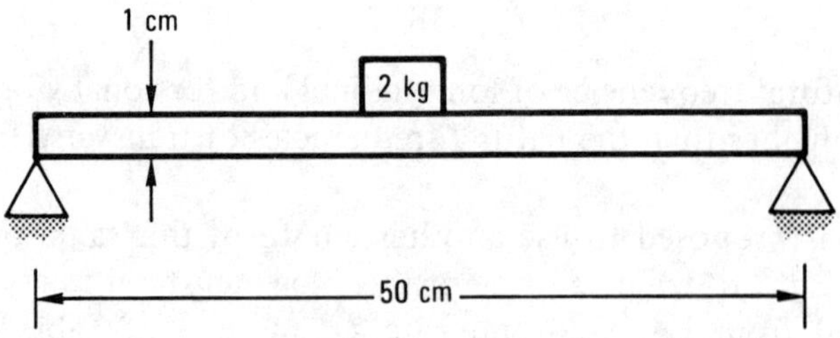

Fig. 8-28. Pinned-pinned beam which supports a mass at midspan.

$$M = \frac{2 \text{ kg}}{980.7 \text{ cm/sec}^2} = 2.04 \times 10^{-3} \text{ kg-sec}^2\text{/cm},$$

$$M_b = mL = 1.42 \times 10^{-4} \text{ kg-sec}^2\text{/cm}.$$

In the absence of the mass, the natural frequency of the beam can be computed from the exact solution given in frame 5 of Table 8-1. The result is:

$$f_1 = 90.4 \text{ Hz}.$$

The natural frequency of the beam with the mass can be calculated from frame 4 of Table 8-8. The result is:

$$f_1 = 16.4 \text{ Hz}.$$

The natural frequency of the beam without the mass can also be calculated from frame 4 of Table 8-8 with M = 0. This gives:

$$f_1 = 90.7 \text{ Hz},$$

which differs from the exact result by only 0.33%.

8.5.3. Laminated Two-Span Beam Example

A laminated, three-layer, two-span beam is shown in Fig. 8-29. The central core is made of aluminum honeycomb. The outer two layers are aluminum plate. The beam is clamped at one end, the other end is free, and there is a pinned support at a distance of one-third of the beam span from the clamped end. The width of the plate is 50 cm. The properties of the beam materials, in the units of frame 3 of Table 3-1 and the notation of Fig. 8-2, are:

$$E_1 = 1.33 \times 10^4 \text{ kg/cm}^2,$$

$$\mu_1 = \frac{7.2 \times 10^{-5} \text{ kg/cm}^3}{980.7 \text{ cm/sec}^2} = 7.34 \times 10^{-8} \text{ kg-sec}^2\text{/cm}^4,$$

$$d_1 = 0.75 \text{ cm},$$

$$E_2 = 7 \times 10^5 \text{ kg/cm}^2,$$

$$\mu_2 = \frac{2.7 \times 10^{-3} \text{ kg/cm}^3}{980.7 \text{ cm/sec}^2} = 2.75 \times 10^{-6} \text{ kg-sec}^2\text{/cm}^4,$$

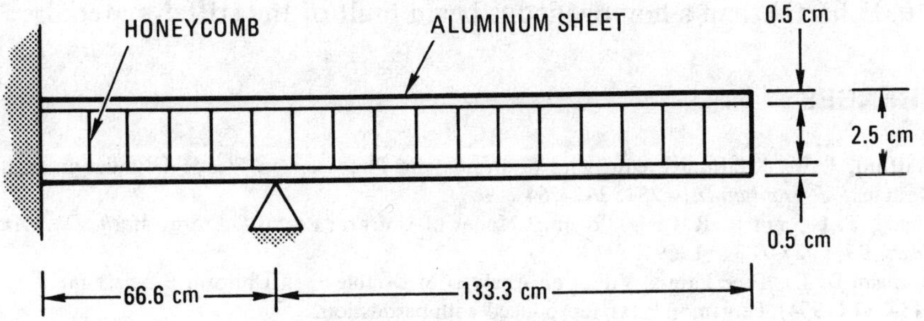

Fig. 8-29. A three-layer, two-span, aluminum honeycomb laminated beam.

$$d_2 = 1.25 \text{ cm},$$
$$L = 200 \text{ cm},$$
$$b = 50 \text{ cm},$$
$$a/L = 0.33.$$

The equivalent homogeneous beam stiffness and beam mass from Eqs. 8-8 and 8-9 are:

$$EI = \tfrac{2}{3}\, bE_1\, (d_1^3 - 0) + \tfrac{2}{3}\, bE_2\, (d_2^3 - d_1^3),$$
$$= 3.59 \times 10^7 \text{ kg-cm}^2,$$
$$m = 2b\mu_1\, (d_1 - 0) + 2b\mu_2\, (d_2 - d_1),$$
$$= \frac{0.140 \text{ kg/cm}}{980.7 \text{ cm/sec}^2} = 1.43 \times 10^{-4} \text{ kg-sec}^2/\text{cm}^2.$$

Figure 8-5(d) and Eq. 8-13 give the fundamental natural frequency:

$$f_1 = \frac{\lambda_1^2}{2\pi L^2}\left(\frac{EI}{m}\right)^{1/2},$$

where

$$\lambda_1 = 2.5.$$

Thus,

$$f_1 = 12.46 \text{ Hz}.$$

If the beam had been composed solely of aluminum, then the result would be:

$$EI = \tfrac{2}{3}\, E_2\, bd_2^3 = 4.56 \times 10^7 \text{ kg-cm}^2,$$

$$m = 2b\mu_2 d_2 = \frac{0.338 \text{ kg/cm}}{980.7 \text{ cm/sec}^2} = 3.44 \times 10^{-4} \text{ kg-sec}^2/\text{cm}^2,$$

and

$$f_1 = 9.05 \text{ Hz}.$$

Thus, by using a laminated construction, a beam can be built which is not only lighter (0.140 kg/cm versus 0.338 kg/cm) but has a higher natural frequency (12.5 versus 9.05 hertz) than a homogeneous beam built of the stiffer material.

REFERENCES

8-1. Stafford, J. W., "Natural Frequencies of Beams and Plates on an Elastic Foundation with a Constant Modulus," *J. Franklin Inst.* 284, 262–264 (1967).

8-2. Chang, T. C., and R. R. Craig, "Normal Modes of Uniform Beams," *J. Eng. Mech. Div., Am. Soc. Civil Engrs.* 95, 1027–1031 (1969).

8-3. Gorman, D. J., "Free Lateral Vibration Analysis of Double-Span Uniform Beams," *Int. J. Mech. Sci.* 16, 345–351 (1974), Pergamon Press, reproduced with permission.

8-4. Moretti, P. M., and R. L. Lowery, "Heat Exchanger Tube Vibration Characteristics in a "No-Flow" Condition," Final Report to Tubular Exchanger Manufacturers Association Experimental Program, School of

Mechanical and Aerospace Engineering, Oklahoma State University, Stillwater, Oklahoma, September 1973.

8-5. Gorman, D. J., and R. K. Sharma, "Vibration Frequencies and Modal Shapes for Multi-Span Beams with Uniformly Spaced Supports," Ottawa University, Ontario, Canada, Report No. Conf-740330-1 (1974).

8-6. Gorman, D. J., *Free Vibration Analysis of Beams and Shafts*, John Wiley, New York, 1975.

8-7. Franklin, R. E., B. M. H. Soper, and R. H. Whittle, "Avoiding Vibration-Induced Tube Failures in Shell and Tube Heat Exchangers," B.N.E.S. Vibration in Nuclear Plant Conference, Paper 3:1, Keswick, U.K., 1978.

8-8. Shaker, F. J., "Effect of Axial Load on Mode Shapes and Frequencies of Beams," Lewis Research Center Report NASA-TN-8109, December 1975.

8-9. Chen, S. S., and M. W. Wambsganss, "Design Guide for Calculating Natural Frequencies of Straight and Curved Beams on Multiple Supports," USAEC Report ANL-CT-74-06, Argonne National Laboratory, June 1974.

8-10. Amba-Rao, C. L., "Effect of End Conditions on the Lateral Frequencies of Uniform Straight Columns," *J. Acoust. Soc. Am.* **42**, 900–901 (1967).

8-11. Huang, T., and D. W. Dareing, "Buckling and Frequencies of Long Vertical Pipes," *J. Eng. Mech. Div., Am. Soc. Civil Engrs.* **95**, 167–181 (1969).

8-12. Laird, W. M., and G. Fauconneau, "Upper and Lower Bounds for the Eigenvalues of Vibrating Beams with Linearly Varying Axial Load," National Aeronautics and Space Administration Report NASA-CR-653, University of Pittsburgh, November 1966.

8-13. Laura, P. A. A., J. L. Pombo, and E. A. Susemihl, "A Note on the Vibrations of a Clamped-Free Beam with a Mass at the Free End," *J. Sound Vib.* **37**, 161–168 (1974).

8-14. Haener, J., "Formulas for the Frequencies Including Higher Frequencies of Uniform Cantilever and Free-Free Beams with Additional Masses at the Ends," *J. Appl. Mech.* **25**, 412 (1958).

8-15. Bhat, B. R., and H. Wagner, "Natural Frequencies of a Uniform Cantilever with a Tip Mass Slender in the Axial Direction," *J. Sound Vib.* **45**, 304–307 (1976).

8-16. Baker, W. E., "Vibration Frequencies for Uniform Beams with Central Masses," *J. Appl. Mech.* **31**, 35–37 (1964).

8-17. Timoshenko, S., D. H. Young, and W. Weaver, Jr., *Vibration Problems in Engineering* (4th ed.), John Wiley, New York, 1974.

8-18. Wang, H. C., and W. J. Worley, "Tables of Natural Frequencies and Nodes for Transverse Vibration of Tapered Beams," National Aeronautics and Space Administration Report NASA-CR-443, University of Illinois, April 1966.

8-19. Mabie, H. H., and C. B. Rogers, "Transverse Vibrations of Double-Tapered Cantilever Beams," *J. Acoust. Soc. Am.* **51**, 1771–1774 (1972).

8-20. Housner, G. W., and W. O. Keightley, "Vibrations of Linearly Tapered Cantilever Beams," *J. Eng. Mech. Div., Am. Soc. Civil Engrs.* **88**, EM2, 95–123 (1962).

8-21. Mabie, H. H., and C. B. Rogers, "Transverse Vibrations of Tapered Cantilever Beams with End Support," *J. Acoust. Soc. Am.* **44**, 1739–1741 (1968).

8-22. Conway, H. D., and J. F. Dubil, "Vibration Frequencies of Truncated-Cone and Wedge Beams," *J. Appl. Mech. Div., Am. Soc. Civil Engrs.* **32**, 932–934 (1965).

8-23. Mabie, H. H., and C. B. Rogers, "Technology Transfer in the Vibration Analysis of Linearly Tapered Cantilever Beams," *J. Eng. Industry* **98**, 1335–1341 (1976); also see, Downs, B., "Transverse Vibrations of Cantilever Beams Having Unequal Breadth and Depth Tapers," *J. Appl. Mech.* **44**, 737–742 (1978).

8-24. Goel, R. P., "Transverse Vibration of Tapered Beams," *J. Sound Vib.* **47**, 1–7 (1976).

8-25. Hibbeler, R. C., "Free Vibration of a Beam Supported by Unsymmetrical Spring-Hinges," *J. Appl. Mech.* **42**, 501–502 (1975).

8-26. Chun, K. R., "Free Vibration of a Beam with One End Spring-Hinged and the Other Free," *J. Appl. Mech.* **39**, 1154–1155 (1972).

8-27. Maurizi, M. J., R. E. Rossi, and J. A. Reyes, "Vibration Frequencies for a Uniform Beam with One End Spring-Hinged and Subjected to a Translational Restraint of the Other End," *J. Sound Vib.* **48**, 565–568 (1976).

8-28. Lee, T. W., "Vibration Frequency for a Uniform Beam with One End Spring-Hinged and Carrying a Mass at the Other Free End," *J. Appl. Mech.* **95**, 813–815 (1973).

8-29. Hess, M. S., "Vibration Frequencies for a Uniform Beam with Central Mass and Elastic Supports," *J. Appl. Mech.* **31**, 556–558 (1964).

8-30. Sankaran, G. V., K. K. Raju, and G. V. Rao, "Vibrations of a Tapered Beam with One End Spring-Hinged and Carrying a Mass at the Other Free End," *J. Appl. Mech.* **42**, 740–744 (1975).

8-31. Cowper, G. R., "The Shear Coefficient in Timoshenko's Beam Theory," *J. Appl. Mech.* **33**, 335–340 (1966).

8-32. Rutenberg, A., "Approximate Natural Frequencies for Coupled Shear Walls," *Earthquake Eng. Struct. Dynam.* **4**, 95–100 (1975).

8-33. Housner, G. W., and A. G. Brody, "Natural Periods of Vibration of Buildings," *J. Eng. Mech. Div., Am. Soc. Civil Engrs.* **89**, 31–65 (1963).

8-34. Rinne, J. E., "Building Code Provisions for Aseismic Design," in *Proceedings of Symposium on Earthquake and Blast Effects on Structures*, Los Angeles, Ca., 1952, pp. 291–305.

8-35. Goodman, L. E., and J. G. Sutherland, "Discussion of Natural Frequencies of Continuous Beams of Uniform Span Length," *J. Appl. Mech.* **18**, 217–218 (1951).

8-36. Hurty, W. C., and M. F. Rubenstein, "On the Effect of Rotatory Inertia and Shear in Beam Vibration," *J. Franklin Inst.* **278**, 124–132 (1964).

8-37. Murty, A. V. K., and R. P. Shimpi, "Vibrations of Laminated Beams," *J. Sound Vib.* **36**, 273–284 (1974).

8-38. Gere, J. M., "Torsional Vibration of Beams of Thin-Walled Open Section," *J. Appl. Mech. Div., Am. Soc. Civil Engrs.* **21**, 381–387 (1954).

8-39. Carr, J. B., "The Torsional Vibration of Uniform Thin-Walled Beams of Open Section," *Aeron. J.* **73**, 672–674 (1969).

8-40. Roark, R. J., *Formulas for Stress and Strain* (4th ed.), McGraw-Hill, New York, 1965, pp. 196–199.

8-41. Rutenberg, A., "Vibration Frequencies for a Uniform Cantilever with Rotational Constraint at a Point," *J. Appl. Mech.* **45**, 422–423 (1978).

9

CURVED BEAMS AND FRAMES

9.1. COMPLETE RINGS

General Case. Curved beams and rings possess extensional modes, flexural (i.e., inextensional) modes, and torsional modes which are analogous to the corresponding modes of straight beams. However, the curvature of the curved beam introduces geometric coupling between these modes, unlike the modes of straight beams which are generally uncoupled. As a result of the curvature-induced coupling, the modes of curved beams and rings are two- or three-dimensional, whereas the modes of straight beams are generally one-dimensional.

The natural frequencies and mode shapes of homogeneous, uniform, unsupported circular rings, shown in Fig. 9-1, are given in Table 9-1. The rings are assumed to be slender, that is, the thickness of the ring is much less than the radius of the ring, so that shear deformation can be neglected and rotary inertia is neglected in comparison with flexural inertia, except in the torsion mode. It is assumed that the torsional and flexural vibrations of the ring are not inertially coupled. This requirement is always met if the cross section of the ring is symmetric about both the x and y axes. Warping of the cross section due to torsion has also been neglected (see Section 8.9).

Natural Frequencies and Mode Shapes. The modes of vibration of complete rings can be classified as either (1) extensional (longitudinal elongation and contraction of the ring along its own axis), (2) torsional (twisting of the ring about its own axis), (3) in-plane flexural (inextensional vibrations in the plane of the ring), or (4) out-of-plane flexural (inextensional vibrations out of the plane of the ring). Out-of-plane flexural vibrations are coupled to torsional vibrations. Generally, only the flexural modes are of practical importance since the fundamental natural frequencies of the torsional and extensional modes are much higher than the fundamental natural frequencies of the flexural modes.

The natural frequencies of the flexure modes in and out of the plane of the ring are only slightly different for rings with circular cross sections ($I_x = I_y$, $C = 2I_x$). The difference between the natural frequencies of the in-plane and out-of-plane flexure modes is only 2.96% for the fundamental mode ($i = 2$) and 1.5% for the second mode ($i = 3$) of slender rings with a circular cross section.

The mode number, i, equals the number of complete modal waves around the circumference of the ring. 2i is the number of equally spaced vibration nodes in the circumference of the ring (Fig. 9-2). $i = 2$ is the lowest flexural mode. $i = 1$ corresponds to rigid body translation for the flexural modes (frames 3 and 4 of Table 9-1). $i = 0$ corresponds to uniform radial expansion for the extensional mode and to uniform twist for the torsion mode (frames 1 and 2 of Table 9-2).

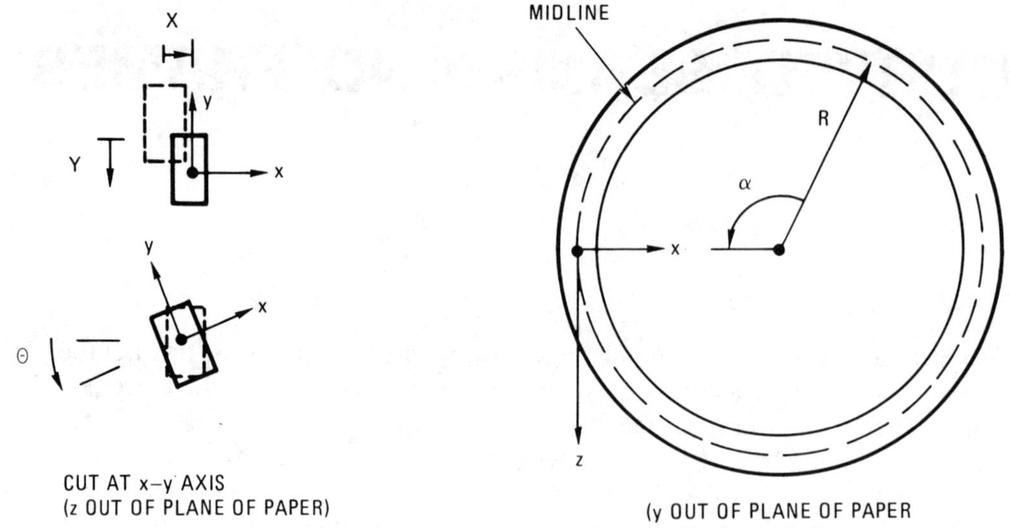

Fig. 9-1. Coordinate system for slender rings. X, Y, and θ are deformations.

Stress and Strain. The relationships between deformation, strain, stress, and the stress resultants will be given separately for each class of vibration modes in Table 9-1. The following notation applies:

ϵ_z = normal strain parallel to z axis (Fig. 9-1),
$\epsilon_{z\theta}$ = shear strain on z face in θ direction,
σ_{zz} = normal stress applied parallel to z axis (Fig. 9-1),
$\sigma_{z\theta}$ = shear stress on z face in θ direction,
x, y = orthogonal distances measured from centroid of cross section (Fig. 9-1),
α = angular coordinate of arcs (Fig. 9-1),
R = mean radius of the ring,
X, Y = deformations parallel to x and y axes, respectively,

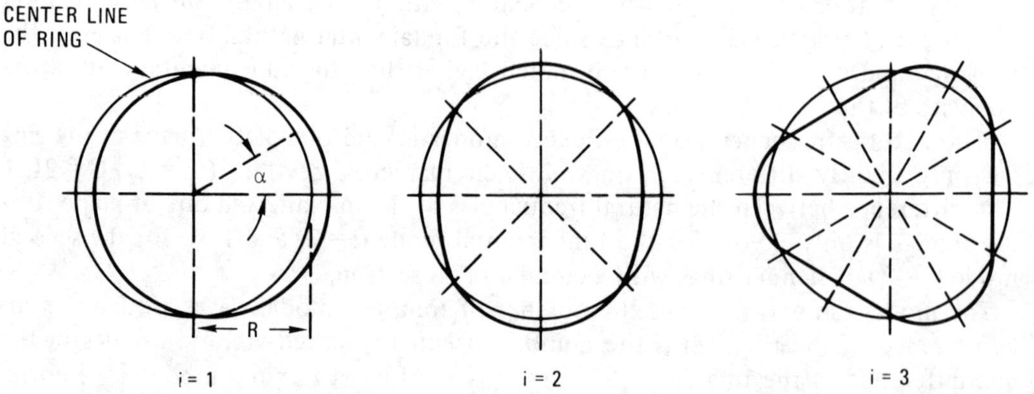

Fig. 9-2. In-plane flexural modes of a ring (frame 3 of Table 9-1) showing the deformation normal to the ring axis for i = 1, 2, and 3. i = 1 is a rigid body translation.

Table 9-1. Complete Rings.

Notation: x = distance from centroid of cross section toward center of ring (Fig. 9-1); y = distance perpendicular to plane of ring (Fig. 9-1); z = distance along ring (Fig. 9-1); θ = angle of twist about z axis (Fig. 9-1); $\tilde{x}, \tilde{y}, \tilde{z}, \tilde{\theta}$ = mode shapes corresponding to deformations parallel to x, y, z axes and rotation about z axis, respectively; m = mass per unit length of ring; C = torsion constant (Table 8-18); E = modulus of elasticity; G = shear modulus $\{E/[2(1+\nu)]\}$; I_x, I_y = area moments of inertia about x and y axes, respectively (Table 5-1); I_{zz} = polar area moment of inertia about z axis (Chapter 5); R = radius to midline of ring; α = angular position about ring; μ = mass density of ring material; ν = Poisson's ratio; see Fig. 9-1; see Table 3-1 for consistent sets of units. Ref. 9-1.

Description	Natural Frequency, f_i (Hz)	Mode Shape
1. Extension Modes i = 0,1,2,3...	$\frac{(1+i^2)^{1/2}}{2\pi R}\left(\frac{E}{\mu}\right)^{1/2}$ (Extension modes have natural frequencies above those of flexure modes for slender rings)	$\begin{bmatrix}\tilde{x}\\ \tilde{y}\\ \tilde{z}\\ \tilde{\theta}\end{bmatrix}_i = \begin{bmatrix}\cos i\alpha\\ 0\\ -i\sin i\alpha\\ 0\end{bmatrix}$
2. Torsion Modes i = 0,1,2,3...	$\frac{1}{2\pi R}\left(\frac{i^2 CG + I_x E}{\mu I_{zz}}\right)^{1/2}$ for circular cross sections: $\frac{(i^2+\nu+1)^{1/2}}{2\pi R}\left(\frac{G}{\mu}\right)^{1/2}$	$\begin{bmatrix}\tilde{x}\\ \tilde{y}\\ \tilde{z}\\ \tilde{\theta}\end{bmatrix}_i = \begin{bmatrix}0\\ \varepsilon^*\\ 0\\ \cos i\alpha\end{bmatrix}$ *$\varepsilon \ll 1$; if i = 0, $\varepsilon = 0$
3. In-Plane Flexure Modes i = 1,2,3...	$\frac{i(i^2-1)}{2\pi R^2 (i^2+1)^{1/2}}\left(\frac{EI_y}{m}\right)^{1/2}$	$\begin{bmatrix}\tilde{x}\\ \tilde{y}\\ \tilde{z}\\ \tilde{\theta}\end{bmatrix}_i = \begin{bmatrix}i\cos i\alpha\\ 0\\ \sin i\alpha\\ 0\end{bmatrix}$
4. Out-of-Plane Flexure Modes i = 1,2,3...	$\frac{i(i^2-1)}{2\pi R^2}\left[\frac{EI_x}{m\left(i^2+\frac{EI_x}{GC}\right)}\right]^{1/2}$ for circular cross sections: $\frac{i(i^2-1)}{2\pi R^2 (i^2+1+\nu)^{1/2}}\left(\frac{EI}{m}\right)^{1/2}$	$\begin{bmatrix}\tilde{x}\\ \tilde{y}\\ \tilde{z}\\ \tilde{\theta}\end{bmatrix}_i = \begin{bmatrix}0\\ \sin i\alpha\\ 0\\ -\frac{i^2}{R}\left(\frac{1+\frac{GC}{EI_x}}{1+i^2\frac{GC}{EI_x}}\right)\sin i\alpha\end{bmatrix}$ $\begin{bmatrix}\tilde{x}\\ \tilde{y}\\ \tilde{z}\\ \tilde{\theta}\end{bmatrix}_i = \begin{bmatrix}0\\ \sin i\\ 0\\ -\frac{i^2}{R}\left(\frac{2+\nu}{i^2+1+\nu}\right)\sin i\alpha\end{bmatrix}$

θ = twist about z axis (Fig. 9-1),
E = modulus of elasticity,
G = shear modulus, $E/[2(1+\nu)]$,
C = torsion constant (Table 8-18),
I_x, I_y = area moments of inertia about x and y axes, respectively (Table 5-1),
N_z = resultant load in z direction,
M_z = bending moment about either x or y axis,
M_θ = twisting moment about z axis.

The following relationships apply to both complete rings and circular arcs (only the non-zero stresses and stress resultants are presented):

1. Extension Modes (frame 1 of Tables 9-1 and 9-2).

$$\epsilon_z = \frac{-X}{R}, \qquad \sigma_{zz} = \frac{-EX}{R},$$

$$N_z = \int_A \sigma_{zz}\, dA = -EA\frac{X}{R}. \tag{9-1}$$

2. Torsion Modes (frame 2 of Table 9-1). Both an r–θ and an x–y coordinate system are used to represent the x–y plane. A circular cross section has been assumed in the strain-deformation and stress-strain relationships.

$$\epsilon_z = \theta\frac{y}{R}, \qquad \sigma_{zz} = E\theta\frac{y}{R},$$

$$\epsilon_{z\theta} = \frac{r}{R}\frac{\partial\theta}{\partial\alpha}, \qquad \sigma_{z\theta} = \frac{Gr}{R}\frac{\partial\theta}{\partial\alpha},$$

$$\mathfrak{M}_z = \int_A \sigma_{zz}\, y\, dA = EI_x\frac{\theta}{R},$$

$$\mathfrak{M}_\theta = \int_A \sigma_{r\theta}\, r\, dA = \frac{GC}{R}\frac{\partial\theta}{\partial\alpha}. \tag{9-2}$$

3. In-Plane Flexure Modes (frame 3 of Table 9-1, frames 2 through 4 of Table 9-2).

$$\epsilon_z = \frac{x}{R^2}\frac{\partial^2 X}{\partial\alpha^2}, \qquad \sigma_{zz} = \frac{Ex}{R^2}\frac{\partial^2 X}{\partial\alpha^2},$$

$$\mathfrak{M}_z = \int_A \sigma_{zz}\, x\, dA = \frac{EI_y}{R^2}\frac{\partial^2 X}{\partial\alpha^2}. \tag{9-3}$$

4. Out-of-Plane Flexure Modes (frame 4 of Table 9-1, frames 5 and 6 of Table 9-2). Both an r–θ and an x–y coordinate system are used to represent the x–y

Table 9-2. Circular Arcs.

Notation: x = distance from centroid of cross section toward geometric center of arc; y = distance perpendicular to plane of arc; z = distance along arc; θ = angle of twist about z axis; $\tilde{x}, \tilde{y}, \tilde{z}, \tilde{\theta}$ = mode shapes corresponding to deformations parallel to x, y, z axes and rotation about z axis, respectively; m = mass per unit length of ring; C = torsion constant (Table 8-18); E = modulus of elasticity; G = shear modulus $\{E/[2(1+\nu)]\}$; I_x, I_y = area moments of inertia about x and y axes, respectively (Table 5-1); R = radius to midline of partial ring; α = angular position about arc; α_0 = angle subtended by arc (radians); μ = mass density of partial ring material; ν = Poisson's ratio; see Fig. 9-1 for coordinate system; see Table 3-1 for consistent sets of units. The formulas in this table are approximate.

Description	Natural Frequencies, f_i (Hz)	Mode Shape
1. In-plane Extension Mode, Pinned-Pinned or Clamped-Clamped Arc 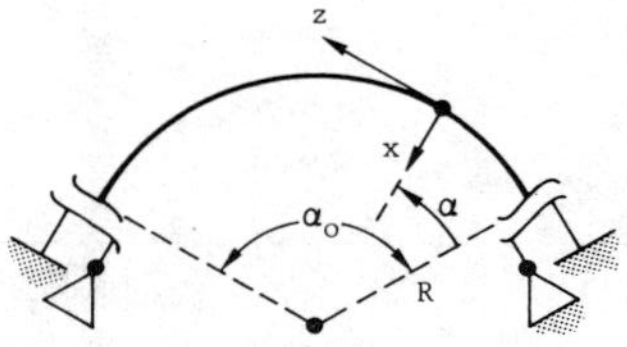 (The frequency of extension modes is well above those of flexure modes for slender arcs.)	Modes with $\tilde{x}$ symmetric and $\tilde{z}$ antisymmetric about midspan: $\frac{0.97}{2\pi R}\left(\frac{E}{\mu}\right)^{1/2}$; i = 0 , $\frac{i}{2\alpha_o R}\left[1+\left(\frac{\alpha_o}{i\pi}\right)^2\right]^{1/2}\left(\frac{E}{\mu}\right)^{1/2}$; i = 2,4,6... Modes with $\tilde{x}$ antisymmetric and $\tilde{z}$ symmetric about midspan: $\frac{i}{2\alpha_o R}\left[1+\left(\frac{\alpha_o}{i\pi}\right)^2\right]^{1/2}\left(\frac{E}{\mu}\right)^{1/2}$; i = 1,3,5...	$\begin{bmatrix}\tilde{x}\\ \tilde{y}\\ \tilde{z}\\ \tilde{\theta}\end{bmatrix}_i = \begin{bmatrix}\cos\frac{i\pi\alpha}{\alpha_o}\\ 0\\ -\frac{i\pi}{\alpha_o}\sin\frac{i\pi\alpha}{\alpha_o}\\ 0\end{bmatrix}$; i = 0,1,2,3... Ref. 9-7
2. In-Plane Flexure Mode, Pinned-Pinned Arc 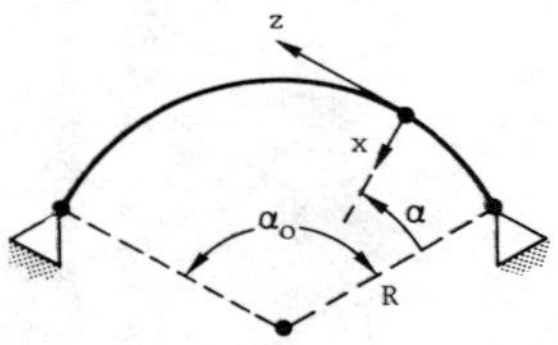	Modes with $\tilde{x}$ antisymmetric and $\tilde{z}$ symmetric about midspan: $\frac{i^2\pi^2}{2\pi(R\alpha_o)^2}\left\{\frac{\left[1-\left(\frac{\alpha_o}{i\pi}\right)^2\right]^2}{1+3\left(\frac{\alpha_o}{i\pi}\right)^2}\right\}^{1/2}\left(\frac{EI_y}{m}\right)^{1/2}$; i = 2,4,6... Modes with $\tilde{x}$ symmetric and $\tilde{z}$ antisymmetric about midspan: $\frac{i^2\pi^2}{2\pi(R\alpha_o)^2}\left\{\frac{\left[1-\left(\frac{\alpha_o}{i\pi}\right)^2\right]^2}{1+\frac{1}{i^2}+2\left(\frac{\alpha_o}{i\pi}\right)^2}\right\}^{1/2}\left(\frac{EI_y}{m}\right)^{1/2}$; i = 3,5,7,...	$\begin{bmatrix}\tilde{x}\\ \tilde{y}\\ \tilde{z}\\ \tilde{\theta}\end{bmatrix}_i = \begin{bmatrix}\sin\frac{i\pi\alpha}{\alpha_o}\\ 0\\ \frac{\alpha_o}{i\pi}\left(1-\cos\frac{i\pi\alpha}{\alpha_o}\right)\\ 0\end{bmatrix}$ Ref. 9-7 i = 2,4,6... $\begin{bmatrix}\tilde{x}\\ \tilde{y}\\ \tilde{z}\\ \tilde{\theta}\end{bmatrix}_i = \begin{bmatrix}\sin\frac{i\pi\alpha}{\alpha_o}-\frac{1}{i}\sin\frac{\pi\alpha}{\alpha_o}\\ 0\\ \frac{-\alpha_o}{i\pi}\left(\cos\frac{i\pi\alpha}{\alpha_o}-\frac{1}{\pi^3}\cos\frac{\pi\alpha}{\alpha_o}\right)\\ 0\end{bmatrix}$; i = 3,5,7,... Ref. 9-7

Table 9-2. Circular Arcs. *(Continued)*

Description	Natural Frequencies, f_i (Hz)	Mode Shape
3. In-Plane Flexure Mode, Clamped-Clamped Arc 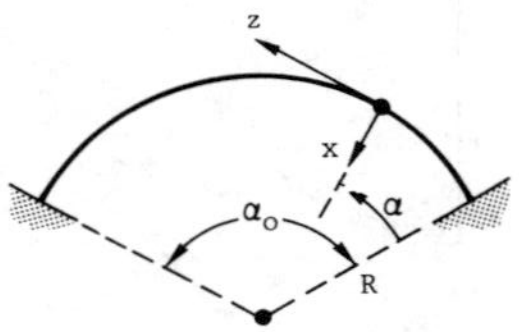	Modes with $\tilde{x}$ antisymmetric and $\tilde{z}$ symmetric about midspan: $\frac{\lambda_i^2}{2\pi(R\alpha_o)^2}\left[\frac{1 - 2\sigma_i^2\left(1 - \frac{2}{\sigma_i\lambda_i}\right)\left(\frac{\alpha_o}{\lambda_i}\right)^2 + \left(\frac{\alpha_o}{\lambda_i}\right)^4}{1 + 5\sigma_i^2\left(1 - \frac{2}{\sigma_i\lambda_i}\right)\left(\frac{\alpha_o}{\lambda_i}\right)^2}\right]^{1/2}\left(\frac{EI_y}{m}\right)^{1/2};$ i = 2,4,6...	$\begin{bmatrix}\tilde{x}\\ \tilde{y}\\ \tilde{z}\\ \tilde{\theta}\end{bmatrix}_i = \begin{bmatrix}\psi_i\\ 0\\ \frac{\alpha_o}{\lambda_i}\left(\frac{1}{\lambda_i^3}\frac{d^3\psi_i}{d\left(\frac{\alpha}{\alpha_o}\right)^3} + 2\sigma_i\right)\\ 0\end{bmatrix}$ i = 2,4,6...
Parameters in frequency and mode shape: λ_i, σ_i, and ψ_i are given in frame 7 of Table 8-1 with $\psi_i = \tilde{y}_i$ and substitution "α/α_o" for "x/L".	Modes with $\tilde{x}$ symmetric and $\tilde{z}$ antisymmetric about midspan: $\frac{\lambda_i^2}{2\pi(R\alpha_o)^2}\left[\frac{1 - 1.82\left(\frac{\alpha_o}{\lambda_i}\right)^2}{1 + \left(\frac{\lambda_1}{\lambda_i}\right)^2 + 2\left(1.63 - \frac{5}{\lambda_i}\right)\left(\frac{\alpha_o}{\lambda_i}\right)^2}\right]^{1/2}\left(\frac{EI_y}{m}\right)^{1/2};$ i = 3,5,7...	$\begin{bmatrix}\tilde{x}\\ \tilde{y}\\ \tilde{z}\\ \tilde{\theta}\end{bmatrix}_i = \begin{bmatrix}\psi_i - \frac{\sigma_i}{\sigma_1}\frac{\lambda_1}{\lambda_i}\psi_1\\ 0\\ \frac{\alpha_o}{\lambda_i^4}\left[\frac{d^3\psi_i}{d\left(\frac{\alpha}{\alpha_o}\right)^3} - \frac{\sigma_i}{\sigma_1}\left(\frac{\lambda_i}{\lambda_1}\right)^3\frac{d^3\psi_1}{d\left(\frac{\alpha}{\alpha_o}\right)^3}\right]\\ 0\end{bmatrix}$ Ref. 9-7 i = 3,5,7...
4. Out-of-Plane Flexure Mode, Pinned-Pinned Arc 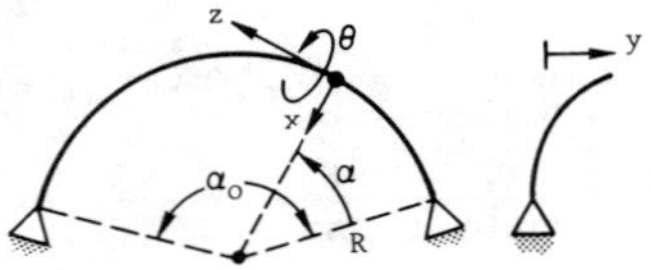	$\frac{\pi^2}{2\pi(R\sigma_o)^2}\left\{\frac{\frac{GC}{EI_x}\left[1 - \left(\frac{\alpha_o}{\pi}\right)^2\right]^2}{\left(\frac{\alpha_o}{\pi}\right)^2 + \frac{GC}{EI_x}}\right\}^{1/2}\left(\frac{EI_x}{m}\right)^{1/2};$ $0 < \frac{EI_x}{GC} < 2$	$\begin{bmatrix}\tilde{x}\\ \tilde{y}\\ \tilde{z}\\ \tilde{\theta}\end{bmatrix} = \begin{bmatrix}0\\ \sin\frac{\pi\alpha}{\alpha_o}\\ 0\\ -\frac{1}{R}\frac{\left(1 + \frac{GC}{EI_x}\right)}{\left[\left(\frac{\alpha_o}{\pi}\right)^2 + \frac{GC}{EI_x}\right]}\sin\frac{\pi\alpha}{\alpha_o}\end{bmatrix}$
Fundamental Mode Only	For circular cross sections: $\frac{\pi^2}{2\pi(R\alpha_o)^2}\left\{\frac{\left[1 - \left(\frac{\alpha_o}{\pi}\right)^2\right]^2}{1 + (1 + \nu)\left(\frac{\alpha_o}{\pi}\right)^2}\right\}^{1/2}\left(\frac{EI_x}{m}\right)^{1/2}$	$\begin{bmatrix}\tilde{x}\\ \tilde{y}\\ \tilde{z}\\ \tilde{\theta}\end{bmatrix} = \begin{bmatrix}0\\ \sin\frac{\pi\alpha}{\alpha_o}\\ 0\\ -\frac{1}{R}\frac{(2 + \nu)}{\left[1 + (1 + \nu)\left(\frac{\alpha_o}{\pi}\right)^2\right]}\sin\frac{\pi\alpha}{\alpha_o}\end{bmatrix}$ Ref. 9-8

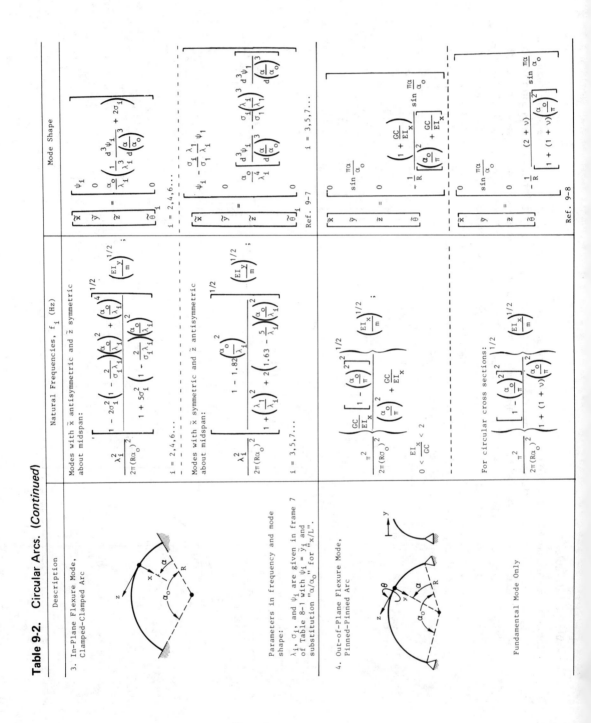

5. Out-of-Plane Flexure Mode, Clamped-Clamped Arc 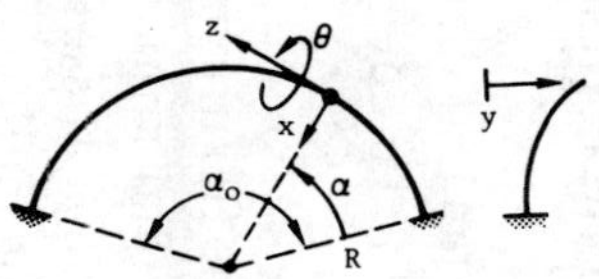 Parameters in Frequency and Mode Shape: $\beta = \left(\frac{\alpha_o}{\pi}\right)^4 - 2.492 \left(\frac{\alpha_o}{\pi}\right)^2 + 5.139$ ψ is given by $\psi = \tilde{y}_1$ in frame 7 of Table 8-1 with "α/α_o" substituted for "x/L", Ref. 9-8.	$\frac{\pi^2}{2\pi(R\alpha_o)^2} \left[\frac{3.586\left(\frac{\alpha_o}{\pi}\right)^2 + 1.246 \frac{GC}{EI_x}\beta}{\left(\frac{\alpha_o}{\pi}\right)^2 + 1.246 \frac{GC}{EI_x}}\right]^{1/2} \left(\frac{EI_x}{m}\right)^{1/2}$ $0 < \frac{EI_x}{GC} < 2$	$\begin{bmatrix}\tilde{x}\\ \tilde{y}\\ \tilde{z}\\ \tilde{\theta}\end{bmatrix} = \begin{bmatrix}0\\ \psi\\ 0\\ \dfrac{-\left(1 + \frac{GC}{EI_x}\right)\frac{\psi}{R}}{0.8025\left(\frac{\alpha_o}{\pi}\right)^2 + \frac{GC}{EI_x}}\end{bmatrix}$
	For circular cross sections: $\frac{\pi^2}{2\pi(R\alpha_o)^2} \left[\frac{3.586\left(\frac{\alpha_o}{\pi}\right)^2 + \frac{1.246}{1 + \nu}\beta}{\left(\frac{\alpha_o}{\pi}\right)^2 + \frac{1.246}{1 + \nu}}\right]^{1/2} \left(\frac{EI_x}{m}\right)^{1/2}$	$\begin{bmatrix}\tilde{x}\\ \tilde{y}\\ \tilde{z}\\ \tilde{\theta}\end{bmatrix} = \begin{bmatrix}0\\ \psi\\ 0\\ \dfrac{-(2 + \nu)\psi}{R\left[1 + 0.8025(1 + \nu)\left(\frac{\alpha_o}{\pi}\right)^2\right]}\end{bmatrix}$
6. Out-of-Plane Flexure Mode, Clamped-Pinned Arc 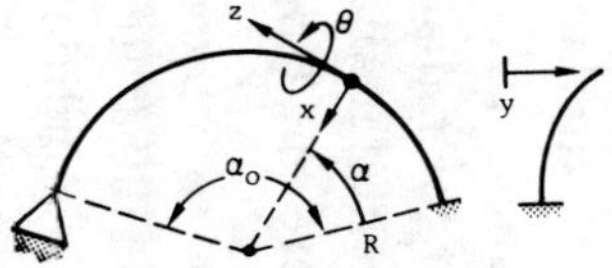 Parameters in Frequency and Mode Shape: $\beta = \left(\frac{\alpha_o}{\pi}\right)^4 - 2.332 \left(\frac{\alpha_o}{\pi}\right)^2 + 2.440$ ψ is given by $\psi = \tilde{y}_1$ in frame 6 of Table 8-1 with "α/α_o" substituted for "x/L", Ref. 9-8.	$\frac{\pi^2}{2\pi(R\alpha_o)^2} \left[\frac{1.080\left(\frac{\alpha_o}{\pi}\right)^2 + 1.166 \frac{GC}{EI_x}\beta}{\left(\frac{\alpha_o}{\pi}\right)^2 + 1.166 \frac{GC}{EI_x}}\right]^{1/2} \left(\frac{EI_x}{m}\right)^{1/2}$ $0 < \frac{EI_x}{GC} < 2;$	$\begin{bmatrix}\tilde{x}\\ \tilde{y}\\ \tilde{z}\\ \tilde{\theta}\end{bmatrix} = \begin{bmatrix}0\\ \psi\\ 0\\ \dfrac{-\left(1 + \frac{GC}{EI_x}\right)\frac{\psi}{R}}{0.8576\left(\frac{\alpha_o}{\pi}\right)^2 + \frac{GC}{EI_x}}\end{bmatrix}$
	For circular cross sections: $\frac{\pi^2}{2\pi(R\alpha_o)^2} \left[\frac{1.080\left(\frac{\alpha_o}{\pi}\right)^2 + \frac{1.166}{1 + \nu}\beta}{\left(\frac{\alpha_o}{\pi}\right)^2 + \frac{1.166}{1 + \nu}}\right]^{1/2} \left(\frac{EI_x}{m}\right)^{1/2}$	$\begin{bmatrix}\tilde{x}\\ \tilde{y}\\ \tilde{z}\\ \tilde{\theta}\end{bmatrix} = \begin{bmatrix}0\\ \psi\\ 0\\ \dfrac{-(2 + \nu)\psi/R}{1 + 0.8576(1 + \nu)\left(\frac{\alpha_o}{\pi}\right)^2}\end{bmatrix}$

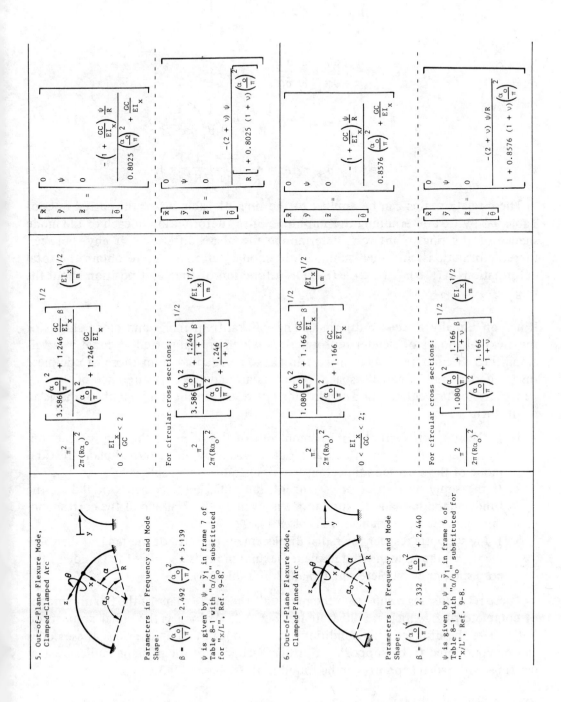

plane. A circular cross section has been assumed in the strain-deformation and stress-strain relationships.

$$\epsilon_z = \frac{\theta y}{R} - \frac{y}{R}\frac{\partial^2 Y}{\partial \alpha^2}, \qquad \sigma_{zz} = E\epsilon_z,$$

$$\epsilon_{z\theta} = \frac{r}{R}\left(\frac{\partial \theta}{\partial \alpha} + \frac{1}{R}\frac{\partial Y}{\partial \alpha}\right), \qquad \sigma_{z\theta} = G\epsilon_{z\theta},$$

$$\mathfrak{M}_z = \int_A \sigma_{zz}\, y\, dA = \frac{EI_y}{R}\left(\theta - \frac{1}{R}\frac{\partial^2 Y}{\partial \alpha^2}\right),$$

$$\mathfrak{M}_\theta = \int_A \sigma_{r\theta}\, r\, dA = \frac{GC}{R}\left(\frac{\partial \theta}{\partial \alpha} + \frac{1}{R}\frac{\partial Y}{\partial \alpha}\right). \tag{9-4}$$

These relationships can be applied to the rings of Table 9-1 or the partial rings of Table 9-2 by first determining the amplitude of the deformation in each of the mode shapes of the ring of interest, determining the phase difference, if any, between these deformations, and then summing the modal deformations to obtain the total deformations $X(t, \alpha)$, $Y(t, \alpha)$, $\theta(t, \alpha)$ as a function of time and position about the ring.

Rings on Equally Spaced Supports. The natural frequencies and mode shapes of the flexural modes of slender uniform rings which are supported at equal intervals along their circumference can, in certain cases, be adapted from those of complete rings. If there are 2i equally spaced supports around the circumference of the ring which do not prevent rotation of the ring but resist certain components of displacement, then:

1. If the supports restrict displacement out of the plane of the ring (y), the fundamental out-of-plane flexural mode is given by the i out-of-plane flexural mode of the corresponding complete ring (frame 4 of Table 9-1).
2. If the supports restrict *both* tangent and radial displacements (x and z), the fundamental in-plane flexural mode is given by the 2i mode of the corresponding complete ring (frame 3 of Table 9-1).
3. If the supports restrict the radial displacement (z) but not the tangent displacement (x), the fundamental in-plane flexural mode is given by the i mode of the corresponding complete ring (frame 2 of Table 9-1).

These relationships result from the nature of the mode shape of the flexural modes of complete rings, which are illustrated in Fig. 9-2. As long as the radial supports do not violate the displacement conditions at the nodes of complete rings, these nodes can be considered to be supports. Further discussion of the in-plane modes of slender rings with radial supports can be found in Refs. 9-2 and 9-3.

Thick Rings. The analysis of thick rings must incorporate the effects of shear deformation and rotary inertia (Ref. 9-3). It is shown in Ref. 9-4 that the in-plane flexural vibrations of a thick circular ring, taking into account shear deformation and rotary inertia, are given by:

$$\frac{f_i\big|_{\text{rotary + shear}}}{f_i\big|_{\text{frame 3 of Table 9-1}}} = \left[\frac{1}{1 + i^2\gamma + \dfrac{I_y}{AR^2}\dfrac{(i^2 - 1)^2}{i^2 + 1}\dfrac{1}{1 + i^2\gamma}}\right]^{1/2},$$

where i = 1, 2, 3, . . . , and

$$\gamma = \frac{I_y}{AR^2}\frac{E}{G}\frac{1}{K}.$$

A is the area of the ring cross section, E is the modulus of elasticity, G is the shear modulus $\{E/[2(1 + \nu)]; \nu = \text{Poisson's ratio}\}$, I_y is the area moment of inertia of the cross section, K is the shear coefficient (Table 8-14), and R is the radius of the ring. The out-of-plane vibration of thick rings is discussed in Ref. 9-5.

Elliptical Rings. The natural frequencies in hertz of slender, uniform elliptical rings vibrating in their own planes are given by:

$$f_i = \frac{\lambda_i^2}{2\pi b^2}\left(\frac{EI_y}{m}\right)^{1/2}; \qquad i = 1, 2, 3, \ldots,$$

where E is the modulus of elasticity, I_y is the area moment of inertia of the ring cross section about the axis perpendicular to the plane of the ring, and m is the mass per unit length of ring. The overall length of the ellipse (twice the major axis) is 2a, and the overall width of the ellipse (twice the minor axis) is 2b. The dimensionless frequency parameters, λ_i, are functions of the ratio a/b. For the fundamental in-plane flexure mode (Ref. 9-6):

$\frac{a}{b}$	1.0	1.1	1.2	1.4	1.7	2.0	2.5	3.0
λ	1.638	1.558	1.481	1.342	1.167	1.028	0.8528	0.7271

9.2. CIRCULAR ARCS

General Case. A circular arc is a segment of a circular ring. The natural frequencies and mode shapes of homogeneous, slender, circular arcs with both clamped-clamped and pinned-pinned ends are given in Table 9-2. The torsion mode is not included in Table 9-2. The arcs are assumed to be slender, that is, the thickness of the arc is much less than the radius of the arc, so that shear deformation is negligible and rotary inertia can be neglected in comparison with flexural inertia. It is assumed that the torsional and flexural vibrations of the ring are not inertially coupled. This requirement is always met if the cross section of the arc is symmetric about both the x and y axes. Warping of the cross section due to torsion has been neglected (see Section 8.9).

Boundary Conditions. The boundary conditions applied at the ends of the circular arcs are:

In-Plane Modes

Pinned-pinned:

$$Z = X = \frac{\partial^2 X}{\partial z^2} = 0, \quad \alpha = 0, \theta_0,$$

$$Y = \theta = 0, \qquad 0 \leqslant \alpha \leqslant \alpha_0;$$

Clamped-clamped:

$$Z = X = \frac{\partial X}{\partial z} = 0, \qquad \alpha = 0, \alpha_0,$$

$$Y = \theta = 0, \qquad 0 \leqslant \alpha \leqslant \alpha_0;$$

Out-of-Plane Modes

Pinned-pinned:

$$Y = \frac{\partial^2 Y}{\partial z^2} = \theta = \frac{\partial^2 \theta}{\partial z^2} = 0, \qquad \alpha = 0, \alpha_0,$$

$$X = Z = 0, \qquad 0 \leqslant \alpha \leqslant \alpha_0;$$

Clamped-clamped:

$$Y = \frac{\partial Y}{\partial z} = \theta = \frac{\partial \theta}{\partial z} = 0, \qquad \alpha = 0, \alpha_0,$$

$$X = Z = 0, \qquad 0 \leqslant \alpha \leqslant \alpha_0.$$

X, Y, and Z are the deformations parallel to the x, y, and z axes, respectively (Fig. 9-1). θ is the rotation about the z axis. Note that for the out-of-plane modes, the term pinned support refers to a support which prevents displacement and rigid body rotation ($\theta = 0$ at $\alpha = 0, \alpha_0$) but which cannot support an out-of-plane moment. These boundary conditions do not permit motion of the arc parallel to its own axis at the supports.

Natural Frequencies and Mode Shapes. The natural frequencies given in Table 9-2 are approximate formulas which were generated using the Raleigh technique with curved beam theory and shallow shell theory. The mode shapes were adapted from the mode shapes of straight beams; they do not always satisfy all the boundary conditions. Nevertheless, these formulas have proven to provide very good estimates of exact numerical solutions. The formulas in frames 4, 5, and 6 were adapted from Ref. 9-8 by neglecting the effect of rotary inertia and warping of the cross section. The formulas for the out-of-plane modes (frames 4, 5, and 6 of Table 9-2) are considerably simplified if the arcs have circular cross section, since for a circular cross section $C = 2I_x$, $G = E/[2(1 + \nu)]$, and thus $GC/(EI_x) = 1/(1 + \nu)$. The formulas presented in Table 9-2 are generally within a few percent of the exact result for the lower modes of slender arcs with $0° < \alpha \leqslant 180°$.

Analysis has shown that the modes (other than purely torsional modes) of circular arcs can be divided into three classes: (1) extensional modes (extension and contraction of the ring along its own axis with deformation in the plane of the ring), (2) in-plane flexure modes (inextensional vibration in the plane of the ring), and (3) out-of-plane flexure modes (coupled flexural and torsional inextensional vibrations out of the plane of the ring). These divisions apply to complete rings as well (Section 9.1). The natural frequencies of the extensional modes are generally much higher than the natural frequencies of the lower flexural modes.

The approximate solutions of Table 9-2 suggest that the extensional and flexure modes do not interact. This is not always the case. Numerical solutions (Fig. 9-3) show some interaction between the in-plane flexural mode and the extensional mode for relatively thick rings. The "breathing" extensional mode, corresponding to i = 1 in frame 1 of Table 9-2, is particularly prone to interaction with flexural modes (Ref. 9-7).

The curvature of the arc results in coupling of the in-plane transverse deflection (X) and longitudinal deflection of the arc parallel to its own axis (Z). This coupling

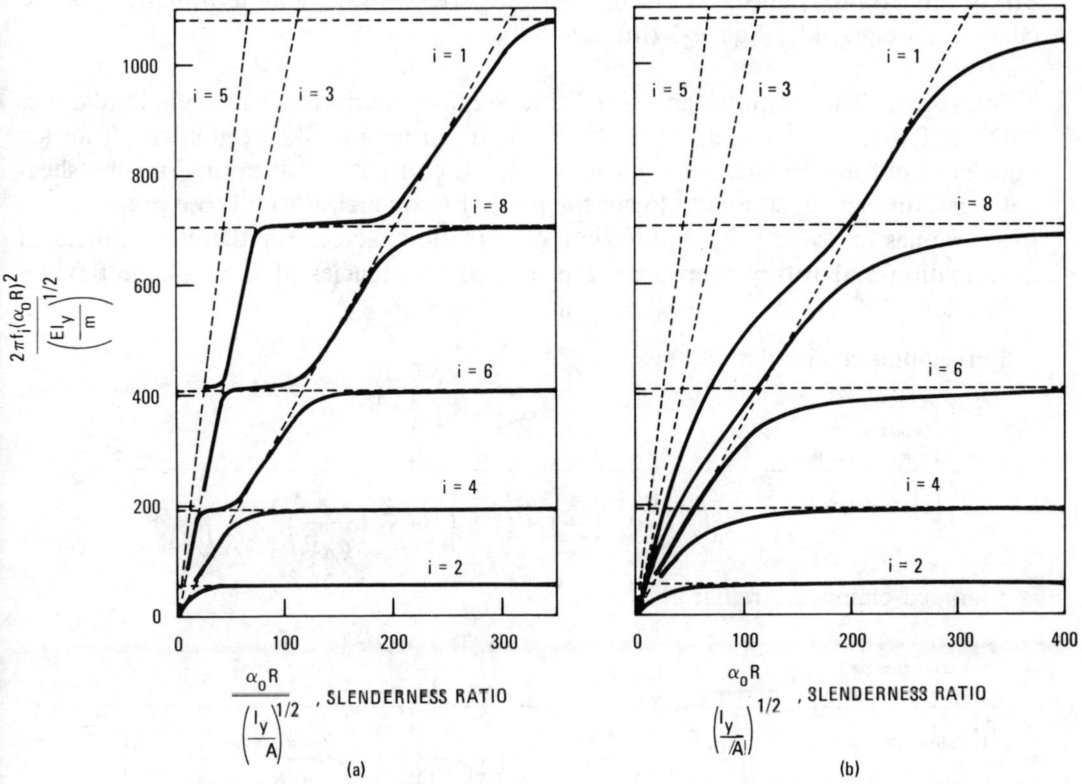

Fig. 9-3. Comparison of approximate natural frequencies (- - - -) (Table 9-2, frames 1 and 3) of a fixed-fixed 90° arc with exact solutions (———): (a) without shear deformation and rotary inertia and (b) with shear deformation and rotary inertia; antisymmetric in-plane modes only. Horizontal lines are flexural modes; diagonal lines are extensional modes (Ref. 9-10).

has considerable influence on the first in-plane flexural mode. A curved beam cannot support an in-plane flexural mode corresponding to the first (i = 1) transverse vibration mode of an analogous straight beam if the boundaries of the arc prevent longitudinal deflection of the arc parallel to its own axis. Modes corresponding to the higher flexural modes of straight beams can exist in the arc because these modes do not require longitudinal motion at the boundaries (although the mode shapes of the odd modes, i = 3, 5, 7, . . . , must generally be modified from that of straight beams to meet this condition). If the ratio of the rise of the arc is a small fraction of the characteristic dimension of the cross section

$$\frac{R\left[1 - \cos\left(\frac{\alpha_0}{2}\right)\right]}{\left(\frac{I_y}{A}\right)^{1/2}} \ll 1,$$

then the natural frequency of the shallow arc can be predicted from the natural frequencies of straight beams as given in Table 8-1 (Ref. 9-9). α_0, I_y, and R are defined in Table 9-2. A is the area of the cross section.

Stress and Strain. The relationships between stress, strain, and deformation for circular arcs are given by Eqs. 9-1 through 9-4.

Thick Arcs. The formulas given in Table 9-2 were derived for slender circular arcs with $(\alpha_0 R)/(I_y/A)^{1/2}$ or $(\alpha_0 R)/(I_x A)^{1/2}$ on the order of 100 or greater. If an arc does not conform to these assumptions, then the effects of rotary inertia and shear deformation can substantially lower the natural frequencies from those predicted by the formulas in Table 9-2. Approximate correction factors for the effects of shear deformation and rotary inertia on the natural frequencies of the in-plane flexural modes are (Ref. 9-10):

Pinned-pinned circular arc:

$$\frac{f_i\Big|_{\text{rotary inertia + shear deformation}}}{f_i\Big|_{\text{Table 9-2, frame 2}}} = \frac{\frac{\alpha_0 R}{i\pi}\left(\frac{A}{I_y}\right)^{1/2}}{\left\{\left(\frac{\alpha_0 R}{i\pi}\right)^2\left(\frac{A}{I_y}\right) + \Omega + \left[1 + \Omega\left(\frac{i\pi}{\alpha_0 R}\right)^2\left(\frac{I_y}{A}\right)\right]^{-1}\right\}^{1/2}}.$$

Clamped-clamped circular arcs:

$$\frac{f_i\Big|_{\text{rotary inertia + shear deformation}}}{f_i\Big|_{\text{Table 9-2, frame 3}}} = \frac{\frac{\alpha_0 R}{i\pi}\left(\frac{A}{I_y}\right)^{1/2}}{\left\{\left(\frac{\alpha_0 R}{i\pi}\right)^2\left(\frac{A}{I_y}\right) + \gamma_i^2\Omega + \gamma_i^2\left[1 + \Omega\left(\frac{i\pi}{\alpha_0 R}\right)^2\left(\frac{I_y}{A}\right)\right]^{-1}\right\}^{1/2}},$$

where

$$\Omega = \frac{E}{KG}.$$

K is the shear coefficient (Table 8-14), G is the shear modulus $\{E/[2(1+\nu)]$, ν = Poisson's ratio$\}$, and A is the area of the cross section of the arc. γ_i is the ratio of the natural frequency of a clamped-clamped arc to the natural frequency of a pinned-pinned arc in the mode of interest, with shear deformation and rotary inertia neglected. These frequencies are given in frames 2 and 3, respectively, of Table 9-2. All other quantities in the above equations are defined in Table 9-2.

The rotary inertia associated with twisting of the arc about its own axis can have a large effect on the natural frequencies of the out-of-plane flexural mode if the torsional stiffness of the arc is relatively small compared with the flexural stiffness, i.e., $EI_x/(GC) > 2$. The influence of rotary inertia on the out-of-plane flexural modes is discussed in Refs. 9-10 and 9-11. The natural frequencies of noncircular arcs are discussed in Ref. 9-12.

9.3. RIGHT-ANGLE AND U BENDS

The natural frequencies of slender right-angle and U bends with intermediate supports are shown in Figs. 9-4 and 9-5. The intermediate supports are pinned to allow rotation about any axis and to prevent transverse motion at the support, but the in-

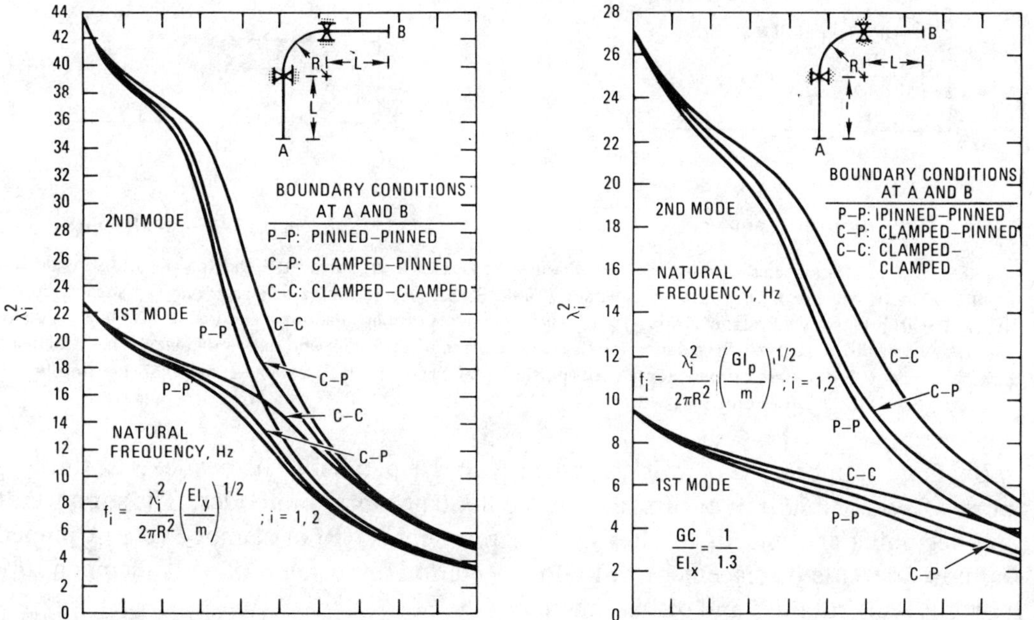

Fig. 9-4. Natural frequencies of the first two flexural modes of a supported right-angle bend for vibration in and out of the plane of the bend. C = torsion constant (Table 8-18). E = modulus of elasticity. G = shear modulus $\{E/[2(1+\nu)]\}$, where ν = Poisson's ratio. I_x, I_y = area moments of inertia about x and y axes (Table 5-1); y axis is perpendicular to the plane of the paper, x axis is in the plane of the paper and perpendicular to the local beam axis, as in Fig. 9-1. I_p = area polar moment of inertia of the beam about its own axis; m = mass per unit length of beam. See Table 3-1 for consistent sets of units (Ref. 9-13).

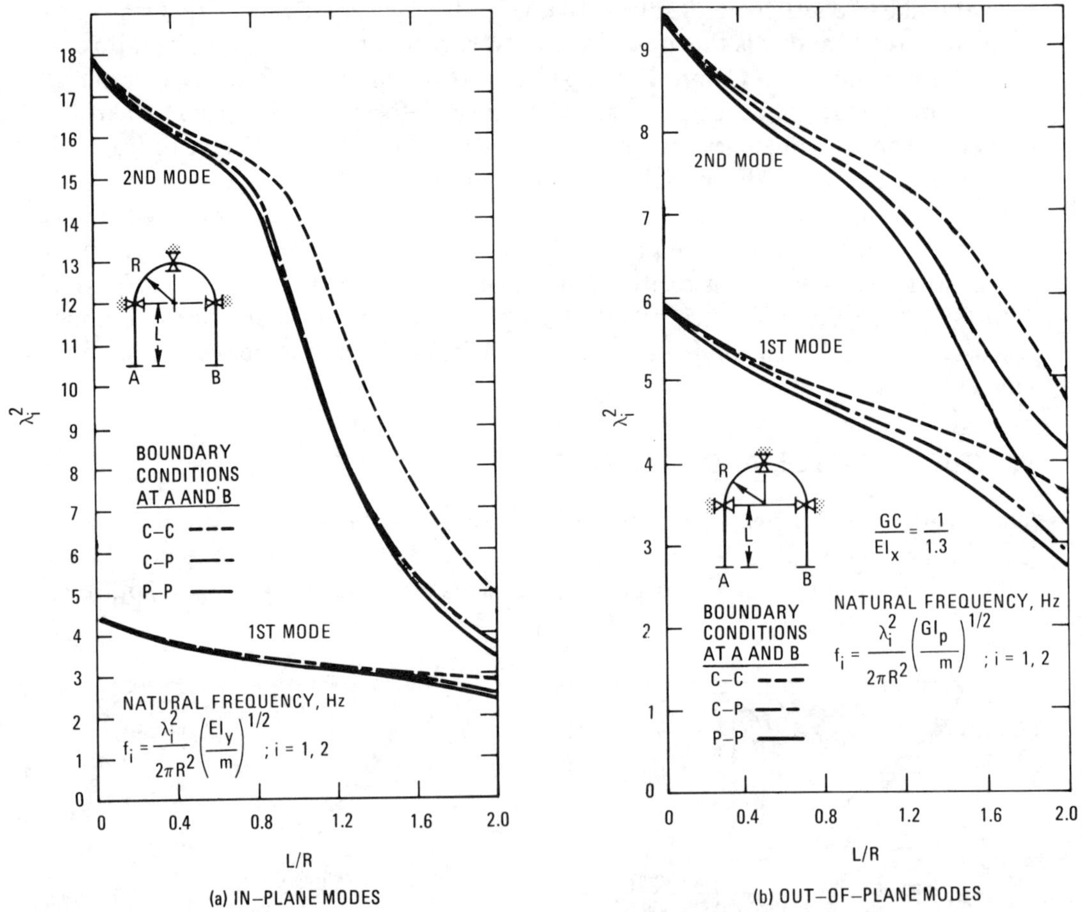

Fig. 9-5. Natural frequencies of the first two flexural modes of a supported U bend for vibration in and out of the plane of the bend. C = torsion constant (Table 8-18). E = modulus of elasticity. G = shear modulus $\{E/[2(1 + \nu)]\}$, where ν = Poisson's ratio. I_x, I_y = area moments of inertia about x and y axes (Table 5-1); y axis is perpendicular to the plane of the paper, x axis is in the plane of the paper and perpendicular to the local beam axis, as in Fig. 9-1. I_p = area polar moment of inertia of the beam about its own axis; m = mass per unit length of beam. See Table 3-1 for consistent sets of units (Ref. 9-13).

termediate supports do permit the beam to move parallel to its own axis at the support. This axial motion occurs only in the in-plane vibration modes. The supports at the beginning and end of the beams are either pinned (P) or clamped (C). A pinned support prevents displacement but allows rotation about any axis. A clamped support prevents rotation and displacement.

Shear deformation, warping of the cross section due to torsion, and the possibility of inertia coupling of rotation and displacement have been neglected in Figs. 9-4 and 9-5. Rotary inertia has been neglected in Figs. 9-4(a) and 9-5(a). The rotary inertia associated with the beam twisting about its own axis has been included in Figs. 9-4(b) and 9-5(b); these figures are valid only for $GC/(EI_x) = 1.3$. If the beam has circular cross section, then $C = 2I_x$ and $GC/(EI_x) = 1/(1 + \nu)$, where ν = Poisson's ratio. Thus, Figs. 9-4(b) and 9-5(b) are valid for circular rods or tubes composed of materials with a Poisson's ratio of $\nu = 0.3$.

9.4. SUPPORTED HELICES

A multiply supported helix, shown in Fig. 9-6, can arise in helically coiled heat exchangers where a tube is coiled through a series of support plates. A closed form solution is not available for a multiply supported helix. However, if the supports are spaced uniformly about the helix and the helix angle is small, then the natural frequencies of the helix can be approximated from the natural frequencies of rings in cases where there are an even number of supports per turn.

The natural frequencies of complete rings for in-plane and out-of-plane flexural vibration are compared with the results of finite element analysis of multiply supported helices of various lengths with six supports per turn in Fig. 9-7. The mode i = 3 is used in the ring calculation since this mode gives six uniformly spaced nodes about the ring circumference, which corresponds to six supports per turn. The helix angle is 5 deg and the ratio of the helix radius (R) to the radius (r) of the circular rod used to represent the helix is R/r = 89. Thus, the helix is slender. The other helix parameters were E = m = 1, R = 100, r = 1.128, and ν = 0.3.

The symbols in Fig. 9-7 correspond to various boundary conditions at the ends of the helix and at the intermediate supports as listed in Table 9-3. Only the lowest natural frequencies of modes which were predominantly in the plane of the helical coil (x–z plane) and perpendicular to the y axis are given in Fig. 9-7. Figure 9-7 suggests:

1. The out-of-plane natural frequencies of slender helices with even numbers of supports per turn may be approximated from the natural frequencies of complete rings.
2. The in-plane natural frequencies of rings can adequately represent the in-plane natural frequencies of helices if the helix is permitted to move tangent to its own axis (z axis) at some of the supports.

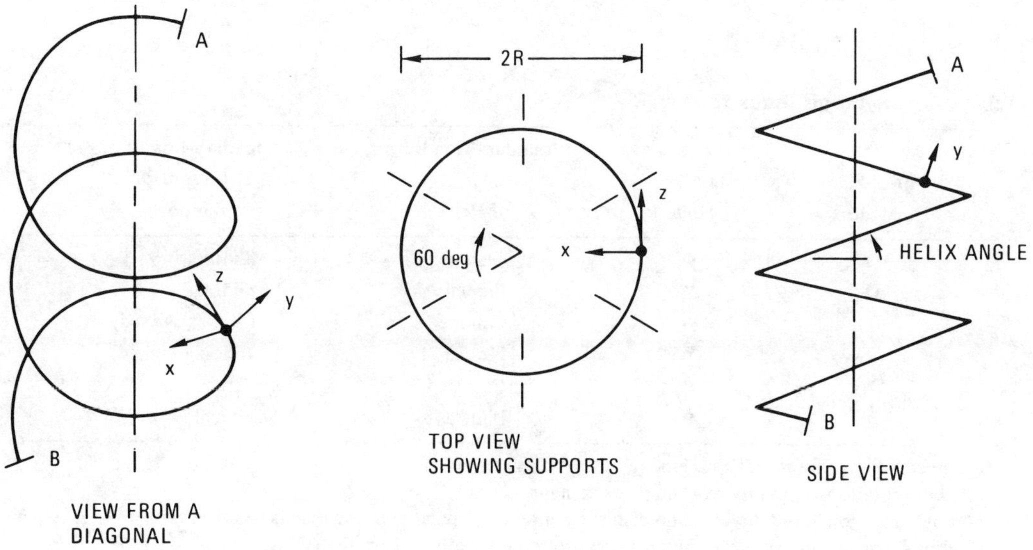

Fig. 9-6. A $2\frac{1}{2}$ turn helix with six equally spaced supports per turn.

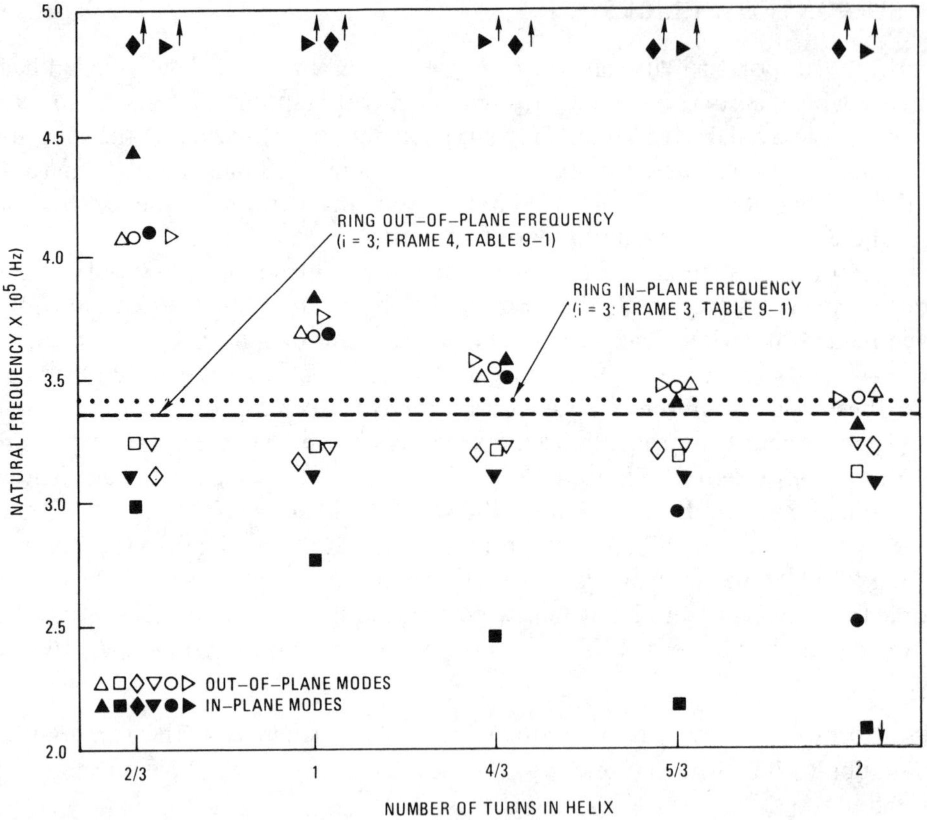

Fig. 9-7. Comparison of the fundamental natural frequency of a helix with six supports per turn with the natural frequencies of rings. The data points are numerical results for helices with various boundary conditions (Table 9-3) (author's result).

Table 9-3. Symbols Index for Fig. 9-7.[a]

Out-of-Plane Mode	In-Plane Mode	Boundary Conditions at Ends of Helix[b]	Boundary Conditions at Intermediate Supports
○	●	Clamped	Sliding
□	■	Pinned	Sliding
△	▲	Clamped	(Sliding-Pinned)
▽	▼	Pinned	(Sliding-Pinned)
▷	▶	Clamped	Pinned
◇	◆	Pinned	Pinned

[a]Clamped = no rotation or displacement,
Pinned = rotation about any axis, no displacement,
Sliding = rotation about any axis, no displacement except parallel to local helix axis (z axis),
(Sliding-Pinned) = intermediate supports are alternately sliding, then pinned, etc.
[b]At A and B in Fig. 9-6.

3. If none of the helix supports permit motion parallel to the local helix axis, then the in-plane fundamental frequencies will be substantially above that predicted by the ring formulas.
4. If all intermediate supports permit unrestricted motion parallel to the local helix axis, then a low-frequency, in-plane "unwinding" mode will arise which has a natural frequency below that of the fundamental in-plane ring frequency.

9.5. PORTAL FRAMES

A portal frame consists of a cap beam supported by two leg beams which are welded to the cap beam. The natural frequency of portal frames vibrating in the fundamen-

Table 9-4A. Portal Frame, Symmetric Mode.

Notation: E = modulus of elasticity; I = area moment of inertia about neutral axis; m = mass per unit length of beam; vertical legs are identical; see Table 3-1 for consistent sets of units (Ref. 9-14)

Portal Frame, First Symmetric In-Plane Mode	Natural Frequency, f (Hz)
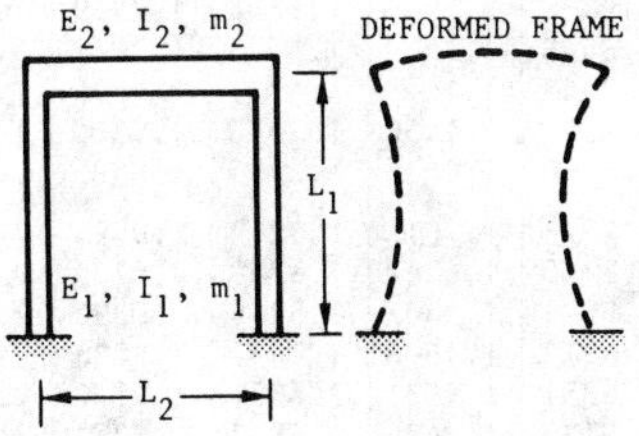	$f = \dfrac{\lambda^2}{2\pi L_1^2}\left(\dfrac{E_1 I_1}{m_1}\right)^{1/2}$

$$\lambda = \lambda\left[\left(\frac{E_1 I_1}{E_2 I_2}\frac{m_2}{m_1}\right)^{1/4}\frac{L_2}{L_1},\ \left(\frac{E_2 I_2}{E_1 I_1}\right)^{3/4}\left(\frac{m_2}{m_1}\right)^{1/4},\ \text{Boundary Conditions at Feet}\right]$$

$\left(\frac{m_2}{m_1}\right)^{1/4}\left(\frac{E_2 I_2}{E_1 I_1}\right)^{3/4}$	$\left(\frac{E_1 I_1}{E_2 I_2}\frac{m_2}{m_1}\right)^{1/4}\frac{L_2}{L_1}$ = 10.0	8.0	6.0	4.0	2.0	1.5	1.2	1.0	0.8	0.4	0.2	0.1
Pinned feet 10.0	0.3710	0.4543	0.5909	0.8596	1.6507	2.1684	2.6722	3.1416	3.6237	3.8519	3.8891	3.9074
8.0	0.3790	0.4637	0.6022	0.8735	1.6676	2.1845	2.6840	3.1416	3.5954	3.8355	3.8802	3.9028
6.0	0.3897	0.4767	0.6182	0.8939	1.6937	2.2097	2.7023	3.1416	3.5569	3.8097	3.8660	3.8952
4.0	0.4051	0.4958	0.6429	0.9270	1.7394	2.2546	2.7348	3.1416	3.4997	3.7637	3.8390	3.8804
2.0	0.4290	0.5273	0.6862	0.9911	1.8411	2.3582	2.8082	3.1416	3.4003	3.6578	3.7690	3.8392
1.5	0.4374	0.5387	0.7030	1.0181	1.8904	2.4102	2.8438	3.1416	3.3612	3.6054	3.7297	3.8143
1.2	0.4430	0.5467	0.7150	1.0383	1.9302	2.4531	2.8725	3.1416	3.3328	3.5628	3.6950	3.7910
1.0	0.4471	0.5525	0.7240	1.0540	1.9633	2.4894	2.8961	3.1416	3.3110	3.5275	3.6642	3.7693
0.8	0.4515	0.5589	0.7340	1.0720	2.0037	2.5346	2.9246	3.1416	3.2864	3.4845	3.6240	3.7394
0.4	0.4614	0.5735	0.7577	1.1173	2.1214	2.6729	3.0041	3.1416	3.2259	3.3622	3.4903	3.6244
0.2	0.4670	0.5819	0.7720	1.1466	2.2150	2.7948	3.0621	3.1416	3.1877	3.2706	3.3663	3.4907
0.1	0.4699	0.5865	0.7799	1.1636	2.2793	2.8921	3.0985	3.1416	3.1658	3.2121	3.2731	3.3666
Clamped feet (Shown) 10.0	0.3814	0.4665	0.6057	0.8780	1.6760	2.1999	2.7172	3.2231	3.9269	4.6367	4.6850	4.7070
8.0	0.3897	0.4767	0.6183	0.8941	1.6973	2.2223	2.7386	3.2408	3.9269	4.6167	4.6745	4.7015
6.0	0.4007	0.4902	0.6356	0.9172	1.7296	2.2567	2.7720	3.2682	3.9269	4.5859	4.6576	4.6924
4.0	0.4156	0.5093	0.6611	0.9532	1.7847	2.3170	2.8312	3.3166	3.9268	4.5321	4.6260	4.6748
2.0	0.4374	0.5388	0.7031	1.0185	1.9008	2.4510	2.9664	3.4258	3.9268	4.4138	4.5454	4.6265
1.5	0.4445	0.5488	0.7183	1.0443	1.9540	2.5159	3.0334	3.4788	3.9268	4.3580	4.5012	4.5976
1.2	0.4493	0.5557	0.7289	1.0631	1.9958	2.5683	3.0884	3.5213	3.9267	4.3138	4.4629	4.5709
1.0	0.4527	0.5606	0.7367	1.0773	2.0295	2.6119	3.1347	3.5564	3.9267	4.2779	4.4293	4.5462
0.8	0.4563	0.5659	0.7452	1.0932	2.0696	2.6652	3.1921	3.5988	3.9267	4.2351	4.3861	4.5125
0.4	0.4641	0.5776	0.7647	1.1316	2.1790	2.8207	3.3655	3.7176	3.9267	4.1186	4.2481	4.3871
0.2	0.4684	0.5842	0.7759	1.1551	2.2575	2.9457	3.5157	3.8052	3.9266	4.0361	4.1276	4.2489
0.1	0.4707	0.5876	0.7819	1.1683	2.3066	3.0330	3.6335	3.8605	3.9266	3.9856	4.0415	4.1281

Table 9-4B. Portal Frame, Asymmetric Mode.

Notation: E = modulus of elasticity; I = area moment of inertia about neutral axis; m = mass per unit length of beam; vertical legs are identical; see Table 3-1 for consistent sets of units (Ref. 9-14)

Portal Frame, First Asymmetric In-Plane Mode	Natural Frequency, f (Hz)
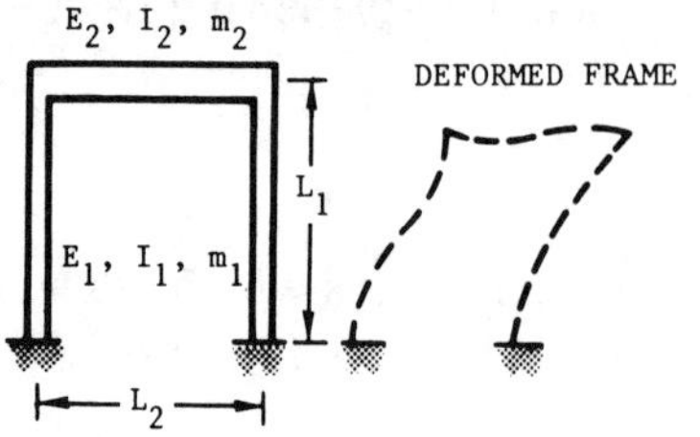	$f = \frac{\lambda^2}{2\pi L_1^2}\left(\frac{E_1 I_1}{m_1}\right)^{1/2}$

$\lambda = \lambda\,[L_1/L_2,\ m_1/m_2,\ E_1I_1/(E_2I_2)$, Boundary Conditions at Feet]

		Pinned Feet						Clamped Feet (Shown)					
		L_1/L_2						L_1/L_2					
$\frac{m_1}{m_2}$	$\frac{E_1I_1}{E_2I_2}$	0.25	0.75	1.5	3.0	6.0	12.0	0.25	0.75	1.5	3.0	6.0	12.0
0.25	0.25	0.6964	0.9520	1.1124	1.2583	1.3759	1.4586	0.9953	1.3617	1.6003	1.8270	2.0193	2.1612
	0.75	0.6108	0.8961	1.0764	1.2375	1.3649	1.4530	0.9030	1.2948	1.5544	1.7999	2.0051	2.1543
	1.5	0.5414	0.8355	1.0315	1.2093	1.3491	1.4449	0.8448	1.2323	1.5023	1.7649	1.9853	2.1442
	3.0	0.4695	0.7562	0.9635	1.1610	1.3201	1.4292	0.7968	1.1648	1.4329	1.7096	1.9504	2.1253
	6.0	0.4014	0.6663	0.8737	1.0870	1.2702	1.4001	0.7547	1.1056	1.3573	1.6350	1.8946	2.0917
	12.0	0.3405	0.5757	0.7711	0.9881	1.1925	1.3496	0.6966	1.0626	1.2905	1.5525	1.8179	2.0372
0.75	0.25	0.8947	1.1740	1.3168	1.4210	1.4882	1.5271	1.2873	1.7014	1.9262	2.0994	2.2156	2.2852
	0.75	0.7867	1.1088	1.2776	1.3998	1.4773	1.5217	1.1715	1.6242	1.8779	2.0733	2.2026	2.2788
	1.5	0.6983	1.0368	1.2281	1.3707	1.4617	1.5136	1.0979	1.5507	1.8218	2.0390	2.1843	2.2696
	3.0	0.6061	0.9413	1.1516	1.3203	1.4327	1.4982	1.0373	1.4698	1.7454	1.9838	2.1516	2.2521
	6.0	0.5186	0.8314	1.0485	1.2414	1.3822	1.4694	0.9851	1.3981	1.6601	1.9072	2.0983	2.2207
	12.0	0.4400	0.7195	0.9284	1.1336	1.3024	1.4191	0.9137	1.3455	1.5832	1.8201	2.0235	2.1694
1.5	0.25	1.0300	1.2964	1.4103	1.4826	1.5243	1.5469	1.4941	1.9006	2.0860	2.2090	2.2819	2.3221
	0.75	0.9085	1.2280	1.3707	1.4616	1.5136	1.5415	1.3652	1.8214	2.0390	2.1842	2.2695	2.3159
	1.5	0.8079	1.1514	1.3203	1.4326	1.4982	1.5336	1.2823	1.7444	1.9837	2.1515	2.2521	2.3070
	3.0	0.7021	1.0482	1.2414	1.3821	1.4694	1.5182	1.2141	1.6583	1.9070	2.0983	2.2206	2.2900
	6.0	0.6011	0.9279	1.1335	1.3024	1.4191	1.4897	1.1570	1.5808	1.8198	2.0234	2.1693	2.2596
	12.0	0.5103	0.8042	1.0061	1.1921	1.3391	1.4395	1.0807	1.5235	1.7399	1.9368	2.0964	2.2095
3.0	0.25	1.1597	1.3898	1.4719	1.5189	1.5442	1.5573	1.7022	2.0612	2.1963	2.2756	2.3190	2.3416
	0.75	1.0275	1.3202	1.4326	1.4981	1.5336	1.5519	1.5649	1.9834	2.1515	2.2520	2.3070	2.3356
	1.5	0.9161	1.2412	1.3821	1.4694	1.5182	1.5440	1.4752	1.9063	2.0982	2.2206	2.2899	2.3268
	3.0	0.7977	1.1333	1.3024	1.4191	1.4896	1.5287	1.4015	1.8185	2.0233	2.1693	2.2595	2.3101
	6.0	0.6838	1.0058	1.1921	1.3391	1.4395	1.5002	1.3425	1.7382	1.9366	2.0964	2.2094	2.2803
	12.0	0.5809	0.8732	1.0603	1.2277	1.3596	1.4503	1.2695	1.6781	1.8561	2.0111	2.1381	2.2310
6.0	0.25	1.2691	1.4516	1.5083	1.5388	1.5545	1.5627	1.8889	2.1727	2.2635	2.3128	2.3385	2.3517
	0.75	1.1304	1.3821	1.4694	1.5181	1.5440	1.5573	1.7501	2.0980	2.2206	2.2899	2.3268	2.3457
	1.5	1.0112	1.3023	1.4191	1.4896	1.5287	1.5494	1.6576	2.0228	2.1693	2.2595	2.3101	2.3370
	3.0	0.8827	1.1919	1.3391	1.4395	1.5002	1.5341	1.5817	1.9358	2.0963	2.2095	2.2802	2.3205
	6.0	0.7578	1.0601	1.2277	1.3595	1.4502	1.5057	1.5244	1.8550	2.0110	2.1380	2.2309	2.2909
	12.0	0.6443	0.9217	1.0936	1.2477	1.3704	1.4558	1.4659	1.7939	1.9310	2.0539	2.1604	2.2421
12.0	0.25	1.3488	1.4882	1.5283	1.5492	1.5600	1.5654	2.0336	2.2410	2.3011	2.3326	2.3486	2.3568
	0.75	1.2073	1.4190	1.4896	1.5287	1.5494	1.5600	1.8990	2.1692	2.2595	2.3101	2.3370	2.3509
	1.5	1.0835	1.3391	1.4395	1.5002	1.5341	1.5521	1.8075	2.0960	2.2094	2.2802	2.3206	2.3422
	3.0	0.9481	1.2276	1.3595	1.4502	1.5057	1.5369	1.7323	2.0105	2.1379	2.2309	2.2909	2.3259
	6.0	0.8153	1.0935	1.2477	1.3703	1.4557	1.5085	1.6785	1.9303	2.0538	2.1603	2.2421	2.2964
	12.0	0.6938	0.9518	1.1124	1.2583	1.3760	1.4586	1.6364	1.8692	1.9745	2.0770	2.1720	2.2477

tal asymmetric mode is given in Table 9-4B. The beams comprising these frames are assumed to be uniform and sufficiently slender so that shear deformation, axial deformation, and rotary inertia can be neglected. The two legs of the frame are assumed to be identical. The analysis of non-slender frames is discussed in Ref. 9-15. All vibrations are in the plane of the frame, and frames with both clamped feet and pinned feet are considered. The natural frequency parameter (λ) of the symmetric mode can be expressed as a function of two dimensionless groups plus the boundary condition at the feet. The natural frequency parameter for the asymmetric mode is a function of three dimensionless parameter groups plus the boundary conditions at the feet.

If the top beam in the portal frame is rigid (Fig. 9-8) and the legs are slender and uniform, although not necessarily identical to each other, then the fundamental natural frequency of the frame can be found approximately using the Rayleigh technique. The approximate fundamental natural frequency of the frame shown in Fig. 9-8 for vibrations in its own plane is:

$$f_1 = \frac{1}{2\pi}\left[\frac{12 \sum E_i I_i}{L^3 (M + 0.37 \sum M_i)}\right]^{1/2} \text{ Hz}, \tag{9-5}$$

where M is the mass of the cap beam, M_i is the mass of the i leg, and E_i and I_i are the modulus of elasticity and the area moment of inertia of the i leg. The summation (Σ) refers to the sum over the number of legs. There must be at least two legs so that the cap is always parallel to a line running between the feet of the frame. The mode shape of deformation of the legs of the frame in Fig. 9-8 is:

$$\tilde{y} = \left(1 - \frac{2x}{L}\right)^3 - 3\left(1 - \frac{2x}{L}\right) + 2.$$

If the legs of the frame are pinned to the rigid cap and clamped at the feet, then the fundamental natural frequency of the frame can be adapted from frame 3 of Table

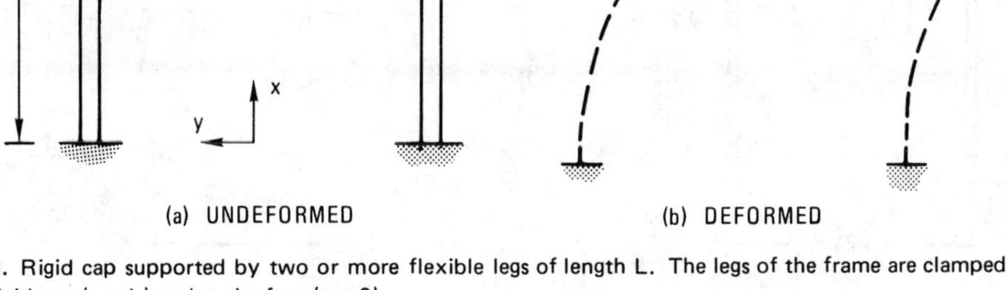

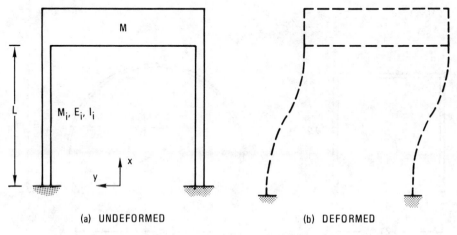

Fig. 9-8. Rigid cap supported by two or more flexible legs of length L. The legs of the frame are clamped both at the rigid cap (x = L) and at the feet (x = 0).

8-18. Additional discussion of a portal frame with a rigid cap can be found in Ref. 9-16.

The in-plane natural frequencies and mode shapes of two beam frames are discussed in Ref. 9-17.

9.6. EXAMPLE

Two designs for frames used to support a tent are shown in Fig. 9-9. Both frames are composed of 3.7-m lengths of aluminum tubing, 1 cm in outside diameter and 2 mm thick. Both frames have pinned feet. Using the units of frame 4 of Table 3-1, the parameters describing the aluminum tubing are:

$$E = 6.89 \times 10^{11} \text{ dyn/cm}^2,$$

$$\mu = 2.77 \text{ g/cm}^3,$$

$$I_y = \frac{\pi}{4}(r_0^4 - r_i^4) = 0.464 \text{ cm}^4 \text{ (frame 24 of Table 5-1)},$$

$$m = \mu 2\pi r t = 1.74 \text{ g/cm (frame 27 of Table 5-1)}.$$

Out-of-plane motion of the frames is prevented by the tent fabric.

The fundamental in-plane natural frequency of the circular ring frame can be calculated from frame 2 of Table 9-2 with the following additional parameters:

$$i = 2,$$

$$\alpha_0 = 225 \text{ deg} = 3.93 \text{ rad},$$

$$R = 95.0 \text{ cm}.$$

These parameters give:

$$f = 7.98 \text{ Hz}.$$

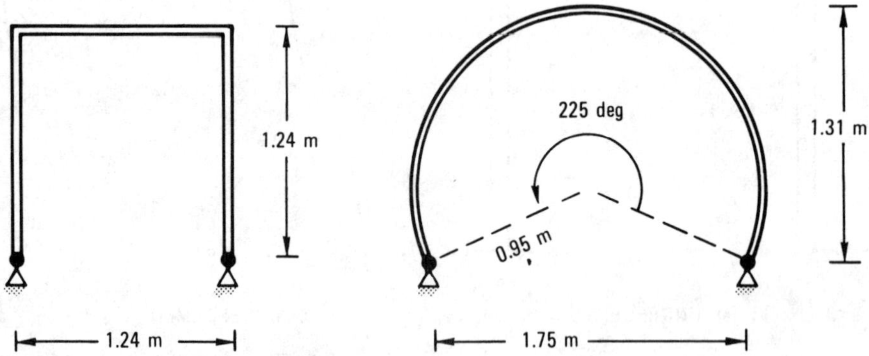

Fig. 9-9. Two frames composed of aluminum tubing.

The fundamental in-plane natural frequency of the square frame can be found from Table 9-4B using the following parameters:

$$\frac{m_1}{m_2} = \frac{E_1 I_1}{E_2 I_2} = \frac{L_1}{L_2} = 1,$$

$$L_1 = 124 \text{ cm}.$$

Interpolation from Table 9-4B gives:

$$\lambda = 1.26,$$

$$f = 7.04 \text{ Hz}.$$

Thus, the circular frame has a significantly higher natural frequency than the square frame. In addition, the circular frame encloses a larger area (2.09 m^2 versus 1.54 m^2) than the square frame.

REFERENCES

9-1. Love, A. E., "A Treatise on the Mathematical Theory of Elasticity," (4th ed.) (first published in 1927 by Cambridge University Press), Dover Press, New York, 1944, pp. 451–454.

9-2. Mallik, A. K., and D. J. Mead, "Free Vibration of Thin Circular Rings on Periodic Radial Supports," *J. Sound Vib.* **54**, 13–27 (1977).

9-3. Rao, S. S., and V. Sundararajan, "In-Plane Flexural Vibrations of Circular Rings," *J. Appl. Mech.* **36**, 620–625 (1969).

9-4. Kirkhope, J., "Simple Frequency Expression for the In-Plane Vibration of Thick Circular Rings," *J. Acoust. Soc. Am.* **59**, 86–88 (1976).

9-5. Kirkhope, J., "Out-of-Plane Vibration of Thick Circular Ring," *J. Eng. Mech.* **102**, 239–247 (1976).

9-6. Sato, K., "Free Flexural Vibrations of an Elliptical Ring in its Plane," *J. Acoust. Soc. Am.* **57**, 113–115 (1975).

9-7. Veletsos, A. S., *et al.*, "Free In-Plane Vibration of Circular Arches," *J. Eng. Mech.* **98**, 311–329 (1972).

9-8. Culver, C. G., "Natural Frequencies of Horizontally Curved Beams," *J. Struct.* **93**, ST2, 189–203 (1967).

9-9. Reissner, E., "Note on the Problem of Vibrations of Slightly Curved Bars," *J. Appl. Mech.* **21**, 195–196 (1954).

9-10. Austin, W. J., and A. S. Veletsos, "Free Vibration of Arches Flexible in Shear," *J. Eng. Mech.* **98**, 735–753 (1973).

9-11. Ojalvo, U., "Coupled Twisting-Bending Vibrations of Incomplete Elastic Rings," *Int. J. Mech. Sci.* **4**, 53–72 (1962).

9-12. Romanelli, E., and P. A. Laura, "Fundamental Frequencies of Non-Circular Elastic, Hinged Arcs," *J. Sound Vib.* **24**, 17–22 (1972).

9-13. Lee, L. S. S., "Vibrations of an Intermediately Supported U-Bend Tube," *J. Eng. Industry* **97**, 23–32 (1975).

9-14. Rieger, N. F., and H. McCallion, "The Natural Frequencies of Portal Frames–I," *Int. J. Mech. Sci.* **7**, 253–261 (1965).

9-15. Yang, Y. T., and C. T. Sun, "Axial-Flexural Vibration of Frameworks Using Finite-Element Approach," *J. Acoust. Soc. Am.* **53**, 137–146 (1973).

9-16. Anderson, G. L., "Natural Frequencies of Two Cantilevers Joined by a Rigid Connector at Their Free Ends," *J. Sound Vib.* **57**, 403–412 (1978).

9-17. Chang, C. H., "Vibration of Frames with Inclined Members," *J. Sound Vib.* **56**, 201–214 (1978).

10

MEMBRANES

General Case. A membrane is a thin structure which conforms to a surface in space. The bending rigidity of a membrane is zero. Thus, a membrane can support only tensile loads. Soap films and drum heads often approach the theoretical ideal of a membrane. A uniform homogeneous membrane is described by its mass per unit surface area (γ) and the tension it bears per unit length of boundary (S).

Natural Frequencies and Mode Shapes. The natural frequencies and mode shapes of flat, uniform, homogeneous membranes with fixed boundaries (displacement is zero along the boundary) are given in Table 10-1 for vibration perpendicular to the plane of the membrane. The mode shapes of membranes, unlike beams, are described by two spatial coordinates since the membrane is two-dimensional. The nodal lines can be used to characterize the mode shape, as shown in Fig. 10-1. A nodal line gives the same conditions as a fixed boundary. Thus, the higher modes of membranes with simple geometries such as a rectangle or circle can be used to generate the natural frequencies of right triangles and circular sectors (frames 4, 5, 11, and 12 of Table 10-1).

The mode shapes of uniform membranes with fixed boundaries are orthogonal over the membrane

$$\int_A \tilde{z}_k(x, y)\, \tilde{z}_j(x, y)\, dx\, dy = 0, \qquad k \neq j, \tag{10-1}$$

where $\tilde{z}_k$ and $\tilde{z}_j$ are the mode shapes of any two modes of the membrane and A is the area of the membrane.

There is a unique natural frequency for each vibration mode of a membrane; however, two or more of these frequencies may have the same value. For example, the natural frequency f_{12} equals the natural frequency f_{21} of a square membrane. Note that the natural frequency parameter λ for the fundamental mode of all relatively compact membranes–square, circle, equilateral triangle, etc.–is between 1 and 2 and that λ increases with the fineness of the membrane.

The natural frequency of a star-shaped membrane is discussed in Ref. 10-8.

Membrane Analogy. There exists an analogy between the vibration of polygonal membranes and similarly shaped simply supported plates. A solution of the plate equation for a plate with straight edges and zero moment about those edges also solves the membrane equation for a similarly shaped membrane with fixed edges.

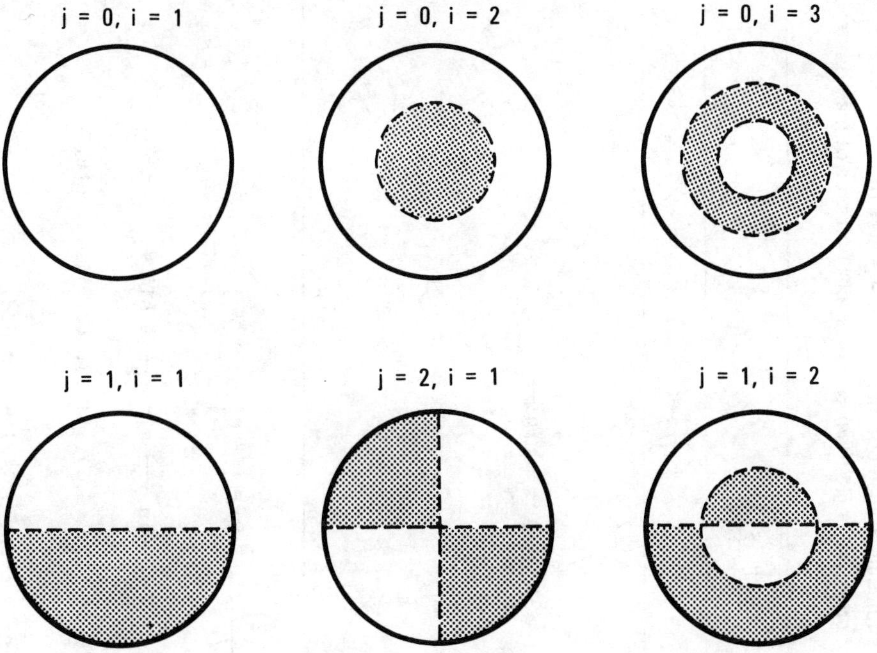

Fig. 10-1. Mode shapes of a circular membrane (frame 9 of Table 10-1).

It is possible to relate the natural frequencies of the membrane to the natural frequencies of the plate (Ref. 10-1). This transformation is effected by:

$$\lambda_{ij}\Big|_{\text{membrane}} = \frac{A^{1/2}}{\pi a}\lambda_{ij}\Big|_{\text{plate}}, \tag{10-2}$$

where λ_{ij} is the nondimensional frequency parameter, A is the area of the shape of interest (Table 10-1), and a is the characteristic length used in the formula for the natural frequency of the plate (Tables 11-4, 11-9, and 11-10). For example, the general solution for the natural frequencies of a simply supported rectangular plate can be found from frame 1 of Table 10-1 using Eq. 10-2.

$$f_{ij}\Big|_{\text{plate}} = \frac{\pi^2}{2\pi a^2}\left[i^2 + j^2\left(\frac{a}{b}\right)^2\right]\left[\frac{Eh^3}{12\gamma(1-\nu^2)}\right]^{1/2}; \qquad i = 1, 2, 3, \ldots, \quad j = 1, 2, 3, \ldots$$

where the symbols are as defined in Table 11-4. This expression can be verified by comparison with frame 16 of Table 11-4. The mode shapes of the membrane and the analogous plate are identical.

It must be cautioned that the analogy of Eq. 10-2 is valid only for polygonal shapes, i.e., shapes composed of a series of straight edges such as triangles, rectangles, and pentagons, and the analogy only applies to membranes with fixed edges and plates with simply supported edges.

Table 10-1. Membranes.

Notation: a, b = lengths; i, j = integers; r = radius; $\tilde{z}$ = mode shape for vibration perpendicular to plane of membrane; A = area of membrane; $\mathcal{J}_j$ = Bessel function of first kind and j order; S = tension per unit length of edge; β = angle (deg); γ = mass per unit area of membrane; θ = angle (radians); consistent sets of units are given in Table 3-1

Natural Frequency (hertz), $f_{ij} = \frac{\lambda_{ij}}{2}\left(\frac{S}{\gamma A}\right)^{1/2}$

Description	Area, A	λ_{ij} and Remarks
1. Rectangle 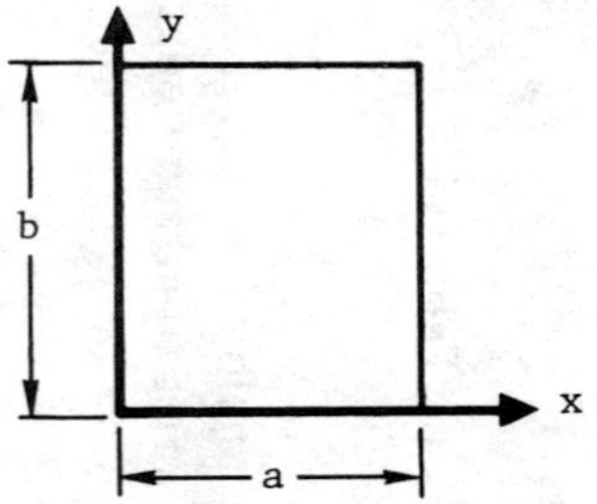	ab	$\lambda_{ij} = \left(i^2 \frac{b}{a} + j^2 \frac{a}{b}\right)^{1/2}$ Mode shape: $\tilde{z}_{ij} = \sin \frac{i\pi x}{a} \sin \frac{j\pi y}{b}$ $i = 1,2,3...$ $j = 1,2,3...$ Ref. 10-7, p. 311
2. Isosceles Right Triangle 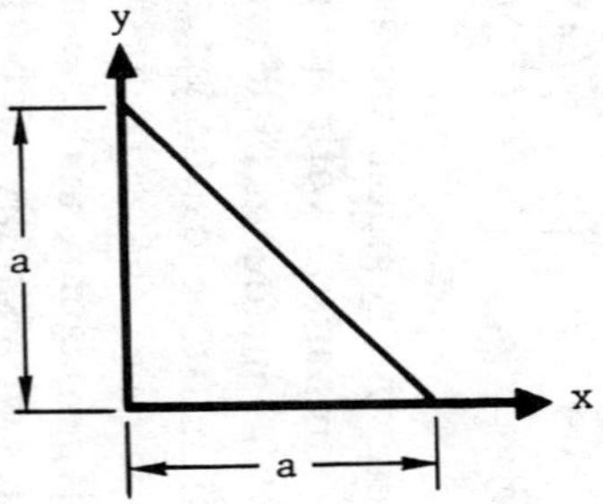	$\frac{1}{2} a^2$	$\lambda_{ij} = \frac{\left(i^2 + j^2\right)^{1/2}}{2^{1/2}}$ for $i,j > 0$ but $i \neq j$ Mode shape:* $\tilde{z}_{ij} = \sin \frac{i\pi x}{a} \sin \frac{j\pi y}{a} \pm \sin \frac{j\pi x}{a} \sin \frac{i\pi y}{a}$ $(i,j) = (1,2), (2,1), (3,1), (1,3), (2,3), ...$ etc. * − if i + j = even; + if i + j = odd. Ref. 10-7, p. 318

<table>
<tr><td>3. Isosceles Triangle
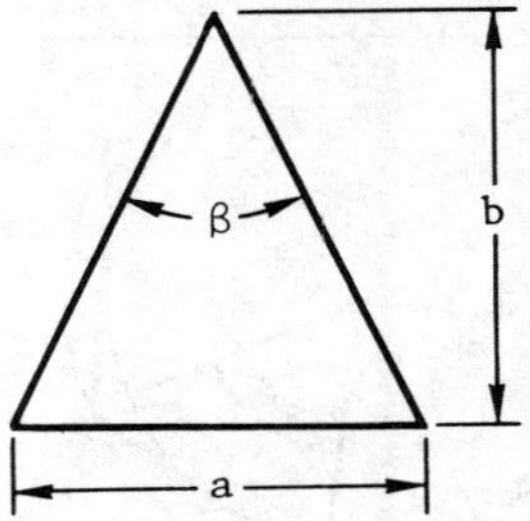
</td><td>$\frac{1}{2}$ ab</td><td>Fundamental mode:

$\frac{a}{b}$ =	0.347	0.518	0.684	0.845	1.0	1.15	1.29	1.41
β (deg) =	20	30	40	50	60	70	80	90
λ =	1.782	1.631	1.564	1.528	1.520	1.527	1.548	1.581

Ref. 10-1.</td></tr>
<tr><td>4. Parallelogram
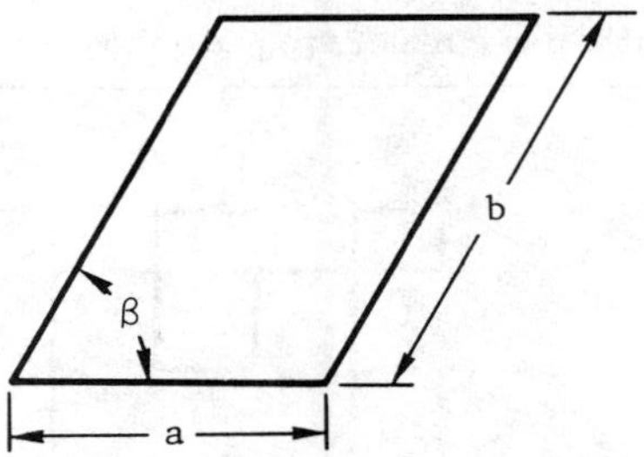
</td><td>ab sin β</td><td>Fundamental mode:

$\frac{b}{a}$	β (deg)			
	90	75	60	45
1	1.414	1.429	1.478	1.579
1.5	1.472	1.489	1.546	1.662
2	1.581	1.602	1.670	1.810
3	1.826	1.853	1.943	2.127

Ref. 10-2; higher modes are also given in Ref. 10-2.</td></tr>
<tr><td>5. L Shape
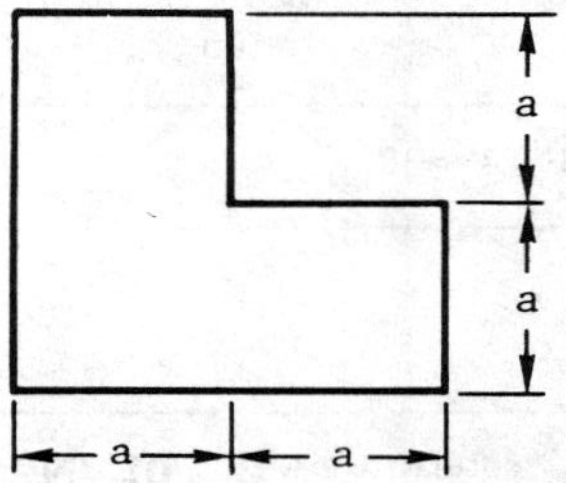
</td><td>$3a^2$</td><td>

Mode number =	1	2	3	4	5
λ =	1.717	2.150	2.449	2.999	3.124

Ref. 10-3.</td></tr>
</table>

Table 10-1. Membranes. *(Continued)*

Natural Frequency (hertz), $f_{ij} = \frac{\lambda_{ij}}{2}\left(\frac{S}{\gamma A}\right)^{1/2}$

Description	Area, A	λ_{ij} and Remarks
6. H Shape 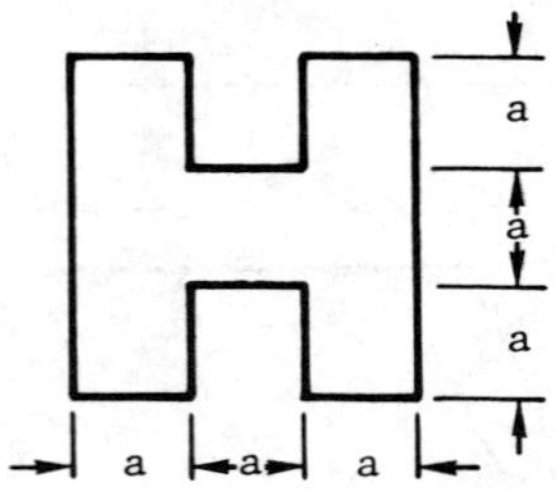	$7a^2$	Mode number = 1, 2, 3 λ = 6.556, 12.06, 16.63 Ref. 10-3.
7. Regular Polygon with N Equal Sides 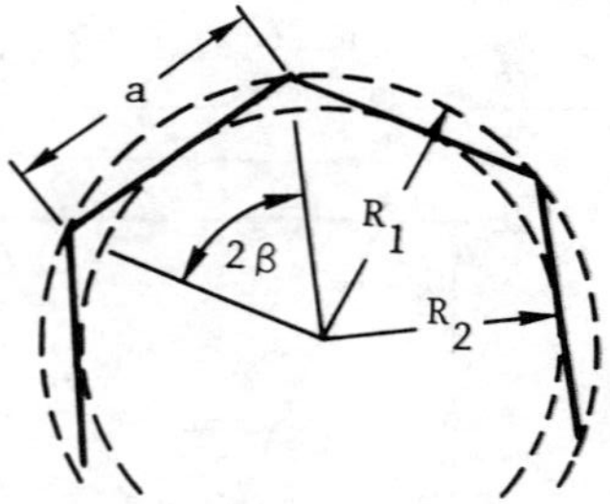	$\frac{N}{4} a^2 \cot \beta$	Fundamental mode: (see table below) Ref. 10-4

N	λ
4	1.414
5	1.385
6	1.372
7	1.366
8	1.362

8. Arbitrary Continuous Boundary

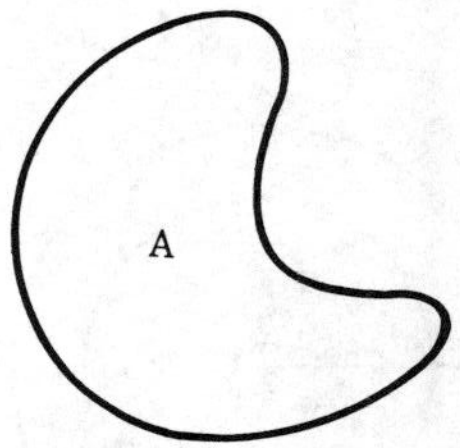

Area: A

Lower bound on λ for i^{th} mode:

$$\lambda_i = 1.357\ i^{1/2}; \quad i = 1,2,3...$$

Refs. 10-5, 10-6

9. Circle

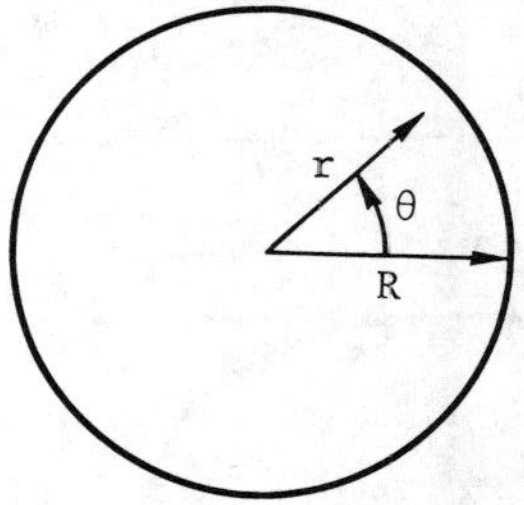

Area: πR^2

	j			
i	0	1	2	3
1	1.357	2.162	2.897	3.600
2	3.114	3.958	4.749	5.507
3	4.882	5.740	6.556	7.343
4	6.653	7.517	8.348	9.153

Mode shape: $\tilde{z}_{ij} = J_j\left(\pi^{1/2}\ \lambda_{ji}\ \frac{r}{R}\right) \cos j\theta$

i = number of nodal radii

j = number of nodal diameters

Ref. 10-7

10. Circular Sector

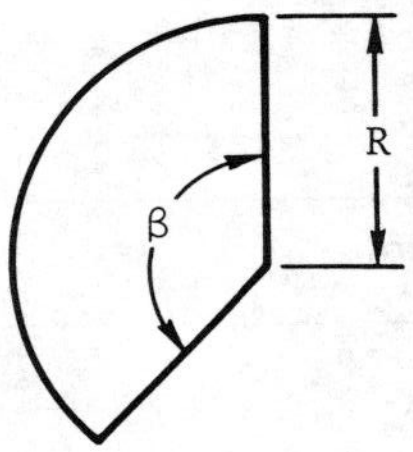

Area: $\pi R^2 \left(\frac{\beta}{360}\right)$

Fundamental mode:

β (deg) =	20	30	45	60	90	180
λ =	1.724	1.618	1.514	1.470	1.448	1.529

Ref. 10-7, pp. 332-334.

Table 10-1. Membranes. (*Continued*)

Natural Frequency (hertz), $f_{ij} = \frac{\lambda_{ij}}{2}\left(\frac{S}{\gamma A}\right)^{1/2}$

Description	Area, A	λ_{ij} and Remarks
11. Ellipse $\frac{x^2}{a^2} + \frac{y^2}{b^2} = 1$ 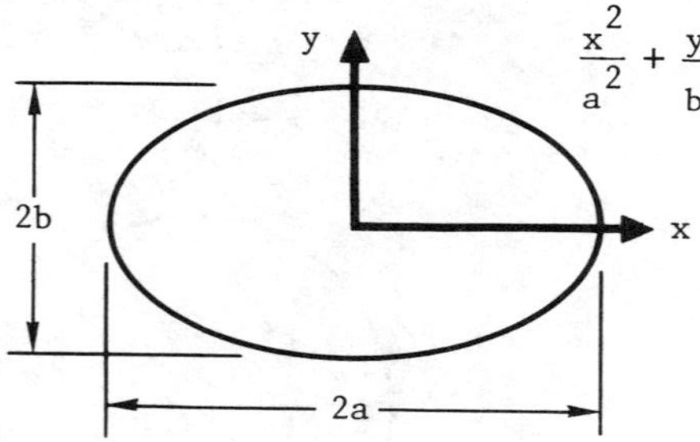 	πab	Fundamental mode: $\lambda = 2.405 \left[\frac{\frac{b}{a} + \frac{a}{b}}{2\pi}\right]^{1/2}$ Ref. 10-8
12. Annulus 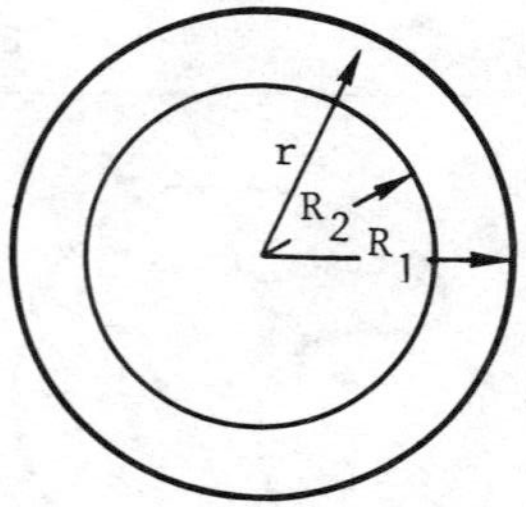 	$\pi(R_2^2 - R_1^2)$	For narrow annuli, $R_2/R_1 > 0.5$. $\lambda_{ij} = \frac{1}{\pi^{1/2}}\left[4j^2\left(\frac{R_1 - R_2}{R_1 + R_2}\right) + i^2\pi^2\left(\frac{R_1 + R_2}{R_1 - R_2}\right)\right]^{1/2}$ $i = 1,2,3\ldots \quad ; \quad j = 0,1,2\ldots$ Mode Shape: $\tilde{z}_{ij} = \sin\left[\frac{i\pi(r - R_2)}{R_1 - R_2}\right] \cos j\theta$ Other solutions can be adapted from the higher modes of a circular membrane.

13. Parabolic Section $x^2 = a^2\left(1 - \frac{y}{2^{1/2}a}\right)$ 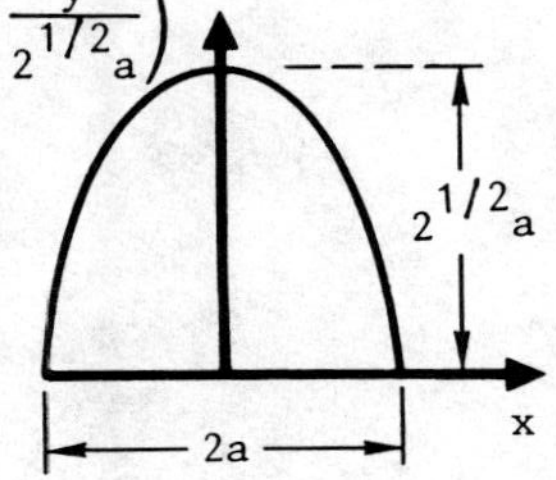	$\frac{4\sqrt{2}}{3}a^2$	Approximate solution, fundamental mode only: $\lambda = \frac{4.6701}{\pi}$ Ref. 10-8

REFERENCES

10-1. Conway, H. D., and K. A. Karnham, "The Free Flexural Vibration of Triangular, Rhombic and Parallelogram Plates and Some Analogies," *Int. J. Mech. Sci.* **7,** 811–816 (1965).

10-2. Durvasula, S., "Natural Frequencies and Modes of Skew Membranes," *J. Acoust. Soc. Am.* **44,** 1636–1646 (1968).

10-3. Milsted, M. G., and J. R. Hutchinson, "Use of Trigonometric Terms in the Finite Element with Application to Vibrating Membranes," *J. Sound Vib.* **32,** 327–346 (1974).

10-4. Shahady, P. A., R. Pasarelli, and P. A. A. Laura, "Application of Complex Variable Theory to the Determination of the Fundamental Frequency of Vibrating Plates," *J. Acoust. Soc. Am.* **42,** 806–809 (1967).

10-5. Pnueli, D., "Lower Bound to the nth Eigenvalue of the Helmholtz Equation over Two Dimensional Regions of Arbitrary Shape," *J. Appl. Mech.* **36,** 630–631 (1969).

10-6. Pnueli, D., "Lower Bounds to the Gravest and All Higher Frequencies of Homogeneous Vibrating Plates of Arbitrary Shape," *J. Appl. Mech.* **42,** 815–820 (1975).

10-7. Rayleigh, J. W. S., *The Theory of Sound*, Vol. 1, Dover Publications, New York, 1945 (first published in 1894), pp. 332–345.

10-8. Mazumdar, J., "Transverse Vibration of Membranes of Arbitrary Shape by the Method of Constant-Deflection Contours," *J. Sound Vib.* **27,** 47–57 (1973).

11

PLATES

11.1. GENERAL CASE

General Assumptions. A plate is a two-dimensional sheet of elastic material which lies in a plane. Plates, unlike membranes, possess bending rigidity as a result of their thickness and the elasticity of the plate material. During transverse vibration, plates deform primarily by flexing perpendicular to their own plane. Common examples of plates are windows, walls, and computer cards.

The general assumptions used in the analysis of plates in this chapter are:

1. The plates are flat and have constant thickness.
2. The plates are composed of a homogeneous, linear elastic, isotropic material.
3. The plates are thin. The thickness of each plate is less than about $\frac{1}{10}$ the minimum lateral plate dimension.
4. The plates deform through flexural deformation. The deformations are small in comparison with the thickness of the plate. Normals to the midsurface of the undeformed plate remain straight and normal to the midplane during deformation. Rotary inertia and shear deformation are neglected.
5. The in-plane load on the plates is zero.

These assumptions apply to all the results presented in Chapter 11 unless an exception is specifically noted.

Stress and Strain. As the plate deflects, the midsurface of the plate (halfway between the top and bottom surfaces) remains unstressed. At all other points there is a biaxial state of stress. Normals to the midsurface remain straight and normal to the midsurface as the plate deflects. The stresses and strains are proportional to the distance from the midsurface. The maximum stresses and strains are at the surface of the plate. For a plate lying in the x–y plane, the normal strains (ϵ_x and ϵ_y) and shear strain (ϵ_{xy}) in the plane of the plate are:

$$\epsilon_x = -z\frac{\partial^2 Z}{\partial x^2}, \qquad \epsilon_y = -z\frac{\partial^2 Z}{\partial y^2}, \qquad \epsilon_{xy} = -2z\frac{\partial^2 Z}{\partial x \partial y}. \tag{11-1}$$

The out-of-plane strains are zero:

$$\epsilon_{xz} = \epsilon_{yz} = \epsilon_{zz} = 0,$$

where z is the distance perpendicular to the plate midsurface and Z is the transverse deflection of the plate midsurface. These strains are associated with the following stresses for a homogeneous isotropic material:

$$\sigma_{xx} = \frac{E}{1-\nu^2}(\epsilon_x + \nu\epsilon_y), \qquad \sigma_{yy} = \frac{E}{1-\nu^2}(\epsilon_y + \nu\epsilon_x),$$

$$\sigma_{xy} = G\,\epsilon_{xy}, \tag{11-2}$$

$$\sigma_{xz} = \sigma_{yz} = \sigma_{zz} = 0.$$

The first subscript refers to the normal to the face on which the stress acts; the second subscript is the direction of the stress. σ_{xx} and σ_{yy} are the normal stresses; σ_{xy} is the shear stress. These stresses are shown in Fig. 11-1. Note the σ_{zz} is neglected in the thin plate theory. E is the elastic modulus, G is the shear modulus, and ν is Poisson's ratio. The bending moments, $\mathfrak{M}_x$ and $\mathfrak{M}_y$, and twisting moment, $\mathfrak{M}_{xy}$, in the plate are:

$$\mathfrak{M}_x = -\int_{-h/2}^{h/2} \sigma_{xx}\, z dz = \frac{Eh^3}{12(1-\nu^2)}\left(\frac{\partial^2 Z}{\partial x^2} + \nu\frac{\partial^2 Z}{\partial y^2}\right),$$

$$\mathfrak{M}_y = -\int_{-h/2}^{h/2} \sigma_{yy}\, z dz = \frac{Eh^3}{12(1-\nu^2)}\left(\frac{\partial^2 Z}{\partial y^2} + \nu\frac{\partial^2 Z}{\partial x^2}\right),$$

$$\mathfrak{M}_{xy} = -\int_{-h/2}^{h/2} \sigma_{xy}\, z dz = \frac{Eh^3}{12(1+\nu)}\frac{\partial^2 Z}{\partial x \partial y}. \tag{11-3}$$

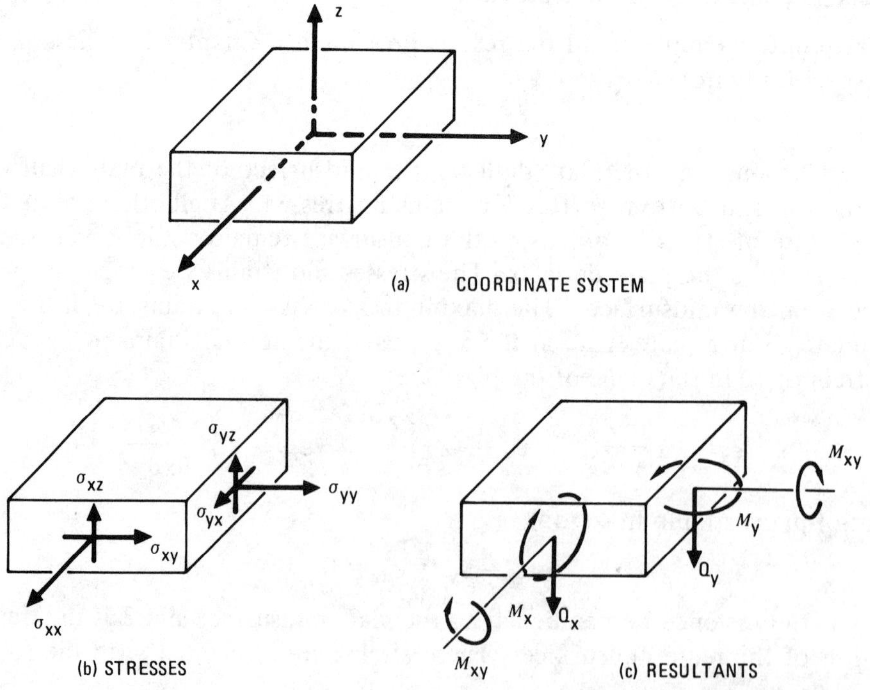

Fig. 11-1. Coordinate system, stresses, and resultants for a plate with rectangular coordinates.

The integrals are taken through the plate thickness, h. The transverse shearing forces are the resultant of the transverse shear stresses and can be related to the deformations in the plate by:

$$Q_x = -\int_{-h/2}^{h/2} \sigma_{xz}\, dz = \frac{Eh^3}{12(1-\nu^2)} \frac{\partial}{\partial x}\left(\frac{\partial^2 Z}{\partial x^2} + \frac{\partial^2 Z}{\partial y^2}\right),$$

$$Q_y = -\int_{-h/2}^{h/2} \sigma_{yz}\, dz = \frac{Eh^3}{12(1-\nu^2)} \frac{\partial}{\partial y}\left(\frac{\partial^2 Z}{\partial x^2} + \frac{\partial^2 Z}{\partial y^2}\right). \qquad (11\text{-}4)$$

Although the out-of-plane shear deformations ϵ_{xz}, ϵ_{yz} and the associated shear stresses σ_{xz}, σ_{yz} (or σ_{rz} and $\sigma_{\theta z}$ in polar coordinates) have been neglected in the thin plate theory, equilibrium considerations dictate that the integral of these shear stresses over the thickness cannot be zero. The out-of-plane shear stresses will generally have a parabolic distribution over the plate thickness, maximum at the mid-surface and zero at the edges.

In polar coordinates the strain-deformation relationships for a plate lying in the r–θ plane are:

$$\epsilon_r = -z\frac{\partial^2 Z}{\partial r^2}, \qquad \epsilon_\theta = -\frac{z}{r}\frac{\partial Z}{\partial r} - \frac{z}{r^2}\frac{\partial^2 Z}{\partial \theta^2},$$

$$\epsilon_{r\theta} = -\frac{z}{r}\frac{\partial^2 Z}{\partial r \partial \theta} - rz\frac{\partial}{\partial r}\left(\frac{1}{r^2}\frac{\partial Z}{\partial \theta}\right), \qquad \epsilon_{rz} = \epsilon_{\theta z} = \epsilon_{zz} = 0. \qquad (11\text{-}5)$$

The stress-strain relationships in polar coordinates for a homogeneous isotropic material are:

$$\sigma_{rr} = \frac{E}{1-\nu^2}(\epsilon_r + \nu\epsilon_\theta), \qquad \sigma_{\theta\theta} = \frac{E}{1-\nu^2}(\epsilon_\theta + \nu\epsilon_r),$$

$$\sigma_{r\theta} = G\epsilon_{r\theta}, \qquad \sigma_{rz} = \sigma_{\theta z} = \sigma_{zz} = 0. \qquad (11\text{-}6)$$

The moments in the plate in polar coordinates are:

$$\mathfrak{M}_r = -\int_{-h/2}^{h/2} \sigma_{rr}\, z dz = \frac{Eh^3}{12(1-\nu^2)}\left[\frac{\partial^2 Z}{\partial r^2} + \nu\left(\frac{1}{r}\frac{\partial Z}{\partial r} + \frac{1}{r^2}\frac{\partial^2 Z}{\partial \theta^2}\right)\right],$$

$$\mathfrak{M}_\theta = -\int_{-h/2}^{h/2} \sigma_{\theta\theta}\, z dz = \frac{Eh^3}{12(1-\nu^2)}\left(\frac{1}{r}\frac{\partial Z}{\partial r} + \frac{1}{r^2}\frac{\partial^2 Z}{\partial \theta^2} + \nu\frac{\partial^2 Z}{\partial r^2}\right),$$

$$\mathfrak{M}_{r\theta} = -\int_{-h/2}^{h/2} \sigma_{r\theta}\, z dz = \frac{Eh^3}{12(1+\nu)}\frac{\partial}{\partial r}\left(\frac{1}{r}\frac{\partial Z}{\partial \theta}\right). \qquad (11\text{-}7)$$

The transverse shearing forces in polar coordinates are:

$$Q_r = -\int_{-h/2}^{h/2} \sigma_{rz}\, dz = \frac{Eh^3}{12(1-\nu^2)}\frac{\partial}{\partial r}\left(\frac{\partial^2 Z}{\partial r^2} + \frac{1}{r}\frac{\partial Z}{\partial r} + \frac{1}{r^2}\frac{\partial^2 Z}{\partial \theta^2}\right),$$

$$Q_\theta = -\int_{-h/2}^{h/2} \sigma_{\theta z}\, dz = \frac{Eh^3}{12(1-\nu^2)} \frac{1}{r} \frac{\partial}{\partial \theta} \left(\frac{\partial^2 Z}{\partial r^2} + \frac{1}{r}\frac{\partial Z}{\partial r} + \frac{1}{r^2}\frac{\partial^2 Z}{\partial \theta^2} \right). \tag{11-8}$$

Natural Frequencies, Mode Shapes, and Deformation. The mode shapes and natural frequencies of plates are functions of the two integer indices i and j. Each of these indices may be associated with the number of flexural half-waves in one of the two plate dimensions. For each i and j for which a vibration mode exists, there is a natural frequency and an associated mode shape. If the plate vibrates freely, then the total transverse deformation is the sum of the modal deformations:

$$Z = \sum_i \sum_j A_{ij} \tilde{z}_{ij} \sin (2\pi f_{ij} t + \phi_{ij}). \tag{11-9}$$

$\tilde{z}_{ij}$ is the mode shape of the ij mode, and the modal amplitude A_{ij} and the modal phase angle ϕ_{ij} are determined by the means used to set the plate in motion. The natural frequency of the plate is:

$$f_{ij} = \frac{\lambda_{ij}^2}{2\pi a^2} \left[\frac{Eh^3}{12\gamma(1-\nu^2)} \right]^{1/2}; \qquad i = 1, 2, 3, \ldots, \quad j = 1, 2, 3, \ldots, \tag{11-10}$$

where λ_{ij} is a dimensionless parameter which is generally a function of the mode indices (i, j), Poisson's ratio (ν), the plate geometry, and the boundary conditions on the plate. γ is the mass per unit area of the plate and a is a characteristic dimension of the plate. Once the total displacement Z has been determined, then Eqs. 11-1 through 11-8 may be used to determine the stresses, strains, and resultant forces in the plate.

Additional information concerning the solutions presented in the tables can be found by consulting the references in the tables and in the text. Many of the pre-1966 references in this chapter can also be found in Ref. 11-1.

Orthogonality. A general orthogonality principle does not exist for the mode shapes of vibration of thin plates. However, the mode shapes of a thin uniform plate can be shown to be orthogonal,

$$\int_A \tilde{z}_{ij} \tilde{z}_{mn}\, dA = 0 \qquad \text{if} \quad i \neq m, \quad j \neq n,$$

where A is the area of the plate, if (1) the natural frequencies of the modes are well separated and (2) either (a) the edges of the plate are clamped or (b) the portion of the edges which are not clamped are straight segments, as those of a rectangular or polygonal plate, and these segments are simply supported (Ref. 6-13, p. 183).

The modes of plates which do not fall into these classes cannot generally be shown to be orthogonal because of the complexity of the rigorous boundary conditions. (The boundary conditions associated with simply supported and free edges of plates are functions of the Poisson's ratio of the plate material as well as the displacement of the plate.) However, this lack of a general proof of orthogonality has not proven to be a serious practical limitation. Many of the modes of plates for which orthogonality cannot be proven in general can be shown to be orthogonal in specific cases.

Moreover, many modes of plates with simply supported or free boundaries cannot be expressed in closed form. In order to facilitate numerical solution, the boundary conditions associated with simply supported and free edges are often simplified by neglecting some of the terms in the rigorous boundary conditions. Orthogonality can be proven for these approximate boundary conditions. Additional discussion of plate orthogonality can be found in Section 11.3.

Sandwich Plates. A laminated sandwich plate is shown in Fig. 11-2. The plate consists of uniform layers of material which are glued together. Each layer is assumed to be homogeneous and isotropic. The plate is symmetric about the midsurface. The analysis of sandwich plates is considerably more difficult than the analysis of homogeneous plates because large shear strains can develop between the layers (Refs. 11-84, 11-85). However, if the sandwich plate is slender, that is, the thickness of the plate is small compared with typical lateral dimensions and the distance between vibration nodes, then it is reasonable to assume that normals to the midsurface remain normal during vibration. Using this assumption, the natural frequencies of slender sandwich plates can be computed using the formulas developed for homogeneous plates by defining the sandwich plate equivalent stiffness,

$$\frac{Eh^3}{12} = \sum_{k} 2E_k \int_{d_k}^{d_{k+1}} z^2\,dz,$$

$$= \frac{2}{3} \sum_{k=0,1,2,\ldots} E_k (d_{k+1}^3 - d_k^3), \qquad (11\text{-}11)$$

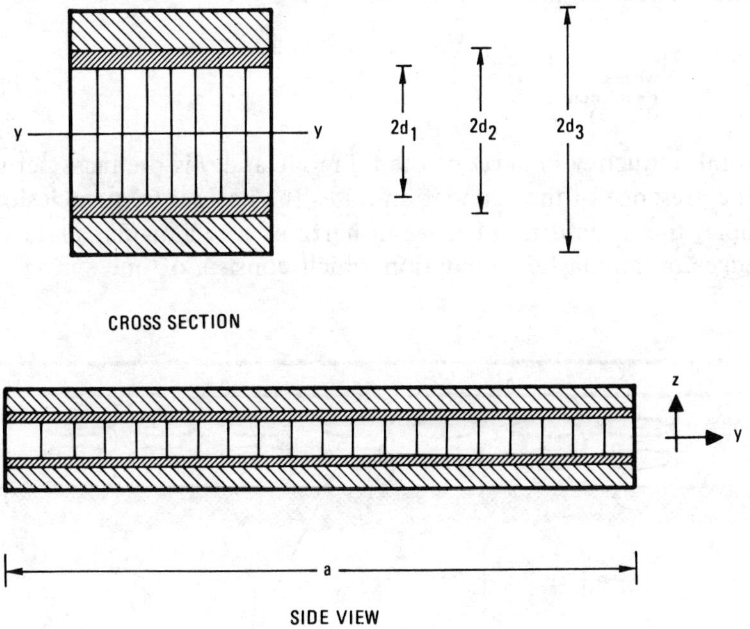

Fig. 11-2. Sandwich plate.

and equivalent mass per unit area,

$$\gamma = 2 \sum \mu_k \int_{d_k}^{d_{k+1}} dz,$$

$$= 2 \sum_{k=0,1,2} \mu_k (d_{k+1} - d_k) = 2 \sum_{k=0,1,2,\ldots} \gamma_k. \qquad (11\text{-}12)$$

The distance from the plate midsurface to the seam between the k and the k + 1 layers is d_k ($d_0 = 0$). E_k and μ_k are the modulus of elasticity and the density of the k layer which has mass per unit area γ_k. The natural frequencies of sandwich plates can be found from the formulas for the natural frequencies of homogeneous plates by substituting equivalent stiffness and equivalent mass per unit area for the stiffness ($Eh^3/12$) and mass per unit area (γ) of the homogeneous plate. However, it should be noted that the effects of shear deformation and rotary inertia are generally more important in the analysis of sandwich plates than comparable uniform plates.

Plates on Elastic Foundations. A plate on an elastic foundation is shown in Fig. 11-3. The elastic foundation could consist of a pattern of springs, as in Fig. 11-3, or simply an elastic pad. The foundation modulus, E_f, is defined as the ratio of the pressure applied to a unit area of foundation to the deformation that pressure produces. E_f has units of pressure per length. $E_f = k_f/\Delta^2$ for the foundation shown in Fig. 11-3 with springs of spring constant k_f located on a square grid of side Δ.

The natural frequencies of a uniform thin plate on elastic foundations with constant modulus can easily be adapted from the natural frequencies of the plate in the absence of the foundation provided the boundary conditions of the plate are independent of the foundation modulus (Ref. 11-17):

$$f_{ij} = \left(f_{ij}^2 \Big|_{\substack{\text{without}\\ \text{foundation}\\ (E_f=0)}} + \frac{E_f}{4\pi^2\gamma} \right)^{1/2}, \qquad i = 1, 2, 3, \ldots, \quad j = 1, 2, \ldots. \qquad (11\text{-}13)$$

f_{ij} is the natural frequency in hertz of the i, j mode and γ is the mass per unit area of the plate. The presence of the foundation raises the natural frequencies of the plate.

For example, the natural frequencies in hertz of a rectangular plate with simply supported edges on an elastic foundation which consists of one spring, with spring

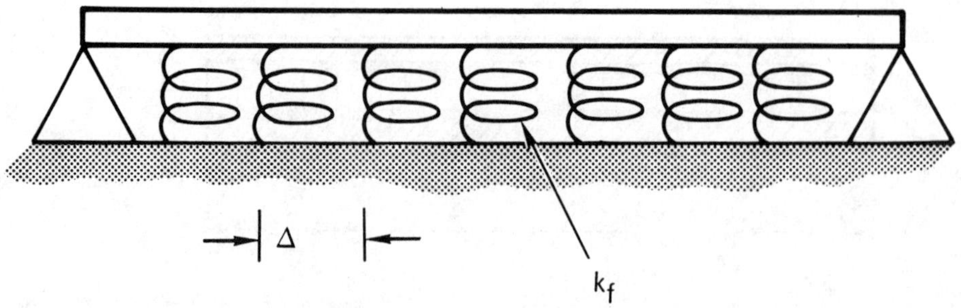

Fig. 11-3. Side view of a plate on an elastic foundation.

constant k_f, per Δ^2 area of the plate (Fig. 11-3) are:

$$f_{ij} = \left\{ \frac{\pi^2}{4a^4} \left[i^2 + j^2 \left(\frac{a}{b}\right)^2 \right]^2 \frac{Eh^3}{12\gamma(1-\nu^2)} + \frac{k_f}{4\gamma\pi^2\Delta^2} \right\}^{1/2}$$

$$i = 1, 2, 3, \ldots, j = 1, 2, 3 \ldots \qquad (11\text{-}14)$$

where a is the width of the plate and b is its length.

The mode shapes of plates on elastic foundations are unchanged by the presence of the foundation (Ref. 11-17).

Membrane Analogy. There exists an analogy between the vibration of pentagonal membranes and similarly shaped simply supported plates. A solution of the plate equation for a plate with straight edges and zero moment about those edges also solves the membrane equation for a similarly shaped membrane with fixed edges. It is possible to relate the membrane solution to the plate solution (Ref. 10-1). This transformation is effected by:

$$\lambda_{ij}\Big|_{\text{plate}} = \frac{\pi a}{A^{1/2}} \lambda_{ij}\Big|_{\text{membrane}}, \qquad (11\text{-}15)$$

where A is the area of the shape of interest (Table 10-1) and a is the characteristic length used in the formula for the natural frequency of the plate. For example, the general solution for the natural frequencies of a simply supported rectangular plate can be found from frame 2 of Table 10-1:

$$f_{ij}\Big|_{\text{plate}} = \frac{\pi^2}{2\pi a^2} \left[i^2 + j^2 \left(\frac{a}{b}\right)^2 \right] \left[\frac{Eh^3}{12\gamma(1-\nu^2)} \right]^{1/2}; \quad \begin{matrix} i = 1, 2, 3, \ldots, \\ j = 1, 2, 3, \ldots, \end{matrix} \qquad (11\text{-}16)$$

where the symbols are defined in Table 11-10. This expression can be verified by comparison with frame 16 of Table 11-10.

The mode shapes of the plate and the analogous membrane are identical. Thus, the mode shape of the simply supported rectangular plate is:

$$\tilde{z}_{ij}(x, y) = \sin \frac{i\pi x}{a} \sin \frac{j\pi y}{b}, \qquad (11\text{-}17)$$

where the origin of the coordinate system is at one of the corners of the plate with the x axis parallel to the sides of length b and the y axis parallel to the sides of length a.

It must be cautioned that the analogy of Eq. 11-15 is valid only for polygonal shapes, that is, shapes composed of a series of straight edges such as triangles, rectangles, and pentagons, and the analogy applies only to membranes with fixed edges (Table 10-1) and plates with simply supported edges.

11.2. CIRCULAR, ANNULAR, AND ELLIPTICAL PLATES

Tables. The natural frequencies of circular plates are given in Table 11-1. Evidently, the mode shapes of circular plates can be expressed in closed form only for

Table 11-1. Circular Plates.

Notation: a = radius of plate; h = thickness of plate; i = number of nodal diameters; j = number of nodal circles, not counting the boundary; k = torsion spring constant per unit length of edge; r = a radius; C = clamped edge; E = modulus of elasticity; F = free edge; $\mathcal{I}$ = modified Bessel function of first kind; $\mathcal{J}$ = Bessel function of first kind; $\mathcal{K}$ = modified Bessel function of second kind; M = mass; N = load per unit length of edge; S = simply supported edge; $\mathcal{Y}$ = Bessel function of second kind; β = angle (degrees); γ = mass per unit area of plate (μh for a plate of material with density μ); ν = Poisson's ratio; see Table 3-1 for consistent sets of units.

Natural Frequency (hertz), $f_{ij} = \frac{\lambda_{ij}^2}{2\pi a^2}\left[\frac{Eh^3}{12\gamma(1-\nu^2)}\right]^{1/2}$; i=0,1,2...; j=0,1,2...

Description	λ_{ij}^2	Mode Shape, $\tilde{z}_{ij}$, and Remarks

1. Free Edge

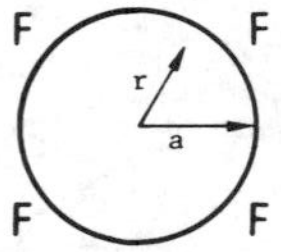

λ_{ij}^2:

j	i = 0	1	2	3
0	*	*	5.253	12.23
1	9.084	20.52	35.25	52.91
2	38.55	59.86	83.9	111.3
3	87.80	119.0	154.0	192.1

(λ_{41} = 21.6)
ν = 0.33; Ref. 11-2

Transcendental equation for λ is given in Ref. 11-3

*Rigid body mode.

Radii of Nodal Circles, r/a:

j	i = 0	1	2	3
1	0.680	0.781	0.822	0.847
2	0.841 0.391	0.871 0.497	0.890 0.562	0.925 0.605
3	0.893 0.591 0.257	0.932 0.643 0.351	0.936 0.678 0.414	0.939 0.704 0.460

ν = 0.33; Ref. 11-2

2. Simply Supported Edge

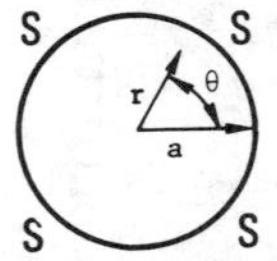

λ_{ij}^2:

j	i = 0	1	2
0	4.977	13.94	25.65
1	29.76	48.51	70.14
2	74.20	102.8	134.3
3	138.3	176.8	218.2

ν = 0.3; Ref. 11-4

Transcendental Equation for λ_{ij} (Ref. 11-5):

$$\frac{J_{i+1}(\lambda)}{J_i(\lambda)} + \frac{I_{i+1}(\lambda)}{I_i(\lambda)} = \frac{2\lambda}{1-\nu}$$

Mode Shape:

$$\left[J_i\left(\frac{\lambda r}{a}\right) - \frac{J_i(\lambda)}{I_i(\lambda)} I_i\left(\frac{\lambda r}{a}\right)\right] \cos i\theta; \quad i=0,1,2...$$

Radii of Nodal Circles, r/a:

j	i = 0	1	2
0	1	1	1
1	1 0.441	1 0.550	1 0.613
2	1 0.644 0.279	1 0.692 0.378	1 0.726 0.443
3	1 0.736 0.469 0.204	1 0.765 0.528 0.288	1 0.787 0.520 0.348

ν = 0.3
Ref. 11-4

3. Clamped Edge

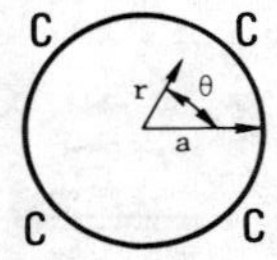

	i			
j	0	1	2	3
0	10.22	21.26	34.88	51.04
1	39.77	60.82	84.58	111.0
2	89.10	120.1	153.8	190.3
3	158.2	199.1	242.7	289.2

Ref. 11-6 ($\lambda_{41} = 69.67$)

Transcendental equation for λ_{ij}:

$$J_i(\lambda)I_{i+1}(\lambda) + I_i(\lambda)J_{i+1}(\lambda) = 0$$

Ref. 11-1

λ is independent of ν.

Mode Shape:

$$\left[J_i\left(\frac{\lambda r}{a}\right) - \frac{J_i(\lambda)}{I_i(\lambda)} I_i\left(\frac{\lambda r}{a}\right)\right] \cos i\theta$$

i=0,1,2...

Radii of Nodal Circles, r/a:

	i			
j	0	1	2	3
0	1	1	1	1
1	1 0.379	1 0.490	1 0.559	1 0.606
2	1 0.583 0.255	1 0.640 0.350	1 0.679 0.414	1 0.708 0.462
3	1 0.688 0.439 0.191	1 0.721 0.497 0.272	1 0.746 0.540 0.330	1 0.765 0.574 0.375

Ref. 11-4

4. Clamped Along Part of Edge, Simply Supported Along Remainder

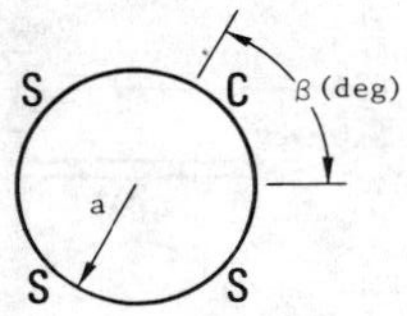

Fundamental Mode Only, Bounds for λ_{11}:

β	Upper Bound	Lower Bound
0	--	4.862
22.5	5.871	5.842
45.0	6.350	6.335
67.5	6.880	6.864
90.0	7.508	7.480
112.5	8.231	8.162
135.0	9.120	8.880
157.5	9.855	9.126
180.0	10.21	--

$\nu = 0.25$; Ref. 11-7

Approximate Transcendental Equation for λ_{11} (Ref. 11-8):

$$2\lambda \left[\frac{J_1(\lambda)}{J_0(\lambda)} + \frac{I_1(\lambda)}{I_0(\lambda)}\right]^{-1} = 1 - \nu + \frac{1}{\ln(\sin \beta/2)}$$

5. Clamped at Center, Free Along Edge

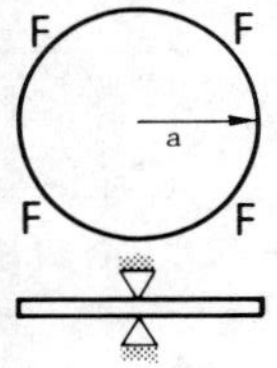

Polar Symmetric Modes (i=0):

j	λ^2
0	3.752
1	20.91
2	61.2
3	120.6
4	199.9
5	298.2
6	416.6

$\nu = 1/3$; Ref. 11-2

In higher modes values of λ are separated by π.

Table 11-1. Circular Plates. *(Continued)*

Natural Frequency (hertz), $f_{ij} = \frac{\lambda_{ij}^2}{2\pi a^2}\left[\frac{Eh^3}{12\gamma(1-\nu^2)}\right]^{1/2}$; i=0,1,2...; j=0,1,2...

Description	λ_{ij}^2	Mode Shape, $\tilde{z}_{ij}$, and Remarks
6. Clamped at Center, Simply Supported Along the Edge 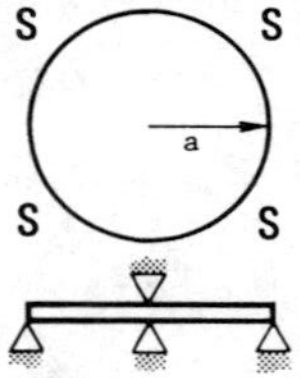 	Polar Symmetric Modes (i=0): j | λ^2 0 | 14.8 1 | 49.4 ν = 0.3; Ref. 11-9	Transcendental Equation for Polar Symmetric Modes (Ref. 11-9): $(1-\nu)\left\{[I_0(\lambda) - J_0(\lambda)]\left[Y_1(\lambda) + \frac{2}{\pi}K_1(\lambda)\right]\right.$ $\left. + [J_1(\lambda) + I_1(\lambda)]\left[Y_0(\lambda) + \frac{2}{\pi}K_0(\lambda)\right]\right\}$ $- 2\lambda\left[I_0(\lambda)Y_0(\lambda) + \frac{2}{\pi}J_0(\lambda)K_0(\lambda)\right] = 0$
7. Clamped at Center, Clamped Along the Edge 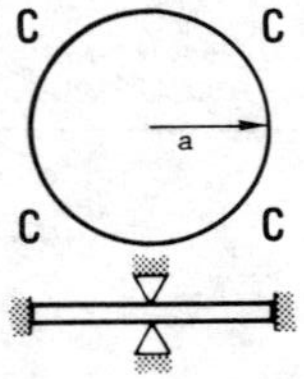 	Polar Symmetric Modes (i=0): j | λ^2 0 | 22.7 1 | 61.9 Ref. 11-9	Transcendental Equation for Polar Symmetric Modes (Ref. 11-9): $[J_0(\lambda) - I_0(\lambda)]\left[Y_1(\lambda) + \frac{2}{\pi}K_1(\lambda)\right]$ $- [J_1(\lambda) + I_1(\lambda)]\left[Y_0(\lambda) + \frac{2}{\pi}K_0(\lambda)\right] = 0$ λ is independent of ν.
8. Simple Support at Arbitrary Radius 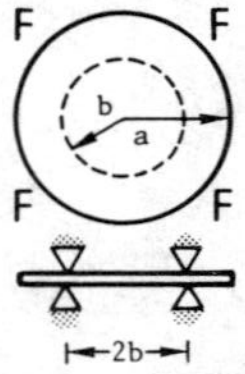 	Fundamental mode: $\frac{b}{a}$ | λ^2 0 | 3.75 0.2 | 4.5 0.4 | 6.7 0.6 | 8.8 0.8 | 7.5 1.0 | 5.0 ν = 1/3; Ref. 11-10	

9. Edge Simply Supported and Elastically Restrained by Torsion Springs

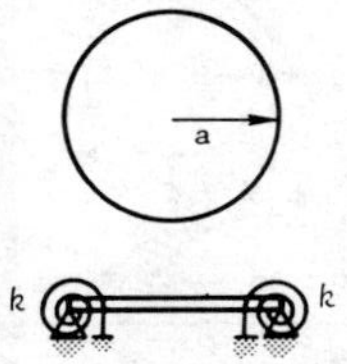

$\frac{Eh^3}{12k(1 - \nu^2)a}$	λ^2 i=0 j=0	λ^2 i=0 j=1	λ^2 i=1 j=0
∞	10.2	39.7	21.2
1	10.2	39.7	21.2
0.1	10.0	39.1	20.9
0.01	8.76	35.2	18.6
0.001	6.05	30.8	15.0
0	4.93	29.7	13.9

$\nu = 0.3$; Ref. 11-11

k = moment/radian-unit length of edge. Also see Ref. 11-12.

10. Free Edge, Point Mass at Center

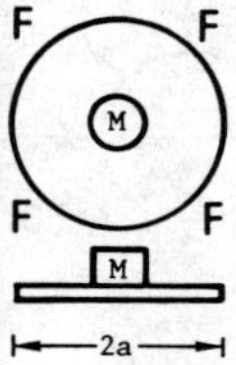

First Symmetric Mode (i=0, j=1):

$\frac{M}{\gamma\pi a^2}$	λ^2
0.0	9.0
0.05	8.2
0.1	7.5
0.2	6.8
0.4	5.9
0.6	5.5
1.0	4.9
1.4	4.7
∞	3.7

$\nu = 0.3$; Ref. 11-13

Higher modes are discussed in Ref. 11-13.

11. Simply Supported Edge, Point Mass at Center

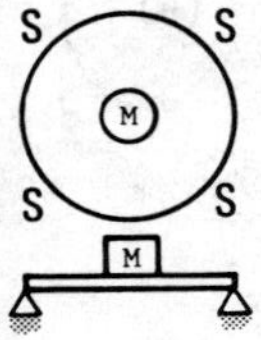

Approximate formula for First Symmetric Mode (i=j=0) with $M > \gamma\pi a^2$:

$$\lambda^2 = \frac{4}{(3 + 4\nu + \nu^2)^{1/2}}\left(\frac{\gamma\pi a^2}{M}\right)^{1/2}$$

Ref. 11-14

Table 11-1. Circular Plates. *(Continued)*

Natural Frequency (hertz), $f_{ij} = \frac{\lambda_{ij}^2}{2\pi a^2}\left[\frac{Eh^3}{12\gamma(1-\nu^2)}\right]^{1/2}$; i=0,1,2...; j=0,1,2...

Description	λ_{ij}^2	Mode Shape, $\tilde{z}_{ij}$, and Remarks

12. Clamped Edge, Point Mass at Center

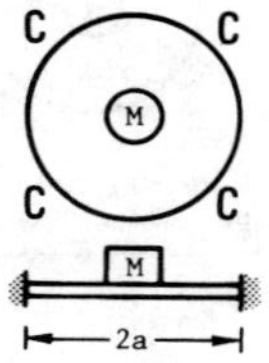

Fundamental Mode (i=j=0):

$\frac{M}{\gamma\pi a^2}$	λ^2
0.0	10.2
0.05	9.0
0.1	8.1
0.2	6.9
0.4	5.4
0.6	4.75
1.0	3.8
1.4	3.3

Ref. 11-15

λ is independent of ν. An approximate solution for $M > \gamma\pi a^2$ is (Ref. 11-14):

$$\lambda^2 = 4\left(\frac{\gamma\pi a^2}{M}\right)^{1/2}$$

13. Simply Supported Edge, Uniform Radial Edge Load

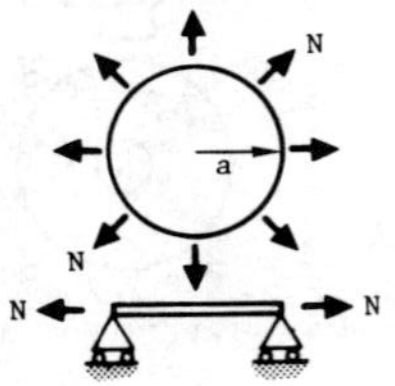

$\frac{12(1-\nu^2)Na^2}{4.2\ Eh^3}$	λ^2 i=0 j=0	λ^2 i=1 j=0
2.0	8.55	17.47
1.5	7.81	16.55
1.0	6.99	15.57
0.5	6.05	14.55
0.25	5.52	13.98
0.0	4.94	13.47
-0.25	4.27	12.86
-0.5	3.46	12.23
-1.0	0.0	10.95

ν = 0.3, Ref. 11-16

N = force per unit length of edge, positive outward. Higher modes are given in Ref. 11-16. i=1, j=0 is second mode. Also see Ref. 11-12.

14. Clamped Edge, Uniform Radial Edge Load

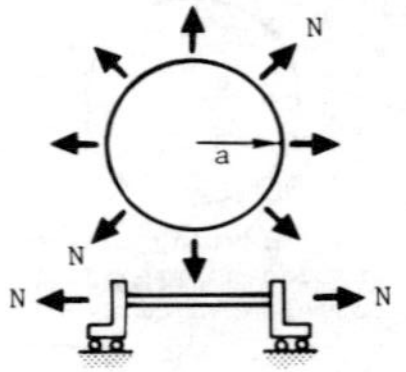

$\frac{12(1-\nu^2)Na^2}{14.68\ Eh^3}$	λ^2 i=0 j=0	λ^2 i=1 j=0
2.0	17.37	30.61
1.5	15.92	28.51
1.0	14.30	26.41
0.5	12.44	24.00
0.25	11.39	22.81
0.0	10.21	21.25
-0.25	8.91	19.61
-0.5	7.28	17.94
-1.0	0.0	14.31

Ref. 11-16

N = force per unit length of edge, positive outward. Higher modes are given in Ref. 11-16. λ is independent of ν. i=1, j=0 is second mode. Also see Ref. 11-2.

the three elementary boundary conditions (i.e., free, simply supported, or clamped). The natural frequencies of annular plates in all nine possible combinations of elementary boundary conditions are given in Table 11-2. The fundamental natural frequencies of elliptical plates are given in Table 11-3.

Mode Shapes and Natural Frequencies. The plate equation in polar coordinates has the following general solution for the transverse vibration of a circular plate which has elementary, polar symmetric boundary conditions:

$$Z(r, \theta, t) = \left[a_i \mathcal{J}_i\left(\frac{\lambda r}{a}\right) + b_i \mathcal{Y}_i\left(\frac{\lambda r}{a}\right) + c_i \mathcal{I}_i\left(\frac{\lambda r}{a}\right) + d_i K_i\left(\frac{\lambda r}{a}\right)\right] \cos i\theta \sin 2\pi f t;$$

$$i = 0, 1, 2, 3, \ldots, \qquad (11\text{-}18)$$

where the natural frequency, f, is a function of the dimensionless parameter λ:

$$f = \frac{\lambda^2}{2\pi a^2}\left[\frac{Eh^3}{12\gamma(1-\nu^2)}\right]^{1/2}.$$

The plate lies in the r–θ plane, where r is a radius from the center of the plate and θ is the angle of that radius. Z is the deformation of the midsurface of the plate perpendicular to the plane of the plate. The plate has thickness h, modulus of elasticity E, Poisson's ratio ν, radius a, and mass per unit area γ. $\mathcal{J}_i$ and $\mathcal{Y}_i$ are Bessel functions of the first and second kinds, respectively, and i order. $\mathcal{I}_i$ and K_i are modified Bessel functions of the first and second kinds, respectively, and i order. a_i, b_i, c_i, and d_i are constants which are determined to within an arbitrary constant by the boundary conditions and mode number. Since the functions $\mathcal{Y}$ and K approach infinity as r approaches zero, the terms b_i and d_i are set equal to zero in the analysis of circular plates; they are retained in the analysis of annular plates.

The dimensionless frequency parameter λ is generally a function of the boundary conditions on the plate, the plate geometry (for example, the ratio of the inner to outer diameter of annular plates), and in many cases Poisson's ratio:

$$\lambda = \lambda\,(\text{boundary conditions, geometry, Poisson's ratio}).$$

λ is determined by setting the determinant of the characteristic matrix to zero to impose the desired boundary conditions on the general solution. The result is an infinite number of solutions for λ for each i. These solutions are numbered λ_{ij} (i = 0, 1, 2, 3, . . . ; j = 0, 1, 2, 3, . . .). Once λ_{ij} is found, the associated modal constants a_{ij}, etc., are determined to within an arbitrary constant by the response matrix. These considerations imply that the mode shape of a circular plate can generally be expressed as

$$\tilde{z}_{ij}(r, \theta) = \left[a_{ij}\mathcal{J}_i\left(\frac{\lambda_{ij} r}{a}\right) + b_{ij}\mathcal{I}_i\left(\frac{\lambda_{ij} r}{a}\right)\right] \cos i\theta; \quad \begin{array}{l} i = 0, 1, 2, 3, \ldots, \\ j = 0, 1, 2, 3, \ldots, \end{array} \qquad (11\text{-}19)$$

Table 11-2. Annular Plates.

Notation: a = outside radius; b = inside radius; h = plate thickness; i = number of nodal diameters; j = number of nodal circles, not counting boundary circles; C = clamped edge; E = modulus of elasticity; F = free edge; S = simply supported edge; γ = mass per unit area of plate; ν = Poisson's ratio; consistent sets of units are given in Table 3-1. This table was adapted from Refs. 11-19 and 11-20.

Natural Frequency (hertz), $$f_{ij} = \frac{\lambda_{ij}^2}{2\pi a^2}\left[\frac{Eh^3}{12\gamma(1-\nu^2)}\right]^{1/2}; \quad i=0,1,2,\cdots; \; j=0,1,2,\cdots$$

Description	λ_{ij}^2

1. Free-Free

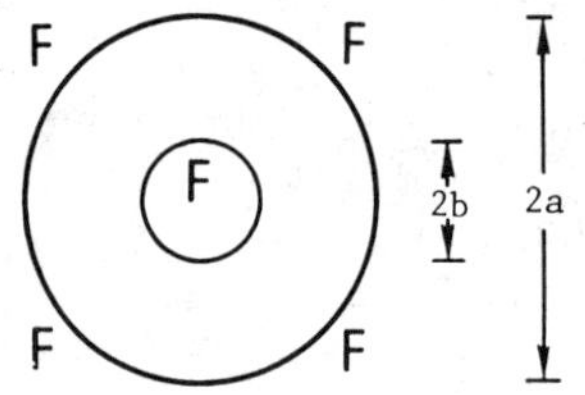

i	j	b/a 0.1	0.3	0.5	0.7
2	0	5.30	4.91	4.28	3.57
0	1	8.77	8.36	9.32	13.2
3	0	12.4	12.26	11.4	9.86
1	1	20.5	18.3	17.2	22.0
2	1	34.9	33.0	31.1	37.8

$\nu = 0.3$

2. Free-Simply Supported

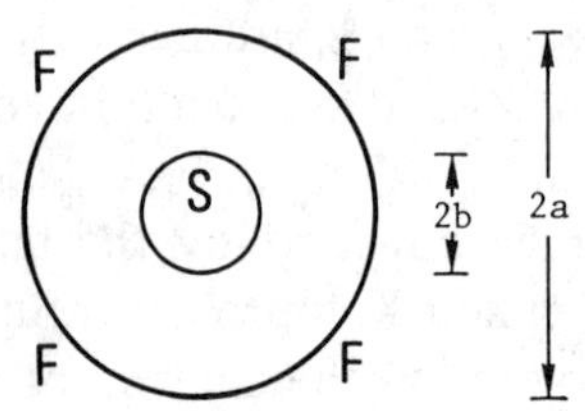

i	j	b/a 0.1	0.3	0.5	0.7
0	0	3.45	3.42	4.11	6.18
1	0	2.30	3.32	4.86	8.34
2	0	5.42	6.08	7.98	13.4
3	0	12.4	12.6	14.0	20.5
0	1	20.8	31.6	61.0	170

$\nu = 0.3$

3. Simply Supported-Free

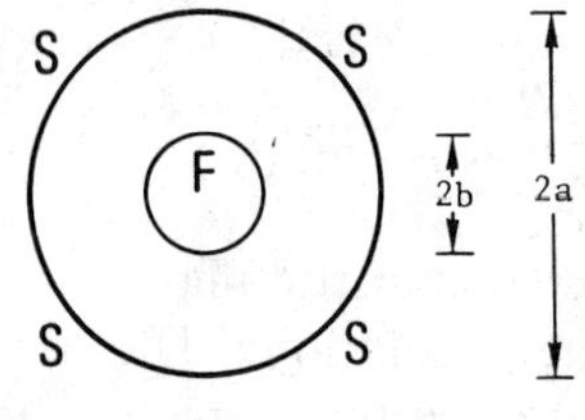

i	j	b/a 0.1	0.3	0.5	0.7
0	0	4.86	4.66	5.07	6.94
1	0	13.9	12.8	11.6	13.3
2	0	25.4	24.1	22.3	24.3
3	0	40.0	38.8	35.7	37.2
0	1	29.4	36.9	65.8	175

$\nu = 0.3$

Table 11-2. Annular Plates. *(Continued)*

Natural Frequency (hertz), $f_{ij} = \frac{\lambda_{ij}^2}{2\pi a^2}\left[\frac{Eh^3}{12\gamma\,(1 - \nu^2)}\right]^{1/2}$; i=0,1,2,···; j=0,1,2,···

Description	λ_{ij}^2

4. Free-Clamped

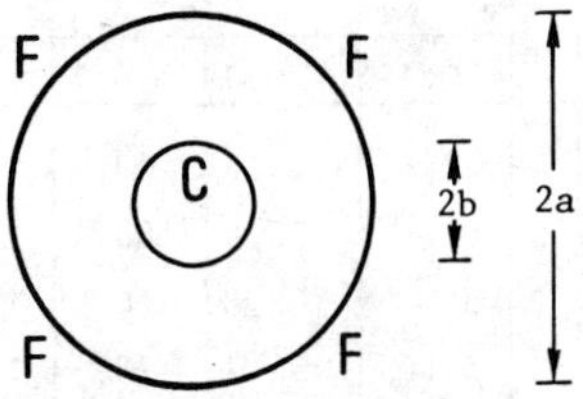

		b/a			
i	j	0.1	0.3	0.5	0.7
0	0	4.23	6.66	13.0	37.0
1	0	3.14	6.33	13.3	37.5
2	0	5.62	7.95	14.7	39.3
3	0	12.4	13.3	18.5	42.6
0	1	25.3	42.6	85.1	239

$\nu = 0.3$

5. Clamped-Free

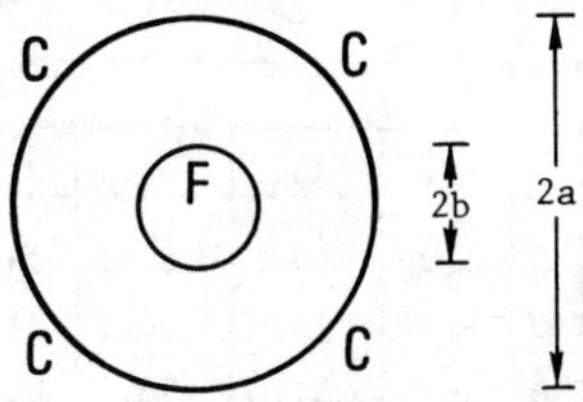

		b/a			
i	j	0.1	0.3	0.5	0.7
0	0	10.2	11.4	17.7	43.1
1	0	21.1	19.5	22.1	45.3
2	0	34.5	32.5	32.0	51.5
3	0	51.0	49.1	45.8	61.3
0	1	39.5	51.7	93.8	253

$\nu = 0.3$

6. Simply Supported-Simply Supported

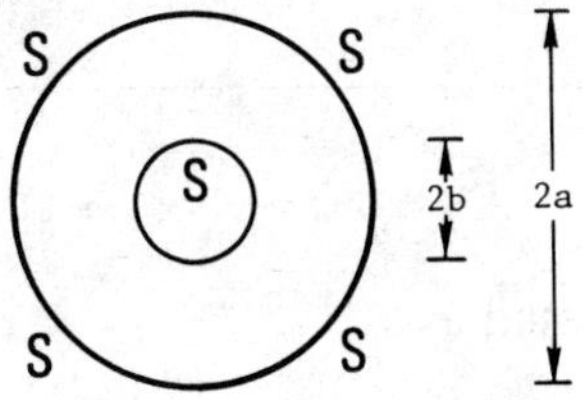

		b/a			
i	j	0.1	0.3	0.5	0.7
0	0	14.5	21.1	40.0	110
1	0	16.7	23.3	41.8	112
2	0	25.9	30.2	47.1	116
3	0	40.0	42.0	56.0	122
0	1	51.70	81.8	159.0	439

λ is independent of ν.

7. Simply Supported-Clamped

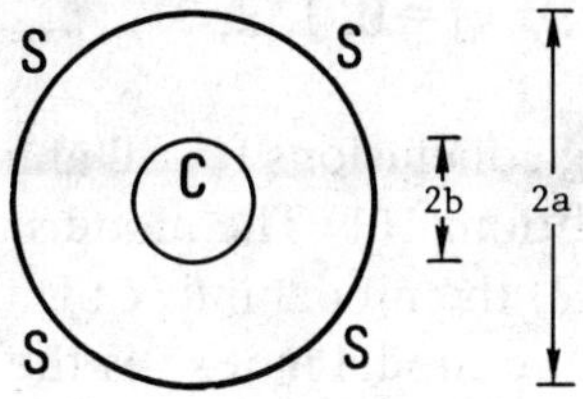

		b/a			
i	j	0.1	0.3	0.5	0.7
0	0	17.8	29.9	59.8	168
1	0	19.0	31.4	61.0	170
2	0	26.0	36.2	64.6	172
3	0	40.0	45.4	71.0	177
0	1	60.1	100	198	552

λ is independent of ν.

Table 11-2. Annular Plates. (*Continued*)

Natural frequency (hertz), $f_{ij} = \dfrac{\lambda_{ij}^2}{2\pi a^2}\left[\dfrac{Eh^3}{12\gamma(1-\nu^2)}\right]^{1/2}$; i=0,1,2,···; j=0,1,2,···

Description	λ_{ij}^2

8. Clamped-Simply Supported

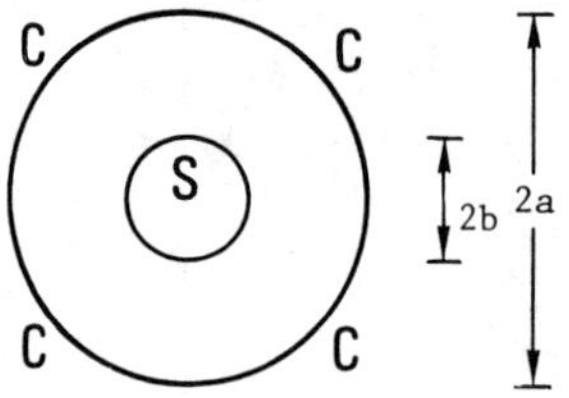

i	j	b/a 0.1	0.3	0.5	0.7
0	0	22.6	33.7	63.9	175
1	0	25.1	35.8	65.4	175
2	0	35.4	42.8	70.0	178
3	0	51.0	54.7	78.1	185
0	1	65.6	104	202	558

λ is independent of ν.

9. Clamped-Clamped

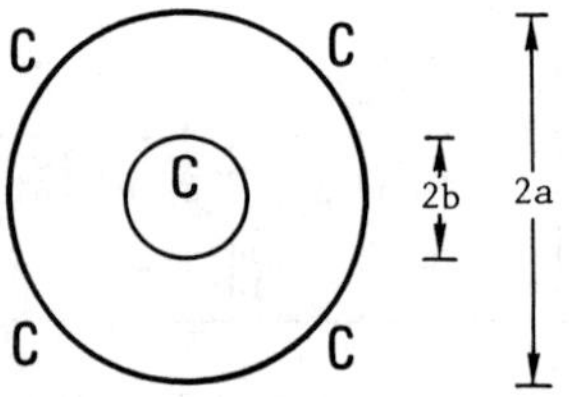

i	j	b/a 0.1	0.3	0.5	0.7
0	0	27.3	45.2	89.2	248
1	0	28.4	45.6	90.2	249
2	0	36.7	51.0	93.3	251
3	0	51.2	60.0	99.0	256
0	1	75.3	125	246	686

λ is independent of ν.

and the mode shape of an annular plate can be expressed as

$$\tilde{z}_{ij}(r,\theta) = \left[a_{ij}\mathcal{J}_i\left(\frac{\lambda_{ij}r}{a}\right) + b_{ij}\mathcal{Y}_i\left(\frac{\lambda_{ij}r}{a}\right) + c_{ij}\mathcal{I}_i\left(\frac{\lambda_{ij}r}{a}\right) + d_{ij}\mathcal{K}_i\left(\frac{\lambda_{ij}r}{a}\right)\right] \cos i\theta;$$

$$i = 0, 1, 2, 3, \ldots, \quad j = 0, 1, 2, 3, \ldots. \qquad (11\text{-}20)$$

The form of these mode shapes leads to the following conclusions for all annular and circular plates with polar symmetric boundary conditions: (1) The mode shapes are separable into a function of r and a function of θ; (2) the modal index i is the number of nodal diameters in the mode shape; and (3) the modal index j is the number of nodal rings in the mode shape, not counting modal rings enforced by boundary conditions. For example, the relationship between the modal indexes i and j and the mode shape of a simply supported circular plate is shown in Fig. 11-4.

Table 11-3. Elliptical Plates.

Notation: a = semi-minor axis; b = semi-major axis; h = plate thickness; C = clamped edge; F = free edge; E = modulus of elasticity; S = simply supported edge; γ = mass per unit area of plate; ν = Poisson's ratio; consistent sets of units are given in Table 3-1.

$$\text{Natural Frequency (hertz), } f_i = \frac{\lambda_i^2}{2\pi a^2}\left[\frac{Eh^3}{12\gamma(1-\nu^2)}\right]^{1/2};\ i=1,2,3$$

Description	λ^2

1. Free

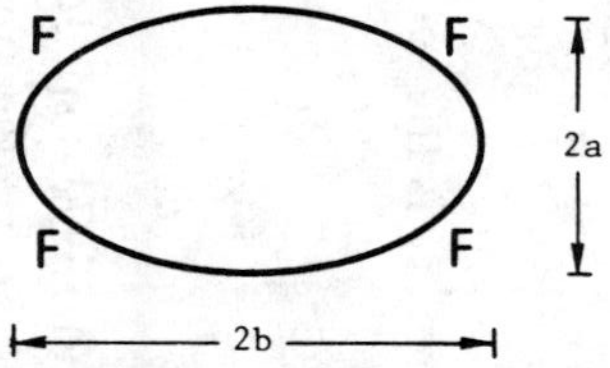

$\frac{b}{a}$ =	1.00	1.10	1.20	1.40	1.70	2.0	2.5	3.0	5.0	7.0
λ_1 =	2.315	2.189	2.056	1.806	1.510	1.291	1.038	0.8865	0.5207	0.3719
λ_2 =	3.002	2.883	2.811	2.733	2.673	2.635	2.590	2.558	2.487	2.454
λ_3 =	4.673	4.443	4.222	3.802	3.249	2.807	2.272	1.902	1.146	0.8180

ν = 0.3; Ref. 11-21.

2. Simply Supported Edge

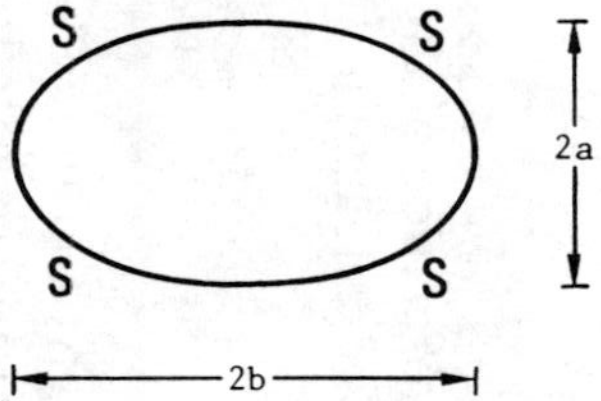

$\frac{b}{a}$ =	1.0	1.1	1.2	1.4	1.7	2.0	2.5	3.0	5.0	10.0	20.0
λ_1^2 =	4.865	4.454	4.157	3.773	3.463	3.292	3.128	3.027	2.846	2.750	2.725

ν = 0.25; Ref. 11-22.

Higher modes are presented in Ref. 11-22.

3. Clamped

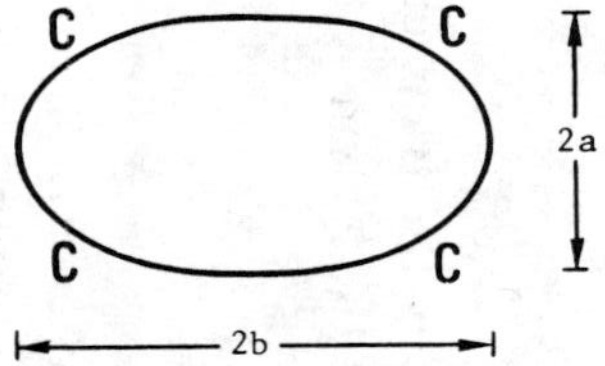

$\frac{b}{a}$ =	1.0	1.1	1.2	1.5	2.0	3.0	5.0
λ_1^2 =	10.22	9.350	8.726	7.567	6.937	6.521	6.354

$$\lambda_1^2 \approx 6.263\left[1+\frac{2}{3}\left(\frac{a}{b}\right)^2+\left(\frac{a}{b}\right)^4\right]^{1/2}$$

$$\frac{\lambda_2^2}{\lambda_1^2} \approx 1.815;\ \text{Ref. 11-23.}$$

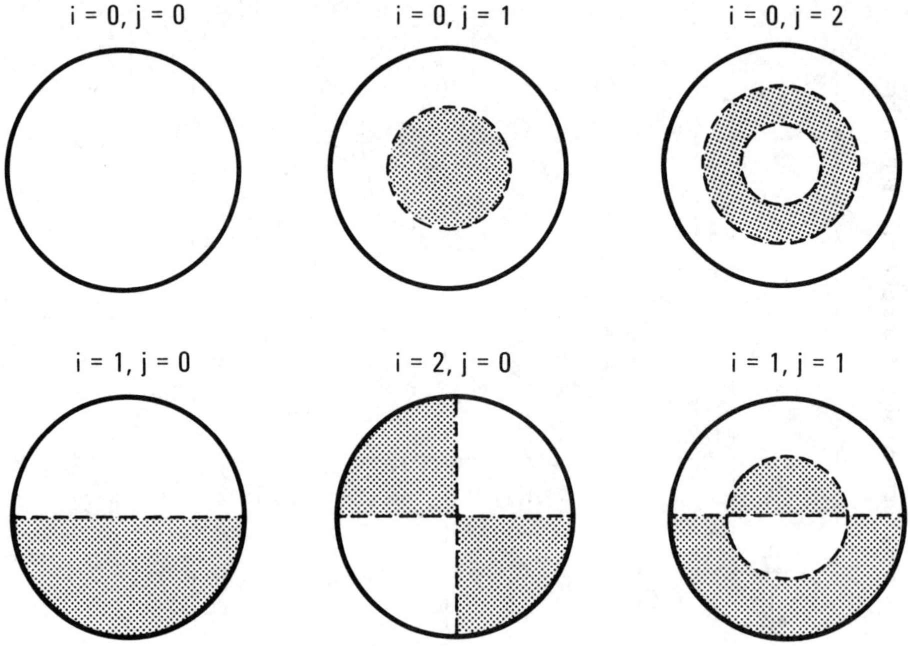

Fig. 11-4. Nodal lines for a circular plate whose edge is simply supported.

Boundary Conditions. The boundary conditions associated with the circular and annular plates are:

$$\text{Free Edge:}\quad \mathcal{M}_r = \mathcal{V}_r = 0$$

$$\text{Simply Supported Edge:}\quad Z = \mathcal{M}_r = 0$$

$$\text{Clamped Edge:}\quad Z = \frac{\partial Z}{\partial r} = 0$$

$$\text{Simply Supported Edge with a Torsion Spring:}\quad Z = 0$$

$$\mathcal{M}_r = k \frac{\partial Z}{\partial r}$$

Z is the transverse deformation of the plate. $\mathcal{M}_r$ is the moment given in Eq. 11-7. $\mathcal{V}_r$ is the Kelvin–Kirchoff edge reaction:

$$\mathcal{V}_r = Q_r + \frac{1}{r}\frac{\partial \mathcal{M}_{r\theta}}{\partial \theta},$$

where Q_r and $\mathcal{M}_{r\theta}$ are given in Eqs. 11-8 and 11-7, respectively. A clamped edge permits neither displacement nor rotation. A free edge supports neither bending moment nor shear. A simply supported edge is the two-dimensional generalization of the pinned joint discussed in Section 8.1. However, both the free and the simply supported boundary conditions for plates involve the Poisson's ratio of the plate, and they are considerably more complex than the equivalent beam boundary condition.

Sectorial Plates. A sectorial plate is formed from a circle or annulus with boundaries along two radii. Solutions for sectorial plates can be easily adapted from the solutions for circular and annular plates with elementary polar symmetric boundary conditions if (1) the angle (α) enclosed by the sectorial plate is an integer submultiple of 360 degrees,

$$\alpha = \frac{360}{j}, \qquad j = 2 \text{ or } 3 \text{ or } 4, \ldots$$

and (2) the radial (straight) boundaries of the sectorial plate are simply supported. The sectorial plates which satisfy both these conditions have natural frequencies and mode shapes identical to the associated complete circular (frames 1, 2, and 3 of Table 11-1) or annular (Table 11-3) plate with the following choices of modal indices,

$$j = 0, 1, 2, \ldots \qquad \text{(number of nodal circles)},$$

$$i = n, 2n, 3n, \ldots \qquad \text{(number of nodal diameters)},$$

where

$$n = \frac{360}{2\alpha} = \text{an integer} \qquad (\alpha \text{ in degrees}),$$

so that the nodal lines along the radii of the complete plate align with the boundaries of the sectorial plate.

The natural frequencies of sectorial plates with clamped radial boundaries are discussed in Ref. 11-18.

Poisson's Ratio Effect. The natural frequency parameters (λ_{ij}) given in Tables 11-1, 11-2, and 11-3 are functions of Poisson's ratio except for the clamped plate (frames 3, 11, and 13 of Table 11-1), the clamped-clamped annulus (frame 9 of Table 11-2), and the clamped ellipse (frame 3 of Table 11-3). Generally, the natural frequency parameters change only slightly over the range of Poisson's ratios of practical importance. For example, the natural frequency parameter as a function of Poisson's ratio for the fundamental mode of a circular plate with a simply supported edge is (Ref. 11-12):

$\nu = 0.0$	0.25	0.30	0.50,
$\lambda^2 = 4.452$	4.872	4.947	5.227.

Vogel and Skinner (Ref. 11-20) found that the natural frequencies of annular plates varied only slightly with large variations in Poisson's ratio. They found that the natural frequency of the annular plates given in Table 11-2 varied, at most, between 3% and 5% as Poisson's ratio was varied between 0.0 and 0.5, which is the maximum possible variation in Poisson's ratio for an isotropic material. The natural frequency of the fundamental mode of a elliptical plate with an axis ratio (a/b) of 2 varies only 7% as Poisson's ratio is varied between 0.0 and 0.5 (Ref. 11-21). Thus, Poisson's ratio is generally a parameter of secondary importance in the calculation of the natural frequencies of these plates.

Rotary Inertia and Shear Deformation. Rotary inertia is the inertia associated with local rotation of the plate as it flexes. Shear deformation is deformation due to transverse shearing of the cross section under dynamic shear loads. These effects are neglected in Tables 11-1, 11-2, and 11-3. Rotary inertia and shear deformation will tend to lower the natural frequencies from those given in the tables. The importance of rotary inertia and shear deformation generally increases with the ratio of plate thickness to plate diameter and with increasing mode number; these effects are generally insignificant for plates whose thickness is less than $\frac{1}{10}$ of the plate diameter for vibrations in the fundamental mode. The effects of rotary inertia and shear deformation on the natural frequency of circular plates are discussed in Refs. 11-24, 11-25, and 11-26. Extensive tables of the natural frequencies of annular plates, including shear deformation and rotary inertia, are given in Ref. 11-19.

Orthotropic Plates. An isotropic plate is a plate composed of a material whose properties are independent of direction. The plates in Tables 11-1, 11-2, and 11-3 are assumed to be isotropic. Two material constants (E and ν) are required to describe the linear stress-strain relationship of an isotropic plate. An orthotropic plate is a plate whose material properties are symmetric about three mutually perpendicular axes. Four material constants are required to describe the linear stress-strain relationship of an orthotropic plate. Common orthotropic materials are wood and fiber-reinforced composites. The solutions available for orthotropic plates are generally limited to cases where the orthotropic material properties and the plate possess the same symmetries. Woo, Kirmser, and Huang (Refs. 11-27, 11-28) have shown that cylindrical orthotropy implies the existence of a singularity at the center of a circular orthotropic plate. The singularity can be removed by introducing a small isotropic core. The natural frequencies of orthotropic circular plates are discussed in Refs. 11-27 through 11-30. The natural frequencies of annular orthotropic plates are discussed in Ref. 11-31.

Other Effects. Spinning circular disks are discussed in Refs. 11-32, 11-33 and 11-34. Variable-thickness circular plates are discussed in Refs. 11-34 through 11-38. Extensive tables for the natural frequencies of tapered annular plates are given in Ref. 11-39. The natural frequencies of annular plates with reinforced edges are given in Ref. 11-40.

11.3. RECTANGULAR PLATES

Exact Natural Frequencies. The natural frequencies of the first six modes of rectangular plates for all 21 possible combinations of the three elementary boundary conditions on the four edges of the plates are given in Table 11-4. This table was adapted from the numerical results of Leissa (Ref. 11-41).

The dimensionless frequency parameter λ_{ij} of rectangular plates is generally a function of the boundary conditions applied at the edges of the plate, the aspect

Table 11-4. Rectangular Plates.

Notation: a = length of plate; b = width of plate; h = thickness of plate; i = number of half-waves in mode shape along horizontal axis; j = number of half-waves in mode shape along vertical axis; C = clamped edge; E = modulus of elasticity; F = free edge; S = simply supported edge; γ = mass per unit area of plate (μh for a plate of a material with density μ); ν = Poisson's ratio; see Table 3-1 for consistent sets of units. This table was adapted from Ref. 11-41, copyright Academic Press Inc (London) Ltd, reproduced with permission.

Natural Frequency (hertz) $f_{ij} = \frac{\lambda_{ij}^2}{2\pi a^2}\left[\frac{Eh^3}{12\gamma(1-\nu^2)}\right]^{1/2}$; i=1,2,3...; j=1,2,3...

Description	λ_{ij}^2 and (ij)

1. Free-Free - Free-Free

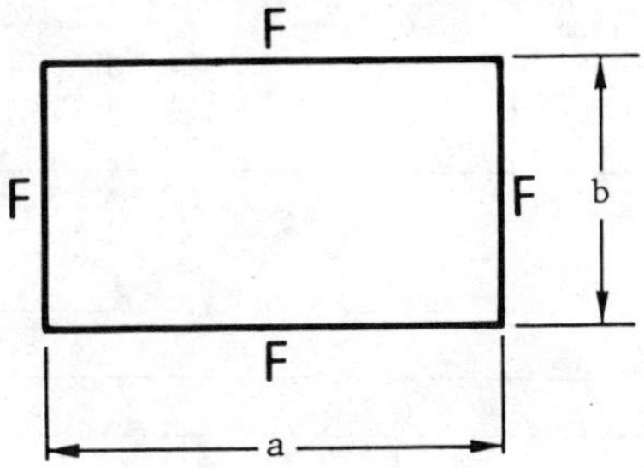

a/b	Mode Sequence 1	2	3	4	5	6
0.4	3.463 (13)	5.288 (22)	9.622 (14)	11.44 (23)	18.79 (15)	19.10 (24)
2/3	8.946 (22)	9.602 (13)	20.74 (23)	22.35 (31)	25.87 (14)	29.97 (32)
1.0	13.49 (22)	19.79 (13)	24.43 (31)	35.02 (32)	35.02 (23)	61.53 (41)
1.5	20.13 (22)	21.60 (31)	46.65 (32)	50.29 (13)	58.20 (41)	67.49 (23)
2.5	21.64 (31)	33.05 (22)	60.14 (41)	71.48 (32)	117.5 (51)	119.4 (42)

ν = 0.3; (ij) = (11), (12), (21) are rigid body modes. Also see Ref. 11-82.

2. Simply Supported-Free - Free-Free

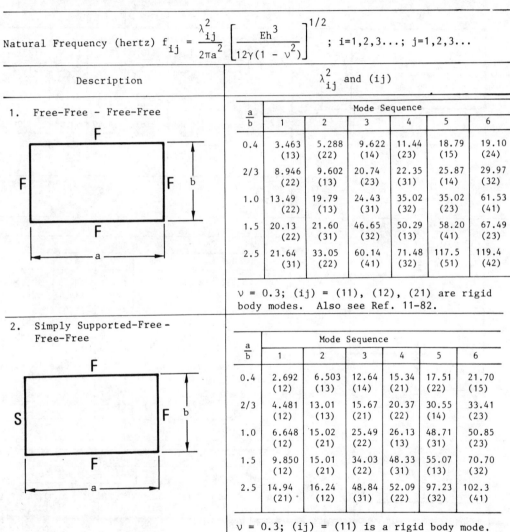

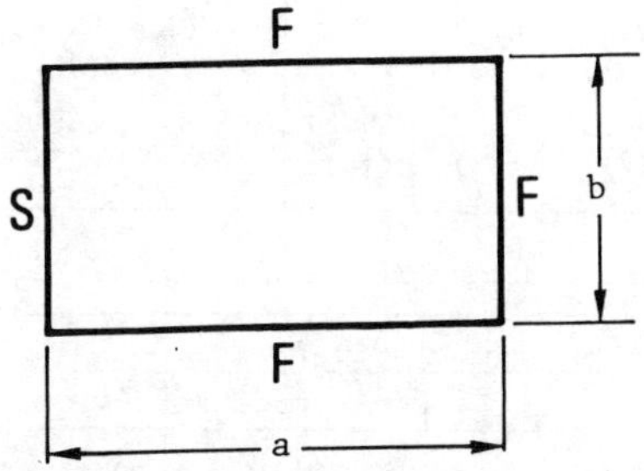

a/b	Mode Sequence 1	2	3	4	5	6
0.4	2.692 (12)	6.503 (13)	12.64 (14)	15.34 (21)	17.51 (22)	21.70 (15)
2/3	4.481 (12)	13.01 (13)	15.67 (21)	20.37 (22)	30.55 (14)	33.41 (23)
1.0	6.648 (12)	15.02 (21)	25.49 (22)	26.13 (13)	48.71 (31)	50.85 (23)
1.5	9.850 (12)	15.01 (21)	34.03 (22)	48.33 (31)	55.07 (13)	70.70 (32)
2.5	14.94 (21)	16.24 (12)	48.84 (31)	52.09 (22)	97.23 (32)	102.3 (41)

ν = 0.3; (ij) = (11) is a rigid body mode.

Table 11-4. Rectangular Plates. *(Continued)*

Natural Frequency (hertz), $f_{ij} = \dfrac{\lambda_{ij}^2}{2\pi a^2}\left[\dfrac{Eh^3}{12\gamma(1-\nu^2)}\right]^{1/2}$; i=1,2,3...; j=1,2,3...

Description	λ_{ij}^2 and (ij)

3. Clamped-Free-Free-Free

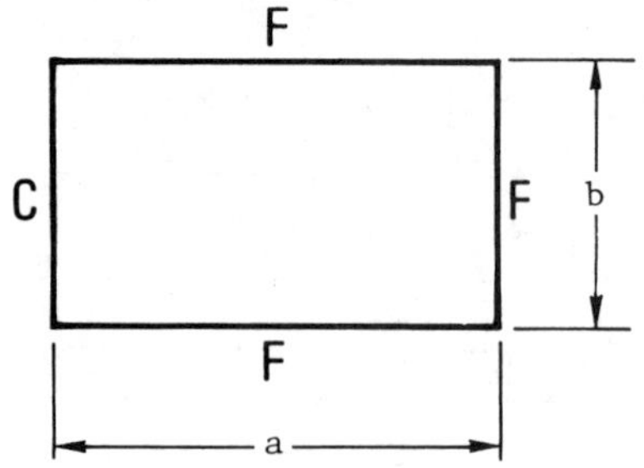

a/b	Mode Sequence 1	2	3	4	5	6
0.40	3.511 (11)	4.786 (12)	8.115 (13)	13.88 (14)	21.64 (21)	23.73 (22)
2/3	3.502 (11)	6.406 (12)	14.54 (13)	22.04 (21)	26.07 (22)	31.62 (14)
1.0	3.492 (11)	8.525 (12)	21.43 (21)	27.33 (13)	31.11 (22)	54.44 (23)
1.5	3.477 (11)	11.68 (12)	21.62 (21)	39.49 (22)	53.88 (13)	61.99 (31)
2.5	3.456 (11)	17.99 (12)	21.56 (21)	57.46 (22)	60.58 (31)	106.5 (32)

$\nu = 0.3$

4. Simply Supported-Free-Free-Simply Supported

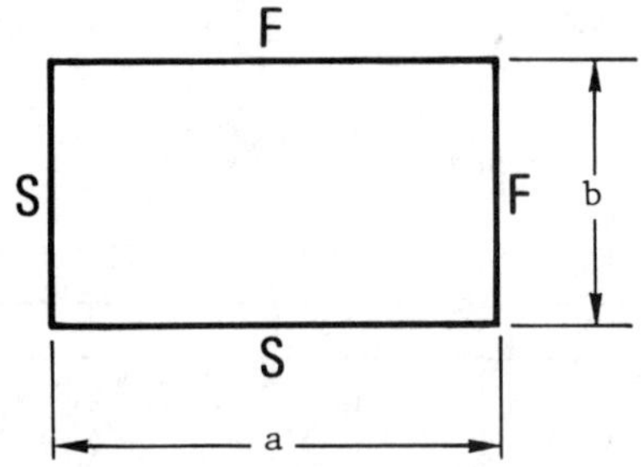

a/b	Mode Shape 1	2	3	4	5	6
0.4	1.320 (11)	4.743 (12)	10.36 (13)	15.87 (21)	18.93 (14)	20.17 (22)
2/3	2.234 (11)	9.575 (12)	16.76 (21)	24.66 (13)	27.06 (22)	44.17 (23)
1.0	3.369 (11)	17.41 (12)	19.37 (21)	38.29 (22)	51.32 (13)	53.74 (31)
1.5	5.026 (11)	21.54 (21)	37.72 (12)	55.49 (31)	60.88 (22)	99.39 (32)
2.5	8.251 (11)	29.65 (21)	64.76 (31)	99.21 (12)	118.3 (41)	126.1 (22)

$\nu = 0.3$

5. Simply Supported-Free-Simply Supported-Free

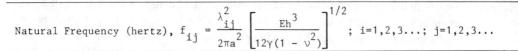

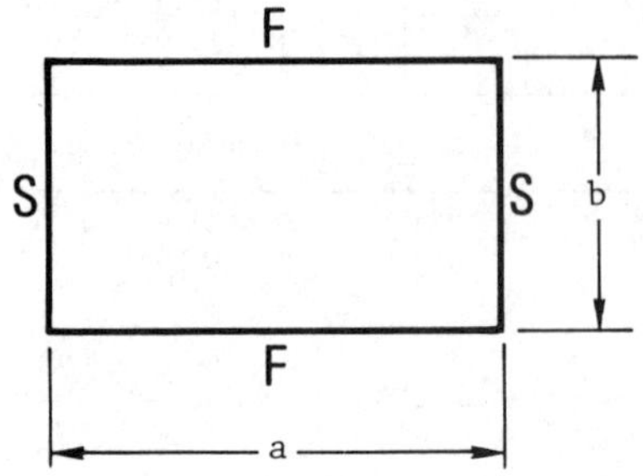

a/b	Mode Sequence 1	2	3	4	5	6
0.4	9.760 (11)	11.04 (12)	15.06 (13)	21.71 (14)	31.18 (15)	39.24 (21)
2/3	9.698 (11)	12.98 (12)	22.95 (13)	39.11 (21)	40.36 (14)	42.69 (22)
1.0	9.631 (11)	16.14 (12)	36.73 (13)	38.95 (21)	46.74 (22)	70.74 (23)
1.5	9.558 (11)	21.62 (12)	38.72 (21)	54.84 (22)	65.79 (13)	87.63 (31)
2.5	9.484 (11)	33.62 (12)	38.36 (21)	75.20 (22)	86.97 (31)	130.4 (32)

$\nu = 0.3$

Table 11-4. Rectangular Plates. *(Continued)*

Natural Frequency (hertz), $f_{ij} = \frac{\lambda_{ij}^2}{2\pi a^2}\left[\frac{Eh^3}{12\gamma(1-\nu^2)}\right]^{1/2}$; i=1,2,3...; j=1,2,3...

Description	λ_{ij}^2 and (ij)

6. Clamped-Free - Simply Supported-Free

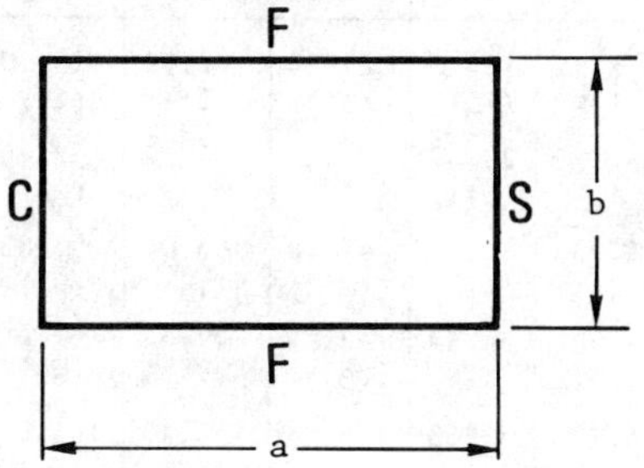

a/b	Mode Sequence 1	2	3	4	5	6
0.4	15.38 (11)	16.37 (12)	19.66 (13)	25.55 (14)	34.51 (15)	46.44 (16)
2/3	15.34 (11)	17.95 (12)	26.73 (13)	43.19 (14)	49.84 (21)	53.01 (22)
1.0	15.29 (11)	20.67 (12)	39.78 (13)	49.73 (21)	56.62 (22)	77.37 (14)
1.5	15.22 (11)	25.71 (12)	49.55 (21)	64.01 (22)	68.13 (13)	103.7 (31)
2.5	15.13 (11)	37.29 (12)	49.23 (21)	83.33 (22)	103.1 (31)	143.7 (32)

$\nu = 0.3$

7. Clamped-Free - Free-Simply Supported

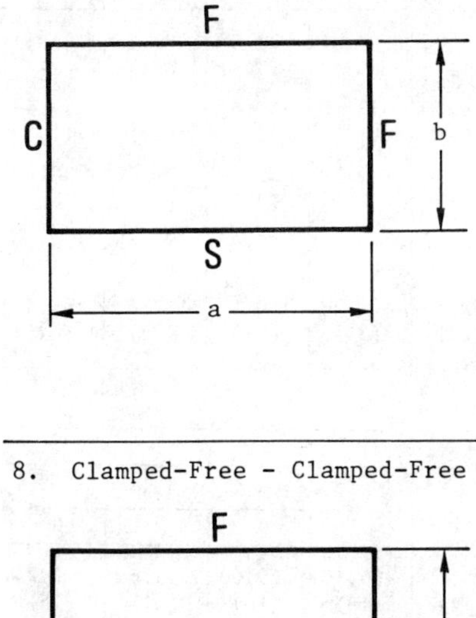

a/b	Mode Sequence 1	2	3	4	5	6
0.4	3.854 (11)	6.420 (12)	11.58 (13)	19.77 (14)	22.52 (21)	26.02 (22)
2/3	4.425 (11)	10.91 (12)	22.96 (21)	25.70 (13)	32.43 (22)	48.47 (23)
1.0	5.364 (11)	19.17 (12)	24.77 (21)	43.19 (22)	53.00 (13)	64.05 (31)
1.5	6.931 (11)	27.29 (21)	38.59 (12)	64.25 (22)	67.47 (31)	108.0 (32)
2.5	10.10 (11)	35.16 (21)	74.99 (31)	99.93 (12)	127.7 (22)	135.5 (41)

$\nu = 0.3$

8. Clamped-Free - Clamped-Free

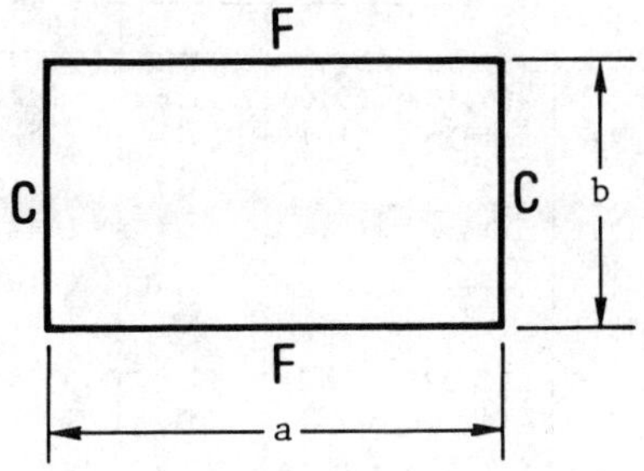

a/b	Mode Sequence 1	2	3	4	5	6
0.4	22.35 (11)	23.09 (12)	25.67 (13)	30.63 (14)	38.69 (15)	49.86 (16)
2/3	22.31 (11)	24.31 (12)	31.70 (13)	46.82 (14)	61.57 (21)	64.34 (22)
1.0	22.27 (11)	26.53 (12)	43.66 (13)	61.47 (21)	67.55 (22)	79.90 (14)
1.5	22.21 (11)	30.90 (12)	61.30 (21)	70.96 (13)	74.26 (22)	118.3 (23)
2.5	22.13 (11)	41.69 (12)	61.00 (21)	92.38 (22)	119.9 (31)	157.8 (32)

$\nu = 0.3$

Table 11-4. Rectangular Plates. (*Continued*)

Natural Frequency (hertz), $f_{ij} = \frac{\lambda_{ij}^2}{2\pi a^2}\left[\frac{Eh^3}{12\gamma(1-\nu^2)}\right]^{1/2}$; i=1,2,3...; j=1,2,3...

Description	λ_{ij}^2 and (ij)

9. Clamped-Free - Free-Clamped

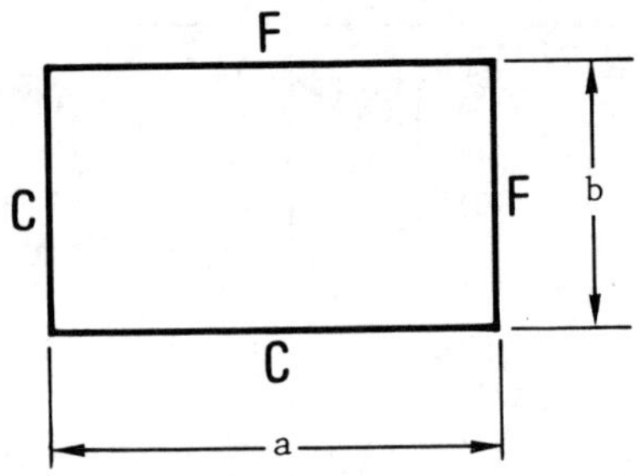

a/b	Mode Sequence 1	2	3	4	5	6
0.4	3.986 (11)	7.155 (12)	13.10 (13)	21.84 (14)	22.90 (21)	26.50 (22)
2/3	4.985 (11)	13.29 (12)	23.38 (21)	30.26 (13)	34.24 (22)	52.40 (23)
1.0	6.942 (11)	24.03 (21)	26.68 (12)	47.78 (22)	63.04 (13)	65.83 (31)
1.5	11.22 (11)	29.90 (21)	52.62 (12)	68.09 (31)	77.04 (22)	117.9 (32)
2.5	24.91 (11)	44.72 (21)	81.88 (31)	136.5 (41)	143.1 (12)	165.6 (22)

$\nu = 0.3$

10. Simply Supported-Free-Simply Supported-Simply Supported

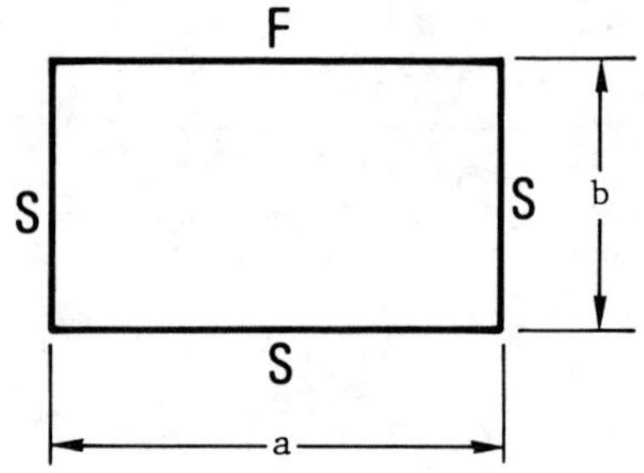

a/b	Mode Sequence 1	2	3	4	5	6
0.4	10.13 (11)	13.06 (12)	18.84 (13)	27.56 (14)	39.34 (15)	39.61 (21)
2/3	10.67 (11)	18.30 (12)	33.70 (13)	40.13 (21)	48.41 (22)	57.59 (14)
1.0	11.68 (11)	27.76 (12)	41.20 (21)	59.07 (22)	61.86 (13)	90.29 (31)
1.5	13.71 (11)	43.57 (21)	47.86 (12)	81.48 (22)	92.69 (31)	124.56 (13)
2.5	18.80 (11)	50.54 (21)	100.23 (31)	100.2 (12)	147.6 (22)	169.1 (41)

$\nu = 0.3$

11. Simply Supported-Free - Simply Supported-Clamped

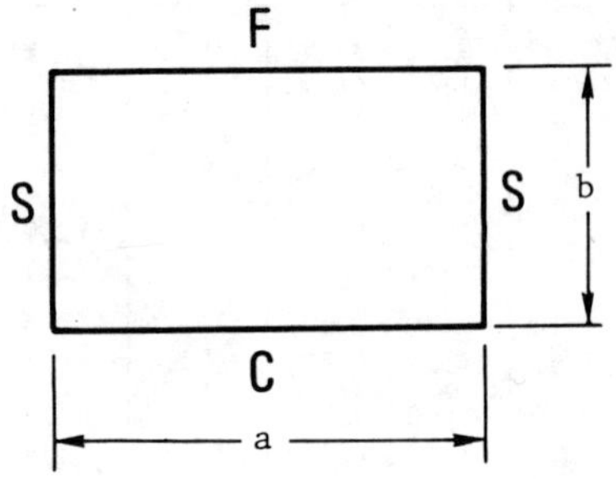

a/b	Mode Sequence 1	2	3	4	5	6
0.4	10.19 (11)	13.60 (12)	20.10 (13)	29.62 (14)	39.64 (21)	42.24 (15)
2/3	10.98 (11)	20.34 (12)	37.96 (13)	40.27 (21)	49.73 (22)	64.19 (14)
1.0	12.69 (11)	33.07 (12)	41.70 (21)	63.01 (22)	72.40 (13)	90.61 (31)
1.5	16.82 (11)	45.30 (21)	61.02 (12)	92.31 (22)	93.83 (31)	141.8 (32)
2.5	30.63 (11)	58.08 (21)	105.5 (31)	149.46 (12)	173.1 (41)	182.8 (22)

$\nu = 0.3$

Table 11-4. Rectangular Plates. (*Continued*)

$$\text{Natural Frequency (hertz), } f_{ij} = \frac{\lambda_{ij}^2}{2\pi a^2}\left[\frac{Eh^3}{12\gamma(1 - \nu^2)}\right]^{1/2} \text{; i=1,2,3...; j-1,2,3...}$$

Description	λ_{ij}^2 and (ij)

12. Clamped-Free - Clamped-Simply Supported

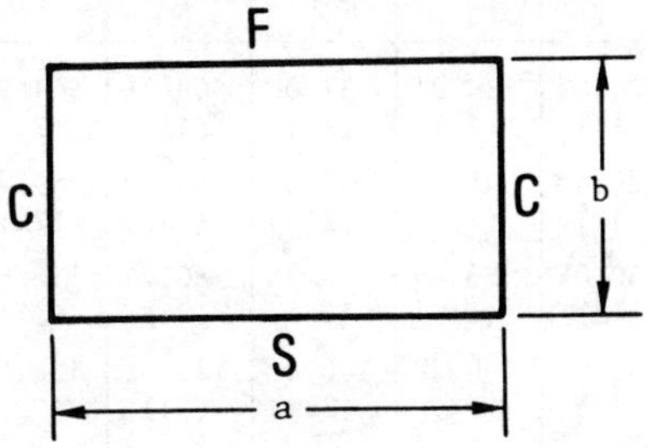

$\frac{a}{b}$	Mode Sequence					
	1	2	3	4	5	6
0.4	22.54 (11)	24.30 (12)	28.34 (13)	35.35 (14)	45.71 (15)	59.56 (16)
2/3	22.86 (11)	27.97 (12)	40.68 (13)	62.31 (21)	62.70 (14)	68.68 (22)
1.0	23.46 (11)	35.61 (12)	63.13 (21)	66.81 (13)	77.50 (22)	109.0 (23)
1.5	24.78 (11)	53.73 (12)	64.96 (21)	97.26 (22)	124.5 (31)	127.9 (13)
2.5	28.56 (11)	70.56 (21)	114.0 (12)	130.8 (31)	159.5 (22)	210.3 (41)

$\nu = 0.3$

13. Clamped-Free - Simply Supported-Simply Supported

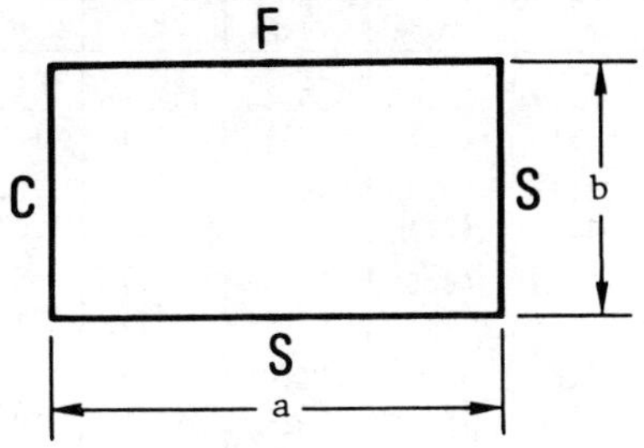

$\frac{a}{b}$	Mode Sequence					
	1	2	3	4	5	6
0.4	15.65 (11)	17.95 (12)	22.90 (13)	30.89 (14)	42.11 (15)	50.22 (21)
2/3	16.07 (11)	22.45 (12)	36.70 (13)	50.70 (21)	57.91 (22)	59.84 (14)
1.0	16.87 (11)	31.14 (12)	51.63 (21)	64.04 (13)	67.65 (22)	101.2 (23)
1.5	18.54 (11)	50.44 (12)	53.72 (21)	88.80 (22)	108.2 (31)	126.1 (13)
2.5	23.07 (11)	59.97 (21)	112.0 (12)	115.1 (31)	153.2 (22)	189.5 (41)

$\nu = 0.3$

14. Clamped-Free - Simply Supported-Clamped

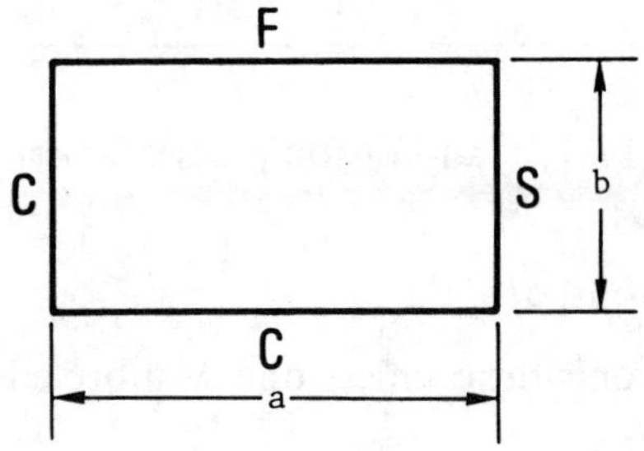

$\frac{a}{b}$	Mode Sequence					
	1	2	3	4	5	6
0.4	15.70 (11)	18.37 (12)	23.99 (13)	32.81 (14)	44.86 (15)	50.25 (21)
2/3	16.29 (11)	24.20 (12)	40.70 (13)	50.82 (21)	59.07 (22)	66.26 (14)
1.0	17.62 (11)	36.05 (12)	52.07 (21)	71.19 (22)	74.35 (13)	106.3 (31)
1.5	21.04 (11)	55.18 (21)	63.18 (12)	99.01 (22)	109.2 (31)	150.9 (13)
2.5	33.58 (11)	66.61 (21)	119.9 (31)	150.8 (12)	187.6 (22)	193.2 (41)

$\nu = 0.3$

Table 11-4. Rectangular Plates. *(Continued)*

Natural Frequency (hertz), $f_{ij} = \frac{\lambda_{ij}^2}{2\pi a^2}\left[\frac{Eh^3}{12\gamma(1-\nu^2)}\right]^{1/2}$; i=1,2,3...; j=1,2,3...

Description	λ_{ij}^2 and (ij)

15. Clamped-Free - Clamped-Clamped

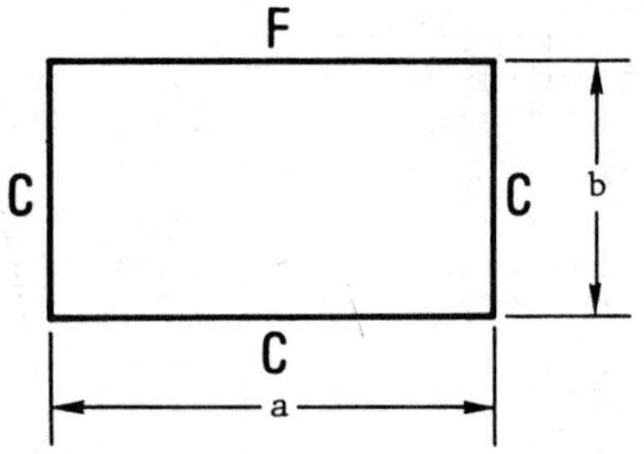

a/b	Mode Sequence 1	2	3	4	5	6
0.4	22.58 (11)	24.62 (12)	29.24 (13)	37.06 (14)	48.28 (15)	61.92 (21)
2/3	23.02 (11)	29.43 (12)	44.36 (13)	62.42 (21)	68.89 (14)	69.70 (22)
1.0	24.02 (11)	40.04 (12)	63.49 (21)	76.76 (13)	80.71 (22)	116.8 (23)
1.5	26.73 (11)	65.92 (12)	66.22 (21)	106.8 (22)	125.4 (31)	152.5 (13)
2.5	37.66 (11)	76.41 (21)	135.2 (31)	152.5 (12)	193.0 (22)	213.7 (41)

$\nu = 0.3$

16. Simply Supported-Simply Supported - Simply Supported-Simply Supported

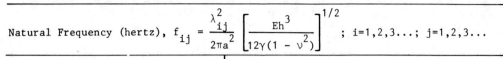

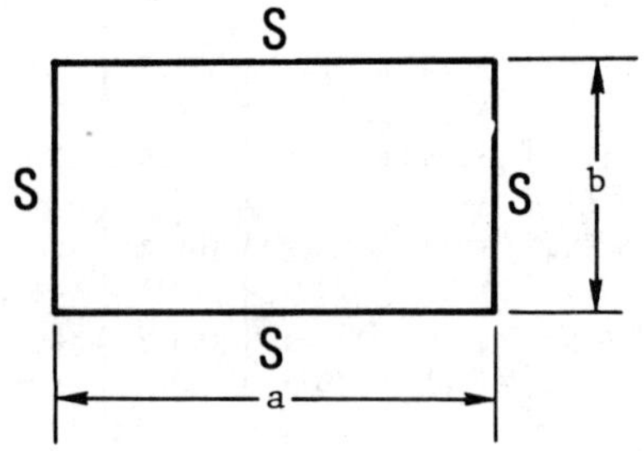

a/b	Mode Sequence 1	2	3	4	5	6
0.4	11.45 (11)	16.19 (12)	24.08 (13)	35.14 (14)	41.06 (21)	45.80 (22)
2/3	14.26 (11)	27.42 (12)	43.86 (21)	49.35 (13)	57.02 (22)	78.96 (23)
1.0	19.74 (11)	49.35 (21)	49.35 (12)	78.96 (22)	98.70 (31)	98.70 (13)
1.5	32.08 (11)	61.69 (21)	98.70 (12)	111.0 (31)	128.30 (22)	177.7 (32)
2.5	71.56 (11)	101.16 (21)	150.5 (31)	219.6 (41)	256.6 (12)	286.2 (22)

$$\lambda_{ij}^2 = \pi^2\left[i^2 + j^2\left(\frac{a}{b}\right)^2\right]$$

ratio of the plate (a/b for the plate shown in Fig. 11-13), and in some cases Poisson's ratio (ν):

$$\lambda_{ij} = \lambda_{ij}\,(\text{boundary conditions, a/b}, \nu).$$

It has been found that λ_{ij} is independent of Poisson's ratio unless one or more edges of the rectangular plate are free.

Approximate Natural Frequencies. If the mode shape of a rectangular plate is approximated by the mode shape of single beams (Table 8-1) along the x and y axes,

Table 11-4. Rectangular Plates. *(Continued)*

Natural Frequency (hertz), $f_{ij} = \frac{\lambda_{ij}^2}{2\pi a^2}\left[\frac{Eh^3}{12\gamma(1-\nu^2)}\right]^{1/2}$; i=1,2,3...; j=1,2,3...

Description	λ_{ij}^2 and (ij)

17. Simply Supported-Simply Supported - Simply Supported-Clamped

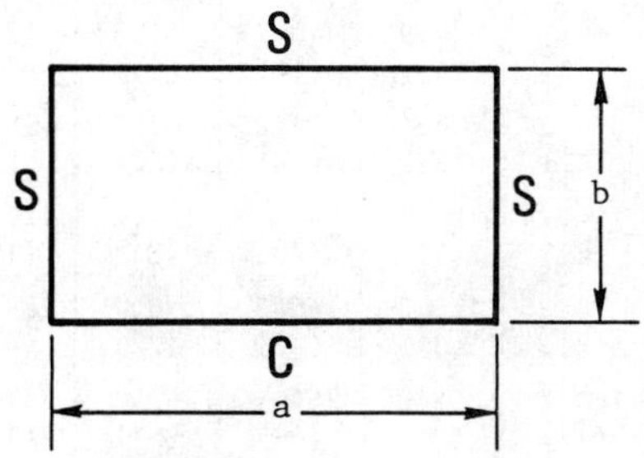

a/b	Mode Sequence 1	2	3	4	5	6
0.4	11.75 (11)	17.19 (12)	25.92 (13)	37.83 (14)	41.21 (21)	46.36 (22)
2/3	15.58 (11)	31.07 (12)	44.56 (21)	55.39 (13)	59.46 (22)	83.61 (23)
1.0	23.65 (11)	51.67 (21)	58.65 (12)	86.13 (22)	100.3 (31)	113.2 (13)
1.5	42.53 (11)	69.00 (21)	116.3 (31)	121.0 (12)	147.6 (22)	184.1 (41)
2.5	103.9 (11)	128.3 (21)	172.4 (31)	237.3 (41)	320.8 (12)	323.0 (51)

λ is independent of ν.

18. Clamped-Simply Supported - Simply Supported-Clamped

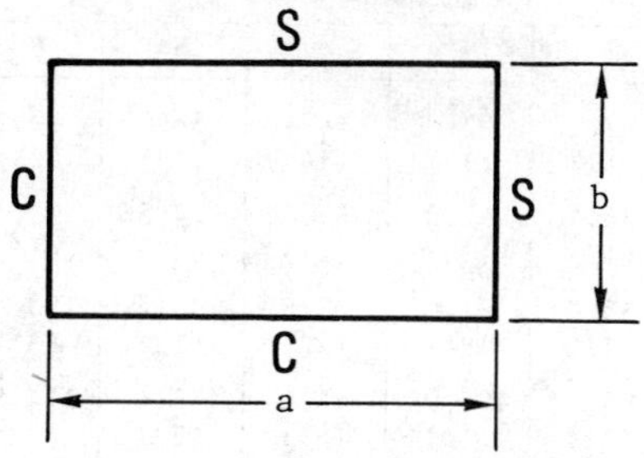

a/b	Mode Sequence 1	2	3	4	5	6
0.4	16.85 (11)	21.36 (12)	29.24 (13)	40.51 (14)	51.46 (21)	55.12 (15)
2/3	19.95 (11)	34.02 (12)	54.37 (21)	57.52 (13)	67.82 (22)	90.07 (14)
1.0	27.06 (11)	60.54 (21)	60.79 (12)	92.86 (22)	114.6 (13)	114.7 (31)
1.5	44.89 (11)	76.55 (21)	122.3 (12)	129.4 (31)	152.6 (22)	202.7 (41)
2.5	105.3 (11)	133.5 (21)	182.7 (31)	253.2 (41)	321.6 (12)	344.5 (51)

λ is independent of ν.

then approximate closed form solutions for the natural frequencies can be developed using the energy (Rayleigh) technique. This gives an approximate natural frequency, in hertz, of the form (Ref. 11-61):

$$f_{ij} = \frac{\pi}{2}\left[\frac{G_1^4}{a^4} + \frac{G_2^4}{b^4} + \frac{2J_1J_2 + 2\nu(H_1H_2 - J_1J_2)}{a^2b^2}\right]^{1/2}\left[\frac{Eh^3}{12\gamma(1-\nu^2)}\right]^{1/2}$$

$$i = 1, 2, 3 \ldots, \quad j = 1, 2, 3 \ldots . \qquad (11\text{-}21)$$

The dimensionless parameters G, H, and J, given in Table 11-5, are functions of the indices i and j and the boundary conditions on the plate. a is the length of the plate,

Table 11-4. Rectangular Plates. *(Continued)*

Natural Frequency (hertz), $f_{ij} = \frac{\lambda_{ij}^2}{2\pi a^2}\left[\frac{Eh^3}{12\gamma(1-\nu^2)}\right]^{1/2}$; i=1,2,3...; j=1,2,3...

Description	λ_{ij}^2 and (ij)

19. Simply Supported-Clamped - Simply Supported-Clamped

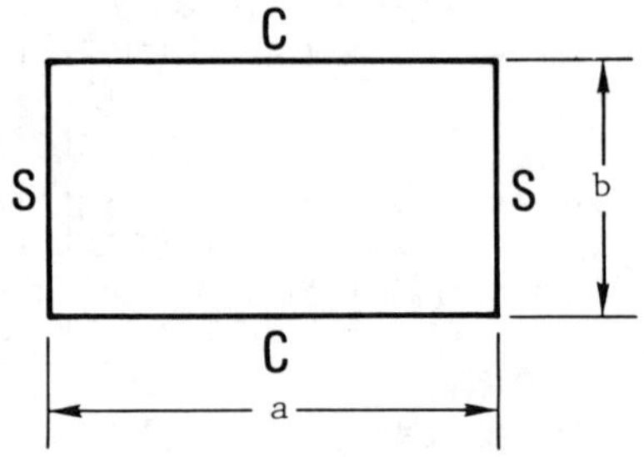

a/b	Mode Sequence 1	2	3	4	5	6
0.4	12.13 (11)	18.36 (12)	27.97 (13)	40.75 (14)	41.38 (21)	47.00 (22)
2/3	17.37 (11)	35.34 (12)	45.43 (21)	62.05 (13)	62.31 (22)	88.80 (23)
1.0	28.95 (11)	54.74 (21)	69.32 (12)	94.59 (22)	102.2 (31)	129.1 (13)
1.5	56.35 (11)	78.98 (21)	123.2 (31)	146.3 (12)	170.1 (22)	189.1 (41)
2.5	145.5 (11)	164.7 (21)	202.2 (31)	261.1 (41)	342.1 (51)	392.9 (12)

λ is independent of ν.

20. Clamped-Simply Supported - Clamped-Clamped

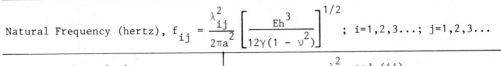

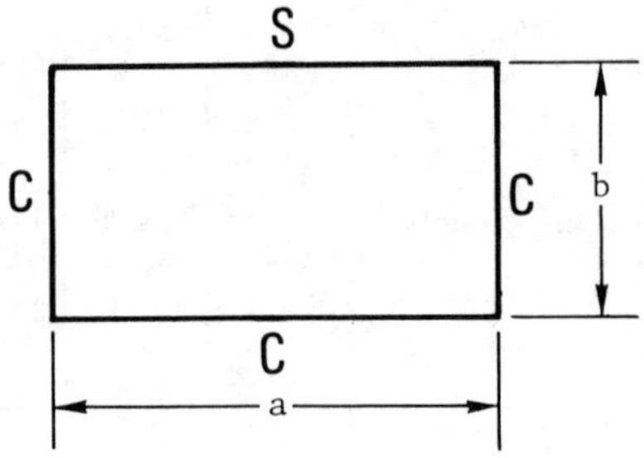

a/b	Mode Sequence 1	2	3	4	5	6
0.4	23.44 (11)	27.02 (12)	33.80 (13)	44.13 (14)	58.03 (15)	62.97 (21)
2/3	25.86 (11)	38.10 (12)	60.33 (13)	65.62 (21)	77.56 (22)	92.15 (14)
1.0	31.83 (11)	63.35 (12)	71.08 (21)	100.8 (22)	116.4 (13)	130.4 (31)
1.5	48.17 (11)	85.51 (21)	124.0 (12)	144.0 (31)	158.4 (22)	214.8 (32)
2.5	107.1 (11)	139.7 (21)	194.4 (31)	270.5 (41)	322.6 (12)	353.4 (22)

λ is independent of ν.

b is the plate width, h is the plate thickness, E is the modulus of elasticity, ν is Poisson's ratio, and γ is the mass per unit area of the plate. Consistent sets of units are given in Table 3-1.

The approximate natural frequencies predicted by Eq. 11-21 are directly analogous to the exact solutions of Table 11-4. The approximate natural frequencies can be expected to be within 5% of the exact solution with significantly better accuracy for plates with either clamped or simply supported edges.

Table 11-4. Rectangular Plates. *(Continued)*

Natural Frequency (hertz), $f_{ij} = \frac{\lambda_{ij}^2}{2\pi a^2}\left[\frac{Eh^3}{12\gamma(1-\nu^2)}\right]^{1/2}$; i=1,2,3...; j=1,2,3...

Description	λ_{ij}^2 and (ij)

21. Clamped-Clamped - Clamped-Clamped

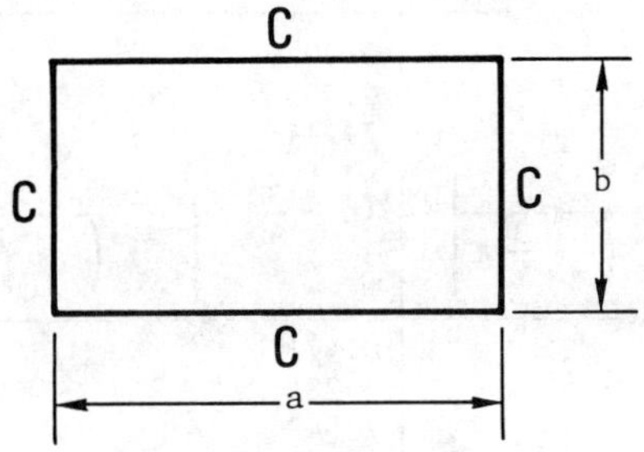

$\frac{a}{b}$	Mode Sequence 1	2	3	4	5	6
0.4	23.65 (11)	27.82 (12)	35.45 (13)	46.70 (14)	61.55 (15)	63.10 (21)
2/3	27.01 (11)	41.72 (12)	66.14 (21)	66.55 (13)	79.85 (22)	100.9 (14)
1.0	35.99 (11)	73.41 (21)	73.41 (12)	108.3 (22)	131.6 (31)	132.2 (13)
1.5	60.77 (11)	93.86 (21)	148.8 (12)	149.74 (31)	179.7 (22)	226.9 (41)
2.5	147.80 (11)	173.9 (21)	221.5 (31)	291.9 (41)	384.7 (51)	394.4 (12)

λ is independent of ν.

Mode Shapes. A general closed-form solution does not exist for vibration of a rectangular plate with various elementary boundary conditions on each of the four edges. However, it has been found that the mode shapes of rectangular plates lying in the x–y plane (Fig. 11-5) can be well approximated by a series of beam modes in the separable form of the variables:

$$\tilde{z}_{ij}(x, y) = \sum_m \sum_n a_{mn}^{ij}\, \tilde{z}_m(x)\, \tilde{z}_n(y),$$

where $\tilde{z}_{ij}(x, y)$ is the ij mode shape of the plate. $\tilde{z}_m$ and $\tilde{z}_n$ are beam mode shapes (Table 8-1) which satisfy the appropriate boundary conditions on the edges y = 0, b and x = 0, a, respectively, and a_{mn}^{ij} are constants which generally can be found to within an arbitrary constant using the Rayleigh-Ritz procedure (Ref. 11-42). The above series is often dominated by a single term and the plate mode shape can be expressed in separable variable form:

$$\tilde{z}_{ij}(x, y) \approx \tilde{x}_i(x)\tilde{y}_j(y), \qquad i, j = 1, 2, 3, \ldots. \qquad (11\text{-}22)$$

This approximation was employed in the derivation of Eq. 11-21. In general, the modal indices i and j describe the number of flexure half-waves along the x (horizontal) and y (vertical) axes, respectively, as can be seen for beam modes in Fig. 8-4. If two opposite sides of the rectangular plate are free (frames 1, 2, 3, 5, 6, and 8 of Table 11-4), the plate can possess a fundamental beam-like mode such that $\tilde{y}_1(y) \approx 1$. This is the limiting case of one flexural half-wave along the vertical axis as that wave becomes increasingly flat.

Table 11-5. Coefficients in Approximate Formulas for Isotropic and Orthotropic Rectangular Plates.

Boundary Conditions on Opposite Edges	Mode Index[(a)] (n)	$G^{(a)}$	$H^{(a)}$	$J^{(a)}$
1. Free-Free	1	0	0	0
	2	0	0	1.216
	3	1.506	1.248	5.017
	n (n > 3)	$n - \frac{3}{2}$	$\left(n - \frac{3}{2}\right)\left[1 - \frac{2}{\left(n - \frac{3}{2}\right)\pi}\right]$	$\left(n - \frac{3}{2}\right)^2\left[1 + \frac{6}{\left(n - \frac{3}{2}\right)\pi}\right]$
2. Simply Supported-Free	1	0	0	0.3040
	2	1.25	1.165	2.756
	3	2.25	4.346	7.211
	n (n > 1)	$n - \frac{3}{4}$	$\left(n - \frac{3}{4}\right)^2\left[1 - \frac{1}{\left(n - \frac{3}{4}\right)\pi}\right]$	$\left(n - \frac{3}{4}\right)^2\left[1 + \frac{3}{\left(n - \frac{3}{4}\right)\pi}\right]$
3. Clamped-Free	1	0.597	-0.0870	0.471
	2	1.494	1.347	3.284
	3	2.500	4.658	7.842
	n (n > 2)	$n - \frac{1}{2}$	$\left(n - \frac{1}{2}\right)^2\left[1 - \frac{2}{\left(n - \frac{1}{2}\right)\pi}\right]$	$\left(n - \frac{1}{2}\right)^2\left[1 + \frac{2}{\left(n - \frac{1}{2}\right)\pi}\right]$
4. Simply Supported-Simply Supported	1	1	1	J = H
	2	2	4	
	3	3	9	
	n	n	n^2	
5. Clamped-Simply Supported	1	1.25	1.165	J = H
	2	2.25	4.346	
	3	3.25	9.528	
	n	$n + \frac{1}{4}$	$\left(n + \frac{1}{4}\right)^2\left[1 - \frac{1}{\left(n + \frac{1}{4}\right)\pi}\right]$	
6. Clamped-Clamped	1	1.506	1.248	J = H
	2	2.5	4.658	
	3	3.5	10.02	
	n (n > 1)	$n + \frac{1}{2}$	$\left(n + \frac{1}{2}\right)^2\left[1 - \frac{2}{\left(n + \frac{1}{2}\right)\pi}\right]$	

[(a)] $G = G_1$, $H = H_1$, $J = J_1$, and mode index = i when boundary conditions are applied to sides of length b. $G = G_2$, $H = H_2$, $J = J_2$, and mode index = j when boundary conditions are applied to sides of length a. Ref. 11-61.

The one-term modal expansion of Eq. 11-22 is the exact solution form for rectangular plates which have two opposite sides simply supported and elementary boundary conditions (i.e., clamped, simply supported, or free) on the other two sides. This occurs in frames 5, 10, 11, 16, 17, and 19 of Table 11-4. The mode shape for

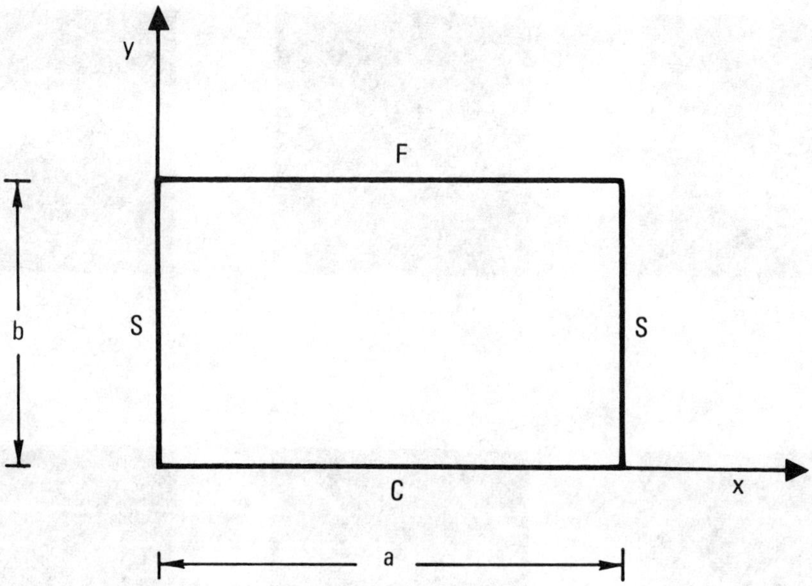

Fig. 11-5. Coordinate systems for rectangular simply supported-free-simply supported-clamped plate.

these frames can be expressed exactly as (Ref. 11-41):

$$\tilde{z}_{ij}(x, y) = \left(\sin \frac{i\pi x}{a}\right) \tilde{y}_j(y).$$

The function $\tilde{y}_j(y)$ is similar to, but generally not identical with, the beam mode shape which satisfies the boundary conditions at $y = 0$ and $y = b$. The coordinate system is shown in Fig. 11-5. This equation implies that the exact mode shape for a completely simply supported rectangular plate (frame 16 of Table 11-4) is:

$$\tilde{z}_{ij}(x, y) = \sin \frac{i\pi x}{a} \sin \frac{j\pi y}{b}, \qquad i, j = 1, 2, 3, \ldots.$$

Note that i is the number of flexure half-waves in the mode shape along the x (horizontal) axis and j is the number of flexure half-waves along the y (vertical) axis. This relationship holds for all 21 frames of Table 11-4, although the amplitude of the waves along each axis may not be equal as with the completely simply supported rectangular plate.

The one-term modal expansion (Eq. 11-22) is not exact for rectangular plates without two opposite sides simply supported. For example, experimental measurements of the first six modes of a free-square plate (frame 1 of Table 11-4) are shown in Fig. 11-6. The modes corresponding to $i = 1, j = 3$ and $i = 3, j = 1$ have considerably different form due to differences in the sign of higher-order terms in the modal expansion. The $i = 1, j = 3$ mode is approximately

$$\tilde{z}_{13} \approx \cos \frac{\pi x}{a} \cos \frac{3\pi y}{a} - \cos \frac{3\pi x}{a} \cos \frac{\pi y}{a},$$

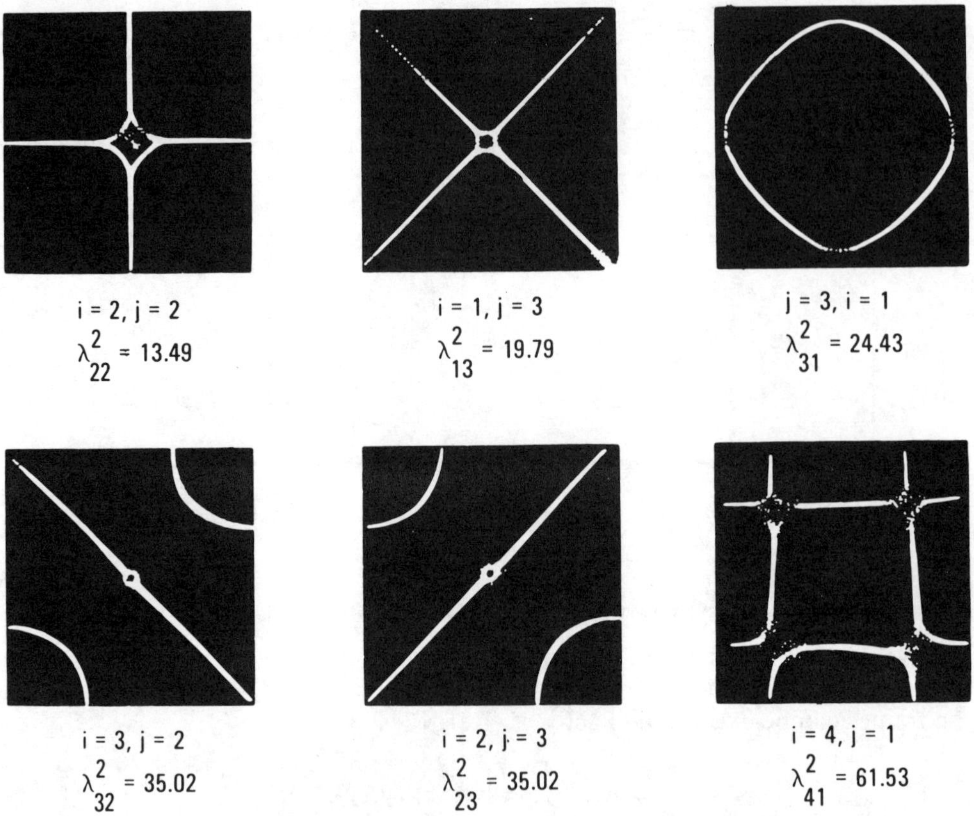

Fig. 11-6. Experimentally determined mode shapes of a completely free square plate (Ref. 11-43).

and the i = 3, j = 1 mode is approximately

$$\tilde{z}_{31} \approx \cos\frac{\pi x}{a}\cos\frac{3\pi y}{a} + \cos\frac{3\pi x}{a}\cos\frac{\pi y}{a}.$$

Clearly, in these cases at least two terms are required to adequately differentiate these two mode shapes. Thus, the one-term mode expansion should be used with considerable caution when approximating the mode shapes of plates without two opposite sides simply supported. Additional discussion of the natural frequencies and mode shapes of a completely free plate can be found in Ref. 11-82.

The mode shape indexes i and j can be associated with the number of half-waves in the plate along the x axis (horizontal axis) and y axis (vertical axis), respectively, as the plate vibrates. The intertwining of the deformations along the two axes can produce complex and fanciful geometric nodal patterns as shown in Fig. 11-6. Vibration modes need not exist for all possible integers i and j. For example, the modes of a completely free plate (Fig. 11-6, frame 1 of Table 11-4) corresponding to (ij) = (11), (12), and (21) are rigid body modes since the index 1 corresponds to a rigid body mode in a completely free plate.

Boundary Conditions. The boundary conditions associated with rectangular plates are:

$$\text{Free Edge:}\quad \mathcal{M} = \mathcal{V} = 0$$

$$\text{Simply Supported Edge:}\quad Z = \mathcal{M} = 0$$

$$\text{Clamped Edge:}\quad Z = \frac{\partial Z}{\partial z} = 0$$

Z is the transverse deflection of the plate. The moment $\mathcal{M}$ is $\mathcal{M}_x$ if the edge is parallel to the y axis or $\mathcal{M}_y$ if the edge is parallel to the x axis. $\mathcal{M}_x$ and $\mathcal{M}_y$ are defined in Eq. 11-3. $\mathcal{V}$ is the Kelvin-Kirchoff edge reaction:

$$\mathcal{V}_x = Q_x + \frac{\partial \mathcal{M}_{xy}}{\partial y},$$

$$\mathcal{V}_y = Q_y + \frac{\partial \mathcal{M}_{xy}}{\partial x}.$$

$\mathcal{V}_x$ is employed in the boundary condition if the edge is parallel to the y axis and $\mathcal{V}_y$ is employed if the edge is parallel to the x axis. Q_x and Q_y are defined in Eq. 11-4. The simply supported edge is the two-dimensional equivalent of the pinned joint discussed in Chapter 8. A free edge supports neither moment nor shear. A clamped edge permits neither rotation nor translation.

These boundary conditions are uniformly applied along the edges of the plates in Table 11-4. For example, the boundary conditions of the simply supported-free-simply supported-clamped plate shown in Fig. 11-5 (also frame 11 of Table 11-4) are:

$$Z = \mathcal{M}_x = 0 \qquad 0 < y < a; \quad x = 0, \quad x = b,$$

$$Z = \frac{\partial Z}{\partial y} = 0 \qquad 0 < x < b; \quad y = 0,$$

$$\mathcal{M}_y = \mathcal{V}_y = 0 \qquad 0 < x < b; \quad y = a.$$

The boundary conditions for the simply supported edge of a thin plate of arbitrary shape can be expressed as (Ref. 6-13, p. 182)

$$Z = 0 \quad \text{and} \quad \frac{\partial^2 Z}{\partial n^2} + \frac{\nu}{R}\frac{\partial Z}{\partial n} = 0$$

and for a free edge as

$$\nabla^2 Z - (1-\nu)\left(\frac{1}{R}\frac{\partial Z}{\partial n} + \frac{\partial^2 Z}{\partial s^2}\right) = 0,$$

and

$$\frac{\partial}{\partial n}\nabla^2 Z + (1-\nu)\frac{\partial}{\partial s}\left(\frac{\partial^2 Z}{\partial n \partial s} - \frac{1}{R}\frac{\partial Z}{\partial s}\right) = 0.$$

Z is the transverse deformation of the midsurface. n is the local coordinate normal to the edge, s is the local coordinate tangent to the edge, and R is the curvature of the edge. Note that both sets of boundary conditions involve Poisson's ratio, ν. As a result, the natural frequencies and mode shapes of thin plates with simply supported and free edges are functions of Poisson's ratio. These boundary conditions are considerably more complex than the analogous boundary conditions for beams. If the plate edge is composed of a series of straight line segments, a rectangular or polygonal plate for example, and the local effect at corners is neglected, then $R = \infty$ and terms proportional to $1/R$ can be neglected in the boundary conditions. This considerably simplifies the simply supported boundary condition by removing the Poisson's ratio dependence.

Two results of the general complexity of the boundary conditions for simply supported and free edges of plates are that (1) orthogonality of the mode shapes of plates cannot be proven in general for plates with free edges or nonpolygonal simply supported plates, and (2) approximate solutions generally do not satisfy the rigorous boundary condition at a free edge. Usually, some terms in the free edge boundary condition are neglected.

Poisson's Ratio Effect. The natural frequency parameter λ_{ij} is a function of Poisson's ratio only for rectangular plates with one or more free edges; this arises in 15 of the 21 possible combinations of simple boundary conditions listed in Table 11-4. Generally, the natural frequency parameter changes only slightly over the range of Poisson's ratios of practical importance. For example, the natural frequency parameter for a simply supported-free-simply supported-free plate (frame 5 of Table 11-4) for the fundamental mode as a function of aspect ratio and Poisson's ratio is (Ref. 11-41):

λ_{11}^2

$\frac{a}{b}$	$\nu = 0.0$	$\nu = 0.3$	$\nu = 0.5$
0.4	9.870	9.760	9.451
1.0	9.870	9.631	9.079
2.5	9.870	9.484	8.704

Generally, λ_{ij}^2 decreases with increasing Poisson's ratio (Ref. 11-41).

Rotary Inertia and Shear Deformation. Rotary inertia is the inertia associated with local rotation of the plate as it flexes. Shear deformation is deformation due to transverse shearing of the cross section under dynamic shear loads. These effects are neglected in Table 11-4. Rotary inertia and shear deformation will tend to lower the natural frequencies from those given in the tables. The importance of rotary inertia and shear deformation generally increases with the ratio of plate thickness to plate length or width and with increasing mode number; these effects are generally insignificant for plates whose thickness is less than $\frac{1}{10}$ the minimum of the plate length or width for vibrations in the fundamental mode. The effects of rotary inertia and shear deformation on rectangular plates are discussed in Refs. 11-44 and 11-45.

Orthotropic Plates. An isotropic plate is a plate composed of a material whose properties are independent of direction. The plates in Table 11-4 are assumed to be isotropic. Two material constants (E and ν) are required to describe the linear stress-strain relationship of an isotropic plate. An orthotropic plate is a plate whose material properties are symmetric about three mutually perpendicular axes. Common orthotropic materials are wood and fiber-reinforced composites. Four material constants are required to describe the linear stress-strain relationship of a thin orthotropic plate.

The stress-strain relationships for a thin orthotropic plate whose orthotropic axes align with the x–y axes are:

$$\sigma_x = \frac{1}{1 - \nu_x \nu_y} (E_x \epsilon_x + \nu_y E_x \epsilon_y),$$

$$\sigma_y = \frac{1}{1 - \nu_x \nu_y} (E_y \epsilon_y + \nu_y E_x \epsilon_x),$$

$$\sigma_{xy} = G\epsilon_{xy}, \qquad (11\text{-}23)$$

where E_x and E_y are the elastic moduli for stresses and strains in the x and y directions, ν_x and ν_y are the associated Poisson's ratios, and G is the shear modulus. It can be shown using a symmetry argument that

$$\nu_y E_x = \nu_x E_y;$$

so there are only four independent orthotropic constants for a thin orthotropic plate. In the isotropic case, $E_x = E_y = E$, $\nu_x = \nu_y = \nu$, and $G = E/[2(1 + \nu)]$ and the above equations simplify to Eq. 11-2. In plate analysis it is convenient to define the four orthotropic constants as follows:

$$D_x = \frac{E_x h^3}{12(1 - \nu_x \nu_y)}, \qquad D_y = \frac{E_y h^3}{12(1 - \nu_x \nu_y)},$$

$$D_k = \frac{Gh^3}{12}, \qquad D_{xy} = D_x \nu_y + 2D_k. \qquad (11\text{-}24)$$

If the orthotropic axes align with the plate axes of a uniform rectangular plate which spans x = 0 to x = a and y = 0 to y = b, then an approximate expression for the natural frequencies, in hertz, of the plate is (Ref. 11-61):

$$f_{ij} = \frac{\pi}{2\gamma^{1/2}} \left[\frac{G_1^4 D_x}{a^4} + \frac{G_2^4 D_y}{b^4} + \frac{2H_1H_2D_{xy}}{a^2b^2} + \frac{4D_k(J_1J_2 - H_1H_2)}{a^2b^2} \right]^{1/2};$$

$$i = 1, 2, 3, \ldots, \quad j = 1, 2, 3 \ldots. \qquad (11\text{-}25)$$

The dimensionless parameters G, H, and J are functions of the indices i and j and the boundary conditions on the plate. These parameters are given in Table 11-5. γ is the mass per unit area of the plate. Consistent sets of units are given in Table 3-1.

The above expression was derived using the energy (Rayleigh) technique and approximating the plate mode shape with the mode shape of single beams (Table 8-1) along the x and y axes. It can be expected to be within 5% of the exact result and it reduces to Eq. 11-21 for the case of an isotropic material. Additional discussion of the natural frequencies of orthotropic plates can be found in Refs. 11-47, 11-48, 11-11-49, 11-61, and 11-86.

Point Supported Plates. A point support is a boundary condition applied at a point on the plate which prevents deflection of the plate at that point but which places no restriction on rotation of the plate about that point. The natural frequencies of various plates with point supports are given in Table 11-6.

Tapered Plates. The natural frequencies of tapered rectangular plates are given in Table 11-7. Many approximate solutions are presented in Ref. 11-4. The natural frequencies of plates with an exponential variation in thickness are discussed in Ref. 11-60.

In-Plane Loads. The natural frequencies of a uniform rectangular plate subject to normal in-plane loads (Fig. 11-7) can be related to the natural frequencies of the same plate without loads by the following approximate expression (Ref. 11-61):

$$f_{ij}\Big|_{\substack{\text{In-plane}\\ \text{loads } N_1,\, N_2}} = \left(f_{ij}^2\Big|_{\substack{\text{No in-plane}\\ \text{loads,}\\ N_1 = N_2 = 0}} + \frac{N_1 J_1}{4\gamma a^2} + \frac{N_2 J_2}{4\gamma b^2} \right)^{1/2},$$

where f_{ij} is the natural frequency of the ij mode in hertz and γ is the mass per unit area of the plate. The dimensionless coefficients J_1 and J_2 are functions of the mode number and the boundary conditions applied on the sides of lengths b and a, respectively, and are given in Table 11-5. The natural frequencies for the case of no in-plane loads can be calculated from Table 11-4 or Eq. 11-21. This equation applies to isotropic plates and to orthotropic plates if the orthotropic axes align with the plate axes.

N_1 and N_2 are the in-plane loads per unit length of edge. As shown in Fig. 11-7, N_1 acts over the edges of length b and N_2 acts over the edges of length a. If N_1 and N_2 are tensile ($N_1 > 0$, $N_2 > 0$), then the in-plane loads increase the natural frequencies of the plate. If N_1 and N_2 are compressive ($N_1 < 0$, $N_2 < 0$), then the in-plane loads decrease the natural frequencies of the plate. Sufficiently large compressive loads will cause the plate to buckle as the fundamental natural frequency goes to zero. The mode shape is unaffected by in-plane loads to the order of this approximation.

Other discussions of the effect of in-plane loads on the natural frequencies of rectangular plates can be found in Refs. 11-61 and 11-62.

Table 11-6. Point Supported Rectangular Plates.

Notation: a = width of plate; b = length of plate; h = thickness of plate; C = clamped edge; E = modulus of elasticity; S = simply supported edge; γ = mass per unit area of plate (μh for a plate of material of density μ); ν = Poisson's ratio; X denotes a point support which prevents displacement but places no restriction on rotation; unmarked edges of plates are free; see Table 3-1 for consistent sets of units

$$\text{Natural Frequency (hertz)}, \; f_i = \frac{\lambda_i^2}{2\pi a^2}\left[\frac{Eh^3}{12\gamma(1-\nu^2)}\right]^{1/2}; \quad i = 1,2,3,\ldots$$

Description	λ_i^2 and Remarks

1. Rectangular Plate, Corner Supports

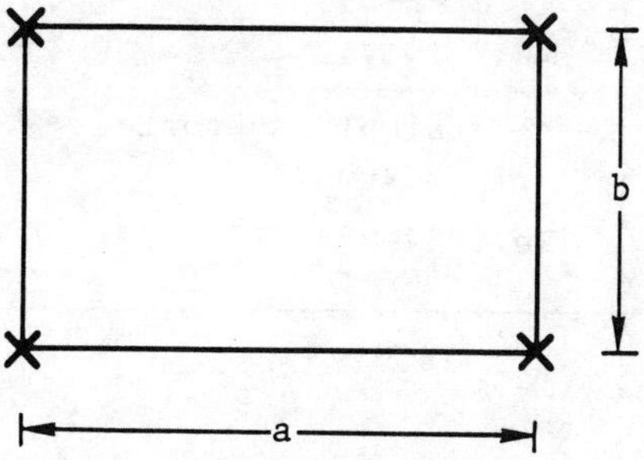

$\frac{a}{b}$	λ_1^2	λ_2^2
1.0	7.12	15.8
1.5	8.92	21.5
2.0	9.29	27.5
2.5	9.39	35.5

$\nu = 0.3$; Ref. 11-50

2. Square Plate, Four Point Supports

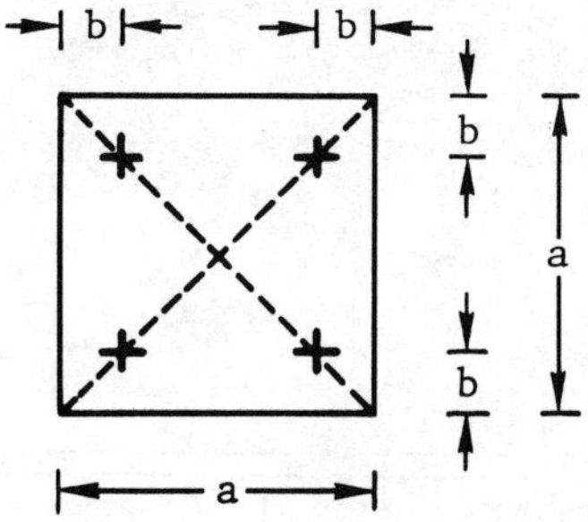

$\frac{b}{a}$	λ_1^2	λ_2^2	λ_3^2
0.0	7.14	15.79*	19.69
0.1	12.89	19.69	23.97*
0.2	19.69	23.13	32.56
0.3	19.31	19.72	24.30*
0.4	13.35	14.06*	16.83
0.5	11.34	13.47	19.69

*Repeated values.

$\nu = 0.3$; Ref. 11-51.

Note: b/a = 0.5 gives a single point support at center of plate.

Table 11-6. Point Supported Rectangular Plates. (*Continued*)

Natural Frequency (hertz), $f_i = \dfrac{\lambda_i^2}{2\pi a^2}\left[\dfrac{Eh^3}{12\gamma(1-\nu^2)}\right]^{1/2}$; $i = 1,2,3,\ldots$

Description	λ_i^2 and Remarks

3. Square Plate with n Equally Spaced Point Supports on Edges

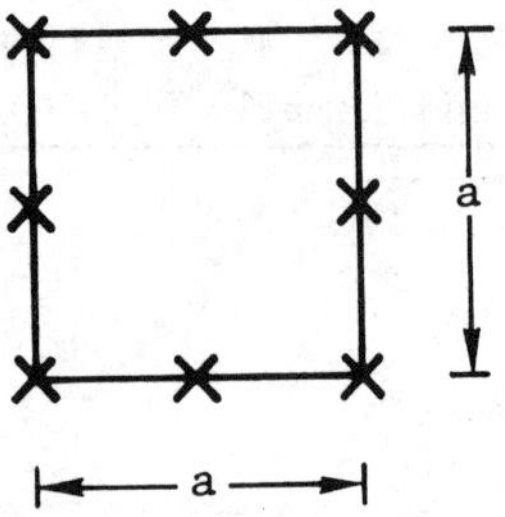

(n = 3 shown)

n	λ_1^2
3	18.20
5	19.64
7	19.71
9	19.73
∞*	19.74

*∞ = simply supported edges

Ref. 11-83

4. Square Plate, Supports at Midpoints of Sides

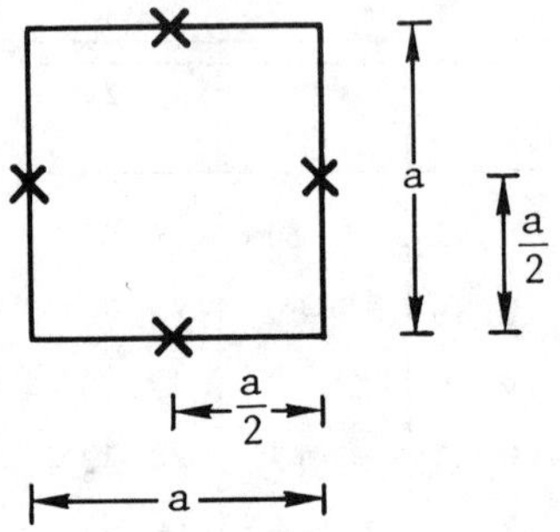

$\lambda_1^2 = 13.5$

$\nu = 0.3$

Ref. 11-52

5. N-Bay Plate

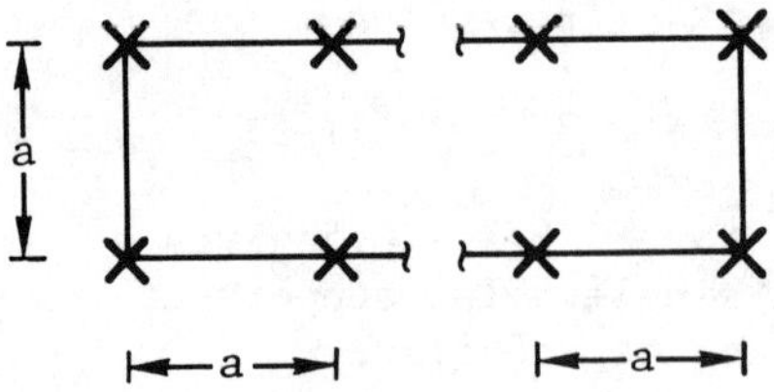

Number of Bays	λ_1^2	λ_2^2	λ_3^2
1	7.18	16.30	16.30
2	16.27	16.76	33.28
3	24.41	25.41	28.39
4	33.02	33.41	37.20
5	41.41	41.86	45.43

Ref. 11-53. 1 bay = square plate.

Table 11-6. Point Supported Rectangular Plates. *(Continued)*

Natural Frequency (hertz), $f_i = \dfrac{\lambda_i^2}{2\pi a^2}\left[\dfrac{Eh^3}{12\gamma(1-\nu^2)}\right]^{1/2}$; i = 1,2,3,...

Description	λ_i^2 and Remarks
6. Rectangular Plate With Simply Supported Edges With a Central Support 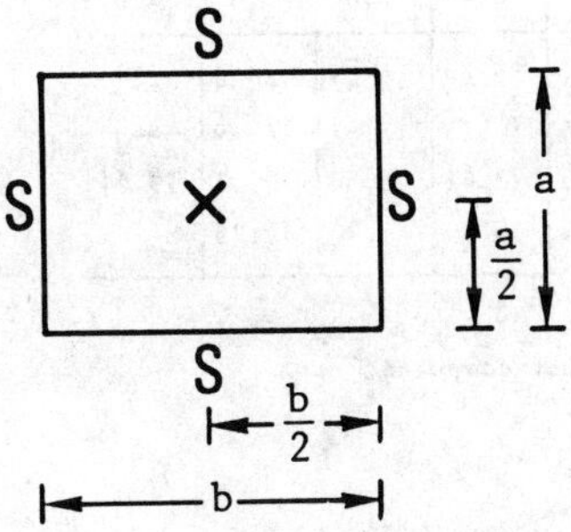	$\frac{a}{b}$ = 1.0 1.5 2.0 λ_1^2 = 52.6 73.1 91.1 Ref. 11-54 (Ref. 11-52 gives λ_1^2 = 53.4 for a/b = 1.0 and ν = 0.3.)
7. Square Plate With Two Simply Supported Sides and a Support at Corner 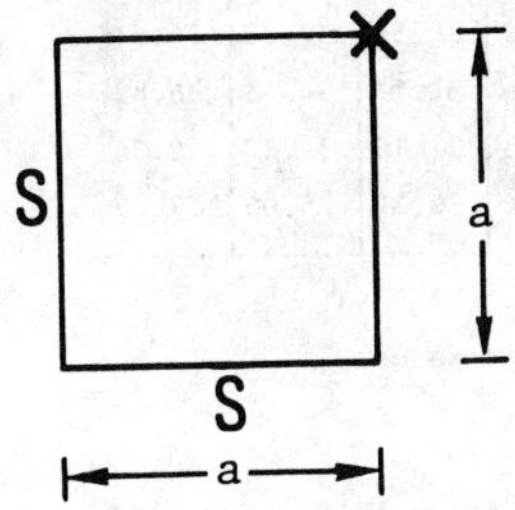	λ_1^2 = 9.00 ν = 0.3 Ref. 11-55
8. Square Plate With Two Clamped Sides and a Support at Corner 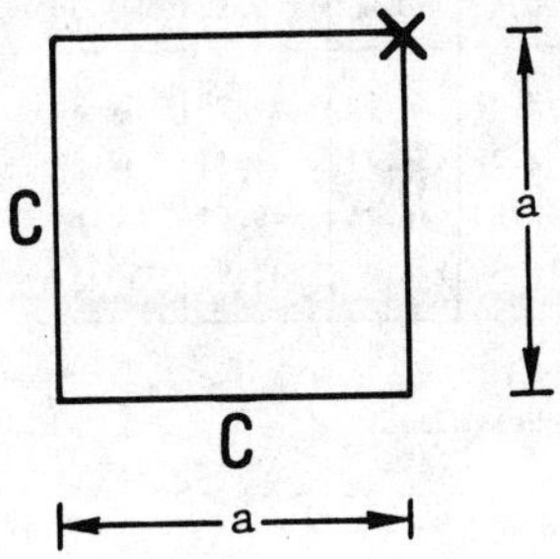	λ_1^2 = 13.7 Ref. 11-55 ν = 0.3

Table 11-7. Linearly Tapered Rectangular Plates.

Notation: a = length of plate; b = width of plate; h = thickness of plate; h_0 = thickness of plate along y axis; C = clamped edge; E = modulus of elasticity; S = simply supported edge; μ = density of material; ν = Poisson's ratio; plates are linearly tapered along x axis; see Table 3-1 for consistent sets of units

Fundamental Natural Frequency (hertz), $f = \dfrac{\lambda^2}{2\pi a^2}\left[\dfrac{E h_o^2}{12\mu(1-\nu^2)}\right]^{1/2}$ Taper: $h = h_o\left(1 + \dfrac{\alpha x}{a}\right)$

Description	λ^2

1. Simply Supported

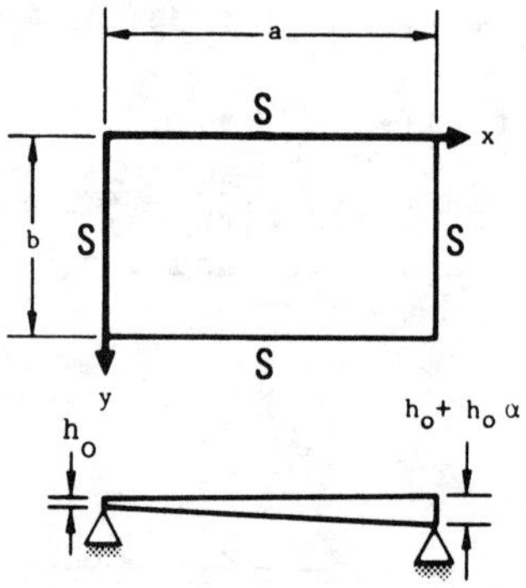

$\frac{a}{b}$	α = 0.0	0.2	0.4	0.6	0.8	1.0
0.25	--	11.51	12.48	13.42	14.34	--
0.50	--	13.55	14.73	15.87	16.99	--
1.0	19.74	21.69	23.61	25.50	27.36	29.21
1.5	32.08	35.22	38.30	41.30	44.25	--

ν = 0.3; Refs. 11-56 and 11-57.

-- denotes value not computed.

2. Simply Supported-Clamped-Simply Supported-Clamped

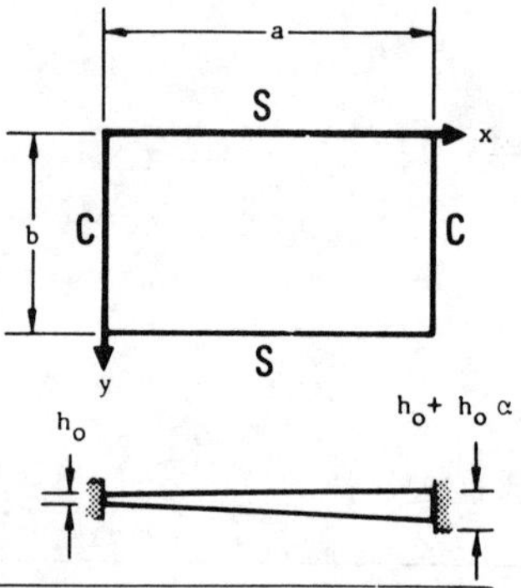

$\frac{a}{b}$	α = 0.0	0.2	0.4	0.6	0.8	1.0
0.5	23.82	26.15	28.40	30.58	32.73	34.81
2/3	25.04	27.50	29.86	32.17	34.41	36.62
1.0	28.95	31.80	34.53	37.19	39.80	42.36
2.0	54.75	60.11	65.24	70.21	75.05	79.79

ν = 0.3; Ref. 11-58.

See Ref. 11-58 for higher modes.

3. Clamped

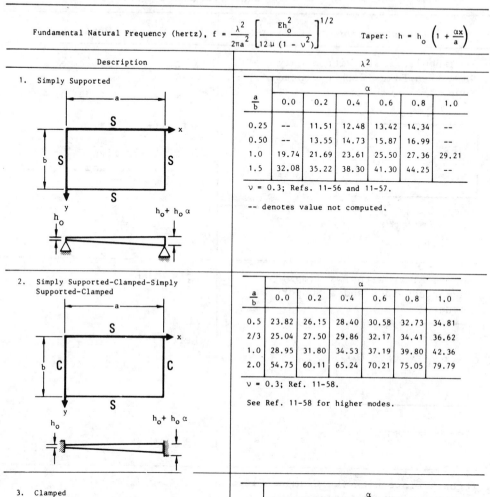

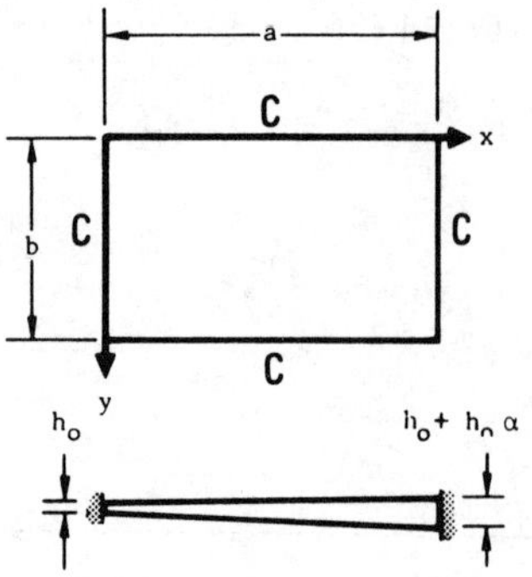

$\frac{a}{b}$	α = 0.0	0.2	0.4	0.6	0.8	1.0
0.5	24.59	27.00	29.32	31.58	33.90	35.96
2/3	27.02	29.67	32.22	34.71	37.13	39.52
1.0	36.00	39.52	42.93	46.24	49.47	52.64
2.0	98.33	107.8	116.6	124.9	132.9	140.5

ν = 0.3; Ref. 11-59.

See Ref. 11-59 for higher modes.

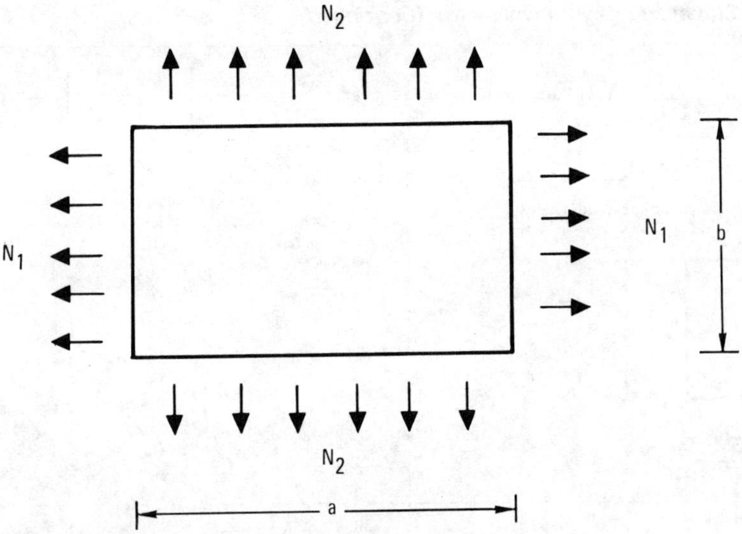

Fig. 11-7. Rectangular plate with normal in-plane loads.

Table 11-8. Square Plates with Openings.

Notation: a = length of one side of plate; h = thickness of plate; C = clamped edge; E = modulus of elasticity; S = simply supported edge; γ = mass per unit area of plate (μh for a plate of material of density μ); ν = Poisson's ratio; the edges of the openings are free; see Table 3-1 for consistent sets of units.

Fundamental Natural Frequency (hertz), $f = \dfrac{\lambda^2}{2\pi a^2}\left[\dfrac{Eh^3}{12\gamma(1-\nu^2)}\right]^{1/2}$

Description	λ^2 and Remarks
1. Simply Supported Plate, Round Opening 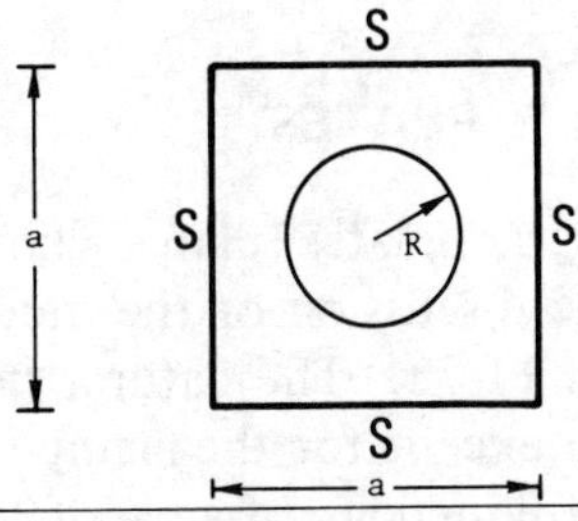	$\frac{2R}{a}$ = 0, 0.05, 0.10, 0.15, 0.20, 0.25, 0.3 λ^2 = 19.9, 19.75, 19.5, 19.4, 19.3, 19.35, 19.5 ν = 0.3; Ref. 11-63
2. Simply Supported Plate, Square Opening 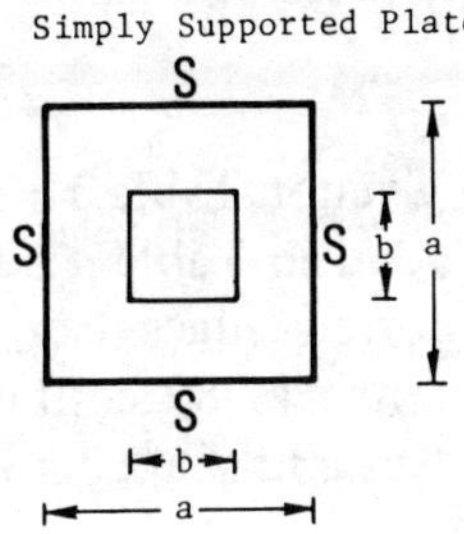	$\frac{b}{a}$ = 0, 1/6, 1/3, 1/2 λ^2 = 19.63, 19.48, 21.45, 26.05 ν = 0.0; Ref. 11-64 (λ^2 is 7% to 10% lower for ν = 0.3.)

Table 11-8. Square Plates with Openings. *(Continued)*

Fundamental Natural Frequency (hertz), $f = \frac{\lambda^2}{2\pi a^2}\left[\frac{Eh^3}{12\gamma(1-\nu^2)}\right]^{1/2}$

Description	λ^2 and Remarks
3. Clamped Plate, Square Opening 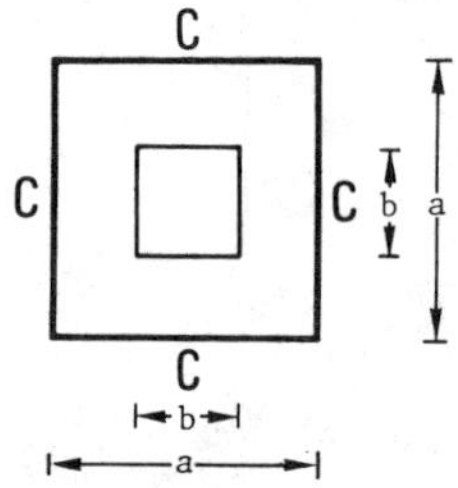 	$\frac{b}{a}$ = 0 1/6 1/3 1/2 λ^2 = 34.85 35.80 43.25 62.40 ν = 0.0; Ref. 11-64. (λ^2 is about 7% to 10% lower for ν = 0.3.)
4. Clamped Plate, Circular Opening 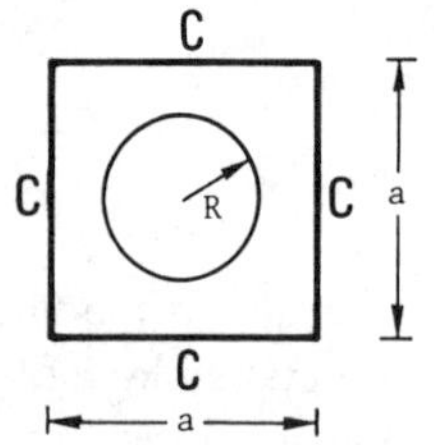 	$\frac{2R}{a}$ = 0.0 0.05 0.10 0.15 0.20 0.25 λ^2 = 36.0 35.5 35.1 35.2 35.7 36.7 ν = 0.3; Ref. 11-63 (See Ref. 11-65 for a rectangular plate.)

Plates With Openings. The fundamental natural frequencies of square plates with square and circular openings are given in Table 11-8. Cracked plate natural frequencies are discussed in Ref. 11-66. The natural frequencies of composite orthotropic plates with square openings are discussed in Ref. 11-86.

11.4. PARALLELOGRAM, TRIANGULAR, AND OTHER PLATES

Parallelogram Plates. The natural frequencies of parallelogram plates (plates whose opposite sides are parallel) are given in Table 11-9. Some discussion of the mode shapes of these plates is contained in Refs. 11-70 through 11-72. The natural frequencies of these plates are independent of Poisson's ratio except for the clamped-free-free-free plate of frame 1 of Table 11-9. The natural frequencies of a parallelogram plate on point supports is discussed in Ref. 11-51.

Other Plates. The natural frequencies of triangular plates are given in Table 11-10. The fundamental natural frequencies of other plates can be found in Table 11-11. Additional solutions for polygonal plates with simply supported edges can be adapted from Table 10-1 using the membrane analogy. Reference 11-89 contains solutions for the natural frequencies of plates with clamped edges and shaped as an I, L, or X.

Table 11-9. Parallelogram Plates.

Notation: a = length of first axis; b = length of second axis; h = thickness of plate; C = clamped edge; E = modulus of elasticity; F = free edge; S = simply supported edge; γ = mass per unit area of plate (μh for a plate of material with density μ); β = angle (degrees); ν = Poisson's ratio; see Table 3-1 for consistent sets of units

Natural Frequency (hertz), $f_i = \frac{\lambda_i^2}{2\pi a^2}\left[\frac{Eh^3}{12\gamma(1-\nu^2)}\right]^{1/2}$; i=1,2,3...

Description	λ_i^2

1. Clamped-Free - Free-Free

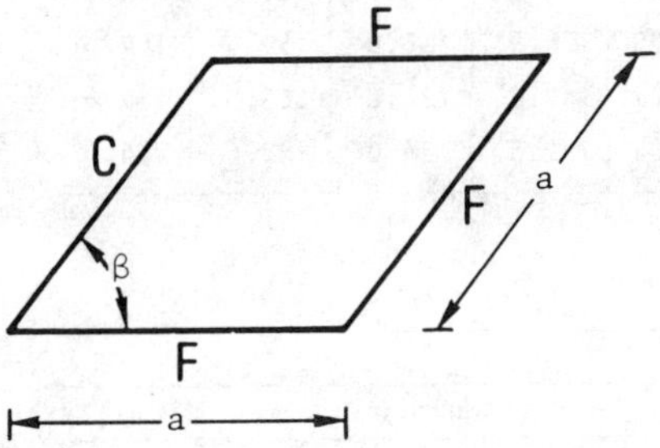

β	Mode Number 1	Mode Number 2
90	3.492	8.525
75	3.601	8.872
60	3.961	10.19
45	4.824	13.75

ν = 0.3; Refs. 11-70, 11-41.

2. Simply Supported Rhombus

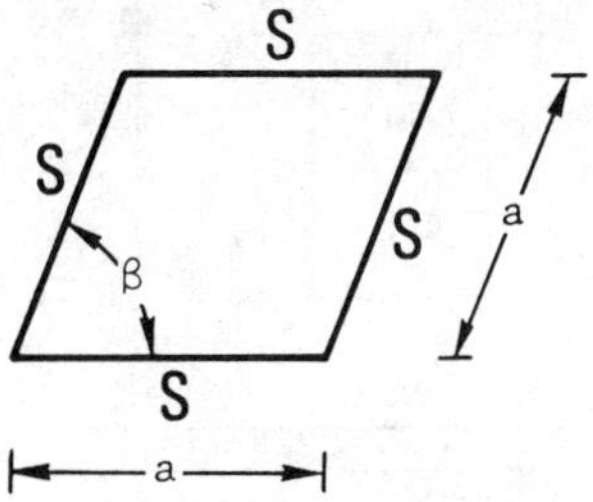

β	Mode Number					
	1	2	3	4	5	6
90	19.74	49.35	49.35	78.96	98.70	98.70
70	21.82	49.04	60.07	79.94	108.5	116.9
60	24.91	52.67	71.79	83.92	122.9	123.1
45	34.79	66.36	100.5	107.3	141.0	168.3

Ref. 11-71

3. Simply Supported Parallelogram

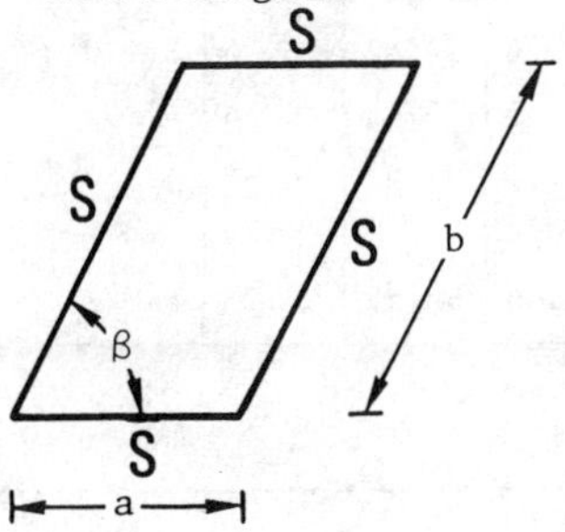

β	Mode Number					
	a/b = 2/3		a/b = 1/2		a/b = 1/3	
	1	2	1	2	1	2
90	14.26	27.42	--	--	--	--
70	15.82	29.44	13.76	21.44	12.31	15.72
60	18.15	32.48	15.89	23.96	14.35	17.88
45	25.69	42.17	22.87	32.05	21.06	24.95

Ref. 11-71

Table 11-9. Parallelogram Plates. *(Continued)*

Natural Frequency (hertz), $f_i = \dfrac{\lambda_i^2}{2\pi a^2}\left[\dfrac{Eh^3}{12\gamma(1-\nu^2)}\right]^{1/2}$; i=1,2,3...

Description	λ_i^2

4. Clamped-Simply Supported - Simply Supported-Simply Supported

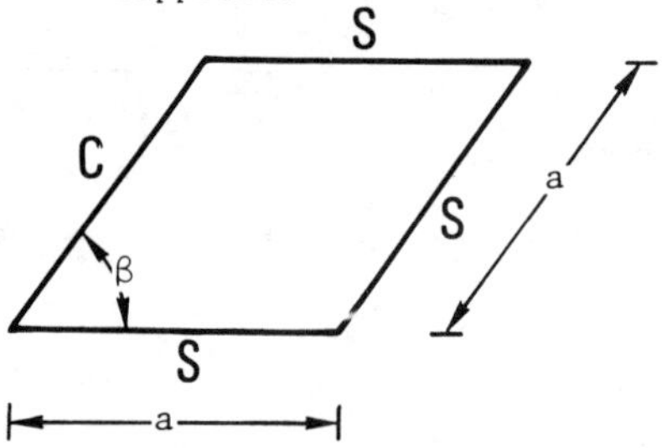

β	Mode Number 1	2	3	4	5	6
90	23.65	51.68	58.65	86.15	100.3	113.2
70	26.47	54.43	68.55	88.16	114.9	130.1
60	30.79	59.72	82.37	94.06	133.0	136.1
50	38.80	69.99	105.9	108.0	151.4	164.8
45	45.18	78.27	117.3	126.8	163.8	189.5
40	54.24	90.01	132.5	154.0	182.6	224.0

Ref. 11-72

5. Clamped-Simply Supported - Clamped-Simply Supported

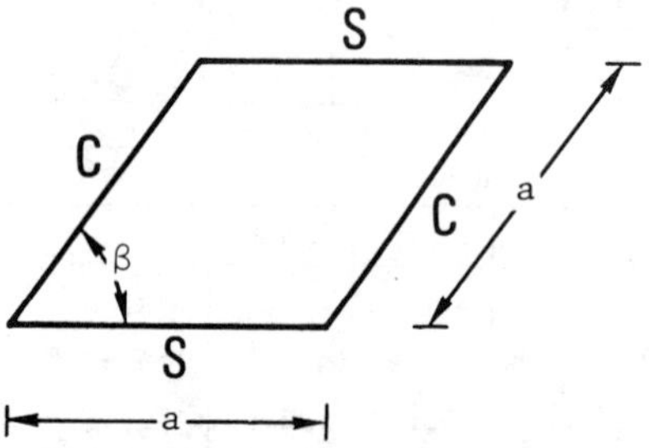

β	Mode Number 1	2	3	4	5	6
90	28.95	54.76	69.33	94.61	102.3	129.1
70	32.33	58.77	79.35	95.73	119.4	138.66
60	37.47	65.12	94.15	101.7	139.2	145.1
50	46.91	76.89	116.5	120.4	161.9	173.7
45	54.38	86.26	127.3	142.1	176.7	200.4
40	65.04	99.68	144.1	171.7	194.4	236.9

Ref. 11-72

6. Clamped-Simply Supported - Clamped-Simply Supported

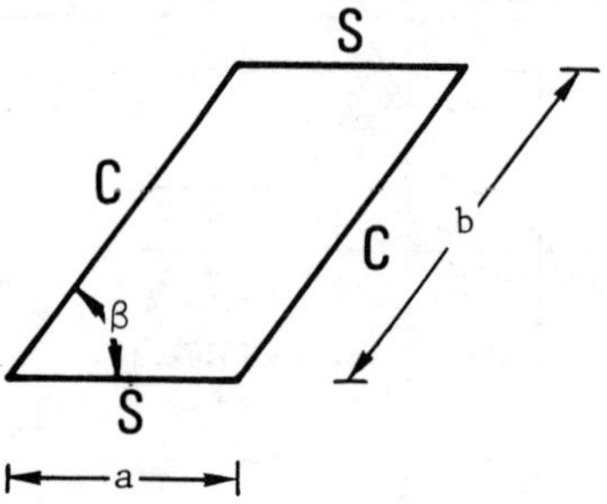

	Mode Number					
	a/b = 2/3			a/b = 1/2		
β	1	2	3	1	2	3
90	25.04	35.11	54.76	23.82	28.96	39.10
75	26.71	36.94	56.65	25.45	30.66	40.91
60	32.77	43.59	63.81	31.41	36.91	47.53
50	41.35	52.92	73.99	--	--	--
45	48.18	60.26	82.06	46.59	52.73	64.22

Ref. 11-72

Table 11-9. Parallelogram Plates. *(Continued)*

$$\text{Natural Frequency (hertz)}, \quad f_i = \frac{\lambda_i^2}{2\pi a^2}\left[\frac{Eh^3}{12\gamma(1-\nu^2)}\right]^{1/2}; \quad i=1,2,3\ldots$$

Description	λ_i^2

7. Clamped-Simply Supported - Simply Supported-Clamped Rhombic

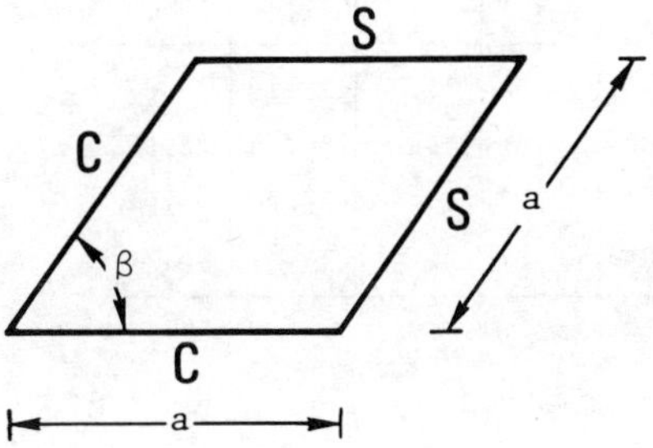

β	Mode Number 1	2	3	4	5	6
140	61.61	100.5	144.1	165.8	194.4	239.8
135	51.45	87.45	127.3	137.2	175.7	203.3
130	44.28	78.28	113.9	117.8	161.9	177.7
120	35.23	66.94	88.94	102.2	144.0	146.4
110	30.33	61.27	73.90	95.54	126.6	136.0
90	27.05	60.54	60.80	92.86	114.6	114.7
70	30.07	61.45	73.44	95.63	126.6	135.5
60	34.73	67.10	88.29	102.2	144.8	145.7
50	43.30	78.41	114.4	116.2	162.2	177.7
45	50.09	87.55	127.8	134.9	175.7	203.3
40	59.70	100.5	144.3	163.8	194.4	239.8

Ref. 11-72

8. Clamped-Clamped-Simply Supported-Clamped

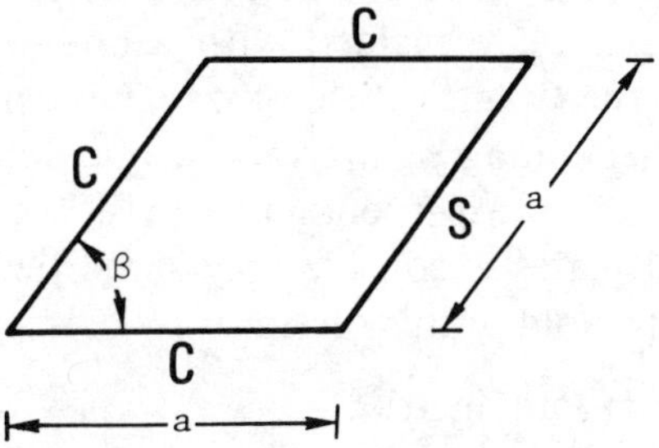

β	Mode Number 1	2	3	4	5	6
90	31.83	63.34	71.08	100.9	116.4	130.4
70	35.47	66.89	82.53	103.5	132.6	148.5
60	40.97	73.51	98.40	110.7	153.1	155.5
50	51.02	86.24	125.3	127.1	172.7	188.5
45	58.95	96.43	138.2	149.0	187.5	216.1
40	70.15	110.8	156.1	179.6	207.3	254.6

Ref. 11-72

9. Clamped Rhombic

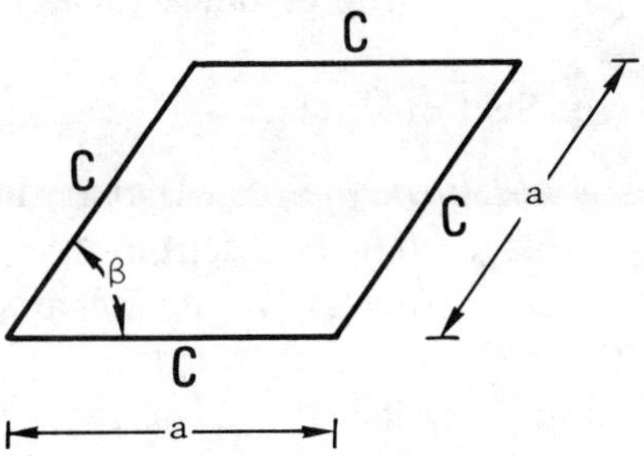

β	Mode Number 1	2	3	4	5	6
90	35.99	73.41	73.41	108.3	131.6	132.2
70	40.05	74.65	88.32	111.6	145.3	155.5
60	46.14	81.20	105.5	119.5	165.7	165.8
45	65.93	106.6	149.0	158.9	199.4	231.9

Ref. 11-73

Table 11-9. Parallelogram Plates. (*Continued*)

Natural Frequency (hertz), $f_i = \frac{\lambda_i^2}{2\pi a^2}\left[\frac{Eh^3}{12\gamma(1-\nu^2)}\right]^{1/2}$; i=1,2,3...

Description	λ_i^2
10. Clamped	

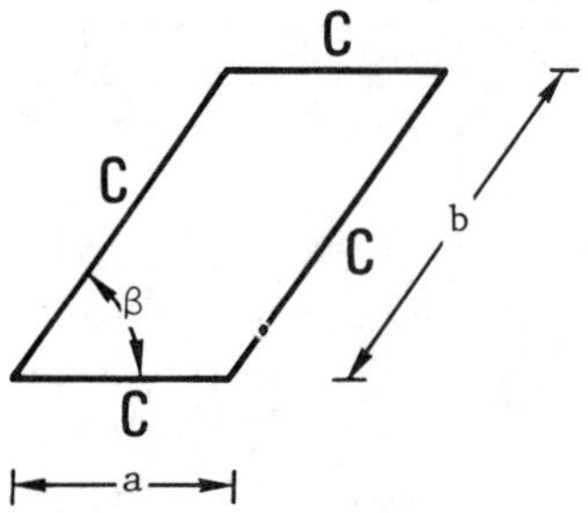

	Mode Number					
	a/b = 2/3		a/b = 1/2		a/b = 1/3	
β	1	2	1	2	1	2
90	27.01	41.72	24.58	31.84	23.20	25.87
70	30.19	45.44	27.62	35.10	26.18	28.92
60	35.01	51.04	32.22	39.98	30.70	33.52
45	50.79	69.19	47.37	56.06	45.71	48.76

Ref. 11-73

11.5. GRILLAGES AND STIFFENED PLATES

A grillage is a rectangular network of interlocking beams, welded at their intersections. A stiffened rectangular plate is a rectangular plate melded with a one- or two-dimensional grillage. Both grillages and stiffened plates differ from uniform plates in that their properties are directional; their bending rigidity about one axis is not necessarily the same as the bending rigidity about a perpendicular axis. The analysis of the vibrations of grillages and stiffened plates is considerably simplified under the following conditions which are assumed to hold in this section:

1. The stiffeners are uniform. All parallel stiffeners are identical.
2. The spacing between stiffeners is uniform so that the plate or grillage possesses two perpendicular planes of symmetry which are normal to its own plane.
3. The spacing between stiffeners is small compared with the distance between vibration nodes.
4. The general assumptions of plate theory hold (see Section 11.1).

Smeared analysis of grillages and stiffened plates is based on the idea that if the stiffeners are closely spaced, the stiffened plate or grillage behaves essentially as a uniform orthotropic plate. If the properties of discrete stiffeners can be smeared over the surface to create an equivalent uniform orthotropic plate, then orthotropic plate solutions can be employed. This smeared approach has the advantage of generality and allows the considerable body of literature on orthotropic plates to be tapped once expressions are developed for the properties of the equivalent orthotropic plate.

Table 11-10. Triangular Plates.

Notation: a, b = lengths; h = thickness of plate; C = clamped edge; E = modulus of elasticity; F = free edge; S = simply supported edge; γ = mass per unit area of plate (μh for a plate of material with density μ); ν = Poisson's ratio; see Table 3-1 for consistent sets of units

$$\text{Natural Frequency (hertz)},\ f_i = \frac{\lambda_i^2}{2\pi a^2}\left[\frac{Eh^3}{12\gamma(1-\nu^2)}\right]^{1/2};\ i = 1,2,3,\ldots$$

Description	λ_i^2

1. Clamped-Free-Free Isosceles Triangle

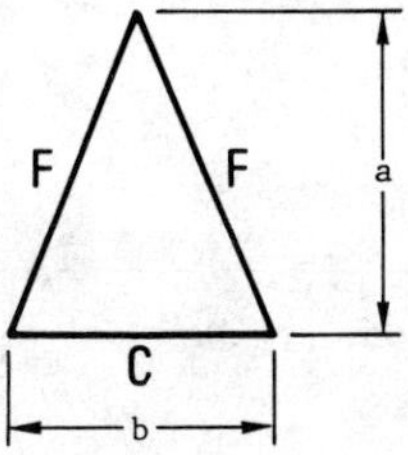

a/b	Mode Number 1	2	3	4
1	7.149	30.80	61.13	148.8
2	7.122	30.72	90.11	259.4
4	7.080	30.65	157.7	493.4
7	7.068	30.64	266.0	853.6

ν = 0.30, Ref. 11-74

2. Clamped-Free-Free Right Triangle

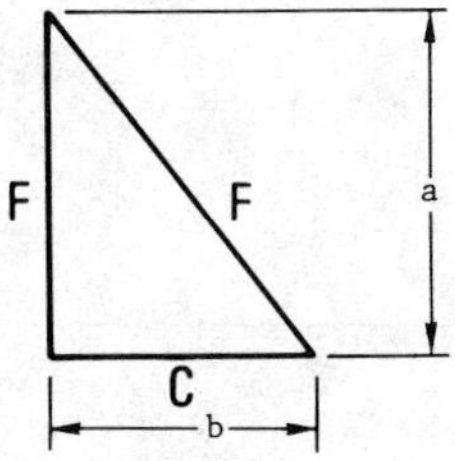

a/b	Mode 1	2
2	5.887	25.40
4	6.617	28.80
7	6.897	30.28

ν = 0.30, Ref. 11-74

3. Simply Supported Isosceles Triangle

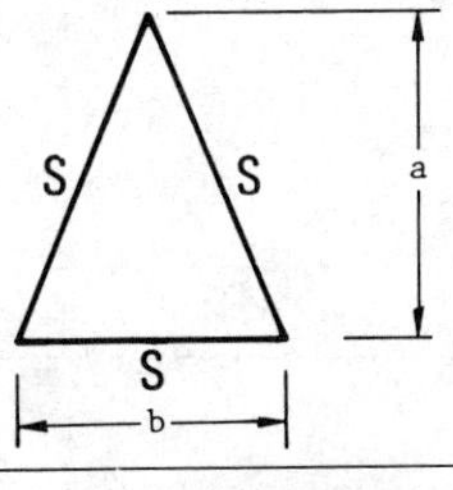

a/b	Mode Number 1	2	3	4	5
0.5	24.69	49.43	64.30	84.19	99.19
2/3	30.98	67.31	77.08	116.1	131.7
1.0	45.85	102.9	121.0	177.6	199.9
1.5	73.66	146.8	196.8	238.7	321.0

Ref. 11-75

4. Simply Supported Asymmetric Triangle

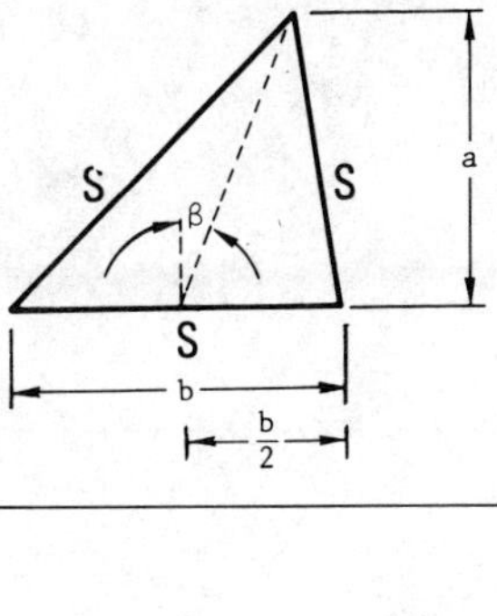

Fundamental mode:

a/b	β (deg) 0	10	20	30	45
0.5	24.69	24.78	25.06	25.64	27.78
1.0	45.85	46.28	47.71	50.57	60.22
1.5	73.66	74.64	77.85	84.21	105.1

Ref. 11-76

See Ref. 11-76 for higher modes.

Table 11-10. Triangular Plates. (*Continued*)

$$\text{Natural Frequency (hertz)},\ f_i = \frac{\lambda_i^2}{2\pi a^2}\left[\frac{Eh^3}{12\gamma(1-\nu^2)}\right]^{1/2};\ i = 1,2,3,\ldots$$

Description	λ_i^2
5. Simply Supported Right Triangle 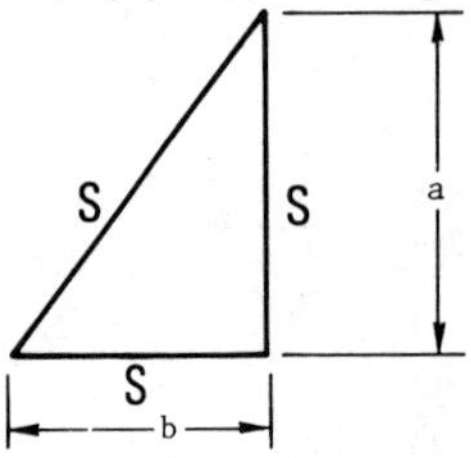	For a = b, $\lambda_{ij}^2 = \pi^2(i^2 + j^2)$; i,j > 0 but i ≠ j, (i,j) = (1,2),(2,1),(3,1),(1,3),(2,3)... etc. For $b = (3^{1/2}/3)a$, λ^2 = 92.11 in fundamental mode (Ref. 11-1).
6. Clamped-Simply Supported-Simply Supported Isosceles Triangle 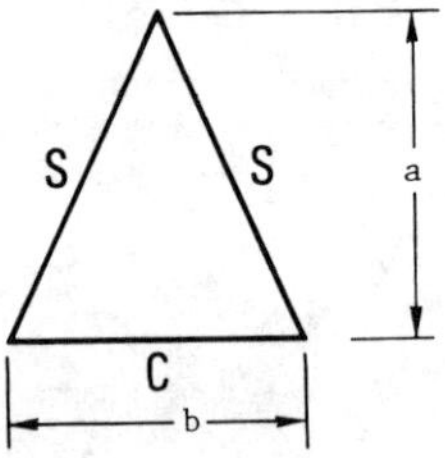	$\frac{a}{b}$ = 0.25 0.5 1.0 1.5 2.0 λ_1^2 = 31.0 34.5 55.0 87.0 120.0 Ref. 11-77
7. Clamped-Clamped-Simply Supported Isosceles Triangle 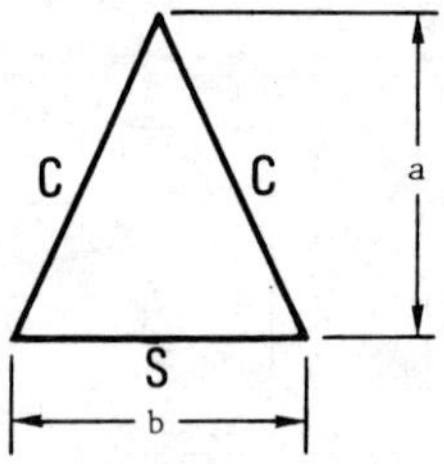	Fundamental mode only: $\frac{a}{b}$ = 0.1340 0.2887 0.5 0.866 1.866 λ^2 = 20.36 26.30 36.80 61.20 166.8 Ref. 11-78
8. Clamped Isosceles Triangle 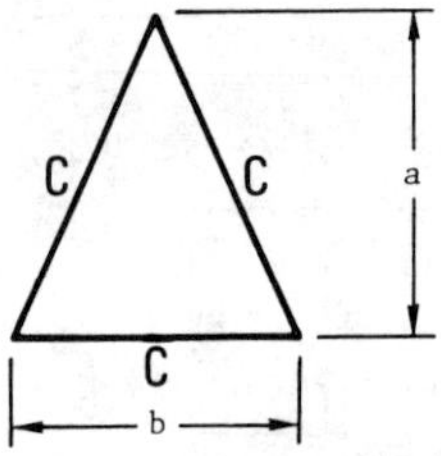	Fundamental mode only: $\frac{a}{b}$ = 0.5 0.866 1.866 λ^2 = 46.8 74.4 186.0 Ref. 11-78

Table 11-11. Other Plates.

Notation: a = length; h = thickness of plate; E = modulus of elasticity; S = simply supported edge; γ = mass per unit area of plate (μh for a plate of material with density μ); ν = Poisson's ratio; see Table 3-1 for consistent sets of units. λ^2 is independent of ν.

Fundamental Natural Frequency (hertz), $f = \frac{\lambda^2}{2\pi a^2}\left[\frac{Eh^3}{12\gamma(1-\nu^2)}\right]^{1/2}$

Description	λ^2

1. Simply Supported Symmetric Trapezoid

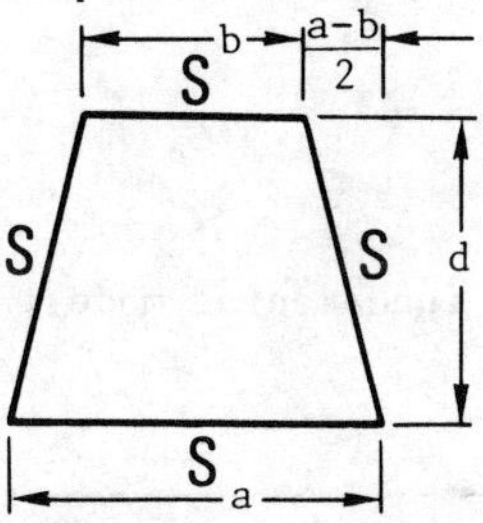

d/a	b/a = 0.0	0.2	0.4	0.6	0.8	1.0
0.5	98.78	76.50	63.18	55.97	51.85	49.35
2/3	69.70	55.09	44.70	38.38	34.54	32.08
1.0	45.85	37.75	30.79	25.64	22.13	19.74
1.5	32.74	28.04	23.64	19.72	16.58	14.26

Ref. 11-75

2. Simply Supported Unsymmetric Trapezoid

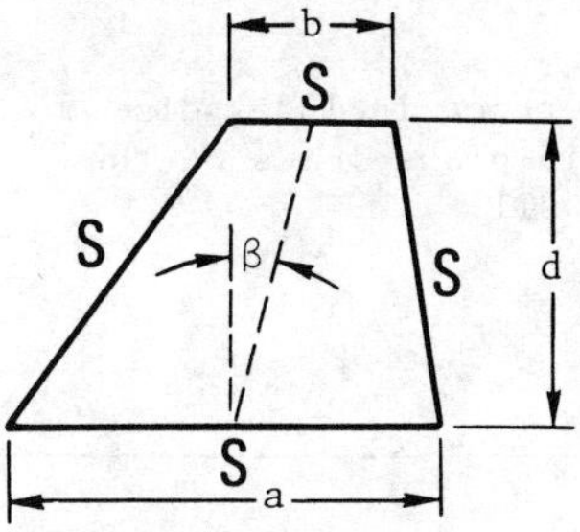

	d/a = 0.5		1.0		1.5	
β	b/a = 0.4	b/a = 0.8	b/a = 0.4	b/a = 0.8	b/a = 0.4	b/a = 0.8
10	63.42	52.00	31.20	22.37	24.05	16.88
20	64.26	52.47	32.50	23.18	25.39	17.86
30	66.02	53.44	35.14	24.87	28.02	19.85
45	72.81	56.94	44.06	30.97	36.59	26.81

Ref. 11-76

3. Simply Supported Regular Polygon with n Sides

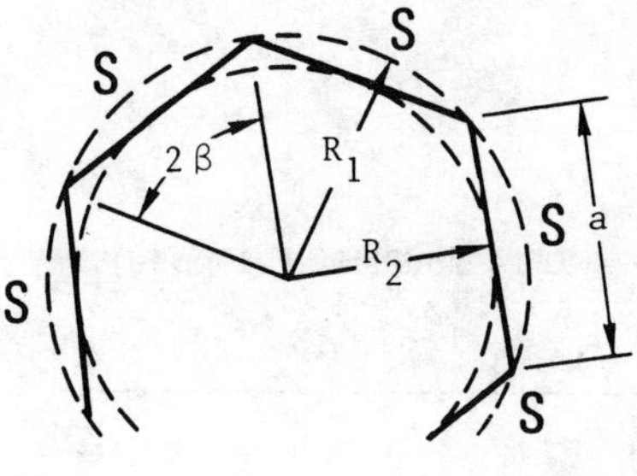

Number of sides =	4	5	6	7	8
λ^2 =	19.74	11.01	7.152	5.068	3.794

[$R_1 = a/(2 \sin \beta)$; $R_2 = a/(2 \tan \beta)$]

Ref. 11-79

Table 11-11. Other Plates. *(Continued)*

$$\text{Natural Frequency (hertz)},\ f_i = \frac{\lambda_i^2}{2\pi a^2}\left[\frac{Eh^3}{12\gamma(1-\nu^2)}\right]^{1/2}$$

Description	λ^2
4. Simply Supported Plate with Arbitrarily Shaped Edge 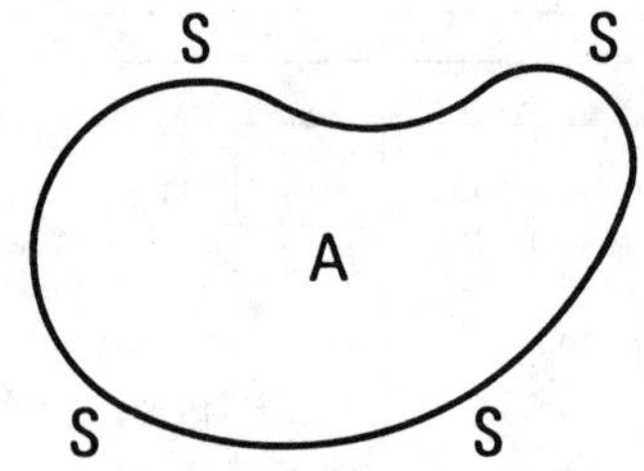 	Lower bound for λ_i: $\lambda_i^2 = 4.977i$; i=1,2,3 The characteristic length is defined as $a = \frac{A^{1/2}}{\pi^{1/2}}$. A = area of plate. i = mode number (i = 1 is fundamental mode). $\nu = 0.3$ Ref. 11-80.
5. Clamped Regular Polygon with n Sides 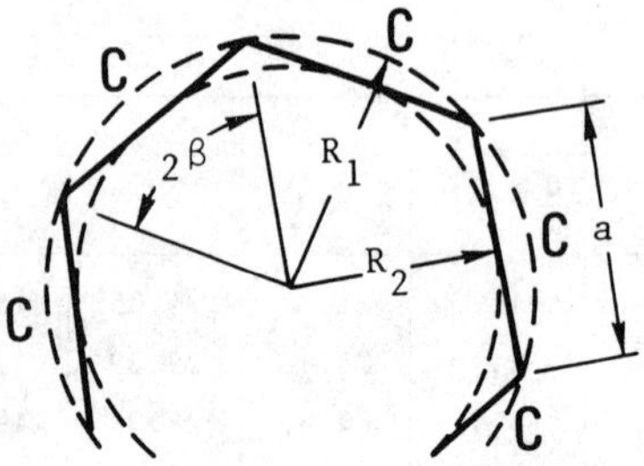 	Number of sides = 4 5 6 7 8 λ^2 = 35.08 19.71 12.81 9.081 6.787 [$R_1 = a/(2 \sin \beta)$; $R_2 = a/(2 \tan \beta)$] Fundamental Mode. Ref. 11-81. This reference includes discussion of effect of in-plane loads on the natural frequency as well.
6. Clamped Plate with Arbitrarily Shaped Edge 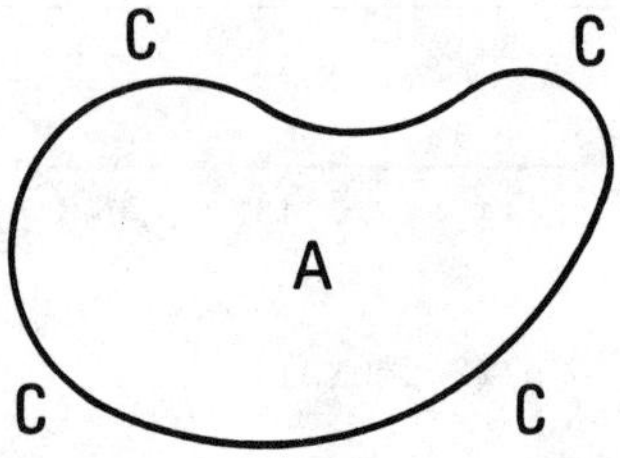 	Lower bound for λ_i: $\lambda_i^2 = 10.22i$; i=1,2,3... The characteristic length is defined as $a = \frac{A^{1/2}}{\pi^{1/2}}$. A = area of plate i = mode number (i = 1 is fundamental mode). Ref. 11-80.

Table 11-12. Equivalent Orthotropic Constants of Grillages and Stiffened Plates.

Notation: a = width; b = length; h = thickness of plate; C = torsion constant (Table 8-18); E = modulus of elasticity; I = moment of inertia about neutral axis (Table 5-1, midsurface = neutral axis if plate or grillage is symmetric about neutral axis); H = height; μ = mass density; ν = Poisson's ratio. Subscripts: a = stiffener of length a; b = stiffener of length b. Consistent sets of units are given in Table 3-1. See Eq. 11-25 for applications. These solutions were adapted from Ref. 11-87. $D_k \approx D_{xy}/2$.

Description	D_x	D_y	D_{xy}
1. Grillage 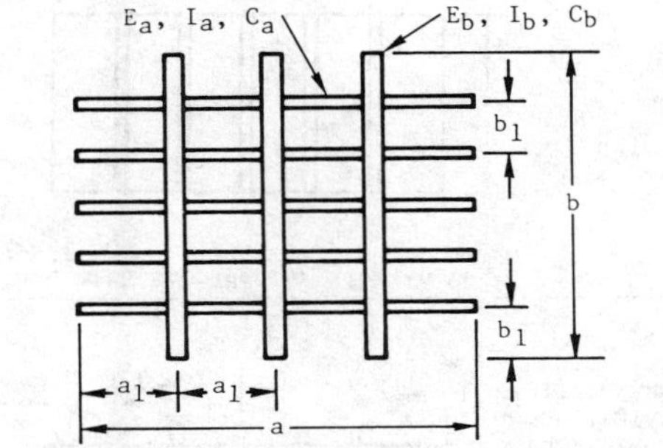 GRILLAGE IS SYMMETRIC ABOUT MIDSURFACE	$\frac{E_a I_a}{b_1}$	$\frac{E_b I_b}{a_1}$	$\frac{E_a C_a}{2b_1} + \frac{E_b C_b}{2a_1}$
2. Plate with Stiffeners in One Direction 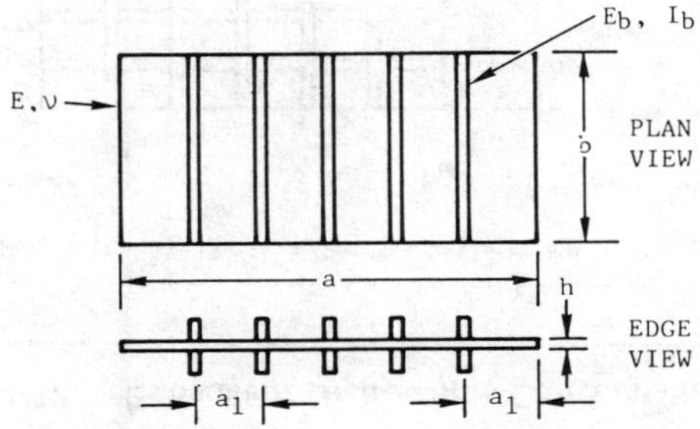 PLATE IS SYMMETRIC ABOUT MIDSURFACE	$\frac{Eh^3}{12(1 - \nu^2)}$	$\frac{Eh^3}{12(1 - \nu^2)} + \frac{E_b I_b}{a_1}$ I_b = moment of inertia of stiffener alone with respect to midsurface.	$\frac{Eh^3}{12(1 - \nu^2)}$

Table 11-12. Equivalent Orthotropic Constants of Grillages and Stiffened Plates. *(Continued)*

Description	D_x	D_y	D_{xy}
3. Plate with Stiffeners in Two Directions 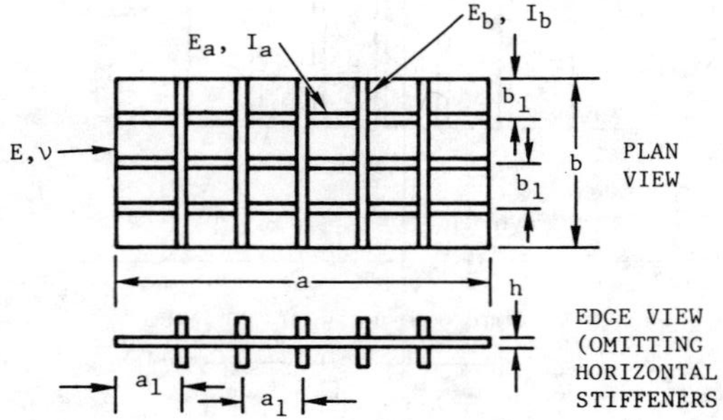	$\frac{Eh^3}{12(1-\nu^2)} + \frac{E_a I_a}{b_1}$ I_a = moment of inertia of stiffener alone with respect to midsurface.	$\frac{Eh^3}{12(1-\nu^2)} + \frac{E_b I_b}{a_1}$ I_b = moment of inertia of stiffener alone with respect to midsurface.	$\frac{Eh^3}{12(1-\nu^2)}$
4. Plate with Rectangular Ribs in One Direction 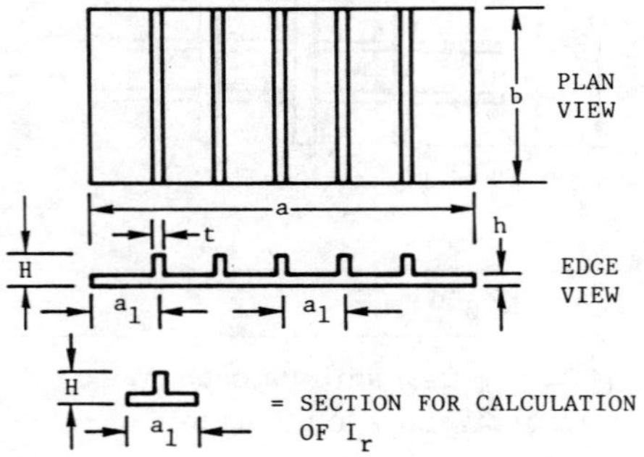	$\frac{Ea_1 h^3}{12\left[a_1 - t + \left(\frac{h}{H}\right)^3 t\right]}$	$\frac{EI_r}{a_1}$ I_r = see first column.	$\frac{Eh^3}{12(1+\nu)} + \frac{C_r E}{a_1}$ C_r = torsion constant of rib alone.

5. Corrugated Plate 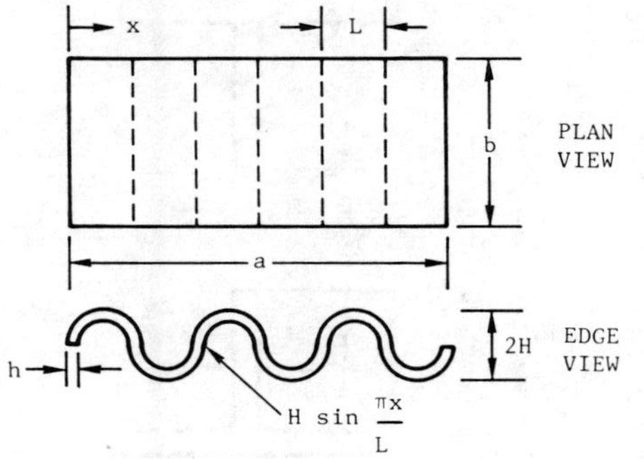	$\dfrac{Eh^3}{12(1 - \nu^2)\left(1 + \dfrac{\pi^2 H^2}{4L^2}\right)}$	$\dfrac{EHh^2}{2}\left[1 - \dfrac{0.81}{1 + \dfrac{5}{2}\left(\dfrac{H}{2L}\right)^2}\right]$	$\dfrac{Eh^3}{12(1 + \nu)}\left(1 + \dfrac{\pi^2 H^2}{4L^2}\right)$
6. Fiber-Reinforced Plate 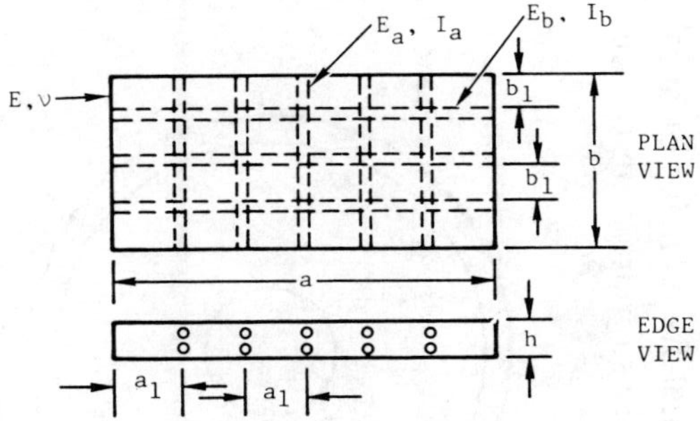 PLATE IS SYMMETRIC ABOUT MIDSURFACE.	$\dfrac{E}{1 - \nu^2}\left[\dfrac{h^3}{12} + \left(\dfrac{E_a}{E} - 1\right)\dfrac{I_a}{b_1}\right]$ I_a = moment of inertia of fibers of length a with respect to midsurface.	$\dfrac{E}{1 - \nu^2}\left[\dfrac{h^3}{12} + \left(\dfrac{E_b}{E} - 1\right)\dfrac{I_b}{a_1}\right]$ I_b = moment of inertia of fibers of length b with respect to midsurface.	$(D_x D_y)^{1/2}$

Equivalent orthotropic constants are given in Table 11-12 for grillages and stiffened plates. These constants were developed by matching the stiffness of grillages and stiffened plates with that of an orthotropic plate. Note that the moment of inertia of the stiffeners in Table 11-12 is calculated about the neutral axis, which will be the midsurface if the grillage or stiffened plate is symmetric about the midsurface. If the stiffeners of a plate such as shown in frame 2 or 3 of Table 11-12 are rectangular, then the moment of inertia of a stiffener is found from the moment of inertia of two rectangles in space separated by one-half the plate thickness from the midsurface (see Section 5.1).

Approximate solutions for the natural frequency and mode shape of rectangular orthotropic plates are given by Eq. 11-25 (p. 267) with the constants of Tables 11-5 and 11-12. The mass per unit area, γ, in Eq. 11-25 now represents the average mass per unit area, i.e., the total mass divided by the area ab. Other solutions for orthotropic plates having two opposite sides simply supported are discussed in Ref. 11-48.

Solutions for simply supported and clamped grillages developed using the discrete approach are given in Refs. 11-68 and 11-88. Other literature on vibration of grillages is surveyed in Ref. 11-67 and discussed in Ref. 11-69. Also see the discussion of orthotropic plates in Section 11.3.

11.6. EXAMPLE

An aluminum annular plate, shown in Fig. 11-8, is used as a baffle in a cylindrical pressure vessel. It is generally necessary to know both the in-plane loads and the local stiffness of the pressure vessel in order to calculate the natural frequency of the baffle. However, if in-plane loads can be neglected, then upper and lower bounds for the fundamental natural frequency of the baffle can be calculated from the fundamental natural frequencies of annular plates which are free on the inside and clamped and simply supported on the outside.

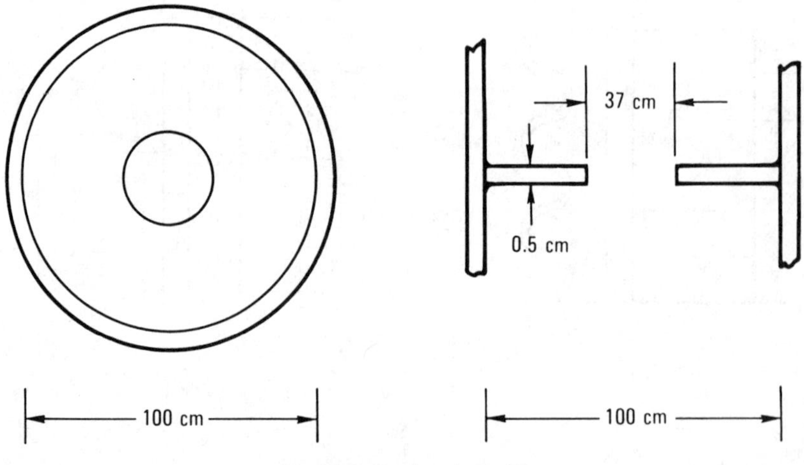

Fig. 11-8. An annular baffle.

The parameters describing the annular plate in terms of the symbols of Table 11-2 and the units of frame 3 of Table 3-1 are:

$$\left.\begin{aligned} a &= 50 \text{ cm} \\ b &= 18.5 \text{ cm} \end{aligned}\right\} b/a = 0.37 = 0.30 + 0.07,$$

$$h = 0.5 \text{ cm},$$

$$E = 7.03 \times 10^5 \text{ kg/cm}^2,$$

$$\mu = \frac{2.7 \times 10^{-3} \text{ kg/cm}^3}{980.7 \text{ cm/sec}^2} = 2.75 \times 10^{-6} \text{ kg-sec}^2/\text{cm}^4,$$

$$\gamma = \mu h = 1.415 \times 10^{-6} \text{ kg-sec}^2/\text{cm}^3,$$

$$\nu = 0.3.$$

For the fundamental mode, bounds for λ^2 are estimated from frames 3 and 5 of Table 11-2 using linear interpolation. A lower bound for λ^2 is found from frame 3 of Table 11-2 for a simply supported-free annular plate:

$$\lambda^2 = 4.66 + \frac{0.07}{0.2}(5.07 - 4.66),$$

$$= 4.80.$$

An upper bound for λ^2 is found from frame 5 of Table 11-2:

$$\lambda^2 = 11.4 + \frac{0.07}{0.2}(17.7 - 11.4),$$

$$= 13.6.$$

The bounds for the fundamental natural frequency of the annular plate are found using the formula presented in Table 11-2:

$$f = \frac{\lambda^2}{2\pi a^2}\left[\frac{Eh^3}{12\gamma(1-\nu^2)}\right]^{1/2}.$$

The result is

$$23.0 < f < 65.3 \text{ hertz}.$$

REFERENCES

11-1. Leissa, A. W., "Vibration of Plates," NASA-SP-160, Office of Technology Utilization, NASA, Washington D.C., 1969 (out of print).

11-2. Colwell, R. C., and H. C. Hardy, "The Frequencies and Nodal Systems of Circular Plates," *Phil. Mag.* **24**(7), 1041–1055 (1937).

11-3. Airey, J., "The Vibration of Circular Plates and Their Relation to Bessel Functions," *Proc. Phys. Soc.* (*London*) **23**, 225–232 (1911).

11-4. Gontkevich, V. S., *Natural Vibrations of Plates and Shells*, A. P. Filippov (ed.), Nauk Dumka (Kiev), 1964 (translated by Lockheed Missiles and Space Company, Sunnyvale, Calif.) (see also Ref. 11-1).

11-5. Prescott, T., *Applied Elasticity*, Dover Publications, New York, 1961 (originally published by Langmons, Green and Company, 1924).

11-6. Carrington, H., "The Frequencies of Vibration of Flat Circular Plates Fixed at the Circumference," *Phil. Mag.* **50(6)**, 1261–1264 (1925).
11-7. Bartlett, C. C., "The Vibration and Buckling of a Circular Plate Clamped on Part of Its Boundary and Simply Supported on the Remainder," *Quart. J. Mech. Appl. Math.* **16**, 431–440 (1963).
11-8. Noble, B., "The Vibration and Buckling of a Circular Plate Clamped on Part of Its Boundary and Simply Supported on the Remainder," *Proc. 9th Midwest Conf. Solid and Fluid Mech.*, August 1965.
11-9. Sakharov, I. E., "Dynamic Stiffness in the Theory of Axisymmetric Vibrations of Circular and Annular Plates," *Izv. An SSSR, OTN, Mekh. i Mashin* **5**, 90–98 (1959).
11-10. Bodine, R. Y., "The Fundamental Frequencies of a Thin Flat Plate Simply Supported Along a Circle of Arbitrary Radius," *J. Appl. Mech.* **26**, 666–668 (1959).
11-11. Kantham, C. L., "Bending and Vibration of Elastically Restrained Circular Plates," *J. Franklin Inst.* **265**, 483–491 (1958).
11-12. Laura, P. A. A., J. C. Paloto, and R. D. Santos, "A Note on the Vibration of a Circular Plate Elastically Restrained Against Rotation," *J. Sound Vib.* **41**, 177–180 (1975).
11-13. Roberson, R. E., "Transverse Vibrations of a Free Circular Plate Carrying Concentrated Mass," *J. Appl. Mech.* **18**, 280–282 (1951).
11-14. Harris, C. M., and C. E. Crede (eds.), *Shock and Vibration Handbook*, McGraw-Hill, New York, 1961, p. 1–15.
11-15. Roberson, R. E., "Vibrations of a Clamped Circular Plate Carrying Concentrated Mass," *J. Appl. Mech.* **18**, 349–352 (1951).
11-16. Wah, T., "Vibration of Circular Plates," *J. Acoust. Soc. Am.* **34**, 275–281 (1962).
11-17. Stafford, J. W., "Natural Frequencies of Beams and Plates on an Elastic Foundation with a Constant Modulus," *J. Franklin Inst.* **284**, 262–264 (1967).
11-18. Ben-Amoz, M., "Note on Deflections and Flexural Vibrations of Clamped Sectorial Plates," *J. Appl. Mech.* **26**, 136–137 (1959).
11-19. Rao, S. S., and A. S. Prasad, "Vibrations of Annular Plates Including the Effects of Rotary Inertia and Transverse Shear Deformation," *J. Sound Vib.* **42**, 305–324 (1975).
11-20. Vogel, S. M., and D. W. Skinner, "Natural Frequencies of Transversely Vibrating Uniform Annular Plates," *J. Appl. Mech.* **32**, 926–931 (1965).
11-21. Sato, K., "Free-Flexural Vibrations of an Elliptical Plate with Free Edge," *J. Acoust. Soc. Am.* **54**, 547–550 (1973).
11-22. Leissa, A. W., "Vibration of a Simply Supported Elliptical Plate," *J. Sound Vib.* **6**, 145–148 (1967); also see Sato, K., "Free-Flexural Vibrations of an Elliptical Plate with Simply Supported Edge," *J. Acoust. Soc. Am.* **52**, 919–922 (1972).
11-23. McNitt, R. P., "Free Vibration of a Clamped Elliptical Plate," *J. Aerospace Sci.* **29**, 1124–1125 (1962).
11-24. Reissner, E., "On Axi-Symmetrical Vibrations of Circular Plates of Uniform Thickness Including the Effects of Shear Deformation and Rotary Inertia," *J. Acoust. Soc. Am.* **26**, 252–253 (1954).
11-25. Mindlin, R. D., and H. Deresiewicz, "Thickness-Shear and Flexural Vibrations of a Circular Disk," *J. Appl. Phys.* **25**, 1320–1332 (1954).
11-26. Mindlin, R. D., "Influence of Rotary Inertia and Shear on Flexural Motions of Isotropic Plates," *J. Appl. Mech.* **18**, 31–38 (1951).
11-27. Woo, H. H., P. G. Kirmser, and C. L. Huang, "Vibration of an Orthotropic Plate with an Isotropic Core," *AIAA J.* **11**, 1421–1422 (1973).
11-28. Kirmser, P. G., C. L. Huang, and H. K. Woo, "Vibration of Cylindrically Orthotropic Circular Plates," *AIAA J.* **10**, 1690–1691 (1972).
11-29. Rao, K. S., K. Ganapathi, and G. V. Rao, "Vibration of Cylindrically Orthotropic Circular Plates," *J. Sound Vib.* **36**, 433–434 (1973).
11-30. Minkarah, I. A., and W. H. Hoppmann, "Flexural Vibrations of Cylindrically Aeolotropic Circular Plates," *J. Acoust. Soc. Am.* **36**, 470–475 (1964).
11-31. Ramaiah, G. K., and K. Vijayakumar, "Natural Frequencies of Polar Orthotropic Annular Plates," *J. Sound Vib.* **26**, 517–531 (1973).
11-32. Eversman, W., and R. O. Dodson, Jr., "Free Vibration of a Centrally Clamped Spinning Circular Disk," *AIAA J.* **7**, 2010–2012 (1969).
11-33. Krauter, A. T., and P. Z. Bulkeley, "Effect of Central Clamping on Transverse Vibrations of Spinning Membrane Disks," *J. Appl. Mech.* **37**, 1037–1042 (1970).
11-34. Kennedy, W., and D. Gorman, "Vibration Analysis of Variable Thickness Discs Subjected to Centrifugal and Thermal Stresses," *J. Sound Vib.* **53**, 83–101 (1977).
11-35. Conway, H. D., "Some Special Solutions for the Flexural Vibration of Discs of Varying Thickness," *Ingr.-Arch.* **26**, 408–410 (1958).

11-36. Conway, H. D., E. C. H. Becker, and J. F. Dubil, "Vibration Frequencies of Tapered Bars and Circular Plates," *J. Appl. Mech.* **31** (2), 329–331 (1964).
11-37. Thurston, E. G., and Y. T. Tsui, "On the Lowest Flexural Resonant Frequency of a Circular Disk of Linearly Varying Thickness Driven at Its Center," *J. Acoust. Soc. Am.* **27**, 926–929 (1955).
11-38. Jain, R. K., "Vibrations of Circular Plates of Variable Thickness under an Inplane Force," *J. Sound Vib.* **23**, 407–414 (1972).
11-39. Ramaiah, G. K., and K. Vijayakumar, "Vibrations of Annular Plates with Linear Thickness Profiles," *J. Sound Vib.* **40**, 293–298 (1975).
11-40. Stuart, R., and J. F. Carney, III, "Vibration of Annular Plates with Edge-Beams," *AIAA J.* **12**, 5–6 (1974).
11-41. Leissa, A. W., "The Free Vibration of Rectangular Plates," *J. Sound Vib.* **31**, 257–293 (1973).
11-42. Gorman, D. J., "Free Vibration Analysis of Cantilever Plates by the Method of Superposition," *J. Sound Vib.* **49**, 453–467 (1976).
11-43. Waller, M. D., "Vibrations of Free Square Plates," *Proc. Phys. Soc. (London)* **51**, 831–844 (1939).
11-44. Mindlin, R. D., A. Schacknow, and H. Deresiewicz, "Flexural Vibrations of Rectangular Plates," *J. Appl. Mech.* **23**, 430–436 (1956).
11-45. Kristiansen, V. R., W. Soedel, and J. F. Hamilton, "An Investigation of Scaling Laws for Vibrating Beams and Plates with Special Attention to the Effects of Shear and Rotary Inertia," *J. Sound Vib.* **20**, 113–122 (1972).
11-46. Hearmon, R. F. S., "The Fundamental Frequency of Rectangular Wood and Plywood Plates," *Proc. Phys. Soc. (London)* **58**, 78–92 (1946).
11-47. Hearmon, R. F. S., "The Frequency of Flexural Vibration of Rectangular Orthotropic Plates with Clamped or Supported Edges," *J. Appl. Mech.* **26**, 537–540 (1959).
11-48. Vijayakumar, K., "Natural Frequencies of Rectangular Orthotropic Plates with a Pair of Parallel Edges Simply Supported," *J. Sound Vib.* **35**, 379–394 (1974).
11-49. Magrab, E., "Natural Frequencies of Elastically Supported Orthotropic Rectangular Plates," *J. Acoust. Soc. Am.* **67**, 79–83 (1977).
11-50. Reed, R. E., Jr., "Comparison of Methods in Calculating Frequencies of Corner-Supported Rectangular Plates," NASA Report NASA TN D-3030, 1965.
11-51. Srinivasan, R. S., and K. Munaswamy, "Frequency Analysis of Skew Orthotropic Point Supported Plates," *J. Sound Vib.* **39**, 207–216 (1975).
11-52. Johns, D. J., and R. Nataroja, "Vibration of a Square Plate Symmetrically Supported at Four Points," *J. Sound Vib.* **25**, 75–82 (1972).
11-53. Petyt, M., and W. H. Mirza, "Vibration of Column-Supported Floor Slabs," *J. Sound Vib.* **21**, 355–364 (1972).
11-54. Nowacki, W., *Dynamics of Elastic Systems*, John Wiley, New York, 1963, p. 228.
11-55. Cox, H. L., "Vibration of Certain Square Plates Having Similar Adjacent Edges," *Quart. J. Mech. Appl. Math.* **8**, 454–456 (1955).
11-56. Appl, F. C., and N. R. Byers, "Fundamental Frequency of Simply Supported Rectangular Plates with Linearly Varying Thickness," *J. Appl. Mech.* **32**, 163–167 (1965).
11-57. Chopra, I., and S. Durvasula, "Natural Frequencies and Modes of Tapered Skew Plates," *J. Sound Vib.* **13**, 935–944 (1971).
11-58. Ashton, J. E., "Natural Modes of Vibration of Tapered Plates," *ASCE, J. Struct. Div.* **95**, 787–790 (1969).
11-59. Ashton, J. E., "Free Vibration of Linearly Tapered Clamped Plates," *ASCE, J. Eng. Mech. Div.* **95**, 497–500 (1969).
11-60. Soni, S. R., and K. S. Rao, "Vibrations of Non-Uniform Rectangular Plates: A Spline Technique Method of Solution," *J. Sound Vib.* **35**, 35–45 (1974).
11-61. Dickinson, S. M., "The Buckling and Frequency of Flexural Vibration of Rectangular, Isotropic and Orthotropic Plates Using Rayleigh's Method," *J. Sound Vib.* **61**, 1-8 (1978).
11-62. Bassily, S. F., and S. M. Dickinson, "Buckling and Lateral Vibrations of Rectangular Plates Subject to In-plane Loads–A Ritz Approach," *J. Sound Vib.* **24**, 219–239 (1972).
11-63. Hegarty, R. F., and T. Ariman, "Elasto-Dynamic Analysis of Rectangular Plates with Circular Holes," *Int. J. Solids Structures* **11**, 895–906 (1975).
11-64. Paramasivam, P., "Free Vibration of Square Plates with Square Openings," *J. Sound Vib.* **30**, 173–178 (1973).
11-65. Anderson, R. G., B. M. Iroms, and O. C. Zienkiewicz, "Vibration and Stability of Plates Using Finite Elements," *Int. J. Solids Structures* **4**, 1031–1055 (1968).
11-66. Stahl, B., and L. M. Kerr, "Vibration and Stability of Cracked Rectangular Plates," *Int. J. Solids Structures* **8**, 69–91 (1972).

11-67. Rao, H. V. S. G., "Vibration Analysis of Grid-Works," *Shock Vib. Digest* **8** (9), 25–30 (1976).

11-68. Chang, P. Y., and F. C. Michelsen, "A Vibration Analysis of Grillage Beams," *J. Ship Research* **13**, 32–39 (1969).

11-69. Armstrong, I. D., "The Natural Frequencies of Grillages," *Int. J. Mech. Sci.* **10**, 43–55 (1968).

11-70. Barton, M. V., "Vibration of Rectangular and Skew Cantilever Plates," *J. Appl. Mech.* **18**, 129–134 (1951).

11-71. Durvasula, S., "Natural Frequencies and Modes of Skew Membranes," *J. Acoust. Soc. Am.* **44**, 1636–1646 (1968).

11-72. Nair, P. S., and S. Durvasula, "Vibration of Skew Plates," *J. Sound Vib.* **26**, 1–19 (1973).

11-73. Durvasula, S., "Natural Frequencies and Modes of Clamped Skew Plates," *AIAA J.* **7**, 1164–1167 (1969).

11-74. Anderson, B. W., "Vibration of Triangular Cantilever Plates by the Ritz Method," *J. Appl. Mech.* **21**, 365–376 (1954).

11-75. Chopra, I., and S. Durvasula, "Vibration of Simply-Supported Trapezoidal Plates, I. Symmetric Trapezoids," *J. Sound Vib.* **19**, 379–392 (1971).

11-76. Chopra, I., and S. Durvasula, "Vibration of Simply Supported Trapezoidal Plates, II. Unsymmetric Trapezoids," *J. Sound Vib.* **20**, 125–134 (1972).

11-77. Cox, H., and B. Klein, "Vibrations of Isosceles Triangular Plates Having the Base Clamped and Other Edges Simply-Supported," *Aeron. Quart.* **7**, 221–224 (1956).

11-78. Ota, T., M. Hamada, and T. Tarumoto, "Fundamental Frequency of Isosceles Rectangular Plates," *Bull. JSME* **4**, 478–481 (1961).

11-79. Shahady, P. A., R. Pasarelli, and P. A. A. Laura, "Application of Complex Variable Theory to the Determination of the Fundamental Frequency of Vibrating Plates," *J. Acoust. Soc. Am.* **42**, 806–809 (1967).

11-80. Pnueli, D., "Lower Bounds to the Gravest and All Higher Frequencies of Homogeneous Vibrating Plates," *J. Appl. Mech.* **42**, 815–820 (1975).

11-81. Laura, P. A. A., and R. Gutierrez, "Fundamental Frequency of Vibration of Clamped Plates of Arbitrary Shape Subjected to a Hydrostatic State of In-Plane Stress," *J. Sound Vib.* **48**, 327–332 (1976).

11-82. Gorman, D. J., "Free Vibration of the Completely Free Rectangular Plate by the Method of Superposition," *J. Sound Vib.* **57**, 437–447 (1978).

11-83. Rao, G. V., "Fundamental Frequency of a Square Panel with Multiple Point Supports on Edges," *J. Sound Vib.* **38**, 271 (1975).

11-84. Venkatesan, S., and V. X. Kunukkasseril, "Free Vibration of Layered Circular Plates," *J. Sound Vib.* **60**, 511–534 (1978).

11-85. Solecki, R., "Oscillations of Rectangular Sandwich Plates with Concentrated Masses," *J. Sound Vib.* **33**, 295–303 (1974).

11-86. Rajamani, A., and R. Prabhakaran, "Dynamic Response of Composite Plates with Cut-Outs, Part I: Simply Supported Plates, Part II: Clamped-Clamped Plates," *J. Sound Vib.* **54**, 549–576 (1977).

11-87. Timoshenko, S., and S. Woinowsky-Krieger, *Theory of Plates and Shells*, 2nd. ed., McGraw-Hill, New York, 1959, pp. 366–371.

11-88. Wah, T. H., "Grillages," in *Shock and Vibration Computer Programs*, W. Pilkey and B. Pilkey (editors), Naval Research Laboratory, Washington, D.C., 1975, pp. 329–338.

11-89. Irie, T., G. Yamada, and Y. Narita, "Free Vibration of Cross-Shaped, I-Shaped and L-Shaped Plates Clamped at All Edges," *J. Sound Vib.* **61**, 571–583 (1978).

12
SHELLS

12.1. GENERAL CASE

General Assumptions. A shell is a sheet of elastic material which conforms to a curved surface, the midsurface of the shell. A shell can be curved about one axis like a cylindrical storage tank or about multiple axes like a wine glass. A shell can be open like a curved roof or closed like a hollow ball. The curvature of the shell couples its flexural and extensional vibrations and considerably complicates the analysis of shells.

The assumptions generally employed in the analysis of shells in this chapter are:

1. The shells have constant thickness.
2. The shell walls are thin. The shell wall is less than 10% of the shell radius.
3. The shells are composed of a linear, elastic, homogeneous isotropic material.
4. There are no loads applied to the shells.
5. The deformations of the shell are small in comparison with the radius of the shell. Straight lines perpendicular to the midsurface of the shell remain straight and perpendicular to the midsurface during deformation.
6. Rotary inertia and shear deformation are neglected.

These assumptions apply to all the solutions in this chapter unless an exception is specifically noted.

Shell Equations. There is not general agreement in the literature on the linear differential equations which describe the deformations of a shell. A number of theories have arisen and are used. Prominent among these are the theories of A. E. H. Love (1888, Refs. 12-1 and 12-2), W. Flügge (1934, Refs. 12-3 and 12-4), L. H. Donnell (1938, Ref. 12-5), K. M. Mushtari (1938, Ref. 12-6), E. Reissner (1941, Ref. 12-7), and J. L. Sanders (1959, Ref. 12-8). The differences among the theories are due to the various assumptions made about the form of small terms and the order of terms which are retained in the analysis. The Donnell and Mushtari shell theories are the simplest of these theories. The Flügge and Sanders shell theories are generally felt to be the most accurate. In some cases the various shell theories predict significantly different results. However, over broad ranges of parameters of engineering importance these theories yield similar results. A summary and comparison of the shell theories can be found in Ref. 12-9.

Solution of the Shell Equations. Each of the shell theories mentioned above describes the motion of the shell in terms of an eighth-order differential equation.

Inertia terms associated with each of the three mutually orthogonal displacements, shown in Fig. 12-1 for a cylindrical shell, are retained in the analysis. If the spatial dependence of each of the deformations can be estimated, then the natural frequencies of the shell can be reduced to the solution of a cubic characteristic polynomial and the relative amplitude of the three displacements can be found from a three-by-three matrix of simultaneous linear equations. Experience has shown that the natural frequencies of shells do not fall in an ascending series with increasing values of the modal index. The lowest-frequency mode of a shell may as well be associated with the modal index i = 9 as i = 1.

Analysis for the natural frequencies and mode shapes of shells is generally much more complex than analysis of beams and plates because:

1. The generality of the shell equations permits a wide variety of mode shapes with vastly different character. For example, some of the solutions which can be obtained from the equations describing a cylindrical shell are (a) the transverse vibration of tubular beams, (b) the longitudinal vibration of tubular

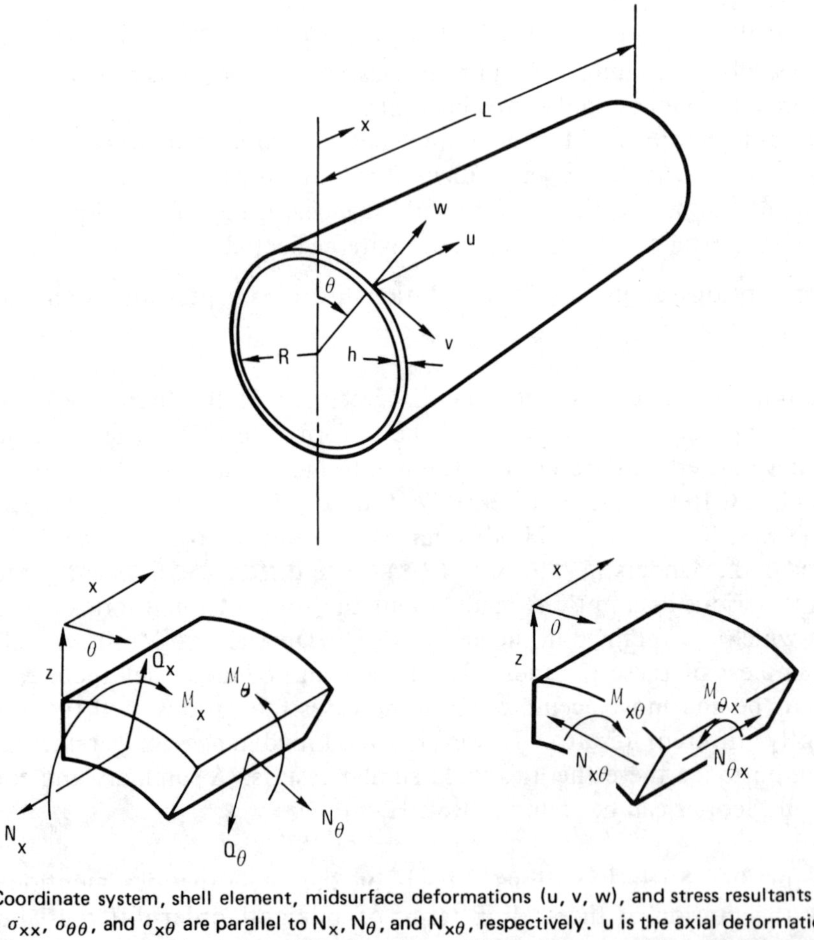

Fig. 12-1. Coordinate system, shell element, midsurface deformations (u, v, w), and stress resultants (N, $\mathcal{M}$, Q). The stresses σ_{xx}, $\sigma_{\theta\theta}$, and $\sigma_{x\theta}$ are parallel to N_x, N_θ, and $N_{x\theta}$, respectively. u is the axial deformation, v is the circumferential deformation, and w is the radial deformation. All quantities are positive as shown.

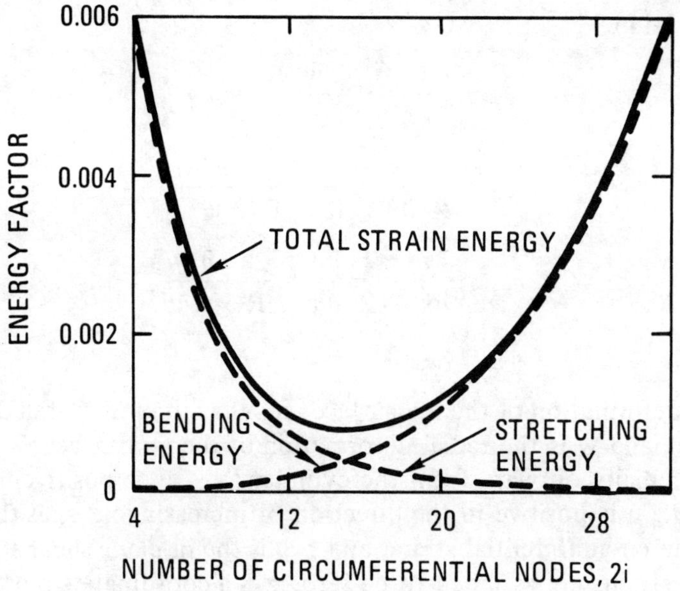

Fig. 12-2. Variation in strain energy in a cylindrical shell with increasing number of circumferential waves. (Ref. 12-10)

beams, (c) the torsional vibration of tubular beams, (d) the flexural in-plane vibration of rings, (e) the extensional in-plane vibration of rings, and (f) vibration modes unique to shells.

2. The curvature of the shell results in coupling of the extensional and flexural deformations of the shell. The deformation of a shell can vary from purely extensional to purely flexural, as shown in Fig. 12-2.

Shell theories must incorporate both flexural and extensional deformations. In contrast, the theory of transverse vibration of beams and plates (Chapters 8 and 11) considers only flexural deformations, since extensional deformation of beams and plates is zero to first (linear) order during transverse vibration. Extensional deformations of shells enter the shell equations as a first-order term. As a result, the natural frequencies of shells are a function of the axial constraint applied at the ends of the shell, whereas the natural frequencies of a transverse vibration of a straight beam are not.

Because of the complexity of the shell equations and their solution, few closed form solutions are available for the natural frequencies and mode shapes of shells. The results presented in this chapter are mostly approximate solutions which are valid for a particular type of mode over a certain range of parameters.

12.2. CYLINDRICAL SHELLS

12.2.1. Strain-Deformation Relationships

For a cylindrical shell, shown in Fig. 12-1, the midsurface deformations u, v, and w produce the following strains in the plane of the shell according to the Donnell shell theory (Ref. 12-11):

$$\epsilon_x = \frac{\partial u}{\partial x} - z\frac{\partial^2 w}{\partial x^2},$$

$$\epsilon_\theta = \frac{1}{R}\frac{\partial v}{\partial \theta} + \frac{w}{R} - \frac{z}{R^2}\frac{\partial^2 w}{\partial \theta^2},$$

$$\epsilon_{x\theta} = \frac{\partial v}{\partial x} + \frac{1}{R}\frac{\partial u}{\partial \theta} - \frac{2z}{R}\frac{\partial^2 w}{\partial x \partial \theta},$$

$$\epsilon_{xz} = \epsilon_{\theta z} = \epsilon_{zz} = 0. \qquad (12\text{-}1)$$

u is the axial deformation of the midsurface, v is the circumferential deformation of the midsurface, and w is the radial deformation of the midsurface. w is positive for deformations radially outward from the cylinder axis. v is positive in the direction of increasing θ. u is positive in the direction of increasing x. ϵ_x is the longitudinal strain, ϵ_θ is the circumferential strain, and $\epsilon_{x\theta}$ is the in-plane shear strain. The out-of-plane shear strains are assumed to be zero. z is a coordinate from the midsurface of the shell, normal to the surface of the shell and directed outward from the cylinder axis. $z = 0$ is the midsurface of the shell. R is the radius of the shell midsurface. Other shell theories produce somewhat more complex strain-deformation relations for the cylindrical shell. For example, the Flügge theory (Ref. 12-4, pp. 208–212) gives:

$$\epsilon_x = \frac{\partial u}{\partial x} - z\frac{\partial^2 w}{\partial x^2},$$

$$\epsilon_\theta = \frac{1}{R}\frac{\partial v}{\partial \theta} - \frac{z}{R(R+z)}\frac{\partial^2 w}{\partial \theta^2} + \frac{w}{R+z},$$

$$\epsilon_{x\theta} = \frac{1}{R+z}\frac{\partial u}{\partial \theta} + \frac{R+z}{R}\frac{\partial v}{\partial x} - \frac{\partial^2 w}{\partial x \partial \theta}\left(\frac{z}{R} + \frac{z}{R+z}\right),$$

$$\epsilon_{xz} = \epsilon_{\theta z} = \epsilon_{zz} = 0. \qquad (12\text{-}2)$$

It can easily be seen that the Flügge relationships reduce to the Donnell equations (Eq. 12-1) if terms of order z are neglected in comparison with terms of order R.

The stresses in the shell are given by the following isotropic stress-strain relationships:

$$\sigma_{xx} = \frac{E}{1-\nu^2}(\epsilon_x + \nu\epsilon_\theta),$$

$$\sigma_{\theta\theta} = \frac{E}{1-\nu^2}(\epsilon_\theta + \nu\epsilon_x),$$

$$\sigma_{x\theta} = \sigma_{\theta x} = \frac{E}{2(1+\nu)}\epsilon_{x\theta},$$

$$\sigma_{xz} = \sigma_{\theta z} = \sigma_{zz} = 0. \qquad (12\text{-}3)$$

E is the modulus of elasticity and ν is Poisson's ratio. σ_{xx}, $\sigma_{\theta\theta}$, and σ_{zz} are normal stresses acting on the x, θ, and z faces of the shell element, respectively. σ_{xz}, $\sigma_{\theta z}$, and $\sigma_{x\theta}$ are shear stresses. The first index is the face on which the stress acts; the second index is the direction of the stress. Note that σ_{zz} and the out-of-plane shear stress have been neglected in the thin shell theory.

The stress resultants in the shell, evaluated according to the Donnell shell theory, are (Ref. 12-4, pp. 213–217):

$$N_x = \int_{-h/2}^{h/2} \sigma_{xx} \left(1 + \frac{z}{R}\right) dz = \frac{Eh}{1 - \nu^2} \left(\frac{\partial u}{\partial x} + \frac{\nu}{R} \frac{\partial v}{\partial \theta} + \nu \frac{w}{R}\right),$$

$$N_\theta = \int_{-h/2}^{h/2} \sigma_{\theta\theta}\, dz = \frac{Eh}{1 - \nu^2} \left(\frac{1}{R} \frac{\partial v}{\partial \theta} + \frac{w}{R} + \nu \frac{\partial u}{\partial x}\right),$$

$$N_{x\theta} = \int_{-h/2}^{h/2} \sigma_{x\theta} \left(1 + \frac{z}{R}\right) dz = \frac{Eh}{2(1 + \nu)} \left(\frac{\partial v}{\partial x} + \frac{1}{R} \frac{\partial u}{\partial \theta}\right),$$

$$N_{\theta x} = \int_{-h/2}^{h/2} \sigma_{\theta x}\, dz = N_{x\theta},$$

$$\mathcal{M}_x = -\int_{-h/2}^{h/2} \sigma_{xx} \left(1 + \frac{z}{R}\right) z dz = \frac{Eh^3}{12(1 - \nu^2)} \left(\frac{\partial^2 w}{\partial x^2} + \frac{\nu}{R^2} \frac{\partial^2 w}{\partial \theta^2}\right),$$

$$\mathcal{M}_\theta = -\int_{-h/2}^{h/2} \sigma_{\theta\theta}\, z dz = \frac{Eh^3}{12(1 - \nu^2)} \left(\frac{1}{R^2} \frac{\partial^2 w}{\partial \theta^2} + \nu \frac{\partial^2 w}{\partial x^2}\right),$$

$$\mathcal{M}_{x\theta} = -\int_{-h/2}^{h/2} \sigma_{x\theta} \left(1 + \frac{z}{R}\right) z dz = \frac{Eh^3}{12\,(1 + \nu)} \frac{1}{R} \frac{\partial^2 w}{\partial x \partial \theta},$$

$$\mathcal{M}_{\theta x} = -\int_{-h/2}^{h/2} \sigma_{\theta x}\, z dz = \mathcal{M}_{x\theta}. \qquad (12\text{-}4)$$

E is the modulus of elasticity, ν is Poisson's ratio, and h is the thickness of the cylinders. N_x, N_θ, and $N_{x\theta}$ are resultant forces per unit length of edge. $\mathcal{M}_x$, $\mathcal{M}_\theta$, and $\mathcal{M}_{x\theta}$ are resultant moments per unit length of edge. Terms of order z/R have been neglected in evaluation of the integrals. These terms are retained in the Flügge shell theory (Ref. 12-4, pp. 208–211).

The transverse shear forces per unit length of edge in the cylindrical shell are the resultant of the transverse shear stresses and can be related to the deformations of the shell by:

$$Q_x = -\int_{-h/2}^{h/2} \sigma_{xz} \left(1 + \frac{z}{R}\right) dz = \frac{Eh^3}{12(1 - \nu^2)} \frac{\partial}{\partial x} \left(\frac{1}{R^2} \frac{\partial^2 w}{\partial \theta^2} + \frac{\partial^2 w}{\partial x^2}\right),$$

$$Q_\theta = -\int_{-h/2}^{h/2} \sigma_{\theta z}\, dz = \frac{Eh^3}{12(1 - \nu^2)} \frac{1}{R} \frac{\partial}{\partial \theta} \left(\frac{1}{R^2} \frac{\partial^2 w}{\partial \theta^2} + \frac{\partial^2 w}{\partial x^2}\right).$$

Terms of order z/R have been neglected in evaluation of the integrals. Although the out-of-plane shear deformations ϵ_{xz}, $\epsilon_{\theta z}$ and the associated shear stresses σ_{xz}, $\sigma_{\theta z}$ have been neglected in the thin shell theory, equilibrium considerations dictate that the integral of these shear stresses over the thickness cannot be zero. The out-of-plane shear stresses will generally have a parabolic distribution over the shell thickness, maximum at the midsurface and zero at the edges.

12.2.2. Infinitely Long Cylindrical Shells

The equations describing the free vibrations of cylindrical shells admit solutions which are independent of the axial coordinate (x) and have the following form:

$$u = A \cos i\theta \cos \omega t,$$
$$v = B \sin i\theta \cos \omega t,$$
$$w = C \cos i\theta \cos \omega t, \qquad i = 0, 1, 2, \ldots \tag{12-5}$$

The midsurface deformations u, v, and w are defined in Fig. 12-1. A, B, and C are constants, with the units of length. i is the number of circumferential waves in the mode shape, ω is the circular frequency of vibration, and t is time. These solutions apply to infinitely long cylinders in the sense that the distance between axial nodes (~L/j, see Fig. 12-3) is much greater than the cylinder radius, so that boundary conditions at the ends of the cylinder do not influence the solution and the solutions do not vary along the length of the cylinder.

If the solution form of Eq. 12-5 is substituted into the Donnell-Mushtari equations of motion, the following matrix equation is generated (Ref. 12-9):

$$\begin{bmatrix} \frac{(1-\nu)}{2} i^2 - \lambda^2 & 0 & 0 \\ 0 & i^2 - \lambda^2 & i \\ 0 & i & 1 + ki^4 - \lambda^2 \end{bmatrix} \begin{bmatrix} A \\ B \\ C \end{bmatrix} = \begin{bmatrix} 0 \\ 0 \\ 0 \end{bmatrix}, \tag{12-6}$$

where

$$\lambda^2 = \frac{\mu(1-\nu^2) R^2 \omega^2}{E}, \tag{12-7}$$

$$k = \frac{h^2}{12R^2}. \tag{12-8}$$

μ is the density of the shell material, h is the wall thickness, and R is the radius of the cylinder midsurface. λ (Eq. 12-7) is the dimensionless parameter which specifies the circular natural frequencies (ω) of the cylindrical shell. The determinant of the matrix on the left-hand side of Eq. 12-6 is set to zero for non-trivial solutions to produce a sixth order polynomial in λ. The three real solutions of this polynomial define the following natural frequencies and mode shapes:

1. *Axial Modes (same result for both Donnell and Flügge shell theories)*

   $B = C = 0,$

$$\lambda^2 = \frac{(1-\nu)}{2} i^2; \quad i = 1, 2, 3, \ldots \tag{12-9}$$

2. *Coupled Radial-Circumferential Modes*
   a. *Donnell Shell Theory* (*Eq. 12-6*)

$$A = 0,$$

$$\frac{B}{C} = \frac{i}{\lambda^2 - i^2},$$

$$\lambda^2 = \tfrac{1}{2} \{(1 + i^2 + ki^4) \mp [(1 + i^2)^2 + 2ki^4(1 - i^2)]^{1/2}\};$$

$$i = 0, 1, 2, 3, \ldots \tag{12-10}$$

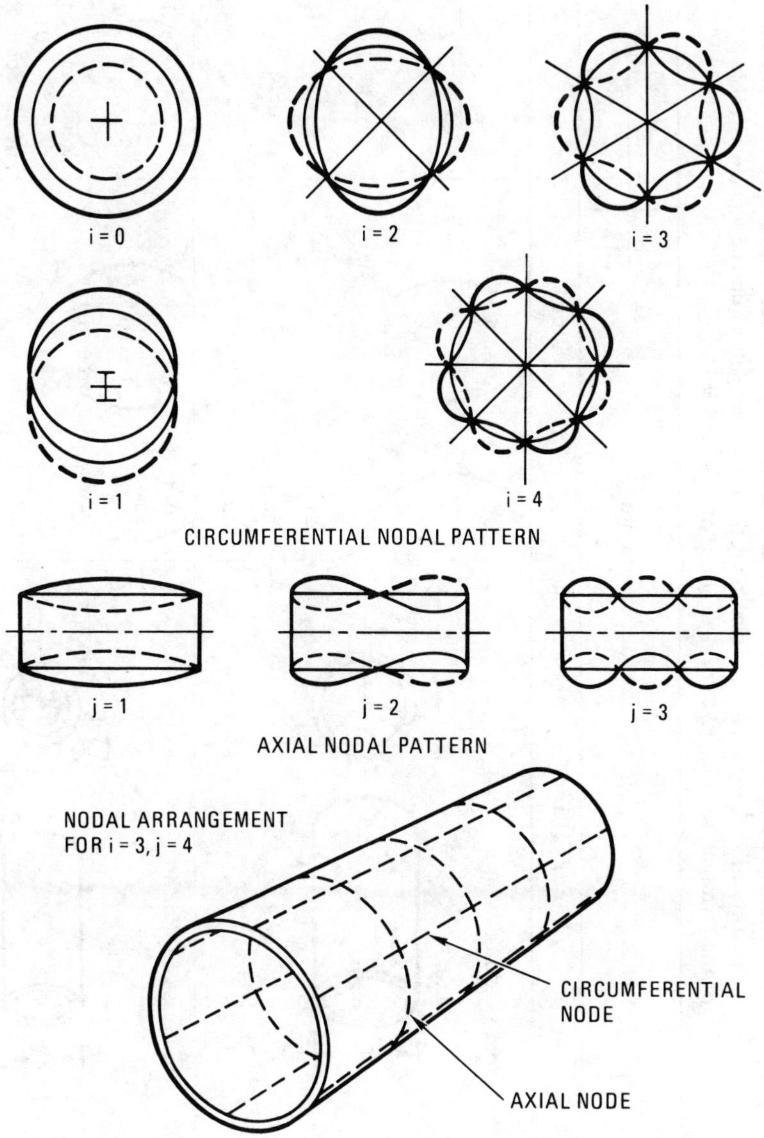

Fig. 12-3. Nodal patterns for a simply supported cylinder without axial constraint.

Table 12-1. Cylindrical Shells of Infinite Length.

Notation: h = cylinder thickness; i = number of circumferential waves in mode shape; j = number of axial half-waves in mode shape; u = deformation parallel to cylinder axis; v = circumferential deformation; w = radial deformation, positive outward from cylinder axis; $\tilde{u}$, $\tilde{v}$, $\tilde{w}$ = mode shapes associated with u, v, and w, respectively; E = modulus of elasticity; R = cylinder radius to midsurface; θ = angle (radians); μ = density of shell material; ν = Poisson's ratio; consistent sets of units are given in Table 3-1; see text for discussion of the derivation of the formulas; see Fig. 12-1 for definitions of u, v, and w.

Natural Frequency (hertz), $f_i = \frac{\lambda_i}{2\pi R}\left[\frac{E}{\mu(1-\nu^2)}\right]^{1/2}$

Description	Mode Shape	λ_i and Remarks
1. Axial Modes 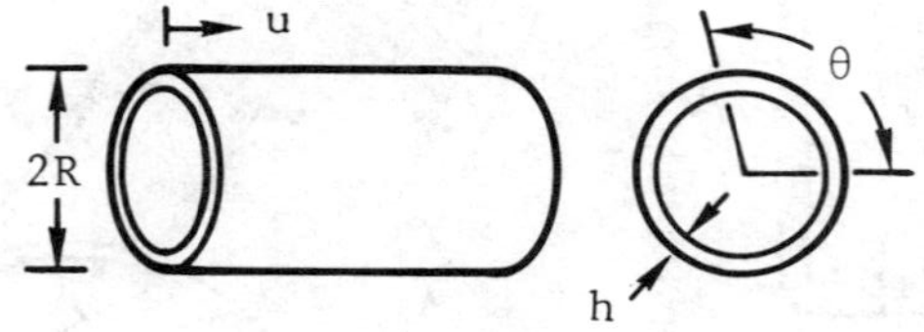	$\begin{pmatrix}\tilde{u}\\ \tilde{v}\\ \tilde{w}\end{pmatrix}_i = \begin{pmatrix}\cos i\theta\\ 0\\ 0\end{pmatrix}$ i=1,2,3...	$\frac{(1-\nu)^{1/2} i}{2^{1/2}}$; i=1,2,3... Shell theory result (Ref. 12-9)
2. Extension Modes 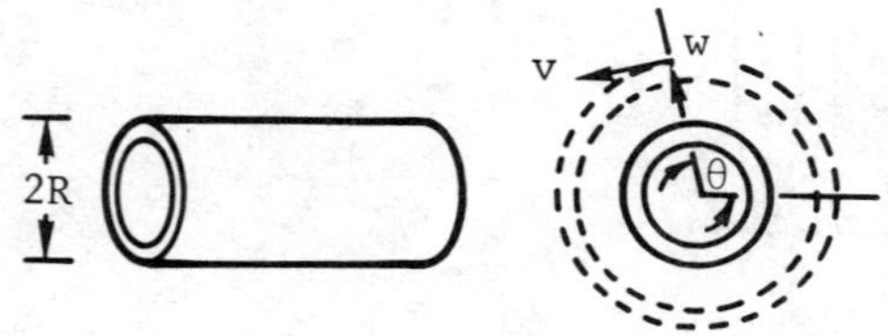 (i=0 Mode Shown)	$\begin{pmatrix}\tilde{u}\\ \tilde{v}\\ \tilde{w}\end{pmatrix}_i = \begin{pmatrix}0\\ i\sin i\theta\\ \cos i\theta\end{pmatrix}$	$\left(1+i^2\right)^{1/2}$; i=0,1,2,... Membrane theory result, exact only for i = 0. Bending stiffness of shell neglected for i = 1,2,3.... (Refs. 12-2, 12-9)

3. Radial-Circumferential Flexural Modes 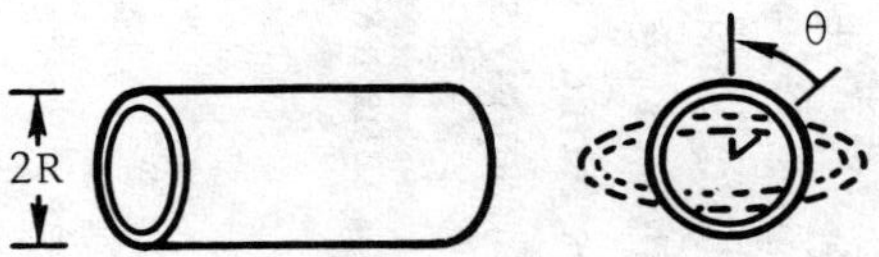 (i=2 Mode Shown)	$\begin{pmatrix} \tilde{u} \\ \tilde{v} \\ \tilde{w} \end{pmatrix}_i = \begin{pmatrix} 0 \\ -(1/i)\cos i\theta \\ \sin i\theta \end{pmatrix}$ i=2,3,4...	$\frac{1}{12^{1/2}} \frac{h}{R} \frac{i(i^2-1)}{\left(1+i^2\right)^{1/2}}$; i=2,3,4... Inextensional theory result (Ref. 12-12). Valid for $ih \ll R$.

b. *Flügge Shell Theory*

$$A = 0,$$

$$\lambda^2 = \tfrac{1}{2}\,\{(1 + i^2 + ki^4) \mp [(1 + i^2)^2 - 2ki^6]^{1/2}\};$$

$$i = 0, 1, 2, 3, \ldots \tag{12-11}$$

The axial mode solution is presented in frame 1 of Table 12-1. Two limiting forms of the radial-circumferential mode solutions are presented in frames 2 and 3 of Table 12-1. The first limiting form, frame 2, was obtained by neglecting the stiffness terms ($k = 0$) in Eq. 12-10 or Eq. 12-11. In this membrane solution it is assumed that the shell deforms by extension only; flexural deformations are neglected. This solution is exact for the $i = 0$ mode even if k is retained in the analysis. The second limiting form, frame 3, can be obtained by neglecting the circumferential extensional deformation of the cylinder (Ref. 12-12). This solution is applicable only to the higher modes, $i = 2, 3, 4, \ldots$. Terms containing k^2 (Eq. 12-8) have been neglected in Eqs. 12-10 and 12-11 since these terms are negligible for thin shells.

12.2.3. Simply Supported Cylindrical Shells Without Axial Constraint

The solution form used in Section 12.2.2 can be adapted for the solution of a simply supported cylindrical shell without axial constraints at its ends. The boundary conditions for this case are:

$$\text{At} \quad x = 0 \quad \text{and} \quad x = L,$$

$$v = w = 0,$$

$$\mathfrak{M}_x = N_x = 0, \tag{12-12}$$

where the deformations v and w and the stress resultants $\mathfrak{M}_x$ and N_x are defined by Eq. 12-4 and shown in Fig. 12-1. These boundary conditions can be approximated by attaching thin, flat, circular diaphragms to the ends of the cylinder. The diaphragms have considerable stiffness in their own plane, thereby restraining the v and w components of the shell displacement at their mutual boundaries. However, the diaphragms, as a result of their thinness, generate negligible bending moment, $\mathfrak{M}_x$, and longitudinal force, N_x, in the shell as the shell deforms and do not restrain axial deformation, u. Thus, the boundary conditions are sometimes termed shear diaphragm boundary conditions, and they are analogous to those of a simply supported beam without axial constraint.

The following solution form for the midsurface deformations satisfies both the equations of motion and the boundary conditions:

$$u = A \cos \frac{j\pi x}{L} \cos i\theta \cos \omega t,$$

$$v = B \sin \frac{j\pi x}{L} \sin i\theta \cos \omega t,$$

$$w = C \sin \frac{j\pi x}{L} \cos i\theta \cos \omega t, \qquad i = 1, 2, 3, \ldots; \quad j = 1, 2, 3, \ldots \tag{12-13}$$

A, B, and C are constants with the units of length which describe the amplitude of the axial (u), circumferential (v), and radial (w) deformations of the shell. i is the number of circumferential waves in the mode shape, i = 1, 2, 3, and j is the number of longitudinal half-waves in the mode shape (Fig. 12-3). If this solution form is substituted into the Donnell shell theory equation of motion, the following matrix equation is produced (Ref. 12-9):

$$\begin{bmatrix} a_{11} & a_{12} & a_{13} \\ a_{21} & a_{22} & a_{23} \\ a_{31} & a_{32} & a_{33} \end{bmatrix} \begin{bmatrix} A \\ B \\ C \end{bmatrix} = 0, \tag{12-14}$$

where

$$a_{11} = -\xi^2 - \frac{(1-\nu)}{2} i^2 + \lambda^2, \qquad a_{22} = -\frac{(1-\nu)}{2} \xi^2 - i^2 + \lambda^2,$$

$$a_{12} = \frac{(1+\nu)}{2} \xi i, \qquad a_{23} = -i,$$

$$a_{31} = -\nu\xi,$$

$$a_{13} = \nu\xi, \qquad a_{32} = i,$$

$$a_{21} = a_{12}, \qquad a_{33} = 1 + k(\xi^2 + i^2)^2 - \lambda^2,$$

where

$$\lambda^2 = \frac{\mu(1-\nu^2) R^2 \omega^2}{E},$$

$$k = \frac{h^2}{12R^2},$$

$$\xi = \frac{j\pi R}{L}, \quad j = 1, 2, 3, \ldots. \tag{12-15}$$

μ is the density of the cylinder material, ν is the Poisson's ratio, L is the cylinder length, h is the wall thickness, R is the radius of the midsurface, ω is the circular frequency of vibration, which is a function of the dimensionless parameter λ, and E is the modulus of elasticity.

If the determinant of the matrix on the left-hand side of Eq. 12-14 is set equal to zero to generate non-trivial solutions and k (Eq. 12-15) is assumed to be small for thin shells, the result is a cubic characteristic polynomial in λ^2 for the natural frequencies of vibration:

$$\lambda^6 - K_2 \lambda^4 + K_1 \lambda^2 - K_0 = 0, \tag{12-16}$$

where the Donnell shell theory gives (Ref. 12-9):

$$K_2 = 1 + \tfrac{1}{2}(3-\nu)(i^2 + \xi^2) + k(i^2 + \xi^2)^2,$$

$$K_1 = \tfrac{1}{2}(1-\nu)\left[(3+2\nu)\xi^2 + i^2 + (i^2+\xi^2)^2 + \frac{3-\nu}{1-\nu}k(i^2+\xi^2)^3\right],$$

$$K_0 = \tfrac{1}{2}(1-\nu)\left[(1-\nu)^2\xi^4 + k(i^2+\xi^2)^4\right]. \qquad (12\text{-}17)$$

For small k, the Flügge shell theory gives a result of the same form as Eq. 12-16 with K_1 and K_2 as defined by Eqs. 12-17 but with the following K_0 (Ref. 12-9):

$$K_0 = \frac{1}{2}(1-\nu)\left[(1-\nu^2)\xi^4 + k(i^2+\xi^2)^4\right]$$
$$+\frac{k}{2}(1-\nu)\left[2(2-\nu)\xi^2 i^2 + i^4 - 2\nu\xi^6 - 6\xi^4 i^2 - 2(4-\nu)\xi^2 i^4 - 2i^6\right]. \qquad (12\text{-}18)$$

k and ξ are defined by Eq. 12-15.

The natural frequencies of a simply supported cylindrical shell are found by solving Eq. 12-16 for λ and then determining the mode shape of the solution from Eq. 12-14. Thus,

$$\frac{A}{C} = \frac{a_{32}a_{23} - a_{33}a_{22}}{a_{22}a_{31} - a_{32}a_{21}},$$

$$\frac{B}{C} = \frac{a_{33}a_{21} - a_{31}a_{23}}{a_{22}a_{31} - a_{32}a_{21}}, \qquad (12\text{-}19)$$

where a_{ij} are defined by Eq. 12-15 and A, B, and C are defined by Eq. 12-13. This procedure results in three natural frequencies and mode shapes for each i and j. Only the lowest of these three solutions is ordinarily of practical importance for $i = 1, 2, 3, \ldots$. Equation 12-16 does not possess closed form solutions except for the case $i = 0$; thus, numerical or approximate techniques must generally be employed for the higher modes.

For the case $i = 0$, which corresponds to modes independent of the circumferential angle θ, the equation of motion for the simply supported cylindrical shell without axial constraint (Eq. 12-16) has the following solutions:

1. *Torsional Mode (both Donnell and Flügge shell theories)*

$$A = C = 0,$$
$$\lambda^2 = \tfrac{1}{2}(1-\nu)\xi^2. \qquad (12\text{-}20)$$

2. *Coupled Axial-Radial Modes*

   a. *Donnell Shell Theory*

$$\frac{A}{C} = \frac{\nu\xi}{\xi^2 - \lambda^2},$$
$$B = 0,$$
$$\lambda^2 = \tfrac{1}{2}\left\{(1+\xi^2+k\xi^4) \mp \left[(1-\xi^2)^2 + 2\xi^2(2\nu^2 + k\xi^2 - k\xi^4)\right]^{1/2}\right\}. \qquad (12\text{-}21)$$

Table 12-2. Simply Supported Cylindrical Shells Without Axial Constraint.

Notation: h = cylinder thickness; i = number of circumferential waves in mode shape; j = number of axial half-waves in mode shape; u = deformation parallel to cylinder axis; v = circumferential deformation; w = radial deformation, positive outward from cylinder axis; $\tilde{u}$, $\tilde{v}$, $\tilde{w}$ = mode shapes associated with u, v, and w, respectively; E = modulus of elasticity; L = length of cylinder; R = cylinder radius to midsurface; θ = angle (radians); μ = density of cylinder material; ν = Poisson's ratio; consistent sets of units are given in Table 3-1; see text for discussion of the derivation of formulas and boundary conditions; see Fig. 12-1 for definition of u, v, and w.

Natural Frequency (hertz), $f_{ij} = \frac{\lambda_{ij}}{2\pi R}\left[\frac{E}{\mu(1-\nu^2)}\right]^{1/2}$

Description	Mode Shape	λ_{ij}
1. Torsion Modes 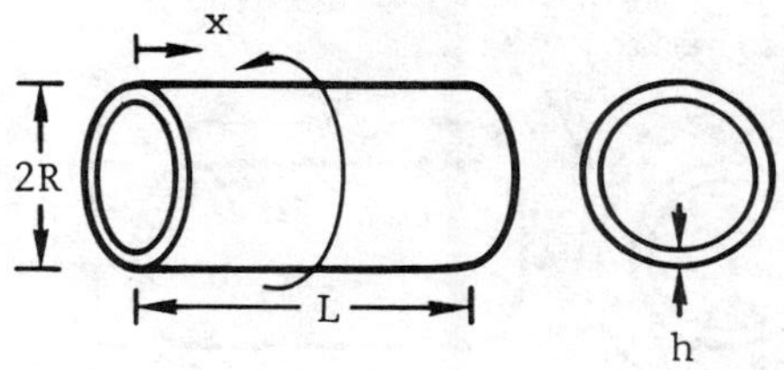	$\begin{pmatrix}\tilde{u}\\ \tilde{v}\\ \tilde{w}\end{pmatrix}_{ij} = \begin{pmatrix}0\\ \sin\frac{j\pi x}{L}\\ 0\end{pmatrix}$ i = 0 j = 1,2,3...	$\frac{(1-\nu)^{1/2}}{2^{1/2}}\frac{j\pi R}{L}$ i = 0 j = 1,2,3... Donnell shell theory result.
2. Axial Modes 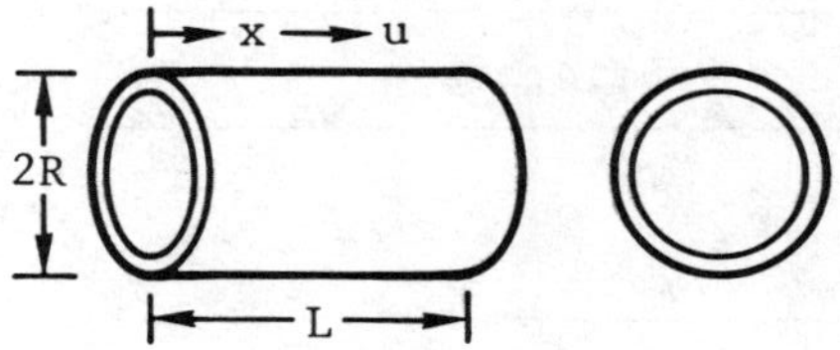	$\begin{pmatrix}\tilde{u}\\ \tilde{v}\\ \tilde{w}\end{pmatrix}_{ij} = \begin{pmatrix}\cos\frac{j\pi x}{L}\\ 0\\ 0\end{pmatrix}$ i = 0 j = 1,2,3...	For long cylinders, L/(jR) > 8, $\lambda_{ij} = j\pi(1-\nu^2)^{1/2}\frac{R}{L}$ i = 0 j = 1,2,3... Beam theory result (see Chapter 8 for longitudinal vibration of free-free beam). See text for shell theory for L ≃ jR.

Table 12-2. Simply Supported Cylindrical Shells Without Axial Constraint. *(Continued)*

Natural Frequency (hertz), $f_{ij} = \frac{\lambda_{ij}}{2\pi R}\left[\frac{E}{\mu(1-\nu^2)}\right]^{1/2}$

Description	Mode Shape	λ_{ij}
3. Radial Mode 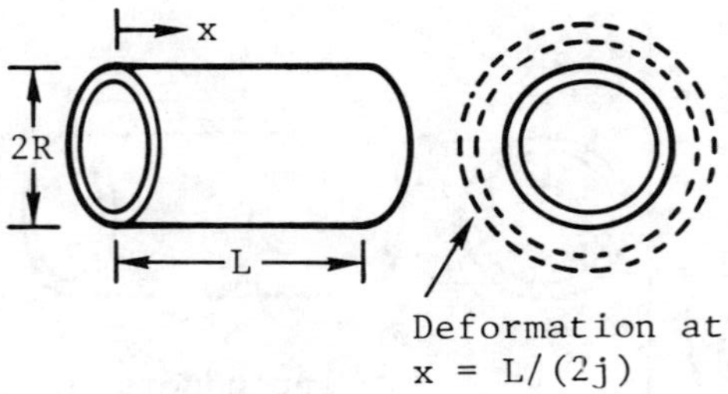	$\begin{pmatrix}\tilde{u}\\ \tilde{v}\\ \tilde{w}\end{pmatrix}_{ij} = \begin{pmatrix}0\\ 0\\ \sin\frac{j\pi x}{L}\end{pmatrix}$ $i = 0$ $j = 1,2,3\ldots$	For long cylinders, $L/(jR) > 8$, $\lambda_{ij} = 1$ $i = 0$ $j = 1,2,3\ldots$ λ_{ij} is independent of j for $L >> jR$. See text for shell theory for $L \approx jR$.
4. Bending Modes 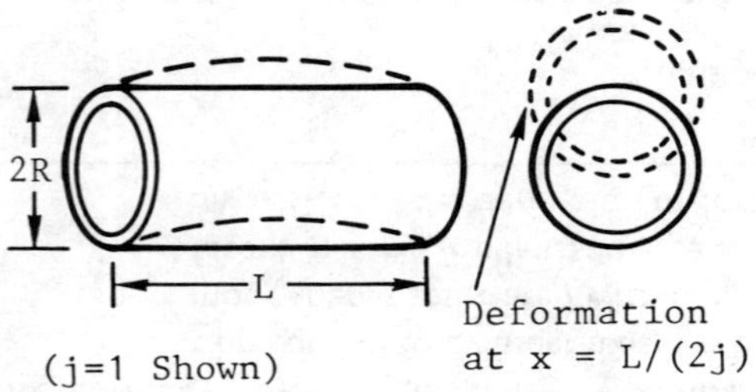 (j=1 Shown)	For long cylinders, $L >> jR$; cylinder axis deforms as $\sin j\pi x/L$. See Chapter 8 for transverse deflection of simply supported beams. See text for shell theory.	For long cylinders, $L/(jR) > 8$, $\frac{j^2\pi^2(1-\nu^2)^{1/2}}{2^{1/2}}\frac{R^2}{L^2}$ $i = 1$ $j = 1,2,3\ldots$ Beam theory result (Chapter 8). See text for shell theory for $L \approx jR$.

5. Radial-Axial Modes 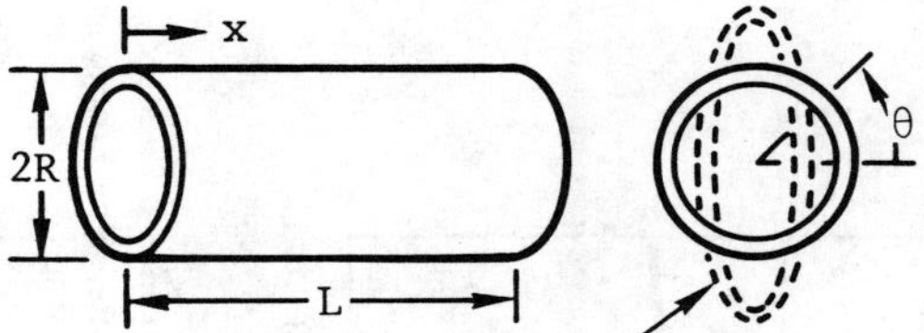 Deformation at x = L/(2j) for i=2	$\tilde{u} = A \cos i\theta \cos \frac{j\pi x}{L}$ $\tilde{v} = B \sin i\theta \sin \frac{j\pi x}{L}$ $\tilde{w} = C \cos i\theta \sin \frac{j\pi x}{L}$ i = 2,3,4... j = 1,2,3... See text for A, B, C.	$\dfrac{\left\{(1-\nu^2)\left(\frac{j\pi R}{L}\right)^4 + \left(\frac{h^2}{12R^2}\right)\left[i^2 + \left(\frac{j\pi R}{L}\right)^2\right]^4\right\}^{1/2}}{\left(\frac{j\pi R}{L}\right)^2 + i^2}$ i = 2,3,4... j = 1,2,3... Inertia associated with u and v is neglected. Note that i=2, j=1 is not generally the lowest frequency mode of these modes. This Donnell shell theory result is not generally as accurate as the Flügge theory (see text).

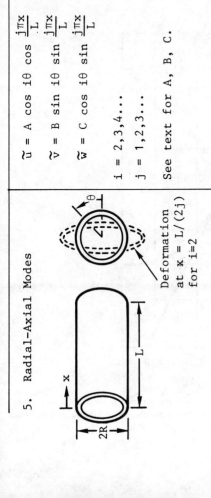

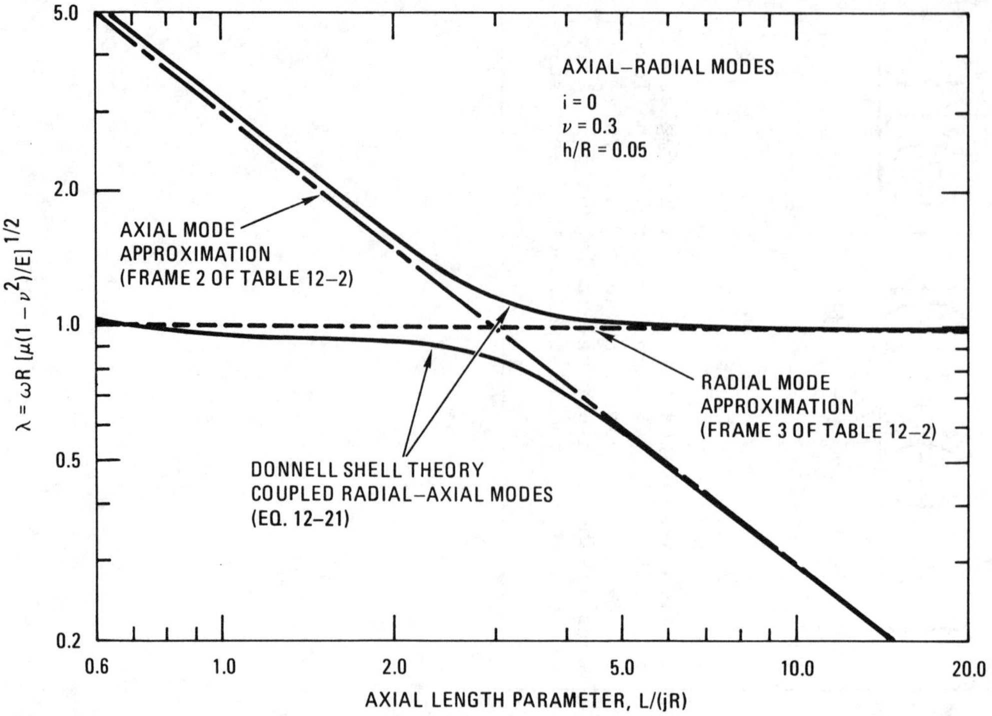

Fig. 12-4. Natural frequencies of axial-radial modes (i = 0) for a simply supported cylindrical shell without axial constraint.

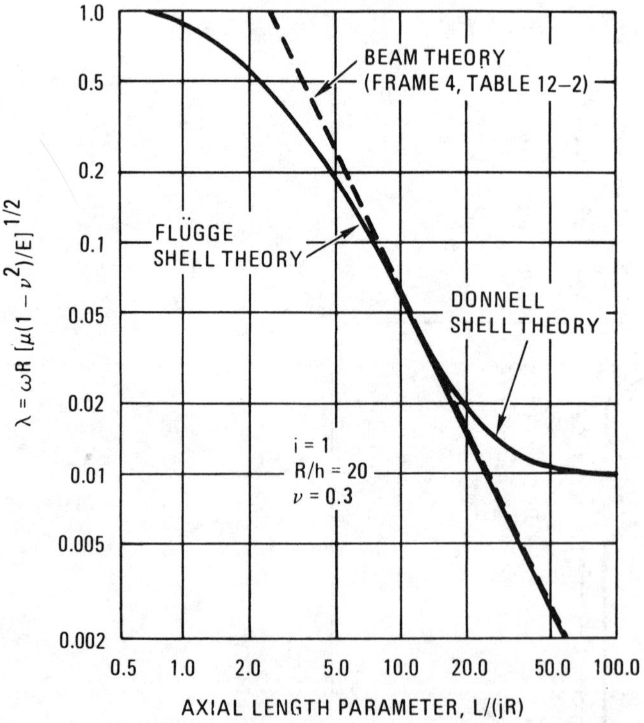

Fig. 12-5. Comparison of beam theory with Flügge and Donnell shell theories for lowest-frequency parameter for i = 1. Flügge shell theory is nearly identical to the exact result in this example. The Flügge and Donnell theories give the same result for $0.5 < L/(jR) < 10$. (Adapted in part from Ref. 12-9.)

b. *Flügge Shell Theory*

$$B = 0,$$

$$\lambda^2 = \frac{1}{2}\{(1+\xi^2+k\xi^4) \mp [(1-\xi^2)^2 + 4\nu^2\xi^2 - 2k\xi^6]^{1/2}\}. \tag{12-22}$$

k and ξ are defined by Eq. 12-15. The torsional mode is presented in frame 1 of Table 12-2. The coupled axial-radial modes given by Eqs. 12-21 and 12-22 decouple into an axial mode and a radial mode for long cylinders, $L/(jR) \gg 1$. These uncoupled modes are presented in frames 2 and 3 of Table 12-2. In Fig. 12-4, the coupled modes are compared with the uncoupled modes.

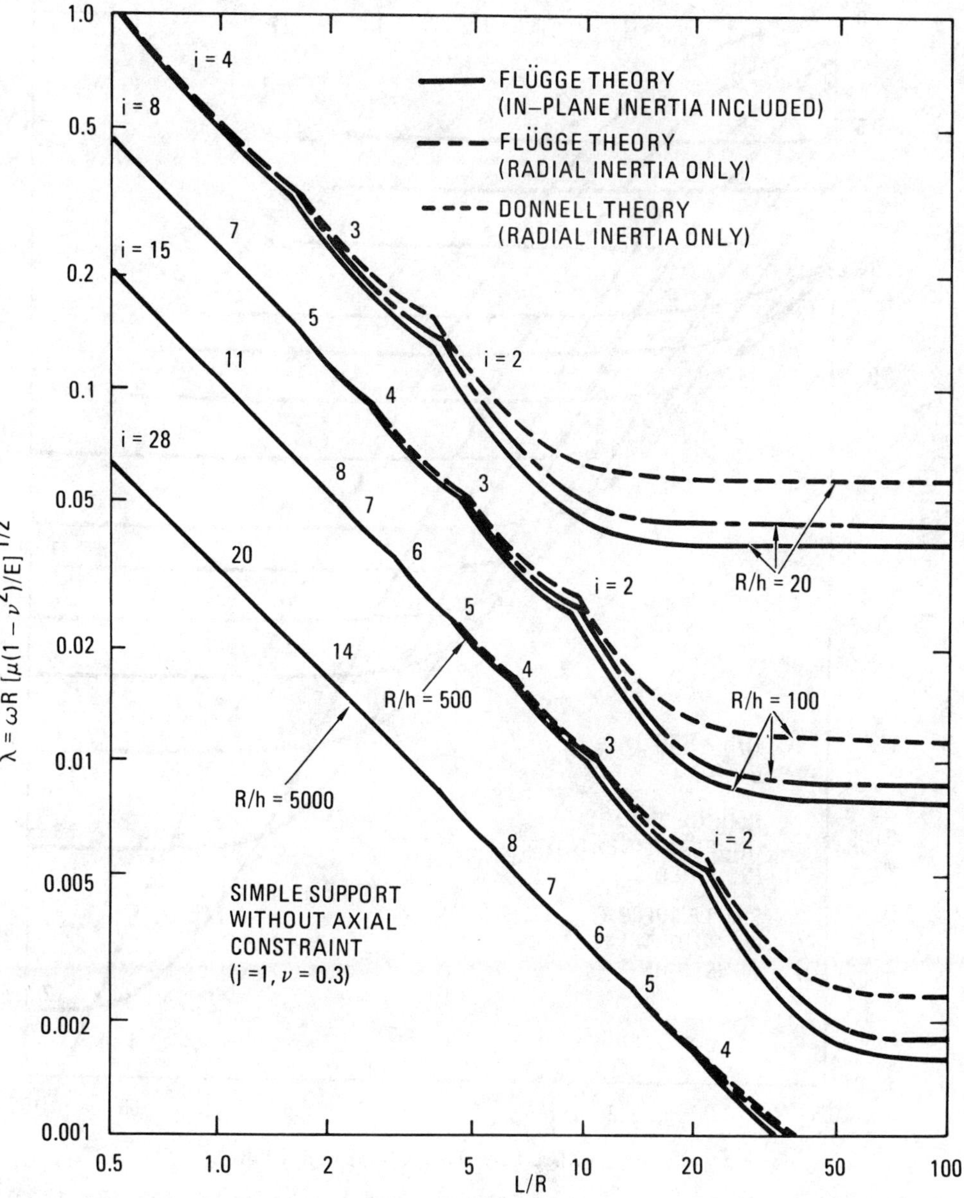

Fig. 12-6. Frequency envelope for a cylindrical shell, comparing results of Flügge and Donnell theories (with and without in-plane inertia) for simple support without axial constraint. (Ref. 12-13).

The deformation of a slender $[L/(jR) > 8]$ cylindrical shell in the $i = 1$ mode is the same as the deformation of a slender pinned-pinned beam. In Fig. 12-5, the beam theory approximation to the $i = 1$ shell mode is compared with two shell theories. Note that for $L/(jR) > 10$, the beam theory provides a better approximation to the exact result (nearly identical to the Flügge shell theory) than does the Donnell theory.

For higher circumferential modes, $i = 2, 3, 4, \ldots$, the deformation of the shell is dominated by radial motion as shown in Fig. 12-3. For these modes it is reasonable

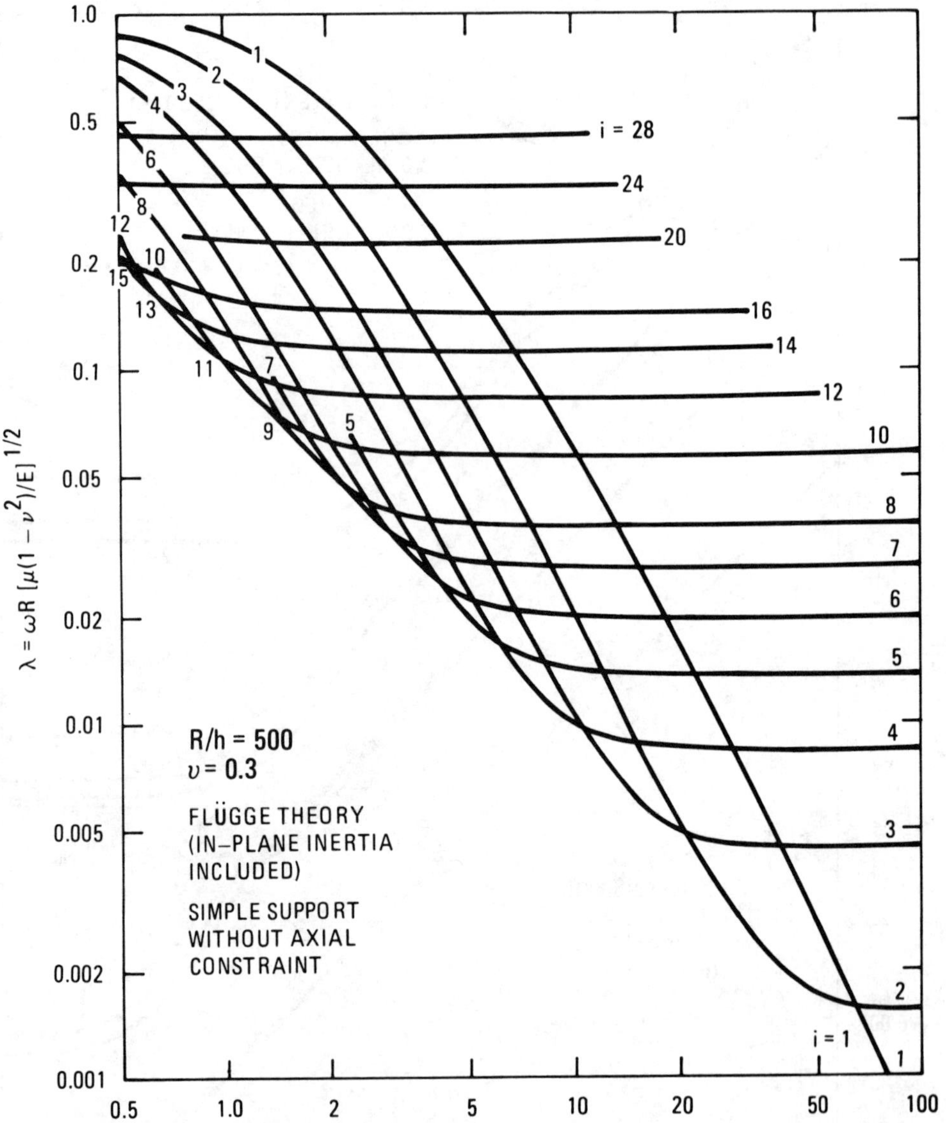

Fig. 12-7. Portion of frequency spectrum for a cylindrical shell. λ is given by Eq. 12-15. (Ref. 12-13)

to retain only the inertia term associated with radial deformation (w) in the equation of motion and neglect the inertia terms associated with the in-plane (u and v) deformations. If the in-plane inertia terms are neglected by deleting λ^2 in the terms a_{11} and a_{22} in Eq. 12-14, the result is a simple closed form expression for the natural frequencies of the cylinder:

$$\lambda^2 = \frac{2K_0}{(1 - \nu)(\xi^2 + i^2)^2}; \quad i = 2, 3, 4, \ldots, \tag{12-23}$$

where ξ is given by Eq. 12-15 and K_0 is given by Eq. 12-17 for the Donnell theory and by Eq. 12-18 for the Flügge theory. Equation 12-23 is presented in frame 5 of Table 12-2 for the Donnell theory. Leissa's results (Ref. 12-9) suggest that using Eq. 12-23 instead of numerical solutions for Eq. 12-16 results in an overprediction of the natural frequencies. The degree of the overprediction decreases with increasing i. For $i = 2$, the error incurred by using Eq. 12-23 instead of Eq. 12-16 can be as high as 13% for both the Donnell and Flügge shell theories, but the error drops to 6% for $i = 3$ and to 3% for $i = 4$. For long cylinders, $L/(jR) > 20$, the Donnell shell theory tends to overpredict the natural frequencies; the Flügge shell theory gives a more accurate result for long shells. This can be seen in Fig. 12-6, which gives the envelope of the lowest-frequency modes for $i = 2, 3, 4, \ldots$ with and without consideration of the in-plane inertia terms. Note that the lowest-frequency mode does not generally correspond to $i = 2$, as shown in Fig. 12-7.

12.2.4. Other Boundary Conditions

Solution Types. Cylindrical shells that have boundary conditions independent of θ will possess vibration modes analogous to the modes of a simply supported cylindrical shell without axial constraint (Table 12-2). The beam-like modes (the $i = 0$ torsion mode, the $i = 0$ axial mode, and the $i = 1$ transverse bending mode) of long shells, $L/(jR) > 8$, can be analyzed using the classical beam theory given in Chapter 8. Beam theory may also be used for shorter shells if the effects of rotary inertia and shear deformation are taken into account. Certain other modes of long shells can be described independently of the boundary conditions on the ends of the shell (Table 12-1). However, most of the higher modes of cylindrical shells involving circumferential deformation can be described only by shell theory. Approximate formulas have been developed to describe these shell modes.

A number of approximate formulas for the natural frequencies and mode shapes of cylindrical shells with various boundary conditions are presented in the following paragraphs. The accuracy of these approximate formulas generally increases with the circumferential mode number, i, and the effective shell length, $L/(jR)$.

General Solution Form. A general form for the deformations of a freely vibrating cylindrical shell (Fig. 12-1) is:

$$u = A\tilde{\phi}'(x/L) \cos i\theta \cos \omega t,$$

$$v = B\tilde{\phi}_j(x/L) \sin i\theta \cos \omega t,$$

$$w = C\tilde{\phi}_j(x/L)\cos i\theta \cos \omega t,$$
$$i = 0, 1, 2, 3, \ldots$$
$$j = 1, 2, 3, \ldots \qquad (12\text{-}24)$$

A, B, and C are constants which describe the amplitude of the axial (u), tangential (v), and radial (w) deformations of the shell midsurface. ω is the circular natural frequency of vibration $2\pi f$. i is the number of circumferential waves in the mode shape and j is the number of longitudinal half-waves in the mode shape. The deformations and the coordinate system are shown in Fig. 12-1. The axial modal functions $\tilde{\phi}(x/L)$ are chosen to satisfy the desired boundary conditions at $x = 0$ and $x = L$; the beam modal functions of Table 8-1 are generally employed. For a simply supported cylinder without axial constraints, the mode shape of pinned-pinned beam $\tilde{\phi}_j = \sin j\pi x/L$ would be used. Substituting this mode shape into Eq. 12-24 yields the exact solution form for these boundary conditions (Eq. 12-13).

If Eq. 12-24 is used in conjunction with the Flügge shell theory and the Rayleigh-Ritz technique, the following characteristic equation is produced for the natural frequencies and mode shapes of a cylindrical shell (Ref. 12-14):

$$\begin{bmatrix} a_{11} - \alpha_2\lambda^2 & a_{12} & a_{13} \\ a_{12} & a_{22} - \lambda^2 & a_{23} \\ a_{13} & a_{23} & a_{33} - \lambda^2 \end{bmatrix} \begin{Bmatrix} A \\ B \\ C \end{Bmatrix} = 0. \qquad (12\text{-}25)$$

Setting the determinant of the characteristic matrix to zero for non-trivial solutions results in the following characteristic polynomial and mode shapes:

$$\begin{aligned} &\alpha_2\lambda^6 - \lambda^4(a_{11} + \alpha_2 a_{22} + \alpha_2 a_{33}) \\ &\quad - \lambda^2(a_{12}^2 + a_{13}^2 + \alpha_2 a_{23}^2 - \alpha_2 a_{22} a_{33} - a_{11} a_{33} - a_{11} a_{22}) \\ &\quad + a_{12}^2 a_{33} + a_{23}^2 a_{11} + a_{13}^2 a_{22} - a_{11} a_{22} a_{33} - 2 a_{12} a_{23} a_{13} = 0, \end{aligned} \qquad (12\text{-}26)$$

$$\frac{A}{C} = \frac{(a_{33} - \lambda^2)(a_{22} - \lambda^2) - a_{23}^2}{a_{12} a_{23} - a_{22} a_{13}},$$

$$\frac{B}{C} = \frac{a_{12}(a_{33} - \lambda^2) - a_{13} a_{23}}{a_{22} a_{13} - a_{23} a_{12}}, \qquad (12\text{-}27)$$

where

$$\begin{aligned} \lambda &= \omega R[\mu(1 - \nu^2)/E]^{1/2}, \\ a_{11} &= \beta_j^2 + \tfrac{1}{2}(1 + k)(1 - \nu) i^2 \alpha_2, \\ a_{12} &= -\nu i \beta_j \alpha_1 - \tfrac{1}{2}(1 - \nu) i \beta_j \alpha_2, \\ a_{13} &= -\nu \beta_j \alpha_1 + k\beta_j[-\beta_j^2 + \tfrac{1}{2}(1 - \nu) i^2 \alpha_2], \\ a_{22} &= i^2 + \tfrac{1}{2}(1 + 3k)(1 - \nu)\beta_j^2 \alpha_2, \\ a_{23} &= i + k i \beta_j^2[\nu\alpha_1 + \tfrac{3}{2}(1 - \nu)\alpha_2], \\ a_{33} &= 1 + k[\beta_j^4 + (i^2 - 1)^2 + 2\nu i^2 \beta_j^2 \alpha_1 + 2(1 - \nu) i^2 \beta_j^2 \alpha_2], \end{aligned}$$

$$\beta_j = \underline{\lambda}_j R/L,$$

$$k = \frac{h^2}{12R^2}.$$

R is the radius of the cylinder to the midsurface. h is the thickness of the shell, and ω is the circular natural frequency which is specified by the dimensionless parameter λ. L is the axial length of the cylindrical shell. $\underline{\lambda}_j$ are the dimensionless beam frequency parameters given in the second column of Table 8-1, which correspond to the desired axial mode shape, $\tilde{\phi}_j$.

α_1 and α_2 are integrals of the mode shape:

$$\alpha_1 = -\frac{1}{L}\int_0^L \tilde{\phi}_j''(x)\,\tilde{\phi}_j(x)\,dx, \qquad (12\text{-}28)$$

Table 12-3. Formulas for Parameters α_1 and α_2 (Eqs. 12-28 and 12-29) Using Beam Mode Shapes of Table 8-1.

Boundary Conditions	Formula for α_1 [a]	Formula for α_2 [a]
1. Free-free	$\frac{\sigma_j}{\underline{\lambda}_j}(\sigma_j\underline{\lambda}_j - 2)$	$\frac{\sigma_j}{\underline{\lambda}_j}(\sigma_j\underline{\lambda}_j + 6)$
2. Free-pinned	$\frac{\sigma_j}{\underline{\lambda}_j}(\sigma_j\underline{\lambda}_j - 1)$	$\frac{\sigma_j}{\underline{\lambda}_j}(\sigma_j\underline{\lambda}_j + 3)$
3. Clamped-free	$\frac{\sigma_j}{\underline{\lambda}_j}(\sigma_j\underline{\lambda}_j - 2)$	$\frac{\sigma_j}{\underline{\lambda}_j}(\sigma_j\underline{\lambda}_j + 2)$
4. Clamped-pinned	$\frac{\sigma_j}{\underline{\lambda}_j}(\sigma_j\underline{\lambda}_j - 1)$	$\frac{\sigma_j}{\underline{\lambda}_j}(\sigma_j\underline{\lambda}_j - 1)$
5. Clamped-clamped	$\frac{\sigma_j}{\underline{\lambda}_j}(\sigma_j\underline{\lambda}_j - 2)$	$\frac{\sigma_j}{\underline{\lambda}_j}(\sigma_j\underline{\lambda}_j - 2)$
6. Pinned-pinned, sliding-sliding, sliding-pinned	1/2	1/2

[a] $\underline{\lambda}_j$ and σ_j are given in Table 8-1 in columns 2 and 4, respectively. $\underline{\lambda}_j = \lambda_j$ of Table 8-1.

$$\alpha_2 = \frac{1}{L} \int_0^L [\tilde{\phi}_j'(x)]^2 \, dx. \tag{12-29}$$

The primes (′) in Eqs. 12-28, 12-29, and 12-24 denote differentiation with respect to the dimensionless spanwise parameter $\underline{\lambda}_j$ x/L (see Section 8.1.2). Using the beam mode shapes of Table 8-1 for the axial mode shape $\tilde{\phi}$, α_1 and α_2 can be evaluated as shown in Table 12-3. Numerical values for α_1 and α_2 are given in Table 12-4. Note that in Table 12-3 the parameter $\underline{\lambda}_j$ is identical to the λ_i of Table 8-1, the underbar being added to avoid confusion between the beam and shell frequency parameters.

The natural frequencies and mode shapes of the cylindrical shell can be found by choosing a beam mode shape which satisfies the desired boundary conditions at x = 0 and x = L, solving Eq. 12-26 numerically for λ, and solving Eq. 12-27 for the ratios between A, B, and C. The result will be three natural frequencies and mode shapes for each i and j, but only the lowest of the three frequencies is ordinarily of practical importance.

Simplified General Solution. Sharma and Johns (Refs. 12-14 and 12-15) found that Eq. 12-26 could be considerably simplified if the circumferential strain ϵ_θ and the shear strain $\epsilon_{x\theta}$ of the midsurface of the shell (i.e., at z = 0) were neglected. These assumptions lead to the following midsurface deformations,

$$u = -\frac{A}{i^2} R \frac{d\tilde{\phi}_j}{dx} \cos i\theta \cos \omega t,$$

Table 12-4. Parameters α_1 and α_2 (Eqs. 12-28, 12-29, and Table 12-3) for Beam Modes of Table 8-1.

Mode	α_1 / α_2	j = 1	2	3	4	5
1. Free-free	α_1	0.5499	0.7467	0.8180	0.8585	0.8843
	α_2	2.2116	1.7662	1.5456	1.4244	1.3473
2. Free-pinned	α_1	0.7467	0.8585	0.9021	0.9251	0.9394
	α_2	1.7662	1.4244	1.2938	1.2247	1.1819
3. Clamped-free	α_1	-0.2441	0.6033	0.7440	0.8182	0.8585
	α_2	1.3219	1.4712	1.2529	1.1820	1.1415
4. Clamped-pinned	α_1	0.7467	0.8585	0.9021	0.9251	0.9394
	α_2	0.7467	0.8585	0.9021	0.9251	0.9394
5. Clamped-clamped	α_1	0.5499	0.7467	0.8180	0.8585	0.8843
	α_2	0.5499	0.7467	0.8180	0.8585	0.8843
6. Pinned-pinned, sliding-sliding, sliding-pinned	α_1	0.5	0.5	0.5	0.5	0.5
	α_2	0.5	0.5	0.5	0.5	0.5

$$v = -\frac{A}{i}\tilde{\phi}_j \sin i\theta \cos \omega t,$$

$$w = A\tilde{\phi}_j \cos i\theta \cos \omega t,$$

$$i = 2, 3, 4, \ldots = \text{number of circumferential waves},$$

$$j = 1, 2, 3, \ldots = \text{number of axial half-waves}, \tag{12-30}$$

and the higher circumferential mode natural frequencies in hertz,

$$f_{ij} = \frac{\lambda_{ij}}{2\pi R}\left[\frac{E}{\mu(1-\nu^2)}\right]^{1/2}, \qquad i = 2, 3, 4, \ldots, \quad j = 1, 2, 3, \ldots, \tag{12-31}$$

where E is the modulus of elasticity, R is the radius of the shell to the midsurface, μ is the density of the shell material, ν is Poisson's ratio, and A is a constant. The dimensionless frequency parameter λ_{ij} is given by (Ref. 12-15):

$$\lambda_{ij}^2 = \frac{\beta_j^4 + ki^2\beta_j^2[\beta_j^2 i^2 + 2\nu i^2(i^2-1)\alpha_1 + 2(1-\nu)(i^2-1)^2\alpha_2] + ki^4(i^2-1)^2}{\beta_j^2\alpha_2 + i^2(i^2+1)}, \tag{12-32}$$

where

$$\beta_j = \underline{\lambda}_j R/L,$$
$$k = h^2/(12R^2),$$

and α_1 and α_2 are given by Eqs. 12-28 and 12-29, respectively, or Tables 12-3 and 12-4. $\underline{\lambda}_j$ is the natural frequency parameter given in Table 8-1 for the desired beam mode shape. For example, the clamped-free beam modes and $\underline{\lambda}_1 = 1.8751$ would be used for a clamped-free cylindrical shell with axial constraint in the mode j = 1.

Since the circumferential and in-plane shear strains of the midplane have been neglected, Eqs. 12-30 through 12-32 can be applied successfully only to the higher circumferential modes (i = 2, 3, 4, . . .) of shells. These equations can be expected to yield an accurate result in higher circumferential modes for long shells, $L/(jR) > 8$, since the vibrations are dominated by bending rather than stretching deformation.

For a cylinder of infinite length, $L/(jR) \gg 10$, Eq. 12-32 reduces to the exact inextensional result given by frame 3 of Table 12-1. Equation 12-32 is compared with the Donnell and Flügge shell theories, neglecting in-plane inertia, in Fig. 12-8. The Flügge theory curve is the most accurate of the curves in Fig. 12-8. Note that Eq. 12-32 is superior to the Donnell theory for long cylinders, $L/(jR) > 8$, but both the Donnell and Flügge theories are superior to Eq. 12-32 for short cylinders.

Sharma (Ref. 12-16) has developed a second approximate expression for the natural frequency parameter λ_{ij} (Eq. 12-31) which generally provides a better (i.e., lower) estimate of the natural frequencies than Eq. 12-32 but still does not require numerical solution of a cubic equation as does Eq. 12-26. The expression for λ_{ij} is:

$$\lambda_{ij}^2 = \frac{(a_{11}a_{22}a_{33} + 2a_{12}a_{13}a_{23} - a_{11}a_{23}^2 - a_{22}a_{13}^2 - a_{33}a_{12}^2)\,i^4}{(a_{11}a_{22} - a_{12}^2)(\beta_j^2\alpha_2 + i^4 + i^2)},$$

$$i = 2, 3, 4, \ldots = \text{number of circumferential waves},$$

$$j = 1, 2, 3, \ldots = \text{number of axial half-waves}. \tag{12-33}$$

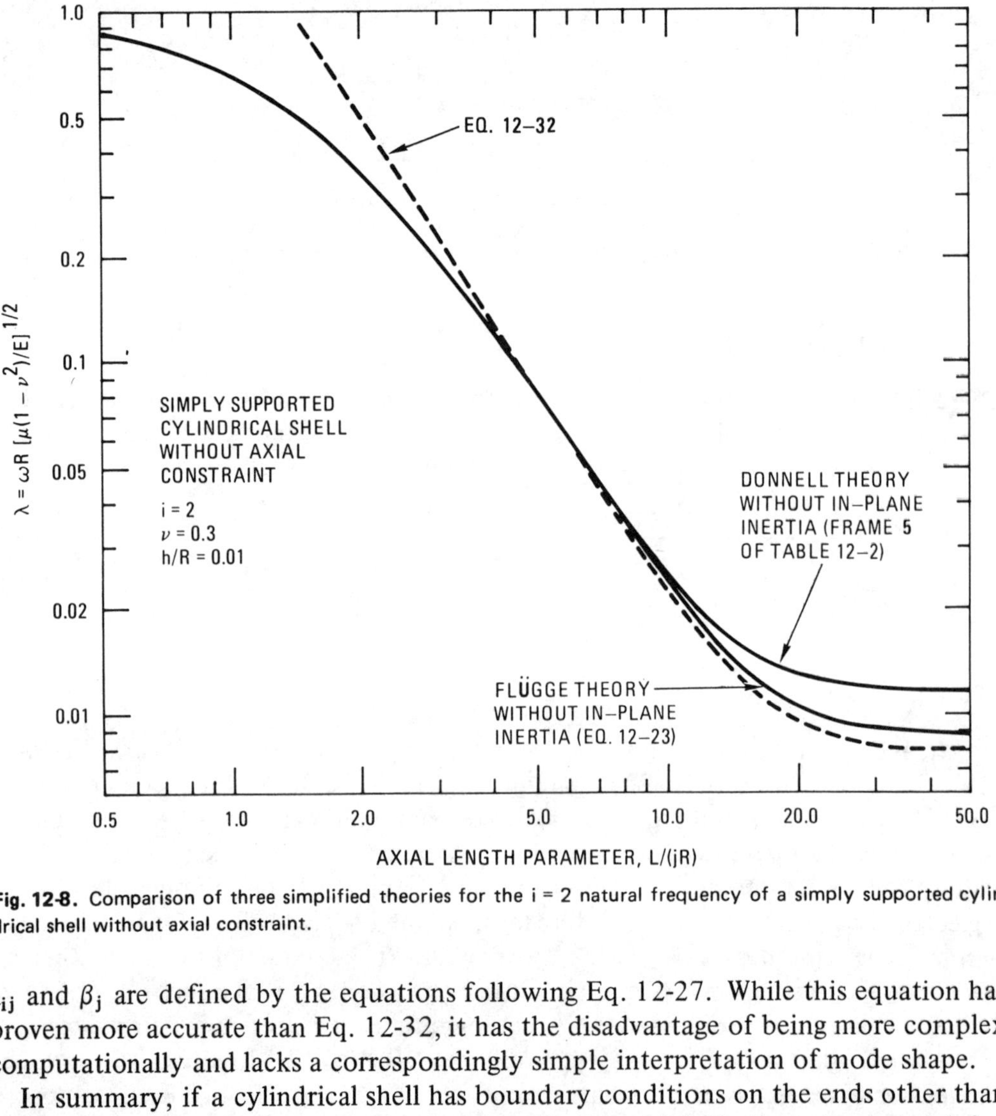

Fig. 12-8. Comparison of three simplified theories for the i = 2 natural frequency of a simply supported cylindrical shell without axial constraint.

a_{ij} and β_j are defined by the equations following Eq. 12-27. While this equation has proven more accurate than Eq. 12-32, it has the disadvantage of being more complex computationally and lacks a correspondingly simple interpretation of mode shape.

In summary, if a cylindrical shell has boundary conditions on the ends other than simply supported without axial constraint (Section 12.2.3), then the natural frequencies and mode shapes of the shell can often be approximately determined by either (1) application of Eq. 12-26, 12-32, or 12-33, or (2) beam theory for certain modes corresponding to i = 0 or i = 1. The application of Eqs. 12-26, 12-32, and 12-33 to cylindrical shells with various boundary conditions is illustrated in the following paragraphs.

Free-Free Boundary Conditions. The boundary conditions that describe a cylindrical shell which is completely free at both ends are (Ref. 12-9):

$$N_x = \mathfrak{M}_x = 0,$$

$$N_{x\theta} + \frac{\mathfrak{M}_{x\theta}}{R} = 0,$$

$$Q_x = \frac{1}{R}\frac{\partial \mathfrak{M}_{x\theta}}{\partial \theta} = 0 \quad \text{at} \quad x = 0, L.$$

Q_x is the integral of σ_{xz} over the thickness of the shell.

The higher modes ($i = 2, 3, 4, \ldots$) of a free-free cylindrical shell can be grouped by the axial (x) behavior of the mode shape. If the mode does not vary axially or there is a linear axial variation, then the mode corresponds roughly to $j = 0$. Inextensional vibration theory provides a good approximation to these modes.

There are two sets of inextensional modes corresponding to $j = 0$. The first set, analyzed by Rayleigh, has the following mode shape and circular natural frequency ω (Ref. 12-18):

$$u = 0,$$

$$v = -\frac{A}{i}\sin i\theta \cos \omega t,$$

$$w = A \cos i\theta \cos \omega t,$$

$$\omega = \frac{i(i^2 - 1)}{(i^2 + 1)^{1/2} R}\left(\frac{h^2}{12R^2}\right)^{1/2}\left[\frac{E}{\mu(1 - \nu^2)}\right]^{1/2},$$

$$i = 2, 3, 4, \ldots = \text{number of circumferential waves.} \tag{12-34}$$

These modes do not vary axially. The second set of modes corresponding to $j = 0$ have linear axial variation in the deformations. For these modes Love found (Ref. 12-2, pp. 543–547):

$$u = \frac{AR}{i^2}\cos i\theta \cos \omega t,$$

$$v = \frac{A(x - L/2)}{i}\sin i\theta \cos \omega t,$$

$$w = A(x - L/2)\cos i\theta \cos \omega t,$$

$$\omega^2 = \frac{i^2(i^2 - 1)^2}{(i^2 + 1)R^2}\,\frac{h^2}{12R^2}\,\frac{E}{\mu(1 - \nu^2)}\,\frac{1 + \dfrac{24(1 - \nu)R^2}{i^2 L^2}}{1 + \dfrac{12R^2}{i^2(i^2 + 1)L^2}},$$

$$i = 2, 3, 4, \ldots = \text{number of circumferential waves,} \tag{12-35}$$

where the axial coordinate x spans the cylinder from $x = 0$ to $x = L$. A is a constant with the units of length. The natural frequencies of the Rayleigh and Love modes are nearly the same for long cylinders, $L/R > 8$. The principal difference between the two modes is that the ends of the cylinders "toe" in or out in the Love modes, while the cylinder flexes uniformly along its length in the Rayleigh modes. Both the Rayleigh and Love modes of free-free cylindrical shells have been observed experimentally, as shown in Fig. 12-9 and discussed in Ref. 12-19.

The higher modes of the free-free cylindrical shell ($j = 1, 2, 3, \ldots; i = 2, 3, 4, \ldots$)

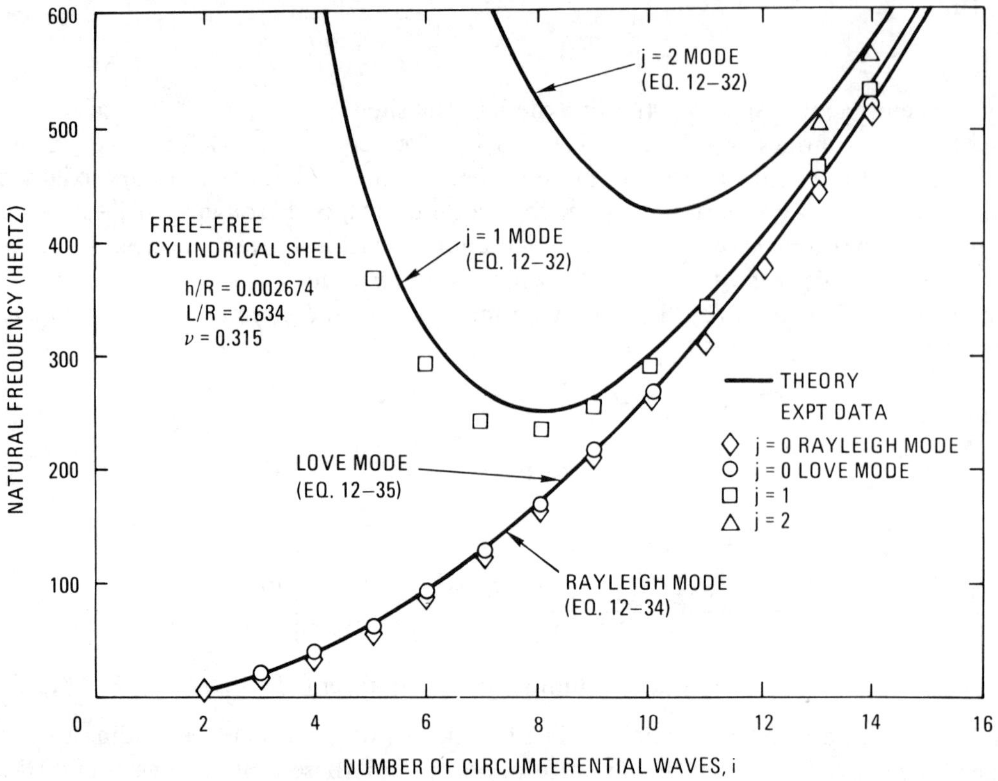

Fig. 12-9. Comparison of analytical results with the experimental results of Sewall and Naumann (Ref. 12-17) for a free-free cylindrical shell. $E = 10^7$ lb/in.^{2}; R = 9.538 in.

can be analyzed using the free-free beam modes in Eq. 12-26, 12-32, or 12-33. Equations 12-26, 12-32, and 12-33 are compared with Warburton's numerical results in Table 12-5. Note that the accuracy of these equations increases with increasing L/(jR) and i.

Clamped-Free Boundary Conditions. The boundary conditions for a cylindrical shell which is clamped at one end and free at the other end are (Ref. 12-9):

$$u = v = w = \frac{\partial w}{\partial x} = 0 \qquad \text{at} \quad x = 0,$$

$$\left.\begin{aligned} N_x &= N_{x\theta} + \frac{\mathcal{M}_{x\theta}}{R} = 0 \\ \mathcal{M}_x &= Q_x + \frac{1}{R}\frac{\partial \mathcal{M}_{x\theta}}{\partial \theta} = 0 \end{aligned}\right\} \qquad \text{at} \quad x = L.$$

The higher modes (i = 2, 3, 4, . . .) of a clamped-free cylindrical shell can be analyzed either through numerical solution of Eq. 12-26 or by means of the simplified solutions, Eq. 12-32 or 12-33, using the clamped-free beam modes. Both of

Table 12-5. Comparison of the Numerical Results of Warburton (Ref. 12-19) with Analytical Results for the Natural Frequency Parameter of a Free-Free Steel Cylindrical Shell

h/R = 0.002, ν = 0.3

	i=2			i=6		
	j=1	j=3	j=5	j=1	j=3	j=5
L/R	61.5	143	224	22.3	50.7	79.1
λ^2(Ref. 12-19)	4×10^{-6}	4×10^{-6}	4×10^{-6}	4×10^{-4}	4×10^{-4}	4×10^{-4}
λ^2(Eq. 12-32)	4.159×10^{-6}	4.027×10^{-6}	4.178×10^{-6}	4.005×10^{-4}	4.003×10^{-4}	4.003×10^{-4}
λ^2(Eq. 12-33)	4.106×10^{-6}	4.156×10^{-6}	4.039×10^{-6}	4.005×10^{-4}	4.003×10^{-4}	4.001×10^{-4}
λ^2(Eq. 12-26)	4.111×10^{-6}	4.027×10^{-6}	4.043×10^{-6}	4.005×10^{-4}	4.002×10^{-4}	4.001×10^{-4}
L/R	21.9	50.8	79.8	8.30	19.1	30.0
λ^2(Ref. 12-19)	1×10^{-4}	1×10^{-4}	1×10^{-4}	4.8×10^{-4}	4.8×10^{-4}	4.8×10^{-4}
λ^2(Eq. 12-32)	1.107×10^{-4}	1.118×10^{-4}	1.120×10^{-4}	4.886×10^{-4}	4.894×10^{-4}	4.887×10^{-4}
λ^2(Eq. 12-33)	1.006×10^{-4}	1.037×10^{-4}	1.026×10^{-4}	4.858×10^{-4}	4.835×10^{-4}	4.819×10^{-4}
λ^2(Eq. 12-26)	1.006×10^{-4}	1.037×10^{-4}	1.025×10^{-4}	4.857×10^{-4}	4.835×10^{-4}	4.818×10^{-4}
L/R	12.1	28.0	44.0	5.46	12.6	19.8
λ^2(Ref. 12-19)	0.001	0.001	0.001	8×10^{-4}	8×10^{-4}	8×10^{-4}
λ^2(Eq. 12-32)	0.001151	0.001178	0.001180	8.477×10^{-4}	8.546×10^{-4}	8.527×10^{-4}
λ^2(Eq. 12-33)	0.001080	0.001055	0.001038	8.283×10^{-4}	8.177×10^{-4}	8.099×10^{-4}
λ^2(Eq. 12-26)	0.001079	0.001053	0.001036	8.282×10^{-4}	8.175×10^{-4}	8.098×10^{-4}
L/R	2.57	5.59	8.67	2.37	5.44	8.52
λ^2(Ref. 12-19)	0.2	0.2	0.2	0.01	0.01	0.01
λ^2(Eq. 12-32)	0.4174	0.5762	0.6223	0.01238	0.01299	0.01316
λ^2(Eq. 12-33)	0.1623	0.1124	0.1693	0.01088	0.01068	0.01050
λ^2(Eq. 12-26)	0.2275	0.2356	0.2266	0.01089	0.01068	0.01050

these solutions are compared with experimentally measured natural frequencies in Table 12-6. Note that the analytical solutions overpredict the experimental results.

Clamped-Clamped Boundary Conditions with Axial Constraint. The boundary conditions for cylindrical shells which have both ends clamped and constrained against axial motion are:

$$u = v = w = \frac{\partial w}{\partial x} = 0 \qquad \text{at} \quad x = 0, L.$$

The natural frequencies of these shells can be estimated from Eq. 12-26, 12-32, or 12-33 using the clamped-clamped beam modes. Two other characteristic polynomials of the same form as Eq. 12-26 for clamped-clamped axially constrained cylindrical shells can be found in Refs. 12-20 and 12-11 (pp. 299–302). The experimentally measured frequencies of Koval and Cranch (Ref. 12-21) for a clamped-clamped cylindrical shell are compared with the analytical predictions of Eqs. 12-26, 12-32, and 12-33 in Table 12-7.

Table 12-6. Comparison of Experimental and Analytical Results for a Clamped-Free Cylindrical Shell

L = 502 mm, R = 63.5 mm, h = 1.63 mm, E = 2.1 × 10^{11} N/m^2, ν = 0.28, μ = 7.8 × 10^3 kg/m^3 (Ref. 12-15).

		Natural Frequency (hertz)				
i	Source	j=1	j=2	j=3	j=4	j=5
2	Experimental	293.0	827.0	1894	--	--
	Eq. 12-32	320.9	1095	2922	5566	8922
	Eq. 12-33	319.4	1018	2358	3762	4948
	Eq. 12-26	319.5	1020	2399	3963	5471
3	Experimental	760.0	886.0	1371	2155	3208
	Eq. 12-32	769.9	941.7	1638	2883	4596
	Eq. 12-33	769.8	930.6	1514	2413	3431
	Eq. 12-26	769.9	930.4	1515	2428	3486
4	Experimental	1451	1503	1673	2045	2713
	Eq. 12-32	1466	1527	1756	2284	3149
	Eq. 12-33	1466	1525	1730	2156	2775
	Eq. 12-26	1465	1524	1730	2159	2783
5	Experimental	2336	2384	2480	2667	2970
	Eq. 12-32	2367	2410	2519	2755	3169
	Eq. 12-33	2367	2410	2514	2723	3058
	Eq. 12-26	2367	2409	2513	2723	3059
6	Experimental	3429	3476	3546	3667	3880
	Eq. 12-32	3470	3509	3589	3734	3970
	Eq. 12-33	3470	3509	3587	3725	3936
	Eq. 12-26	3470	3508	3587	3724	3936

12.2.5. Effect of Axial Constraint

Forsberg (Ref. 12-22) found that the difference in the natural frequencies of the higher circumferential modes, i = 2, 3, 4, . . . , of a simply supported cylindrical shell without axial constraint and a similar clamped-clamped cylindrical shell with axial constraint was primarily due to the addition of the axial constraint rather than the clamping of the ends, as can be seen in Fig. 12-10. This suggests that in many cases the axial constraint may be more important than the constraint against local rotation in calculation of the natural frequencies.

12.2.6. Effect of Loads on Natural Frequencies of Cylindrical Shells

Only a few general results exist for the effect of external loads on the natural frequencies of cylindrical shells. The cylindrical shell shown in Fig. 12-11 is subject to a uniform axial tension $\overline{N}_x$ per unit length of edge and a uniform hoop tension $\overline{N}_\theta$ per unit axial length. $\overline{N}_x$ and $\overline{N}_\theta$ are positive if they produce tensile stresses in the

Table 12-7. Comparison of Experimental Results of Koval and Cranch (Ref. 12-21) with Analytical Results for a Clamped-Clamped Cylindrical Shell with Axial Constraint

L = 12 in., h = 0.01 in., ν = 0.3, R = 3 in., E = 30 X 10^6 lb/in.^{2}

j	Source	Natural Frequency (hertz)						
		i=4	i=5	i=6	i=7	i=8	i=9	i=10
1	Experimental	700	552[a]	525	592[a]	720	885	1095[a]
	Eq. 12-32	957.2	663.7	570.0	605.5	719.8	881.3	1075
	Eq. 12-33	797.4	594.9	540.7	593.7	715.0	879.2	1074
	Eq. 12-26	797.8	594.9	540.7	593.7	715.0	879.2	1074
2	Experimental	1620	1210	980	856[a]	900	995	1140[a]
	Eq. 12-32	2596	1704	1241	1021	966.2	1028	1167
	Eq. 12-33	1877	1360	1063	927.0	918.0	1004	1155
	Eq. 12-26	1884	1362	1063	927.0	917.9	1004	1155
3	Experimental	--	--	1650	1359	1350	1278[a]	1325
	Eq. 12-32	5051	3301	2342	1793	1497	1382	1400
	Eq. 12-33	3077	2312	1801	1482	1316	1278	1342
	Eq. 12-26	3114	2321	1831	1482	1316	1278	1342
4	Experimental	--	--	--	1960	1765	--	1690
	Eq. 12-32	8279	5425	3828	2873	2292	1961	1813
	Eq. 12-33	4203	3299	2627	2155	1850	1687	1645
	Eq. 12-26	4370	3329	2636	2159	1851	1687	1645
5	Experimental	--	--	--	--	--	2300	2100
	Eq. 12-32	12240	8062	5686	4241	3326	2747	2406
	Eq. 12-33	5162	4230	3461	2877	2458	2186	2043
	Eq. 12-26	5369	4296	3484	2886	2462	2187	2044

[a] Result is average of two values.

shell and negative if they produce compressive stresses in the shell. These loads could result from internal pressure ($\overline{N}_x = pR/2$, $\overline{N}_\theta = pR$, where p is the internal pressure), rotation of the cylinder about its axis ($\overline{N}_x = 0$, $\overline{N}_\theta = \mu h \omega^2 R^2$, where ω is the frequency of rotation in radians per second), or applied mechanical loads.

For a simply supported cylindrical shell without axial constraints, it is possible to derive a closed form expression for the effect of the loads $\overline{N}_x$ and $\overline{N}_\theta$ on the natural frequencies if either of the following sets of assumptions is made: (1) the Donnell shell theory is used and the in-plane inertia terms are neglected; or (2) the Flügge shell theory is used and certain terms which are ordinarily small are neglected (Refs. 12-9, 12-22). Under either of these sets of assumptions, the natural frequencies of a loaded simply supported shell without axial constraint are related to the natural frequencies of the unloaded shell by (Ref. 12-9):

$$\omega_{ij}^2\Big|_{\substack{\text{loads,}\\ \overline{N}_x,\, \overline{N}_\theta}} = \omega_{ij}^2\Big|_{\substack{\text{no loads,}\\ \overline{N}_x = \overline{N}_\theta = 0}} + \frac{\overline{N}_x j^2 \pi^2}{\mu h L^2} + \frac{\overline{N}_\theta i^2}{\mu h R^2}. \tag{12-36}$$

ω_{ij} is the natural frequency of vibration in radians per second ($2\pi f_{ij}$). The remaining terms are defined in Table 12-2. The ω_{ij} for no load can be obtained from the Donnell or Flügge theory (frame 5 of Table 12-2, Eq. 12-16 or 12-23). It can be seen

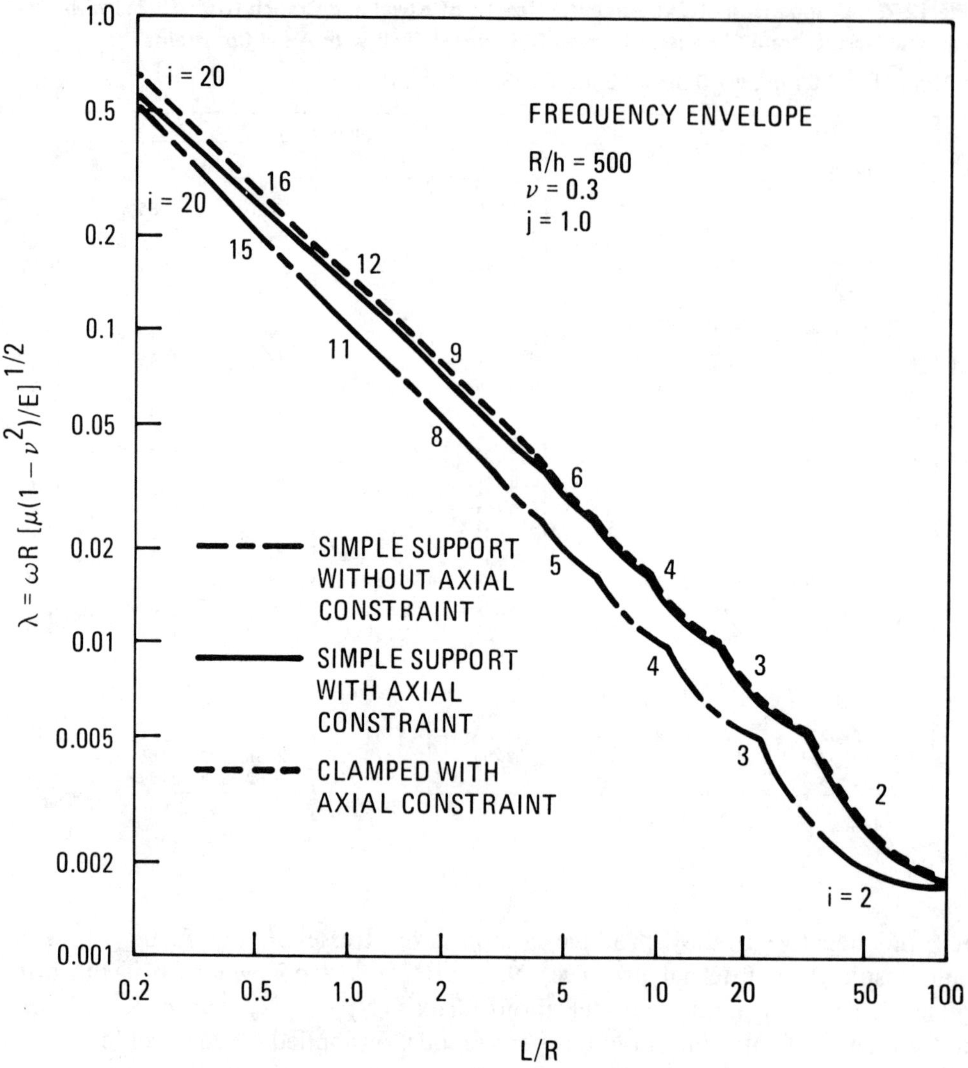

Fig. 12-10. Lowest-frequency modes for cylindrical shells with and without axial constraint. (Adapted from Ref. 12-22)

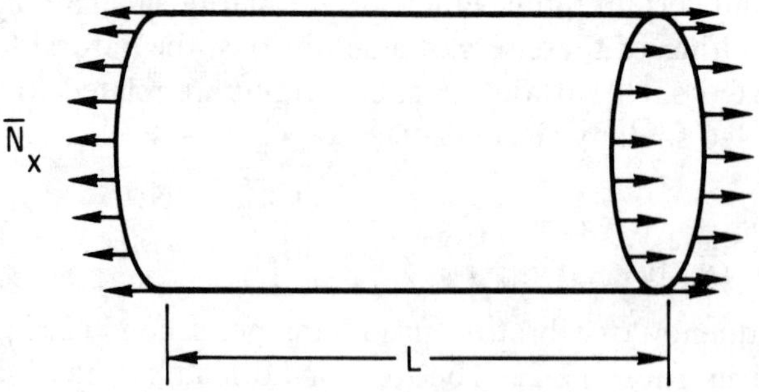

Fig. 12-11. Cylindrical shell subject to uniform axial load.

from Eq. 12-36 that tensile loads ($\overline{N}_x > 0$, $\overline{N}_\theta > 0$) increase the natural frequencies of the shell. Compressive loads ($\overline{N}_x < 0$, $\overline{N}_\theta < 0$) decrease the natural frequencies of the shell and ultimately cause buckling as ω_{ij} approaches zero. The mode shapes of a simply supported cylindrical shell without axial constraint are unaffected by the loads $\overline{N}_x$ and $\overline{N}_\theta$.

Evidently, there are no general closed form results available for the effect of loads on the natural frequencies of cylindrical shells with other boundary conditions. The relationship between the onset of buckling and natural frequency suggests that the effect of axial load on the natural frequency may be approximated by:

$$\frac{\omega_{ij}|_{\overline{N}_x}}{\omega_{ij}|_{\overline{N}_x=0}} = \left(1 + \frac{\overline{N}_x}{|\overline{N}_{xbij}|}\right)^{1/2}, \tag{12-37}$$

where $\overline{N}_{xbij}$ is the value of the load which will produce buckling of the ij mode. This equation holds exactly for simply supported cylindrical shells without axial constraint.

Other discussions of the effect of loads on the natural frequencies of cylindrical shells may be found in Refs. 12-21 and 12-23 through 12-25.

12.3. CYLINDRICALLY CURVED PANELS

The natural frequencies and mode shapes of cylindrically curved panels, shown in Fig. 12-12, can be adapted from the natural frequencies and mode shapes of the cylindrical shells given in the previous section if the edge of the cylinder at $\theta = 0$ and $\theta = \theta_0$ are simply supported without circumferential constraint:

$$u = w = N_\theta = M_\theta = 0 \qquad \text{along} \quad \theta = 0, \theta_0. \tag{12-38}$$

These boundary conditions are shown in Fig. 12-13. For these boundary conditions, the shell equations have solutions of the form:

$$u = A\tilde{\phi}'(x/L) \sin i\theta \cos \omega t,$$

$$v = B\tilde{\phi}(x/L) \cos i\theta \cos \omega t,$$

$$w = C\tilde{\phi}(x/L) \sin i\theta \cos \omega t, \tag{12-39}$$

where A, B, and C are constants which specify the amplitude of the axial (u), tangential (v), and radial (w) deformation of the panel midsurface. The function $\tilde{\phi}$, determines the longitudinal mode shape and is chosen to satisfy the boundary conditions at $x = 0$, L. In order to meet the boundary conditions along the edges $\theta = 0$, θ_0, the parameter i is allowed to take on noninteger values:

$$i = \frac{n\pi}{\theta_0}, \tag{12-40}$$

where $n = 1, 2, 3, \ldots$, n is the number of circumferential half-waves in the mode shape and θ_0 is the angle subtended by the panel in radians.

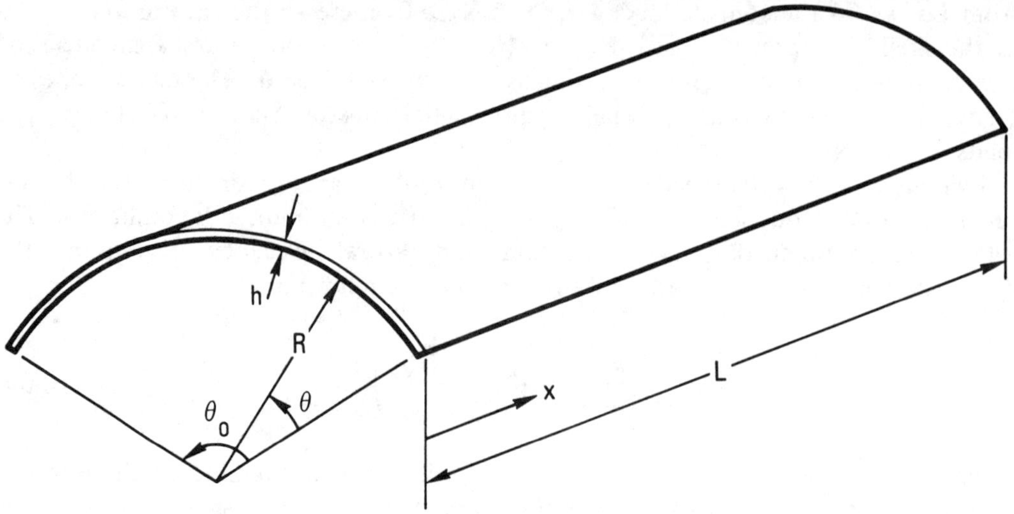

Fig. 12-12. Cylindrically curved panel. R is the radius to the midplane of the shell.

The solution form of Eq. 12-39 is identical to the general solution form for cylindrical shells given by Eq. 12-24 in Section 12.2 but with rotation of the circumferential coordinate θ by $\pi/2$. Thus, the solutions for the natural frequencies of cylindrical shells given in Section 12.2 can be directly applied to cylindrically curved panels which have the boundary conditions of Eq. 12-38 along the edges $\theta = 0, \theta_0$ by substituting the value of i given by Eq. 12-40. The mode shapes, stresses, strains, and stress resultants of Section 12.2 are also directly applicable if the circumferential coordinate θ in the mode shape is first rotated by $\pi/2$.

The results of Leissa (Ref. 12-9) indicate that if the in-plane inertia terms are neglected in the analysis of long, cylindrically curved panels with $i < 1$, the natural frequency may be substantially underestimated. Thus, it is advisable to numerically solve the cubic Eq. 12-16 or 12-26 for the natural frequencies of cylindrically curved panels with small i rather than using one of the simplified formulas in Section 12.2.

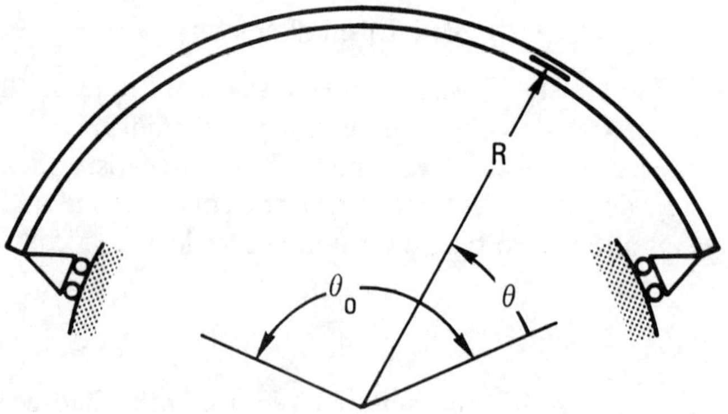

Fig. 12-13. End view of cylindrically curved panel with simply supported edges without circumferential constraint at $\theta = 0, \theta_0$.

Specialized approximate formulas for cylindrically curved panels have been developed by Sewall (Ref. 12-26), whose formulas were recast by Webster (Ref. 12-27), and by Szechenyi (Ref. 12-54). These formulas apply to cylindrically curved panels which are as shown in Fig. 12-12 with the following two sets of boundary conditions: (1) all edges are simply supported without constraint on in-plane (u and v) deformation, and (2) all edges are clamped and the in-plane (u and v) deformation is zero at the edges.

For these sets of boundary conditions, the natural frequencies of the panels may be expressed in terms of the natural frequencies of a flat plate with the same physical dimensions and analogous boundary conditions:

$$\omega_{ij}\Big|_{\substack{\text{curved}\\ \text{panel}}} = \left[\omega_{ij}^2\Big|_{\substack{\text{flat plate}\\ \text{width} = R\theta\\ \text{length} = L^0}} + \frac{\alpha_{ij} E h}{R^2 \gamma (1 - \nu^2)}\right]^{1/2} \qquad i, j = 1, 2, 3, \ldots, \tag{12-41}$$

where

ω_{ij} = circular natural frequency with appropriate boundary conditions (radians/second); there are 2π radians in 1 cycle,

i = number of half-waves along circumferential ($R\theta$) axis,

j = number of half-waves along longitudinal (x) axis,

E = modulus of elasticity,

h = plate thickness,

R = radius to midsurface of the curved panel (Fig. 12-13),

γ = mass per unit area of panel, μh,

α_{ij} = dimensionless constant which is a function of the boundary conditions (see below),

ν = Poisson's ratio.

Consistent sets of units are given in Table 3-1. The natural frequencies of the appropriate flat plates with length a = L and width $b = R\theta_0$ and clamped or simply supported edges can be found from either frames 21 and 16 of Table 11-4, respectively, or from Eq. 11-21. The dimensionless constants α_{ij} are given by:

$$\alpha_{ij} = \begin{cases} \dfrac{\left(\dfrac{j\pi}{L}\right)^4 (1 - \nu^2)}{\left[\left(\dfrac{j\pi}{L}\right)^2 + \left(\dfrac{i\pi}{R\theta_0}\right)^2\right]^2} & \text{simply supported edges; Refs. 12-26, 12-27, 12-54} \\[3ex] \dfrac{\lambda_i^4\lambda_j^4\left\{1 - \left(\dfrac{\nu\beta_j}{\lambda_j^2}\right)^2 - \left(\dfrac{\beta_i}{\lambda_i^2}\right)^2\right\} + \left(\dfrac{1-\nu}{2}\right)\beta_i\beta_j\left[\dfrac{b^2}{L^2}(\lambda_j^4 - \nu^2\beta_j^2) + \dfrac{L^2}{b^2}(\lambda_i^4 - \beta_i^2)\right] + (\nu\beta_i\beta_j)^2}{\lambda_i^4\lambda_j^4 + \left(\dfrac{1-\nu}{2}\right)\beta_i\beta_j\left(\dfrac{b^2}{L^2}\lambda_j^4 + \dfrac{L^2}{b^2}\lambda_i^4\right) - \nu(\beta_i\beta_j)^2} & \text{clamped edges, Refs. 12-26, 12-27} \end{cases}$$

where $\beta_j = \sigma_j\lambda_j(\sigma_j\lambda_j - 2)$; $\beta_i = \sigma_i\lambda_i(\sigma_i\lambda_i - 2)$, $b = R\theta_0$, and λ and σ are given in frame 7 of Table 8-1. An alternate expression for α_{ij} for clamped edges is given in Ref. 12-54. It can be seen from Eq. 12-41 and the above equations that the curvature raises the natural frequencies of a curved panel above that of an analogous flat plate. Equation 12-41 can be expected to be most accurate for the lower modes of thin panels with $\theta_0 < 1$ radian.

Szechenyi (Ref. 12-54) has found that the natural frequencies of a cylindrically curved panel with simply supported edges without in-plane constraints but with in-plane loads can be found from Eq. 12-41 with two additional terms:

$$\omega_{ij}^2\Big|_{\substack{\text{simply supported}\\ \text{curved panel with}\\ \text{in-plane loads}}} = \omega_{ij}^2\Big|_{\substack{\text{simply supported}\\ \text{curved panel with-}\\ \text{out in-plane loads}\\ \text{(Eq. 12-41)}}} + \frac{N_x}{\gamma}\left(\frac{j\pi}{L}\right)^2 + \frac{N_\theta}{\gamma}\left(\frac{i\pi}{R\theta_0}\right)^2.$$

N_x is the longitudinal in-plane load per unit length of edge and N_θ is the circumferential load per unit length of edge. (See Figs. 12-11 and 11-7.) N_x and N_θ are positive for tensile loads and negative for compressive loads. This equation is identical to the corresponding equation for a simply supported cylindrical shell, Eq. 12-36.

If edges of the cylindrically curved panel $x = 0$ and $x = L$ are free, then certain modes can be analyzed by modeling the cylindrically curved panel as a circular arc. Solutions for the natural frequencies and mode shapes of circular arcs are discussed in Section 9.2.

12.4. CONICAL SHELLS

The natural frequencies of thin conical shells can be analyzed in the same manner as those of cylindrical shells. The equations of motion are solved by assuming displacement functions of the form:

$$u = \sum_{i=0}^{\infty} F_{ij}(s) \cos i\theta \cos \omega t,$$

$$v = \sum_{i=0}^{\infty} G_{ij}(s) \sin i\theta \cos \omega t,$$

$$w = \sum_{i=0}^{\infty} H_{ij}(s) \cos i\theta \cos \omega t, \qquad j = 0, 1, 3, \ldots,$$

where s is the coordinate from the apex of the cone, θ is the circumferential angle, i is the number of circumferential waves in the mode shape, and j is the number of longitudinal half-waves in the mode shape.

The natural frequencies and mode shapes of several complete conical shells and frustrums of conical shells are given in Table 12-8. The solutions presented in this table, with the exception of frame 1, may also be found in Ref. 12-9.

As the conical angle becomes increasingly shallow ($\alpha \sim 0$), the natural frequencies of a frustrum of a conical shell must approach the natural frequencies of a similar cylindrical shell. This suggests that the natural frequencies of shallow frustrums of conical shells may be calculated from the natural frequencies of cylindrical shells using an average radius $\overline{R} = (R_1 + R_2)/2$, where R_1 and R_2 are the radii of the mid-planes of the ends of the conical shell. This approximation is discussed in Refs. 12-32 and 12-33 and appears to generally yield a result within about 20% of the true natural frequency if the ratio of the radii of the ends of the conical shell is less than the ratio $R_2/R_1 < 1.5$, where R_2 is the larger radius.

Table 12-8. Conical Shells.

Notation: h = thickness of the shell; i = number of circumferential waves in the mode shape; j = number of longitudinal half-waves in the mode shape; s = distance from apex; u = deformation parallel to s; v = circumferential deformation; w = radial deformation; $\tilde{u}$, $\tilde{v}$, $\tilde{w}$ = mode shapes associated with u, v, and w, respectively; C = clamped edge; E = modulus of elasticity; F = free edge; R = radius of cone base; S = simply supported edge; α = half angle of cone; θ = angle; μ = density of shell material; ν = Poisson's ratio; consistent sets of units are given in Table 3-1.

Description	Mode Shape	Natural Frequency (hertz)	λ
1. Complete Cone With Free Base 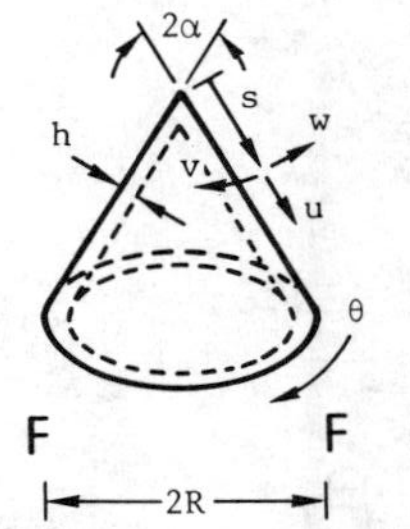	Lowest Frequency Modes; Radial-Circumferential Modes: $\begin{pmatrix}\tilde{u}\\ \tilde{v}\\ \tilde{w}\end{pmatrix} = \begin{pmatrix}0\\ \varepsilon\\ F_j(s)\cos i\theta\end{pmatrix}$ i=2,3,4... j=1,2,3... $\varepsilon << 1$ Tip of cone does not move, $F_j(0) = 0$.	$\frac{\lambda_{ij}}{2\pi R}\left(\frac{E}{\mu}\right)^{1/2}$ $\nu = 0.3$ Ref. 12-55 See Ref. 12-28 for axisymmetric (higher frequency) modes.	$0.8745\left(\frac{h}{R}\right)^{7/8}(\sin\alpha)^{1/4}\left[i^4\nu^2 + \frac{\beta_j^4\sin^4\alpha}{2}(1-4\nu^2\sin^2\alpha) - 2\nu(i-2)(i-3)(2i-5)\right]^{1/2}$ i=2,3,4...; j=1,2,3... ($\beta_1 = 1.875$; $\beta_2 = 4.694$; $\beta_j = (j - 1/2)\pi$, $j > 2$)
2. Complete Cone With Clamped Base 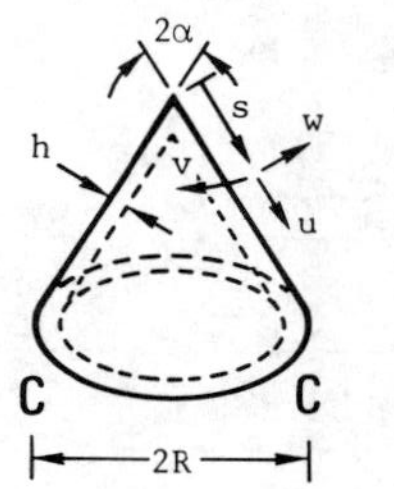	Axisymmetric Modes: $\begin{pmatrix}\tilde{u}\\ \tilde{v}\\ \tilde{w}\end{pmatrix} = \begin{pmatrix}F_j(s)\\ 0\\ G_j(s)\end{pmatrix}$ j=1,2,3... i=0	$\frac{\lambda_j}{2\pi R}\left(\frac{E}{\mu}\right)^{1/2}$ $\nu = 0.3$ Ref. 12-28	See table below

$\frac{12(1-\nu^2)}{\tan^4\alpha}\left(\frac{R}{h}\right)^2$	j = 1	2	3	4
0.1	1,049	15,830	79,410	250,200
0.4	266.9	3,964	19,860	62,570
1.0	110.4	1,591	7,954	25,040
4.0	32.07	404.8	1,999	6,275
10	16.33	167.5	808.2	2,521
100	6.096	24.69	93.37	269.0
1000	3.574	7.648	18.58	40.76
100,000	1.802	2.431	3.449	4.604

Table 12-8. Conical Shells. *(Continued)*

Description	Mode Shape	Natural Frequency (hertz)	λ
3. Frustrum of a Cone With Free Edges 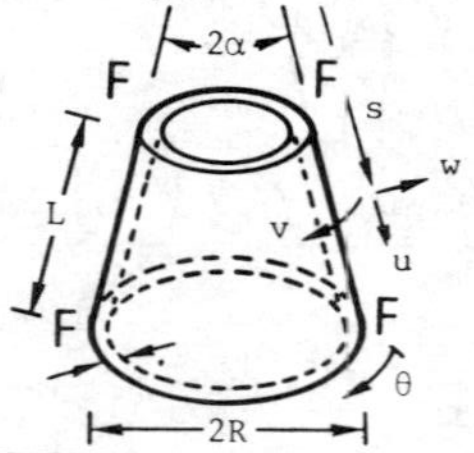	Inextensional Shell Modes: $\begin{pmatrix}\tilde{u}\\ \tilde{v}\\ \tilde{w}\end{pmatrix}_i = \begin{pmatrix}\sin\alpha\cos\alpha\cos i\theta\\ \left(1 + B\frac{s}{S_2}\right) i\cos\alpha\sin i\theta\\ i^2 - \sin^2\alpha + Bi^2\frac{s}{S_2}\cos i\theta\end{pmatrix}_i$ i=2,3,4... j=0 B = constant S_2 = R/sin α	$\frac{\lambda_i}{2\pi R}\left[\frac{E}{\mu(1-\nu^2)}\right]^{1/2}$ Refs. 12-9, 12-29 i=2,3,4...	$\frac{h}{12^{1/2}R}\frac{i^2-i}{(i^2+1)^{1/2}}\left(i+1-4\sin\frac{3\alpha}{2}\right)$ i = 2,3,4... ; α < 60° (Note similarity of this result to frame 3 of Table 12-1.) See Ref. 12-29 for other modes.
4. Frustrum of a Cone With Clamped-Free Edges 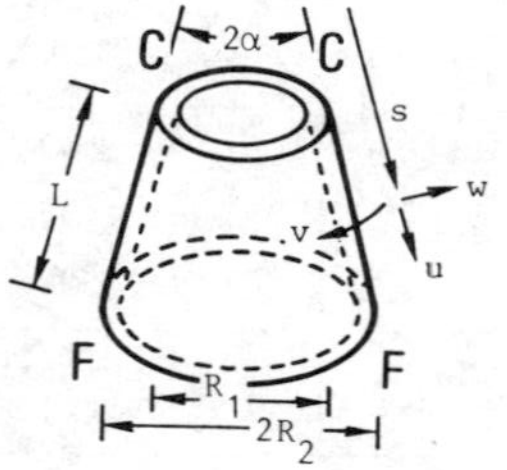	Fundamental Axisymmetric Mode: $\begin{pmatrix}\tilde{u}\\ \tilde{v}\\ \tilde{w}\end{pmatrix} = \begin{pmatrix}F(s)\\ 0\\ G(s)\end{pmatrix}$ i=j=0	$\frac{\lambda}{2\pi R_2}\left(\frac{E}{\mu}\right)^{1/2}$ ν = 0.3 Ref. 12-9	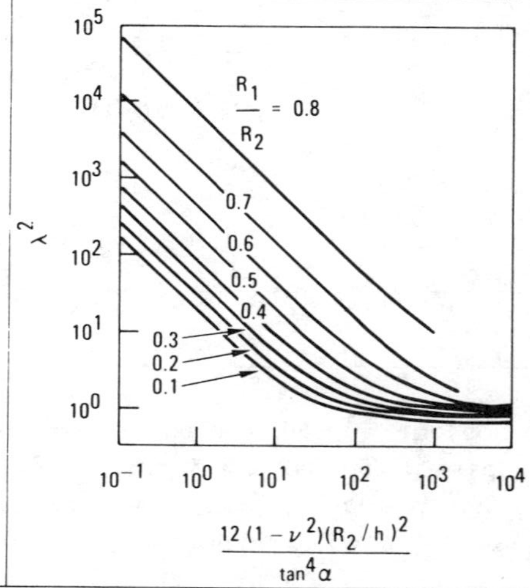

5. Frustrum of a Cone With Simply-Supported Edges Without Axial Constraint

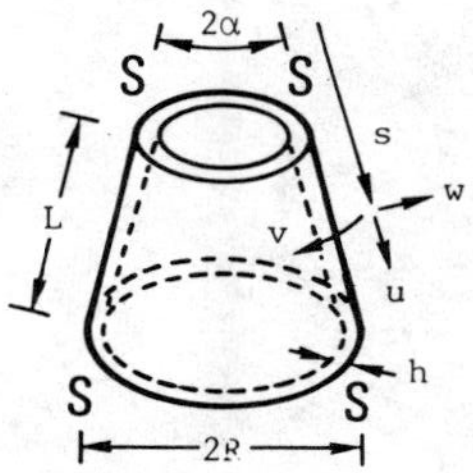

$$\begin{pmatrix}\tilde{u}\\ \tilde{v}\\ \tilde{w}\end{pmatrix}_i = \begin{bmatrix}\cos\dfrac{\pi(s-S_1)}{L}\cos i\theta\\ \sin\dfrac{\pi(s-S_1)}{L}\sin i\theta\\ \sin\dfrac{\pi(s-S_1)}{L}\cos i\theta\end{bmatrix}_i$$

$$S_1 = \frac{R}{\sin\alpha} - L$$

i=2,3,4...
j=1

$$\frac{\lambda_i}{2\pi L}\left[\frac{E}{\mu(1-\nu^2)}\right]^{1/2}$$

Ref. 12-30

λ_i of lowest frequency mode and (i)

90 − α (deg)	h/R 0.03	0.01	0.005	0.001
5	--	0.141 (2)	0.0967 (3)	0.0448 (6)
20	0.479 (3)	0.287 (5)	0.199 (5)	0.0950 (9)
40	0.652 (3)	0.386 (5)	0.282 (6)	0.138 (10)
60	0.776 (3)	0.479 (5)	0.350 (6)	0.172 (9)
80	0.891 (2)	0.553 (3)	0.432 (4)	0.229 (6)

$\nu = 0.3$

6. Frustrum of a Cone With Clamped Edges

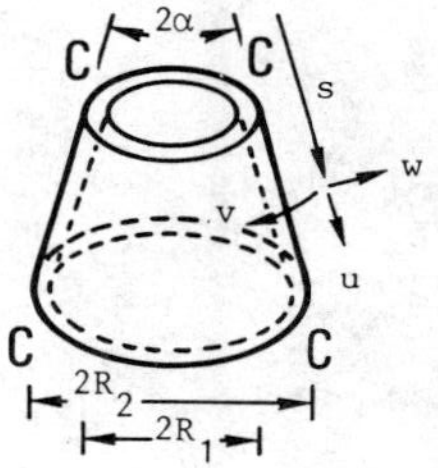

Torsional Modes:

$$\begin{pmatrix}\tilde{u}\\ \tilde{v}\\ \tilde{w}\end{pmatrix}_j = \begin{bmatrix}0\\ J_1(\lambda_j s/S_1)\\ +\ BY_1(\lambda_j s/S_1)\\ 0\end{bmatrix}$$

j=1,2,3...
i=0

B = constant

J_1, Y_1 = Bessel functions of first and second kind, respectively

$$S_1 = \frac{R_1}{\sin\alpha}$$

$$\frac{\lambda_j}{2\pi S_1}\left[\frac{E}{2(1+\nu)\mu}\right]^{1/2}$$

i=1,2...

$S_1 = R_1/\sin\alpha$

Ref. 12-31

Solution also applicable to annular disks (i.e., α = 90°).

R_2/R_1	$\left(\frac{R_2}{R_1}-1\right)\lambda_j$ j=1	j=2	j=3	j=4	j=5
1	3.142	6.283	9.425	12.566	15.708
2	3.197	6.312	9.444	12.581	15.720
5	3.389	6.445	9.541	12.657	15.782
20	3.667	6.749	9.830	12.920	16.020
50	3.760	6.887	9.99	13.10	16.21

λ_j is independent of α.

12.5. SPHERICAL SHELLS

Complete Shells. The natural frequencies and mode shapes of the axisymmetric modes (i.e., modes independent of the circumferential angle θ) of thin spherical shells are given in Table 12-9. The modes are due to extensional deformation alone; the bending stiffness of the shell wall has been neglected. This is a reasonable approximation for thin shells since a complete spherical shell cannot support an inextensional bending mode (Ref. 12-18, p. 423). The modes must be either purely extensional or coupled bending-extensional. If the effect of the bending stiffness of the shell wall is incorporated in the analysis, the natural frequencies of the non-torsional modes corresponding to frames 1 and 3 of Table 12-9 can be obtained from the solution of the following cubic equation in λ^2 (Ref. 12-36):

$$
\begin{aligned}
&2\lambda^6 k_s k_1 (k_r k_1 - c_r c_1)/(1-\nu) - \lambda^4 \{(k_r k_1 - c_r c_1)[r + 4k_s(1+\nu)/(1-\nu)] \\
&\quad + k_1[\xi(k_1 + c_1) + c_r + k_r + 2k_s(k_1 + k_r)(r/(1-\nu) - 1)]\} \\
&\quad + \lambda^2\{(\xi c_1 + c_r)(1+\nu)(2-r) + k_r[r(r-3-\nu) + 2(1+\nu) \\
&\quad \cdot ((r-2)k_s + 1)] + k_1[2k_s r(r+4\nu)/(1-\nu) + r(r+\xi+\nu) \\
&\quad + (1+3\nu)(\xi - 2k_s) - (1-\nu)]\} - (r-2)[r(r-2) + 2k_s(1+\nu) \\
&\quad \cdot (r - 1 + \nu) + (1-\nu^2)(\xi + 1)] = 0. \qquad (12\text{-}42)
\end{aligned}
$$

The torsional natural frequencies corresponding to frame 2 of Table 12-8, including the effect of the bending stiffness of the shell wall, can be obtained from the following quadratic equation in λ^2 (Ref. 12-36):

$$
\begin{aligned}
&4\lambda^4 k_s(k_r k_1 - c_r c_1)/(1-\nu) - 2\lambda^2[\xi(k_1 + c_1) + c_r + k_r + k_s(k_1 + k_r)(r-2)] \\
&\qquad + (1-\nu)(r-2)[\xi + 1 + (r-2)k_s] = 0, \qquad (12\text{-}43)
\end{aligned}
$$

where

$$
\begin{aligned}
\lambda &= \omega R[(1-\nu^2)\mu/E]^{1/2}, \\
k &= h^2/(12R^2), \\
\xi &= 1/k, \\
k_1 &= 1 + k, \\
k_r &= 1 + 1.8k, \\
c_1 &= 2k, \\
c_r &= 2, \\
k_s &\approx 1.2, \\
r &= i(i+1); \qquad i = 0, 1, 2, \ldots
\end{aligned}
$$

R is the radius of the shell to the midsurface, h is the thickness of the shell, and ω is the circular natural frequency of the shell which is determined by the nondimensional frequency parameter λ. i is the modal index. The natural frequencies pre-

Table 12-9. Thin Spherical Shells—Axisymmetric Modes

Notation: i = integer index; f_i = natural frequency of i mode (hertz); h = thickness of the shell; v = circumferential deformation; u = in-plane deformation perpendicular to u; w = radial deformation; $\tilde{u}, \tilde{v}, \tilde{w}$ = mode shapes associated with u, v, and w, respectively; E = modulus of elasticity; P_i, P_i^1 = Legendre polynomials; R = radius of the sphere to the midsurface; μ = density of the shell material; ν = Poisson's ratio; ϕ = angle; formulas neglect bending stiffness and are valid for $ih \ll R$; see text for formulas taking into account bending stiffness; see Table 3-1 for consistent sets of units.

Natural Frequency (hertz), $f_i = \frac{\lambda_i}{2\pi R}\left[\frac{E}{\mu(1-\nu^2)}\right]^{1/2}$; $i=0,1,2,\ldots$

Description	Mode Shape	λ_i
Complete Spherical Shell 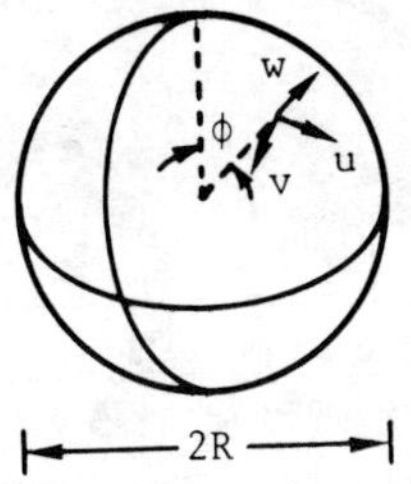	Fundamental Radial Mode: $\begin{pmatrix}\tilde{u}\\ \tilde{v}\\ \tilde{w}\end{pmatrix} = \begin{pmatrix}0\\ 0\\ 1\end{pmatrix}$ $i=0$	$\left[\frac{2(1+\nu)}{1+h^2/(12R^2)}\right]^{1/2}$ Ref. 12-34
	Torsional Modes: $\begin{pmatrix}\tilde{u}\\ \tilde{v}\\ \tilde{w}\end{pmatrix}_i = \begin{pmatrix}--\\ -j \csc \phi P_i^1(\cos\phi)\\ 0\end{pmatrix}$ $i=1,2,3\ldots$ $(i < j)$ $j=1,2,3\ldots$	$\left[\frac{(1-\nu)(i^2+i-2)}{2+5h^2/(6R^2)}\right]^{1/2}$ $i=1,2,3\ldots$ λ is independent of j. Ref. 12-34
	Radial - Tangential Modes: $\begin{pmatrix}\tilde{u}\\ \tilde{v}\\ \tilde{w}\end{pmatrix}_i = \begin{pmatrix}AP_i^1(\cos\phi)\\ P_i(\cos\phi)\\ 0\end{pmatrix}$ $i=0,1,2,3\ldots$ $A = 2(1+\nu)/\left\{(i^2+i-3-\nu) \mp \left[(3+\nu-i(i+1))^2 + 4(i^2+i) \times (1+\nu)^2\right]^{1/2}\right\}$	$\frac{1}{2^{1/2}}\left\{(i^2+i+1+3\nu) \mp \left[(i^2+i+1+3\nu)^2 - 4(1-\nu^2)(i^2+i-2)\right]^{1/2}\right\}^{1/2}$ $i=0,1,2,3\ldots$ Ref. 12-35 (Note: There is only one real root for $i=0$.)

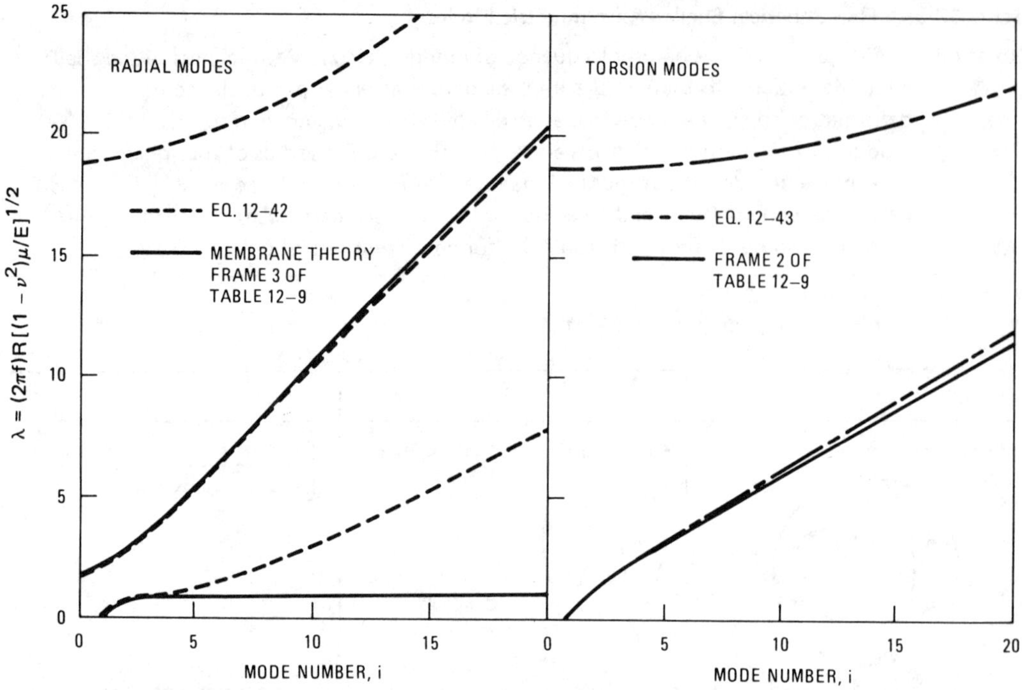

Fig. 12-14. Natural frequency parameters of complete spherical shells. $h/R = 0.1$, $\nu = 0.3$. The dashed lines represent the more accurate theories. (Ref. 12-36)

dicted by Eqs. 12-42 and 12-43 are more accurate than the formulas given in Table 12-9 if the parameter ih becomes of the same order as R, as shown in Fig. 12-14.

Deep Spherical Shell Segments. There has been considerable discussion of the natural frequencies of segments of spherical shells in the literature (Refs. 12-34 (pp. 602–607) and 12-37 through 12-43). However, theoretical and computational difficulties have not permitted the formulation of a simple general expression for the natural frequencies of deep spherical shell segments. One exception is the inextensional (bending) modes of a free open spherical shell. Lord Rayleigh found that thin, open spherical shells, $0 < \phi < \phi_0$, having a free edge at $\phi = \phi_0$ possess inextensional modes which are very similar to the higher circumferential modes of cylindrical shells (Eq. 12-24). The mode shapes are (Ref. 12-18, pp. 419–432)

$$\begin{pmatrix} \tilde{\theta} \\ \tilde{\phi} \\ \tilde{r} \end{pmatrix}_i = \begin{pmatrix} -\sin\phi \left(\tan\dfrac{\phi}{2}\right)^i \cos i\theta \\ -\left(\tan\dfrac{\phi}{2}\right)^i \sin i\theta \\ R(i+\cos\phi)\left(\tan\dfrac{\phi}{2}\right)^i \cos i\theta \end{pmatrix}, \qquad (12\text{-}44)$$

and the natural frequencies are

$$f_i = \frac{\lambda_i}{2\pi R}\left(\frac{E}{\mu}\right)^{1/2}; \qquad i = 2, 3, 4, \ldots \qquad (12\text{-}45)$$

$\tilde{\theta}$ is the circumferential mode shape, $\tilde{\phi}$ is the polar angle mode shape, and $\tilde{r}$ is the radial mode shape. f_i is the natural frequency in hertz. R is the radius to the mid-surface of the shell, E is the modulus of elasticity, and μ is the density of the shell material. i is the number of circumferential waves in the mode shape. ϕ is the angle measured from the pole (Table 12-9) and θ is the circumferential angle. The dimensionless parameter λ_i is a function of the angle subtended by the shell segment:

$$\lambda_i^2 = \frac{(i^2 - 1)^2 i^2}{3(1+\nu)} \left(\frac{h}{R}\right)^2 \frac{g_1(i, \phi_0)}{g_2(i, \phi_0)}, \tag{12-46}$$

where

$$g_1(i, \phi_0) = \frac{1}{8}\left[\frac{\left(\tan\frac{\phi_0}{2}\right)^{2i-2}}{i-1} + \frac{2\left(\tan\frac{\phi_0}{2}\right)^{2i}}{i} + \frac{\left(\tan\frac{\phi_0}{2}\right)^{2i+2}}{i+1}\right],$$

$$g_2(i, \phi_0) = \int_0^{\phi_0} \left(\tan\frac{\phi}{2}\right)^{2i} [(i+\cos\phi)^2 + 2(\sin\phi)^2] \sin\phi \, d\phi.$$

Some numerically evaluated values of g_1/g_2 are:

g_1/g_2

Mode Index	ϕ_0 (degrees)				
i	10	30	60	90	120
2	34.32	1.569	0.3070	0.1907	0.3101
3	13.23	0.6382	0.1363	0.09412	0.1720
4	7.153	0.3535	0.07838	0.05626	0.1056
5	4.506	0.2257	0.05100	0.03724	0.07036

For a hemisphere, $[\phi_0 = \pi/2, \tan(\phi_0/2) = 1]$ and Eq. 12-46 becomes

$$\lambda_i^2 = \frac{1}{12(1+\nu)} \left(\frac{h}{R}\right)^2 \frac{(i^3 - i)(2i^2 - 1)}{g_2(i, \phi_0 = \pi/2)}, \qquad i = 2, 3, 4, \ldots, \tag{12-47}$$

where

$$g_2(i, \phi_0 = \pi/2) = \begin{cases} 1.52961 & i = 2 \\ 1.88156 & i = 3. \\ 2.29609 & i = 4 \end{cases}$$

Equations 12-45 and 12-47 have been experimentally verified for the case of a thin hemispherical shell with a free edge (Ref. 12-38) and agree with shallow shell results for small ϕ_0 (Ref. 12-44).

Shallow Spherical Shell Segments. A simplified shell theory, called shallow shell theory, has been developed to describe the vibrations of shallow curved shells, such as that shown in Fig. 12-15. For shallow shell theory to be valid, the rise of the shell

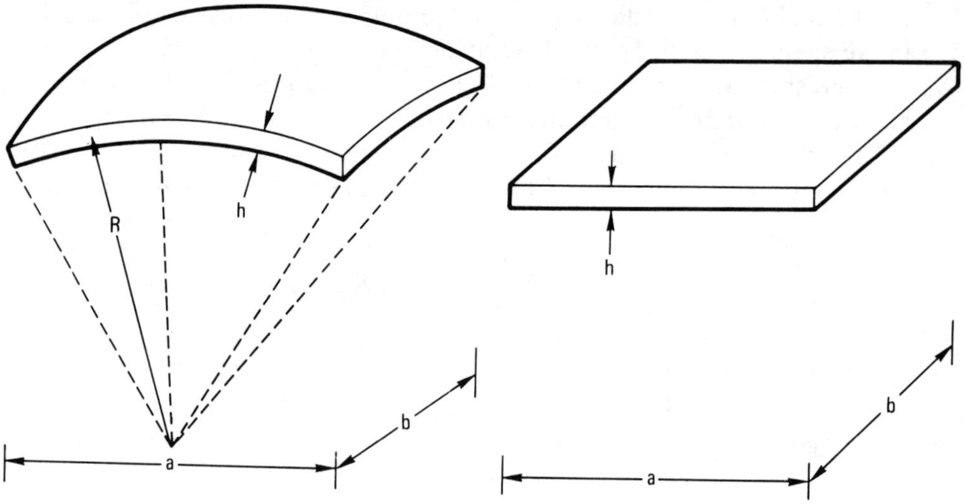

Fig. 12-15. A shallow spherical shell segment with a rectangular plan and a similar flat plate.

must be less than about $\frac{1}{8}$ the typical lateral projection of the shell and the shell must be thin. The deformation during vibration of these shells is dominated by vibration perpendicular to the shell surface. Thus, it is reasonable to retain only the inertia associated with normal deformation in the analysis. Using shallow shell theory, Soedel (Ref. 12-45) has shown that a powerful analogy exists between the natural frequencies and mode shapes of flat plates and those of similar shallow spherical shell segments.

If a flat plate and a segment of a shallow spherical shell have (1) the same thickness, (2) the same homogeneous isotropic material, (3) analogous, homogeneous boundary conditions (i.e., clamped edge, free edge, simply supported edge), and (4) the same projected lateral dimensions (i.e., they result in shadows of the same dimensions on a plane surface due to a light source at infinity), then (1) the mode shapes governing the deformation normal to the surface are identical and (2) the natural frequencies are related by:

$$\omega_{ij}\Big|_{\substack{\text{shallow}\\ \text{spherical}\\ \text{shell}}} = \left(\omega_{ij}^2\Big|_{\substack{\text{flat}\\ \text{plate}}} + \frac{E}{\mu R^2}\right)^{1/2}. \tag{12-48}$$

ω_{ij} is the circular natural frequency in the ij mode. E is the modulus of elasticity, μ is the material density, and R is the radius to the midsurface of the shallow, spherical shell segment. For a flat plate, $1/R = 0$.

Using this analogy, it is possible to relate the natural frequencies of flat plates of various shapes, such as rectangular, annular, and circular, to the natural frequencies of similar shallow, spherical shell segments. For example, the natural frequencies of a flat simply supported plate with lateral dimensions a and b (Fig. 12-15) are:

$$\omega_{ij}^2 = \frac{\pi^4}{12}\left(\frac{i^2}{a^2} + \frac{j^2}{b^2}\right)^2 \frac{Eh^2}{\mu(1-\nu^2)}; \qquad i, j = 1, 2, 3, \ldots .$$

Therefore, the natural frequencies of a similar shallow curved shell segment with projected dimensions a and b are:

$$\omega_{ij} = \left[\frac{\pi^4}{12}\left(\frac{i^2}{a^2} + \frac{j^2}{b^2}\right)^2 \frac{Eh^2}{\mu(1 - \nu^2)} + \frac{E}{\mu R^2}\right]^{1/2}; \qquad i, j = 1, 2, 3, \ldots.$$

This agrees exactly with the solution in the literature (Ref. 12-46). Equation 12-48 has also been found to agree very well with seminumerical solutions for the fundamental symmetric mode of clamped and free shallow spherical shells with circular plan forms (Ref. 12-47).

Despite the power of the analogy of Eq. 12-48, it may lead to errors in certain cases. In the derivation of this equation, the transverse and the longitudinal deformations are assumed to be intimately coupled. Inextensional, i.e., purely flexural, deformations are not incorporated in the analysis. While this is appropriate for the fundamental symmetric mode of shallow shells with fixed edges, it is of questionable accuracy for the higher modes of certain shallow shells, particularly those with free edges.

For example, consider a spherically curved arch with fixed ends. If the arch is modeled as a shallow spherically curved shell with pinned ends, then the predicted natural frequencies are (frame 6 of Table 8-1 and Eq. 12-48):

$$\omega_i\bigg|_{arc} = \left(\frac{i^4\pi^4}{L^4}\frac{EI}{m} + \frac{E}{\mu R^2}\right)^{1/2}; \qquad i = 1, 2, 3, \ldots.$$

If i = 1 and, for slender arches,

$$\frac{\pi^2}{L^4}\frac{EI}{m} << \frac{E}{\mu R^2},$$

the fundamental natural frequency is predicted to be:

$$\omega_1 = \frac{1}{R}\left(\frac{E}{\mu}\right)^{1/2}.$$

This agrees well with the extensional arc mode of frame 1 of Table 9-2 and the results of Reisner (Ref. 9-8). However, as frame 2 of Table 9-2 and Ref. 9-8 indicate, the arch can have inextensional higher modes which have frequencies lower than this mode. The analogy of Eq. 12-48 cannot accurately predict the natural frequencies of these higher modes. Thus, Eq. 12-48 should be applied with caution for the higher modes of shallow shells, particularly those with free edges.

Torsion of Shallow Spherical Caps. Seide (Ref. 12-34, pp. 610–614) has developed approximate formulas for the natural frequencies of the purely torsional modes (modes with only circumferential deformation) of thin spherical caps which subtend $0 < \phi < \phi_0$. $\phi = \phi_0$ is the edge of the cap. The torsional natural frequencies of the cap in hertz are:

$$f_i = \frac{\lambda_i}{2\pi R}\left[\frac{E}{\mu(1 - \nu^2)}\right]^{1/2}; \qquad i = 1, 2, 3, \ldots, \qquad (12\text{-}49)$$

where

$$\lambda_i^2 = \frac{1 - \nu}{2(1 + 5k)} \left[\left(\frac{\alpha_i}{\phi_0} \right)^2 - 2 \right], \qquad k = \frac{h^2}{12R^2}.$$

R is the radius of the cap to the midsurface and h is the cap thickness. E is the modulus of elasticity of the cap material which has Poisson's ratio ν and mass density μ. The angle ϕ_0 is measured in radians. Consistent sets of units are given in Table 3-1.

The dimensionless parameter α_i is a function of the index i and the boundary conditions on the cap:

α_i

Boundary Condition at $\phi = \phi_0$	Mode Number (i)				
	1	2	3	4	i
Free	5.136	8.417	11.62	14.80	$[\mathcal{J}_2(\alpha_i) = 0]$
Fixed	3.832	7.016	10.17	13.32	$[\mathcal{J}_1(\alpha_i) = 0]$

$\mathcal{J}_1$ and $\mathcal{J}_2$ are Bessel functions of the first kind (see Ref. 13-4, p. 409 for higher roots). Equation 12-49 can be expected to be within a few percent of the exact result for thin spherical caps as deep as $\phi_0 = 90°$ ($\pi/2$ radians).

Other Shallow Shells. The natural frequencies of shallow shells, other than spherical shallow shells, are discussed in Refs. 12-48 and 12-49. Additional studies of shallow spherical shells can be found in Refs. 12-46 and 12-50 through 12-53.

REFERENCES

12-1. Love, A. E. H., "The Small Free Vibrations of a Thin Elastic Shell," *Phil. Trans. Roy. Soc. (London), Ser. A.* **179,** 491–549 (1888).

12-2. Love, A. E. H., *A Treatise on the Mathematical Theory of Elasticity*, 4th ed., Dover Publications, New York, 1944.

12-3. Flügge, W., *Statik und Dynamik der Schalen*, Julius Springer, Berlin, 1934 (reprinted by Edwards Brothers, Ann Arbor, Michigan, 1943).

12-4. Flügge, W., *Stresses in Shells*, 2nd ed., Springer-Verlag, New York, 1973.

12-5. Donnell, L. H., "Stability of Thin Walled Tubes under Torsion," NACA Report No. 479, 1933.

12-6. Mushtari, K. M., "On the Stability of Cylindrical Shells Subject to Torsion," *Trudy Kaz. avais, in-ta* **2** (1938) (in Russian).

12-7. Reissner, E., "A New Derivation of the Equations of the Deformation of Elastic Shells," *Am. J. Math.* **63,** 177–184 (1941).

12-8. Sanders, J. L., "An Improved First Approximation Theory for Thin Shells," NASA Report NASA-TR-R24, 1959.

12-9. Leissa, A. W., "Vibrations of Shells," NASA Report NASA-SP-288, Ohio State University, 1973 (out of print).

12-10. Arnold, R. N., and G. B. Warburton, "Flexural Vibrations of Thin Cylindrical Shells Having Freely Supported Ends," *Proc. Roy. Soc. (London), Ser. A* **197,** 238–256 (1949).

12-11. Kraus, H. *Thin Elastic Shells*, John Wiley, New York, 1967, pp. 200–201.

12-12. Dym, C. L., "Some New Results for the Vibrations of Circular Cylinders," *J. Sound Vib.* **29,** 189–205 (1973).

12-13. Forsberg, K., "A Review of Analytical Methods Used to Determine the Modal Characteristics of Cylindrical Shells," NASA Report NASA CR-613, Lockheed Aircraft Company, Ca., September 1966.
12-14. Sharma, C. B., and D. J. Johns, "Vibration Characteristics of a Clamped-Free and Clamped-Ring-Stiffened Circular Cylindrical Shell," *J. Sound Vib.* **14,** 459–474 (1971).
12-15. Sharma, C. B., "Frequencies of Clamped-Free Circular Cylindrical Shells," *J. Sound Vib.* **30,** 525–528 (1973).
12-16. Sharma, C. B., "Simple Linear Formula for Critical Frequencies for Cantilevered Circular Cylindrical Shells," *J. Sound Vib.* **55,** 467–471 (1977).
12-17. Sewall, J. L., and E. C. Naumann, "An Experimental and Analytical Vibration Study of Thin Cylindrical Shells With and Without Longitudinal Stiffeners," NASA Report NASA-TN D-4705, Langley Research Center, September 1968.
12-18. Rayleigh, J. W. S., *Theory of Sound*, Vol. 1, 2nd ed., MacMillan Company, 1894; also Dover Publications, New York, 1945, pp. 385–386.
12-19. Warburton, G. B., "Vibration of Thin Cylindrical Shells," *J. Mech. Eng. Sci.* **7,** 399–407 (1965).
12-20. Arnold, R. N., and G. B. Warburton, "The Flexural Vibrations of Thin Cylinders," *Proc. Inst. Mech. Engineers (Ser. A)* **167,** 62–80 (1953).
12-21. Koval, L. R., and E. T. Cranch, "On The Free Vibrations of Thin Cylindrical Shells Subjected to an Initial Static Torque," *4th U.S. Nat. Congr. Appl. Mech., June 18-21, 1962,* Vol. 1, pp. 107–117.
12-22. Forsberg, K., "Influence of Boundary Conditions on the Modal Characteristics of Thin Cylindrical Shells," *AIAA J.* **2,** 2150–2157 (1964).
12-23. Bozick, W. F., "The Vibration and Buckling Characteristics of Cylindrical Shells Under Axial Load and External Pressure," AFFDL-TR-67-28, May 1967.
12-24. Armenakas, A. E., and G. Herrmann, "Vibrations of Infinitely Long Cylindrical Shells Under Initial Stress," *AIAA J.* **1,** 100–106 (1963).
12-25. Fung, Y. C., E. E. Sechler, and A. Kaplan, "On the Vibration of Thin Cylindrical Shells Under Internal Pressure," *J. Aeronaut. Sci.* **24,** 650–661 (1957).
12-26. Sewall, J. L., "Vibration Analysis of Cylindrically Curved Panels with Simply Supported or Clamped Edges and Comparison with Some Experiments," NASA Report NASA-TN-D-3791, Langley Research Center, 1967.
12-27. Webster, J. J., "Free Vibration of Rectangular Curved Panels," *Int. J. Mech. Sci.* **10,** 571–582 (1968).
12-28. Dreher, J. F., and A. W. Leissa, "Axisymmetric Vibration of Thin Conical Shells," *Proc. 4th Southwestern Conf. on Theoretical and Appl. Mech. (New Orleans, La.), Feb. 29-1 Mar., 1968,* pp. 163–181, also Ref. 12-9.
12-29. Hu, W. C. L., J. F. Gormley, and U. S. Lindholm, "An Experimental Study and Inextensional Analysis of Free-Free Conical Shells," *Int. J. Mech. Sci.* **9,** 123–135 (1967).
12-30. Grigolyuk, E. I., "Small Oscillations of Thin Elastic Shells," *Izv. Akad. Nauk. SSR, O.T.D.*, No. 6, 1956, NASA TT F-25.
12-31. Garnet, H., M. A. Goldberg, and V. L. Salerno, "Torsional Vibrations of Shells of Revolution," *J. Appl. Mech.* **28,** 571–573 (1961).
12-32. Hartung, R. F., and W. A. Loden, "Axisymmetric Vibration of Conical Shells," *J. Spacecraft and Rockets* **7,** 1153–1159 (1970).
12-33. Herrmann, G., and I. Mirsky, "On Vibrations of Conical Shells," *J. Aerospace Sci.* **25,** 451–458 (1958).
12-34. Seide, P., *Small Elastic Deformations of Thin Shells*, Noordhoff International Publishing, Leyden, The Netherlands, 1975, pp. 615–620.
12-35. Baker, W. E., "Axisymmetric Modes of Vibration of Thin Spherical Shells," *J. Acoust. Soc. Am.* **33,** 1749–1758 (1961).
12-36. Wilkinson, J. P., "Natural Frequencies of Closed Spherical Shells," *J. Acoust. Soc. Am.* **38,** 367–368 (1965).
12-37. Shah, A. H., C. V. Ramkrishnan, and S. K. Datta, "Three Dimensional and Shell-Theory Analysis of Elastic Waves in a Hollow Sphere," *J. Appl. Mech.* **38,** 431–444 (1969).
12-38. Hwang, C., "Some Experiments on the Vibration of a Hemispherical Shell," *J. Appl. Mech.* **33,** 817–824 (1966).
12-39. Kalnis, A., "Discussion of 'Some Experiments on the Vibration of a Hemispherical Shell'," *J. Appl Mech.* **34,** 792–794 (1967).
12-40. Eikrem, A. K., and A. G. Dodge, "Natural Frequencies of a Hemispherical Shell," *Exp. Mech.* **12,** 575–579 (1972).
12-41. Kalnins, A., "Effect of Bending on Vibrations of Spherical Shells," *J. Acoust. Soc. Am.* **36,** 74–81 (1964).
12-42. Archer, R. R., "On the Influence of Uniform Stress States on the Natural Frequencies of Spherical Shells," *J. Appl. Mech.* **29,** 502–505 (1962).

12-43. Naghdi, P. M., and A. Kalnins, "On Vibrations of Elastic Spherical Shells," *J. Appl. Mech.* **29**, 65–72 (1962).

12-44. Johnson, M. W., and E. Reissner, "On Inextensional Deformations of Shallow Elastic Shells," *J. Math. Phys.* **34**, 335–346 (1956).

12-45. Soedel, W., "A Natural Frequency Analogy Between Spherically Curved Panels and Flat Plates," *J. Sound Vib.* **29**, 457–461 (1973).

12-46. Nowacki, W., *Dynamics of Elastic Systems*, John Wiley, New York, 1963, pp. 269–272.

12-47. Reissner, E., "On Axi-Symmetrical Vibrations of Shallow Spherical Shells," *Quart. Appl. Math.* **13**, 279–290 (1955).

12-48. Reissner, E., "On Transverse Vibrations of Thin, Shallow Elastic Shells," *Quart. Appl. Math.* **13**, 169–176 (1955).

12-49. Jones, R., and J. Mazumdar, "Transverse Vibrations of Shallow Shells by the Method of Constant-Deflection Contours," *J. Acoust. Soc. Am.* **56**, 1487–1492 (1974).

12-50. Koplik, B., and Y. Yu, "Approximate Solutions for Frequencies of Axisymmetric Vibrations of Spherical Caps," *J. Appl. Mech.* **34**, 785–787 (1967).

12-51. Hoppmann, W. H., II, and C. N. Baronet, "A Study of the Vibrations of Shallow Spherical Shells," *J. Appl. Mech.* **30**, 329–334 (1963).

12-52. Archer, R. R., and J. Famili, "On the Vibration and Stability of Finitely Deformed Spherical Shells," *J. Appl. Mech.* **32**, 116–120 (1965).

12-53. Lock, M. H., J. S. Whittier, and H. A. Malcom, "Transverse Vibrations of a Shallow Spherical Dome," *J. Appl. Mech.* **35**, 402–403 (1968).

12-54. Szechenyi, E., "Approximate Formulas for the Determination of the Natural Frequencies of Stiffened and Curved Panels," *J. Sound Vib.* **14**, 401–418 (1971).

12-55. Jager, E. H., "An Engineering Approach to Calculating the Lowest Natural Frequencies of Thin Conical Shells," *J. Sound Vib.* **63**, 259–264 (1979).

13

FLUID SYSTEMS

13.1. ACOUSTIC CAVITIES

General Case. Acoustic modes in fluid-filled cavities arise from pressure waves which traverse the cavities at the speed of sound and reflect off the boundaries of the cavities. The speed of sound in a fluid (i.e., a gas or a liquid) is generally a function of the nature of the fluid, its temperature, and its pressure. The speed of sound in various media is given in Table 13-1. The natural frequencies of acoustic cavities can be heard in the echoes of a large room or the resonant tones of a pipe organ. Acoustic cavity modes are of practical importance in the design of auditoriums and fluid machinery such as turbines and tubular heat exchangers.

The general assumptions used in the analysis of the acoustic cavities presented in this section are:

1. The fluid contained in the cavity is homogeneous, inviscid, and under a uniform mean pressure.
2. Any mean flow in the cavity is much less than the speed of sound.
3. The cavity is bordered by rigid walls.
4. The fluctuations of pressure in the field are much less than the mean pressure on the fluid.

These assumptions are applicable to the natural frequency analysis of most acoustic cavities of practical importance.

As a sound wave moves through a fluid, the fluid particles are cyclically displaced from their equilibrium positions and the pressure in the fluid rises and then falls. If a homogeneous, inviscid fluid in an acoustic cavity oscillates freely in one of the normal modes of the cavity, the motion of the fluid can be expressed in terms of a velocity potential:

$$\Phi = A \frac{c^2}{\omega} \tilde{\phi} \sin(\omega t + \psi), \qquad (13\text{-}1)$$

where

A = a dimensionless constant which specifies the amplitude of vibration,
c = speed of sound (Table 13-1),
t = time,
$\tilde{\phi} = \tilde{\phi}(x, y, z)$, mode shape of the potential function,
ω = circular natural frequency of vibration ($2\pi f$),
ψ = a phase angle.

Table 13-1. Speed of Sound.

Notation: γ = ratio of specific heat at constant pressure to that at constant volume; p = mean pressure; R = gas constant (Section 16.3); W = molecular weight (Table 16-10); ρ = mass density (fluid); B = isothermal bulk modulus; E = modulus of elasticity; μ = mass density (solid); ν = Poisson's ratio; consistent sets of units are given in Table 3-1

Medium	Speed of Sound, c	Examples and Remarks
1. Gases	$\left(\frac{\gamma p}{\rho}\right)^{1/2}$ or $\left(\frac{\gamma RT}{W}\right)^{1/2}$	Adiabatic Gas Law. c is nearly independent of pressure for most gases. For air, $c = 20.05\ T^{1/2}$ m/sec, where T = temperature in °Kelvin (°C + 273.2); c (air) = 343 m/sec at 20°C. Also see Tables 16-10 and 16-11.
2. Liquids	$\left(\frac{\gamma B}{\rho}\right)^{1/2}$	c is nearly independent of pressure for water. For distilled water, $c = 1403 + 5T - 0.06T^2 + 0.0003T^3$ m/sec. For sea water, $c = 1449 + 4.6T - 0.055T^2 + 0.0003T^3 + (1.39 - 0.012T)(S - 35) + 0.017d$ m/sec, where T = temperature in °C, S = salinity in parts per thousand, d = depth below surface in meters. c (fresh water) = 1481 m/sec at 20°C. c (salt water) ≈ 1500 m/sec at 13°C. Refs. 13-1 (pp. 116–118, 461), 13-2, 13-3. Also see Tables 16-7, 16-8, and 16-9.
3. Solids	Slender solids: $\left(\frac{E}{\mu}\right)^{1/2}$ Bulk solids: $\left[\frac{E(1-\nu)}{\mu(1+\nu)(1-2\nu)}\right]^{1/2}$	Slender solids = beams, bars, etc. c is nearly independent of pressure for most solids. (See table below.)

Material[a]	c (m/sec) Slender Solid	c (m/sec) Bulk Solid
Aluminum	5150	6300
Brass	3500	4700
Copper	3700	5000
Iron	3700	4350
Nickel	4900	5850
Steel	5050	6100
Glass	5200	5600
Lucite	1800	2650
Concrete	–	3100

[a]At 20°C.

The dynamic variation in pressure in the cavity can be expressed in terms of the velocity potential by:

$$p = \rho \frac{\partial \Phi}{\partial t} = A\rho c^2 \tilde{\phi} \cos(\omega t + \psi), \qquad (13\text{-}2)$$

where ρ is the mean density of the fluid. The dynamic variation in fluid velocity is:

$$\vec{u} = -\nabla \Phi, \tag{13-3}$$

where $\vec{u}$ is the vector fluid velocity and ∇ is the gradient operator. For rectangular coordinates (frames 1 through 7 of Table 13-2), Eq. 13-3 implies:

$$u_x = -\frac{\partial \Phi}{\partial x}, \qquad u_y = -\frac{\partial \Phi}{\partial y}, \qquad u_z = -\frac{\partial \Phi}{\partial z}, \tag{13-4}$$

where u_x, u_y, and u_z are the components of fluid velocity in the x, y, and z coordinate directions, respectively. For cylindrical coordinates (frames 8 through 15 of Table 13-2), Eq. 13-3 implies:

$$u_r = -\frac{\partial \Phi}{\partial r}, \qquad u_\theta = -\frac{1}{r}\frac{\partial \Phi}{\partial \theta}, \qquad u_z = -\frac{\partial \Phi}{\partial z}, \tag{13-5}$$

where u_r is the radial component of velocity, u_θ is the circumferential component of velocity, and u_z is the axial component of velocity. The fluid displacements are obtained by integrating the fluid velocities over time.

If the fluid oscillates freely in more than one mode, then the dynamic component of fluid pressure, the fluid velocities, and fluid displacements are simply the sum of the contributions by the individual modes and the total velocity potential function is:

$$\Phi = \sum_{i,j,k} A_{ijk} \frac{c^2}{\omega_{ijk}} \tilde{\phi}_{ijk} \sin(\omega_{ijk} t + \psi_{ijk}), \tag{13-6}$$

where $\tilde{\phi}_{ijk}$ and ω_{ijk} are the mode shapes of the potential functions and the circular natural frequency ($2\pi f_{ijk}$) of the ijk mode. A_{ijk} and ψ_{ijk} specify the amplitude of each mode and its phase angle with respect to other modes. A_{ijk} and ψ_{ijk} are determined by the means used to set the fluid into motion. The three indices i, j, and k are generally required to specify each of the acoustic modes of a three-dimensional cavity.

Boundary Conditions. When a pressure wave traveling through a fluid encounters a boundary, it may be reflected by the boundary, transmitted through the boundary, or partially reflected and partially transmitted, as shown in Fig. 13-1. For a plane wave striking a semi-infinite boundary at right angles to the boundary, the portion of the sound power which is reflected by the boundary is (Ref. 13-1, pp. 128–134):

$$\alpha_r = \left[\frac{\dfrac{\rho_2 c_2}{\rho_1 c_1} - 1}{\dfrac{\rho_2 c_2}{\rho_1 c_1} + 1}\right]^2. \tag{13-7}$$

This equation is exact for a fluid-fluid boundary and approximate for a fluid-solid boundary. ρ is the density, c is the speed of sound, and the subscripts 1 and 2 refer to the incident medium and boundary medium, respectively. If $\rho_2 c_2 >> \rho_1 c_1$, the

Table 13.2. Acoustic Cavities.

Notation: D = diameter; L = length; L_x, L_y, L_z = lengths associated with x, y, and z coordinates, respectively; R = radius; $\mathcal{J}_j$ = Bessel function of first kind and j order; $\mathcal{Y}_j$ = Bessel function of second kind and j order; c = speed of sound (Table 13-1); r = a radius; x, y, z = mutually orthogonal coordinates; α = angle (degrees); θ = angle (radians); see Table 3-1 for consistent sets of units. The pressure, velocity, and displacement of the fluid in the cavity are expressed in terms of $\tilde{\phi}$ using the formulas in the text.

Description	Natural Frequency, f_i (hertz)	Mode Shape of Velocity Potential, $\tilde{\phi}_i$
1. Slender Tube with Both Ends Cpen, L >> D 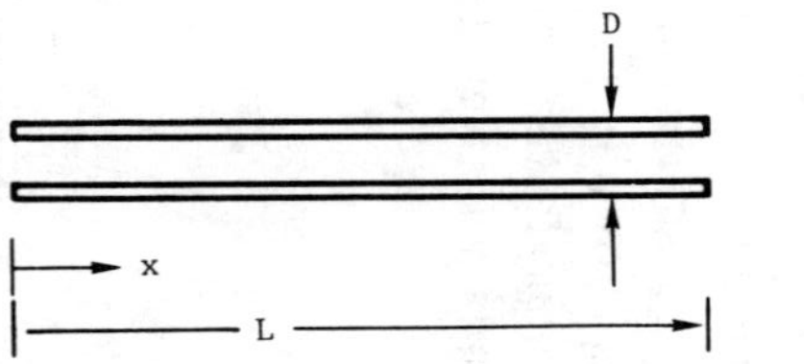	$\frac{ic}{2L}$; i=1,2,3...	$\sin \frac{i\pi x}{L}$, i=1,2,3... Result is independent of the cross section of the tube.
2. Slender Tube with One End Open, One End Closed, L >> D 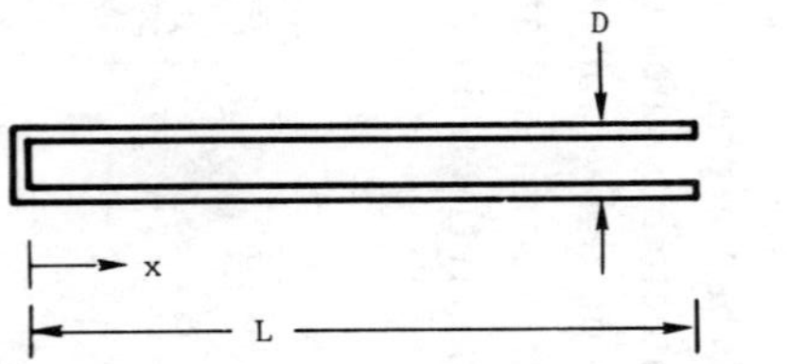	$\frac{ic}{4L}$; i=1,3,5,...	$\cos \frac{i\pi x}{2L}$, i=1,3,5... Result is independent of the cross section of the tube. See text for more accurate formula for short tubes.
3. Slender Tube with Both Ends Closed, L >> D 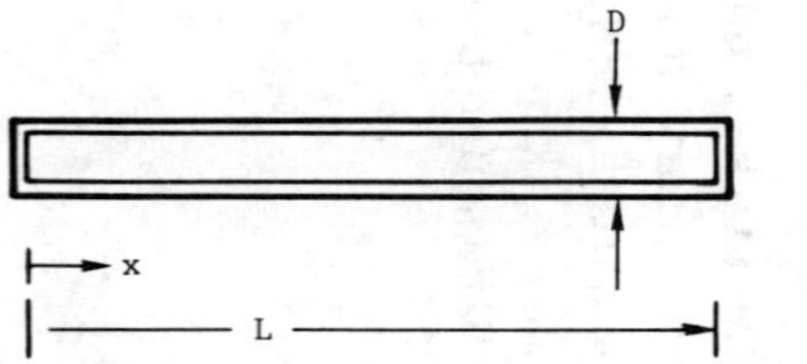	$\frac{ic}{2L}$; i=1,2,3...	$\cos \frac{i\pi x}{L}$, i = 1,2,3 Result is independent of the cross section of the tube.

4. Rectangular Volume Open at Opposite Ends, x = 0, L 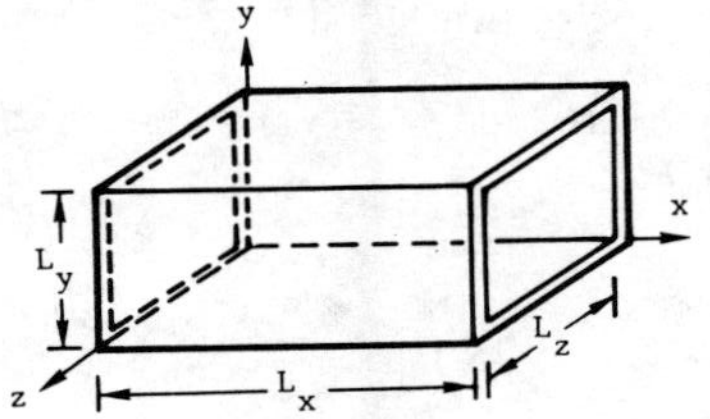	$\frac{c}{2}\left(\frac{i^2}{L_x^2} + \frac{j^2}{L_y^2} + \frac{k^2}{L_z^2}\right)^{1/2}$ i = 0, 1, 2, ... j = 0, 1, 2, ... k = 0, 1, 2, ...	$\sin\frac{i\pi x}{L_x} \cos\frac{j\pi y}{L_y} \cos\frac{k\pi z}{L_z}$ i = 0, 1, 2, ... j = 0, 1, 2, ... k = 0, 1, 2, ...
5. Rectangular Volume Open at One End x = L, Closed at x = 0 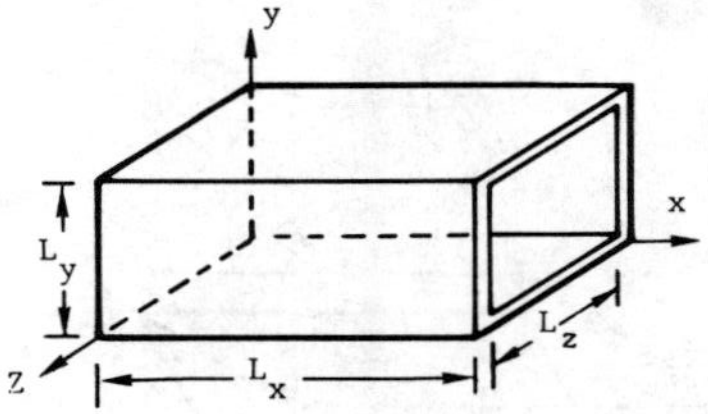	$\frac{c}{2}\left(\frac{i^2}{4L_x^2} + \frac{j^2}{L_y^2} + \frac{k^2}{L_z^2}\right)^{1/2}$ i = 0, 1, 3, 5, ... j = 0, 1, 2, ... k = 0, 1, 2, ...	$\cos\frac{i\pi x}{2L_x} \cos\frac{j\pi y}{L_y} \cos\frac{k\pi z}{L_z}$ i = 0, 1, 3, 5, ... j = 0, 1, 2, ... k = 0, 1, 2, ...
6. Closed Rectangular Volume 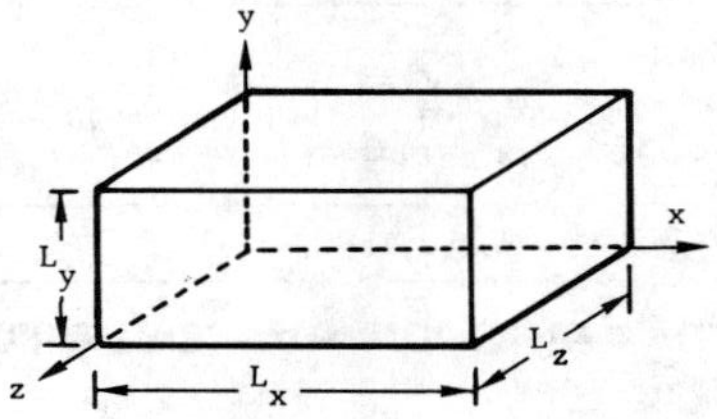	$\frac{c}{2}\left(\frac{i^2}{L_x^2} + \frac{j^2}{L_y^2} + \frac{k^2}{L_z^2}\right)^{1/2}$ i = 0, 1, 2, ... j = 0, 1, 2, ... k = 0, 1, 2, ...	$\cos\frac{i\pi x}{L_x} \cos\frac{j\pi y}{L_y} \cos\frac{k\pi z}{L_z}$ i = 0, 1, 2, ... j = 0, 1, 2, ... k = 0, 1, 2, ...

Table 13-2. Acoustic Cavities. (*Continued*)

Description	Natural Frequency, f_{ijk} (hertz)	Mode Shape of Velocity Potential, $\tilde{\phi}_{ijk}$
7. Volume with Isosceles Right Triangular Cross Section 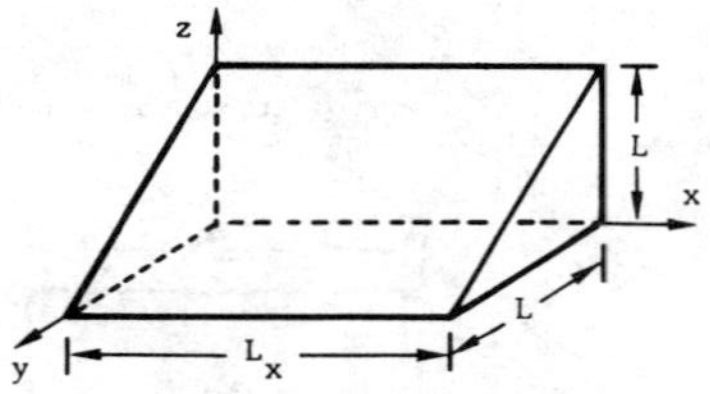 Sides at y = 0, z = 0, and z = -y + L are closed.	If ends at x = 0, L_x are open: $f_{ijk} = \frac{c}{2}\left(\frac{i^2}{L_x^2} + \frac{j^2}{L^2} + \frac{k^2}{L^2}\right)^{1/2}$ i = 0,1,2...	$\tilde{\phi}_{ijk} = \sin\frac{i\pi x}{L_x}\left(\cos\frac{j\pi y}{L}\cos\frac{k\pi z}{L} \pm \cos\frac{j\pi z}{L}\cos\frac{k\pi y}{L}\right)$ - if j + k = even, + if j + k = odd. (j,k) = (0,0), (1,2), (2,1), (3,1), (1,3), (2,3), etc. $j \neq k$ unless j = k = 0
	If end at x = 0 is closed and end at x = L_x is open: $f_{ijk} = \frac{c}{2}\left(\frac{i^2}{4L_x^2} + \frac{j^2}{L^2} + \frac{k^2}{L^2}\right)^{1/2}$ i=0,1,3,5...	$\tilde{\phi}_{ijk} = \cos\frac{i\pi x}{2L_x}\left(\cos\frac{j\pi y}{L}\cos\frac{k\pi z}{L} \pm \cos\frac{j\pi z}{L}\cos\frac{k\pi y}{L}\right)$ - if j + k = even, + if j + k = odd. (j,k) = (0,0), (1,2), (2,1), (3,1), (1,3), (2,3), etc. $j \neq k$ unless j = k = 0
	If ends at x = 0, L_x are closed: $f_{ijk} = \frac{c}{2}\left(\frac{i^2}{L_x^2} + \frac{j^2}{L^2} + \frac{k^2}{L^2}\right)^{1/2}$ i=0,1,2...	$\tilde{\phi}_{ijk} = \cos\frac{i\pi x}{L_x}\left(\cos\frac{j\pi y}{L}\cos\frac{k\pi z}{L} \pm \cos\frac{j\pi z}{L}\cos\frac{k\pi y}{L}\right)$ - if j + k = even, + if j + k = odd. (j,k) = (0,0), (1,2), (2,1), (3,1), (1,3), (2,3), etc. $j \neq k$ unless j = k = 0

8. Right Cylindrical Volume Open at Both Ends

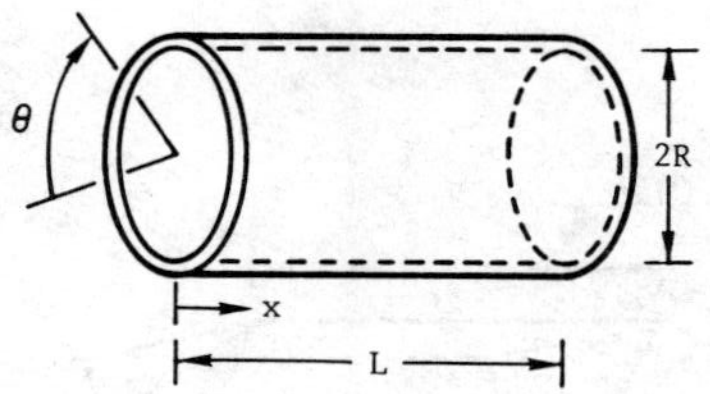

$$\frac{c}{2\pi}\left(\frac{\lambda_{jk}^2}{R^2} + \frac{i^2\pi^2}{L^2}\right)^{1/2}$$

i=0,1,2...
j=0,1,2...
k=0,1,2...

$$\left[J_j'(\lambda_{jk}) = 0\right]$$

$$J_j\left(\lambda_{jk}\frac{r}{R}\right)\sin\frac{i\pi x}{L}\begin{Bmatrix}\sin j\theta \\ \text{or} \\ \cos j\theta\end{Bmatrix}$$

i=0,1,2...
j=0,1,2... = number of nodal diameters
k=0,1,2... = number of nodal circles

	λ_{jk}						
	j						
k	0	1	2	3	4	5	6
0	0	1.8412	3.0542	4.2012	5.3176	6.4156	7.5013
1	3.8317	5.3314	6.7061	8.0152	9.2824	10.5199	11.7349
2	7.0156	8.5363	9.9695	11.3459	12.6819	13.9872	15.2682
3	10.173	11.7060	13.1704	14.5859	15.9641	17.3128	18.6374

$\lambda_{j=0,k} \approx \pi(k + 0.25)$ for $k \geq 3$

Additional values of λ_{jk} are tabulated in Ref. 13-4, pp. 411, 468.

9. Right Cylindrical Volume, Open at One End, Closed at Other End

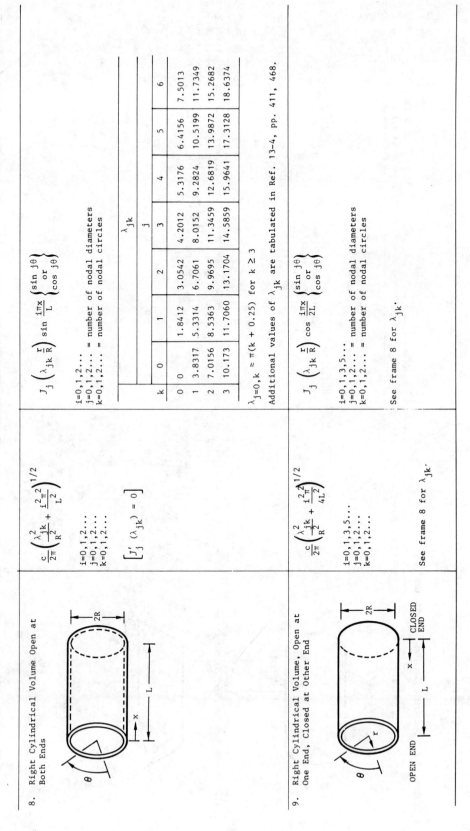

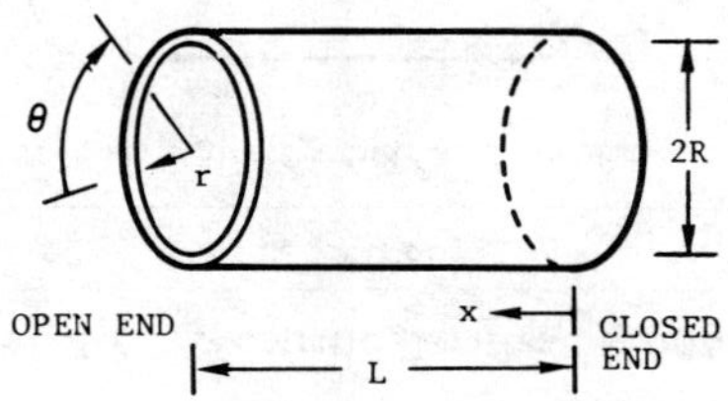

$$\frac{c}{2\pi}\left(\frac{\lambda_{jk}^2}{R^2} + \frac{i^2\pi^2}{4L^2}\right)^{1/2}$$

i=0,1,3,5...
j=0,1,2...
k=0,1,2...

See frame 8 for λ_{jk}.

$$J_j\left(\lambda_{jk}\frac{r}{R}\right)\cos\frac{i\pi x}{2L}\begin{Bmatrix}\sin j\theta \\ \text{or} \\ \cos j\theta\end{Bmatrix}$$

i=0,1,3,5...
j=0,1,2... = number of nodal diameters
k=0,1,2... = number of nodal circles

See frame 8 for λ_{jk}.

Table 13-2. Acoustic Cavities. *(Continued)*

Description	Natural Frequency, f_{ijk} (hertz)	Mode Shape of Velocity Potential, $\tilde{\phi}_{ijk}$
10. Closed Right Cylindrical Volume 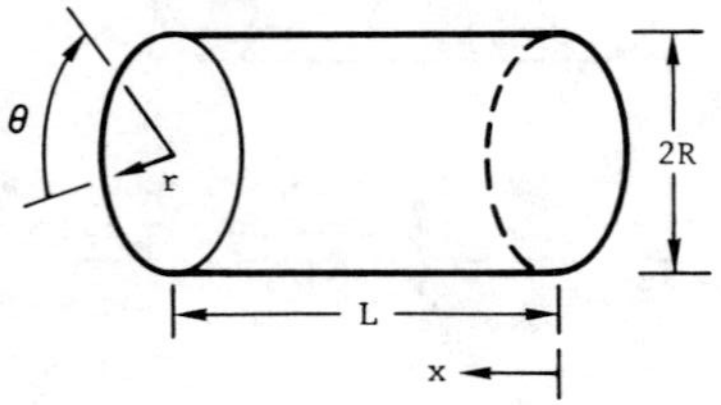	$\frac{c}{2\pi}\left(\frac{\lambda_{jk}^2}{R^2} + \frac{i^2\pi^2}{L^2}\right)^{1/2}$ i=0,1,2... j=0,1,2... k=0,1,2... See frame 8 for λ_{jk}.	$J_j\left(\lambda_{jk}\frac{r}{R}\right)\cos\frac{i\pi x}{L}\begin{Bmatrix}\sin j\theta \\ \text{or} \\ \cos j\theta\end{Bmatrix}$ i=0,1,2... j=0,1,2... = number of nodal diameters k=0,1,2... = number of nodal circles See frame 8 for λ_{jk}.
11. Volume With Circular Sector Cross Section 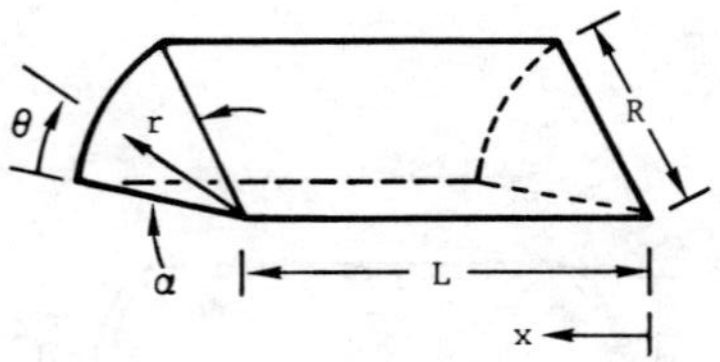	The natural frequencies can be adapted from those in frames 8 through 10 according to the boundary conditions at x = 0,L and the following choices of the index j: $j = \frac{180°n}{\alpha}$; n=0,1,2... $0 < \alpha \leq 360°$ (α in degrees) and i and k are as given in frames 8 through 10. j will be integer only if α is a submultiple of 180n. For non-integer j, λ_{jk} can be interpolated from the table in frame 8.	The mode shapes can be adapted from those in frames 8 through 10 according to the boundary conditions at x = 0,L and j as indicated in the adjacent column. Only the mode shapes with cos jθ dependence are valid. If α is not a submultiple of 180n, the mode shapes will generally involve Bessel functions of fractional order. Ref. 13-5, pp. 299-300.

12. Annular Volume Open at Both Ends, $x = 0, L$

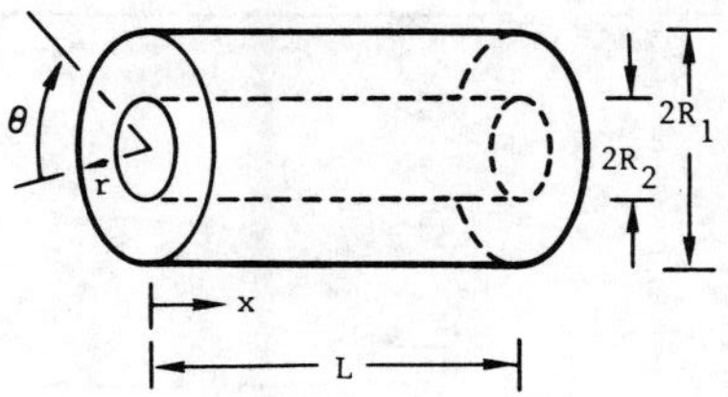

$$\frac{c}{2\pi}\left(\frac{\lambda_{jk}^2}{R_1^2} + \frac{i^2\pi^2}{L^2}\right)^{1/2}$$

i=0,1,2...
j=0,1,2...
k=0,1,2...

$$\left[J_j'\left(\lambda_{jk}\right) Y_j'\left(\frac{R_2}{R_1}\lambda_{jk}\right) - J_j'\left(\frac{R_2}{R_1}\lambda_{jk}\right) Y_j'(\lambda_{jk}) = 0\right]$$

$$G_{jk}(r) \sin\frac{i\pi x}{L}\begin{Bmatrix}\sin j\theta \\ \text{or} \\ \cos j\theta\end{Bmatrix} \qquad R_2 \le r \le R_1$$

i=0,1,2... = number of axial nodes
j=0,1,2... = number of nodal diameters
k=0,1,2... = number of nodal circles

$$G_{jk}(r) = Y_j'(\lambda_{jk})\, J_j\left(\lambda_{jk}\frac{r}{R_1}\right) - J_j'(\lambda_{jk})\, Y_j\left(\lambda_{jk}\frac{r}{R_1}\right)$$

$$G_{jk}(r) \approx \cos\frac{k\pi(r - R_2)}{R_1 - R_2} \quad \text{for } \frac{R_2}{R_1} > 0.8$$

		λ_{jk}						
		j						
$\frac{R_2}{R_1}$	k	0	1	2	3	4	5	6
0.3	0	0	1.5821	2.9685	4.1801	5.3130	6.4147	7.5011
	1	4.7058	5.1374	6.2738	7.7213	9.1526	10.4750	11.7214
0.5	0	0	1.3547	2.6812	3.9577	5.1752	6.3389	7.4622
	1	6.3932	6.5649	7.0626	7.8401	8.8364	9.9858	11.2270

$$\lambda_{jk}^2 \approx \left(\frac{k\pi R_1}{R_1 - R_2}\right)^2 + \left(\frac{2jR_1}{R_1 + R_2}\right)^2, \text{ for } \frac{R_2}{R_1} > 0.5$$

13. Annular Volume Closed at One End, Open at Other End

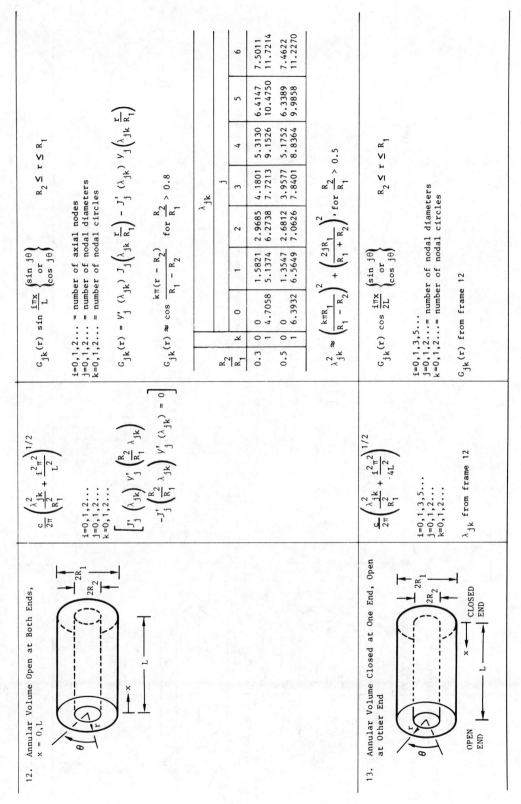

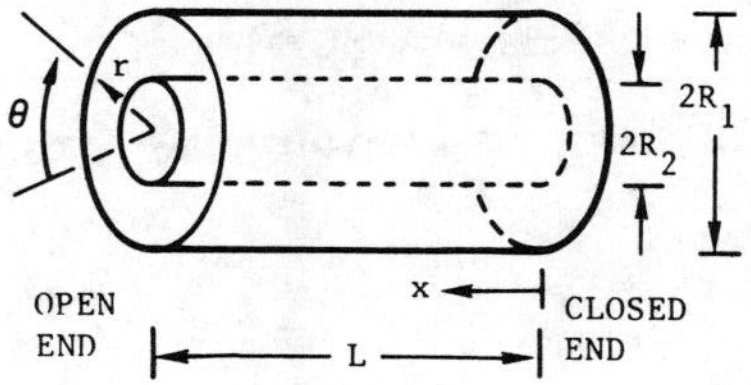

$$\frac{c}{2\pi}\left(\frac{\lambda_{jk}^2}{R_1^2} + \frac{i^2\pi^2}{4L^2}\right)^{1/2}$$

i=0,1,3,5...
j=0,1,2...
k=0,1,2...

λ_{jk} from frame 12

$$G_{jk}(r) \cos\frac{i\pi x}{2L}\begin{Bmatrix}\sin j\theta \\ \text{or} \\ \cos j\theta\end{Bmatrix} \qquad R_2 \le r \le R_1$$

i=0,1,3,5...
j=0,1,2...= number of nodal diameters
k=0,1,2...= number of nodal circles

$G_{jk}(r)$ from frame 12

Table 13-2. Acoustic Cavities. *(Continued)*

Description	Natural Frequency, f_{ijk} (hertz)	Mode Shape of Velocity Potential, $\tilde{\phi}_{ijk}$
14. Closed Annular Volume 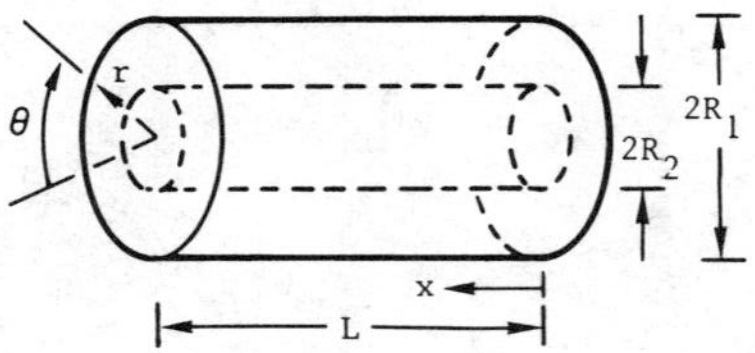	$\frac{c}{2\pi}\left(\frac{\lambda_{jk}^2}{R_1^2} + \frac{i^2\pi^2}{L^2}\right)^{1/2}$ i=0,1,2... j=0,1,2... k=0,1,2... λ_{jk} from frame 12	$G_{jk}(r) \cos \frac{i\pi x}{L} \begin{Bmatrix} \sin j\theta \\ \text{or} \\ \cos j\theta \end{Bmatrix}$ $R_2 \leq r \leq R_1$ i=0,1,2... j=0,1,2... k=0,1,2... $G_{jk}(r)$ from frame 12
15. Annular Segment 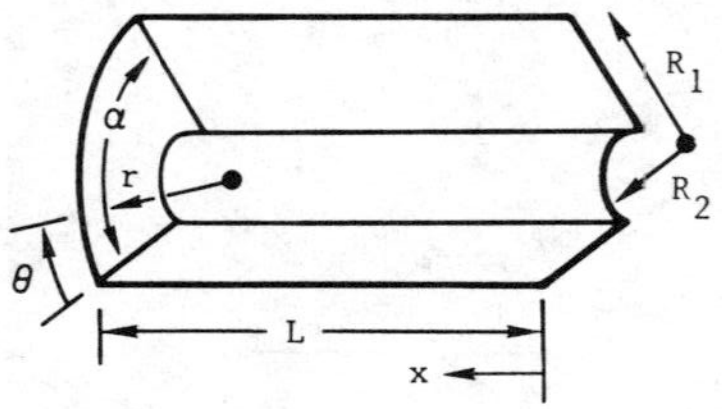	The natural frequencies can be adapted from those in frames 12 through 14 according to the boundary conditions at x = 0,L and the following choices of the index j: $j = \frac{180°n}{\alpha}$; n=0,1,2,... 0 < α ≤ 360° (α in degrees) and i and k are as given in frames 12 through 14. j will be integer only if α is a submultiple of 180n. For non-integer j, λ is obtained by interpolation from the table in frame 12.	The mode shapes can be adapted from those in frames 12 through 14 according to the boundary conditions at x = 0,L and j as indicated in the adjacent column. Only the mode shapes with cos jθ dependence are valid.

<table>
<tr>
<td>16. Closed Spherical Cavity
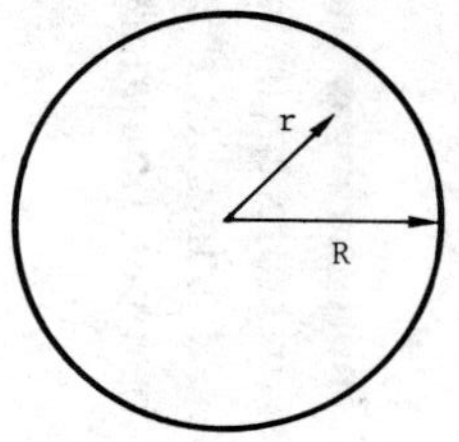
</td>
<td>$\frac{c\lambda_i}{2\pi R}$

i=0,1,2...</td>
<td>Modes symmetric about the center of the sphere:

$\frac{R}{\lambda_i r} \sin \frac{\lambda_i r}{R}$

i=0,1,2... = number of radial nodes

i   0   1   2   3   4   i>4
λ_i 0   4.4934   7.7253   10.9041   14.0662 $\lambda_i \approx \pi(i + 1/2)$

$(\tan \lambda_i = \lambda_i)$

Additional tabulated values can be found in Ref. 13-4, p. 224. Other solutions are discussed in Refs. 13-5 (pp. 264-268), 13-6 (pp. 506-508), and 13-7.</td>
</tr>
<tr>
<td>17. Arbitrary Volume With Closed Boundaries
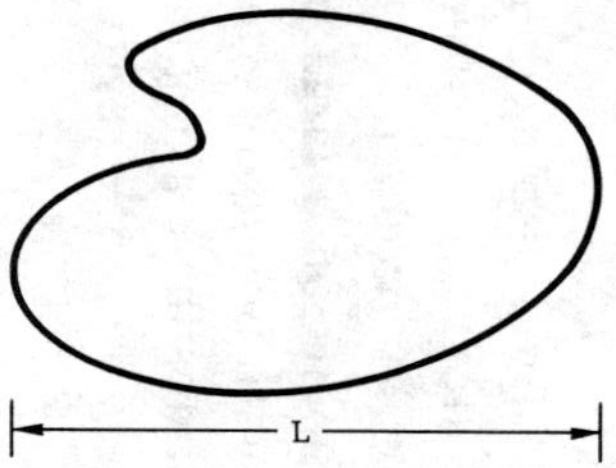
</td>
<td>Fundamental Natural frequency, approximate:

$\frac{c}{2L}$</td>
<td>L is maximum linear dimension.

Note that the natural frequencies of a volume with all open boundaries, for example, a fluid held in place by a membrane wall, are identical to those of the same volume with all closed boundaries.</td>
</tr>
</table>

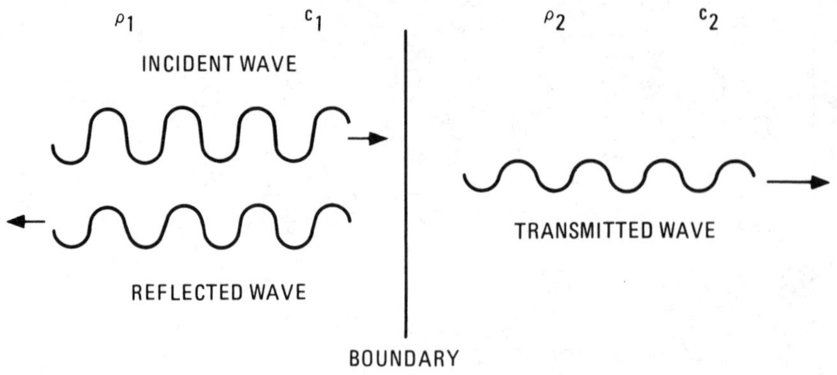

Fig. 13-1. Reflection and transmission of a wave at a boundary.

majority of the incident wave will be reflected by the boundary and the boundary can be considered a rigid wall. For example, for an air/steel interface:

$$\frac{\rho_2 c_2}{\rho_1 c_1} = \frac{8030\ \text{kg/m}^3 \cdot 6100\ \text{m/sec}}{1.2\ \text{kg/m}^3 \cdot 343\ \text{m/sec}} = 1.2 \times 10^5.$$

Thus,

$$\alpha_r = 0.99996 \approx 1$$

and the steel can be considered a rigid boundary at the air/steel interface. Of course, if the boundary consisted only of a very thin sheet of steel instead of a block of steel, these conclusions would not necessarily apply because the compliance of the steel sheet would reduce the reflected wave.

The boundary condition for a rigid boundary is that the fluid velocity normal to the wall must be zero:

$$\hat{n} \cdot \vec{u} = 0 \tag{13-8}$$

or equivalently, using Eqs. 13-2 and 13-3,

$$\hat{n} \cdot \nabla p = 0 \tag{13-9}$$

or

$$\hat{n} \cdot \nabla \Phi = 0. \tag{13-10}$$

$\vec{u}$ is the vector of fluid displacement, $\hat{n}$ is the unit vector normal to the wall, ∇ is the gradient operator, p is the dynamic component of the fluid pressure, and (·) denotes the vector dot product.

If the cavity opens on a large uniform body of fluid, such as the atmosphere, then an approximate boundary condition is that the pressure cavity must equal the pressure in the atmosphere along the open boundary. This implies that the dynamic component of pressure in the cavity must be zero at the open boundary:

$$p = 0 \tag{13-11}$$

or equivalently from Eq. 13-2,

$$\Phi = 0. \tag{13-12}$$

This boundary condition is only approximately achieved in real cavities because local fluid motion will influence the pressure at an open boundary.

Natural Frequencies and Mode Shapes. The natural frequencies and mode shapes of the velocity potential of fluid-filled cavities of various geometries are given in Table 13-2. The cavities are bounded by both rigid walls and open boundaries (Eqs. 13-10 and 13-12), as indicated in each frame. Note that the fundamental natural frequency of a cavity which is completely enclosed by rigid walls is approximately:

$$f_1 = \frac{c}{2L^*} \text{ (hertz)}, \tag{13-13}$$

where L^* is the maximum linear dimension of the cavity.

Two or more modes of a cavity may have the same natural frequency. For example, the three lowest modes of a closed square box [$L_x = L_y = L_z$ in frame 6 of Table 13-2] corresponding to $(i, j, k) = (1, 0, 0)$, $(0, 1, 0)$, and $(0, 0, 1)$ all have the same natural frequency. The natural frequencies of the three-dimensional cavities in Table 13-2 become very closely spaced in the higher modes.

The mode shapes of cavities become increasingly complex in the higher modes. However, very often these mode shapes have a simple geometric interpretation as can be seen in Fig. 13-2, which shows the sign of the radial mode shape in an annular cavity. The pattern suggested by Fig. 13-2 applies to cylindrical cavities as well.

The mode shapes listed in Table 13-2 are orthogonal over the volume of the cavity. That is:

$$\int_V \tilde{\Phi}_{ijk} \tilde{\Phi}_{mno} \, dV = 0 \quad \text{unless} \quad i = m, j = n, k = 0, \tag{13-14}$$

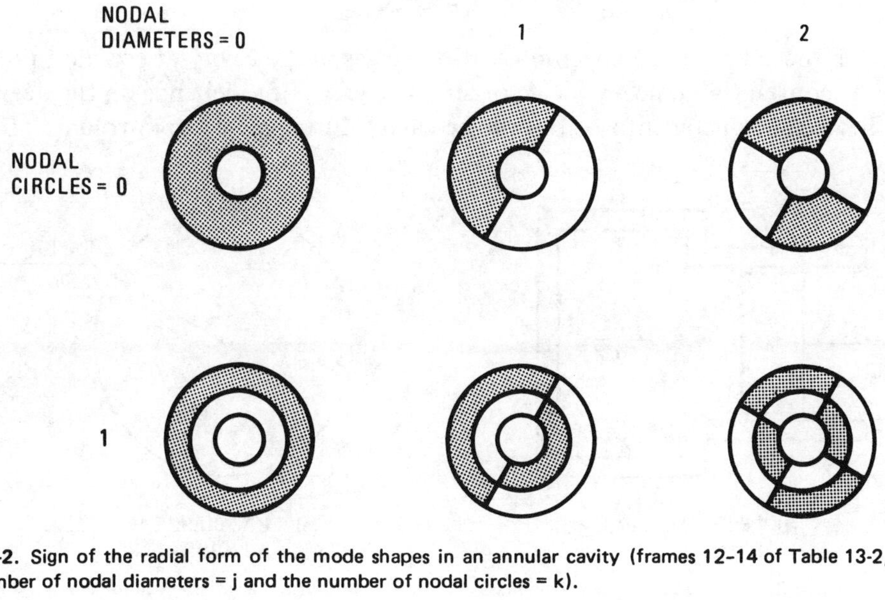

Fig. 13-2. Sign of the radial form of the mode shapes in an annular cavity (frames 12–14 of Table 13-2, where the number of nodal diameters = j and the number of nodal circles = k).

where $\tilde{\phi}$ is a mode shape of the velocity potential of one of the cavities listed in Table 13-2; i, j, k, m, n, and o are integer modal indices; and V is the volume of the cavity.

The natural frequencies and mode shapes of cavities with open boundaries given in Table 13-2 are approximate because the open boundary condition of Eq. 13-12 does not fully model the physics of the fluid behavior at an open boundary. The accuracy of the formulas for open boundary cavities given in Table 13-2 increases with increasing slenderness of the cavity in the direction of the open boundary. Further discussion of open boundaries can be found in Section 13.3.

13.2. BUBBLES AND PISTON-CYLINDER SYSTEMS

General Case. As a piston moves with respect to a cylinder, as in Fig. 13-3(a), the volume of gas in the cylinder cavity will vary, causing the pressure in the cavity to rise or fall. If the mass of fluid contained in the cylinder cavity is constant and the piston moves at a velocity much less than the speed of sound of the gas in the cavity, the relationship between the volume (V) of the cavity and the pressure (p) in the cavity is given by:

$$pV^{\kappa} = \text{constant}. \qquad (13\text{-}15)$$

κ is the dimensionless polytropic constant. Taking the derivative of this expression gives the relationship between an increment in volume and an increment in pressure:

$$dp = -\frac{\kappa p}{V}\, dV. \qquad (13\text{-}16)$$

Thus, as the volume increases the pressure must fall.

The dimensionless polytropic constant can vary within the range

$$1 \leqslant \kappa \leqslant \gamma,$$

where γ is the ratio of the specific heat of the gas in the cavity at constant pressure to that at constant volume. $\gamma = 1.4$ for air. The value of κ depends on the degree to which heat can transfer into or out of the cavity during a change in volume. If heat

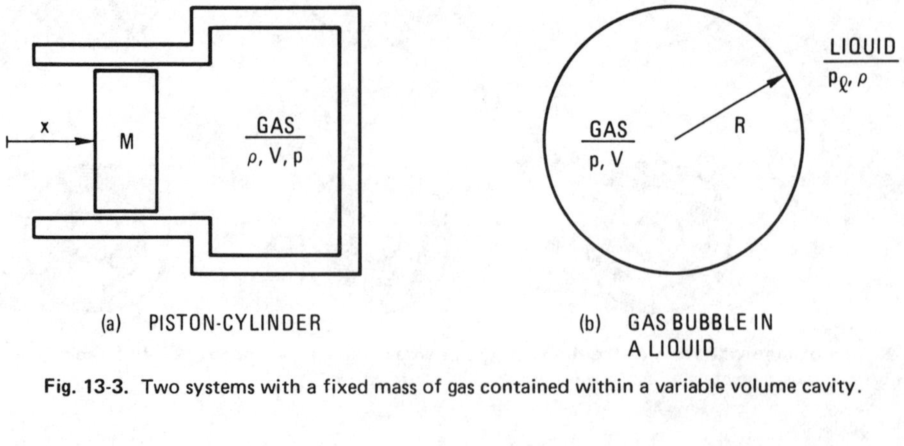

(a) PISTON-CYLINDER (b) GAS BUBBLE IN A LIQUID

Fig. 13-3. Two systems with a fixed mass of gas contained within a variable volume cavity.

can transfer into and out of the cavity sufficiently rapidly to maintain the gas in the cylinder cavity at constant temperature, the process is isothermal and $\kappa = 1.0$. If no heat can transfer into or out of the cylinder, the process is adiabatic and $\kappa = \gamma$. Intermediate cases will have intermediate values of κ.

The compressible gas in a cavity forms an equivalent spring in the vibrations of bubbles and piston-cylinder systems. These systems are discussed in the following paragraphs.

Bubbles. A spherical bubble of gas in a liquid, shown in Fig. 13-3(b), can oscillate in a symmetric radial mode given by:

$$R = R_0 + \delta R \sin \omega t. \tag{13-17}$$

R_0 is the mean radius of the bubble and δR is the amplitude of radial oscillation, which is assumed to be much less than R_0. This mode shape corresponds to alternate compression and expansion of the gas in the bubble. The bubble is analogous to a simple spring-mass system with the spring formed by the compressible gas in the bubble, which reacts against the mass of the fluid surrounding the bubble.

The natural frequency in hertz of the bubble is given by (Ref. 13-8):

$$f = \frac{\omega}{2\pi} = \frac{1}{2\pi R_0 \rho^{1/2}} \left(3\kappa p_0 - \frac{2\sigma}{R_0}\right)^{1/2}, \tag{13-18}$$

where

- p_0 = mean pressure of gas in the bubble (mean pressure of liquid surrounding the bubble = $p_0 - 2\sigma/R_0$),
- R_0 = mean radius of the bubble,
- κ = polytropic constant of the gas in the bubble (Eq. 13-15),
- σ = surface tension constant; $\sigma = 7.27 \times 10^{-2}$ N/m at 20°C for air/water (see Ref. 13-9 for other media),
- ρ = mass density of the liquid surrounding the bubble.

Dissipative effects have been neglected. Consistent sets of units are given in Table 3.1.

The polytropic constant, κ, which should be employed in Eq. 13-18 will depend on the size of the bubble and the nature of the media. Figure 13-4 shows κ as a function of bubble size for an air bubble in water at 20°C. Very small bubbles ($R_0 < 10^{-5}$ m) behave isothermally and $\kappa \approx 1$. Large bubbles behave adiabatically and $\kappa \approx \gamma$. $\gamma = 1.4$ for air.

The effect of surface tension on the bubble natural frequency is of decreasing importance as the size of the bubble increases. For large bubbles ($R_0 > 0.1$ cm for an air bubble in water at 20°C), the bubble dynamics are adiabatic and the effect of surface tension can ordinarily be neglected. For large bubbles the natural frequency in hertz is:

$$f = \frac{(3\gamma p_0)^{1/2}}{2\pi R_0 \rho^{1/2}}. \tag{13-19}$$

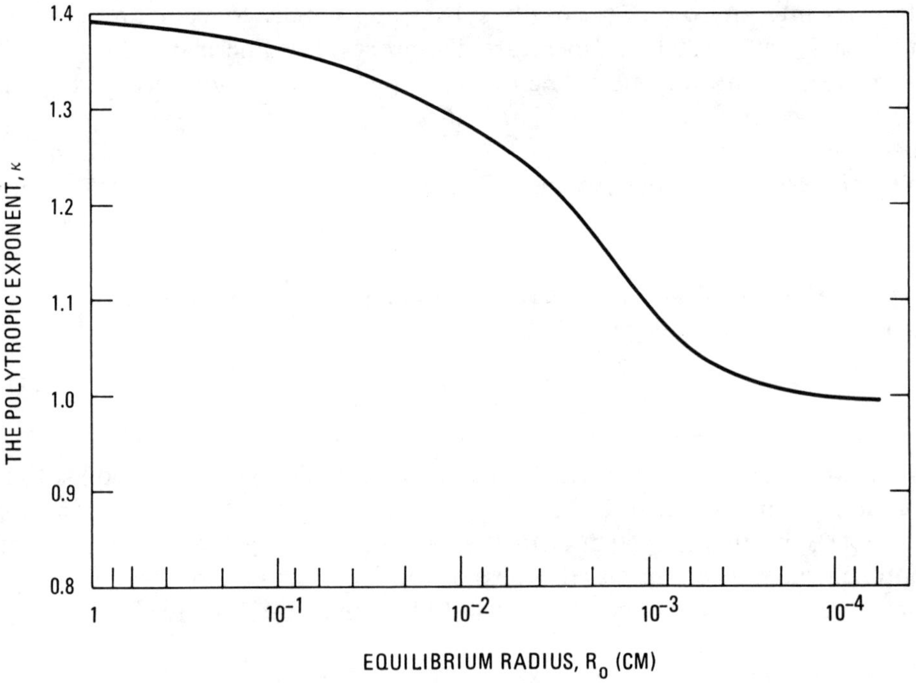

Fig. 13-4. The polytropic constant, κ, as a function of bubble size for an air bubble in water at 20°C (Ref. 13-10).

If a large bubble ascends slowly in a fluid reservoir and if d is the depth of the bubble from the surface of the reservoir, then the pressure in the bubble is proportional to the depth:

$$p_0 \sim d$$

and the bubble radius varies with the depth

$$R_0 \sim \frac{1}{d^{1/3}},$$

where $\sim$ denotes approximately proportional to. Thus, the natural frequency (Eq. 13-19) is proportional to depth to the 5/6 power,

$$f \sim d^{5/6},$$

and the natural frequencies of large bubbles will decrease as they rise to the surface.

Piston-Cylinder. The gas contained in the cylinder cavity shown in Fig. 13-3(a) places the incremental force

$$dF = dpA = \kappa \frac{p_0}{V_0} A^2 \, dx \qquad (13\text{-}20)$$

on the piston of cross-section area A if the piston moves a small amount dx. dp is the incremental increase in pressure and is given by Eq. 13-16. It is assumed that no gas escapes from the cylinder. Since the force applied to the piston is a linear function of its displacement, for small displacements the gas contained in the cylinder cavity can be modeled as a one-dimensional spring. The spring constant is:

$$k = \frac{dF}{dx} = \frac{\kappa p_0 A^2}{V_0}. \tag{13-21}$$

p_0 is the mean gas pressure in the cavity with mean volume V_0. κ is the dimensionless polytropic constant (Eq. 13-15).

The natural frequency of the piston-cylinder system is identical to that of an equivalent spring-mass system (frame 1 of Table 6-1) with the spring constant defined by Eq. 13-21 and mass equal to the piston mass. The natural frequency in hertz for free vibration of the piston in the cylinder is:

$$f = \frac{1}{2\pi}\left(\frac{k}{M}\right)^{1/2} = \frac{1}{2\pi}\left(\frac{\kappa p_0 A^2}{V_0 M}\right)^{1/2}. \tag{13-22}$$

If no heat is transferred into or out of the cavity, then the system is adiabatic,

$$\kappa = \gamma,$$

using frame 1 of Table 13-1,

$$\frac{\gamma p_0}{\rho} = c^2,$$

and the natural frequency in hertz becomes

$$f = \frac{c}{2\pi}\left(\frac{\rho A^2}{M V_0}\right)^{1/2}, \tag{13-23}$$

where

c = speed of sound in the gas (frame 1 of Table 13-1),
A = area of the piston,
ρ = mean density of the gas in the cylinder,
M = mass of the piston,
V_0 = mean volume of the cavity.

Most systems of this type of practical importance are nearly adiabatic.

The analogy between the gas-piston system and an equivalent spring-mass system can also be applied to Helmholtz resonators as shown in the following section.

13.3. HELMHOLTZ RESONATORS

General Case. A Helmholtz resonator is formed by an acoustic cavity with a small opening (neck) to the atmosphere. The system resonates as fluid oscillates back and forth in the neck in response to cyclic pressure fluctuations in the body of the cavity. The fluid in the neck forms the mass of the oscillator and the compressible fluid in the cavity is the spring of the oscillator, as shown in Fig. 13-5. The most familiar Helmholtz resonator is excited by blowing across the neck of an empty soft drink or milk bottle. The resultant low tones occur at the natural frequency

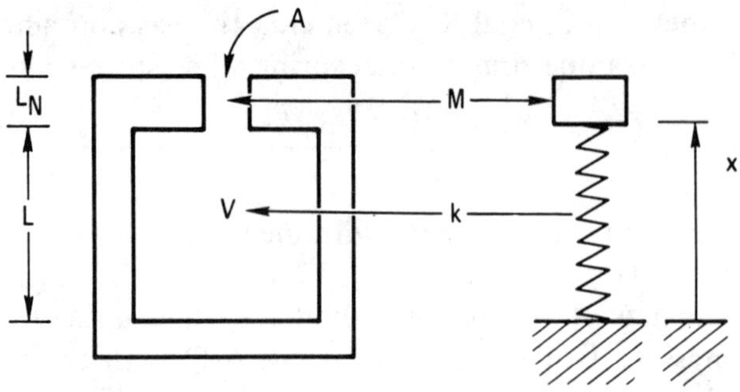

Fig. 13-5. Helmholtz resonator and analogous spring-mass system.

of the Helmholtz resonator formed by the bottle. The practically important property of Helmholtz resonators is their ability to absorb acoustic energy at the natural frequency of the resonator. Helmholtz resonators are often incorporated in chambered mufflers.

The general assumptions in the analysis of Helmholtz resonators are:

1. The cavity walls are rigid.
2. The natural frequency of the resonator is much less than the time required for sound to transverse the resonator cavity. This allows the cavity to be modeled as a one-dimensional spring.
3. The neck has a much smaller cross section than the body of the cavity, so the fluid velocity in the neck is large compared with the fluid velocity in the cavity. This allows the fluid in the neck to be modeled as a mass.
4. No heat is transferred into or out of the cavity during vibration.

If, in addition, the length of the neck is much greater than its diameter, then the mass of fluid in the neck can easily be identified. If the neck is relatively short, the simplified formulas must be modified to include the effective mass in the vicinity of the neck.

Assumption 2, i.e., neglecting the higher frequency modes of the cavity, is employed to focus attention on the Helmholtz mode. The cavity will possess acoustic modes above the Helmholtz mode. If the neck of the cavity is relatively narrow, the natural frequencies and mode shapes of these higher modes can generally be estimated from the modes of the cavity formed by closing the neck (see Table 13-2).

The assumption of no heat transfer (assumption 4) is not required for the modeling which is used in the analysis of Helmholtz resonators. It is employed because most Helmholtz resonator systems of practical importance are nearly adiabatic and the heat transfer is of little importance.

Systems of Helmholtz Resonators. The natural frequencies of systems of Helmholtz resonators are given in Table 13-3. The mode shapes given in this table refer to the mode shape of fluid displacement in the neck of the resonator. In this table it is assumed that the length of the necks of the resonators is much longer than the diameter of the necks and the diameter of the cavities is large compared with the diameter of the necks.

Table 13-3. Helmholtz Resonators.

Notation: c = speed of sound in fluid in cavity (Table 13-1); x = displacement of fluid in neck; $\tilde{x}$ = mode shape associated with displacement of fluid in neck; A = area of neck; L = effective length of neck; V = volume of cavity; see Table 3-1 for consistent sets of units. These formulas are approximate. See text for a discussion of the assumptions and limitations and a more accurate formula for the basic resonator (frame 1).

Geometry	Natural Frequency, f_i (hertz)	Mode Shape, $\tilde{x}_i$, and Remarks
1. Single Vented Resonator 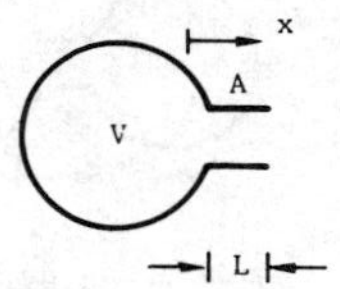	$\frac{c}{2\pi}\left(\frac{A}{VL}\right)^{1/2}$	See text for more accurate formula.
2. Coupled Equal Vented Resonators 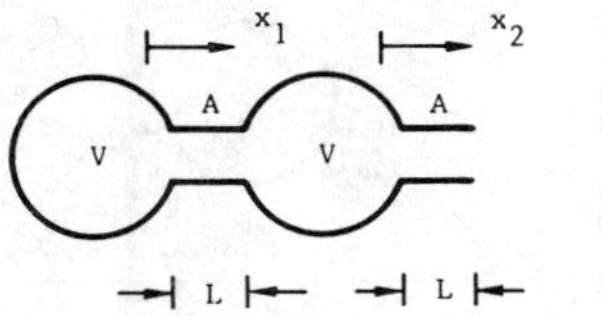	$\frac{(3-5^{1/2})^{1/2}c}{2^{3/2}\pi}\left(\frac{A}{VL}\right)^{1/2}$, $\frac{(3+5^{1/2})^{1/2}c}{2^{3/2}\pi}\left(\frac{A}{VL}\right)^{1/2}$	$\begin{bmatrix}\tilde{x}_1\\ \tilde{x}_2\end{bmatrix} = \begin{bmatrix}1\\ \frac{1+5^{1/2}}{2}\end{bmatrix}, \begin{bmatrix}1\\ \frac{1-5^{1/2}}{2}\end{bmatrix}$
3. Coupled Unequal Vented Resonators 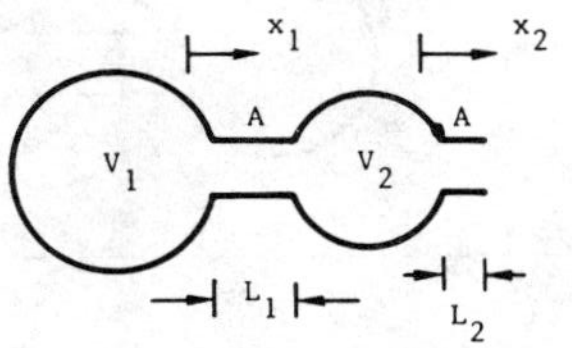	$\frac{cA^{1/2}}{2^{3/2}\pi}\left\{\frac{1}{V_1L_1}+\frac{1}{V_2L_2}+\frac{1}{L_1V_2} \mp \left[\left(\frac{1}{V_1L_1}+\frac{1}{V_2L_2}+\frac{1}{L_1V_2}\right)^2 - \frac{4}{V_1V_2L_1L_2}\right]^{1/2}\right\}^{1/2}$	$\begin{bmatrix}\tilde{x}_1\\ \tilde{x}_2\end{bmatrix}_i = \begin{bmatrix}1\\ 1+\frac{V_2}{V_1}-\frac{L_1V_2}{AC^2}(2\pi f_i)^2\end{bmatrix}$ $i = 1,2$
4. Three Equal Coupled Vented Resonators 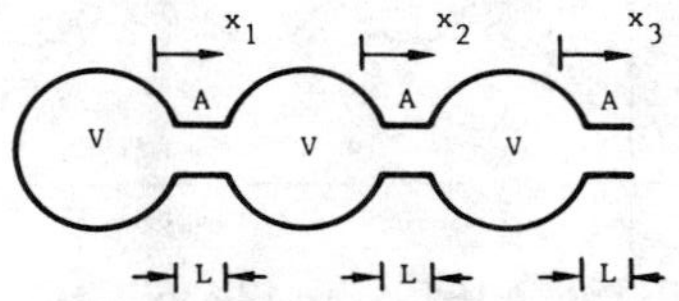	$0.07082\,c\left(\frac{A}{VL}\right)^{1/2}$, $0.1985c\left(\frac{A}{VL}\right)^{1/2}$, $0.2868c\left(\frac{A}{VL}\right)^{1/2}$	$\begin{bmatrix}\tilde{x}_1\\ \tilde{x}_2\\ \tilde{x}_3\end{bmatrix} = \begin{bmatrix}1\\ 1.802\\ 2.247\end{bmatrix}, \begin{bmatrix}1\\ 0.445\\ -0.802\end{bmatrix}, \begin{bmatrix}1\\ -1.247\\ 0.555\end{bmatrix}$

Table 13-3. Helmholtz Resonators. *(Continued)*

Geometry	Natural Frequency, f_i (hertz)	Mode Shape, $\tilde{x}_i$, and Remarks
5. N Equal Coupled Vented Resonators 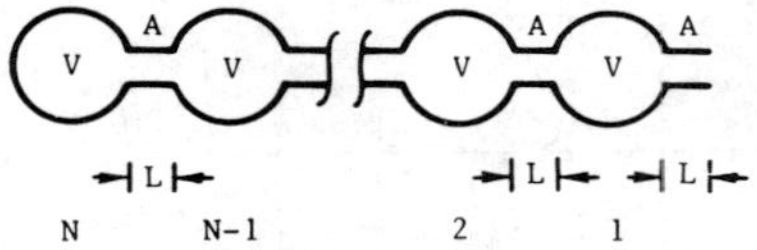	$\frac{\alpha_{N,i}\, c}{2\pi}\left(\frac{A}{VL}\right)^{1/2}$; $i=1,2,\ldots N$ $\alpha_{N,i} = 2 \sin \left[\frac{(2i-1)}{(2N+1)}\frac{\pi}{2}\right]$	--
6. Coupled Equal Resonators 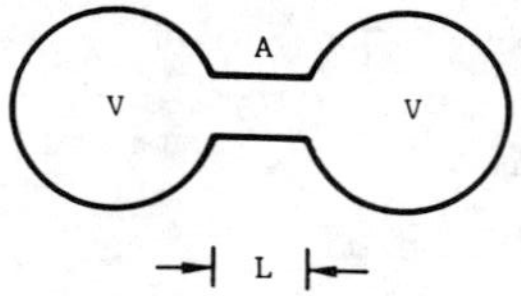	$\frac{c}{2\pi}\left(\frac{2A}{VL}\right)^{1/2}$	--
7. Coupled Unequal Resonators 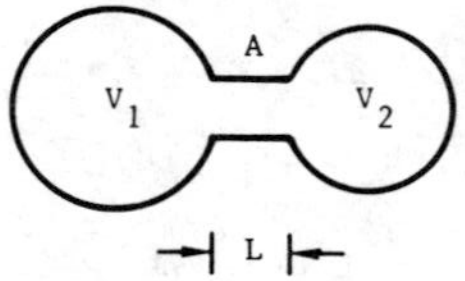	$\frac{c}{2\pi}\left[\left(\frac{A}{L}\right)\left(\frac{1}{V_1}+\frac{1}{V_2}\right)\right]^{1/2}$	--
8. Three Equal Coupled Resonators 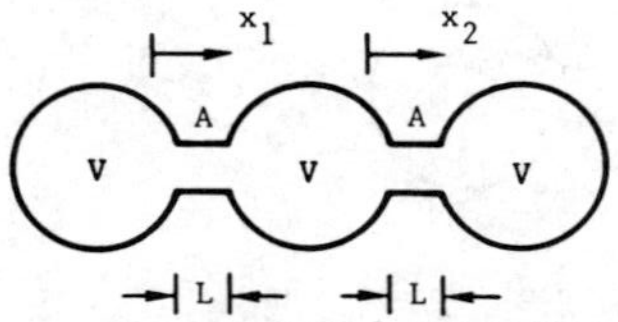	$\frac{c}{2\pi}\left(\frac{A}{VL}\right)^{1/2}$, $\frac{c}{2\pi}\left(\frac{3A}{VL}\right)^{1/2}$	$\begin{bmatrix}\tilde{x}_1\\ \tilde{x}_2\end{bmatrix} = \begin{bmatrix}1\\1\end{bmatrix}, \begin{bmatrix}1\\-1\end{bmatrix}$
9. Three Unequal Coupled Resonators 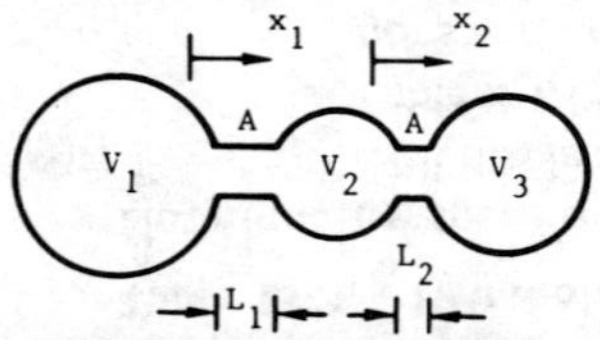	See frame 9 of Table 6-2, with $k_1 = \frac{c^2A^2}{V_1}$, $k_2 = \frac{c^2A^2}{V_2}$, $k_3 = \frac{c^2A^2}{V_3}$, $M_1 = AL_1$, $M_2 = AL_2$.	See frame 9 of Table 6-2.

10. Four Equal Coupled Resonators 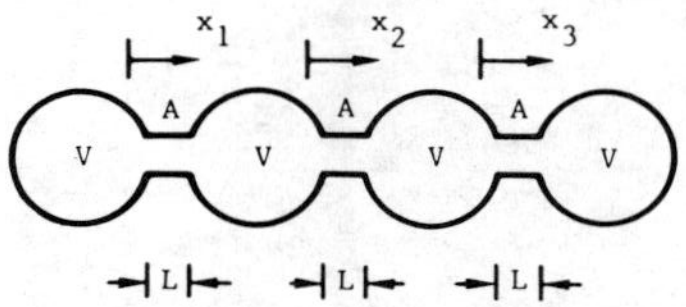	$\frac{(2 - 2^{1/2})^{1/2} c}{2\pi} \left(\frac{A}{VL}\right)^{1/2}, \; 2^{1/2} c \left(\frac{A}{VL}\right)^{1/2},$ $\frac{(2 + 2^{1/2})^{1/2} c}{2\pi} \left(\frac{A}{VL}\right)^{1/2}$	$\begin{bmatrix} \tilde{x}_1 \\ \tilde{x}_2 \\ \tilde{x}_3 \end{bmatrix} = \begin{bmatrix} 1 \\ 2^{1/2} \\ 1 \end{bmatrix}, \begin{bmatrix} 1 \\ 0 \\ -1 \end{bmatrix}, \begin{bmatrix} 1 \\ -2^{1/2} \\ 1 \end{bmatrix}$
11. N Equal Coupled Resonators 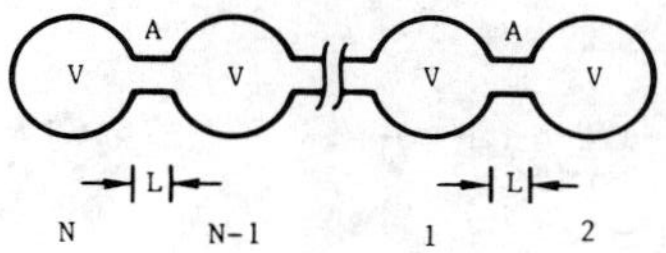	$\frac{\alpha_{N,i}\, c}{2\pi} \left(\frac{A}{VL}\right)^{1/2}, \quad i = 1,2,3,\ldots N$ $\alpha_{N,i} = 2 \sin \left[\frac{i}{(N+1)} \frac{\pi}{2}\right]$	
12. Double-Vented Resonator 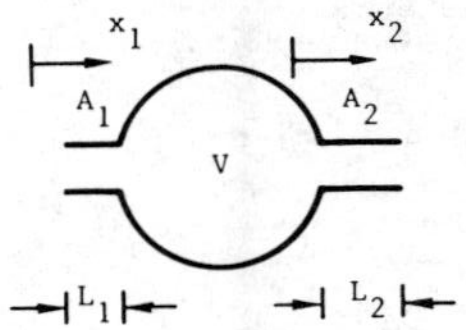	$0, \; \frac{c}{2\pi} \left(\frac{A_1}{L_1 V} + \frac{A_2}{L_2 V}\right)^{1/2}$	$\begin{bmatrix} \tilde{x}_1 \\ \tilde{x}_2 \end{bmatrix} = \begin{bmatrix} 1 \\ -\frac{A_1}{A_2} \end{bmatrix}, \begin{bmatrix} 1 \\ -\frac{L_1}{L_2} \end{bmatrix}$
13. Two Equal, Coupled Double-Vented Resonators 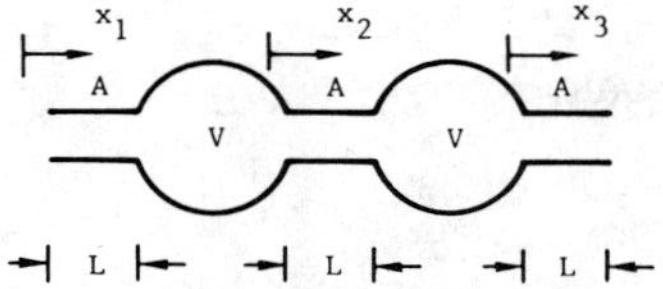	$0, \; \frac{c}{2\pi} \left(\frac{A}{VL}\right)^{1/2}, \; \frac{c}{2\pi} \left(\frac{3A}{VL}\right)^{1/2}$	$\begin{bmatrix} \tilde{x}_1 \\ \tilde{x}_2 \\ \tilde{x}_3 \end{bmatrix} = \begin{bmatrix} 1 \\ 1 \\ 1 \end{bmatrix}, \begin{bmatrix} 1 \\ 0 \\ -1 \end{bmatrix}, \begin{bmatrix} 1 \\ -2 \\ 1 \end{bmatrix}$

Table 13-3. Helmholtz Resonators. *(Continued)*

Geometry	Natural Frequency, f_i (hertz)	Mode Shape, $\tilde{x}_i$, and Remarks
14. Two Unequal, Coupled Double-Vented Resonators 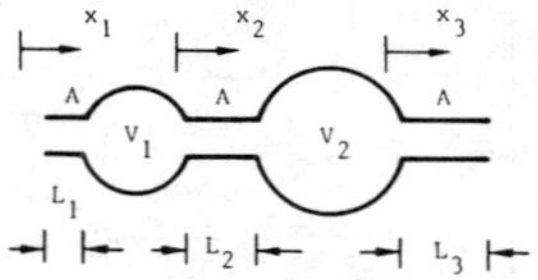	See frame 15 of Table 6-2, with $k_1 = \frac{c^2A^2}{V_1}$, $k_2 = \frac{c^2A^2}{V_2}$, . $M_1 = AL_1$, $M_2 = AL_2$, $M_3 = AL_3$	See frame 15 of Table 6-2.
15. Three Equal Double-Vented Resonators 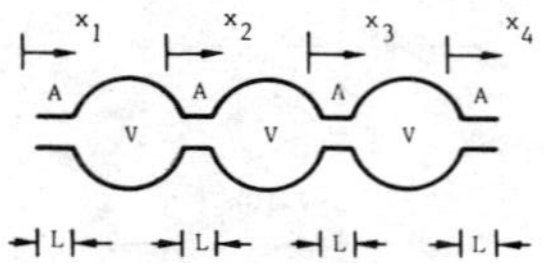	$0,\ \frac{(2 - 2^{1/2})^{1/2} c}{2\pi}\left(\frac{A}{VL}\right)^{1/2},\ \frac{2^{1/2} c}{2\pi}\left(\frac{A}{VL}\right)^{1/2}$ $\frac{(2 + 2^{1/2})^{1/2} c}{2\pi}\left(\frac{A}{VL}\right)^{1/2}$	$\begin{bmatrix}\tilde{x}_1\\ \tilde{x}_2\\ \tilde{x}_3\\ \tilde{x}_4\end{bmatrix} = \begin{bmatrix}1\\1\\1\\1\end{bmatrix}, \begin{bmatrix}1\\-1+2^{1/2}\\1-2^{1/2}\\-1\end{bmatrix}, \begin{bmatrix}1\\-1\\-1\\1\end{bmatrix}, \begin{bmatrix}1\\-1-2^{1/2}\\1+2^{1/2}\\-1\end{bmatrix}$
16. N Equal Double-Vented Resonators 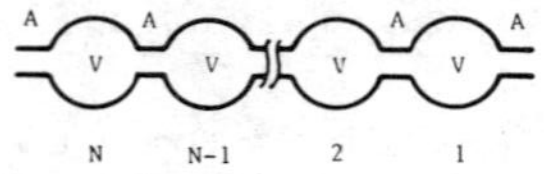	$\frac{\alpha_{N,i}\, c}{2\pi}\left(\frac{A}{VL}\right)^{1/2}$; i=1,2,3...N $\alpha_{N,i} = 2 \sin\left[\frac{(i-1)}{N}\frac{\pi}{2}\right]$	--

The natural frequencies and mode shapes in Table 13-3 are very similar to the analogous spring-mass systems. If the necks of the resonators in a given system have the same cross-sectional area, the resonator cavities may be modeled as springs with a spring constant (Eq. 13-21, frame 1 of Table 13-1) of

$$k = \frac{\rho c^2 A^2}{V} \tag{13-24}$$

for an adiabatic system. The springs are connected to masses of mass

$$M = \rho AL \tag{13-25}$$

which model the fluid in the resonator necks. ρ is the fluid density, c is the speed of sound (frame 1 of Table 13-1), A is the cross-sectional area of the neck of length L, and V is the volume of the cavity. These equations can be derived by noting that the change of pressure in the cavity is proportional to the change in volume of the cavity (Eq. 13-16).

For an adiabatic system,

$$dp = \frac{dF}{A} = -\rho c^2 \frac{dV}{V}.$$

dV is the change of volume of the cavity due to the slug of fluid in the neck entering or leaving the cavity:

$$dV = xA.$$

x is the displacement of the fluid in the neck of the cavity. The change in pressure produces a resultant force on the fluid in the neck which causes the fluid in the neck to accelerate:

$$\rho AL\ddot{x} = dF.$$

Thus, the equation of motion of the fluid in the neck is:

$$(\rho AL)\ddot{x} + \left(\frac{\rho c^2 A^2}{V}\right) x = 0.$$

This equation is identical to the equation of motion of the spring-mass system of frame 1 of Table 6-2 with the substitution of Eqs. 13-24 and 13-25 for the spring constant and the mass of the system, respectively.

Improved Formula for Simple Resonators. The formulas in Table 13-3 neglect the effects of the inertia of fluid in the cavity and the motion of fluid outside the neck. These effects become increasingly important as the ratio of the neck diameter to the neck length increases. If the neck of the resonator is relatively short (diameter $\geqslant$ length) or the diameter of the neck is comparable to the diameter of the cavity, the formulas in Table 13-3 will not yield an accurate result.

Alster (Ref. 13-11) has developed an improved formula for the natural frequency of simple Helmholtz resonators (Fig. 13-5) which takes into account the motion of the fluid in a cavity and outside the neck. The natural frequency in hertz is:

$$f = \frac{c}{2\pi}\left[\frac{A}{1.21\,(V + AL_N)\,L_1}\right]^{1/2}, \tag{13-26}$$

where

$$L_1 = \left\{L_V + L_0 + (L_N + L_0)\left[1 + \frac{L_N + L_0}{2L} + \frac{A\,(L_N + L_0)}{2V} + \frac{A\,(L_N + L_0)}{3V}\,\frac{L_N + L_0}{L}\right]\right\}\frac{V}{V + A\,(L_N + L_0)}\,\frac{L}{L + L_N + L_0},$$

where

c = speed of sound in fluid of the cavity (Table 13-1),
A = cross-sectional area of the neck
L_N = length of the neck (Fig. 13-5),
V = volume of the cavity without the neck,
L = height of the resonator cavity from the bottom to the neck (see Table 13-3 and Fig. 13-5),
L_V = form factor with the units of length (see Eq. 13-27 and Table 13-4),
L_0 = 0.24 r, where r is the radius of the neck. If the neck is not circular, r is a characteristic radius.

L_V is defined as follows:

$$L_V = \frac{A}{VL}\int_0^L \frac{x\left[\int_0^L A_V(x)\,dx\right]dx}{A_V(x)}, \tag{13-27}$$

where x is the coordinate from the bottom of the resonator (Fig. 13-1) and $A_V(x)$ is the cross-sectional area of the cavity at each station x. L_V for various cavities is given in Table 13-4.

Alster (Ref. 13-11) has found that Eq. 13-26 predicts the natural frequency of resonators to within ±5% of the experimentally measured values. For a pipe of length L with one end open and the other closed (frame 2 of Table 13-2), $L_N = 0$, $V = LA$, $L_V = L/3$, and it can be shown that Eq. 13-26 reduces to:

$$f = \frac{c}{4L^{1/2}(L + 4L_0)^{1/2}},$$

where L_0 is defined above. This formula produces a more accurate estimate of the fundamental frequency of relatively short pipes than the corresponding formula in frame 2 of Table 13-2, as can be seen by comparison with experimental data in air for a pipe with a 2.5-cm radius and one open and one closed end (Ref. 13-11).

Length of Pipe, L (cm)	50	40	30	20	10	5
f (experiment) (hertz)	169	211	279	411	778	1380
f (Eq. 13-26) (hertz)	169	210	276	408	775	1415
f (frame 2 of Table 13-2) (hertz)	172	214	286	428	858	1715

Table 13-4. Form Factors for Helmholtz Resonator Cavities.[a]

Description	Form Factor, L_V (Eq. 13-27)
1. Sphere 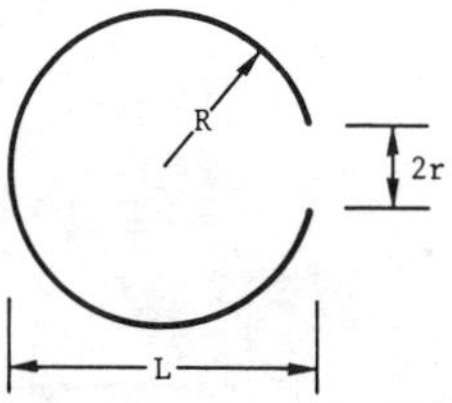	$\frac{r^2}{R}\,\frac{1}{2-\alpha}\left[\frac{1}{3} - \frac{1}{2(\alpha+1)} - \frac{2}{(\alpha+1)^2} + \frac{4}{(\alpha+1)^3}\ln\left(\frac{2}{1-\alpha}\right)\right]$ where $\alpha = \left[1 - \left(\frac{r}{R}\right)^2\right]^{1/2}$
2. Frustrum of a Cone 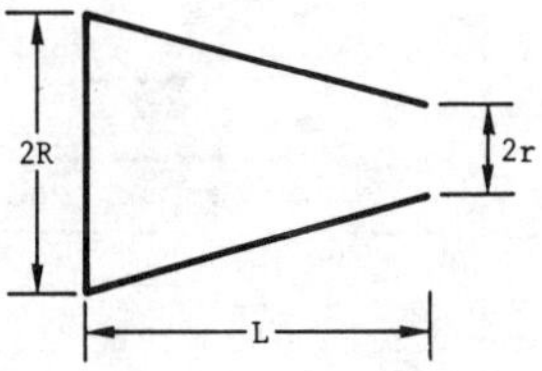	$\frac{L}{\left(\frac{R}{r}\right)^2 + \left(\frac{R}{r}\right) + 1}\left[\frac{1}{3} - \frac{1}{2}\frac{R}{R-r} + \left(\frac{R}{R-r}\right)^3 \cdot \left\{\left(\frac{R}{r}\right) - 1 - \ln\left(\frac{R}{r}\right)\right\}\right]$
3. Prism 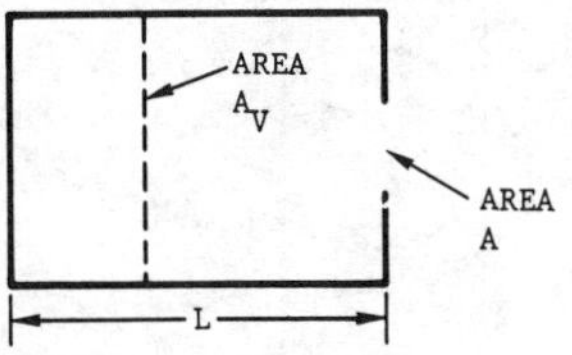	$\frac{L}{3}\,\frac{A}{A_V}$
4. Cylinder with Lateral Hole 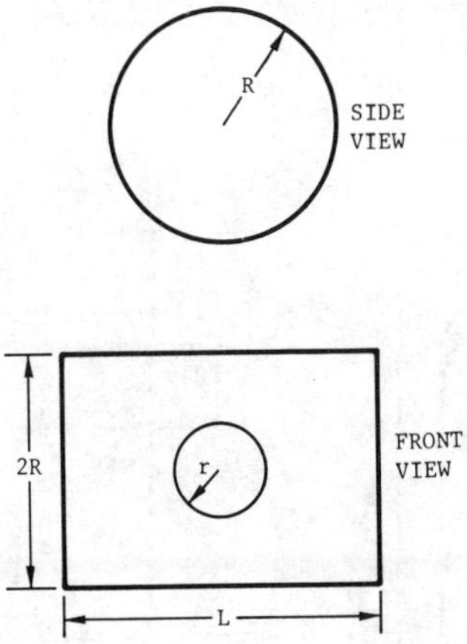	If $r \ll R$ or the resonator has a neck at the opening: $\frac{1.88r^2}{L}$ See Ref. 13-11 for r of the same order as R.
5. Toroid	See Ref. 13-11

(a) Ref. 13-11.

Table 13-5. U-Tubes.

Notation: g = acceleration of gravity; h = height of fluid in tank; u, v = displacement of fluid from mean level; $\tilde{u}$, $\tilde{v}$ = mode shapes associated with u and v; A = cross-sectional area; L = mean length of pipe along center line. The pipes need not have circular cross section. See Table 3-1 for consistent sets of units.

Description	Natural Frequency, f_i (hertz)	Mode Shape
1. Uniform Pipe 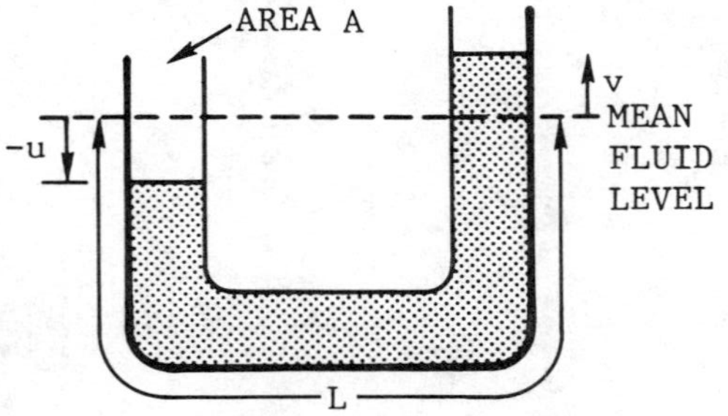	$\frac{1}{2\pi}\left(\frac{2g}{L}\right)^{1/2}$	$\begin{bmatrix}\tilde{u}\\ \tilde{v}\end{bmatrix} = \begin{bmatrix}1\\ -1\end{bmatrix}$ L = mean fluid-filled length of pipe as measured along center line of pipe
2. Tank and Pipe 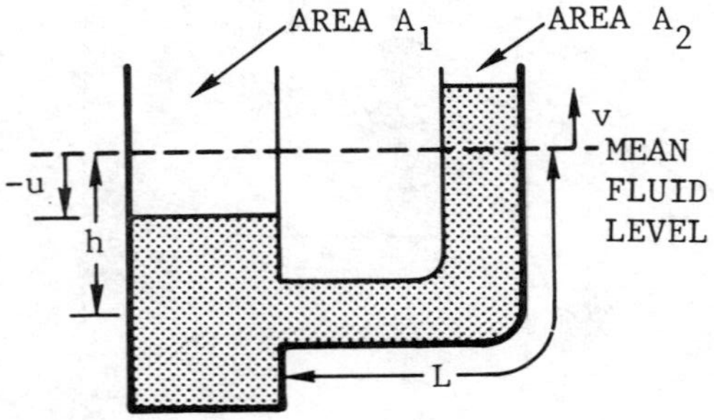	$\frac{1}{2\pi}\left[\frac{g\left(1+\frac{A_2}{A_1}\right)}{h\frac{A_2}{A_1}+L}\right]^{1/2}$	$\begin{bmatrix}\tilde{u}\\ \tilde{v}\end{bmatrix} = \begin{bmatrix}1\\ \frac{-A_1}{A_2}\end{bmatrix}$ L = distance along center line of pipe

<table>
<tr>
<td>3. Two Tanks Connected on Same Level
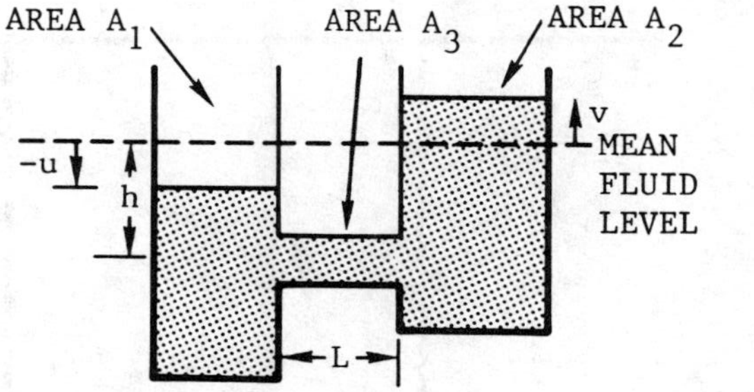
</td>
<td>$\frac{1}{2\pi}\left[\frac{g\left(1+\frac{A_2}{A_1}\right)}{h\left(1+\frac{A_2}{A_1}\right)+\frac{A_2}{A_3}L}\right]^{1/2}$</td>
<td>$\begin{bmatrix}\tilde{u}\\ \tilde{v}\end{bmatrix}=\begin{bmatrix}1\\ \frac{-A_1}{A_2}\end{bmatrix}$
L = length of connecting pipe
Ref. 13-12</td>
</tr>
<tr>
<td>4. Two Tanks Connected on Different Levels
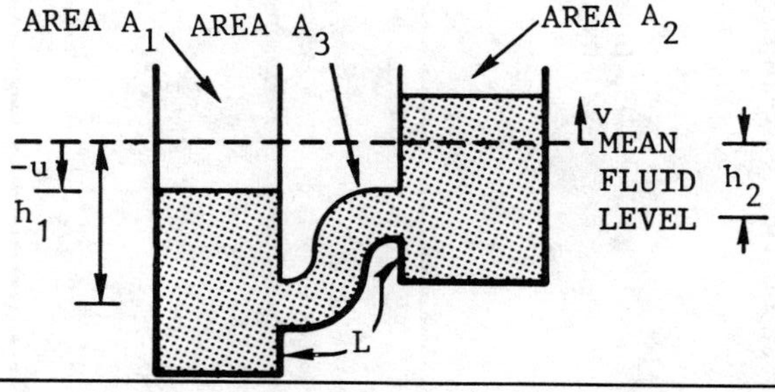
</td>
<td>$\frac{1}{2\pi}\left[\frac{g\left(1+\frac{A_2}{A_1}\right)}{h_2+h_1\frac{A_2}{A_1}+L\frac{A_2}{A_3}}\right]^{1/2}$</td>
<td>$\begin{bmatrix}\tilde{u}\\ \tilde{v}\end{bmatrix}=\begin{bmatrix}1\\ \frac{-A_1}{A_2}\end{bmatrix}$
L = length of connecting pipe</td>
</tr>
</table>

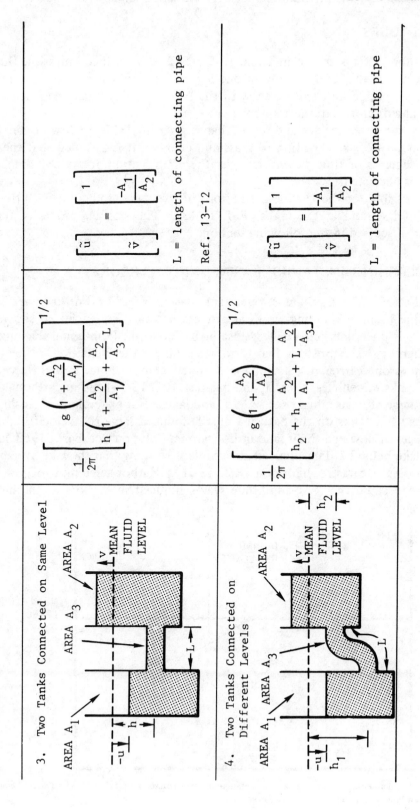

13.4. U-TUBES

The U-tube systems shown in Table 13-5 are filled with incompressible fluid to the mean level. If the level of fluid in one leg of the U-tube is changed and the system is left free, oscillations will occur with the liquid flowing back and forth across the U-tube at a discrete natural frequency.

The natural frequencies and mode shapes given in Table 13-5 were derived from the principles of conservation of mass and energy. It is assumed that the velocity of the elements of fluid is uniform over the cross section at any one time. Viscous friction is neglected.

It is possible for the U-tube oscillations of two tanks to couple with the sloshing modes of the fluid in the tanks (Ref. 13-13). The sloshing modes of fluid-filled tanks are discussed in the following section.

13.5. SLOSHING IN TANKS, BASINS, AND HARBORS

General Case. The liquid in an open tank, basin, or harbor can flow back and forth across the basin in standing waves at discrete natural frequencies, as shown in Fig. 13-6. This phenomenon is called sloshing in liquid-filled tanks and seiching or surging in harbors. The most familiar form of sloshing is the oscillations of tea or coffee in a cup as one carries it. Sloshing is of practical importance in fuel flow problems in fuel tanks of vehicles, aircraft, and missiles (Ref. 13-14) and in earthquake excitation of large storage tanks (Ref. 13-15) and lakes. For example, fuel sloshing in the liquid oxygen tanks on the Saturn I missile induced large roll velocities during the later stages of powered flight, and in Longarone, Italy, on October 9, 1963, landslides into a lake behind a dam induced such violent sloshing that the lake overspilled the dam causing a catastrophic flood (Ref. 13-16). Harbor seiching is of practical importance because the large-amplitude waves induced in harbors as the harbors re-

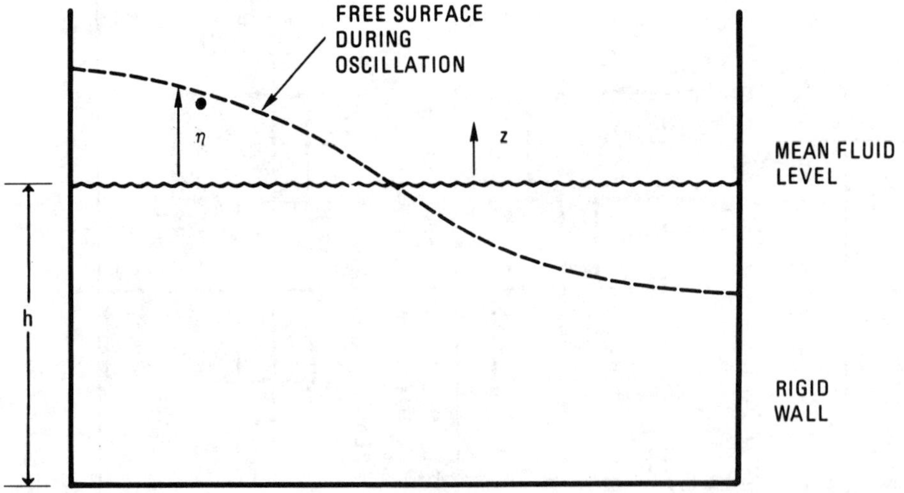

Fig. 13-6. Fundamental sloshing mode in a rectangular basin (frame 1 of Table 13-6). The xy plane is parallel to the mean fluid level and perpendicular to the paper.

spond to low-frequency sea waves or tidal waves can prevent loading and unloading of the boats in the harbors and cause extensive damage in extreme cases (Ref. 13-17).

The general assumptions used in the analysis of the solutions for sloshing liquids in this section are:

1. The liquid is homogeneous, inviscid, irrotational, and incompressible.
2. The boundaries of the basins are rigid.
3. The wave amplitudes are sufficiently small in comparison with the wavelengths and depths to permit nonlinear effects to be neglected.
4. The influence of the surrounding atmosphere is negligible.
5. The influence of surface tension is negligible.

The first assumption, i.e., a homogeneous, inviscid, incompressible liquid, provides an accurate model of most practical situations involving large tanks containing liquids such as water or fuels. The inviscid model generally provides a good first approximation to the natural frequency and mode shape of most real basins and harbors if the Reynolds number of fluid motion, UD/ν, where U is a typical velocity of the sloshing liquid, ν is the kinematic viscosity, and D is a typical fluid displacement amplitude, is of order 1000 or greater so that the influence of boundary layers at the basin walls can be neglected.

The assumption of rigid boundaries is valid so long as the fluctuation in pressure on the basin walls due to sloshing does not include flexing of the basin walls. Of course, real basins have walls which are compliant to some degree. Flexing of the basin walls can have a significant influence on the natural frequencies and mode shapes of sloshing in extreme cases (Refs. 13-16, 13-18).

The influence of surface tension increases as the wave length of the wave decreases. For wave lengths of the order of $\lambda = 2\pi[\sigma/(\rho g)]^{1/2}$ (= 1.7 cm for water in contact with air at 20°C) or smaller, surface tension can substantially influence the wave speed (Ref. 13-29). σ is the surface tension, ρ is the fluid density, λ is the wave length, and g is the acceleration due to gravity. However, for most wave lengths of practical importance, surface tension plays only a minor role in the wave dynamics.

Tanks and Basins of Constant Depth. If the liquid is contained by a continuous rigid wall, the motion of the fluid in the basin can be expressed in terms of a potential function Φ. For free vibrations in one mode in a basin of constant depth, small-amplitude wave theory gives (Ref. 13-17):

$$\Phi = \frac{Ag}{\omega}\,\frac{\cosh \omega(h+z)/c}{\cosh \omega h/c}\,\tilde{\phi}\,\sin(\omega t + \psi), \tag{13-28}$$

where

A = amplitude of the wave,
c = wave speed,
g = acceleration of gravity (Table 3-1),
h = mean depth of liquid in the basin,
t = time,

z = coordinate perpendicular to the mean liquid level, upward (Fig. 13-6),
$\tilde{\phi} = \tilde{\phi}(x, y)$ a dimensionless mode shape (Table 13-6),
ω = circular natural frequency of oscillation ($\omega = 2\pi f$, Table 13-6),
ψ = a phase angle.

The wave speed, c, is a function of the frequency of the wave, the depth of the fluid, and the acceleration of gravity. c is determined by the solution of the following transcendental equation (Ref. 13-17, pp. 24–27):

$$c = \frac{g}{\omega} \tanh \frac{\omega h}{c}. \tag{13-29}$$

This equation can be considerably simplified if the depth of the liquid (h) in the basin is either much less or much greater than the wavelength ($2\pi c/\omega$) of the wave:

If $\frac{\omega h}{c} < \frac{\pi}{10}$ (shallow liquid waves)	If $\frac{\omega h}{c} > \pi$ (deep liquid waves)
$\sinh \frac{\omega h}{c} \approx \frac{\omega h}{c}$	$\sinh \frac{\omega h}{c} \approx \frac{e^{\omega h/c}}{2}$
$\cosh \frac{\omega h}{c} \approx 1$	$\cosh \frac{\omega h}{c} \approx \frac{e^{\omega h/c}}{2}$
$\tanh \frac{\omega h}{c} \approx \frac{\omega h}{c}$	$\tanh \frac{\omega h}{c} \approx 1$

Thus, for shallow liquid waves such that

$$\frac{\omega h}{c} < \frac{\pi}{10},$$

the wave speed (Eq. 13-29) becomes proportional to the square root of the liquid depth:

$$c = (gh)^{1/2}.$$

For deep liquid waves such that

$$\frac{\omega h}{c} > \pi,$$

the wave speed becomes independent of the liquid depth:

$$c = \frac{g}{\omega}.$$

Thus, the wave speed increases with increasing liquid depth until the deep liquid limit is reached. Since the natural frequencies of sloshing waves are proportional to the wave speed, these equations imply that the natural frequencies of sloshing in shallow basins will increase if the liquid depth is increased until the deep liquid limit is reached.

Table 13-6. Waves in Basins of Constant Depth.

Notation: g = acceleration due to gravity (Table 3-1); h = depth of liquid in basin; r = radii; $\mathcal{J}_i$ = Bessel function of first kind and i order; $\mathcal{Y}_i$ = Bessel function of second kind and i order; α = angle (degrees); θ = angle (radians); consistent sets of units are given in Table 3-1. The pressure, velocity, and displacement of the fluid in the basin are expressed in terms of $\tilde{\phi}_{ij}$ using the formulas in the text.

Plan Form	Natural Frequency, f_{ij} (hertz)	Mode Shape of the Potential Function, $\tilde{\phi}_{ij}$ (Eq. 13-28)
1. Rectangular 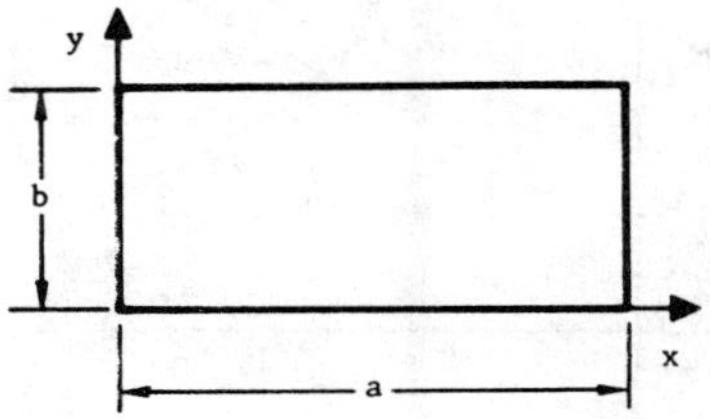	General Case: $f_{ij} = \frac{g^{1/2}}{2\pi^{1/2}}\left[\left(\frac{i^2}{a^2}+\frac{j^2}{b^2}\right)^{1/2} \tanh \pi h\left(\frac{i^2}{a^2}+\frac{j^2}{b^2}\right)^{1/2}\right]^{1/2}$ $i = 0,1,2\ldots$ $j = 0,1,2\ldots$ Shallow Liquid Case, $h\left(\frac{i^2}{a^2}+\frac{j^2}{b^2}\right)^{1/2} < \frac{1}{10}$: $f_{ij} = \frac{(gh)^{1/2}}{2}\left(\frac{i^2}{a^2}+\frac{j^2}{b^2}\right)^{1/2}$ $i = 0,1,2\ldots$ $j = 0,1,2\ldots$ Deep Liquid Case, $h\left(\frac{i^2}{a^2}+\frac{j^2}{b^2}\right)^{1/2} > 1$: $f_{ij} = \frac{g^{1/2}}{2\pi^{1/2}}\left(\frac{i^2}{a^2}+\frac{j^2}{b^2}\right)^{1/4}$ $i = 0,1,2\ldots$ $j = 0,1,2\ldots$	$\cos\frac{i\pi x}{a}\cos\frac{j\pi y}{b}$ $i = 0,1,2\ldots$ $j = 0,1,2\ldots$ Ref. 13-6, p. 440.
2. Isosceles Right Triangle 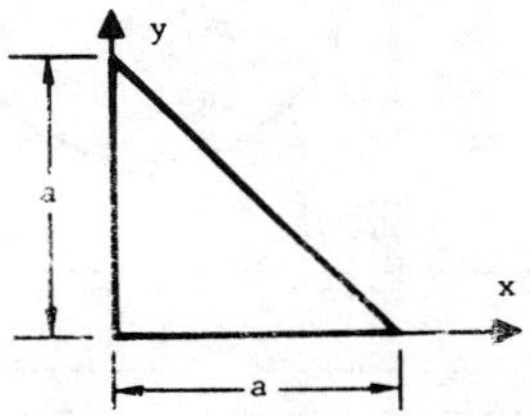	General Case: $f_{ij} = \frac{g^{1/2}}{2\pi^{1/2}a^{1/2}}\left[(i^2+j^2)^{1/2}\tanh\frac{\pi h}{a}(i^2+j^2)^{1/2}\right]^{1/2}$ $(i,j) = (1,2), (2,1), (2,3), (3,2)$, etc. $i,j > 0$, but $i \neq j$ Shallow Liquid Case, $\frac{h}{a}(i^2+j^2)^{1/2} < \frac{1}{10}$: $f_{ij} = \frac{(gh)^{1/2}}{2a}(i^2+j^2)^{1/2}$ Deep Liquid Case, $\frac{h}{a}(i^2+j^2)^{1/2} > 1$: $f_{ij} = \frac{g^{1/2}}{2\pi^{1/2}a^{1/2}}(i^2+j^2)^{1/4}$	$\cos\frac{i\pi x}{a}\cos\frac{j\pi y}{a} \pm \cos\frac{j\pi x}{a}\cos\frac{i\pi y}{a}$ $-$ if $i + j$ = even $+$ if $i + j$ = odd $(i,j) = (1,2), (2,1), (2,3), (3,2)$, etc. $i,j > 0$, but $i \neq j$

Table 13-6. Waves in Basins of Constant Depth. *(Continued)*

Plan Form	Natural Frequency, f_{ij} (hertz)	Mode Shape of the Potential Function, $\tilde{\phi}_{ij}$ (Eq. 13-28)
3. Circular	(see below)	(see below)
4. Circular Sector	(see below)	(see below)

3. Circular

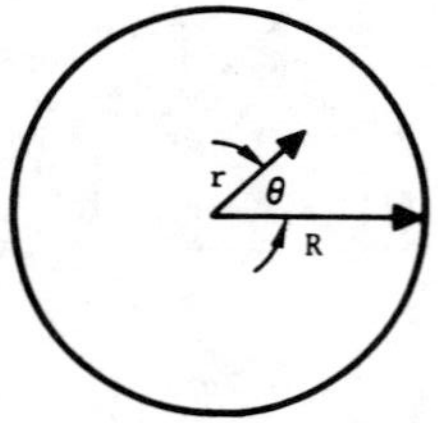

Natural Frequency, f_{ij} (hertz)

General Case:

$$f_{ij} = \frac{1}{2\pi}\left(\frac{\lambda_{ij} g}{R} \tanh \frac{\lambda_{ij} h}{R}\right)^{1/2}$$

$i = 0,1,2....$

$j = 0,1,2....$

Shallow Liquid Case, $\lambda_{ij} \frac{h}{R} < \frac{\pi}{10}$:

$$f_{ij} = \frac{\lambda_{ij}}{2\pi R} (gh)^{1/2} \qquad i = 0,1,2... \quad j = 0,1,2...$$

Deep Liquid Case, $\lambda_{ij} \frac{h}{R} > \pi$:

$$\frac{1}{2\pi}\left(\frac{\lambda_{ij} g}{R}\right)^{1/2} \qquad i = 0,1,2... \quad j = 0,1,2...$$

Mode Shape of the Potential Function, $\tilde{\phi}_{ij}$ (Eq. 13-28)

$$J_i\left(\lambda_{ij}\frac{r}{R}\right)\begin{Bmatrix}\sin i\theta \\ \text{or} \\ \cos i\theta\end{Bmatrix}$$

$i = 0,1,2....$ = number of nodal diameters

$j = 0,1,2....$ = number of nodal circles

λ_{ij}

j \ i	0	1	2	3	4	5	6
0	0	1.8412	3.0542	4.2012	5.3175	6.4156	7.5013
1	3.8317	5.3314	6.7061	8.0152	9.2824	10.5199	11.7349
2	7.0156	8.5363	9.9695	11.3459	12.6819	13.9872	15.2682
3	10.173	11.7060	13.1704	14.5859	15.9641	17.3128	18.6374

$[J_i'(\lambda_{ij}) = 0]$, $\lambda_{i=0,j} \approx \pi\,(j + 0.25)$ for $j \geq 3$.

Additional values of λ_{ij} are tabulated in Ref. 13-4, pp. 411,468. Ref. 13-6, pp. 284-285.

4. Circular Sector

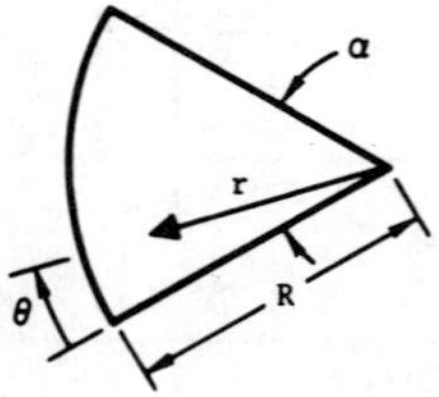

Natural Frequency, f_{ij} (hertz)

The natural frequencies can be adapted from those in frame 3 with the following choices of the index i:

$$i = \frac{180n}{\alpha}\ ; \qquad n = 0,1,2...$$

$0° \leq \alpha \leq 360°$ (α in degrees)

and j as given in frame 3. i will be integer only if α is a submultiple of 180n. λ_{ij} for non-integer i can be interpolated from the table in frame 3.

Mode Shape of the Potential Function, $\tilde{\phi}_{ij}$ (Eq. 13-28)

$$J_i\left(\lambda_{ij}\frac{r}{R}\right) \cos i\theta$$

i = as given in adjacent column

$j = 0,1,2,3...$

λ_{ij} is found from frame 3. For non-integer i, λ_{ij} is obtained by interpolation.

Ref. 13-5, pp. 299-300.

5. Annular

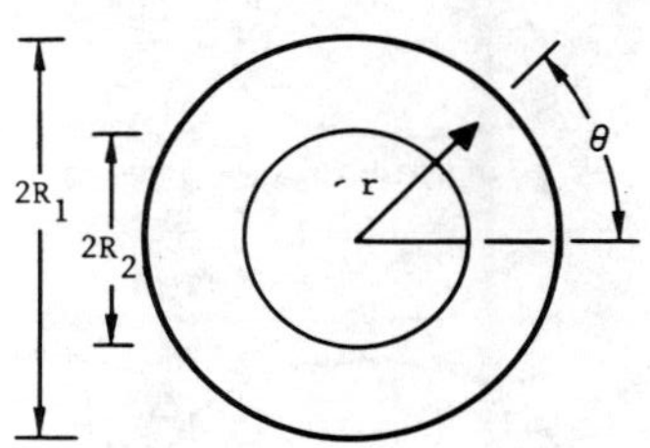

General Case:

$$f_{ij} = \frac{1}{2\pi}\left(\frac{\lambda_{ij} g}{R_1} \tanh \lambda_{ij} \frac{h}{R_1}\right)^{1/2}$$

$i = 0,1,2,...$
$j = 0,1,2...$

Shallow Liquid Case; $\lambda_{ij} \frac{h}{R_1} < \frac{\pi}{10}$:

$$f_{ij} = \frac{\lambda_{ij}}{2\pi R_1} (gh)^{1/2}$$

$i = 0,1,2...$
$j = 0,1,2...$

Deep Liquid Case: $\lambda_{ij} \frac{h}{R_1} > \pi$:

$$f_{ij} = \frac{1}{2\pi}\left(\frac{\lambda_{ij} g}{R_1}\right)^{1/2}$$

$i = 0,1,2...$
$j = 0,1,2...$

$$\tilde{\phi}_{ij} = G_{ij}(r) \begin{Bmatrix} \sin i\theta \\ \text{or} \\ \cos i\theta \end{Bmatrix} \qquad R_2 \le r \le R_1$$

$$G_{ij}(r) = Y_i'(\lambda_{ij}) J_i\left(\lambda_{ij} \frac{r}{R_1}\right) - J_i'(\lambda_{ij}) Y_i\left(\lambda_{ij} \frac{r}{R_1}\right)$$

$$G_{ij}(r) \approx \cos\left[j\pi (r - R_2)/(R_1 - R_2)\right] \text{ for } \frac{R_2}{R_1} > 0.8$$

$\frac{R_2}{R_1}$	j	λ_{ij}, i = 0	1	2	3	4	5	6
0.3	0	0	1.5821	2.9685	4.1801	5.3130	6.4147	7.5011
	1	4.7058	5.1374	6.2738	7.7213	9.1526	10.4750	11.7214
0.5	0	0	1.3547	2.6812	3.9577	5.1752	6.3389	7.4622
	1	6.3932	6.5649	7.0626	7.8401	8.8364	9.9858	11.2270

$$\left(\left.\frac{dG_{ij}}{dr}\right|_{r = R_2} = 0\right)$$

Additional values of λ_{ij} for $i = 1$ and various R_2/R_1 and j are tabulated in Ref. 13-4.

$$\lambda_{ij}^2 \approx \left(\frac{j\pi R_1}{R_1 - R_2}\right)^2 + \left(\frac{2iR_1}{R_1 + R_2}\right)^2 \text{ for } \frac{R_2}{R_1} > 0.5$$

$i = 0,1,2...$ = number of nodal diameters

$j = 0,1,2...$ = number of nodal circles

See Fig. 13-2.

Table 13-6. Waves in Basins of Constant Depth. *(Continued)*

Plan Form	Natural Frequency, f_{ij} (hertz)	Mode Shape of the Potential Function, $\tilde{\phi}_{ij}$ (Eq. 13-28)
6. Annular Sector 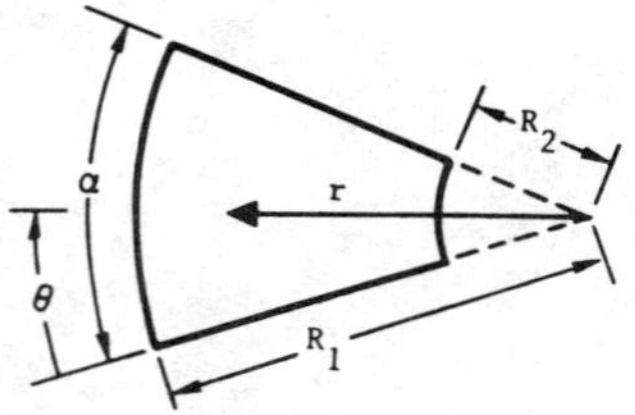	The natural frequencies can be adapted from those in frame 5 with the following choices of the index i: $i = \frac{180n}{\alpha}$; $n = 0,1,2...$ $0 \le \alpha \le 360$ (α in degrees) and j as given in frame 5. i will be integer only if α is a submultiple of 180n. λ_{ij} for non-integer i can be interpolated from the table in frame 5.	G_{ij} (r) cos iθ, see frame 5. i = as given in adjacent column j = 0,1,2... G_{ij} is given in frame 5.
7. Elliptical 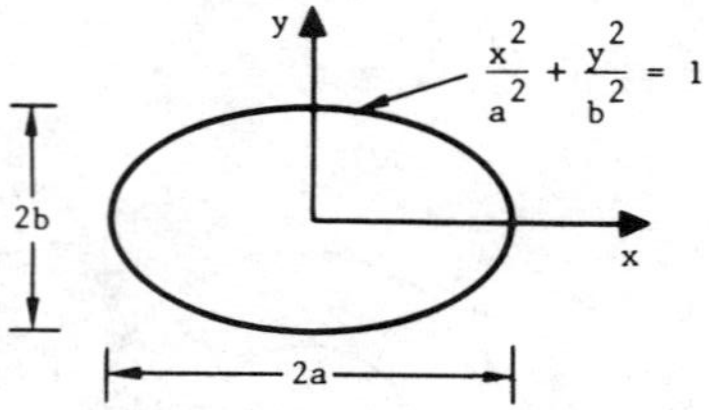	Fundamental Mode Only Shallow Liquid, $\frac{h}{a} < 0.1$: $f = \frac{(gh)^{1/2}}{2\pi a}\left[\frac{18 + 6\left(\frac{b}{a}\right)^2}{5 + 2\left(\frac{b}{a}\right)^2}\right]^{1/2}$ $\frac{b}{a} < 1$	Surface elevation, $\eta = Ax$. A = constant Ref. 13-6, p. 290.
8. Arbitrary Simple Curve 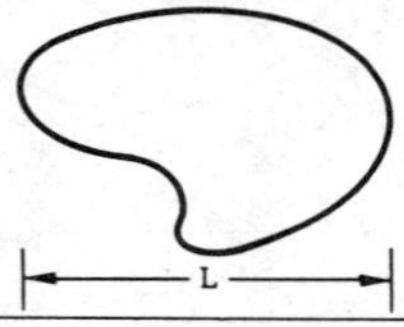	Fundamental Mode Only Approximate formula: $f = \begin{cases} \frac{(gh)^{1/2}}{2L}, & \text{if } \frac{h}{L} < \frac{1}{10} \\ \frac{1}{2}\left(\frac{g}{\pi L}\right)^{1/2}, & \text{if } \frac{h}{L} > 1 \end{cases}$	L = typical maximum lateral dimension

The natural frequencies and potential functions of liquid sloshing in basins of constant depth are given in Table 13-6. The wave form, pressure, and liquid velocities can be expressed in terms of the potential function Φ (Eq. 13-28). For free vibrations in the ij mode, the wave amplitude at the free surface (Fig. 13-6) is:

$$\eta = \frac{1}{g}\frac{\partial \Phi}{\partial t} \quad \text{at} \quad z = 0,$$

$$= A_{ij}\, \tilde{\phi}_{ij} \cos(\omega_{ij}t + \psi_{ij}). \tag{13-30}$$

A_{ij} is the amplitude in the ij mode, which has a circular natural frequency ω_{ij}, a mode shape $\tilde{\phi}_{ij}$, and a phase angle ψ_{ij}. The pressure in the liquid is the sum of a dynamic component and the static head:

$$p = \rho \frac{\partial \Phi}{\partial t} - \rho g z,$$

$$= \rho A g \frac{\cosh \omega (h + z)/c}{\cosh \omega h/c}\, \tilde{\phi}_{ij} \cos(\omega t + \psi_{ij}) - \rho g z. \tag{13-31}$$

The velocity of the liquid in the basin is:

$$\vec{u} = -\nabla \Phi, \tag{13-32}$$

where $\vec{u}$ is the vector fluid velocity and ∇ is the vector gradient operator. In rectangular coordinates (frame 1 of Table 13-6), this equation implies:

$$u_x = -\frac{\partial \Phi}{\partial x}, \quad u_y = -\frac{\partial \Phi}{\partial y}, \quad u_z = -\frac{\partial \Phi}{\partial z}.$$

u_x, u_y, and u_z are the fluid velocities in the x, y, and z directions, respectively. In cylindrical coordinates (frames 3 through 6 of Table 13-6), Eq. 13-32 implies:

$$u_r = -\frac{\partial \Phi}{\partial r}, \quad u_\theta = -\frac{1}{r}\frac{\partial \Phi}{\partial \theta}, \quad u_z = -\frac{\partial \Phi}{\partial z},$$

where u_r is the radial velocity, u_θ is the circumferential velocity, and u_z is the vertical velocity.

For most basins enclosed by a simple closed curve, the fundamental natural frequency in hertz of the basin can be estimated to be:

$$f = \begin{cases} \dfrac{(gh)^{1/2}}{2L}, & \text{if} \quad \dfrac{h}{L} < \dfrac{1}{10}, \\[2ex] \dfrac{1}{2}\left(\dfrac{g}{\pi L}\right)^{1/2}, & \text{if} \quad \dfrac{h}{L} > 1, \end{cases} \tag{13-33}$$

where L is the maximum lateral dimension of the basin, g is the acceleration of gravity, and h is the mean depth of the fluid in the basin. The fundamental natural frequency of the basin decreases with increasing lateral dimension. If the maximum lateral dimension of a basin is of the order of kilometers, such as for a lake, the fundamental natural period of the basin (1/f) will be on the order of minutes or even hours.

The mode shapes of basins become increasingly complex in the higher modes. However, very often these mode shapes have a simple geometric interpretation. For example, the mode shape of waves in an annular basin is identical to that of the annular acoustic cavity shown in Fig. 13-2. The number of nodal diameters is the index i in frame 5 of Table 13-6, and the number of nodal radii is the index j. The pattern suggested by Fig. 13-2 applies to circular basins as well.

Since the equation describing the small amplitude motion of fluid in a basin is linear, the response of the basin to oscillation in several modes with various amplitudes, frequencies, and phases can be found simply by summing the dynamic responses in each of the individual modes. For example, the potential function for multi-modal free vibration becomes:

$$\Phi = \sum_i \sum_j \frac{A_{ij} g}{\omega_{ij}} \frac{\cosh \omega_{ij}(h + z)/c}{\cosh \omega_{ij} h/c} \tilde{\phi}_{ij} \sin(\omega_{ij} t + \psi_{ij}), \tag{13-34}$$

where A_{ij} is the amplitude in the ij mode which has a circular natural frequency ω_{ij}, a phase angle with respect to other modes ψ_{ij}, and a mode shape of the potential function $\tilde{\phi}_{ij}$. The wave amplitude, liquid pressure, and liquid velocity can be determined from this potential by using Eqs. 13-30 through 13-32. Two indices, i and j, are generally required to specify the natural frequencies and mode shapes of waves in basins of constant depth.

The boundary conditions employed in the solutions presented in Table 13-6 are simply that the free surface of the liquid is described by Eq. 13-30 and that the component of fluid velocity normal to the sides and bottom of the basin is zero:

$$\vec{u} \cdot \hat{n} = 0, \tag{13-35}$$

where $\hat{n}$ is the normal outward to the sides and bottom. This boundary condition can be expressed in terms of the potential function as:

$$\nabla \Phi \cdot \hat{n} = 0 \tag{13-36}$$

on the sides and bottom of the basin.

Tanks and Basins of Variable Depth. If the mean depth of liquid in a basin varies over the bottom of the basin, it becomes very difficult to express the potential function (Eq. 13-28), which describes the liquid motion, in closed form. Numerical or approximate solutions are generally required.

Table 13-7 gives the natural frequencies of basins of variable depth. The mode shapes of sloshing in these basins is described only in qualitative terms in Table 13-7 because the mode shapes cannot generally be expressed in terms of simple functions. Additional numerical and experimental solutions are discussed in Refs. 13-14, 13-16, and 13-20. Sloshing can also be analyzed using the NASTRAN computer code (Chapter 15, Ref. 14-56).

By comparing the solutions in Tables 13-6 and 13-7, it can be seen that depth variations do not generally have a large influence on the natural frequencies of sloshing. For example, consider a basin of circular plan form whose maximum depth equals the radius of the plan form. The natural frequency of the fundamental sloshing mode of this basin can be calculated from either frame 3 of Table

Table 13-7. Waves in Basins of Variable Depth.

Notation: H = mean fluid level, R = radius; g = acceleration due to gravity (Table 3-1); graphic symbols: ------ = mean fluid level, ——— = dynamic fluid level during oscillation in the fundamental mode; consistent sets of units are given in Table 3.1.

Description	Natural Frequency, f_i (hertz)

1. Right Circular Canal

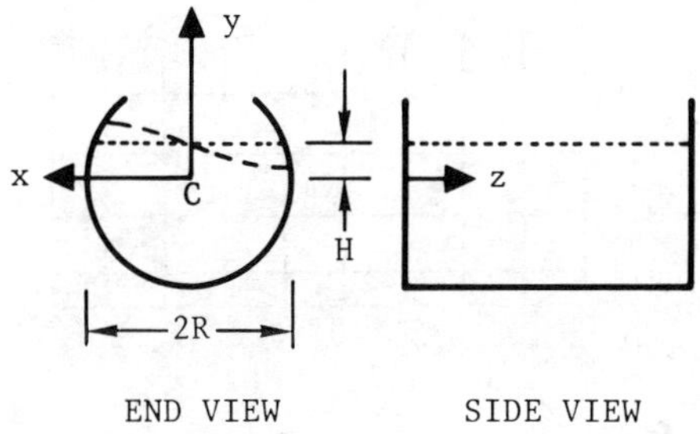

Note: H = height of water above the center C.

Modes independent of z, antisymmetric about the y-z plane:

$$f_i = \frac{\lambda_i^{1/2}}{2\pi}\left(\frac{g}{R}\right)^{1/2} \; ; \quad i = 1,2,3\ldots$$

H/R	λ_1	λ_2	λ_3
-1.0	1.0	6.0	15.0
-0.8	1.045	5.38	10.85
-0.6	1.099	4.97	9.13
-0.4	1.165	4.74	8.33
-0.2	1.249	4.65	7.99
0.0	1.360	4.70	7.96
0.2	1.513	4.91	8.23
0.4	1.742	5.34	8.89
0.6	2.13	6.22	10.28
0.8	3.04	8.42	13.84
1.0	∞	∞	∞

Ref. 13-19

2. Angular Canal

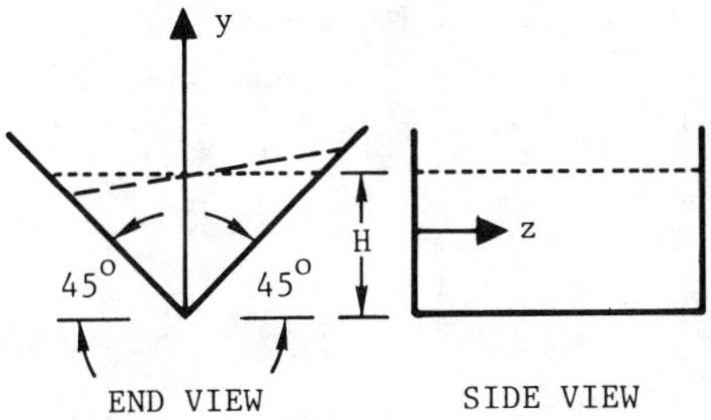

Modes independent of z:

$$f_i = \frac{\lambda_i^{1/2}}{2\pi}\left(\frac{g}{H}\right)^{1/2} \; ; \quad i = 1,2,3\ldots$$

$\lambda_1 = 1.0$

$\lambda_2 = 2.324$

$\lambda_3 = 3.9266$

$\lambda_i = \alpha \tanh \alpha, \; i > 1$

where $\cos 2\alpha \cosh 2\alpha = 1$. Solutions for this equation are given in Section 8.1 (Free-Free Beam). Modes with i = odd are antisymmetric about y-z plane. Modes with i = even are symmetric about y-z plane.

Ref. 13-6, pp. 442-444.

13-6 or frame 3, 4, or 6 of Table 13-7, depending on the contour of the bottom of the basin. The results are as follows:

1. Constant depth bottom (frame 3 of Table 13-6):

$$f = \frac{1.32}{2\pi}\left(\frac{g}{R}\right)^{1/2}$$

Table 13-7. Waves in Basins of Variable Depth. *(Continued)*

Description	Natural Frequency, f_i (hertz)
3. Conical Basin 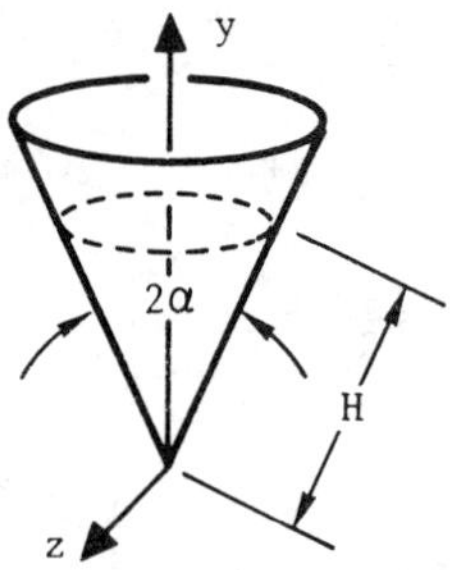 	Fundamental mode (antisymmetric about y-z plane): $f = \frac{\lambda^{1/2}}{2\pi}\left(\frac{g}{H}\right)^{1/2}$ 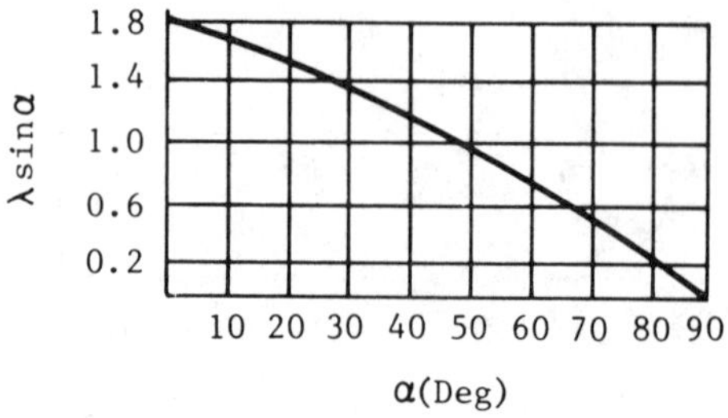 Ref. 13-20
4. Circular Basin with Parabolic Bottom 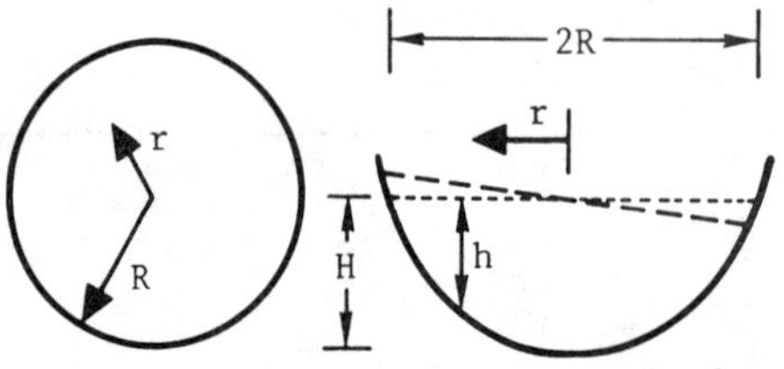 $h = H(1 - r^2/R^2)$ PLAN VIEW SIDE VIEW	For a shallow basin, H << 2R: $f_{ij} = \lambda_{ij}^{1/2} \frac{1}{2\pi} \frac{(gH)^{1/2}}{R},$ where $\lambda_{ij} = i(4j - 2) + 4j\ (j - 1),$ $i = 0,1,2\ldots$ = number of nodal diameters $j = 1,2,3\ldots$ = number of nodal circles (fundamental mode is $i = 1$, $j = 1$, $\lambda = 2$) Ref. 13-6, pp. 291-293; Ref. 13-21.

2. Conical bottom (frame 3 of Table 13-7):

$$f = \frac{1.23}{2\pi}\left(\frac{g}{R}\right)^{1/2}.$$

3. Parabolic bottom (frame 4 of Table 13-7, although the shallow basin assumption is violated):

$$f = \frac{1.41}{2\pi}\left(\frac{g}{R}\right)^{1/2}.$$

Table 13-7. Waves in Basins of Variable Depth. *(Continued)*

Description	Natural Frequency, f_i (hertz)
5. Inclined Right Circular Cylinder 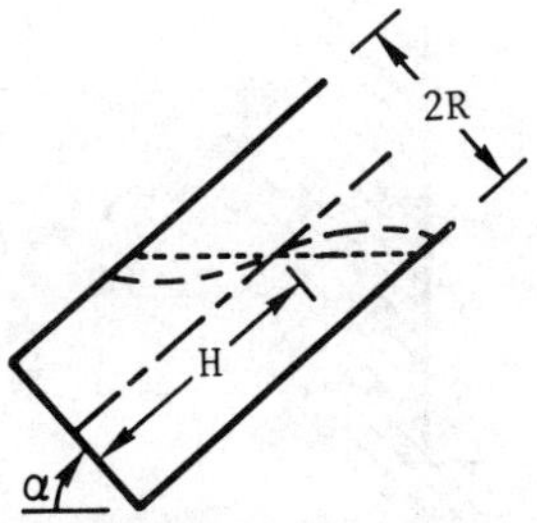 R = RADIUS OF THE CIRCULAR CYLINDER	Fundamental mode as shown: $f = \frac{\lambda^{1/2}}{2\pi}\left(\frac{g}{R}\right)^{1/2}$ 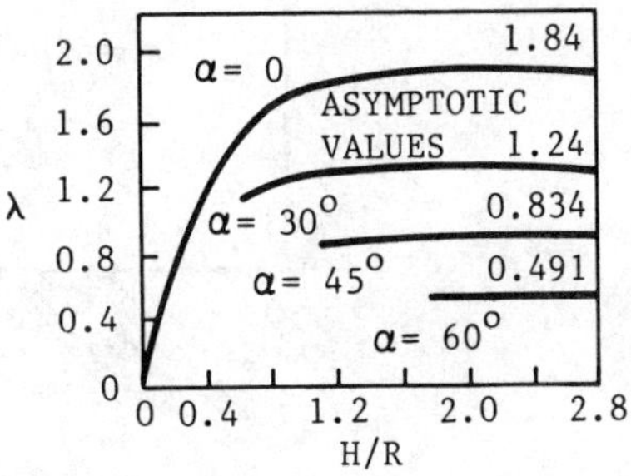 Ref. 13-22
6. Sphere 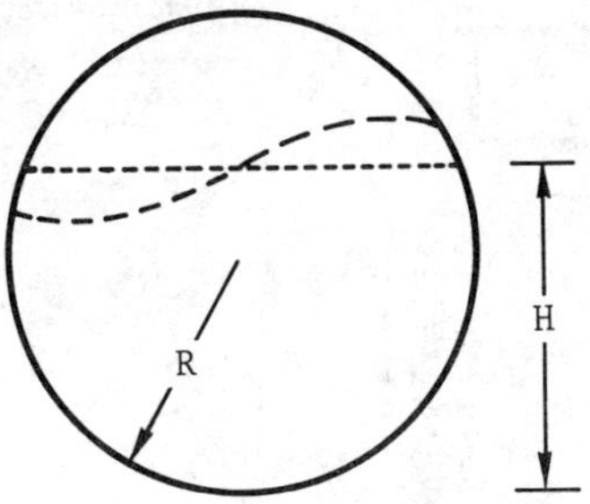 	Fundamental mode as shown: $f = \frac{\lambda^{1/2}}{2\pi}\left(\frac{g}{R}\right)^{1/2}$ 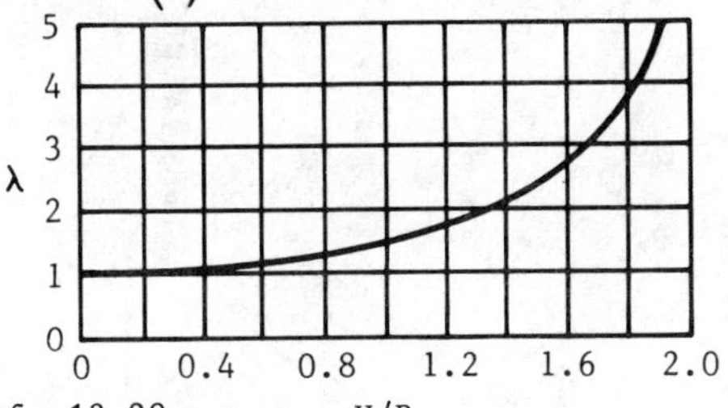 Ref. 13-20

4. Spherical bottom (frame 6 of Table 13-7):

$$f = \frac{1.26}{2\pi}\left(\frac{g}{R}\right)^{1/2}.$$

R is the radius of the plan form. g is the acceleration due to gravity. The difference between the highest and lowest of these natural frequencies is only 14%. Thus, in many cases the natural frequencies of basins with variable depth bottoms can be accurately estimated by computing a mean liquid depth and using the solutions in Table 13-6 for constant depth basins.

Harbors. A harbor is differentiated from the tanks and basins previously discussed by the presence of a channel from the harbor to the open sea, as shown in Fig. 13-7. The channel provides a path for energy to flow between the harbor and the sea. The

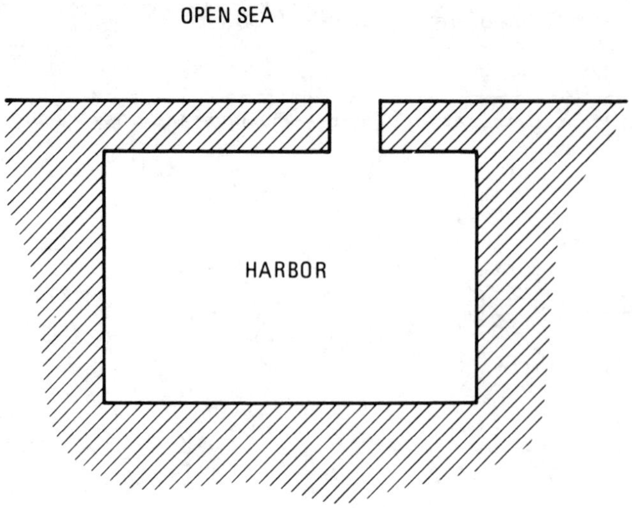

Fig. 13-7. Harbor with a narrow channel to the sea.

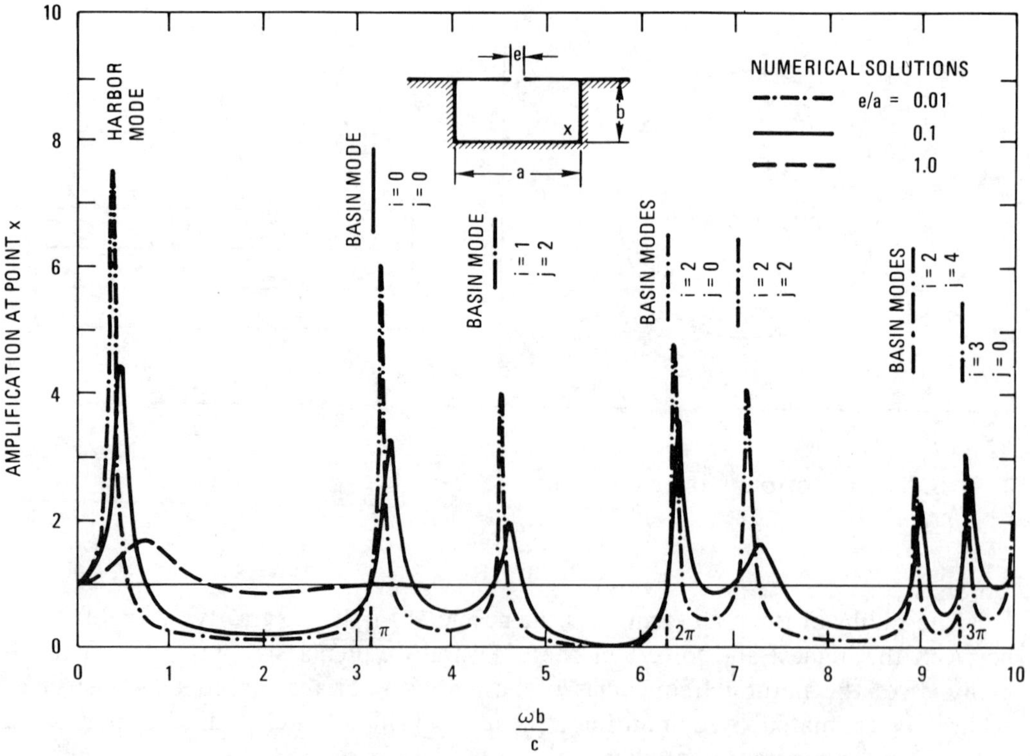

Fig. 13-8. Response of a rectangular harbor with various size openings to sea waves at frequency ω. c = wave speed. a/b = 2. The response is measured at x. The basin natural frequencies are computed from frame 1 of Table 13-6. (Refs. 13-17, 13-23.) From *Estuary and Coastline Hydrodynamics*, A. T. Ippen (ed.), Copyright 1966, McGraw-Hill Book Company, reproduced with permission.

analysis of harbors is considerably complicated by the coupling of the harbor to the sea. Apparently, there are no closed form solutions available for even the simplest harbor geometries which can provide a general and accurate estimate of the lowest natural frequency. However, the following observations can be inferred from the considerable body of numerical solutions and experimental work which have been done on harbors:

1. The presence of the open channel to the sea leads to at least one mode which has a natural frequency below the fundamental frequency of the basin formed by closing the harbor channel. This mode is often called the pumping mode.
2. The higher modes of the harbor can be accurately estimated from the modes of the basin formed by closing the harbor channel if the channel is relatively narrow.

These effects can be seen in Figs. 13-8 and 13-9, which compare experimental data (Fig. 13-9) and numerical solutions (Figs. 13-8 and 13-9) for the response of a harbor to sea waves.

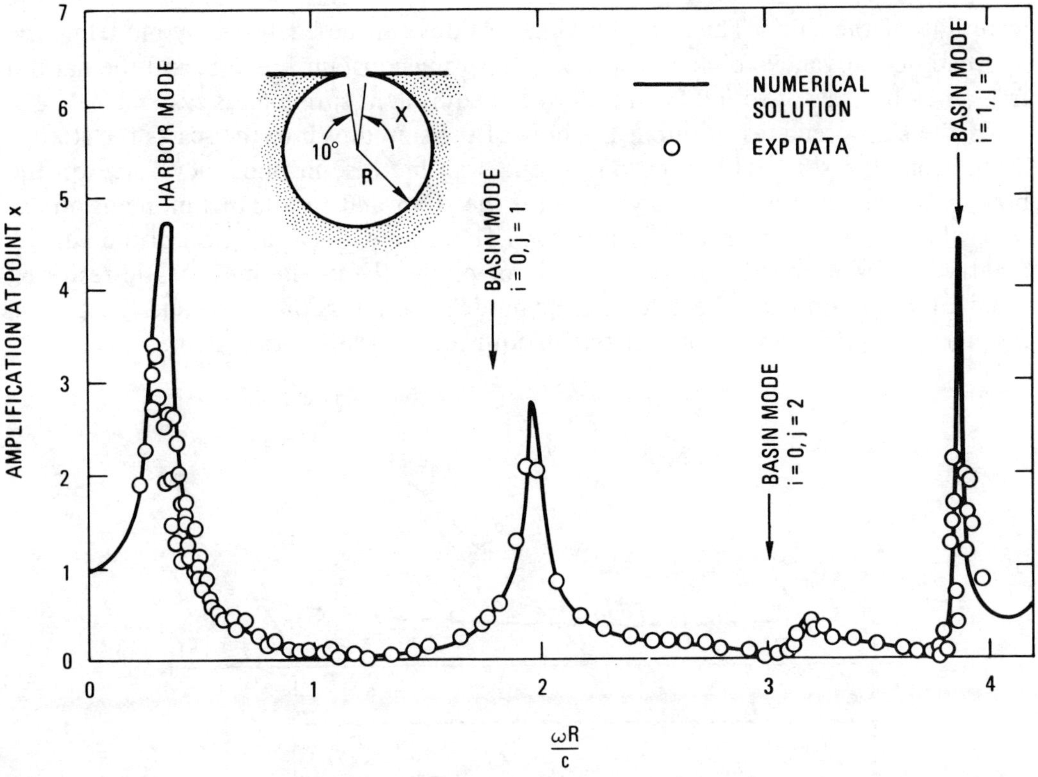

Fig. 13-9. Response at point x of a model of a circular harbor with a 10° opening to sea waves at frequency ω. R = 20.4 cm. c = wave speed. The response is measured at point x. The theoretical values of natural frequencies of a closed circular basin (frame 3 of Table 13-6) are ωR/c = 1.84 (i = 0, j = 1), 3.05 (i = 0, j = 2), and 3.84 (i = 1, j = 0). (Ref. 13-24). From *Journal of Fluid Mechanics*, Copyright 1971, Cambridge University Press, reproduced with permission.

Numerical solution techniques are available for determining the natural frequencies, mode shapes, and, in some cases, damping of an arbitrarily shaped harbor. These numerical techniques are described in Refs. 13-17 and 13-23 through 13-26.

13.6. SHIPS AND FLOATING STRUCTURES

Ships.   Ship motions are a logical starting point for a discussion of the free oscillations of floating structures. The motions of a rigid ship floating in a sea can be described by three mutually orthogonal displacements and three rotations. Natural frequencies can be associated with three of these motions (Fig. 13-10):

Heave: vertical translation perpendicular to the free surface.
Pitch: rotation about the transverse axis.
Roll: rotation about the longitudinal axis.

Heave is the simplest of the ship motions to analyze. If the ship is plunged vertically the distance $-\eta_3$ into the sea, the buoyancy force applied hydrostatically to the ship will increase by the weight of the displaced liquid, $\rho g S \eta_3$, where ρ is the liquid density, g is the acceleration due to gravity, and S is the area enclosed by the waterline of the ship. Thus, the heaving motions can be analyzed by modeling the hydrostatic buoyancy force as a spring with spring constant $k = \rho g S$, and the natural frequency of heave is found from that of the equivalent spring-mass system.

As the ship oscillates in pitch, the bow of the ship dips into the sea as the stern is lifted from the sea. This rotation produces an increase in buoyancy force on the bow and a decrease in buoyancy force on the stern and results in a moment on the ship which opposes the rotation in pitch. This moment is a function of the hull geometry and is proportional to the angle of pitch. Thus, the hydrostatic restoring moment which opposes the pitch can be modeled as a torsion spring and the natural frequency of pitch can be found from an equivalent spring-mass system.

$$\eta_1 = \text{SURGE} \quad \eta_3 = \text{HEAVE} \quad \eta_5 = \text{PITCH}$$
$$\eta_2 = \text{SWAY} \quad \eta_4 = \text{ROLL} \quad \eta_6 = \text{YAW}$$

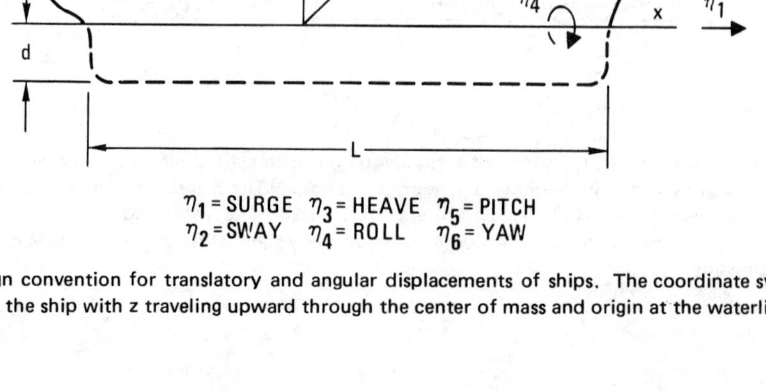

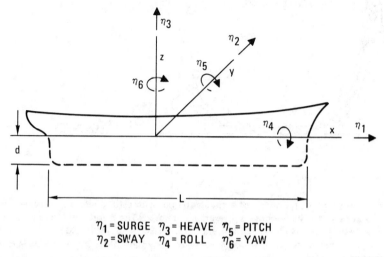

Fig. 13-10. Sign convention for translatory and angular displacements of ships. The coordinate system is fixed with respect to the ship with z traveling upward through the center of mass and origin at the waterline.

Roll motions, like pitch, are rotations. Roll could be analyzed using the same techniques as used for pitch; however, it has proven more useful to approach roll from stability analysis. Nearly all ships have centers of gravity located above the center of buoyancy (centroid of the volume of the ship below the waterline), as shown in Fig. 13-11. For a ship to be stable in roll, the center of buoyancy must move to the lee of the center of gravity as the ship rolls so that the buoyancy-induced moment always counters the roll of the ship. The righting moment in roll is proportional to the angle and can be expressed as:

$$\mathfrak{M} = -Mg\overline{GM} \times \text{angle of pitch (radians)}, \tag{13-37}$$

where M is the mass of the ship, g is the acceleration due to gravity, and $\overline{GM}$ is called the metacentric height. $\overline{GM}$ is typically on the order of one meter for ocean-going ships and can be found by hydrostatic analysis of the ship hull using Eq. 13-37. The above equation can be used to model the hydrostatic restoring moment on the ship as a torsion spring, and the natural frequency in roll can be found from the equivalent spring-mass system.

Table 13-8 gives the natural frequencies of rigid stable ships at rest in a still sea. These ships are assumed to be symmetric fore and aft (about the yz plane) and laterally (about the xz plane) both in terms of the hull geometry and the mass distribution of the ship. This symmetry ensures that the ship motions are not coupled inertially or hydrostatically. Since the natural frequencies of heave and pitch are often very close, any coupling can have a large influence on these two motions. Dissipative effects and coupling in the added mass terms have been neglected in Table 13-8. The approximate formulas given in the third column of Table 13-8 are based on di-

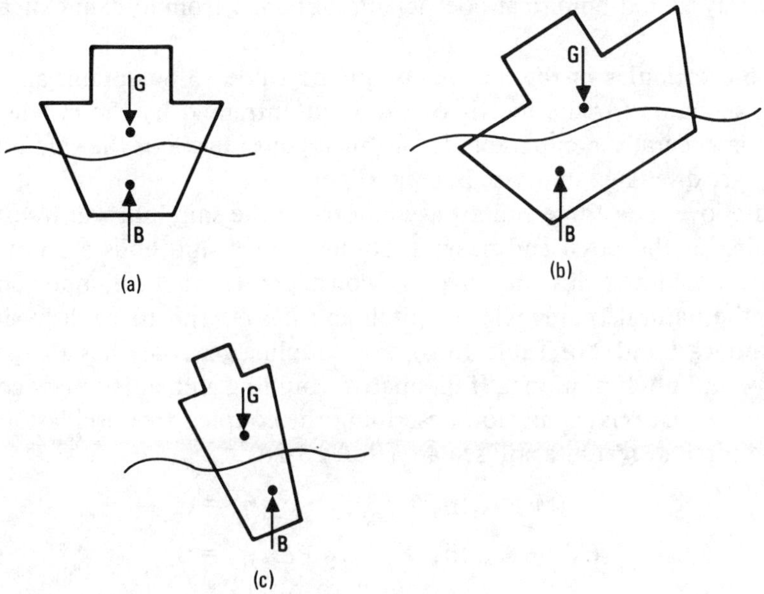

Fig. 13-11. Stability of ship in roll: (a) ship at equilibrium; (b) stable ship (moment tends to return ship to equilibrium); (c) unstable ship (moment tends to increase roll). G = center of gravity, B = center of buoyancy, the centroid of the volume of the hull below the waterline.

Table 13-8. Ships with Lateral and Longitudinal Symmetry.

Notation: b = beam of ship at waterline; b_m = maximum beam at waterline; d = depth of hull; g = acceleration of gravity (Table 3-1); x = fore and aft coordinate; ρ = density of liquid (water); A_z = added mass of liquid in heave; A_{xx}, A_{yy} = polar mass moments of inertia of added mass of liquid about x and y axes, respectively; $\overline{GM}$ = metacentric height (Eq. 13-37); J_{xx}, J_{yy} = polar mass moments of inertia of ship about x and y axes, respectively; L = fore and aft length; M = mass of ship; S = plane area enclosed by waterline. See Fig. 13-10. Solutions are from Ref. 13-27. Consistent sets of units are given in Table 3-1.

Description of Motion	Natural Frequency, f_i (hertz)	Approximate Natural Frequency for Slender Ships, f_i (hertz)
Heave	$\frac{1}{2\pi}\left(\frac{\rho g S}{M + A_z}\right)^{1/2}$	$0.13\left(\frac{g}{d}\right)^{1/2}$
Pitch	$\frac{1}{2\pi}\left[\frac{\rho g \int_L x^2 b(x)\,dx}{J_{yy} + A_{yy}}\right]^{1/2}$	$0.13\left(\frac{g}{d}\right)^{1/2}$
Roll	$\frac{1}{2\pi}\left(\frac{Mg\,\overline{GM}}{J_{xx} + A_{xx}}\right)^{1/2}$	$0.35\left(\frac{g\,\overline{GM}}{b_m^2}\right)^{1/2}$

mensional analysis and empirical coefficients obtained from dynamic measurements on ships.

Each of the formulas in the second column of Table 13-8 contains an added mass term which accounts for the inertia of the fluid entrained by the moving ship hull. This inertia is generally comparable to the analogous inertia of the ship hull. Added mass effects are discussed in more detail in Chapter 14.

As noted above, any force and aft asymmetry in the ship hull will result in hydrostatic coupling of the pitch and heave motions. Since ship hulls are not ordinarily symmetric fore and aft, this coupling is of considerable practical importance. Moreover, since the natural frequencies of pitch and heave tend to be very similar (column 3 of frames 1 and 2 of Table 13-8), this coupling generally has a large influence on the heave and pitch motions. If dissipative coupling and added mass coupling are neglected, the equations of motion describing the coupled free oscillations of a ship in heave and pitch at rest in a still sea are (Ref. 13-28):

$$(M + A_z)\ddot{\eta}_3 + c_{33}\eta_3 + c_{35}\eta_5 = 0, \qquad (13\text{-}38)$$

$$(J_{yy} + A_{yy})\ddot{\eta}_5 + c_{55}\eta_5 + c_{53}\eta_3 = 0, \qquad (13\text{-}39)$$

where η_3 is the heave displacement and η_5 is the angle of pitch, M is the mass of the ship, A_z is the added mass in heave, J_{yy} is the polar mass moment of the ship about the y axis, and A_{yy} is the moment of inertia of the added mass of water in pitch.

The hydrostatic force coefficients c_{33}, c_{55}, c_{35}, and c_{53} are functions of the hull geometry:

$$c_{33} = \rho g S,$$

$$c_{35} = c_{53} = -\rho g \int_L x b(x)\, dx,$$

$$c_{55} = \rho g \int_L x^2\, b(x)\, dx,$$

where ρ is the liquid density, S is the area enclosed by the waterline, and b(x) is the width (beam) of the hull at the waterline at each fore and aft point x. L is the length of the hull at the waterline. If the ship is symmetric fore and aft, then $b(x) = b(-x)$, $c_{35} = c_{53} = 0$, and the system uncouples. If the ship is nearly symmetric, c_{35} and c_{53} will be small.

Equations 13-38 and 13-39 can easily be solved for the coupled natural frequencies and mode shapes by assuming sinusoidal motion and applying the standard matrix techniques. The result is:

$$2\omega_i^2 = (\omega_3^2 + \omega_5^2) \pm \left[(\omega_3^2 - \omega_5^2)^2 + 4\omega_3^2\,\omega_5^2\,\frac{c_{35}\,c_{53}}{c_{33}\,c_{55}}\right]^{1/2}, \tag{13-40}$$

$$\begin{bmatrix}\tilde{\eta}_3\\ \\ \tilde{\eta}_5\end{bmatrix}_i = \begin{bmatrix}\dfrac{(c_{35}/c_{33})\omega_3^2}{\omega_i^2 - \omega_3^2}\\ 1\end{bmatrix}_i \quad \text{or} \quad \begin{bmatrix}1\\ \dfrac{(c_{53}/c_{55})\omega_5^2}{\omega_i^2 - \omega_5^2}\end{bmatrix}_i, \quad i = 1, 2, \tag{13-41}$$

where

$$\omega_3^2 = \frac{c_{33}}{M + A_z}, \qquad \omega_5^2 = \frac{c_{55}}{J_{yy} + A_{yy}}.$$

ω_3 and ω_5 are the circular natural frequencies of the uncoupled oscillation in heave and pitch, respectively. These frequencies are identical to the frequencies given in column 2 of frames 1 and 2 of Table 13-8. If the ship is nearly symmetric fore and aft, then c_{35} and c_{53} are small and the natural frequencies are not much changed by the coupling. However, since the natural frequencies of pitch and heave are generally close, $\omega_3 \approx \omega_5$, the coupling can result in substantial modal interaction (Eq. 13-41).

Buoys and Submarines. Buoys and submarines achieve pendulum stability by having the center of buoyancy above the center of gravity rather than by obtaining a sufficient $\overline{GM}$, as does a surface ship. If the center of buoyancy is above the center of gravity, the buoy has pendulum stability as shown in Fig. 13-12. As the buoy pitches about the center of mass (G), the offset between the center of buoyancy and the center of mass provides a positive righting moment. Buoys are weighted or tethered to a heavy chain to ensure that the center of mass is below the center of buoyancy. Submarines employ buoyancy tanks in the upper portions of the hull to achieve the same effect.

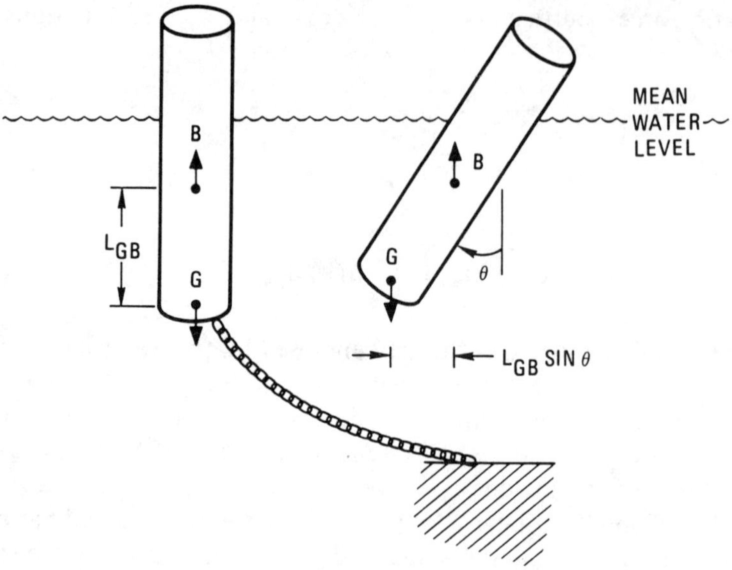

Fig. 13-12. A tethered spar buoy. If the buoy pitches, it rotates about the center of mass (G). B is the center of buoyancy, the centroid of the volume of the buoy below the waterline.

If a buoy pitches to a small angle θ, the righting moment can be expressed as:

$$\mathfrak{M} = -MgL_{GB}\theta,$$

where M is the mass of the buoy, g is the acceleration due to gravity, and L_{GB} is a positive length. If the center of buoyancy does not shift significantly during pitching, the L_{GB} is simply the vertical distance between the center of mass and the center of buoyancy. If the center of buoyancy does move significantly during pitching, the L_{GB}, like the metacentric height $\overline{GM}$ discussed in the previous section, is a proportionality constant.

The natural frequency of the resultant pitching oscillations is:

$$f = \frac{1}{2\pi}\left(\frac{MgL_{GB}}{J_{yy} + A_{yy}}\right)^{1/2} \text{ Hz},$$

where J_{yy} is the polar mass moment of inertia of the buoy about the axis which passes through the center of mass of the buoy and is perpendicular to the plane of rotation. A_{yy} is the polar mass moment of inertia of the water entrained by the rotating buoy about the same axis. A_{yy} is ordinarily of the same order as J_{yy}. Consistent sets of units are given in Table 3-1. Dissipative effects and added mass coupling have been neglected, as has dynamic coupling with the tether.

The heaving motions of a buoy, motion perpendicular to the free surface, can be analyzed by the formula previously given for ships in frame 1 of Table 13-8. Of course, the presence of the tethering chain, as shown in Fig. 13-12, can considerably complicate the analysis.

Other Floating Structures. The natural frequencies of other uncoupled floating structures, such as floats, rafts, and oil rigs, can generally be adapted from the formulas given in Table 13-8. For example, frame 1 of Table 13-8 (column 2) is the exact solution for the heave oscillations of any uncoupled floating structure. If the structure of interest was a semisubmersible oil rig, then area S in the formula refers to the sum of the cross-sectional areas described by the waterline around each leg of the platform.

13.7. EXAMPLE

Consider a rectangular pool of water with dimensions as shown in Fig. 13-13 which forms the cooling water pool for a research reactor. The pool is bounded by concrete walls. The acoustic modes of this pool can be found from the formulas in Table 13-2, and the sloshing modes can be found from the formulas in Table 13-6.

It is useful to calculate the reflection coefficient (Eq. 13-7) before deciding on an acoustic model for the pool boundaries. For the water/concrete boundary,

$$\frac{(\rho c)_{\text{concrete}}}{(\rho c)_{\text{water}}} = \frac{1900\ \text{kg/m}^3 \cdot 3100\ \text{m/sec}}{800\ \text{kg/m}^3 \cdot 1480\ \text{m/sec}} = 5.0,$$

for water at 20°C (Table 13-1), and the reflection coefficient for acoustic waves normal to the boundary (Eq. 13-7) is:

$$\alpha_r = 44\%.$$

This coefficient suggests that substantial acoustic energy will be transmitted into the concrete. This energy will be transmitted through the concrete to the concrete/air interface or concrete/soil interface at the external surfaces of the concrete walls. At the external surfaces the acoustic energy will be reflected back just as longitudinal stress waves are reflected off the ends of a free-free beam. Thus, although the reflection coefficient at the water/concrete boundary is less than 50%, it is doubtful if

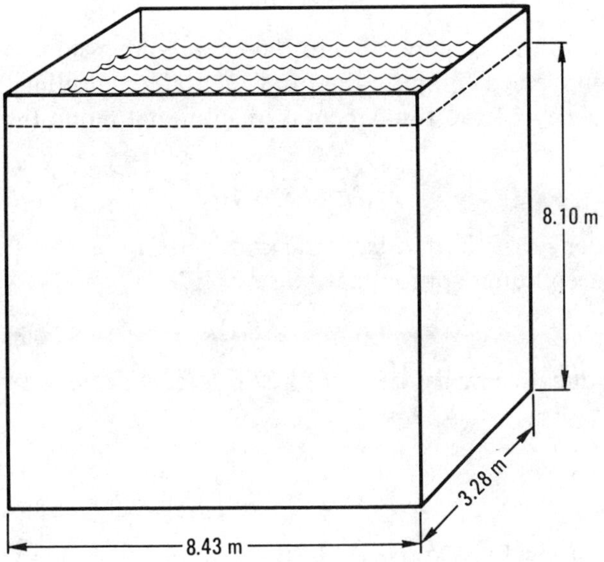

Fig. 13-13. A cooling water tank.

a high percentage of the acoustic energy escapes from the pool and its concrete boundaries.

For preliminary analysis, if the concrete walls are modeled as rigid boundaries and the surface of the pool is an open boundary, the natural frequencies of the pool can be obtained from frame 5 of Table 13-2, where the lengths of the cavity are:

$$L_x = 8.10 \text{ m},$$

$$L_y = 8.43 \text{ m},$$

$$L_z = 3.28 \text{ m}.$$

The natural frequency of the fundamental vertical mode ($i = 1, j = k = 0$) is:

$$f_{100} = 45.7 \text{ Hz}.$$

The natural frequency of the fundamental longitudinal mode ($i = 1, j = 1, k = 0$) is:

$$f_{110} = 99.0 \text{ Hz},$$

and the natural frequency of the fundamental transverse mode ($i = 1, j = 0, k = 1$) is:

$$f_{101} = 230 \text{ Hz},$$

where a speed of sound of $c = 1480$ m/sec has been used in these calculations. While the transmission and reflection off the concrete walls will influence these frequencies, it is reasonable to expect that the tank possesses a number of modes in the 50- to 100-Hz band. Thus, if the reactor in the pool emits acoustic noise in this range, an acoustic resonance may be formed in the pool. Such a resonance has been observed.

The sloshing modes of the pool can be evaluated from frame 1 of Table 13-6 with the lengths defined as follows:

$$a = 8.43 \text{ m},$$
$$b = 3.28 \text{ m},$$
$$h = 8.10 \text{ m}.$$

Since the liquid depth h is comparable to the lateral dimension a, the general case of Table 13-6 is employed to compute the fundamental longitudinal natural frequency. The natural frequency of the fundamental longitudinal mode ($i = 1, j = 0$) of sloshing is:

$$f_{10} = 0.309 \text{ Hz}.$$

The natural frequency of the fundamental transverse mode ($i = 0, j = 1$) can be estimated using the deep liquid approximation since $h > b$:

$$f_{01} = 0.488 \text{ Hz}.$$

An acceleration due to gravity of $g = 9.807 \text{ m/sec}^2$ (Table 3-1) has been used in these calculations.

REFERENCES

13-1. Kinsler, L. E., and A. R. Frey, *Fundamentals of Acoustics*, 2nd ed., John Wiley, New York, 1962.
13-2. Wilson, J., "Speed of Sound in Distilled Water as a Function of Temperature and Pressure," *J. Acoust. Soc. Am.* **31** 1067–1070 (1959).

13-3. Wilson, J., "Speed of Sound in Sea Water as a Function of Temperature, Pressure and Salinity," *J. Acoust. Soc. Am.* **32** 641-644 (1960).
13-4. Abramowitz, M., and I. A. Stegun (eds.), *Handbook of Mathematical Functions*, Dover Publications, Inc., New York, 1970, p. 411, p. 468 (formerly published by U.S. Government Printing Office, Washington, D.C.).
13-5. Rayleigh, J. W. S., *The Theory of Sound*, Vol. 2, 2nd ed., Dover Publications, Inc., New York, 1945.
13-6. Lamb, H., *Hydrodynamics*, 6th ed., Dover Publications, Inc., New York, 1945.
13-7. Morse, P. M., and H. Feshbach, *Methods of Theoretical Physics*, Part II, McGraw-Hill, New York, 1953, pp. 1468-1469.
13-8. Plesset, M. S., and A. Prosperetti, "Bubble Dynamics and Cavitation," in *Annual Review of Fluid Mechanics* (M. Van Dyke, ed.), Annual Reviews, Inc., Palo Alto, Ca., 1977.
13-9. Weast, R. D. (ed.), *Handbook of Chemistry and Physics*, The Chemical Rubber Company, Cleveland, Ohio, 1966.
13-10. Chapman, R. B., and M. S. Plesset, "Thermal Effects in the Free Oscillation of Gas Bubbles," *Trans. ASME, J. Basic Eng.* **93**, 373-376 (1971).
13-11. Alster, M., "Improved Calculation of Resonant Frequencies of Helmholtz Resonators," *J. Sound Vib.* **24**, 64-85 (1972).
13-12. Housner, G. W., and D. E. Hudson, *Applied Mechanics: Dynamics*, 2nd ed., D. Van Nostrand Co., Inc., Princeton, N.J., 1959, p. 128.
13-13. Yeh, G. C. K., "Sloshing of a Liquid in Connected Cylindrical Tanks Owing to U-Tube Free Oscillations," *J. Acoust. Soc. Am.* **40**, 807-812 (1966).
13-14. Abramson, H. N. "Dynamic Behavior of Liquid in Moving Container," *Appl. Mech. Revs.* **16**, 501-506 (1963).
13-15. Epstein, H. I., "Seismic Design of Liquid-Storage Tanks," *Am. Soc. Civil Engrs., J. Struct. Div.* **102**, 1659-1672 (1976).
13-16. Abramson, H. N. (ed.), "The Dynamic Behavior of Liquids in Moving Containers," NASA Report NASA SP-106, Southwest Research Institute, 1966.
13-17. Raichlen, F., "Harbor Resonance," in *Estuary and Coastline Hydrodynamics*, A. T. Ippen (ed.), McGraw-Hill, New York, 1966, pp. 281-340.
13-18. Bauer, H. F., and A. Siekmann, "Dynamical Interaction of a Liquid with the Elastic Structure of a Circular Container," *J. Ingenieur-Archin* **40**, 266-280 (1971).
13-19. Budiansky, B., "Sloshing of Liquids in Circular Canals and Spherical Tanks," *J. Aerospace Sci.* **27**, 161-173 (1960).
13-20. Moiseev, N. N., and A. A. Petrov, "The Calculation of Free Oscillations of a Liquid in a Motionless Container," in *Advances in Applied Mechanics*, G. Kuerti (ed.), Academic Press, New York, 1966, pp. 91-155.
13-21. Goldsbrough, G. R., "The Tidal Oscillations in an Elliptic Basin of Variable Depth," *Proc. Royal Soc. (London), Series A* **130**, 157-167 (1931).
13-22. McNeill, W. A., and J. P. Lamb, "Fundamental Sloshing Frequency for an Inclined, Fluid-Filled Right Circular Cylinder," *J. Spacecraft and Rockets* **7**, 1001-1002 (1970).
13-23. Ippen, A. T., and Y. Goda, "Wave Induced Oscillations in Harbors: The Solution for a Rectangular Harbor Connected to the Open Sea," Report No. 59, Hydrodynamics Lab., M.I.T., July 1963.
13-24. Lee, J. J., "Wave Induced Oscillations in Harbours of Arbitrary Geometry," *J. Fluid Mech.* **45**, 375-394 (1971).
13-25. Lee, J. J., and F. Raichlen, "Oscillations in Harbors with Connected Basins," *J. Waterways, Harbors and Coast Eng. Div., Am. Soc. Civil Engrs.* **91**, 311-331 (1972).
13-26. Miles, J. W., "Harbor Seiching," in *Annual Review of Fluid Mechanics*, M. Van Dyke, W. G. Vincenti, J. V. Wehausen (eds.), Annual Reviews Inc., Palo Alto, Ca., 1974, pp. 17-36.
13-27. Blevins, R. D., *Flow-Induced Vibration*, Van Nostrand Reinhold, New York, 1977, pp. 313-340.
13-28. Salvesen, N., E. O. Tuck, and O. Faltinsen, "Ship Motions and Sea Loads," *Trans. Soc. Naval Architects-Marine Engrs.* **78**, 250-287 (1970).
13-29. Sabersky, R. H., A. J. Acosta, and E. G. Hauptman, *Fluid Flow: A First Course in Fluid Mechanics*, 2nd ed., Macmillan, New York, 1971, pp. 388-389.

14

STRUCTURAL VIBRATIONS IN A FLUID

14.1. INTRODUCTION

Thus far in this book, the effect of a surrounding fluid on structural vibrations has been neglected. The solutions which have been presented are exact only for vibrations in a vacuum, since structural vibrations in a fluid will generally be coupled to motions of the surrounding fluid. The natural frequencies of a structure in a fluid must be found, in general, from a coupled fluid-structural analysis. Fortunately, the effect of a surrounding fluid on the natural frequencies and mode shapes of a structure is not ordinarily significant for relatively compact structures if the fluid density is much less than the average density of the structure. Thus, the surrounding air does not ordinarily affect the natural frequencies or mode shapes of most metallic structures. However, the surrounding water can play a significant role in the free vibrations of marine structures such as ships and pipe lines.

In this chapter techniques are presented for incorporating the effect of a surrounding fluid in the analysis of natural frequencies and mode shapes of structures. The availability and generality of these techniques decrease with increasing complexity of the structure. A number of solutions are available for the analysis of one-dimensional structures such as beams and spring-supported bodies. Few simple solutions are available for multi-dimensional structures such as plates and shells.

14.2. ADDED MASS OF CROSS SECTIONS AND BODIES

14.2.1. General Case

Consider a two-dimensional cross section oscillating in a quiescent fluid as shown in Fig. 14-1. If the structural oscillations are harmonic with amplitude X_0 and circular frequency ω,

$$X = X_0 \sin \omega t, \tag{14-1}$$

then the velocity and acceleration of the structure are:

$$\dot{X} = X_0 \omega \cos \omega t,$$
$$\ddot{X} = -X_0 \omega^2 \sin \omega t. \tag{14-2}$$

The dots (·) denote derivatives with respect to time. The structural motion will induce an oscillating fluid force applied by the surrounding fluid to the structure.

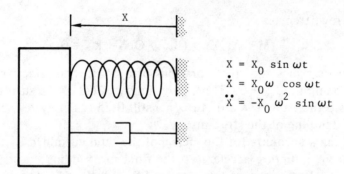

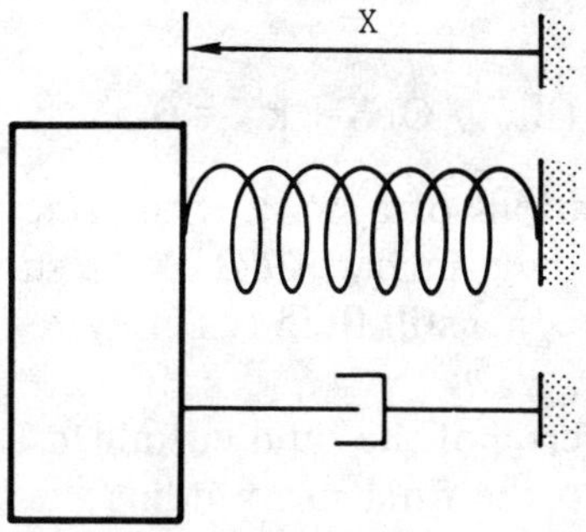

Fig. 14-1. A two-dimensional spring-supported, damped cross section in a still fluid.

This fluid force can generally be expressed in terms of a trigometric series:

$$F = \sum_{i=1}^{N} (b_i \cos i\omega t + a_i \sin i\omega t), \tag{14-3}$$

where a_i and b_i are constants for constant amplitude vibration. If the higher harmonics in the fluid force are neglected, the fluid force on the structure becomes:

$$F = a_1 \sin \omega t + b_1 \cos \omega t. \tag{14-4}$$

By comparing this equation and Eq. 14-2 it can be seen that the force component whose coefficient is b_1 acts in phase with the velocity of the structure and the force component whose coefficient is a_1 acts in phase with the acceleration (and displacement) of the structure. Thus, Eq. 14-4 may be rewritten incorporating Eq. 14-2 as:

$$F = -A_{xx}\ddot{X} - B_{xx}\dot{X}, \tag{14-5}$$

where

$$A_{xx} = \frac{a_1}{X_0 \omega^2}, \quad B_{xx} = -\frac{b_1}{X_0 \omega}.$$

The coefficient A_{xx} is called the added mass. A_{xx} has units of mass or mass per unit length for cross sections. $-A_{xx}\ddot{X}$ is the fluid force applied to the structure due to the inertia of the fluid entrained by the moving structure. This added mass inertia acts with the same sign, frequency, and phase as the inertia of the structural mass. In fact, simple dynamic experiments made on symmetric structures in a still homogeneous fluid cannot be used to differentiate between structural mass and added mass.

For example, the equation of motion of a simple spring-supported symmetric body (Fig. 14-1) vibrating in a still fluid is:

$$M\ddot{X} + C\dot{X} + kX = F, \tag{14-6}$$

where M is the mass of the body, C is the structural damping, k is the spring constant, F is the fluid force applied to the body, and X is the displacement of the body from the equilibrium position. If Eq. 14-5 is used to evaluate the fluid force, Eq. 14-6 becomes

$$M\ddot{X} + C\dot{X} + kX = -A_{xx}\ddot{X} - B_{xx}\dot{X},$$

which can be rewritten as:

$$(M + A_{xx})\ddot{X} + (B_{xx} + C)\dot{X} + kX = 0. \tag{14-7}$$

This equation describes the free vibrations of a body with effective mass $M + A_{xx}$ and effective damping $B_{xx} + C$. Thus, the primary effect of the surrounding fluid on the vibrations of a symmetric structure in a still fluid is simply to increase the effective mass and damping of the structure.

The added mass accounts for the inertia of the fluid entrained by the accelerating structure. As the structure accelerates, the fluid surrounding the structure must accelerate as well. The inertia of the entrained fluid is the added mass. Similarly, the added mass moment of inertia is the moment of inertia of the fluid entrained by the structure as it accelerates in rotation about a fixed point. The added mass and added mass moment of inertia increase the effective mass and effective mass moment of inertia of the structure. If a structure possesses certain symmetries, discussed in the following sections, then the effect of added mass on the natural frequencies of the structure in a still fluid can often be incorporated in the analysis simply by increasing the structure by the added mass and increasing the moment of inertia of the structure by the added mass moment of inertia.

Added mass always decreases the natural frequency of the structure from that which would be measured in a vacuum. The importance of added mass in the dynamic analysis of a particular structure can be estimated from the ratio of the density of the surrounding fluid to the average density of the structure. If a relatively compact structure is much denser than the surrounding fluid, added mass is not likely to have a large influence on the natural frequencies. For example, the density of most tall buildings is approximately 160 kg/m^3 and the density of the air surrounding the building is 1.2 kg/m^3. Since the density of the air is much less than the density of the building, added mass effects are ordinarily negligible in the dynamic analysis of the buildings. However, since the density of water (1000 kg/m^3) is comparable to the density of marine structures such as ships and pipe lines, added mass plays a large role in the dynamic analysis of marine structures.

Added mass can either be measured experimentally or calculated theoretically. Experiments have generally shown that the added mass of a structure vibrating in a still fluid is a function of (1) the geometry of the surface of the structure, including the geometry and relative position of proximate structures and the location of the free surface, if any, (2) the amplitude and direction of vibration, and (3) a Reynolds-number-like parameter:

$$A_{\substack{\text{still}\\\text{fluid}}} = \rho F_e\left(\text{geometry}, \frac{X_0}{D}, \frac{fD^2}{\nu}\right), \tag{14-8}$$

where ρ is the density of the fluid, X_0 is the amplitude of vibration in a given direction relative to the structure, D is a characteristic diameter, f is the frequency of vibration, and ν is the kinematic viscosity of the fluid. Similarly, the added mass force on a stationary structure in an oscillating fluid is a function of (1) the geometry of the structure and proximate structures and the location of the free surface, (2) the amplitude and direction of the fluid oscillations, and (3) the maximum Reynolds number achieved during the oscillations:

$$A_{\substack{\text{oscillating}\\ \text{fluid}}} = \rho F_e \left(\text{geometry}, \frac{U_0}{fD}, \frac{U_0 D}{\nu}\right), \tag{14-9}$$

where U_0 is the amplitude of the fluid velocity which oscillates with frequency f in a given plane.

It can be shown that added mass, unlike fluid drag or fluid damping, can exist in an incompressible, inviscid, irrotational fluid (i.e., ρ = constant, $\nu = 0$). This opens the possibility of evaluating added mass through the mathematical application of potential flow theory. The added mass which is predicted from potential flow theory is only dependent on the geometry of the surface of the structure,

$$A_{\substack{\text{potential}\\ \text{flow}}} = \rho F(\text{geometry}). \tag{14-10}$$

Comparison of experimental and theoretical results (Refs. 14-1, 14-2) suggests that the potential flow theory prediction is within about 10% of the experimentally measured value if the Mach number of the oscillation is small, the amplitude of vibration is small compared with a characteristic diameter of the structure, and the Reynolds-number-like parameter is large:

$$\frac{X_0 (2\pi f)}{c} << 1,$$

$$\frac{X_0}{D} < 1,$$

and

$$\frac{2\pi f D^2}{\nu} > 10000. \tag{14-11}$$

c is the speed of sound in the fluid. Similarly, the potential flow prediction can be expected to yield a good estimate of the added mass force on a stationary structure in an oscillating flow only if

$$\frac{U_0}{c} << 1,$$

$$\frac{U_0}{fD} < 2\pi,$$

and

$$\frac{U_0 D}{\nu} > 10000. \tag{14-12}$$

Outside the ranges specified by Eq. 14-11 or 14-12, the added mass will be a function of the amplitude by Reynolds-number-like parameters given in Eq. 14-8 or 14-9 as well as geometry, and the potential flow prediction of added mass may be in error by as much as 100% (Refs. 14-1 through 14-3). However, since experimental data are not generally available to map out the influence of variables other than geometry, only a single value of added mass for each geometry will be specified in the tables in this chapter.

14.2.2. Two-Dimensional Sections

Table. The added mass of a two-dimensional section with two perpendicular axes of symmetry is completely specified, using potential flow theory, by the added masses for acceleration along each of the axes of symmetry and the added mass moment of inertia for rotation about the intersection of these axes. The added masses of various symmetric cross sections are given in Table 14-1. Added mass is often specified in terms of an added mass coefficient which is defined as the added mass per unit length divided by the area of the cross section. This has not been done in Table 14-1 because certain cross sections which incorporate fins or are composed of thin plates do not lend themselves to this definition.

Coupling and Symmetry. If a cross section is not symmetric, the fluid forces due to added mass will generally couple translation and rotation. For a general cross section [Fig. 14-2(a), for example], the fluid forces on a vibrating section in a still fluid due to added mass may be written in matrix form as:

$$\begin{bmatrix} F_x \\ F_y \\ F_\theta \end{bmatrix}_{\substack{\text{added} \\ \text{mass}}} = - \begin{bmatrix} A_{xx} & A_{xy} & A_{x\theta} \\ A_{xy} & A_{yy} & A_{y\theta} \\ A_{x\theta} & A_{y\theta} & A_{\theta\theta} \end{bmatrix} \begin{bmatrix} \ddot{X} \\ \ddot{Y} \\ \ddot{\theta} \end{bmatrix}, \tag{14-13}$$

where

F_x, F_y = added fluid forces per unit length acting in the x and y directions, respectively,

F_θ = fluid moment per unit length for rotation about the point $x = y = 0$,

$[A]$ = matrix of added masses and added mass moments of inertia per unit length; potential flow theory requires this matrix to be symmetric for vibration in a still fluid,

X, Y = mutually perpendicular displacements,

θ = angle of rotation about the point $x = y = 0$.

Equation 14-13 implies that acceleration in the x direction ($\ddot{X} \neq 0$, $\ddot{Y} = \ddot{\theta} = 0$) of a general cross section will induce not only an added mass force in the x direction,

$$F_x = -A_{xx}\ddot{X}, \tag{14-14}$$

but also a force in the y direction and a moment about the origin

$$F_y = -A_{xy}\ddot{X}, \tag{14-15}$$

$$F_\theta = -A_{x\theta}\ddot{X}. \tag{14-16}$$

Table 14-1. Added Masses of Cross Sections.

Notation: A_{xx}, A_{yy} = added mass per unit length for acceleration along x and y coordinates, respectively; A_{xy} = added mass per unit length along y coordinate as a result of acceleration along x coordinate or vice versa; ρ = mass density of surrounding fluid; see Table 3-1 for consistent sets of units; t = theoretical result, e = experimental result.

Description of Cross Section	Added Mass Per Unit Length (acceleration in direction denoted by ↔)	Added Mass Moment of Inertia Per Unit Length (rotary acceleration about axis through point P, perpendicular to plane of paper)
1. Circle 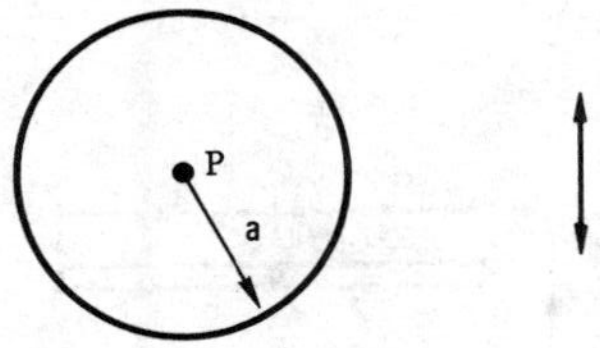	$\rho\pi a^2$ t	0 t
2. Ellipse 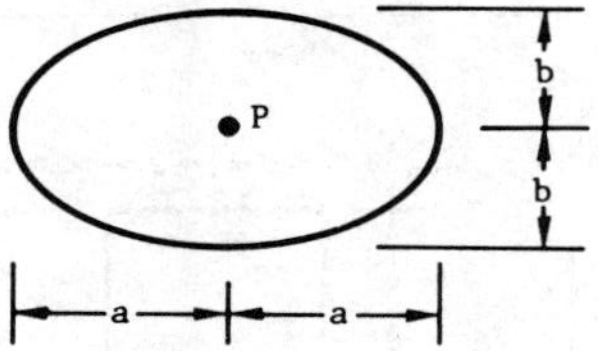	$\rho\pi a^2$ $0 < \frac{a}{b} < \infty$ t	$\rho \frac{\pi}{8}(a^2 - b^2)^2$ t
3. Square 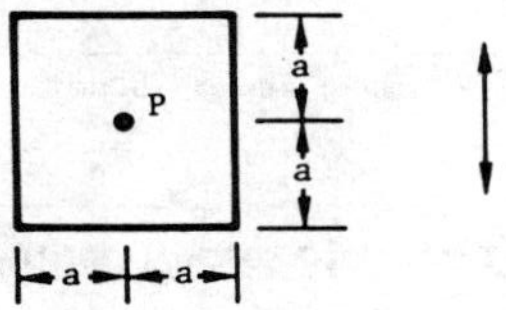	$1.51\pi a^2$ t	$0.234\rho\pi a^4$ t

Table 14-1. Added Masses of Cross Sections. *(Continued)*

Description of Cross Section	Added Mass Per Unit Length (acceleration in direction denoted by ↔)	Added Mass Moment of Inertia Per Unit Length (rotary acceleration about axis through point P, perpendicular to plane of paper)
4. Rectangle	$\alpha \rho \pi a^2$ (see table 4a) t	$\beta_1 \rho \pi a^4$ or $\beta_2 \rho \pi b^4$ (see table 4b) Ref. 14-4 t
5. Thin Plate t<< a	↕ $\rho \pi a^2$ ↔ 0 t	$\rho \frac{\pi}{8} a^4$ t
6. Diamond	$\alpha \rho \pi a^2$ (see table 6a) t	$0.059 \rho \pi a^4$ for a = b only Ref. 14-4 t

4. Rectangle

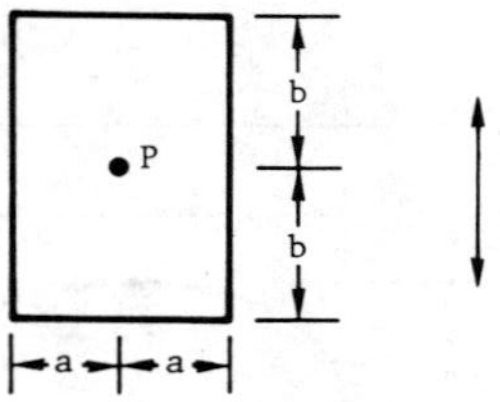

$\alpha \rho \pi a^2$

a/b	α
0.1	2.23
0.2	1.98
0.5	1.70
1.0	1.51
2.0	1.36
5.0	1.21
10.0	1.14
∞	1.0

$\beta_1 \rho \pi a^4$ or $\beta_2 \rho \pi b^4$

a/b	β_1	β_2
0.1	--	0.147
0.2	--	0.15
0.5	--	0.15
1.0	0.234	0.234
2.0	0.15	--
5.0	0.15	--
∞	0.125	--

Ref. 14-4

5. Thin Plate

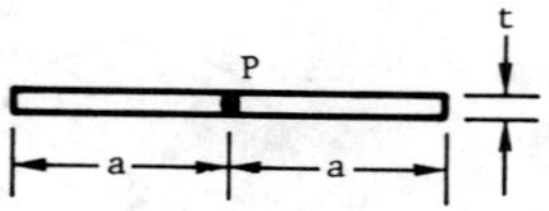

t<< a

6. Diamond

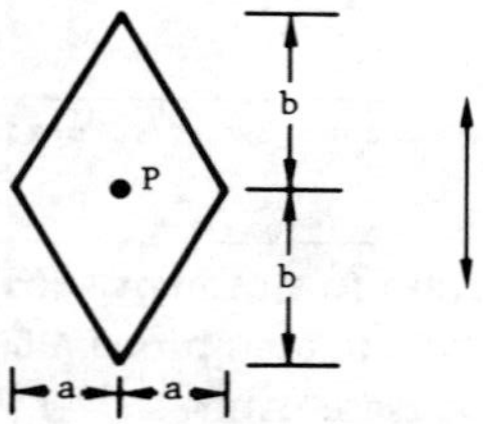

$\alpha \rho \pi a^2$

a/b	α
0.2	0.61
0.5	0.67
1.0	0.76
2.0	0.85

$0.059 \rho \pi a^4$ for a = b only

Ref. 14-4

7. Regular Polygon

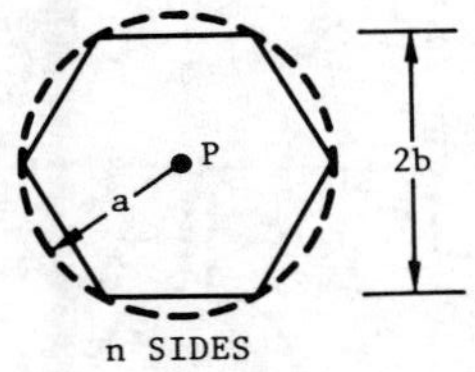

n=6 shown

Motion: ↕ or ↔

$\alpha\rho\pi a^2$

n	α
3	0.654
4	0.787
5	0.823
6	0.867
∞	1.00

n = number of sides

t

$0.055\rho\pi b^4$ for n = 8 only

Refs. 14-4, 14-5

t

8. I Beam

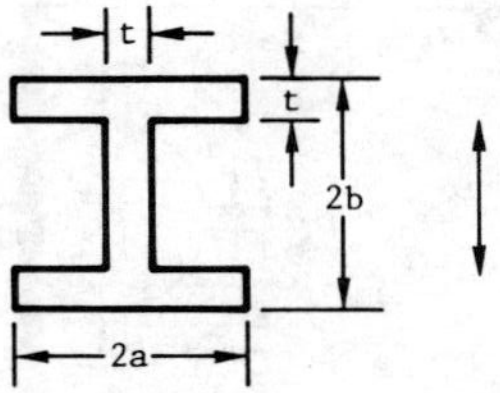

$2.11\rho\pi a^2$

for $\frac{a}{t} = 2.6$

$\frac{b}{t} = 3.6$

e

--

Ref. 14-6

9. Multiple Fins, Equally Spaced

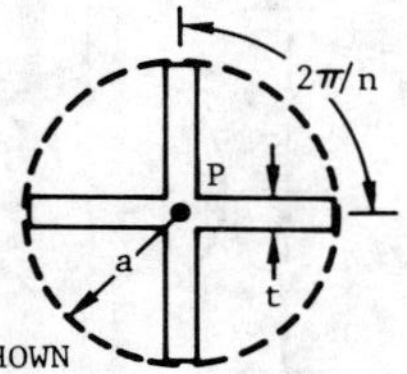

n = 4 SHOWN

t<<a

Motion: ↕ or ↔

$\frac{2\pi\rho a^2}{2^{4/n}}$

n = number of fins

n ≥ 3

See frame 5 for n = 2.

t

$\alpha\rho a^4$

n	α
3	0.533
4	$2/\pi$
∞	$\pi/2$

See frame 5 for n = 2.

Ref. 14-7

t

Table 14-1. Added Masses of Cross Sections. *(Continued)*

Description of Cross Section	Added Mass Per Unit Length (acceleration in direction denoted by ↔)	Added Mass Moment of Inertia Per Unit Length (rotary acceleration about axis through point P, perpendicular to plane of paper)
10. Circle with Two Symmetric Fins 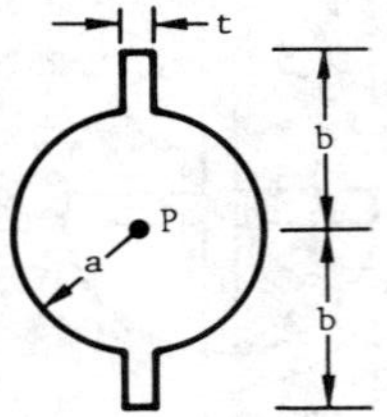 t<<a	↕ $\rho\pi a^2$ ↔ $\rho\pi b^2\left(1 - \frac{a^2}{b^2} + \frac{a^4}{b^4}\right)$	$\rho a^2\left[\frac{2\alpha^2 - \alpha\sin 4\alpha + \frac{1}{2}\sin 2\alpha}{\pi(\sin\alpha)^4} - \frac{\pi}{2}\right]$ where $\sin\alpha = \frac{2ab}{a^2 + b^2}$ and $\frac{\pi}{2} < \alpha < \pi$ Ref. 14-8 (p. 145) t
11. Circle with Multiple Equally Spaced Fins 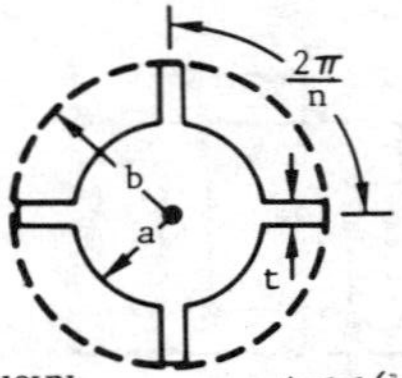 n = 4 SHOWN t<<(b - a)	↕ or ↔ $2\pi\rho b^2\left\{\left[\frac{1 + (a/b)^n}{2}\right]^{4/n} - \frac{1}{2}\left(\frac{a}{b}\right)^2\right\}$ $n \geq 3$ n = number of fins	-- Ref. 14-7
12. Circle with Four Fins t<<(b - a),(c_1 - a) 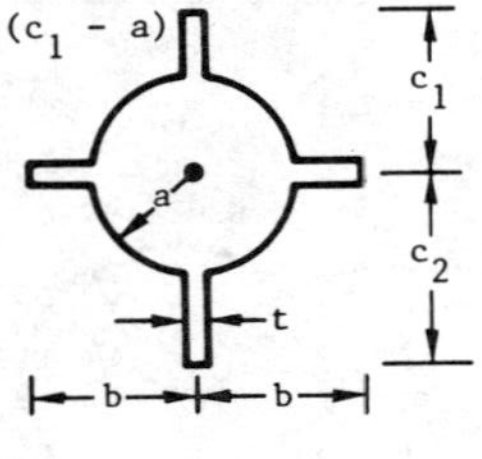	↕ $\pi\rho b^2\left(1 - \frac{a^2}{b^2} + \frac{a^4}{b^4}\right)$ ↔ $\frac{\pi\rho b^2}{4}\left\{\frac{c_1^2}{b^2}\left(1 + \frac{a^4}{c_1^4}\right) + \frac{c_2^2}{b^2}\left(1 + \frac{a^4}{c_2^4}\right) - 2\left(1 + \frac{a^2}{b^2}\right)^2\right.$ $+ 2\left[\left(1 + \frac{a^4}{b^2c_1^2}\right)\left(1 + \frac{c_1^2}{b^2}\right)\right.$ $\left.\left.\cdot\left(1 + \frac{a^4}{b^2c_2^2}\right)\left(1 + \frac{c_2^2}{b^2}\right)\right]^{1/2}\right\}$ See Ref. 14-5 for cross coupling. t	-- Ref. 14-5

13. Finned Square

t<<a

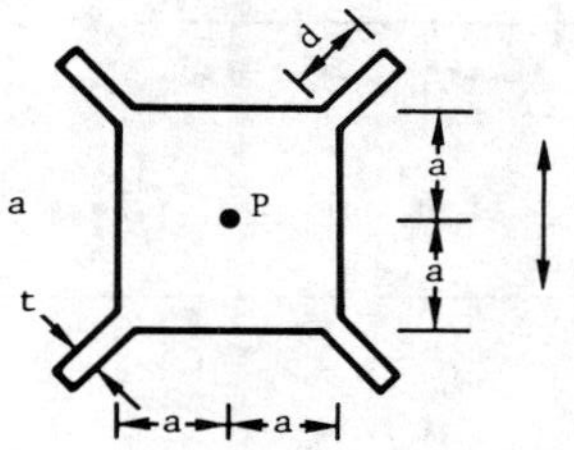

$\alpha \rho \pi a^2$

$\frac{d}{a}$	α
0.05	1.61
0.1	1.72
0.25	2.19

t

$\beta \rho \pi a^4$

$\frac{d}{a}$	β
0.05	0.31
0.1	0.40
0.25	0.69

Ref. 14-4

t

14. Arbitrarily Shaped Cross Section

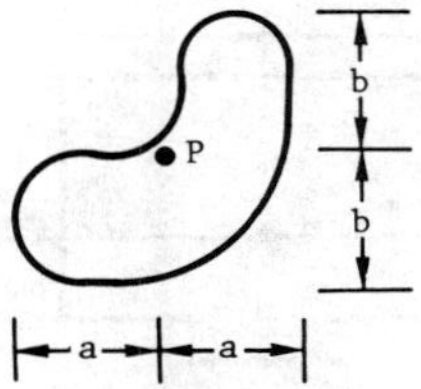

Approximate Formulas:

↕ $\rho \pi a^2$

↔ $\rho \pi b^2$

See Eq. 14-4 for discussion of coupling effects for asymmetric bodies.

Approximate Formula:

$\rho \frac{\pi}{8} \mathrm{Max}\{a^4, b^4\}$

See text for discussion of coupling effects for asymmetric sections.

15. Rotated Cross Section

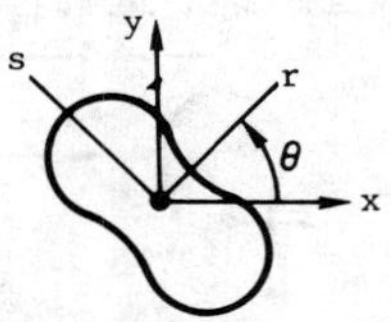

CROSS SECTION IS SYMMETRIC ABOUT r AND s AXES

↕ $\begin{cases} A_{yy} = A_{rr} \sin^2 \theta + A_{ss} \cos^2 \theta \\ A_{xy} = \frac{1}{2}(A_{ss} - A_{rr}) \sin 2\theta \end{cases}$

↔ $\begin{cases} A_{xx} = A_{rr} \cos^2 \theta + A_{ss} \sin^2 \theta \\ A_{xy} = \frac{1}{2}(A_{ss} - A_{rr}) \sin 2\theta \end{cases}$

See Eq. 14-4 for rotation of arbitrary cross section.

t

Unchanged

Ref. 14-9 (p. 24)

t

Table 14-1. Added Masses of Cross Sections. *(Continued)*

Description of Cross Section	Added Mass Per Unit Length (acceleration in direction denoted by ↔)	Added Mass Moment of Inertia Per Unit Length (rotary acceleration about axis through point P, perpendicular to plane of paper)

16. Translated Cross Section

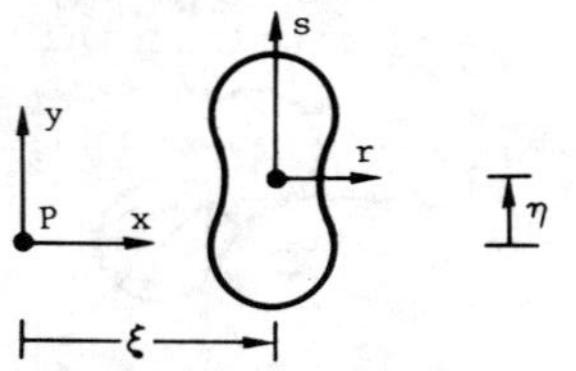

CROSS SECTION IS SYMMETRIC ABOUT r AND s AXES

Added Mass Per Unit Length: Unchanged

Added Mass Moment of Inertia Per Unit Length:

$$A_\theta\Big|_{x=y=0} = A_{rr}\eta^2 + A_{ss}\xi^2 + A_\theta\Big|_{r=s=0}$$

See text for translation of arbitrary cross section.

Ref. 14-9 (p. 24)

17. Floating Rectangle

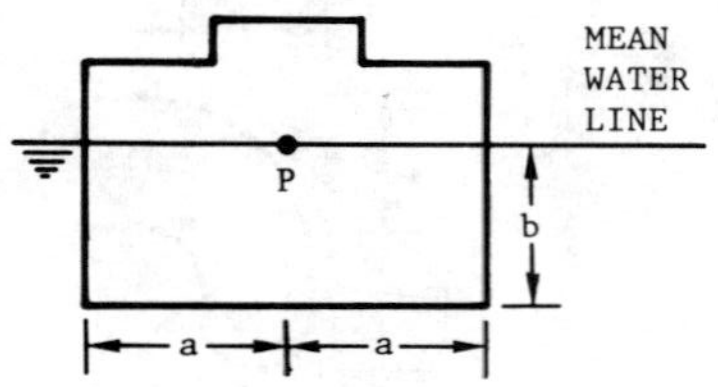

WATER DEPTH>>a,b

Added Mass Per Unit Length:

↕ $\frac{\alpha_1}{2}\rho\pi a^2$

↔ $\frac{\alpha_2}{2}\rho\pi b^2$

a/b	α_1	α_2
0.1	2.23	1.14
0.2	1.98	1.21
0.5	1.70	1.36
1.0	1.51	1.51
2.0	1.36	1.70
5.0	1.21	1.98

t

Added Mass Moment of Inertia Per Unit Length:

$\frac{\beta_1}{2}\rho\pi a^4$ or $\frac{\beta_2}{2}\rho\pi b^4$

a/b	β_1	β_2
0.1	--	0.147
0.2	--	0.15
0.5	--	0.15
1.0	0.234	0.234
2.0	0.15	--
5.0	0.15	--
∞	0.125	--

Free surface remains plane during acceleration.

Refs. 14-10, 14-9 (pp. 169-171) t

18. Symmetric Floating Section

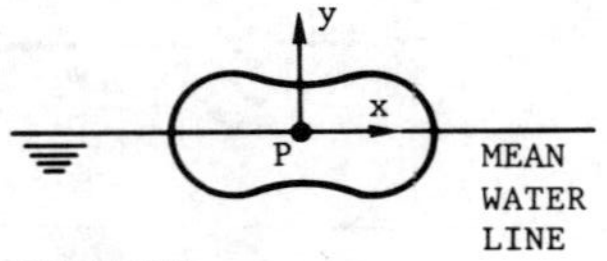

DEEP WATER

SECTION SYMMETRIC ABOUT x AND y AXES

Added Mass Per Unit Length:

↕ or ↔

Added mass is 1/2 that of fully submerged body accelerating in same direction.

See text for discussion.

t

Added Mass Moment of Inertia Per Unit Length:

Added mass moment of inertia is 1/2 that of fully submerged body rotating about same point.

See text for discussion.

Free surface remains plane during acceleration.

Refs. 14-10, 14-9 (pp. 169-171) t

19. Floating Rectangle in Shallow Water

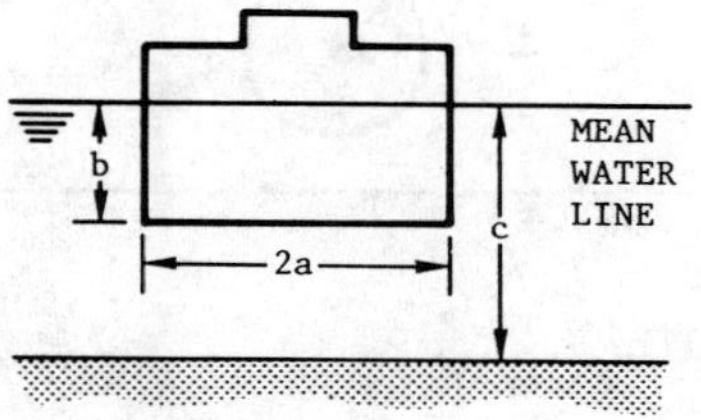

↔ $2\alpha\rho ab$

$\frac{b}{c}$	α			
	a/b			
	0.2	0.5	1.0	2.0
0.2	4.92	2.25	1.29	0.783
0.4	5.43	2.63	1.66	1.16
0.6	6.63	3.56	2.53	2.02
0.8	10.15	6.46	5.23	4.62

--

t

↕ $2\beta\rho ab$

For a=b only

$\frac{c}{b}$	β
1.1	5.52
1.2	3.49
1.5	2.11
3.0	1.35
8.0	1.21
∞	1.19

--

Free surface remains plane during acceleration.

t Refs. 14-10, 14-11

20. Floating Cylinder

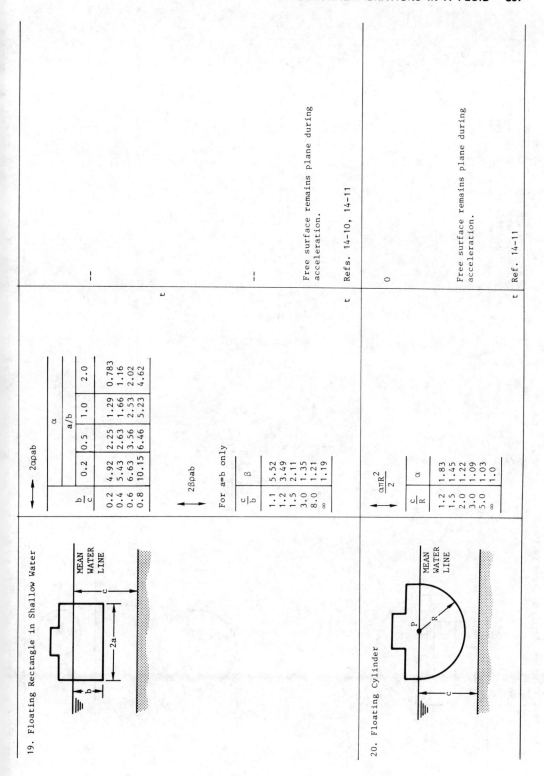

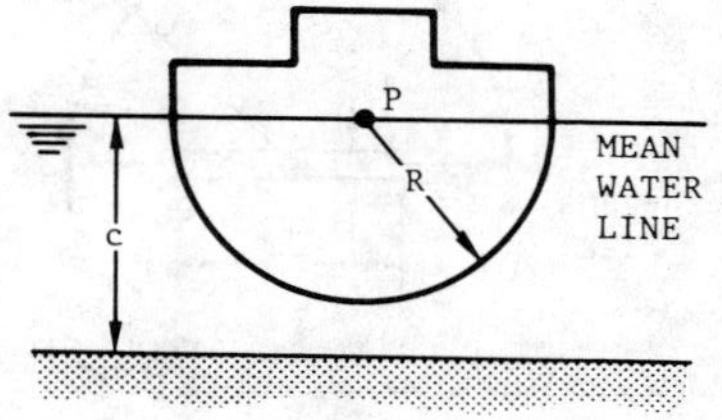

↕ $\frac{\alpha\pi R^2}{2}$

$\frac{c}{R}$	α
1.2	1.83
1.5	1.45
2.0	1.22
3.0	1.09
5.0	1.03
∞	1.0

0

Free surface remains plane during acceleration.

t Ref. 14-11

Table 14-1. Added Masses of Cross Sections. *(Continued)*

Description of Cross Section	Added Mass Per Unit Length (acceleration in direction denoted by ↔)	Added Mass Moment of Inertia Per Unit Length (rotary acceleration about axis through point P, perpendicular to plane of paper)
21. Floating Plate in a Channel 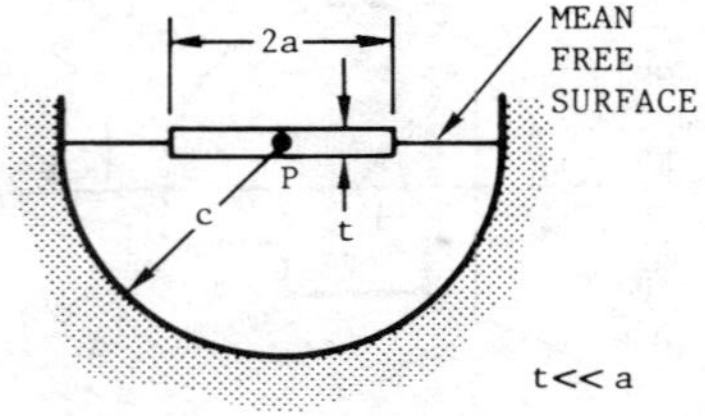 	↕ $\frac{\alpha\rho\pi a^2}{2}$ a/c — α 0 — 1.00 0.2 — 1.04 0.4 — 1.10 0.6 — 1.22 0.8 — 1.60 ↔ 0.0	$\frac{\rho\pi a^4}{16}$ for $c/a = \infty$ Ref. 14-9 (p. 183)
22. Cylinder Within a Pipe 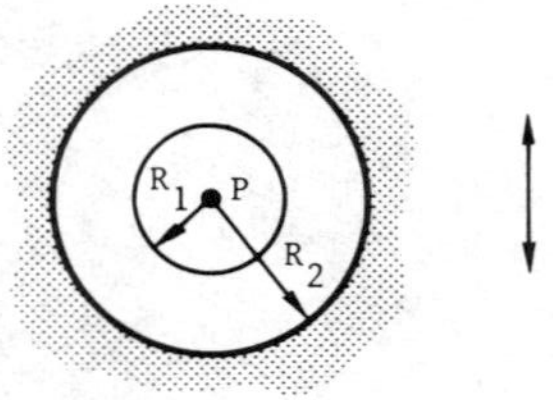 	$\left[\frac{1+\left(\frac{R_1}{R_2}\right)^2}{1-\left(\frac{R_1}{R_2}\right)^2}\right]\rho\pi R_1^2$	0 Ref. 14-12
23. Cylinder in a Channel 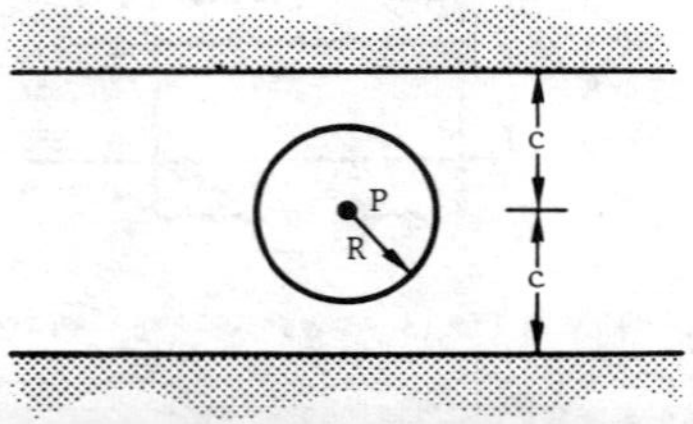 	↕ $\alpha\rho\pi R^2$ α from frame 20	0

24. Rectangle in a Channel

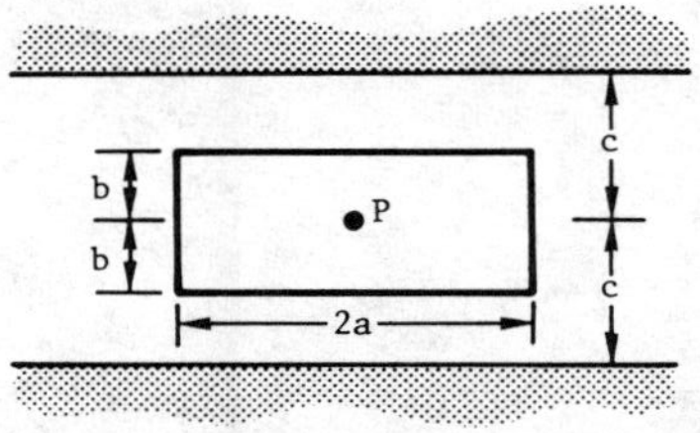

↕ $4\alpha\rho ab$

α from frame 19

↔ $4\beta\rho ab$

β from frame 19

t

--

25. Cylinder Adjacent to a Fixed Cylinder

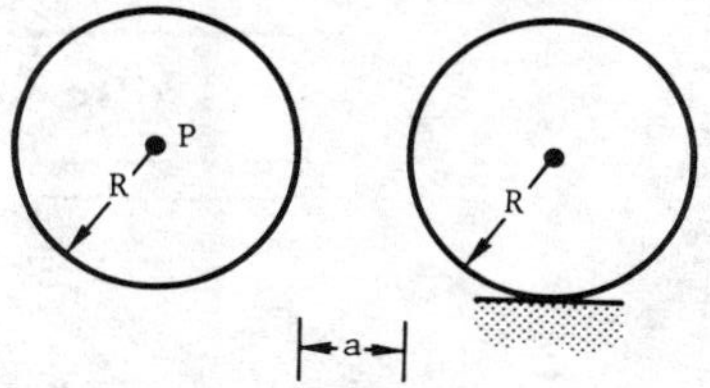

↔ $\alpha\rho\pi^2$

$\frac{a}{R}$	α
∞	1.0
1.2	1.024
0.8	1.044
0.4	1.096
0.2	1.160
0.1	1.224

t

Ref. 14-13 t

26. Cylinder in an Array of Fixed Cylinders

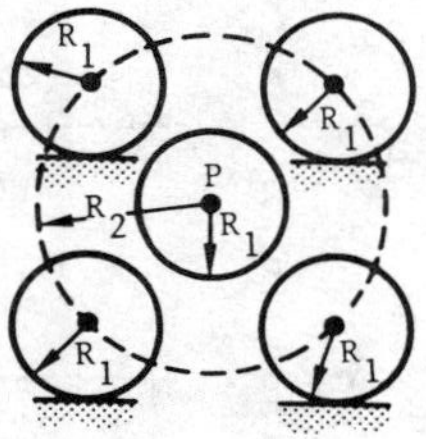

$\alpha\rho\pi R_1^2$

$\frac{R_2}{2R_1}$	α
∞	1.0
1.5	1.4
1.4	1.5
1.3	1.6
1.2	2.0

Approximate values. Result is dependent on specific geometry. See Ref. 14-13 for discussion of coupling effects. e

Ref. 14-15 t

27. Two Rectangles in Tandem

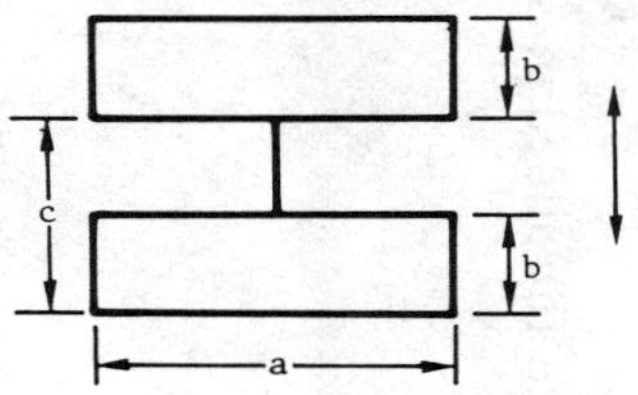

$2\alpha ab$

	α			
	b/a			
$\frac{c}{a}$	0.1	0.2	0.4	1.0
0.5	4.7	2.6	1.3	--
1.0	5.2	3.2	1.7	0.6
1.5	5.8	3.7	2.0	0.7
2.0	6.4	4.0	2.3	0.9
3.0	7.2	4.6	2.5	1.1
4.0	--	4.8	--	--

e

Ref. 14-16

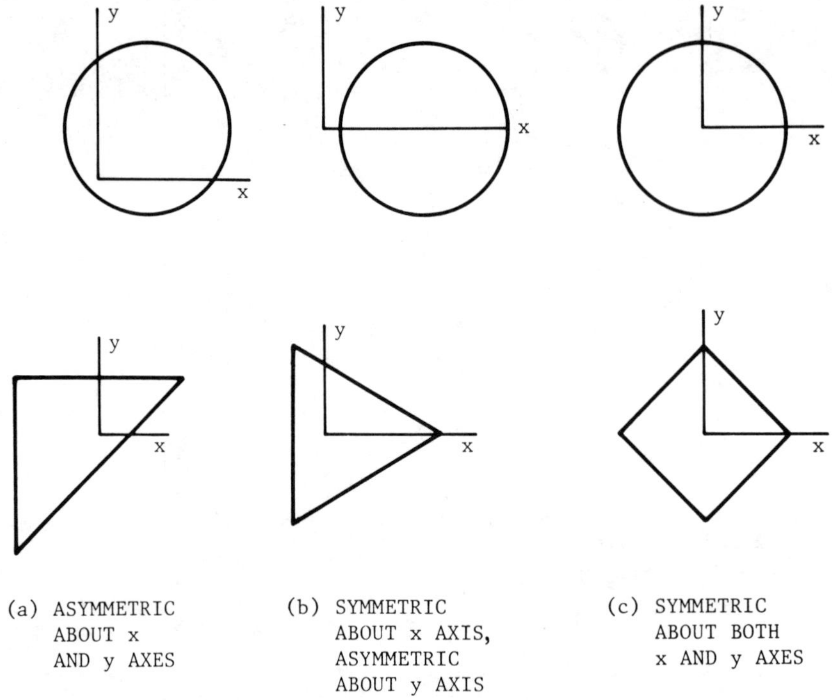

Fig. 14-2. Three degrees of symmetry of cross sections.

A_{xx} and A_{yy} are the added masses for acceleration in the x and y directions, respectively. A_{xx} and A_{yy} have units of mass per unit length. A_{xy} is the added mass translation cross coupling and has units of mass per unit length. $A_{x\theta}$ is the added mass rotation-translation cross coupling and has units of mass.

If the cross section is symmetric about the x axis but asymmetric about the y axis, as in Fig. 14-2(b), the extent of the cross coupling is reduced. x-axis symmetry implies that acceleration in the x direction cannot induce either a force in the y direction or a moment about the origin. Using Eqs. 14-15 and 14-16, x-axis symmetry implies

$$A_{xy} = A_{x\theta} = 0$$

and the added mass forces become:

$$\begin{bmatrix} F_x \\ F_y \\ F_\theta \end{bmatrix}_{\substack{\text{added} \\ \text{mass}}} = - \begin{bmatrix} A_{xx} & 0 & 0 \\ 0 & A_{yy} & A_{y\theta} \\ 0 & A_{y\theta} & A_{\theta\theta} \end{bmatrix} \begin{bmatrix} \ddot{X} \\ \ddot{Y} \\ \ddot{\theta} \end{bmatrix}. \tag{14-17}$$

There are four independent terms in this added mass matrix. If the cross section is symmetric about both the x and y axes as in Fig. 14-2(c), the added mass forces uncouple,

$$\begin{bmatrix} F_x \\ F_y \\ F_z \end{bmatrix}_{\substack{\text{added} \\ \text{mass}}} = -\begin{bmatrix} A_{xx} & 0 & 0 \\ 0 & A_{yy} & 0 \\ 0 & 0 & A_{\theta\theta} \end{bmatrix} \begin{bmatrix} \ddot{X} \\ \ddot{Y} \\ \ddot{\theta} \end{bmatrix}, \tag{14-18}$$

and only three independent diagonal terms remain in the added mass matrix.

Similarly, the forces associated with the inertia of the structural mass can be expressed in terms of a generalized mass matrix for small amplitude vibrations:

$$\begin{bmatrix} F_x \\ F_y \\ F_\theta \end{bmatrix}_{\text{structure}} = -\begin{bmatrix} m & 0 & -my_C \\ 0 & m & mx_C \\ -my_C & mx_C & J_\theta \end{bmatrix} \begin{bmatrix} \ddot{X} \\ \ddot{Y} \\ \ddot{\theta} \end{bmatrix}. \tag{14-19}$$

m is the mass per unit length of the cross section, J_θ is the mass moment of inertia per unit length of the cross section about the origin of the coordinates, and x_C and y_C specify the offset between the center of mass C and the origin of the coordinates (see Fig. 14-3). The mass matrix [M] is diagonal if the center of mass coincides with the origin of the coordinates so that $x_C = y_C = 0$.

The total of the inertia forces is the sum of the inertia forces associated with added mass (Eq. 14-13) and the inertia forces associated with the generalized structural mass (Eq. 14-19):

$$\{F\} = -([M] + [A])\{\ddot{X}\},$$

where {F} is the vector inertia of the structure and the entrained fluid, {X} is the vector displacement of the structure, and [M] and [A] are the mass matrix and added mass matrix, respectively. If both the added mass matrix, [A], and the generalized mass matrix, [M], are diagonal, there is no inertia cross coupling and the analysis is

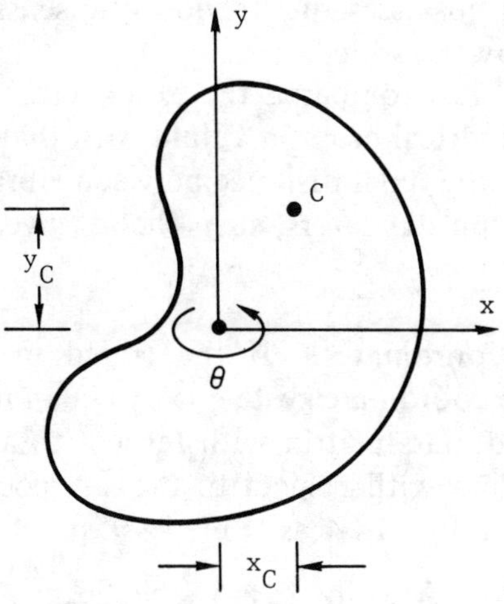

Fig. 14-3. A cross section with an offset between the center of mass, C, and the origin of the coordinates.

considerably simplified. Both the added mass matrix and the generalized mass matrix will be diagonal if (1) the cross-section outline possesses two mutually perpendicular planes of symmetry along the coordinate directions and (2) the center of mass coincides with the origin of the coordinates. If a cross section satisfies both of these criteria, added mass may be simply incorporated in the analysis by (1) increasing the structural mass by the added mass for acceleration along the desired coordinate direction and (2) increasing the mass moment of inertia by the added mass moment of inertia for rotation about the origin.

For example, if a straight uniform beam, such as that discussed in Section 8.1, is symmetric about the plane of transverse vibration, no added mass or structural mass cross coupling can be induced. The effect of added mass on the natural frequencies is simply to increase the structural mass by the added mass. Since the natural frequencies of the beam are inversely proportional to the square root of the mass of the beam, the ratio of the natural frequency of the beam in a vacuum to the beam immersed in a fluid is:

$$\frac{f_{\text{fluid}}}{f_{\text{vacuum}}} = \frac{1}{\left(1 + \dfrac{A}{m}\right)^{1/2}},$$

where A is the added mass per unit length of the beam (Table 14-1) and m is the mass of structural material per unit length. The mode shapes of the uniform beam are unchanged by the added mass.

Strip Theory. In many cases the added mass of slender beams can be approximated using the results presented in Table 14-1 and strip theory. The basic assumptions of strip theory are that (1) the flow over a narrow local section (strip) of the beam is two dimensional and (2) the interaction between adjacent strips is negligible. Under these assumptions the effect of added mass on the transverse vibrations of beams (Section 8.1) is incorporated simply by increasing the mass per unit length of the beam by the appropriate cross-sectional added mass per unit length from Table 14-1, provided the cross section possesses the previously discussed symmetries so that no cross coupling is induced by the added mass.

Landweber (Ref. 14-17) has compared the exact result for the natural frequency of a neutrally buoyant cylindrical beam in a fluid with that predicted by strip theory. His results indicate that if the axial distance between vibration nodes in the beam is more than about three beam diameters, strip theory gives an excellent approximation to the exact result.

Rotated and Translated Coordinates. If the added mass matrix (Eq. 14-13) is known with respect to one coordinate system (x, y) in an infinite fluid region free of other structures, the added mass matrix with respect to another coordinate system (r, s) is rotated by an angle α with respect to the x–y coordinate and has its origin $x = \xi$, and $y = \eta$, as shown in Fig. 14-4, is (Ref. 14-9, p. 24):

$$A_{rr} = A_{xx} \cos^2 \alpha + A_{yy} \sin^2 \alpha + A_{xy} \sin 2\alpha,$$

$$A_{ss} = A_{xx} \sin^2 \alpha + A_{yy} \cos^2 \alpha - A_{xy} \sin 2\alpha,$$

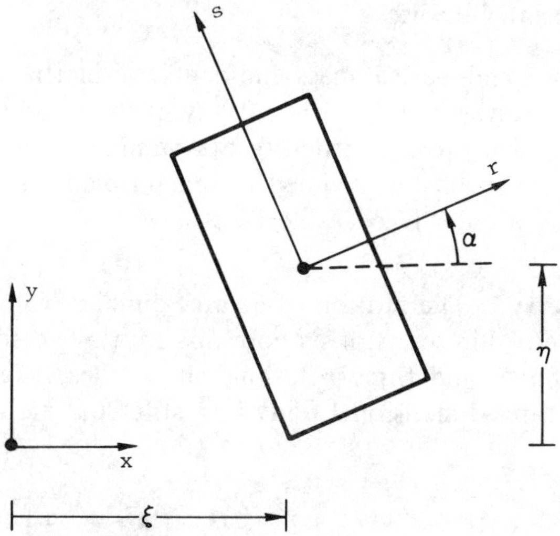

Fig. 14-4. A translated and rotated coordinate frame with origin at $x = \xi$, $y = \eta$.

$$A_{rs} = \tfrac{1}{2}(A_{yy} - A_{xx}) \sin 2\alpha + A_{xy} \cos 2\alpha,$$
$$A_{r\beta} = (A_{xx}\eta - A_{xy}\xi + A_{x\theta}) \cos\alpha + (A_{xy}\eta - A_{yy}\xi + A_{y\theta}) \sin\alpha,$$
$$A_{s\beta} = -(A_{xx}\eta - A_{xy}\xi + A_{x\theta}) \sin\alpha + (A_{xy}\eta + A_{yy}\xi + A_{y\theta}) \cos\alpha,$$
$$A_{\beta\beta} = A_{xx}\eta^2 + A_{yy}\xi^2 - 2A_{xy}\xi\eta + 2(A_{x\theta}\eta - A_{y\theta}\xi) + A_{\theta\theta}, \qquad (14\text{-}20)$$

where θ denotes rotation about an axis through the origin of the x–y system and β denotes rotation about the origin of the r–s system. These relationships are considerably simplified if the original axes (x–y) are axes of symmetry so that

$$A_{xy} = A_{x\theta} = A_{y\theta} = 0.$$

Simplified formulas for this case are presented in Table 14-1.

Floating Cross Sections. The analysis of the added mass of floating cross sections is generally very complex owing to the creation of surface waves as the cross section moves. The analysis is considerably simplified by assuming that the free surface remains plane as the cross section accelerates. This approximation has proven useful [Refs. 14-9 (pp. 169–171), 14-10] and has been incorporated in frames 17 through 21 of Table 14-1.

Inertial Coupling. As a cross section accelerates in a fluid, pressure gradients will be generated in the surrounding fluid. These pressure gradients can exert fluid forces on adjacent structures. If the adjacent structures are not rigid, they will in turn deflect and generate forces on the accelerating cross section. This interaction between adjacent elastic structures is called inertial coupling. Inertial coupling increases as the distance between structures decreases. Inertial coupling can significantly affect the natural frequencies and mode shapes of tubes in closely packed tube arrays if the density of the surrounding fluid is comparable to or larger than the average density of the tubes (Refs. 14-13, 14-18).

14.2.3. Three-Dimensional Bodies

Table. Added masses and added mass moments of inertia of symmetric three-dimensional bodies are given in Table 14-2. Many of these added masses were determined experimentally owing to the difficulty of obtaining three-dimensional theoretical solutions. The free surface is assumed to remain plane as the body accelerates as shown in frames 12 through 14 of Table 14-2.

Coupling and Symmetry. The motion of a three-dimensional rigid body, shown in Fig. 14-5, can be specified in terms of six coordinates: three displacements along the mutually orthogonal axes and three rotations about these axes. The added mass forces on a general three-dimensional body in a still fluid are specified by a 6×6 added mass matrix:

$$\begin{bmatrix} F_1 \\ F_2 \\ F_3 \\ F_4 \\ F_5 \\ F_6 \end{bmatrix} = - \begin{bmatrix} A_{11} & A_{12} & A_{13} & A_{14} & A_{15} & A_{16} \\ A_{12} & A_{22} & A_{23} & A_{24} & A_{25} & A_{26} \\ A_{13} & A_{23} & A_{33} & A_{34} & A_{35} & A_{36} \\ A_{14} & A_{24} & A_{34} & A_{44} & A_{45} & A_{46} \\ A_{15} & A_{25} & A_{35} & A_{45} & A_{55} & A_{56} \\ A_{16} & A_{26} & A_{36} & A_{46} & A_{56} & A_{66} \end{bmatrix} \begin{bmatrix} \ddot{X}_1 \\ \ddot{X}_2 \\ \ddot{X}_3 \\ \ddot{X}_4 \\ \ddot{X}_5 \\ \ddot{X}_6 \end{bmatrix} . \qquad (14\text{-}21)$$

The subscripts refer to the translations ($i = 1, 2, 3$) and rotations ($i = 4, 5, 6$) shown in Fig. 14-5. F_i are the forces ($i = 1, 2, 3$) and moments ($i = 4, 5, 6$) as the body accelerates from rest in a still fluid. Potential flow theory for small-amplitude vibrations in a still, inviscid, irrotational incompressible fluid requires that the added mass matrix be symmetric (Ref. 14-25). Thus, only 21 of the 36 entries in the matrix are independent. (If there is a mean velocity between the body and the fluid, the matrix is generally asymmetric with 36 independent entries.)

The off-diagonal terms in the added mass matrix cross couple all six degrees of freedom of the body. The degree of this cross coupling is reduced if the body possesses one or more planes of symmetry. If the body is symmetric about the x_1–x_3 plane, then 18 of the off-diagonal cross coupling terms must be zero and the added mass matrix becomes (Ref. 14-25):

$$\begin{bmatrix} A_{11} & 0 & A_{13} & 0 & A_{15} & 0 \\ 0 & A_{22} & 0 & A_{24} & 0 & A_{26} \\ A_{13} & 0 & A_{33} & 0 & A_{35} & 0 \\ 0 & A_{24} & 0 & A_{44} & 0 & A_{46} \\ A_{15} & 0 & A_{35} & 0 & A_{55} & 0 \\ 0 & A_{26} & 0 & A_{46} & 0 & A_{66} \end{bmatrix} . \qquad (14\text{-}22)$$

Table 14-2. Added Masses of Bodies.

Notation: ρ = fluid density; consistent sets of units are given in Table 3-1; t = theoretical result, e = experimental result.

<table>
<tr><th>Description of Body</th><th>Added Mass (acceleration in direction denoted by ↔)</th><th>Added Mass Moment of Inertia</th></tr>
<tr>
<td>1. Circular Disk
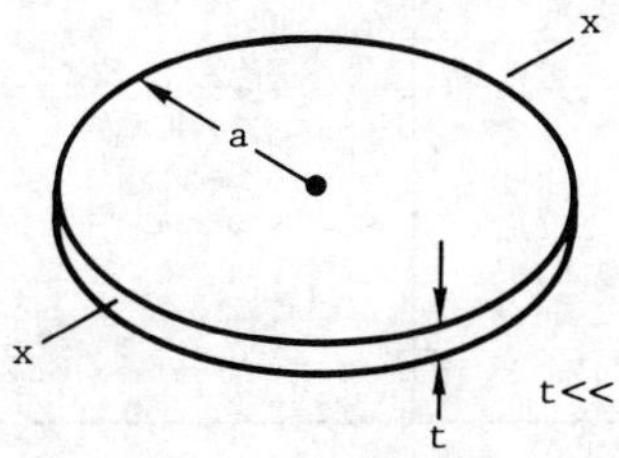</td>
<td>↕ $\frac{8}{3}\rho a^3$
↔ 0.0
t</td>
<td>Rotation about x-x axis:
$0.37\ \rho a^5$
t</td>
</tr>
<tr>
<td>2. Elliptical Disk
t<<a
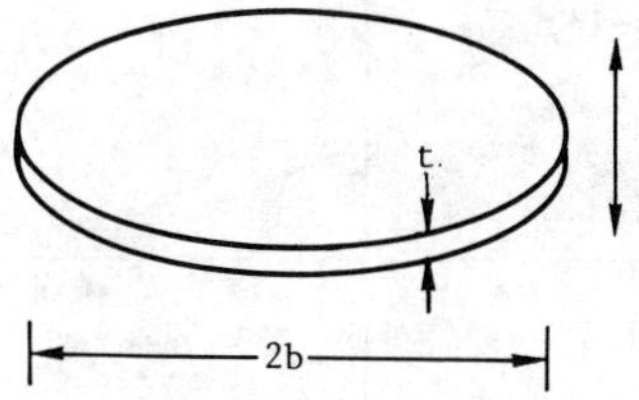a = minor axis
b = major axis</td>
<td>$\alpha \frac{4}{3} \pi \rho a^2 b$
<table>
<tr><th>b/a</th><th>α</th></tr>
<tr><td>∞</td><td>1.0</td></tr>
<tr><td>14.3</td><td>0.991</td></tr>
<tr><td>10.0</td><td>0.984</td></tr>
<tr><td>7.00</td><td>0.972</td></tr>
<tr><td>6.00</td><td>0.964</td></tr>
<tr><td>5.00</td><td>0.952</td></tr>
<tr><td>4.00</td><td>0.933</td></tr>
<tr><td>3.00</td><td>0.900</td></tr>
<tr><td>2.00</td><td>0.826</td></tr>
<tr><td>1.50</td><td>0.758</td></tr>
<tr><td>1.00</td><td>0.637</td></tr>
</table>
t</td>
<td>--
Ref. 14-19</td>
</tr>
<tr>
<td>3. Rectangular Plate
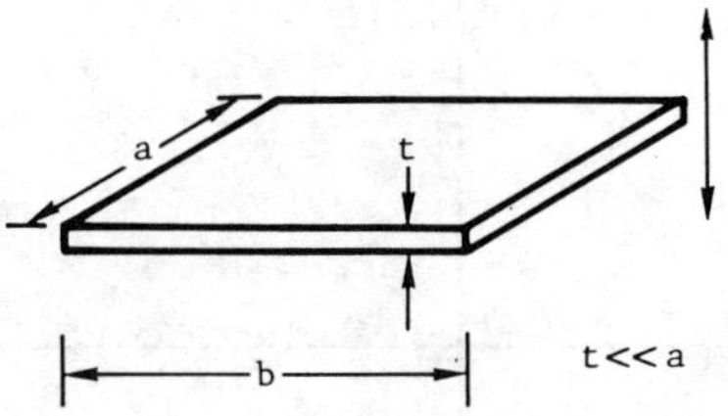t<<a</td>
<td>$\alpha \frac{\pi}{4} \rho a^2 b$
<table>
<tr><th>b/a</th><th>α</th></tr>
<tr><td>1.0</td><td>0.5790</td></tr>
<tr><td>1.25</td><td>0.6419</td></tr>
<tr><td>1.59</td><td>0.7038</td></tr>
<tr><td>2.0</td><td>0.7568</td></tr>
<tr><td>2.5</td><td>0.8008</td></tr>
<tr><td>3.17</td><td>0.8404</td></tr>
<tr><td>4.0</td><td>0.8718</td></tr>
<tr><td>5.0</td><td>0.8965</td></tr>
<tr><td>6.25</td><td>0.9167</td></tr>
<tr><td>8.0</td><td>0.9344</td></tr>
<tr><td>10.0</td><td>0.9469</td></tr>
<tr><td>∞</td><td>1.00</td></tr>
</table>
t</td>
<td>--
Ref. 14-20</td>
</tr>
</table>

Table 14-2. Added Masses of Bodies. *(Continued)*

Description of Body	Added Mass (acceleration in direction denoted by ↔)	Added Mass Moment of Inertia
4. Isosceles Triangular Plate 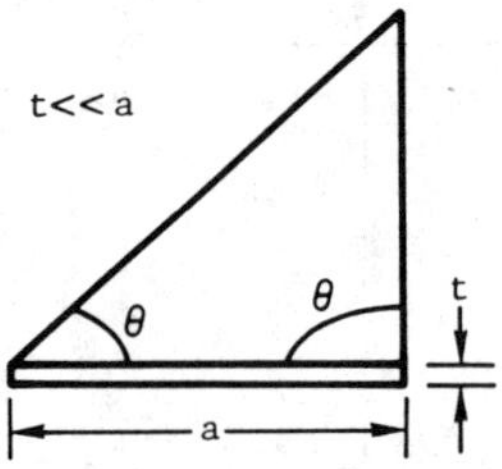	$\frac{\rho a^3}{3\pi}(\tan\theta)^{3/2}$ (motion perpendicular to plane of triangle) e	-- Ref. 14-5
5. Cube 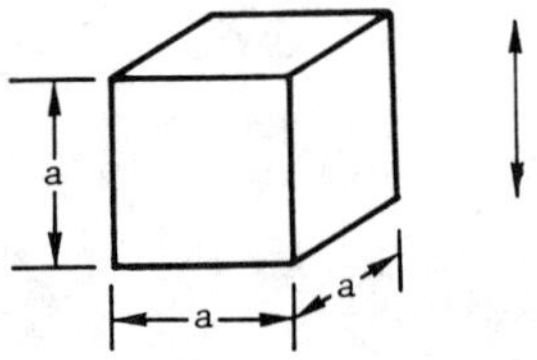	$\alpha\rho a^3$ $\alpha = \begin{cases} 0.67 & \text{Ref. 14-14} \\ 0.70 & \text{Refs. 14-21, 14-16} \end{cases}$ e	--
6. Rectangular Solid 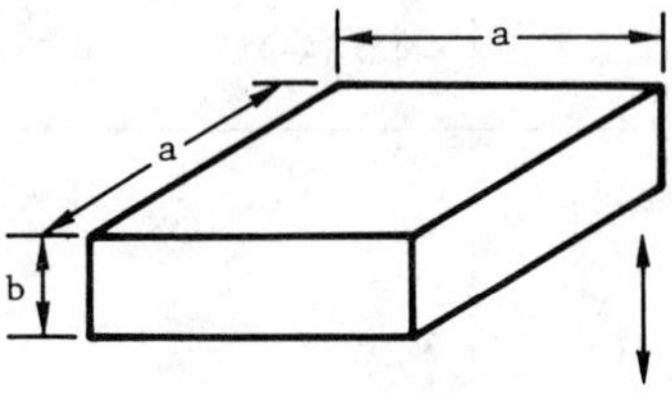	$\alpha\rho a^2 b$ <table><tr><th>b/a</th><th>α</th></tr><tr><td>0.5</td><td>1.32</td></tr><tr><td>0.6</td><td>1.15</td></tr><tr><td>0.8</td><td>0.86</td></tr><tr><td>1.0</td><td>0.70</td></tr><tr><td>1.2</td><td>0.57</td></tr><tr><td>1.6</td><td>0.45</td></tr><tr><td>2.0</td><td>0.35</td></tr><tr><td>2.4</td><td>0.30</td></tr><tr><td>2.8</td><td>0.26</td></tr><tr><td>3.6</td><td>0.22</td></tr></table>e	-- Ref. 14-16
7. Right Circular Cylinder 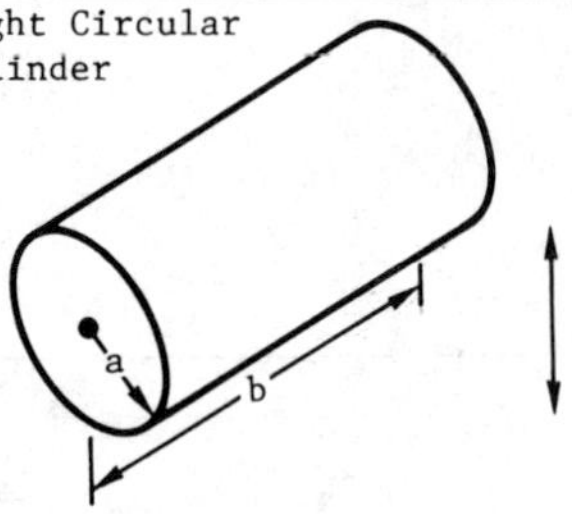	$\alpha\rho\pi a^2 b$ <table><tr><th>b/(2a)</th><th>α</th></tr><tr><td>1.2</td><td>0.62</td></tr><tr><td>2.5</td><td>0.78</td></tr><tr><td>5.0</td><td>0.90</td></tr><tr><td>9.0</td><td>0.96</td></tr><tr><td>∞</td><td>1.00</td></tr></table>(motion perpendicular to cylinder axis) e	-- Ref. 14-4

Table 14-2. Added Masses of Bodies. *(Continued)*

<table>
<tr><th>Description of Body</th><th>Added Mass
(acceleration in direction denoted by ↔)</th><th>Added Mass Moment of Inertia</th></tr>
<tr><td>8. Sphere
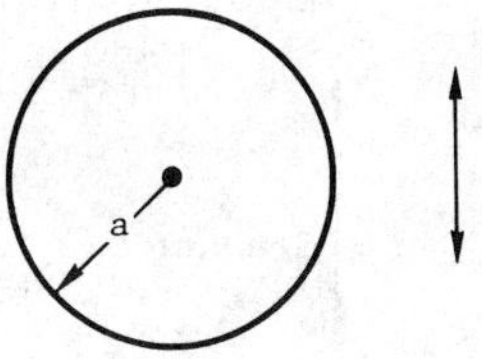
</td><td>$\frac{2}{3}\pi\rho a^3$
t</td><td>0
t</td></tr>
<tr><td>9. Ellipsoid of Revolution
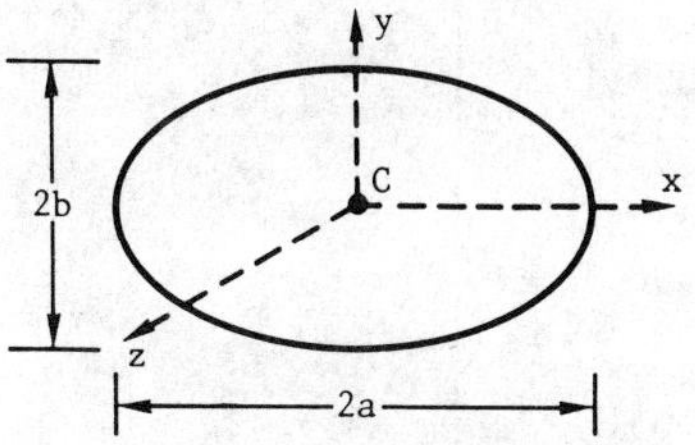

x AXIS IS AXIS OF REVOLUTION</td><td>Acceleration in x direction:
↔ $\alpha\frac{4}{3}\rho\pi ab^2$
or
$\beta\frac{4}{3}\rho\pi b^3$
<table>
<tr><th>a/b</th><th>α</th><th>β</th></tr>
<tr><td>0.01</td><td>--</td><td>0.6348</td></tr>
<tr><td>0.1</td><td>6.148</td><td>0.6148</td></tr>
<tr><td>0.2</td><td>3.008</td><td>0.6016</td></tr>
<tr><td>0.4</td><td>1.428</td><td>0.5712</td></tr>
<tr><td>0.6</td><td>0.9078</td><td>0.5447</td></tr>
<tr><td>0.8</td><td>0.6514</td><td>0.5211</td></tr>
<tr><td>1.0</td><td>0.5000</td><td>0.5000</td></tr>
<tr><td>1.50</td><td>0.3038</td><td>0.4557</td></tr>
<tr><td>2.0</td><td>0.2100</td><td>0.4200</td></tr>
<tr><td>2.5</td><td>0.1563</td><td>0.3908</td></tr>
<tr><td>3.0</td><td>0.1220</td><td>0.3660</td></tr>
<tr><td>5.0</td><td>0.05912</td><td>0.2956</td></tr>
<tr><td>7.0</td><td>0.03585</td><td>0.2510</td></tr>
<tr><td>10.0</td><td>0.02071</td><td>0.2071</td></tr>
<tr><td>∞</td><td>0</td><td>--</td></tr>
</table></td><td>Rotation about z axis or y axis:
$\alpha\frac{4}{15}\pi\rho ab^2(a^2+b^2)$
or
$\beta\frac{8}{15}\pi\rho b^5$
<table>
<tr><th>a/b</th><th>α</th><th>β</th></tr>
<tr><td>0.01</td><td>42.33</td><td>0.2117</td></tr>
<tr><td>0.1</td><td>4.022</td><td>0.2031</td></tr>
<tr><td>0.2</td><td>1.793</td><td>0.1865</td></tr>
<tr><td>0.4</td><td>0.5862</td><td>0.1360</td></tr>
<tr><td>0.6</td><td>0.1843</td><td>0.07519</td></tr>
<tr><td>0.8</td><td>0.03455</td><td>0.02266</td></tr>
<tr><td>1.0</td><td>0.0</td><td>0.0</td></tr>
<tr><td>1.50</td><td>0.09512</td><td>0.2318</td></tr>
<tr><td>2.0</td><td>0.2394</td><td>1.197</td></tr>
<tr><td>2.5</td><td>0.3652</td><td>3.310</td></tr>
<tr><td>3.0</td><td>0.4657</td><td>6.990</td></tr>
<tr><td>5.0</td><td>0.6999</td><td>45.49</td></tr>
<tr><td>7.0</td><td>0.8067</td><td>--</td></tr>
<tr><td>10.0</td><td>0.8835</td><td>--</td></tr>
<tr><td>∞</td><td>1.0</td><td>--</td></tr>
</table></td></tr>
</table>

Table 14-2. Added Masses of Bodies. *(Continued)*

<table>
<tr><th>Description of Body</th><th>Added Mass
(acceleration in direction denoted by ↔)</th><th>Added Mass Moment of Inertia</th></tr>
<tr><td>9. Ellipsoid of Revolution (Continued)</td><td>Acceleration in y or z direction:

$\alpha \frac{4}{3} \rho \pi a b^2$

or

$\beta \frac{4}{3} \rho \pi b^3$

<table>
<tr><th>a/b</th><th>α</th><th>β</th></tr>
<tr><td>0.01</td><td>0.007815</td><td>--</td></tr>
<tr><td>0.1</td><td>0.07480</td><td>0.00748</td></tr>
<tr><td>0.2</td><td>0.1425</td><td>0.02851</td></tr>
<tr><td>0.4</td><td>0.2593</td><td>0.1037</td></tr>
<tr><td>0.6</td><td>0.3552</td><td>0.2131</td></tr>
<tr><td>0.8</td><td>0.4343</td><td>0.3474</td></tr>
<tr><td>1.0</td><td>0.5000</td><td>0.5000</td></tr>
<tr><td>1.50</td><td>0.6221</td><td>0.9331</td></tr>
<tr><td>2.0</td><td>0.7042</td><td>1.408</td></tr>
<tr><td>2.50</td><td>0.7619</td><td>1.905</td></tr>
<tr><td>3.0</td><td>0.8039</td><td>2.412</td></tr>
<tr><td>5.0</td><td>0.8943</td><td>4.472</td></tr>
<tr><td>7.0</td><td>0.9331</td><td>6.532</td></tr>
<tr><td>10.0</td><td>0.9602</td><td>9.602</td></tr>
<tr><td>∞</td><td>1.0</td><td>--</td></tr>
</table>
t</td><td>Rotation about x axis:

0.0

- - - - - - - - - - - - -

See Ref. 14-23 for asymmetric spheroids.

t, Ref. 14-22</td></tr>
<tr><td>10. Quasi-Ellipsoid of Revolution

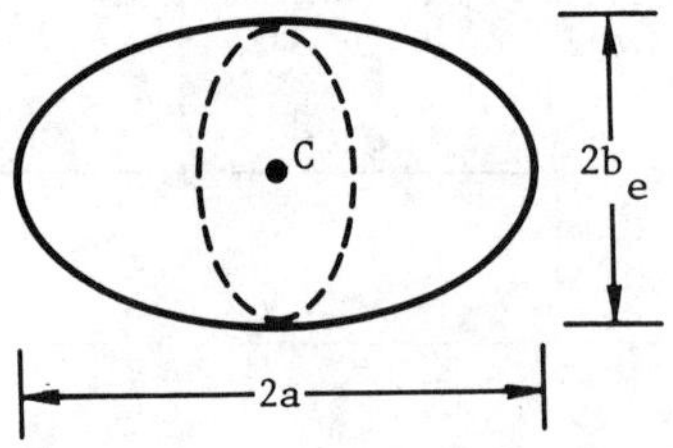
</td><td>For shapes resembling ellipsoids of revolutions, the results of frame 9 can be applied. If the cross section is noncircular, an equivalent radius,

$b_e = A^{1/2}/\pi^{1/2}$,

where A is maximum cross-section area, should be employed.</td><td>See adjacent column.

Ref. 14-23</td></tr>
<tr><td>11. Rotated and Translated Body

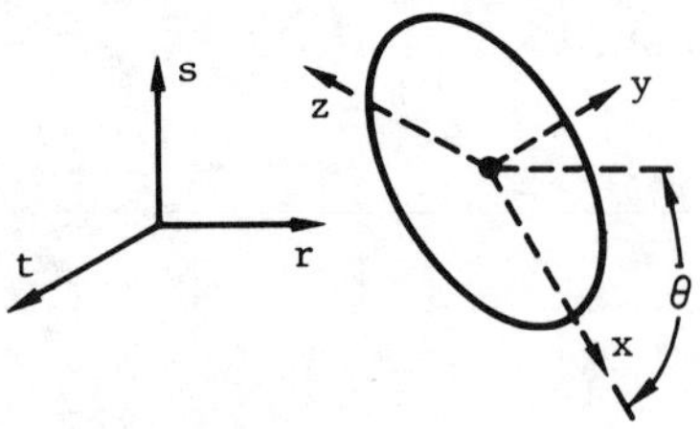
</td><td>For rotation in one plane, see frames 15 and 16 of Table 14-1 or Eq. 14-14.</td><td>See adjacent column.</td></tr>
</table>

Table 14-2. Added Masses of Bodies. *(Continued)*

Description of Body	Added Mass (acceleration in direction denoted by ↔)	Added Mass Moment of Inertia
12. Sphere Near a Free Surface	$\frac{2}{3}\alpha\pi\rho a^3$ (see table for 12 below) e	0 Free surface remains plane during acceleration.
13. Ellipsoid of Revolution Near a Free Surface	$\frac{\alpha 4}{3}\pi\rho ab^2$ for a/b = 2 (see table for 13 below)	Free surface remains plane during acceleration.
14. Floating Body with Three Axes of Symmetry z AXIS PERPENDICULAR TO PAPER. BODY IS SYMMETRIC ABOUT x-y, x-z AND y-z PLANES.	↕ or ↔ Added mass is 1/2 that of fully submerged body. Mean water line is assumed to remain plane during acceleration. t	Added mass moment of inertia is 1/2 that of fully submerged body. Mean water line is assumed to remain plane during rotary acceleration. t
15. Sphere in a Spherical Cavity	$\frac{2}{3}\rho\pi\left[\frac{1+2\left(\frac{a}{b}\right)^3}{1-\left(\frac{a}{b}\right)^3}\right]a^3$ t	0 t, Ref. 14-2

12. Sphere Near a Free Surface

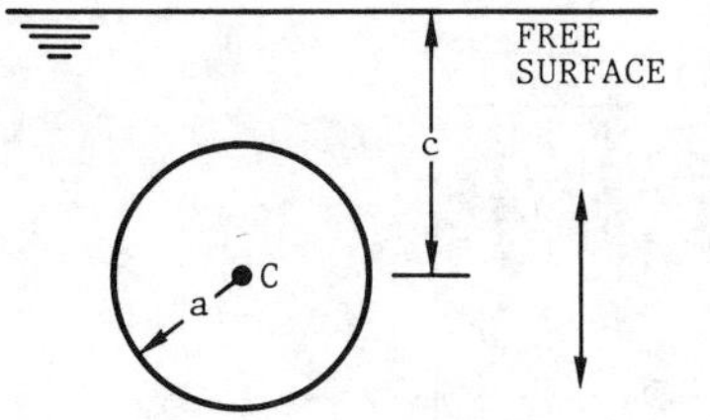

c/2a	α (Ref. 14-6)	α (Ref. 14-24)
0.0	0.50	--
0.50	0.88	0.696
0.75	--	0.942
1.0	1.08	0.976
1.50	--	1.03
2.0	1.18	--
2.5	1.18	--
3.0	1.16	--
4.0	1.05	--
>4.0	1.0	--

13. Ellipsoid of Revolution Near a Free Surface

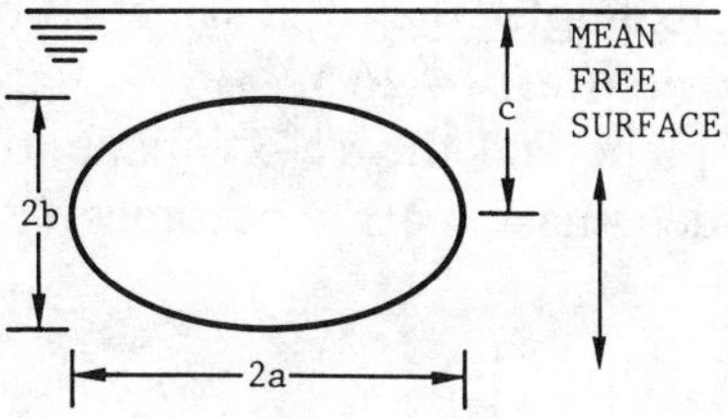

c/2b	α	Ref.
0.0	0.352	Frame 9
1.0	0.913	14-6
2.0	0.915	14-6
∞	0.704	Frame 9

14. Floating Body with Three Axes of Symmetry

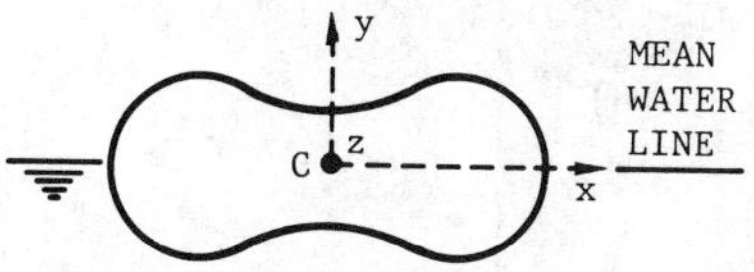

15. Sphere in a Spherical Cavity

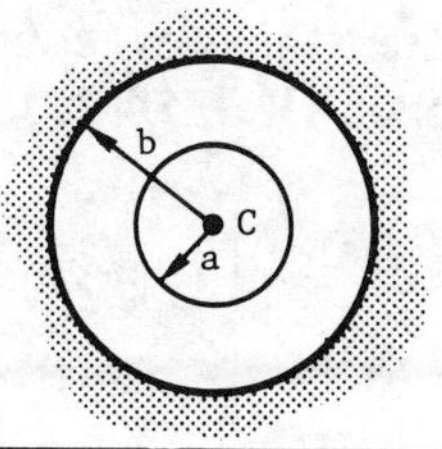

Table 14-2. Added Masses of Bodies. (*Continued*)

Description of Body	Added Mass (acceleration in direction denoted by ↔)	Added Mass Moment of Inertia
16. Two Square Prisims in Tandem	$2\alpha a^2 b$	

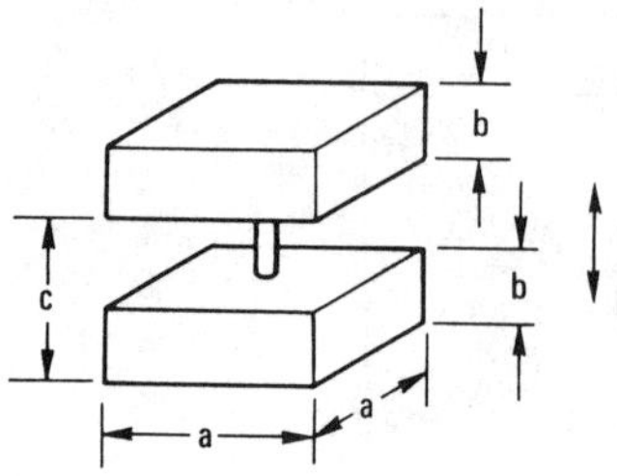

c/a	α, b/a = 0.1	0.2	0.4	1.0
0.5	3.2	1.6	0.75	--
1.0	3.8	2.0	1.1	0.35
1.5	4.4	2.3	1.3	0.45
2.0	4.9	2.7	1.5	0.55
3.0	5.4	2.9	1.6	0.60
4.0	5.6	3.0	1.6	0.70

e Ref. 14-16

There are only 12 independent terms in this matrix. (If the body is in a mean flow parallel to the plane of symmetry, a ship hull for example, then the same 18 off-diagonal terms are zero but the matrix is nonsymmetric, Ref. 14-26.)

If the body is symmetric about both the x_1–x_3 plane and the x_1–x_2 plane, then 26 of the off-diagonal terms are zero and the added mass matrix becomes (Ref. 14-25):

$$\begin{bmatrix} A_{11} & 0 & 0 & 0 & 0 & 0 \\ 0 & A_{22} & 0 & 0 & 0 & A_{26} \\ 0 & 0 & A_{33} & 0 & A_{35} & 0 \\ 0 & 0 & 0 & A_{44} & 0 & 0 \\ 0 & 0 & A_{35} & 0 & A_{55} & 0 \\ 0 & A_{26} & 0 & 0 & 0 & A_{66} \end{bmatrix} . \qquad (14\text{-}23)$$

There are eight independent terms in this matrix.

If the body has three mutually perpendicular planes of symmetry, the x_1–x_3, x_1–x_2, and x_2–x_3 planes, an ellipsoid for example, then the cross coupling is eliminated and only the six independent diagonal terms remain (Ref. 14-25):

$$\begin{bmatrix} A_{11} & 0 & 0 & 0 & 0 & 0 \\ 0 & A_{22} & 0 & 0 & 0 & 0 \\ 0 & 0 & A_{33} & 0 & 0 & 0 \\ 0 & 0 & 0 & A_{44} & 0 & 0 \\ 0 & 0 & 0 & 0 & A_{55} & 0 \\ 0 & 0 & 0 & 0 & 0 & A_{66} \end{bmatrix} . \qquad (14\text{-}24)$$

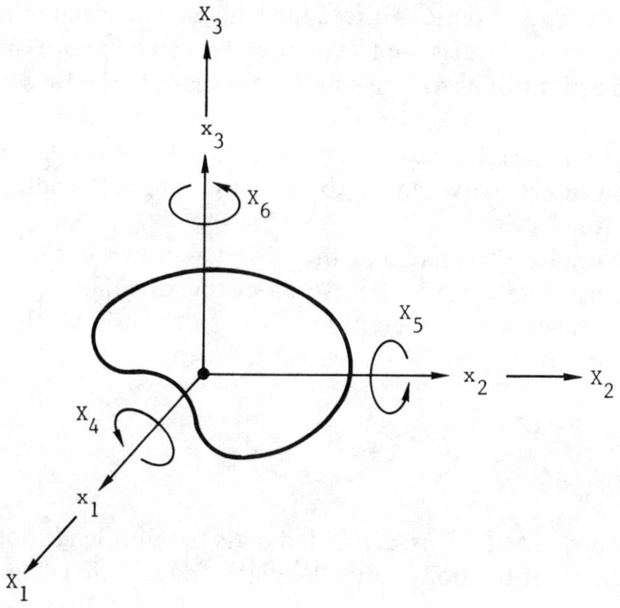

Fig. 14-5. The three rotations and three displacements of a general body.

If, in addition, the body is a body of revolution with the x_1 axis as generator, then the added mass moment of inertia for rotation about the x_1 axis is zero, the remaining two added mass moments of inertia are identical as are two of the added masses,

$$A_{44} = 0, \qquad A_{22} = A_{33}, \qquad A_{55} = A_{66},$$

and only three independent added mass terms remain.

Similarly, the forces associated with the inertia of the structural mass matrix may be expressed in terms of a 6 X 6 generalized mass matrix for small-amplitude vibrations. The generalized mass matrix will be diagonal if the origin of the coordinates coincides with the center of mass and the coordinates are principal coordinates in the sense that all products of inertia are zero for those coordinates (Chapter 5, Ref. 14-8, p. 149).

The total inertia forces are the sum of the inertia forces associated with added mass (Eq. 14-21) and the inertia associated with the generalized structural mass. If both the added mass matrix, [A], and the generalized mass matrix, [M], are diagonal, there is no inertial cross coupling and the analysis is considerably simplified. Both the added mass matrix and the generalized mass matrix will be diagonal if (1) the surface of the body possesses three mutually perpendicular planes of symmetry along the orthogonal axes, (2) the center of mass coincides with the origin of the axes, and (3) the products of inertia of the body are zero. If a body satisfies all of these criteria, for example, an ellipsoid composed of a homogeneous material, the added mass may be simply incorporated in the analysis by (1) increasing the structural mass by the added mass for acceleration along each axis and (2) increasing the mass moments of inertia by the added mass moment of inertia for rotation about each axis.

Strip Theory. In many situations the added masses of slender elongated bodies such as ship hulls, slender bodies, and slender plates can be approximated using strip theory. The basic assumptions of strip theory are that (1) the flow of a narrow local section (strip) over a three-dimensional body is locally two-dimensional and (2) the interaction between adjacent strips is negligible, and therefore the added mass of a slender three-dimensional body can be found by summing the added masses of individual two-dimensional strips.

For example, consider the plate and the cylinder shown in Fig. 14-6. The added mass for vertical acceleration can be estimated by dividing the bodies into two-dimensional strips, neglecting effects due to three-dimensional flow at the ends of the bodies, and summing the added masses of the strips:

$$A = \int_0^b A^S \, dx, \tag{14-25}$$

where A is the added mass, A^S is the added mass per unit length of a strip of width dx, and b is the length of the body. For the plate, frame 5 of Table 14-1 gives:

$$A^S = \frac{\pi}{4} \rho a^2;$$

thus

$$A = \frac{\pi}{4} \rho a^2 b.$$

Similarly, for the cylinder, frame 1 of Table 14-1 gives:

$$A^S = \rho \pi a^2;$$

thus

$$A = \rho \pi a^2 b.$$

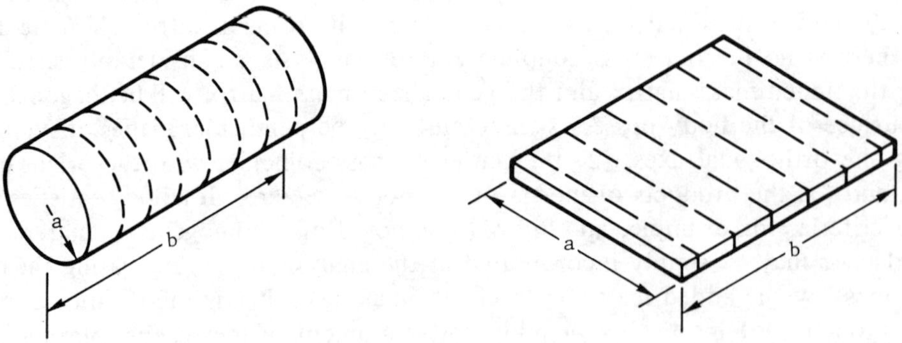

Fig. 14-6. A cylinder and a plate sectioned into strips.

These estimates can be compared with the experimental and theoretical results presented in Table 14-2 as follows:

Cylinder (radius = a, length = b)

b/(2a)	Added Mass/($\rho\pi a^2 b$)	
	Strip Theory	Experiment
1.2	1.0	0.62
2.5	1.0	0.78
5.0	1.0	0.90
9.0	1.0	0.96

Plate (width = a, length = b)

b/a	Added Mass/($\rho\pi a^2 b/4$)	
	Strip Theory	Exact Theory
1.0	1.0	0.5790
2.0	1.0	0.7568
4.0	1.0	0.8718
10.0	1.0	0.9469

As expected, the accuracy of strip theory increases with the slenderness of the body. As the body becomes increasingly squat, strip theory must fail because three-dimensional flow effects play a large role in the added mass of non-slender bodies.

14.3. PLATES

General Case. As an elastic plate vibrates in a fluid, the fluid immediately surrounding the plate is set into motion and the plate will radiate sound waves into the far field just as a loudspeaker radiates sound into a room. In general, the fluid will impose both added mass and damping forces on the plate. The added mass will lower the natural frequency of the plate from that which would be measured in a vacuum, and the damping of the plate will be increased.

Since the natural frequency of an elastic structure is inversely proportional to the square root of the mass of the structure, the effect of added mass on the natural frequencies of the plate can be expressed approximately as:

$$\frac{f_i\big|_{\text{in fluid}}}{f_i\big|_{\text{vacuum}}} = \frac{1}{\left(1 + \frac{A_p}{M_p}\right)^{1/2}} \tag{14-26}$$

if the added mass does not change the mode shape of the plate. M_p is the mass of the plate, i.e., its area times thickness times density of the plate material. A_p is the plate added mass, which is generally a function of the plate geometry, boundary conditions, and mode number. Equation 14-26 applies if only one side of the plate is exposed to the fluid. If both sides of the plate are exposed to the fluid, the added mass is doubled and the term A_p should be replaced with $2A_p$.

By using Eq. 14-26, the problem of determining the effect of fluid on the natural frequencies of the plate is reduced to one of determining its added mass. However, analysis of the added mass of a plate has proven to be an elusive and complex problem. Difficulties arise in the three-dimensional character of the equations describing the fluid motion. While the plate vibrations can be described by two-dimensional modal analysis, the fluid field surrounding the plate is fully three-dimensional. Apparently, the flow field cannot ordinarily be simply modeled even if the fluid viscosity and intermodal coupling are neglected. Generally, the available solutions are either based on simple symmetries, as for circular plates, or rely on numerical solutions, or are experimental results.

One plate solution that is available is the added mass of a disk in an infinite rigid baffle, shown in Fig. 14-7. The baffle is held fixed and the disk moves as a rigid body with respect to the baffle. Both the disk and the baffle are exposed on one side to an inviscid incompressible fluid of density ρ. The added mass of the disk for acceleration perpendicular to the plane of the disk is (Ref. 14-27):

$$A_p = \frac{8}{3}\rho a^3, \tag{14-27}$$

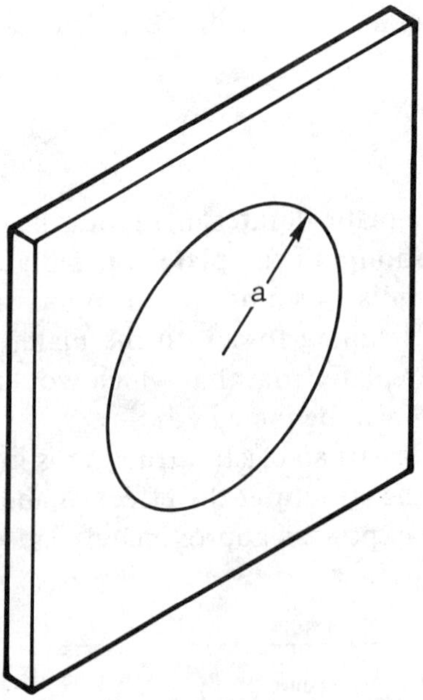

Fig. 14-7. A circular disk in a rigid baffle.

where a is the plate radius. If the fluid surrounds both sides of the disk and baffle, the added mass would be twice this value. Note that the added mass is $2/\pi$ times the mass of fluid contained by the volume described by rotating the disk about a diameter. This suggests that the value of added mass for symmetric plates in the fundamental mode is of the same order as the fluid contained by the volume which is generated by rotating the plate about the longest axis of symmetry. This approximation shows reasonable agreement with experimental data. However, theoretical results show that the plate added mass is a function of the plate boundary conditions and generally declines with increasing mode number.

Circular Plates. The added masses of circular plates with various boundary conditions are given in Table 14-3. The added mass of the rigid baffle (Eq. 14-27) provides a rough estimate of the added mass of these plates. The average added mass of circular plates in the fundamental axisymmetric modes given in frames 2, 3, and 4 of Table 14-3 is 2.43 ρa^3 in comparison with 2.67 ρa^3 for the rigid baffle. Frame 4 of Table 14-3 shows that the added mass decreases in the higher modes of a clamped

Table 14-3. Added Masses of Circular Plates with One Side Exposed to a Fluid.

Notation: a = radius of plate; ρ = density of fluid.

Boundary Conditions	Added Mass, A_P
1. Free (frame 1 of Table 11-1)	Mode with 1 nodal diameter: $\frac{16}{15}\rho a^3$ Mode with 1 nodal circle: $\frac{24}{35}\rho a^3$ Ref. 14-28
2. Free Edge, Central Point Support (frame 5 of Table 11-1)	Axisymmetric mode: $\frac{120}{63}\rho a^3$ Ref. 14-28
3. Simply Supported (frame 2 of Table 11-1)	Fundamental mode: $1.045\ \rho\pi a^3$ Ref. 14-29
4. Clamped (frame 3 of Table 11-1)	Fundamental mode: $0.6689\ \rho\pi a^3$ Mode with 1 nodal diameter: $0.3087\ \rho\pi a^3$ Ref. 14-27

circular plate with one nodal diameter. Similarly, the added masses of the free plate modes with one nodal circle and one nodal diameter (frame 1 of Table 14-3) are well below the added masses of the axisymmetric modes given in frames 2, 3, and 4. This suggests that the added mass of plate modes can generally be expected to decrease with increasing number of nodal lines.

Cantilever Plates. The added masses of slender cantilever plates can be analyzed using strip theory, since it is reasonable to assume the flow over slender plates will be locally two-dimensional. The added masses of cantilever plates as calculated using strip theory are given in Table 14-4. These predictions have shown good agreement with experimental data for slender plates. An aspect ratio correction factor for non-slender plates is presented in Ref. 14-30. An approximate added mass can be predicted using Eq. 14-26 as $2/\pi$ times the volume enclosed by rotating the plate about the longer symmetric axis:

$$2A_p = \left(\frac{2}{\pi}\right)\frac{\pi}{4}\,b^2 a\rho,$$

$$= \rho a b^2 .$$

This is in reasonable agreement with the added mass predicted by strip theory for modes which are primarily translational (modes 1, 3, 6, for example).

Other Rectangular Plates. Greenspon (Ref. 14-31) has proposed the following added mass (Eq. 14-26) for rectangular plates with one side exposed to fluid:

$$A_p = \alpha_{ij}\beta\rho a b^2,$$

where

$$\alpha_{ij} = \frac{\left(\int_A \tilde{z}_{ij}\, dA\right)^2}{2ab\int_A \tilde{z}_{ij}^2\, dA},$$

where $\tilde{z}_{ij}$ is the dimensionless mode shape of the ij mode (Chapter 11) and dA is an element of the plate area A = ab, where a is the width of the plate and b is the length of the plate. ($b \geqslant a$) and β is an aspect ratio dependent factor:

$\frac{a}{b} = 0$	0.1	0.2	0.3	0.4	0.5	0.6	0.7	0.8	0.9	1.0,
$\beta = 0$	0.3	0.42	0.58	0.65	0.72	0.78	0.84	0.92	0.95	1.0.

α_{ij} is a function of the modal indices and the boundary conditions on the plate. α_{ij} is given as follows for the mode shape $\tilde{z}_{ij} = \tilde{x}_i(x)\tilde{y}_j(y)$, where the dimensions of the plate in the x and y directions are a and b, respectively, and $\tilde{x}_i$ and $\tilde{y}_j$ are one-

Table 14-4. Added Masses of Cantilever Plates Submerged in a Fluid.[a]

Notation: ρ = fluid density; i, j = modal indices (frame 3 of Table 11-4).

Mode	i	j	Added Mass, $2A_P$
1. 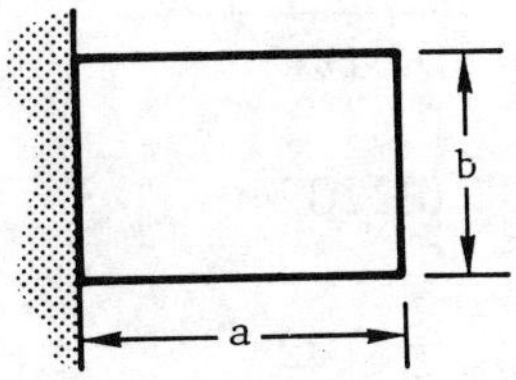	1	1	$\frac{\pi}{4}\rho ab^2$
2. 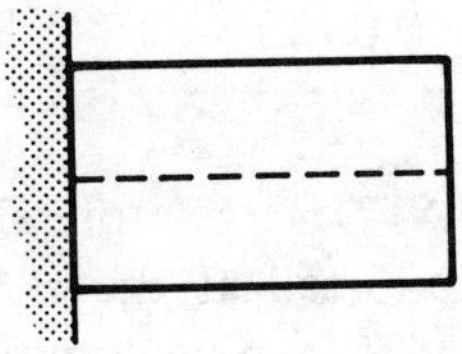	1	2	$\frac{3}{32}\pi\rho ab^2$
3. 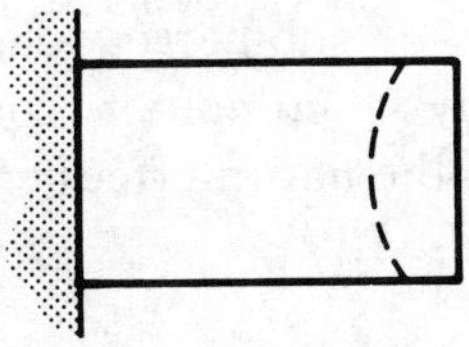	2	1	$\frac{\pi}{4}\rho ab^2$
4. 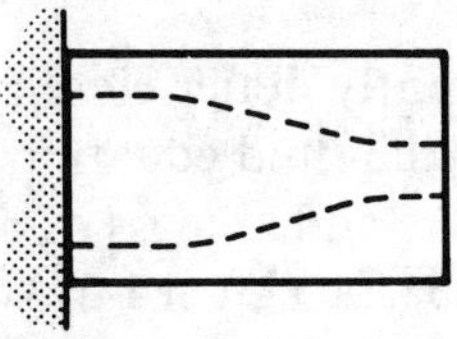	1	3	$0.0803\ \pi\rho ab^2$
5. 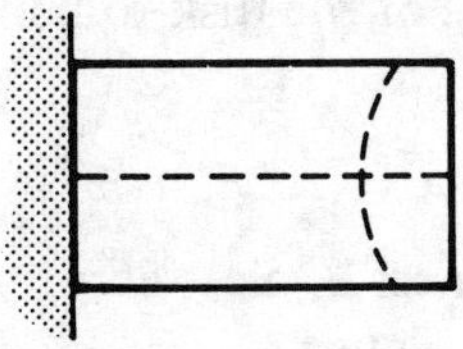	2	2	$\frac{3}{32}\pi\rho ab^2$
6. 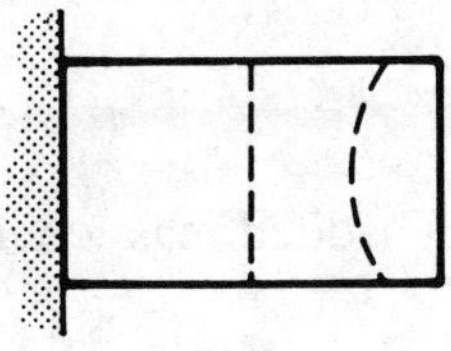	3	1	$\frac{\pi}{4}\rho ab^2$

[a] Ref. 14-30.

dimensional beam modes which satisfy the appropriate boundary conditions:

i	j	α_{ij} C–C–C–C	α_{ij} S–S–S–S
1	1	0.3452	0.8106
	2	0.0	0.0
	3	0.1512	0.2702
	5	0.0962	0.1620
3	1	0.1512	0.2702
	2	0.0	0.0
	3	0.0662	0.0900

C–C–C–C is a completely clamped plate and S–S–S–S is a completely simply supported plate. $\alpha_{11} = 0.2798$ for a plate simply supported at the sides $x = 0$, and clamped along the sides $y = 0, b$. Additional values of α_{ij} can be computed using the formulas in Appendix C.

Some success in the modeling of the vibrations of submerged plates has been achieved using the finite element structural analysis computer codes, such as NASTRAN (Ref. 14-32). Additional discussions of the natural frequencies of fluid-loaded plates can be found in Refs. 14-33 through 14-35.

14.4. SHELLS

The analysis of a shell in contact with a fluid apparently defies closed form solution because of the complexities of the shell equations, the fluid equations, and their interaction. For an order of magnitude estimate of the added mass of a shell, the formula for the added mass of a rigid circular disk in a baffle (Eq. 14-27) can be adapted to the shell problem. An effective mass density for the disk exposed on one side to a fluid can be defined by considering the total mass of the disk to be the sum of the structural mass and the fluid added mass:

$$\mu_e \pi a^2 h = \mu \pi a^2 h + \alpha \pi \rho a^3. \tag{14-28}$$

μ_e is the effective mass density of the disk of thickness h, mass density μ, and radius a which is exposed on one side to a fluid density ρ. Thus,

$$\mu_e = \mu + \frac{\alpha \rho a}{h}.$$

Comparison of Eqs. 14-27 and 14-28 suggests that the added mass coefficient α is:

$$\alpha = \frac{8}{3\pi}.$$

In the case of a shell, the characteristic radius, a, could be defined as the shell radius. For example, for a fluid-filled cylindrical shell, if the length a is defined as the radius of the cylinder, the results of Chen and Rosenberg (Ref. 14-36) for a simply supported end condition imply:

$$\alpha \leqslant \frac{1}{i},$$

where i is the circumferential wave number (see Section 12.2).

A review of the literature on fluid-shell interaction can be found in Ref. 14-37. A few references will also be cited here to provide a starting point for the reader. Discussions of the dynamics of coaxial cylinders coupled by a fluid-filled gap can be found in Refs. 14-38 through 14-43. The vibrations of fluid-filled shells are reviewed in Ref. 14-44 and discussed in Refs. 14-36 and 14-45 through 14-47. The influence of sloshing on the vibrations of partially filled cylindrical shells is examined in Refs. 14-48 through 14-51. The vibrations of shells in an infinite fluid are discussed in Refs. 14-52 through 14-55. The application of the NASTRAN finite element program to coupled fluid-shell vibration is discussed in Ref. 14-56.

14.5. INTERNAL AND EXTERNAL PIPE FLOW

14.5.1. Internal Flow

Fluid flow through a pipe generally depresses the natural frequencies of the pipe. In certain problems involving very high velocity flows through flexible thin-walled pipes, such as those used in the feed lines to rocket motors and water turbines, the decreases in natural frequencies can become important. At very high flow velocities, a pipe can become unstable and either buckle or flail about like an unrestrained garden hose. Reviews of the effect of internal flow on the dynamic response of pipes can be found in Refs. 14-3 (pp. 287–304), 14-57, and 14-58. This section is limited to developing a simple approximate expansion for the change in natural frequency due to steady internal flow in a uniform single-span pipe.

The equation of motion for small, free, transverse vibrations of a uniform pipe without axial load and conveying a steady fluid flow is [Ref. 14-3 (pp. 287–304), 14-57]:

$$EI \frac{\partial^4 Y}{\partial x^4} + \rho A v^2 \frac{\partial^2 Y}{\partial x^2} + 2\rho A v \frac{\partial^2 Y}{\partial x \partial t} + m \frac{\partial^2 Y}{\partial t^2} = 0, \qquad (14\text{-}29)$$

where E is the modulus of elasticity, I is the area moment of inertia of the pipe cross section about the neutral axis, Y is the transverse deflection at each spanwise point x, ρ is the density of the fluid, which travels at average velocity v through the internal area A of the pipe ($A = \pi D^2/4$ for a circular pipe of internal diameter D), and m is the mass per unit length of the pipe plus the fluid in the pipe. If $v = 0$, this equation reduces to the equation of motion for a uniform beam. Generally, Eq. 14-29 can be solved only by considering multiple modes. However, if the flow velocity is much less than that which produces instability, then it is reasonable to assume a single-mode solution of the form:

$$Y(x, t) = y_i(t)\, \tilde{y}_i(x), \qquad (14\text{-}30)$$

where $\tilde{y}_i$ is the mode shape of the i mode in the absence of fluid flow (Chapter 8) and is a function of the dimensionless group $\lambda_i x/L$. L is the pipe span, and λ_i is a dimensionless constant. If Eq. 14-30 is substituted into Eq. 14-29 and the resultant equation is multiplied through by $\tilde{y}_i$ and integrated over the span, an equation of the following form is produced:

$$\ddot{y}_i + 2\zeta_i(2\pi f_i)\dot{y}_i + (2\pi f_i)^2 y_i = 0.$$

The dots ($\cdot$) over the y denote derivatives with respect to time. ζ_i is the fluid damping of the i mode:

$$\zeta_i = \frac{\rho A v}{2\pi f_i m} \frac{\int_0^L \tilde{y}'\tilde{y}\,dx}{\int_0^L \tilde{y}^2\,dx}.$$

This fluid damping is zero for the single-span beams in Table 8-1.

The natural frequency of the i mode of the pipe with fluid flow is related to the corresponding natural frequency at zero flow velocity by:

$$\frac{f_i\big|_{v\neq 0}}{f_i\big|_{v=0}} = \left[1 + \frac{\rho A v^2 L^2 \alpha_i}{EI}\right]^{1/2}, \quad i = 1, 2, 3, \ldots, \tag{14-31}$$

where the dimensionless parameter α_i is only a function of the mode shape:

$$\alpha_i = \frac{L^2}{\lambda_i^4} \frac{\int_0^L \tilde{y}''\tilde{y}\,dx}{\int_0^L \tilde{y}^2\,dx} = \begin{cases} -1/(i\pi)^2 & \text{pinned-pinned} \\ \sigma_i(2 - \sigma_i\lambda_i)/\lambda_i^3 & \text{free-free, clamped-clamped, clamped-free} \\ \sigma_i(1 - \sigma_i\lambda_i)/\lambda_i^3 & \text{clamped-pinned, free-pinned} \end{cases} \tag{14-32}$$

σ_i, λ_i, and the natural frequencies in the absence of fluid flow $[f_i(v = 0)]$ can be found in Table 8-1 for various single-span beams. The primes in Eq. 14-32 refer to differentiation by x.

The approximate solution of Eqs. 14-31 and 14-32 can be expected to be valid only for small shifts in the natural frequencies, although the solution for the pinned-pinned case is in good agreement with a more exact solution [Ref. 14-3 (pp. 287–304)]. Note that the fluid flow generally reduces the natural frequencies of the pipe, since $\alpha < 0$ for single-span pipes of Table 8-1 except for the clamped-free cantilever in the first mode.

14.5.2. Parallel External Flow

As with internal flow, external flow over a pipe parallel to the pipe axis can induce changes in natural frequencies and, at extreme flow velocities, instabilities. Apparently, instabilities induced by a parallel external flow have never been observed in metallic tubes in an industrial environment. Even the changes in natural frequency are quite small. Chen and Wambsganss observed changes in natural frequency of less than 10% for parallel water flow velocities up to 32 m/s over brass rods, which were 1.27 cm in diameter and 1.2 m long (Ref. 14-59). However, a parallel external flow can substantially contribute to the fluid damping, as shown in Ref. 14-3 (pp. 224–227). Reviews of the effect of parallel external flow over circular rods and pipes can be found in Refs. 14-58 through 14-61.

14.6. EXAMPLE

Consider a uniform marine pipe which is supported at intervals along its span. The pipe transports a fluid internally and is exposed to water externally. Since the pipe is symmetric, no cross coupling is introduced by either added mass or the pipe geometry. The effect of the internal fluid mass, for small flow velocities, and the effect of the added mass of externally entrained water can be incorporated in the dynamic analysis of transverse vibrations of the pipe simply by considering the pipe mass per unit length, m, to be the sum of the mass of the fluid carried internally per unit length, m_i, the added mass per unit length, m_A, and the pipe structural mass per unit length, m_s:

$$m = m_s + m_i + m_A .$$

Since the natural frequencies of transverse vibration of the pipe are inversely proportional to the square root of the mass per unit length of the pipe (Table 8-1),

$$f \sim \frac{1}{m^{1/2}},$$

the ratio of the natural frequency of the pipe in a vacuum to the pipe in a fluid is:

$$\frac{f_{\text{fluid}}}{f_{\text{vacuum}}} = \frac{1}{\left(1 + \dfrac{m_A}{m_s + m_i}\right)^{1/2}},$$

where f is the natural frequency in a mode and the pipe is assumed to carry fluid internally (m_i) in both cases. If the mass of internal fluid is negligible ($m_i = 0$), then for a pipe of circular cross section frame 1 of Table 14-1 and frame 24 of Table 5-1 give the corresponding formula:

$$\frac{f_{\text{fluid}}}{f_{\text{vacuum}}} = \frac{1}{\left[1 + \dfrac{\rho}{\mu\left(1 - \dfrac{D_i^2}{D_o^2}\right)}\right]^{1/2}}.$$

ρ is the density of the fluid surrounding the pipe, which is composed of a material of density μ. D_i is the inner diameter of the pipe; D_o is the outer diameter of the pipe.

For a steel pipe ($\mu = 8$ g/cm^3) in water ($\rho = 1$ g/cm^3), this frequency ratio can be expressed as follows:

$$\frac{D_i}{D_o} = 0.1 \quad 0.2 \quad 0.4 \quad 0.6 \quad 0.8 \quad 0.9 \quad 0.95,$$

$$\frac{f_{fluid}}{f_{vacuum}} = 0.943 \quad 0.941 \quad 0.933 \quad 0.915 \quad 0.862 \quad 0.777 \quad 0.662.$$

Note that the decrease in natural frequency is of greatest importance for the thinner-wall pipes because the added mass accounts for a large fraction of the total mass of these pipes.

Similarly, the decrease in pipe frequency due to added mass can be evaluated as a function of the pipe material. For a pipe with a diameter ratio $D_i/D_o = 0.9$ exposed to water ($\rho = 1$ g/cm^3), the reduction in natural frequency due to added mass is as follows:

Pipe Material	Material Density (g/cm^3)	$\frac{f_{fluid}}{f_{vacuum}}$
Copper	8.9	0.792
Steel	8.0	0.777
Aluminum	2.77	0.587
Concrete	2.4	0.559
Acrylic plastic	1.2	0.431

The decrease in natural frequency due to added mass is of greatest importance when using the lighter materials for pipe construction.

For the cantilever steel tube described in Section 8.5.1 ($D_o = 4$ cm, $D_i = 3$ cm), the natural frequencies in water are computed to be

$$\frac{1}{\left\{1 + \dfrac{1.0}{8.0\left[1 - \left(\dfrac{3}{4}\right)^2\right]}\right\}^{1/2}} = 0.882$$

of the natural frequency in air if no water is permitted inside the pipe. Thus, the natural frequencies of the submerged pipe in the first three modes are:

$$f_1 = 30.83 \text{ Hz},$$

$$f_2 = 193.2 \text{ Hz},$$

$$f_3 = 541.0 \text{ Hz}.$$

REFERENCES

14-1. McConnell, K. G., and D. F. Young, "Added Mass of a Sphere in a Bounded Viscous Fluid," *J. Eng. Mech. Div., Am. Soc. Civil Engrs.* **91**, EM4, 147–164 (1965).

14-2. Ackermann, N. L., and A. Arbhabhirama, "Viscous and Boundary Effects on Virtual Mass," *J. Eng. Mech. Div., Am. Soc. Civil Engrs.* **90**, EM4, 123–130 (1964).

14-3. Blevins, R. D., *Flow-Induced Vibration*, Van Nostrand Reinhold, New York, 1977, pp. 134–138.
14-4. Wendel, K., "Hydrodynamische Massen und hydrodynamische Massentragheitmomente," *Jahrbuch Schiffbautechnisches Gesellschaft* 44, 207 (1950). (Available in English as, Wendel, K., "Hydrodynamic Masses and Hydrodynamic Moments of Inertia," David Taylor Model Basin Translation No. 260.)
14-5. Nielsen, J. N., *Missile Aerodynamics*, McGraw-Hill, New York, 1960, pp. 371–373.
14-6. Patton, K. T., "Tables of Hydrodynamic Mass Factors for Translational Motion," ASME Paper 65-WA/UNT-2.
14-7. Bryson, A. E., "Evaluation of the Inertia Coefficients of the Cross Section of a Slender Body," *J. Aeron. Sci.* 21, 424–427 (1954).
14-8. Newman, J. N., *Marine Hydrodynamics*, The MIT Press, Cambridge, Mass., 1977.
14-9. Sedov, L. I., *Two-Dimensional Problems in Hydrodynamics and Aeronautics*, John Wiley, Interscience, New York, 1965.
14-10. Flagg, C. N., and J. N. Newman, "Sway Added-Mass Coefficients for Rectangular Profiles in Shallow Water," *J. Ship Res.* 15, 257–265 (1971).
14-11. Bai, K. J., "The Added Mass of Two-Dimensional Cylinders Heaving in Water of Finite Depth," *J. Fluid Mech.* 81, 85–105 (1977).
14-12. Chen, S. S., M. W. Wambsganss, and J. A. Jendrzejczyk, "Added Mass and Damping of a Vibrating Rod in Confined Viscous Fluid," *J. Appl. Mech.* 98, 325–329 (1976).
14-13. Chen, S. S., "Dynamic Response of Two Parallel Circular Cylinders in a Liquid," *J. Pressure Vessel Tech.* 97, 78–83 (1975).
14-14. Stelson, T. E., and F. T. Mavis, "Virtual Mass and Acceleration in Fluids," *Proc. ASCE* 81, 670-1–670-9 (1955).
14-15. Moretti, P. M., and R. L. Lowery, "Hydrodynamic Inertia Coefficients for a Tube Surrounded by Rigid Tube," *J. Pressure Vessel Tech.* 97 (1975). Also see Chen, S. S., "Vibration of Nuclear Fuel Rod Bundles," *Nucl. Eng. Design*, 35, 399–422 (1975).
14-16. Sarpkaya, T., "Added Masses of Lenses and Parallel Plates," *J. Eng. Mech. Div., Am. Soc. Civil Engrs.* 86, 141–151 (1960).
14-17. Landweber, L., "Vibration of a Flexible Cylinder in a Fluid," *J. Ship Res.* 11, 143–150 (1967).
14-18. Chen, S. S., and J. A. Jendrzejczyk, "Experiments on Fluidelastic Vibration of Cantilevered Tube Bundles," ASME Paper No. 77-DET-96, to be published in *J. Mech. Design*.
14-19. Munk, M. M., "Fluid Mechanics, Part II," in *Aerodynamic Theory*, Vol. I, W. F. Durand (ed.), Julius Springer, Berlin, 1934, p. 302.
14-20. Meyerhoff, W. K., "Added Masses of Thin Rectangular Plates Calculated from Potential Theory," *J. Ship Res.* 14, 100–111 (1970).
14-21. Yu, Y. T., "Virtual Masses of Rectangular Plates and Parallelpipeds in Water," *J. Appl. Phys.* 16, 724–729 (1945).
14-22. Lamb, H., *Hydrodynamics*, 6th ed., Dover Publications, New York, 1945, pp. 152–156, 700–701.
14-23. Munk, M. M., "Aerodynamics of Airships," in *Aerodynamic Theory*, Vol. VI, W. F. Durand (ed.), Julius Springer, Berlin, 1934, pp. 32–36.
14-24. Waugh, J. G., and A. T. Ellis, "Fluid-Free-Surface Proximity Effect on a Sphere Vertically from Rest," *J. Hydronautics* 3, 175–179 (1969).
14-25. Sedov, L. I., *A Course in Continuum Mechanics*, Vol. III, Wolters-Noordhoff Publishing, Groningen, The Netherlands, 1972, pp. 203–207 (translated from Russian).
14-26. Salvesen, N., E. O. Tuck, and O. Faltinsen, "Ship Motions and Sea Loads," *Trans. Soc. Naval Architects Marine Engrs.* 78, 250–287 (1970).
14-27. Lamb, H., "On the Vibrations of an Elastic Plate in Contact with Water," *Proc. Royal Soc. (London), Series A* 98, 205–216 (1921).
14-28. McLachlan, N. W., "The Accession to Inertia of Flexible Discs Vibrating in a Fluid," *Proc. Phys. Soc. (London)* 44, 546–555 (1932).
14-29. Peake, W. H., and E. G. Thurston, "The Lowest Resonant Frequency of a Circular Plate," *J. Acoust. Soc. Am.* 26, 166–168 (March 1954).
14-30. Lindholm, U. S., D. D. Kana, and H. N. Abramson, "Elastic Vibration of Cantilever Plates in Water," *J. Ship Res.* 9, 11–22 (1965).
14-31. Greenspon, J. E., "Vibrations of Cross-Stiffened and Sandwich Plates with Application to Underwater Sound Radiators," *J. Acoust. Soc. Am.* 33, 1485–1497 (1961).
14-32. Marcus, M. S., "A Finite Element Method Applied to the Vibration of Submerged Plates," *J. Ship Res.* 22, 94–99 (1978).
14-33. Junger, M. C., "Normal Modes of Submerged Plates and Shells," in *Symposium on Fluid-Solid Interaction, ASME, Pittsburgh, Pa., 1967*, pp. 79–119.

14-34. Sandman, B. E., "Fluid-Loaded Vibration of an Elastic Plate Carrying a Concentrated Mass," *J. Acoust. Soc. Am.* **61**, 1503–1510 (1977).
14-35. Junger, M. C., and D. Feit, *Sound, Structures and Their Interaction*, The MIT Press, Cambridge, Mass., 1972.
14-36. Chen, S. S., and G. S. Rosenberg, "Free Vibrations of Fluid-Conveying Cylindrical Shells," *J. Eng. Indust.* **96**, 420–426 (1974).
14-37. Chen, S. S., "Flow-Induced Vibrations of Circular Cylindrical Structures, Part I: Stationary Fluids and Parallel Flow," *Shock and Vibration Digest* **9**, No. 10, 25–38 (October 1977).
14-38. Au-Yang, M. K., "Free Vibration of Fluid Coupled Coaxial Cylindrical Shells on Different Lengths," *J. Appl. Mech.* **43**, 480–484 (1976).
14-39. Horvay, G., and G. Bowers, "Influence of Water Mass on the Vibration Modes of a Shell," *J. Fluids Eng.* **97**, 211–216 (1975).
14-40. Krajcinovic, D., "Vibrations of Two Coaxial Cylindrical Shells Containing Fluid," *Nucl. Eng. Des.* **30**, 242–248 (1974).
14-41. Penzes, L. E., and S. K. Bhat, "Generalized Hydrodynamic Effects of a Double Annuli on a Vibrating Cylindrical Shell," *Third International Conference on Structural Mechanics in Reactor Technology*, London, 1975, Paper F2/6.
14-42. Au-Yang, M. K., "Generalized Hydrodynamic Mass for Beam Mode Vibration of Cylinders Coupled by Fluid Gap," *J. Appl. Mech.* **44**, 172–174 (1977).
14-43. Yeh, T. T., and S. S. Chen, "Dynamics of a Cylindrical Shell System Coupled by a Viscous Fluid," *J. Acoust. Soc. Am.* **62**, 262–270 (1977).
14-44. DiMaggio, F. L., "Dynamic Response of Fluid-Filled Shells," *Shock and Vibration Digest* 7, 5–12 (May 1975).
14-45. Kenner, V. H., and W. Goldsmith, "Dynamic Loading of a Fluid-Filled Spherical Shell," *Int. J. Mech. Sci.* **14**, 557–568 (1972).
14-46. Lakis, A. A., and M. P. Paidoussis, "Free Vibration of Cylindrical Shells Partially Filled with Liquid," *J. Sound Vib.* **19**, 1–15 (1971).
14-47. Join, R. K., "Vibration of Fluid-Filled, Orthotropic Cylindrical Shells," *J. Sound Vib.* **37**, 379–388 (1974).
14-48. Abramson, H. N., "Dynamic Behavior of Liquid in Moving Container," *Appl. Mech. Revs.* **16**, 501–506 (1963).
14-49. Abramson, H. N. (ed.), "Dynamic Behavior of Liquid in Moving Containers," NASA Report NASA SP-106, Southwest Research Institute, San Antonio, Texas, 1966.
14-50. Lev, O. E., and B. P. Jain, "Seismic Response of Flexible Liquid Containers," *Conference on Structural Mechanics in Reactor Technology*, San Francisco, 1977, Paper K5/3.
14-51. Bauer, H. F., and J. Siekmann, "Dynamic Interaction of a Liquid with the Elastic Structure of a Circular Cylindrical Container," *J. Ingenieur-Archiv* **40**, 266–280 (1970).
14-52. Chen, F. C., "Axisymmetrical Vibrations of Underwater Hemispherical Shells," *J. Eng. Indust.* **98**, 941–947 (1976).
14-53. Chang, K. Y., and F. L. DiMaggio, "Vibrations of Cylindrical Shells in a Semiinfinite Acoustic Medium," *J. Acoust. Soc. Am.* **49**, 759–766 (1971).
14-54. Sonstegard, D. A., "Effects of a Surrounding Fluid on the Free, Axisymmetric Vibrations of Thin Elastic Spherical Shells," *J. Acoust. Soc. Am.* **45**, 506–510 (1969).
14-55. Lou, Y. K., and T. C. Su, "Free Oscillations of Submerged Spherical Shells," *J. Acoust. Soc. Am.* **63**, 1402–1408 (1978).
14-56. Coppolino, R. N., "A Numerically Efficient Finite Element, Hydroelastic Analysis," NASA-TM-X-3428, NASTRAN: Users' Experiences, Fifth Colloquium, Ames Research Center, Moffett Field, Ca., 5-6 October 1976, pp. 177-206.
14-57. Paidoussis, M. P., and N. T. Issid, "Dynamic Stability of Pipes Conveying Fluid," *J. Sound Vib.* 33, 267–294 (1974).
14-58. Hannoyer, M. J., and M. P. Paidoussis, "Instabilities of Tubular Beams Simultaneously Subjected to Internal and External Flows," *J. Mech. Design* **100**, 328–336 (1978).
14-59. Chen, S. S., and M. W. Wambsganss, "Parallel Flow-Induced Vibration of Fuel Rods," *Nucl. Eng. Design* **18**, 253–278 (1972).
14-60. Paidoussis, M. P., "Dynamics of Flexible Slender Cylinders in Axial Flow," *J. Fluid Mech.* **26**, 717–751 (1966).
14-61. Paidoussis, M. P., "The Dynamical Behavior of Cylindrical Structures in Axial Flow," *Ann. Nucl. Sci. Eng.* **1**, 83–106 (1974).

15

FINITE ELEMENT METHOD

15.1. INTRODUCTION

Many structures of practical importance are too complex to be analyzed by classical techniques. Approximate analysis must be used. There are two general avenues for approximate analysis. The first is to construct approximate solutions to the differential equations of motion describing the structure, using either a series solution or an energy criterion to minimize the error. The second approach to approximate solution is to simulate the structure by decomposing the structure into a series of small but finite elements over which stress and displacements vary monotonically. The equations of motion of the discretized system describe the forces and displacements at the boundary points (nodes) of the finite elements. Thus, complex structures are modeled by the aggregate of simpler structures. This is called the finite element method.

The principal advantage of the finite element method is its generality. It can be used to determine the natural frequencies and mode shapes of any linear, elastic structure. One is limited only by the size of the computer available and the desired accuracy. Moreover, application of the method differs only in the details of input from structure to structure. Thus, as with a foreign language, usage leads to the skills that can be applied to a host of problems.

Of course, the finite element method is not a panacea for structural analysis. First of all, it is purely a numerical technique. The input and the output consist of digits and symbols. The sensitivity of output to a small change in input can be determined only through repetitive runs or reasoning based on experience and classical solutions. Second, the analysis is tied to a middle- or large-sized digital computer. While this will not represent a problem for those associated with an organization which possesses such a computer, it can be a great obstacle for many analysts who work for small firms. Third, even the simplest of the finite element programs has its idiosyncrasies. All too often one attempting to run a finite element program finds himself facing an indecipherable error message. Thus, some degree of consultation is virtually a necessity for running finite element programs, especially for the analyst who has not had previous finite element experience.

Nevertheless, the finite element method has become an indispensable tool for modern analysis. No single technique in the past one-hundred years has had a greater impact on static and dynamic structural analysis than the finite element method. Its usage has increased to the point where entire automobile and aircraft designs are based on the results of finite element analysis. In the following sections, a brief review will be made of some of the principles and applications of the finite element method and a survey of the more widely used finite element programs in the United States will be presented. For more information on the finite element method, the reader should examine Ref. 15-1 or recent issues of the *International Journal for Numerical Methods and Engineering.*

15.2. DISCRETIZATION OF THE STRUCTURE

If a structure can be subdivided into a series of finite elements, the infinite number of degrees of freedom associated with the continuous structure can be reduced to a finite number of degrees of freedom which can be examined individually. This discretation is generally accomplished using the displacement method as follows (Ref. 15-1, p. 11): (1) the structure is divided into a series of elements by imaginary lines, (2) the elements are assumed to be connected only at the intersections of these lines (nodes), (3) the stresses and strains in each element are defined uniquely in terms of the displacements and forces at the nodes, and (4) the mass of the elements is lumped at the nodes. The system of forces and displacements at the nodes results in a series of equations of motion for the displacement of the nodes and hence the displacement of the structure. By solving the resultant series of coupled linear equations, one can determine the stresses, strains, natural frequencies, and mode shapes of the structure.

For example, consider the longitudinal vibrations of the cantilever shown in Fig. 15-1(a). If the cantilever is subdivided into a series of equal elements and the stress over each element is assumed to be approximately constant, each element can be replaced with an equivalent spring-mass, as shown in Fig. 15-1(b). The equivalent spring constant of each element of length L/n is (frame 8 of Table 6-1)

$$k = \frac{EAn}{L}$$

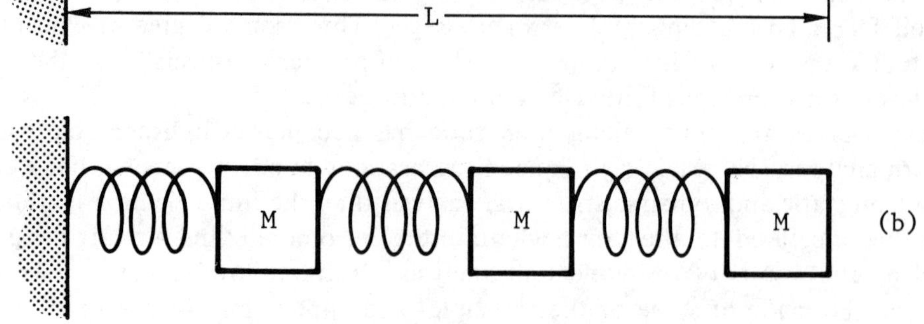

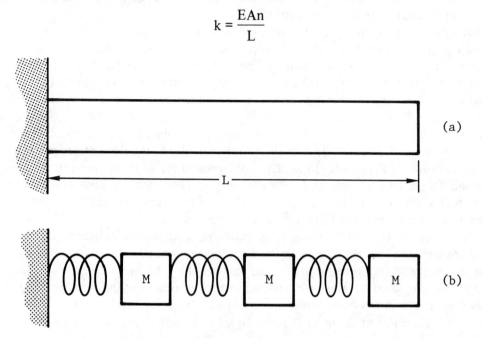

Fig. 15-1. A cantilever beam and spring-mass model for longitudinal vibration.

and the mass of each element is

$$M = \frac{\mu A L}{n}.$$

A is the cross-sectional area of the beam, E is the modulus of elasticity of the beam, L is the length of the beam, n is the number of elements, and μ is the beam density.

The exact solution for the natural frequencies of longitudinal vibration of the beam is (frame 2 of Table 8-16):

$$f_i = \frac{2i - 1}{4L}\left(\frac{E}{\mu}\right)^{1/2} \text{ Hz}, \qquad i = 1, 2, \ldots.$$

The natural frequencies of the equivalent finite element model will depend on how the masses are divided between the nodes. If the mass of each element is assigned to the node to the right of the element, the natural frequencies of the resultant model can easily be evaluated using frame 5 of Table 6-2. For example, for the three mass models shown in Fig. 15-1(b), the natural frequencies are:

	Model	Continuous System
$f_1 L \left(\frac{\mu}{E}\right)^{1/2}$	= 0.2125	0.25
$f_2 L \left(\frac{\mu}{E}\right)^{1/2}$	= 0.5955	0.75
$f_3 L \left(\frac{\mu}{E}\right)^{1/2}$	= 0.8604	1.25

In Fig. 15-2, the natural frequencies of the finite element models are compared with the exact result. Note that (1) the finite element models have a finite number of natural frequencies, whereas the continuous structure has an infinite number, and (2) the accuracy of the finite element models is greatest in the lower modes. The accuracy increases as the number of elements in the model increases. These conclusions hold true for nearly all finite element models. One must generally provide sufficient detail so that the model can accurately simulate the mode of interest and produce an accurate result.

The finite element models of Fig. 15-2 underpredict the natural frequencies. In general, the natural frequencies of the finite element model will approach the exact result as the finite element mesh becomes increasingly fine. The degree of overprediction or underprediction of the finite element model will depend on the manner in which the stiffness and mass matrix of the model are formulated and the nature of the elements employed. For example, if the masses had been divided equally between the nodes in the model of Fig. 15-1, i.e., a mass of one-half element had been assigned to the fixed node, the finite element natural frequencies would be higher and the predictions more accurate.

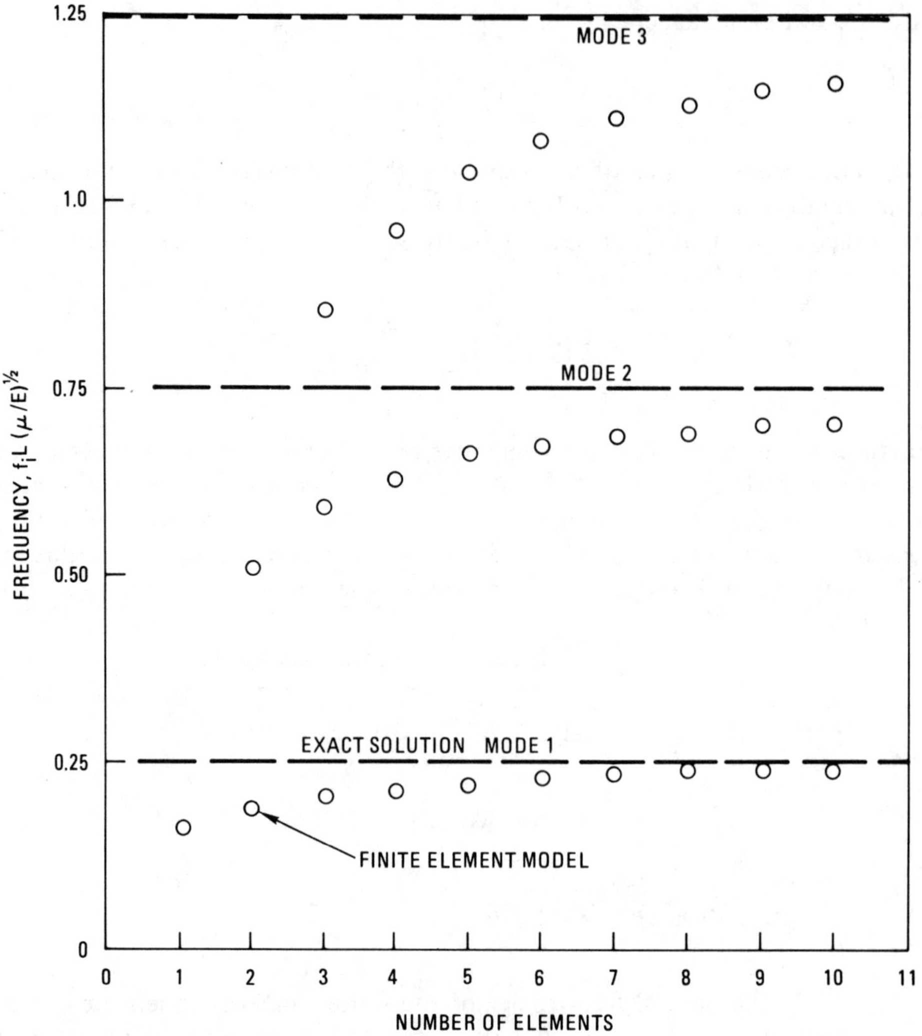

Fig. 15-2. Comparison of natural frequencies of finite element models and exact solution results.

15.3. GENERAL PURPOSE FINITE ELEMENT PROGRAMS

A few of the features of some of the more popular finite element programs in the United States are given in Table 15-1. These are general purpose programs. All are capable of natural frequency and mode shape calculation, time history analysis, and static stress analysis. Table 15-1 does not by any means comprise a complete list of finite element programs which are available. Nevertheless, these programs are proven in the sense that they have been extensively applied for some time, they are being maintained and extended by organizations set up for that purpose, and they are widely available.

The programs listed in Table 15-1 are based on the finite element displacement method and may be summarized as follows (Ref. 15-1):

ANSYS

Capability: Static and dynamic linear and nonlinear structural analysis and heat transfer analysis. Program has plasticity, creep, and large displacement and rotation capability.

Method: Finite element displacement method. Program uses the incremental method of solution accounting for plasticity with isotropic and kinematic hardening. Program uses the wave-front method coupled with an explicit time integration scheme for the solution of the nonlinear equations of motion. Eigenvalues are extracted via Jacobi iteration with Guyan reduction.

Language: FORTRAN

Hardware: Program runs on CDC, IBM, and UNIVAC machines.

Usage: Program has been extensively used in the nuclear industry and indications of its reliability are available.

Developer: John A. Swanson
Swanson Analysis Systems, Inc.
870 Pine View Drive
Elizabeth, PA 15037

Table 15-1. General Purpose Finite Element Programs.

Features	Program[a] ANSYS	MARC	NASTRAN	SAP	STARDYNE
Straight beam, straight pipe, solid and flat plate elements	X	X	X	X	X
Axisymmetric elements	X	X	X	X	O
Curved beam/curved pipe elements	O/X	X/X	O/O	O/X	O/X
Curved shell elements	O	X	O	O	O
Inviscid fluid element	O	O	X	O	O
Buckling analysis	O	X	X	O	O
Shock spectra	X	O	O	X	X
Mesh generation	Yes	Yes	Yes	Some	Some
Nonlinear analysis	Extensive	Extensive	Limited	Limited[b]	None
Pages in manual describing elements, input and output (approximate)	830	820	980	130	560
Proprietary/public	Prop.	Prop.	Public	Public	Prop.
Availability[c]	CDC, W, D	CDC, D	CDC, W	CDC, D	CDC

[a]X = program has this capability; O = program lacks this capability.
[b]Nonlinear capability in MODSAP version.
[c]CDC = Control Data Corporation Cybernet; W = Westinghouse Telecomputer Center, Pittsburgh, PA; D = developer (see text).

Availability: Program and documentation are available from the developer at a fee which is negotiable, or from the Control Data Corporation (CDC) Cybernet or Westinghouse Telecomputer Center, Pittsburgh, PA.

Comment: ANSYS is a large, versatile program which is primarily employed for problems involving geometric or material nonlinearities. Its nonlinear capabilities are comparable to those of MARC.

MARC-CDC

Capability: Static and dynamic linear and nonlinear structural and heat transfer analysis. Program has large deflection, plasticity, creep, and buckling analysis capability. The finite element library includes curved shell elements.

Method: Program uses Von-Mises yield criterion with isotropic and kinematic hardening. The solution technique is the modified Newton-Raphson method coupled with explicit time integration of the nonlinear equations of motion. Shell elements use the Koiter-Sanders shell theory. Eigenvalues are extracted using inverse power iteration. Time integration can be performed by both mode superposition and direct integration using implicit Newmark beta and Houbolt techniques and central difference explicit techniques. Kelvin and Maxwell viscoelastic nonlinear material models are available.

Language: FORTRAN

Hardware: Program runs on CDC, IBM, and UNIVAC machines.

Usage: Program has been used extensively and is found to be reliable in most cases.

Developer: David Hibbet/Pedro V. Marcal
MARC Analysis Research Corporation
105 Medway Street
Providence, RI 02906

Availability: The source or object form of the program along with documentation is available from the developer at a fee which is negotiable, or from Control Data Corporation (CDC) Cybernet.

Comment: MARC is a very sophisticated program which is notable for its extensive nonlinear and shell analysis capabilities. However, its input is complex and some analysts report that more experience is required to become proficient with MARC than with other programs.

NASTRAN

Capability: Static and dynamic structural analysis. Program has buckling, heat transfer, random vibration, and inviscid fluid analysis capability. Program can analyze very large and complex structures and can perform sloshing, acoustic cavity, and coupled fluid-structure dynamic analysis.

Method: Finite element displacement method. Program has capabilities for Guyan reduction. Eigenvalues are extracted using determinant, inverse power, and Givens techniques. Transient analysis is obtained through modal superposition or direct time integration using Newmark beta technique. Random loads are applied using stationary model.

Language: FORTRAN with some machine language subroutines.

Hardware: Program runs on IBM, CDC, and UNIVAC machines. A minimum core of 50,000 single precision words is required.

Usage: Extensive.

Developer: NASA, Goodard Space Flight Center (Management Supervision)
Computer Sciences Corporation
Baltimore Division of Martin-Marietta
The MacNeal-Schwendler Corporation
Bell Aerosystems

Availability: Program is available from Control Data Corporation (CDC) Cybernet; McAuto Computer Network; Westinghouse Telecomputer Center, Pittsburgh, PA; COSMIC (Computer Software Management and Information Center), University of Georgia, Athens, GA 30601; and other sources.

Comment: NASTRAN is the largest general purpose structural analysis program in use today. It is best suited to large linear systems. The large size of the program and its documentation make usage a formidable task. NASTRAN is very expensive for small problems. An enhanced version is available from the MacNeal-Schwendler Corporation, Los Angeles, California.

SAP

Capability: Static and dynamic structural analysis.

Method: Eigenvalues are extracted by determinant search or subspace iteration solution. Transient response by modal superposition or direct integration using Wilson theta technique.

Language: FORTRAN

Hardware: Program runs on IBM, CDC, and UNIVAC machines.

Usage: Extensive

Developer: Klaus-Jürgen Bathe/Edward L. Wilson/Fred E. Peterson
Earthquake Engineering Research Center
University of California
Berkeley, CA 94720

Availability: Program is available from NISEE, University of California, Berkeley, CA 94720.

Cost of the basic program, SAP IV, is $250. Several enhanced versions exist. The program is also available from Control Data Corporation (CDC) Cybernet.

Comment: SAP is best suited to small- or moderate-sized linear system analysis. The program is user oriented and can be easily modified to suit the needs of a particular organization. It is one of the simplest and easiest to use of the general programs available. A number of derivative programs are available.

STARDYNE

Capability: Static and dynamic structural analysis. Program has random vibration capability.

Method: Finite element displacement method. Program has capabilities for Guyan reduction. Eigenvalues are extracted using inverse power, Givens, and Householder techniques. Transient analysis is obtained using modal superposition or direct integration using Newmark beta and Wilson theta techniques. Random vibration analysis employs stationary model.

Language: FORTRAN

Hardware: Program runs on CDC machines.

Usage: Extensive

Developer: Richard Rosen
Mechanics Research Inc.
9841 Airport Boulevard
Los Angeles, CA 90045

Availability: Program is available from Control Data Corporation (CDC) Cybernet.

Comment: STARDYNE is best suited to small- or middle-sized linear systems. The program is notable for its reliability, economy, and ease of use.

The five programs described above have greatly varying capabilities and ease of usage. ANSYS and MARC are known for their nonlinear capabilities, NASTRAN for large system analysis, and SAP and STARDYNE for their economy and ease of use. The nonlinear capabilities of ANSYS and MARC come at the cost of increased computer time and complexity of input. The ability of NASTRAN to handle systems with many thousands of degrees of freedom comes at the cost of inefficiency when analyzing small systems. The ease of usage of SAP and STARDYNE comes at the cost of a lack of nonlinear capability. No one program can efficiently satisfy the needs of all users. It is probably most efficient, in terms of man time and machine time, to employ the simplest program which will solve a given problem. Thus, one would not employ MARC or NASTRAN for a problem that can be solved on SAP or STARDYNE. Most large organizations faced with a variety of analysis problems offer a mix of finite element codes to their members so that an optimum program can be matched to the problem.

The abstracts of the five programs presented here should not be taken as absolute descriptions. These programs are constantly being upgraded and their documentation is constantly being improved. The capabilities of the latest versions can be obtained from the sources listed in the abstracts presented here.

Many other general purpose finite element programs are available. Among these are STRUDL DYNAL, COSA, ASKA, and NISA. Discussions of these programs and many other shock and vibration codes can be found in Ref. 15-2.

REFERENCES

15-1. Zienkiewicz, O. C., and Y. K. Cheung, *The Finite Element Method in Structural and Continuum Mechanics*, McGraw-Hill, New York, 1967.

15-2. Pilkey, W., and B. Pilkey (eds.), *Shock and Vibration Computer Programs, Reviews and Summaries*, The Shock and Vibration Information Center, Naval Research Laboratory, Washington, D.C., 1975.

16

PROPERTIES OF SOLIDS, LIQUIDS, AND GASES

16.1. SOLIDS

The density, modulus of elasticity, and in some cases Poisson's ratio for metals, wood, plastics, and glass are given in Tables 16-1 through 16-6. All the properties are referenced to room temperature, 20°C (68°F), and atmospheric pressure unless otherwise noted. The material properties of solids are substantially independent of pressure. It can be seen from Tables 16-1 and 16-2 that the modulus of elasticity of metals declines with increasing temperature. When the temperature approaches the melting temperature, the modulus of elasticity will approach zero as the metal begins to creep.

The modulus of elasticity of a material can be determined by both static and dynamic methods. In the static method, the modulus of elasticity is obtained from the slope of the stress-strain curve extrapolated back to zero strain using tensile test data. In the most common dynamic method, the modulus of elasticity is obtained from the natural frequency of a sound wave traversing a homogeneous bar of known dimensions. Ordinarily, the modulus of elasticity obtained by this dynamic method is about 10% greater than that obtained from the static method. The dynamically measured modulus of elasticity is probably the more valid estimate for most vibration analyses.

Poisson's ratio, being somewhat more difficult to measure and of less practical importance than the modulus of elasticity, is not known with two-digit accuracy for many materials. However, a good estimate of Poisson's ratio for most engineering materials is $\nu = 0.3$. Poisson's ratio only occasionally varies more than ±20% from this value, although there are certain exceptions such as lead ($\nu \approx 0.45$) and beryllium ($\nu \approx 0.01$ to 0.06). Common steels have $\nu = 0.29$ and aluminum alloys generally have $\nu = 0.33$. The maximum theoretical value of Poisson's ratio for an isotropic material is $\nu = 0.5$. The minimum theoretical value is $\nu = -1$; however, in reality, materials with negative ν are unknown (Ref. 16-1).

The density of iron, steel, and nickel-iron-chromium alloys rarely varies more than ±10% from $\rho = 8$ g/cm^3 (0.29 lb/in.^{3}) because the densities of iron, nickel, and chromium,

Material	Density
Iron	7.86 g/cm^3
Nickel	8.91 g/cm^3
Chromium	7.20 g/cm^3

are within about 10% of 8 g/cm^3.

Table 16-1. Modulus of Elasticity and Poisson's Ratio of Various Alloys.[a]

	304 or 316 Stainless Steel[b]		Ni-Fe-Cr Alloy 800		Ni-Cr-Fe-Cb Alloy 718	
Temp.	E[c]	ν	E[d]	ν	E[d]	ν
(°C)	(Pa × 10^{-9})	--	(Pa × 10^{-9})	--	(Pa × 10^{-9})	--
-100	204	--	207	0.336	--	--
25	194	0.265	196	0.339	200	0.293
50	193	0.267	194	0.339	198	0.291
100	190	0.272	193	0.342	196	0.286
150	187	0.276	190	0.347	193	0.283
200	184	0.280	186	0.353	190	0.279
250	181	0.284	183	0.355	188	0.276
300	177	0.288	179	0.357	185	0.273
350	173	0.292	175	0.360	182	0.272
400	168	0.295	169	0.362	179	0.271
450	164	0.299	164	0.365	176	0.271
500	159	0.302	159	0.369	173	0.272
550	154	0.306	156	0.368	170	0.274
600	149	0.309	152	0.373	167	0.277
650	144	0.313	146	0.377	163	0.282
700	139	0.317	140	0.381	159	--
750	133	0.321	133	0.388	155	--
800	128	0.324	127	0.396	--	--
(°F)	(psi × 10^{-6})	--	(psi × 10^{-6})	--	(psi × 10^{-6})	--
-200	29.7	--	30.0	0.335	--	--
50	28.1	0.265	28.4	0.339	29.0	0.293
100	28.0	0.266	28.2	0.339	28.8	0.292
200	27.6	0.271	27.7	0.341	28.4	0.287
300	27.2	0.276	27.1	0.347	28.0	0.283
400	26.6	0.281	26.6	0.353	27.6	0.279
500	26.1	0.285	26.1	0.355	27.1	0.275
600	25.4	0.289	25.4	0.357	26.7	0.273
700	24.8	0.293	24.8	0.361	26.3	0.271
800	24.1	0.297	24.1	0.363	25.8	0.271
900	23.3	0.301	23.3	0.367	25.3	0.271
1000	22.5	0.305	22.5	0.367	24.8	0.273
1100	21.7	0.309	21.7	0.372	24.2	0.277
1200	20.9	0.313	20.9	0.377	23.6	0.282
1300	20.0	0.317	20.1	0.381	23.0	--
1400	19.2	0.322	19.2	0.389	22.3	--
1500	18.3	0.327	18.3	0.398	--	--

(a) Density: $\rho \approx 7.97\ g/cm^3$ ($0.288\ lb/in.^3$).

(b) Also 310, 321, 347, and other austenitic steels.

(c) Measured statically.

(d) Measured dynamically.

Table 16-1. Modulus of Elasticity and Poisson's Ratio of Various Alloys.[a] (*Continued*)

Temp.	Ni-Cr-Fe Alloy 600 E	Ni-Cr-Fe Alloy 600 ν	2-1/4 Cr-1 Mo E[b]	2-1/4 Cr-1 Mo ν	A-286 (Alloy 660) E	A-286 (Alloy 660) ν
(°C)	(Pa × 10^{-9})	--	(Pa × 10^{-9})	--	(Pa × 10^{-9})	--
25	218	0.291	216	0.260	201	0.306
50	216	0.293	214	0.260	199	0.307
100	213	0.298	209	0.262	196	0.309
150	210	0.303	205	0.265	192	0.312
200	207	0.308	201	0.268	188	0.314
250	205	0.313	198	0.272	184	0.316
300	202	0.318	194	0.276	181	0.318
350	199	0.323	191	0.280	177	0.321
400	195	0.329	187	0.283	173	0.323
450	191	0.334	183	0.287	169	0.325
500	187	0.340	179	0.290	165	0.327
550	183	0.346	174	0.293	161	0.329
600	179	0.352	168	0.295	157	0.332
650	175	0.358	162	0.297	--	--
700	170	0.365	155	0.297	--	--
750	165	0.372	147	0.296	--	--
800	161	0.379	--	--	--	--
(°F)	(psi × 10^{-6})	--	(psi × 10^{-6})	--	(psi × 10^{-6})	--
50	31.8	0.290	31.4	0.260	29.2	0.306
100	31.4	0.292	31.1	0.260	29.0	0.307
200	30.9	0.298	30.4	0.262	28.5	0.309
300	30.4	0.303	29.7	0.265	27.9	0.312
400	30.0	0.308	29.1	0.268	27.3	0.314
500	29.6	0.314	28.6	0.272	26.6	0.317
600	29.2	0.320	28.0	0.277	26.0	0.319
700	28.6	0.325	27.4	0.281	25.4	0.321
800	28.0	0.331	26.8	0.285	24.8	0.324
900	27.4	0.338	26.1	0.289	24.1	0.326
1000	26.7	0.344	25.4	0.293	23.5	0.329
1100	26.0	0.351	24.5	0.295	22.8	0.331
1200	25.3	0.358	23.6	0.297	22.2	0.333
1300	24.6	0.365	22.4	0.297	--	--
1400	23.9	0.373	21.1	0.296	--	--
1500	23.1	0.381	--	--	--	--

[a] Density: $\rho \approx 7.97\ g/cm^3$ ($0.288\ lb/in.^3$).

[b] Dynamically measured modulus; statically measured modulus is lower.

The properties of wood vary with the moisture content, temperature, and direction of the applied stress. The values given in Table 16-5 apply for 12% moisture content, room temperature, and stress applied parallel to the fibers in the wood. Moisture content is the weight of the water contained in the wood, expressed as a percentage of the weight of the oven-dry wood [weight obtained after prolonged drying at 100°C (212°F)]. Air-dry wood has 12% to 15% moisture content, whereas

Table 16-2. Modulus of Elasticity of Steels.[a]

Temp.	Medium and Low Carbon, C ≤ 0.3, 3-1/2 Ni	High Carbon, C > 0.3	C-Mo-Low Cr, Cr ≤ 3	Intermediate Chromium 5 Cr to 9 Cr	Straight Chromium, 12 Cr, 17 Cr, 27 Cr
(°C)	(Pa × 10^{-9})	(Pa × 10^{-9})	(Pa × 10^{-9})	(Pa × 10^{-9})	(Pa × 10^{-9})
-200	207	214	214	203	212
-100	202	210	210	194	208
25	191	206	206	189	201
100	190	203	203	187	197
150	188	200	200	185	195
200	185	195	197	182	191
250	182	190	194	179	187
300	176	185	190	176	181
350	172	179	185	173	174
400	166	--	--	--	165
(°F)	(psi × 10^{-6})	(psi × 10^{-6})	(psi × 10^{-6})	(psi × 10^{-6})	(psi × 10^{-6})
-325	30.0	31.0	31.0	29.4	30.8
-100	29.0	30.4	30.4	28.1	29.8
70	27.9	29.9	29.9	27.4	29.2
200	27.7	29.5	29.5	27.1	28.7
300	27.4	29.0	29.0	26.8	28.3
400	27.0	28.3	28.6	26.4	27.7
500	26.4	27.4	28.0	26.0	27.0
600	25.7	26.7	27.4	25.4	26.0
700	24.8	25.4	26.6	24.9	24.8
800	23.4	--	--	--	23.1

(a) Poisson's ratio: $\nu \approx 0.29$; density: $\rho \approx 7.97$ g/cm^3 (0.288 lb/in.^{3}).

green wood has 40% to 100%. The density of the wood will change with moisture content both because of the additional weight of the water and the tendency of wood to swell with absorption of moisture. The changes in density with moisture are generally small, with the density increasing with increasing moisture content but at a lower rate. As a result of the grain in wood, the properties of wood vary with the direction of the applied load. Ordinarily, the maximum allowable stress parallel to the fibers is about four times the maximum allowable stress perpendicular to the fibers.

The properties of the plastics given in Table 16-6 should only be taken as rough guides because of the variation between types of many subspecies, such as filled and reinforced plastics, which are available.

Tables 16-1 through 16-6 were assembled in part from Refs. 16-2 through 16-6. These references can be consulted for other information, such as yield strength. In addition, one can usually obtain material properties data from manufacturers and suppliers.

Table 16-3. Modulus of Elasticity and Density of Nonferrous Alloys.[a]

	Modulus of Elasticity						
	Aluminum Alloys						
Temp.	3003, 3004, 6063	5052, 5154, 5454, 5456	5083, 5086	6061	2014, 2024	Copper	Unalloyed Titanium
(°C)	(Pa × 10^{-9})	(Pa × 10^{-9})	(Pa × 10^{-9})	(Pa × 10^{-9})	(Pa × 10^{-9})	(Pa × 10^{-9})	(Pa × 10^{-9})
-200	76.5	77.9	78.6	76.5	81.4	117	--
-100	72.7	74.1	74.8	72.7	77.0	115	--
25	68.9	70.3	71.0	68.9	72.4	110	107
100	65.8	67.0	68.6	65.8	72.2	107	103
150	62.7	62.1	65.5	62.7	68.3	106	99.3
200	57.7	55.7	60.4	57.7	63.8	104	95.5
250	--	--	--	--	--	102	91.8
300	--	--	--	--	--	98.9	87.5
350	--	--	--	--	--	95.8	83.2
400	--	--	--	--	--	--	79.3
(°F)	(psi × 10^{-6})	(psi × 10^{-6})	(psi × 10^{-6})	(psi × 10^{-6})	(psi × 10^{-6})	(psi × 10^{-6})	(psi × 10^{-6})
-325	11.1	11.3	11.4	11.1	11.8	17.0	--
-100	10.4	10.6	10.7	10.4	11.0	16.5	--
70	10.0	10.2	10.3	10.0	10.5	16.0	15.5
200	9.6	9.8	10.0	9.6	10.3	15.6	15.0
300	9.1	9.0	9.5	9.1	9.9	15.4	14.4
400	8.3	8.0	8.7	8.3	9.2	15.1	13.8
500	--	--	--	--	--	14.7	13.2
600	--	--	--	--	--	14.2	12.5
700	--	--	--	--	--	13.7	11.8
800	--	--	--	--	--	--	11.2
Density							
g/cm^3	2.74[b]	2.66	--	2.71	2.80	8.97	4.54
lb/in.^{3}	0.099	0.096	--	0.098	0.101	0.324	0.164

[a] Poisson's ratio ≈ 0.33 for aluminum alloys.

[b] Density = 2.71 g/cm^3 (0.098 lb/in.^{3}) for 6063.

Table 16-4. Modulus of Elasticity and Density of Various Metals.

Metal	Symbol	Density $\frac{g}{cm^3}$	Density $\frac{lb}{in.^3}$	Modulus of Elasticity Pa × 10^{-9}	Modulus of Elasticity psi × 10^{-6}
Aluminum	Al	2.70	0.0975	69	10.0
Antimony	Sb	6.62	0.239	78	11.3
Arsenic	As	5.73	0.207	76	11.0
Barium	Ba	3.60	0.13	12	1.8
Beryllium	Be	1.82	0.066	290	42.0
Bismuth	Bi	9.80	0.354	32	4.6
Brass, admiralty	--	8.50	0.307	110	16.0
Brass, aluminum	--	8.33	0.301	110	16.0
Brass, naval	--	8.41	0.304	100	15.0
Brass, red	--	8.44	0.305	120	17.0
Bronze, aluminum	--	7.39	0.267	120	17.0
Bronze, leaded	--	8.83	0.319	120	17.0
Bronze, phosphor	--	8.86	0.320	120	17.0
Boron	B	2.30	0.083	410	60.0
Cadmium	Cd	8.66	0.313	55	8.0
Calcium	Ca	1.55	0.056	21	3.0

Table 16-4. Modulus of Elasticity and Density of Various Metals. *(Continued)*

Metal	Symbol	Density		Modulus of Elasticity	
		$\frac{g}{cm^3}$	$\frac{lb}{in.^3}$	Pa × 10^{-9}	psi × 10^{-6}
Carbon	C	2.21	0.080	4.8	0.7
Chromium	Cr	7.20	0.260	25	36.0
Cobalt	Co	8.86	0.320	21	30.0
Copper	Cu	8.97	0.324	110	16.0
30% cupro nickel	--	8.94	0.323	150	22.0
20% cupro nickel	--	8.94	0.323	140	20.0
10% cupro nickel	--	8.91	0.322	120	18.0
Gallium	Ga	5.98	0.216	6.9	1.0
Germanium	Ge	5.31	0.192	79	11.4
Gold	Au	19.3	0.698	74	12.0
Hafnium	Hf	13.1	0.473	140	20.0
Indium	In	7.31	0.264	11	1.6
Iridium	Ir	22.5	0.813	520	75.0
Iron, pure	Fe	7.86	0.284	210	29.8
Iron, gray cast, C.E.[a] = 4.8	--	7.86	0.284	55	8.0
Iron, gray cast, C.E.[a] = 4.0	--	7.86	0.284	92	13.3
Iron, gray cast, C.E.[a] = 3.3	--	7.86	0.284	140	20.0
Lanthanum	La	6.17	0.223	34	5.0
Lead	Pb	11.3	0.409	18	2.6
Lithium	Li	0.53	0.019	12	1.7
Magnesium	Mg	1.74	0.0628	45	6.5
Manganese	Mn	7.42	0.268	160	23.0
Molybdenum	Mo	10.2	0.369	280	40.0
Nickel	Ni	8.91	0.322	210	30.0
Nickel-Copper (80-20)	--	8.39	0.303	210	31.0
Niobium	Nb	8.58	0.310	100	15.0
Osmium	Os	22.5	0.813	550	80.0
Palladium	Pd	12.0	0.434	120	18.0
Phorosphorus	P	1.82	0.0658	--	--
Platinum	Pt	21.5	0.775	140	21.0
Potassium	K	0.86	0.031	3.4	0.5
Rhenium	Re	21.0	0.76	520	70.0
Rhodium	Rh	12.5	0.450	370	54.0
Ruthenium	Ru	12.2	0.441	470	68.0
Selenium	Se	4.82	0.174	58	8.4
Silicon	Si	2.33	0.084	110	16.0
Silver	Ag	10.5	0.379	76	11.0
Sodium	Na	0.97	0.035	9.0	1.3
Tantalum	Ta	16.6	0.600	190	27.0
Tellurium	Te	6.23	0.225	41	6.0
Thallium	Tl	11.9	0.428	8.3	1.2
Thorium	Th	11.7	0.422	79	11.0
Tin	Sn	7.31	0.264	41	6.0
Titanium	Ti	4.54	0.164	120	16.8
Tungsten	W	19.3	0.697	340	50.0
Uranium	U	19.0	0.687	200	29.7
Vanadium	V	6.00	0.217	130	19.0
Zinc	Zn	7.14	0.258	83	12.0
Zirconium	Zr	6.37	0.230	76	11.0

[a] C.E. = % carbon + 0.33 (% silicon + % phosphorus).

Table 16-5. Modulus of Elasticity and Density of Woods.[a]

Wood	Density		Modulus of Elasticity	
	$\frac{g}{cm^3}$	$\frac{lb}{in.^3}$	Pa × 10^{-9}	psi × 10^{-6}
Hardwoods				
Ash, white	0.59	0.021	12	1.7
Ash, black	0.48	0.017	9.0	1.3
Basswood	0.37	0.014	10	1.5
Beech	0.63	0.023	12	1.7
Birch, yellow	0.60	0.022	13	1.9
Cherry, black	0.53	0.019	9.7	1.4
Chestnut	0.45	0.016	8.3	1.2
Cottonwood, eastern	0.39	0.014	8.3	1.2
Elm, American	0.50	0.018	9.0	1.3
Elm, rock	0.63	0.023	10	1.5
Hickory, shagbark	0.71	0.026	14	2.1
Locust	0.69	0.025	14	2.1
Mahogany, C. American	0.51	0.026	10	1.5
Maple, black	0.57	0.021	12	1.7
Maple, sugar	0.63	0.023	12	1.7
Oak, red	0.63	0.023	11	1.6
Oak, white	0.66	0.024	11	1.6
Poplar, yellow	0.43	0.016	10	1.5
Sweetgum	0.50	0.018	9.7	1.4
Tupelo, black	0.51	0.018	8.3	1.2
Walnut, black	0.56	0.020	11	1.6
Softwoods				
Cedar, western	0.35	0.013	7.6	1.1
Cypress	0.46	0.017	9.7	1.4
Douglas fir	0.48	0.017	12	1.8
Fir, balsam	0.36	0.013	7.6	1.1
Hemlock, western	0.45	0.016	10	1.5
Larch	0.52	0.019	12	1.8
Pine, eastern white	0.37	0.013	8.3	1.2
Pine, ponderosa	0.41	0.015	9	1.3
Pine, southern	0.52	0.019	12	1.8
Pine, western	0.39	0.014	10	1.5
Redwood	0.38	0.014	8.3	1.2
Spruce, eastern	0.40	0.014	9	1.3
Spruce, sitka	0.40	0.014	10	1.5
Tropical				
Balsa	0.14	0.005	3.5	0.5
Ebony	1.08	0.039	19	2.7
Lemon wood	0.78	0.028	16	2.3
Lignum Vitae	1.09	0.039	--	--
Mahogany, C. American	0.50	0.018	10	1.5
Teak	0.65	0.023	12	1.7
Composition Board				
Hardwood, fibrous tempered	1.0–1.3	0.04–0.05	5–8	0.8–1.2
Particle board	0.4–1.3	0.01–0.05	1–7	0.15–1.0

[a] Dry wood: moisture content = 12%.

Table 16-6. Density and Modulus of Elasticity of Plastics, Glass, and Fibers.

Material	Density		Modulus of Elasticity	
	$\frac{g}{cm^3}$	$\frac{lb}{in.^3}$	$Pa \times 10^{-9}$	$psi \times 10^{-6}$
Thermoplastic Plastics				
Cellulose acetate	1.3	0.047	0.7-2	0.1-0.3
Cellulose acetate butyrate	1.2	0.043	0.7-2	0.1-0.3
Cellulose nitrate	1.4	0.051	1-3	0.2-0.4
Ethyl cellulose	1.1	0.040	0.7-3	0.1-0.5
Methyl methacrylate	1.2	0.043	2-3	0.3-0.5
Nylon 6,6	1.1	0.040	1-3	0.2-0.4
Vinyl chloride	1.4	0.049	2-3	0.3-0.5
Polyethylene	0.9-1.0	0.033-0.036	0.6-1	0.1-0.2
Polymethylmethacrylate	1.2	0.043	3.1	0.45
Polypropylene	0.9	0.033	0.7-1	0.1-0.2
Polystyrene	1.0-1.1	0.036-0.04	3	0.4-0.5
Polytetrafluoroethylene (Teflon)	2.1-2.2	0.076-0.079	0.4	0.06
ABS	1.0-1.2	0.036-0.043	1.4-3	0.2-0.4
Thermosetting Plastics				
Epoxy resin	1.1-1.2	0.04-0.043	1-4	0.2-0.6
Phenolic resin	1.3	0.047	2-3	0.3-0.5
Polyester resin	1.1-1.5	0.04-0.054	2-4	0.3-0.6
Laminates				
E-glass fiber/epoxy	1.8-2.2	0.065-0.079	20-60	3-8
E-glass fiber/phenolic	1.7-1.9	0.061-0.069	10-30	2-5
E-glass fiber/polyester	1.5-1.9	0.054-0.069	7-30	1-4
Glass				
Glass sheet	2.3-2.6	0.083-0.094	70	10
Fibers				
E-glass	2.54	0.0918	72	10.5
S-glass	2.49	0.0900	85.5	12.4
D-glass	2.16	0.078	52	7.5
Boron (on tungsten)	2.6	0.094	410	60
Graphite	1.4-1.6	0.051-0.058	170-340	25-50
Quartz	2.2	0.079	70	10

16.2. LIQUIDS

The density, kinematic viscosity, and speed of sound of fresh water and sea water at atmospheric pressure are given in Tables 16-7 and 16-8, respectively. In Table 16-9 these properties are given for various liquids at 20°C (68°F) and atmospheric pressure. The density of most liquids is nearly independent of pressure and temperature for moderate pressures and temperatures below the boiling point. The kinematic vis-

Table 16-7. Properties of Fresh Water.[a]

Temperature	Density	Kinematic Viscosity	Speed of Sound
(°C)	$\left(\frac{kg}{m^3}\right)$	$\left(\frac{m^2}{sec} \times 10^7\right)$	$\left(\frac{m}{sec}\right)$
0	999.9	17.87	1403
5	1000.0	15.19	1427
10	999.7	13.07	1447
15	999.1	11.40	1465
20	998.2	10.04	1481
25	997.1	8.930	1495
30	995.7	8.009	1507
35	994.1	7.237	1517
40	992.2	6.580	1526
45	990.3	6.018	1534
50	988.1	5.534	1541
55	985.7	5.113	1546
60	983.2	4.745	1552
65	980.5	4.421	1553
70	977.8	4.133	1555
75	974.9	3.878	1555
80	971.8	3.650	1555
85	968.7	3.445	1552
90	965.4	3.260	1550
95	961.9	3.093	1547
100	958.4	2.940	1543
(°F)	$\left(\frac{lb}{ft^3}\right)$	$\left(\frac{ft^2}{sec} \times 10^5\right)$	$\left(\frac{ft}{sec}\right)$
32	62.42	1.924	4603
40	62.42	1.663	4672
50	62.38	1.408	4748
60	62.34	1.209	4814
70	62.27	1.052	4871
80	62.19	0.9264	4919
90	62.11	0.8233	4960
100	62.00	0.7381	4995
110	61.84	0.6671	5057
120	61.73	0.6064	5049
130	61.55	0.5550	5071
140	61.38	0.5106	5091
150	61.19	0.4722	5100
160	61.01	0.4385	5101
170	60.79	0.4023	5100
180	60.57	0.3828	5195
190	60.35	0.3595	5090
200	60.13	0.3386	5089
212	59.83	0.3165	5062

[a]At one atmosphere pressure.

Table 16-8. Properties of Sea Water.[a]

Temperature	Density	Kinematic Viscosity	Speed of Sound
(°C)	$\left(\frac{kg}{m^3}\right)$	$\left(\frac{m^2}{sec} \times 10^6\right)$	$\left(\frac{m}{sec}\right)$
0	1028.	1.83	1449.
5	1028.	1.56	1471.
10	1027.	1.35	1490.
15	1026.	1.19	1507.
20	1025.	1.05	1521.
25	1023.	0.946	1534.
30	1022.	0.853	1546.
(°F)	$\left(\frac{lb}{ft^3}\right)$	$\left(\frac{ft^2}{sec} \times 10^5\right)$	$\left(\frac{ft}{sec}\right)$
32	64.18	1.97	4754.
40	64.16	1.71	4818.
50	64.11	1.46	4889.
60	64.04	1.26	4949.
70	63.95	1.11	5002.
80	63.85	0.983	5047.
90	63.72	0.877	5086.

[a]At one atmosphere pressure. Salinity = 35 parts per thousand.

Table 16-9. Properties of Various Liquids.[a]

Liquid	Density		Kinematic Viscosity		Speed of Sound	
	$\frac{kg}{m^3}$	$\frac{lb}{ft^3}$	$\frac{m^2}{sec} \times 10^7$	$\frac{ft^2}{sec} \times 10^5$	$\frac{m}{sec}$	$\frac{ft}{sec}$
Alcohol (ethyl)	790	49.3	15	1.6	1150	3770
Benzene	878	54.8	7.4	0.80	1322	4340
Gasoline	670	42	4.6	0.5	–	–
Glycerin	1260	78.7	11900	1280	1980	6500
Kerosene	804	50.2	23	2.5	1450	4770
Mercury	13600	849	1.2	0.13	1450	4760
Oil, caster	950	59	10000	1090	1540	5050
Oil, crude[b]	850	53	70	7.5	–	–
Oil, SAE 30	920	57	730	78	1290	4240
Turpentine	870	54	17	1.9	1250	4100
Water (fresh)	998.2	62.27	10.0	1.07	1481	4859
Water (sea)	1026	64.05	12.6	1.36	1500	4920

[a]Pressure = 1 atm; temperature = 20°C (68°F), except sea water at 13°C (55°F).
[b]Specific gravity = 0.86.

cosity of most liquids decreases sharply with increasing temperature, as shown in Table 16-7. Kinematic viscosity is nearly independent of pressure for most liquids at moderate pressures, below the boiling point; however, extreme pressures can induce significant increases in viscosity (Ref. 16-7, p. 430). Similarly, high pressures can significantly increase the speed of sound in a liquid. For example, the speed of sound in fresh water as a function of pressure at 20°C (68°F) is (Ref. 16-8):

Pressure (atm):	1	272	544	816
Speed of sound (m/sec):	1483	1529	1575	1621

Unfortunately, there is no general theory of fluid behavior available for predicting the properties of liquids. All techniques for estimating the properties of liquids are based to a greater or lesser degree on experimental data. Techniques for estimating the properties of liquids are given in Ref. 16-7. Data for the speed of sound in fresh water and sea water at various temperatures and pressures are given in Refs. 16-8, 16-9, and 16-10. Data for the viscosity and surface tension of a number of liquids are given in Refs. 16-11 and 16-12. Reference 16-17 is a guide to the literature on thermodynamic properties of solids, liquids, and gases.

16.3. GASES

The kinetic molecular theory of gases provides a general theory for the prediction of the density and speed of sound of gases at low pressures. The kinetic theory of gases results in the ideal gas equation, which can be used to relate the density of a gas, ρ, to the pressure, P, absolute temperature, T, and molecular weight, W, of the gas:

$$\rho = \frac{PW}{RT}. \tag{16-1}$$

R is the universal gas constant. The value of R depends on the units chosen for P, T, and ρ:

Units for ρ	Units for P	Units for T	R
$\frac{\text{kilograms}}{\text{meter}^3}$	Pascals	°Kelvin	$\frac{8315 \text{ Pascal-m}^3}{\text{kg-mole-°K}}$
$\frac{\text{grams}}{\text{liter}}$	Atmospheres	°Kelvin	$\frac{0.08206 \text{ atm-liter}}{\text{g-mole-°K}}$
$\frac{\text{slugs}}{\text{foot}^3}$	$\frac{\text{pounds}}{\text{foot}^2}$	°Rankine	$\frac{49720 \text{ lb-ft}}{\text{slug-mole-°R}}$

The molecular weight, W, always is expressed on a per mole basis and so is independent of the units used for the other parameters in Eq. 16-1. The molecular weight of various gases can be found in Table 16-10. The absolute temperature scales are degrees Kelvin (°K) and degrees Rankine (°R):

$$°K = °C + 273.16,$$

$$°R = °F + 459.69, \tag{16.2}$$

Table 16-10. Properties of Various Gases.

Gas	Molecular Formula	Molecular Weight, W	Ratio of Specific Heats,[a] γ
Air	(b)	28.97	1.400
Argon	Ar	39.94	1.667
Butane	C_4H_{10}	58.12	1.09
Carbon dioxide	CO_2	44.01	1.285
Carbon monoxide	CO	28.01	1.399
Ethane	C_2H_6	30.07	1.183
Ethylene	C_2H_4	28.05	1.208
Helium	He	4.003	1.667
Hydrogen	H_2	2.016	1.404
Methane	CH_4	16.04	1.32
Neon	Ne	20.18	1.667
Nitrogen	N_2	28.02	1.400
Octane	C_8H_{18}	114.2	1.044
Oxygen	O_2	32.00	1.395
Propane	C_3H_8	44.10	1.124
Steam	H_2O	18.02	1.329

[a] At 27°C (80°F), Ref. 16-16.

[b] 78.03 N_2, 20.99 O_2, 0.94 Ar, 0.03 CO_2, 0.01 other (percent by volume).

where °C is the temperature in degrees Celsius and °F is the temperature in degrees Fahrenheit. The pressure corresponding to one standard atmosphere is:

$$1 \text{ atm} = 101325 \text{ Pa},$$
$$= 1.01325 \times 10^6 \text{ dynes/cm}^2,$$
$$= 14.6959 \text{ lb/in.}^2.$$

Further discussion of units can be found in Chapter 3.

The equivalent molecular weight of a mixture of gases is:

$$W_{\text{Mixture}} = \sum_{i=1}^{N} x_i W_i,$$

where there are N components to a gas mixture, each component having a molecular weight W_i (Table 16-10) and representing mole fraction x_i of the total gas mixture ($\Sigma x_i = 1$). The mole fraction x_i, or equivalently the volume fraction, is the number of moles of component i per mole of the gas mixture. One mole is a mass of gas equal to the molecular weight of that gas in grams. For example, air at ordinary temperatures has mole fractions of approximately 78% nitrogen, 21% oxygen, and

1% argon. Thus, from the above equation and Table 16-10, the molecular weight of air is:

$$W_{air} = 0.78(28.02) + 0.21(32.00) + 0.01(39.94),$$
$$= 28.97,$$

which agrees with the value in Table 16-10.

As an example, the density of air at atmospheric pressure and temperature (20°C, 68°F) can be calculated as follows:

$$P = 101300 \text{ Pa } (2116 \text{ lb/ft}^2),$$
$$W = 28.97/\text{mole (Table 16-10)},$$
$$T = 293°\text{K } (527.7°\text{R}).$$

Using Eq. 16-1 and the appropriate value of R, above, gives:

$$\rho = 1.20 \text{ kg/m}^3 \; (0.00234 \text{ slug/ft}^3; 0.0752 \text{ lb/ft}^3).$$

The speed of sound can also be computed from the ideal gas theory. If the motion of sound waves is modeled as an adiabatic reversible process, it can be shown that the speed of waves in a gas is (Eq. 16-1, Ref. 16-13, p. 47):

$$c = \left(\frac{\gamma P}{\rho}\right)^{1/2},$$
$$= \left(\frac{\gamma RT}{W}\right)^{1/2}, \qquad (16\text{-}3)$$

where

T = absolute temperature,
R = universal gas constant,
W = molecular weight of the gas,
γ = ratio of the specific heat of the gas at constant pressure to the specific heat of the gas at constant volume (dimensionless).

The value of the universal gas constant, R, depends on the units employed in Eq. 16-3:

Units for c	Units for T	R
$\frac{\text{meters}}{\text{second}}$	°Kelvin	$\frac{8315 \text{ m}^2}{\text{sec}^2\text{-mole-}°\text{K}}$
$\frac{\text{feet}}{\text{second}}$	°Rankine	$\frac{49720 \text{ ft}^2}{\text{sec}^2\text{-mole-}°\text{R}}$

The molecular weight, W, is always expressed on a per mole basis and so is independent of the units used for other parameters. The molecular weight of various gases can be found in Table 16-10. The ratio of specific heats, γ, generally varies only within narrow limits. The kinetic theory of gases predicts that for a monotonic gas, such as helium or argon (Ref. 16-13, p. 42),

$$\gamma = 1.667,$$

and for a diatomic gas, such as nitrogen or oxygen,

$$\gamma = 1.400.$$

In general, the larger the molecule, the lower the value of γ; however, $\gamma = 1$ is the theoretical lower limit. γ can be expected to vary slightly with temperature. γ values for various gases are given in Table 16-10. Since none of the quantities in Eq. 16-3 are directly dependent on pressure, the speed of sound of a gas at low pressures will be substantially independent of pressure.

The adiabatic sound velocity of Eq. 16-3 suggests that the speed of sound is independent of the amplitude and the frequency of the sound wave. This has been found to be in good agreement with experimental data in nearly all practical cases as long as the fluctuating pressures associated with the sound wave are much less than the mean pressure. However, sound waves with very high frequencies may be transmitted isothermally rather than adiabatically because the larger thermal gradients and shorter distances between maximums and minimums in the high-frequency waves can permit significant heat transfer. The isothermal sound velocities can be obtained by setting $\gamma = 1$ in Eq. 16-3 and thus are somewhat lower than the adiabatic sound velocities.

As an example, the speed of sound in air at room temperature (20°C, 68°F) is computed from Eq. 16-3 as follows:

$$\gamma = 1.400 \text{ (from Table 16-10)},$$
$$T = 293°\text{K} \ (527.7°\text{R}),$$
$$W = 28.97/\text{mole (Table 16-10)}.$$

Using Eq. 16-3 and the appropriate value for R, above, yields:

$$c = 343.2 \text{ m/sec (1126 ft/sec)}.$$

It should be noted that Eqs. 16-1 and 16-3 are based on ideal gas behavior, and these equations are exact only in the limit of low pressures for real gases. While these equations provide accurate estimates for most common gases up to pressures of several atmospheres and temperatures well above the saturation temperature, they must fail at very high pressures. The degree of error and the onset of large deviations can be determined only on a case-by-case basis. Methods for analyzing the properties of non-ideal gases are discussed in Ref. 16-7, pp. 46-82.

At low pressures the absolute viscosity of gases, μ, is independent of pressure and increases with temperature. For example, the viscosity of helium is approximately:

$$\mu = 3.953 \times 10^{-7}\, T_K^{0.687}\ \frac{\text{N-sec}}{\text{m}^2},$$

$$= 5.516 \times 10^{-9}\, T_R^{0.687}\ \frac{\text{lb-sec}}{\text{ft}^2},$$

where T_K is the number of degrees in the temperature in degrees Kelvin, T_R is the number of degrees in the temperature in degrees Rankine, and the units of frame 6 of Table 3-1 have been employed in the second line of this equation. Thus, the viscosity of helium at room temperature (20°C) is 1.96×10^{-5} N-sec/m² (4.09×10^{-7} lb-sec/ft²). The kinematic viscosity, ν, is determined by dividing the viscosity by the density of the gas:

$$\nu = \frac{\mu}{\rho}. \tag{16-4}$$

Consistent sets of units are given in Table 3-1. Since the density of a gas at constant pressure decreases with increasing temperature (Eq. 16-1), the kinematic viscosity of a gas at constant pressure will rise sharply with increasing temperature.

Table 16-11. Properties of Air.[a]

Metric Units

Temp. (°C)	Density $\left(\frac{kg}{m^3}\right)$	Kinematic Viscosity $\left(\frac{m^2}{sec} \times 10^5\right)$	Ratio of Specific Heats, γ	Speed of Sound $\left(\frac{m}{sec}\right)$
-50	1.582	1.01	1.402	299.7
-40	1.514	1.04	1.401	306.2
-30	1.452	1.10	1.401	312.7
-20	1.395	1.17	1.401	319.1
-10	1.342	1.22	1.401	325.3
0	1.292	1.32	1.401	331.4
5	1.269	1.36	1.401	334.4
10	1.247	1.41	1.401	337.4
15	1.225	1.47	1.401	340.4
20	1.204	1.51	1.401	343.3
25	1.184	1.56	1.401	346.3
30	1.165	1.60	1.400	349.1
35	1.146	1.63	1.400	351.9
40	1.127	1.66	1.400	354.7
50	1.109	1.76	1.400	360.3
60	1.060	1.86	1.399	365.7
70	1.029	1.97	1.399	371.2
80	0.9996	2.07	1.399	376.6
90	0.9721	2.20	1.398	381.7
100	0.9461	2.40	1.397	386.9
120	0.8979	2.60	1.396	396.9
140	0.8545	2.56	1.394	406.6
160	0.8150	2.84	1.393	416.5
180	0.7790	3.12	1.392	425.5
200	0.7461	3.39	1.390	434.5
220	0.7159	3.67	1.387	443.1
240	0.6880	3.95	1.386	451.8
260	0.6622	4.23	1.385	460.4

(a) At atmospheric pressure, 101325 Pa; c, γ, and the product $\rho\nu$ are nearly independent of pressure at low pressures.

Table 16.11. Properties of Air[a] (*Continued*)

British Units

Temp. (°F)	Density $\left(\frac{lb}{ft^3}\right)$	Kinematic Viscosity, ν $\left(\frac{ft^2}{sec} \times 10^4\right)$	Ratio of Specific Heats, γ	Speed of Sound, c $\left(\frac{ft}{sec}\right)$
-100	0.1103	0.907	1.402	930.1
-80	0.1045	0.998	1.402	955.6
-60	0.09929	1.09	1.402	980.5
-40	0.09456	1.12	1.401	1004
-20	0.09026	1.19	1.401	1028
0	0.08633	1.26	1.401	1051
10	0.08449	1.31	1.401	1062
20	0.08273	1.36	1.401	1074
30	0.08104	1.42	1.401	1085
40	0.07942	1.46	1.401	1096
50	0.07786	1.52	1.401	1106
60	0.07636	1.58	1.401	1117
70	0.07492	1.64	1.401	1128
80	0.07353	1.69	1.400	1138
90	0.07219	1.74	1.400	1149
100	0.07090	1.79	1.400	1159
120	0.06846	1.89	1.400	1180
140	0.06617	2.01	1.399	1200
160	0.06404	2.12	1.399	1220
180	0.06204	2.25	1.399	1239
200	0.06016	2.40	1.398	1258
250	0.05592	2.80	1.396	1304
300	0.05224	3.06	1.394	1348
350	0.04901	3.36	1.391	1390
400	0.04616	3.65	1.389	1431
500	0.04135	4.56	1.385	1510

(a) At atmospheric pressure, 14.6959 lb/in.^{2}; c, γ, and the product $\rho\nu$ are nearly independent of pressure at low pressures.

Procedures for estimating the viscosity of gases from thermodynamic principles are outlined in Ref. 16-7, pp. 395-428. Numerous data on the viscosity of gases are given in Refs. 16-11, 16-12, and 16-14. Table 16-11 gives the kinematic viscosity and other properties of air at atmospheric pressure. Additional discussion of viscosity can be found in Ref. 16-15. Reference 16-17 is a guide to the literature on thermodynamic properties of solids, liquids, and gases. Properties of various common gases are given in Ref. 16-18.

REFERENCES

16-1. Fung, Y. C., *Foundations of Solid Mechanics*, Prentice-Hall, Englewood Cliffs, N.J., 1965, p. 353.
16-2. Baumeister, T. (ed.), *Standard Handbook for Mechanical Engineers*, 7th ed., McGraw-Hill, New York, 1967.

16-3. *The ASME Boiler and Pressure Vessel Code*, The American Society of Mechanical Engineers, New York.
16-4. "1978 Materials Selector," *Materials Engineering*, **86**, No. 6 (Nov. 1977), Reinhold Publishing Co., Stanford, Conn.
16-5. *Metals Handbook*, 8th ed., American Society for Metals, Metals Park, Novelty, Ohio, 1961.
16-6. American Institute of Timber Construction, *Timber Construction Manual*, John Wiley, New York, 1974.
16-7. Reid, R. C., and T. K. Sherwood, *The Properties of Gases and Liquids, Their Estimation and Correlation*, 2nd ed., McGraw-Hill, New York, 1966.
16-8. Wilson, J., "Speed of Sound in Distilled Water as a Function of Temperature and Pressure," *J. Acoust. Soc. Am.* **31**, 1067–1070 (1959).
16-9. Wilson, J., "Speed of Sound in Sea Water as a Function of Temperature, Pressure and Salinity," *J. Acoust. Soc. Am.* **32**, 641–644 (1960).
16-10. Kinsler, L. E., and A. R. Frey, *Fundamentals of Acoustics*, 2nd ed., John Wiley, New York, 1962.
16-11. Weast, R. C. (ed.), *Handbook of Chemistry and Physics*, The Chemical Rubber Company, Cleveland, Ohio, 1978.
16-12. Perry, J. H., *Chemical Engineers' Handbook*, 4th ed., McGraw-Hill, New York, 1963.
16-13. Shapiro, A. H., *The Dynamics and Thermodynamics of Compressible Fluid Flow*, Vol. 1, The Ronald Press, New York, 1953.
16-14. Keenan, J. H., and J. Kaye, *Gas Tables*, John Wiley, New York, 1948.
16-15. Schlichting, H., *Boundary Layer Theory*, 6th ed., McGraw-Hill, New York, 1968, pp. 7–9.
16-16. Van Wylen, J., and R. E. Sonntag, *Fundamentals of Classical Thermodynamics*, John Wiley, New York, 1973, p. 683.
16-17. Touloukian, Y. S. (ed.), "Thermophysical Properties, Research Literature, Retrieval Guide," 2nd ed., Plenum Press, New York, 1967.
16-18. Hilsenrath, J., *et al.*, *Tables of Thermodynamic and Transport Properties*, Pergamon Press, New York, 1960.

Appendix A

THE RELATIONSHIP BETWEEN STATIC DEFLECTION AND FUNDAMENTAL NATURAL FREQUENCY

General Case. As noted in Chapter 4, the natural frequency of a structure is the result of the exchange of kinetic and potential energy within the structure. The kinetic energy is associated with the motion of structural mass, and the potential energy is associated with the strain energy stored in the elastic structure during deformation. One measure of the strain energy stored in a structure is the static deflection of the structure under the acceleration of gravity. Thus, it is reasonable to believe that a relationship exists between the static deflection of a structure under its own weight and the natural frequency of the structure. This relationship can be used to estimate the fundamental natural frequency of complex structures whose static deflection is known.

For example, the static deflection of the mass in the spring-mass system shown in Fig. A-1 (a) due to gravity is (Eq. 4-9):

$$\delta_s = \frac{Mg}{k}, \tag{A-1}$$

where M is the mass, g is the acceleration due to gravity, and k is the spring constant. The natural frequency of this system is (Eq. 4-6 or frame 1 of Table 6-2):

$$f = \frac{1}{2\pi}\left(\frac{k}{M}\right)^{1/2} \tag{A-2}$$

Incorporating Eq. A-1 into Eq. A-2 to eliminate the spring constant k gives:

$$f = \frac{1}{2\pi}\left(\frac{Mg}{\delta_s M}\right)^{1/2}$$

or

$$f = \frac{1}{2\pi}\left(\frac{g}{\delta_s}\right)^{1/2} \text{ Hz.} \tag{A-3}$$

Equation A-3 allows the natural frequency of the structure to be expressed solely in terms of the acceleration due to gravity and the maximum static deflection, δ_s, that gravity produces. Of course, it is not necessary to limit the application of Eq. A-3 to systems which vibrate in the vertical plane. All that is required is an estimate of the maximum static deflection of the structure produced by a uniform 1-g acceleration field applied in the plane of vibration.

The accuracy of Eq. A-3 depends on the degree to which the static deformation under gravity conforms to the mode shape of vibration. Equation A-3 is exact for the simple spring-mass system shown in Fig. A-1 (a) and generally underestimates the natural frequency of more complex structures.

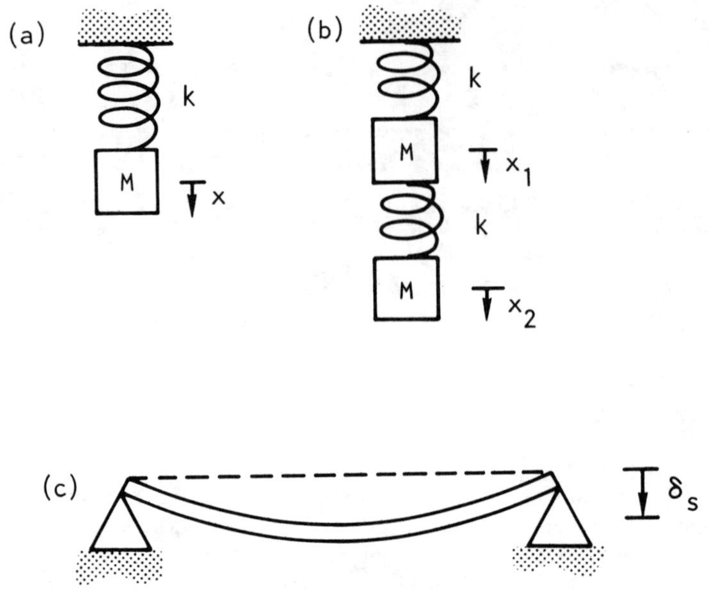

Fig. A-1. Elastic systems.

Two-Spring, Two-Mass System Example. Consider the system shown in Fig. A-1 (b). The maximum static deflection of this system under gravity is the deflection of the lower mass, which is easily computed to be:

$$\delta_s = \frac{2\ Mg}{k} + \frac{Mg}{k} = \frac{3\ Mg}{k}.$$

Using Eq. A-3, the fundamental natural frequency predicted from this deflection is

$$f = \frac{1}{2\pi}\frac{1}{3^{1/2}}\left(\frac{k}{M}\right)^{1/2} \text{ Hz},$$

$$= \frac{0.5774}{2\pi}\left(\frac{k}{M}\right)^{1/2} \text{ Hz},$$

and the mode shape, predicted from the static deflection, is

$$\begin{pmatrix}\tilde{x}_1\\ \tilde{x}_2\end{pmatrix} = \begin{pmatrix}1\\ 1.5\end{pmatrix}.$$

The exact solution for the fundamental natural frequency of this system is (frame 2 of Table 6-2):

$$f = \frac{0.6180}{2\pi}\left(\frac{k}{M}\right)^{1/2} \text{ Hz},$$

and the exact mode shape in the fundamental mode is

$$\begin{pmatrix}\tilde{x}_1\\ \tilde{x}_2\end{pmatrix} = \begin{pmatrix}1\\ 1.618\end{pmatrix}.$$

Thus, Eq. A-3 underestimates the natural frequency of this system by about 6.6%. Equation A-3 cannot be used to estimate the natural frequency of the second, higher, mode of this system.

Beam Example. Consider the slender, uniform pinned-pinned beam shown in Fig. A-1(c). The static deflection of this beam at midspan due to gravity is (Ref. A-1):

$$\delta_s = \frac{5}{384}\frac{mgL^4}{EI} = \frac{1}{(2.960)^4}\frac{mgL^4}{EI},$$

where m is the mass per unit length of the beam, E is the modulus of elasticity, I is the area moment of inertia of the cross section, and L is the length of the beam. Using this result, the fundamental natural frequency of the beam can be estimated from Eq. A-3 to be:

$$f = \frac{(2.960)^2}{2\pi L^2}\left(\frac{EI}{m}\right)^{1/2} \text{ Hz.}$$

The exact result (Table 8-1) has the same form as this equation but with the factor 2.960 replaced by the factor π. Thus, the approximation of Eq. A-3 underestimates the fundamental natural frequency of the beam by about 11%.

Plate Example. It has been found that the maximum static deflection of a thin, uniform elliptical plate with a clamped edge due to its own weight is (Ref. A-2):

$$\delta_s = \frac{\mu g h a^4 b^4}{8(3a^4 + 2a^2b^2 + 3b^4)}\frac{12(1-\nu^2)}{Eh^3}.$$

h is the plate thickness, μ is the plate density, ν is Poisson's ratio, E is the modulus of elasticity, and a and b are the major and minor axes of the ellipse. Using this result, Eq. A-3 predicts that the fundamental natural frequency of the ellipse is:

$$f = \frac{2.828}{2\pi}\left(\frac{3a^4 + 2a^2b^2 + 3b^4}{\mu h a^4 b^4}\right)^{1/2}\left[\frac{Eh^3}{12(1-\nu^2)}\right]^{1/2} \text{ Hz.}$$

A more nearly exact analysis of the fundamental natural frequency of this plate gives (Ref. A-3):

$$f = \frac{3.612}{2\pi}\left(\frac{3a^4 + 2a^2b^2 + 3b^4}{\mu h a^4 b^4}\right)^{1/2}\left[\frac{Eh^3}{12(1-\nu^2)}\right]^{1/2} \text{ Hz.}$$

Thus, Eq. A-3 underestimates the fundamental natural frequency of the plate by about 22%.

The work of Jones (Ref. A-2) and Johns (Ref. A-4) suggests that the fundamental natural frequency of thin uniform plates can be accurately estimated by introducing a correction factor into Eq. A-3 to compensate for the underprediction:

$$f = \frac{1.277}{2\pi}\left(\frac{g}{\delta_s}\right)^{1/2} \text{ Hz.} \qquad \text{(A-4)}$$

This equation has successfully predicted the fundamental natural frequencies of plates of a variety of shapes and boundary conditions to within 3% of the exact result (Ref. A-2).

REFERENCES

A-1. Roark, R., *Formulas for Stress and Strain,* 4th ed., McGraw-Hill, New York, 1965, p. 106.

A-2. Jones, R., "An Approximate Expression for the Fundamental Frequency of Vibration of Elastic Plates," *J. Sound Vib.* **38**, 503–504 (1975).

A-3. Mazumdar, J., "Transverse Vibration of Elastic Plates by the Method of Constant Deflection," *J. Sound Vib.* **18**, 147–155 (1971).

A-4. Johns, D. J., "Comments on 'An Approximate Expression for the Fundamental Frequency of Vibration of Elastic Plates'," *J. Sound Vib.* **41**, 385–387 (1975).

Appendix B
LITERATURE REVIEWS

From time to time certain journal articles, reports, or books survey large sectors of the vibration literature and present an array of solutions which are valuable to both the researcher and engineer. Some of these general sources are listed below:

Subject	*Reference*
Vibration Computer Programs	Pilkey, W., and B. Pilkey (eds), *Shock and Vibration Computer Programs*, Shock and Vibration Information Center, Naval Research Laboratory, Washington, D.C., 1975.
Beams	1. Gorman, D. J., *Free Vibration Analysis of Beams and Shafts*, John Wiley, New York, 1975. 2. Wagner, H., and V. Ramamurli, "Beam Vibrations–A Review," *Shock and Vibration Digest* **9**(9), 17–24 (1977).
Plates	1. Leissa, A. W., "Vibration of Plates," NASA Report SP-160, 1969, Ohio State University (out of print). 2. Leissa, A. W., "Recent Research in Plate Vibrations: Classical Theory," *Shock and Vibration Digest* **9**(10), 13–24 (1977). 3. Leissa, A. W., "Recent Research in Plate Vibrations," *Shock and Vibration Digest* **10**(12), 21–35 (1978).
Shells	Leissa, A. W., "Vibration of Shells," NASA Report SP-288, 1973, Ohio State University (out of print).
Cables	Soler, A. I., "On the Dynamics of Cables and Cable Systems," *Shock and Vibration Digest* **5**(3), 2–10 (1973).
Turbine Blades	Rao, J. S., "Natural Frequencies of Turbine Blading–A Survey," *Shock and Vibration Digest* **5**(10), 3–16 (1973).
Ship Hulls	Jensen, J. J., and N. F. Madsen, "A Review of Ship Hull Vibration," *Shock and Vibration Digest* **9**(4), 13–22 (1977).
Bridges	Huang, T., "Vibration of Bridges," *Shock and Vibration Digest* **8**(4), 61–76 (1976).
Liquid Sloshing	Abramson, H. N., "Dynamic Behavior of Liquid in Moving Container," *Applied Mechanics Reviews* **16**(7), 501–506 (1963).

Appendix C

FORMULAS FOR INTEGRALS CONTAINING MODE SHAPES OF SINGLE SPAN BEAMS

The formulas in this appendix have been taken from Ref. C-1. These formulas have proven very useful in the analysis of dynamic systems whose modes can be described in terms of the mode shapes of single span beams (Table 8-1). The following notation is employed:

ϕ_n, ψ_n = nth beam mode of one of the beams (Table 8-1),
σ_n = nondimensional parameter of column 4 of Table 8-1,
$\beta_n = \lambda_n/L$, where λ_n is the nondimensional frequency parameter given in column 2 of Table 8-1 and L is the span of the beam,
x = coordinate along the span of the beam.

REFERENCE

C-1. Felgar, R. P., "Formulas for Integrals Containing Characteristic Functions of a Vibrating Beam," University of Texas Circular No. 14, Bureau of Engineering Research, Austin, Tex., 1950.

1. $$\int_0^L \phi_n(x)\,dx = \frac{1}{\beta_n^4}\left[\frac{d^3\phi_n}{dx^3}\right]_0^L$$

Clamped-free $$\int_0^L \phi_n(x)\,dx = \frac{2\sigma_n}{\beta_n}$$

Clamped-pinned $$\int_0^L \phi_n(x)\,dx = \frac{1}{\beta_n}\left[(-1)^{n+1}\sqrt{\sigma_n^2+1} - \sqrt{\sigma_n^2-1} + 2\sigma_n\right]$$

Clamped-clamped $$\int_0^L \phi_n(x)\,dx = \frac{2\sigma_n}{\beta_n}\left[1-(-1)^n\right]$$

Free-pinned $$\int_0^L \phi_n(x)\,dx = \frac{1}{\beta_n}\left[(-1)^n\sqrt{\sigma_n^2+1} - \sqrt{\sigma_n^2-1}\right]$$

Free-free $$\int_0^L \phi_n(x)\,dx = 0$$

2. $\int_0^L \frac{d\phi_n}{dx}\,dx = [\phi_n]_0^L$

Clamped-free $\int_0^L \frac{d\phi_n}{dx}\,dx = (-1)^{n+1}2$

Clamped-pinned $\int_0^L \frac{d\phi_n}{dx}\,dx = 0$

Clamped-clamped $\int_0^L \frac{d\phi_n}{dx}\,dx = 0$

Free-pinned $\int_0^L \frac{d\phi_n}{dx}\,dx = -2$

Free-free $\int_0^L \frac{d\phi_n}{dx}\,dx = -2[(-1)^n + 1]$

3. $\int_0^L \frac{d^2\phi_n}{dx^2}\,dx = \left[\frac{d\phi_n}{dx}\right]_0^L$

Clamped-free $\int_0^L \frac{d^2\phi_n}{dx^2}\,dx = (-1)^{n+1}2\sigma_n\beta_n$

Clamped-pinned $\int_0^L \frac{d^2\phi_n}{dx^2}\,dx = \beta_n[(-1)^n\sqrt{\sigma_n^2+1} - \sqrt{\sigma_n^2-1}]$

Clamped-clamped $\int_0^L \frac{d^2\phi_n}{dx^2}\,dx = 0$

Free-pinned $\int_0^L \frac{d^2\phi_n}{dx^2}\,dx = \beta_n[(-1)^{n+1}\sqrt{\sigma_n^2+1} - \sqrt{\sigma_n^2-1} + 2\sigma_n]$

Free-free $\int_0^L \frac{d^2\phi_n}{dx^2}\,dx = 2\sigma_n\beta_n[1-(-1)^n]$

4. $\int_0^L \frac{d^3\phi_n}{dx^3}\,dx = \left[\frac{d^2\phi_n}{dx^2}\right]_0^L$

Clamped-free $\int_0^L \frac{d^3\phi_n}{dx^3}\,dx = -2\beta_n^2$

Clamped-pinned $\int_0^L \frac{d^3\phi_n}{dx^3}\,dx = -2\beta_n^2$

Clamped-clamped $\int_0^L \frac{d^3\phi_n}{dx^3}\,dx = -2\beta_n^2[(-1)^n + 1]$

Free-pinned $\int_0^L \frac{d^3\phi_n}{dx^3}\,dx = 0$

Free-free $\int_0^L \frac{d^3\phi_n}{dx^3}\,dx = 0$

5. $$\int_0^L \phi_n^2\,dx = \frac{1}{4}\left[\frac{3}{\beta_n^4}\phi_n\frac{d^3\phi_n}{dx^3} + x\phi_n^2 - \frac{2x}{\beta_n^4}\frac{d\phi_n}{dx}\frac{d^3\phi_n}{dx^3} - \frac{1}{\beta_n^4}\frac{d\phi_n}{dx}\frac{d^2\phi_n}{dx^2} + \frac{x}{\beta_n^4}\left(\frac{d^2\phi_n}{dx^2}\right)^2\right]_0^L$$

Clamped-free $\int_0^L \phi_n^2\,dx = L$

Clamped-pinned $\int_0^L \phi_n^2\,dx = L$

Clamped-clamped $\int_0^L \phi_n^2\,dx = L$

Free-pinned $\int_0^L \phi_n^2\,dx = L$

Free-free $\int_0^L \phi_n^2\,dx = L$

6. $$\int_0^L \left(\frac{d\phi_n}{dx}\right)^2 dx = \frac{1}{4}\left[3\phi_n\frac{d\phi_n}{dx} + x\left(\frac{d\phi_n}{dx}\right)^2 - 2x\phi_n\frac{d^2\phi_n}{dx^2} - \frac{1}{\beta_n^4}\frac{d^2\phi_n}{dx^2}\frac{d^3\phi_n}{dx^3} + \frac{x}{\beta_n^4}\left(\frac{d^3\phi_n}{dx^3}\right)^2\right]_0^L$$

Clamped-free $\int_0^L \left(\frac{d\phi_n}{dx}\right)^2 dx = \sigma_n\beta_n(2 + \sigma_n\beta_n L)$

Clamped-pinned $\int_0^L \left(\frac{d\phi_n}{dx}\right)^2 dx = \sigma_n\beta_n(\beta_n L\sigma_n - 1)$

Clamped-clamped $\int_0^L \left(\frac{d\phi_n}{dx}\right)^2 dx = \sigma_n\beta_n(\sigma_n\beta_n L - 2)$

Free-pinned $\int_0^L \left(\frac{d\phi_n}{dx}\right)^2 dx = \sigma_n\beta_n(\sigma_n\beta_n L + 3)$

Free-free $\int_0^L \left(\frac{d\phi_n}{dx}\right)^2 dx = \sigma_n\beta_n(\sigma_n\beta_n L + 6)$

7. $$\int_0^L \left(\frac{d^2\phi_n}{dx^2}\right)^2 dx = \frac{1}{4}\left[3\,\frac{d^2\phi_n}{dx^2}\,\frac{d\phi_n}{dx} + x\left(\frac{d^2\phi_n}{dx^2}\right)^2 - 2x\,\frac{d\phi_n}{dx}\,\frac{d^3\phi_n}{dx^3} - \phi_n\,\frac{d^3\phi_n}{dx^3} + x\beta_n^4\phi_n^2\right]_0^L$$

Clamped-free $$\int_0^L \left(\frac{d^2\phi_n}{dx^2}\right)^2 dx = \beta_n^4 L$$

Clamped-pinned $$\int_0^L \left(\frac{d^2\phi_n}{dx^2}\right)^2 dx = \beta_n^4 L$$

Clamped-clamped $$\int_0^L \left(\frac{d^2\phi_n}{dx^2}\right)^2 dx = \beta_n^4 L$$

Free-pinned $$\int_0^L \left(\frac{d^2\phi_n}{dx^2}\right)^2 dx = \beta_n^4 L$$

Free-free $$\int_0^L \left(\frac{d^2\phi_n}{dx^2}\right)^2 dx = \beta_n^4 L$$

8. $$\int_0^L \left(\frac{d^3\phi_n}{dx^3}\right)^2 dx = \frac{1}{4}\left[3\,\frac{d^2\phi_n}{dx^2}\,\frac{d^3\phi_n}{dx^3} + x\left(\frac{d^3\phi_n}{dx^3}\right)^2 - 2\beta_n^4 x\phi_n\,\frac{d^2\phi_n}{dx^2} - \beta_n^4\phi_n\,\frac{d\phi_n}{dx} + x\beta_n^4\left(\frac{d\phi_n}{dx}\right)^2\right]_0^L$$

Clamped-free $$\int_0^L \left(\frac{d^3\phi_n}{dx^3}\right)^2 dx = \sigma_n\beta_n^5(\sigma_n\beta_n L + 2)$$

Clamped-pinned $$\int_0^L \left(\frac{d^3\phi_n}{dx^3}\right)^2 dx = \sigma_n\beta_n^5(\sigma_n\beta_n L + 3)$$

Clamped-clamped $$\int_0^L \left(\frac{d^3\phi_n}{dx^3}\right)^2 dx = \sigma_n\beta_n^5(\sigma_n\beta_n L + 6)$$

Free-pinned $$\int_0^L \left(\frac{d^3\phi_n}{dx^3}\right)^2 dx = \sigma_n\beta_n^5(\sigma_n\beta_n L - 1)$$

Free-free $$\int_0^L \left(\frac{d^3\phi_n}{dx^3}\right)^2 dx = \sigma_n\beta_n^5(\sigma_n\beta_n L - 2)$$

9. $$\int_0^L \phi_n\,\frac{d\phi_n}{dx}\,dx = \frac{1}{2}\,[\phi_n^2]_0^L$$

Clamped-free $$\int_0^L \phi_n\,\frac{d\phi_n}{dx}\,dx = 2$$

Clamped-pinned $$\int_0^L \phi_n\,\frac{d\phi_n}{dx}\,dx = 0$$

Clamped-clamped $\int_0^L \phi_n \frac{d\phi_n}{dx} dx = 0$

Free-pinned $\int_0^L \phi_n \frac{d\phi_n}{dx} dx = -2$

Free-free $\int_0^L \phi_n \frac{d\phi_n}{dx} dx = 0$

10. $$\int_0^L \phi_n \frac{d^2\phi_n}{dx^2} dx = \frac{1}{4}\left[\phi_n \frac{d\phi_n}{dx} - x\left(\frac{d\phi_n}{dx}\right)^2 + 2x\phi_n \frac{d^2\phi_n}{dx^2} + \frac{1}{\beta_n^4} \frac{d^2\phi_n}{dx^2} \frac{d^3\phi_n}{dx^3} - \frac{x}{\beta_n^4}\left(\frac{d^3\phi_n}{dx^3}\right)^2\right]_0^L$$

Clamped-free $\int_0^L \phi_n \frac{d^2\phi_n}{dx^2} dx = \sigma_n\beta_n(2 - \sigma_n\beta_n L)$

Clamped-pinned $\int_0^L \phi_n \frac{d^2\phi_n}{dx^2} dx = \sigma_n\beta_n(1 - \sigma_n\beta_n L)$

Clamped-clamped $\int_0^L \phi_n \frac{d^2\phi_n}{dx^2} dx = \sigma_n\beta_n(2 - \sigma_n\beta_n L)$

Free-pinned $\int_0^L \phi_n \frac{d^2\phi_n}{dx^2} dx = \sigma_n\beta_n(1 - \sigma_n\beta_n L)$

Free-free $\int_0^L \phi_n \frac{d^2\phi_n}{dx^2} dx = \sigma_n\beta_n(2 - \sigma_n\beta_n L)$

11. $$\int_0^L \phi_n \frac{d^3\phi_n}{dx^3} dx = \frac{1}{2\beta_n^4}\left[\left(\frac{d^3\phi_n}{dx^3}\right)^2\right]_0^L$$

Clamped-free $\int_0^L \phi_n \frac{d^3\phi_n}{dx^3} dx = -2\sigma_n^2\beta_n^2$

Clamped-pinned $\int_0^L \phi_n \frac{d^3\phi_n}{dx^3} dx = \beta_n^2[(-1)^n \sqrt{\sigma_n^4 - 1} - \sigma_n^2]$

Clamped-clamped $\int_0^L \phi_n \frac{d^3\phi_n}{dx^3} dx = 0$

Free-pinned $\int_0^L \phi_n \frac{d^3\phi_n}{dx^3} dx = \beta_n^2[\sigma_n^2 - (-1)^n \sqrt{\sigma_n^4 - 1}]$

Free-free $\int_0^L \phi_n \frac{d^3\phi_n}{dx^3} dx = 0$

12. $$\int_0^L \frac{d\phi_n}{dx}\frac{d^2\phi_n}{dx^2}\,dx = \frac{1}{2}\left[\left(\frac{d\phi_n}{dx}\right)^2\right]_0^L$$

Clamped-free $$\int_0^L \frac{d\phi_n}{dx}\frac{d^2\phi_n}{dx^2}\,dx = 2\sigma_n^2\beta_n^2$$

Clamped-pinned $$\int_0^L \frac{d\phi_n}{dx}\frac{d^2\phi_n}{dx^2}\,dx = \beta_n^2[\sigma_n^2 - (-1)^n\sqrt{\sigma_n^4 - 1}]$$

Clamped-clamped $$\int_0^L \frac{d\phi_n}{dx}\frac{d^2\phi_n}{dx^2}\,dx = 0$$

Free-pinned $$\int_0^L \frac{d\phi_n}{dx}\frac{d^2\phi_n}{dx^2}\,dx = \beta_n^2[(-1)^n\sqrt{\sigma_n^4 - 1} - \sigma_n^2]$$

Free-free $$\int_0^L \frac{d\phi_n}{dx}\frac{d^2\phi_n}{dx^2}\,dx = 0$$

13. $$\int_0^L \frac{d\phi_n}{dx}\frac{d^3\phi_n}{dx^3}\,dx = \frac{1}{4}\left[\frac{d\phi_n}{dx}\frac{d^2\phi_n}{dx^2} - x\left(\frac{d^2\phi_n}{dx^2}\right)^2 + 2x\frac{d\phi_n}{dx}\frac{d^3\phi_n}{dx^3} + \phi_n\frac{d^3\phi_n}{dx^3} - \beta_n^4 x\phi_n^2\right]_0^L$$

Clamped-free $$\int_0^L \frac{d\phi_n}{dx}\frac{d^3\phi_n}{dx^3}\,dx = -\beta_n^4 L$$

Clamped-pinned $$\int_0^L \frac{d\phi_n}{dx}\frac{d^3\phi_n}{dx^3}\,dx = -\beta_n^4 L$$

Clamped-clamped $$\int_0^L \frac{d\phi_n}{dx}\frac{d^3\phi_n}{dx^3}\,dx = -\beta_n^4 L$$

Free-pinned $$\int_0^L \frac{d\phi_n}{dx}\frac{d^3\phi_n}{dx^3}\,dx = -\beta_n^4 L$$

Free-free $$\int_0^L \frac{d\phi_n}{dx}\frac{d^3\phi_n}{dx^3}\,dx = -\beta_n^4 L$$

14. $$\int_0^L \frac{d^2\phi_n}{dx^2}\frac{d^3\phi_n}{dx^3}\,dx = \frac{1}{2}\left[\left(\frac{d^2\phi_n}{dx^2}\right)^2\right]_0^L$$

Clamped-free $$\int_0^L \frac{d^2\phi_n}{dx^2}\frac{d^3\phi_n}{dx^3}\,dx = -2\beta_n^4$$

Clamped-pinned $$\int_0^L \frac{d^2\phi_n}{dx^2}\frac{d^3\phi_n}{dx^3}\,dx = -2\beta_n^4$$

Clamped-clamped $\int_0^L \frac{d^2\phi_n}{dx^2}\frac{d^3\phi_n}{dx^3}\,dx = 0$

Free-pinned $\int_0^L \frac{d^2\phi_n}{dx^2}\frac{d^3\phi_n}{dx^3}\,dx = 0$

Free-free $\int_0^L \frac{d^2\phi_n}{dx^2}\frac{d^3\phi_n}{dx^3}\,dx = 0$

15. $$\int_0^L \phi_n\phi_m\,dx = \frac{1}{\beta_n^4 - \beta_m^4}\left[\phi_m \frac{d^3\phi_n}{dx^3} - \phi_n \frac{d^3\phi_m}{dx^3} - \frac{d\phi_m}{dx}\frac{d^2\phi_n}{dx^2} + \frac{d\phi_n}{dx}\frac{d^2\phi_m}{dx^2}\right]_0^L \qquad m \neq n$$

Clamped-free $\int_0^L \phi_n\phi_m\,dx = 0$

Clamped-pinned $\int_0^L \phi_n\phi_m\,dx = 0$

Clamped-clamped $\int_0^L \phi_n\phi_m\,dx = 0$

Free-pinned $\int_0^L \phi_n\phi_m\,dx = 0$

Free-free $\int_0^L \phi_n\phi_m\,dx = 0$

16. $$\int_0^L \frac{d\phi_n}{dx}\frac{d\phi_m}{dx}\,dx = \frac{1}{\beta_n^4 - \beta_m^4}\left[\beta_n^4\phi_n \frac{d\phi_m}{dx} - \beta_m^4\phi_m \frac{d\phi_n}{dx} - \frac{d^2\phi_m}{dx^2}\frac{d^3\phi_n}{dx^3} + \frac{d^2\phi_n}{dx^2}\frac{d^3\phi_m}{dx^3}\right]_0^L$$

$m \neq n$

Clamped-free $$\int_0^L \frac{d\phi_n}{dx}\frac{d\phi_m}{dx}\,dx = \frac{4\beta_m\beta_n}{\beta_n^4 - \beta_m^4}\,[(-1)^{m+n}(\sigma_m\beta_n^3 - \sigma_n\beta_m^3) - \beta_m\beta_n(\sigma_n\beta_n - \sigma_m\beta_m)]$$

Clamped-pinned $$\int_0^L \frac{d\phi_n}{dx}\frac{d\phi_m}{dx}\,dx = \frac{4\beta_m^2\beta_n^2}{\beta_n^4 - \beta_m^4}\,(\sigma_m\beta_m - \sigma_n\beta_n)$$

Clamped-clamped $$\int_0^L \frac{d\phi_n}{dx}\frac{d\phi_m}{dx}\,dx = \frac{4\beta_m^2\beta_n^2(\sigma_m\beta_m - \sigma_n\beta_n)}{\beta_n^4 - \beta_m^4}\,[(-1)^{m+n} + 1]$$

Free-pinned $$\int_0^L \frac{d\phi_n}{dx}\frac{d\phi_m}{dx}\,dx = \frac{4\beta_m\beta_n}{\beta_n^4 - \beta_m^4}\,(\sigma_m\beta_n^3 - \sigma_n\beta_m^3)$$

Free-free $\displaystyle\int_0^L \frac{d\phi_n}{dx}\frac{d\phi_m}{dx}\,dx = \frac{4\beta_m\beta_n(\sigma_m\beta_n^3 - \sigma_n\beta_m^3)}{\beta_n^4 - \beta_m^4}[(-1)^{m+n} + 1]$

17. $$\int_0^L \frac{d^2\phi_n}{dx^2}\frac{d^2\phi_m}{dx^2}\,dx = \frac{1}{\beta_n^4 - \beta_m^4}\left[\beta_n^4\frac{d\phi_n}{dx}\frac{d^2\phi_m}{dx^2} - \beta_m^4\frac{d\phi_m}{dx}\frac{d^2\phi_n}{dx^2} - \beta_n^4\phi_n\frac{d^3\phi_m}{dx^3} + \beta_m^4\phi_m\frac{d^3\phi_n}{dx^3}\right]_0^L \qquad m \neq n$$

Clamped-free $\displaystyle\int_0^L \frac{d^2\phi_n}{dx^2}\frac{d^2\phi_m}{dx^2}\,dx = 0$

Clamped-pinned $\displaystyle\int_0^L \frac{d^2\phi_n}{dx^2}\frac{d^2\phi_m}{dx^2}\,dx = 0$

Clamped-clamped $\displaystyle\int_0^L \frac{d^2\phi_n}{dx^2}\frac{d^2\phi_m}{dx^2}\,dx = 0$

Free-pinned $\displaystyle\int_0^L \frac{d^2\phi_n}{dx^2}\frac{d^2\phi_m}{dx^2}\,dx = 0$

Free-free $\displaystyle\int_0^L \frac{d^2\phi_n}{dx^2}\frac{d^2\phi_m}{dx^2}\,dx = 0$

18. $$\int_0^L \frac{d^3\phi_n}{dx^3}\frac{d^3\phi_m}{dx^3}\,dx = \frac{1}{\beta_n^4 - \beta_m^4}\left[\beta_n^4\frac{d^2\phi_n}{dx^2}\frac{d^3\phi_m}{dx^3} - \beta_m^4\frac{d^2\phi_m}{dx^2}\frac{d^3\phi_n}{dx^3} - \beta_m^4\beta_n^4\phi_m\frac{d\phi_n}{dx} + \beta_m^4\beta_n^4\phi_n\frac{d\phi_m}{dx}\right]_0^L \qquad m \neq n$$

Clamped-free $\displaystyle\int_0^L \frac{d^3\phi_n}{dx^3}\frac{d^3\phi_m}{dx^3}\,dx = \frac{4\beta_m^3\beta_n^3}{\beta_n^4 - \beta_m^4}[(-1)^{m+n}\beta_m\beta_n(\sigma_n\beta_n + \sigma_m\beta_m) + \sigma_m\beta_n^3 - \sigma_n\beta_m^3]$

Clamped-pinned $\displaystyle\int_0^L \frac{d^3\phi_n}{dx^3}\frac{d^3\phi_m}{dx^3}\,dx = \frac{4\beta_m^3\beta_n^3}{\beta_n^4 - \beta_m^4}(\sigma_m\beta_n^3 - \sigma_n\beta_m^3)$

Clamped-clamped $\displaystyle\int_0^L \frac{d^3\phi_n}{dx^3}\frac{d^3\phi_m}{dx^3}\,dx = \frac{4\beta_m^3\beta_n^3(\sigma_m\beta_n^3 - \sigma_n\beta_m^3)}{\beta_n^4 - \beta_m^4}[(-1)^{m+n} + 1]$

Free-pinned $\displaystyle\int_0^L \frac{d^3\phi_n}{dx^3}\frac{d^3\phi_m}{dx^3}\,dx = \frac{4\beta_m^4\beta_n^4}{\beta_n^4 - \beta_m^4}(\sigma_m\beta_m - \sigma_n\beta_n)$

Free-free $\displaystyle\int_0^L \frac{d^3\phi_n}{dx^3}\frac{d^3\phi_m}{dx^3}\,dx = \frac{4\beta_m^4\beta_n^4(\sigma_m\beta_m - \sigma_n\beta_n)}{\beta_n^4 - \beta_m^4}[(-1)^{m+n} + 1]$

19. $$\int_0^L \phi_n \frac{d\phi_m}{dx} dx = \frac{\beta_m^4}{\beta_n^4 - \beta_m^4}\left[\phi_n\phi_m - \frac{1}{\beta_m^4}\frac{d\phi_n}{dx}\frac{d^3\phi_m}{dx^3} + \frac{1}{\beta_m^4}\frac{d^2\phi_n}{dx^2}\frac{d^2\phi_m}{dx^2} - \frac{1}{\beta_m^4}\frac{d\phi_m}{dx}\frac{d^3\phi_n}{dx^3}\right]_0^L$$

$m \neq n$

Clamped-free $$\int_0^L \phi_n \frac{d\phi_m}{dx} dx = \frac{4\beta_m^2}{\beta_n^4 - \beta_m^4}[\beta_n^2 - (-1)^{m+n}\beta_m^2]$$

Clamped-pinned $$\int_0^L \phi_n \frac{d\phi_m}{dx} dx = \frac{\beta_m\beta_n}{\beta_n^4 - \beta_m^4}[-(-1)^{m+n}(\beta_n^2 + \beta_m^2)\sqrt{(\sigma_n^2 + 1)(\sigma_m^2 + 1)}$$
$$+ (-1)^n(\beta_n^2 - \beta_m^2)\sqrt{(\sigma_n^2 + 1)(\sigma_m^2 - 1)}$$
$$- (-1)^m(\beta_n^2 - \beta_m^2)\sqrt{(\sigma_n^2 - 1)(\sigma_m^2 + 1)}$$
$$+ (\beta_n^2 + \beta_m^2)\sqrt{(\sigma_n^2 - 1)(\sigma_m^2 - 1)} + 4\beta_m\beta_n]$$

Clamped-clamped $$\int_0^L \phi_n \frac{d\phi_m}{dx} dx = \frac{4\beta_n^2\beta_m^2}{\beta_n^4 - \beta_m^4}[1 - (-1)^{m+n}]$$

Free-pinned $$\int_0^L \phi_n \frac{d\phi_m}{dx} dx = \frac{\beta_m\beta_n}{\beta_n^4 - \beta_m^4}\Big[-(-1)^{m+n}(\beta_n^2 + \beta_m^2)\sqrt{(\sigma_n^2 - 1)(\sigma_m^2 + 1)}$$
$$- (-1)^n(\beta_n^2 - \beta_m^2)\sqrt{(\sigma_n^2 + 1)(\sigma_m^2 - 1)}$$
$$+ (-1)^m(\beta_n^2 - \beta_m^2)\sqrt{(\sigma_n^2 - 1)(\sigma_m^2 + 1)}$$
$$+ (\beta_n^2 + \beta_m^2)\sqrt{(\sigma_n^2 - 1)(\sigma_m^2 - 1)} + \frac{4\beta_m^3}{\beta_n}\Big]$$

Free-free $$\int_0^L \phi_n \frac{d\phi_m}{dx} dx = \frac{4\beta_m^4}{\beta_n^4 - \beta_m^4}[1 - (-1)^{m+n}]$$

20. $$\int_0^L \phi_n \frac{d^2\phi_m}{dx^2} dx = \frac{1}{\beta_n^4 - \beta_m^4}\left[\beta_m^4\left(\phi_m\frac{d\phi_n}{dx} - \phi_n\frac{d\phi_m}{dx}\right) - \frac{d^2\phi_n}{dx^2}\frac{d^3\phi_m}{dx^3} + \frac{d^2\phi_m}{dx^2}\frac{d^3\phi_n}{dx^3}\right]_0^L$$

$m \neq n$

Clamped-free $$\int_0^L \phi_n \frac{d^2\phi_m}{dx^2} dx = \frac{4\beta_m^2(\sigma_n\beta_n - \sigma_m\beta_m)}{\beta_n^4 - \beta_m^4}[(-1)^{m+n}\beta_m^2 + \beta_n^2]$$

Clamped-pinned $$\int_0^L \phi_n \frac{d^2\phi_m}{dx^2} dx = \frac{4\beta_m^2\beta_n^2}{\beta_n^4 - \beta_m^4}(\sigma_n\beta_n - \sigma_m\beta_m)$$

Clamped-clamped $$\int_0^L \phi_n \frac{d^2\phi_m}{dx^2} dx = \frac{4\beta_m^2\beta_n^2(\sigma_n\beta_n - \sigma_m\beta_m)}{\beta_n^4 - \beta_m^4}[1 + (-1)^{m+n}]$$

Free-pinned $$\int_0^L \phi_n \frac{d^2\phi_m}{dx^2} dx = \frac{4\beta_m^4}{\beta_n^4 - \beta_m^4}(\sigma_n\beta_n - \sigma_m\beta_m)$$

Free-free $$\int_0^L \phi_n \frac{d^2\phi_m}{dx^2} dx = \frac{4\beta_m^4(\sigma_n\beta_n - \sigma_m\beta_m)}{\beta_n^4 - \beta_m^4}[1 + (-1)^{m+n}]$$

21. $$\int_0^L \phi_n \frac{d^3\phi_m}{dx^3}\,dx = \frac{\beta_m^4}{\beta_n^4 - \beta_m^4}\left[\frac{d\phi_m}{dx}\frac{d\phi_n}{dx} - \phi_n \frac{d^2\phi_m}{dx^2} - \phi_m \frac{d^2\phi_n}{dx^2} + \frac{1}{\beta_m^4}\frac{d^3\phi_n}{dx^3}\frac{d^3\phi_m}{dx^3}\right]_0^L$$

$m \neq n$

Clamped-free $$\int_0^L \phi_n \frac{d^3\phi_m}{dx^3}\,dx = \frac{4\beta_m^3\beta_n\sigma_m\sigma_n}{\beta_n^4 - \beta_m^4}\left[(-1)^{m+n}\beta_m^2 - \beta_n^2\right]$$

Clamped-pinned $$\int_0^L \phi_n \frac{d^3\phi_m}{dx^3}\,dx = \frac{\beta_m^3\beta_n^3}{\beta_n^4 - \beta_m^4}\Big[(-1)^{m+n}(\beta_m^2 + \beta_n^2)\sqrt{(\sigma_n^2+1)(\sigma_m^2+1)}$$
$$+ (-1)^m(\beta_n^2 - \beta_m^2)\sqrt{(\sigma_m^2+1)(\sigma_n^2-1)}$$
$$+ (-1)^n(\beta_n^2 - \beta_m^2)\sqrt{(\sigma_m^2-1)(\sigma_n^2+1)}$$
$$+ (\beta_n^2 + \beta_m^2)\sqrt{(\sigma_m^2-1)(\sigma_n^2-1)} - 4\sigma_m\sigma_n\beta_n^2\Big]$$

Clamped-clamped $$\int_0^L \phi_n \frac{d^3\phi_m}{dx^3}\,dx = \frac{4\beta_m^3\beta_n^3\sigma_m\sigma_n}{\beta_n^4 - \beta_m^4}\left[(-1)^{m+n} - 1\right]$$

Free-pinned $$\int_0^L \phi_n \frac{d^3\phi_m}{dx^3}\,dx = \frac{\beta_m^3\beta_n}{\beta_n^4 - \beta_m^4}\Big[(-1)^{m+n}(\beta_n^2 + \beta_m^2)\sqrt{(\sigma_m^2+1)(\sigma_n^2+1)}$$
$$- (-1)^m(\beta_n^2 - \beta_m^2)\sqrt{(\sigma_m^2+1)(\sigma_n^2-1)}$$
$$- (-1)^n(\beta_n^2 - \beta_m^2)\sqrt{(\sigma_m^2-1)(\sigma_n^2+1)}$$
$$+ (\beta_n^2 + \beta_m^2)\sqrt{(\sigma_m^2-1)(\sigma_n^2-1)} - 4\sigma_m\sigma_n\beta_m^2\Big]$$

Free-free $$\int_0^L \phi_n \frac{d^3\phi_m}{dx^3}\,dx = \frac{4\beta_m^5\beta_n\sigma_m\sigma_n}{\beta_n^4 - \beta_m^4}\left[(-1)^{m+n} - 1\right]$$

22. $$\int_0^L \frac{d\phi_n}{dx}\frac{d^2\phi_m}{dx^2}\,dx = \frac{\beta_m^4}{\beta_n^4 - \beta_m^4}\left[\phi_m \frac{d^2\phi_n}{dx^2} - \frac{d\phi_m}{dx}\frac{d\phi_n}{dx} + \phi_n \frac{d^2\phi_m}{dx^2} - \frac{1}{\beta_m^4}\frac{d^3\phi_m}{dx^3}\frac{d^3\phi_n}{dx^3}\right]_0^L$$

$m \neq n$

Clamped-free $$\int_0^L \frac{d\phi_n}{dx}\frac{d^2\phi_m}{dx^2}\,dx = \frac{4\beta_m^3\beta_n\sigma_m\sigma_n}{\beta_n^4 - \beta_m^4}\left[\beta_n^2 - (-1)^{m+n}\beta_m^2\right]$$

Clamped-pinned $$\int_0^L \frac{d\phi_n}{dx}\frac{d^2\phi_m}{dx^2}\,dx = \frac{\beta_m^3\beta_n}{\beta_n^4 - \beta_m^4}\Big[-(-1)^{m+n}(\beta_n^2 + \beta_m^2)\sqrt{(\sigma_m^2+1)(\sigma_n^2+1)}$$
$$- (-1)^n(\beta_n^2 - \beta_m^2)\sqrt{(\sigma_m^2-1)(\sigma_n^2+1)}$$
$$- (-1)^m(\beta_n^2 - \beta_m^2)\sqrt{(\sigma_m^2+1)(\sigma_n^2-1)}$$
$$- (\beta_n^2 + \beta_m^2)\sqrt{(\sigma_m^2-1)(\sigma_n^2-1)} + 4\sigma_m\sigma_n\beta_n^2\Big]$$

Clamped-clamped $$\int_0^L \frac{d\phi_n}{dx}\frac{d^2\phi_m}{dx^2}\,dx = \frac{4\beta_m^3\beta_n^3\sigma_m\sigma_n}{\beta_n^4 - \beta_m^4}\left[1 - (-1)^{m+n}\right]$$

Free-pinned $$\int_0^L \frac{d\phi_n}{dx}\frac{d^2\phi_m}{dx^2}\,dx = \frac{\beta_m^3\beta_n}{\beta_n^4-\beta_m^4}\,[-(-1)^{m+n}(\beta_n^2+\beta_m^2)\sqrt{(\sigma_m^2+1)(\sigma_n^2+1)}$$
$$+(-1)^m(\beta_n^2-\beta_m^2)\sqrt{(\sigma_m^2+1)(\sigma_n^2-1)}$$
$$+(-1)^n(\beta_n^2-\beta_m^2)\sqrt{(\sigma_m^2-1)(\sigma_n^2+1)}$$
$$-(\beta_n^2+\beta_m^2)\sqrt{(\sigma_n^2-1)(\sigma_m^2-1)}+4\sigma_m\sigma_n\beta_m^2]$$

Free-free $$\int_0^L \frac{d\phi_n}{dx}\frac{d^2\phi_m}{dx^2}\,dx = \frac{4\beta_m^5\beta_n\sigma_n\sigma_m}{\beta_n^4-\beta_m^4}\,[1-(-1)^{m+n}]$$

23. $$\int_0^L \frac{d\phi_n}{dx}\frac{d^3\phi_m}{dx^3}\,dx = \frac{\beta_m^4}{\beta_n^4-\beta_m^4}\left[\frac{d\phi_m}{dx}\frac{d^2\phi_n}{dx^2}-\frac{d\phi_n}{dx}\frac{d^2\phi_m}{dx^2}+\phi_n\frac{d^3\phi_m}{dx^3}-\phi_m\frac{d^3\phi_n}{dx^3}\right]_0^L \qquad m\neq n$$

Clamped-free $$\int_0^L \frac{d\phi_n}{dx}\frac{d^3\phi_m}{dx^3}\,dx = 0$$

Clamped-pinned $$\int_0^L \frac{d\phi_n}{dx}\frac{d^3\phi_m}{dx^3}\,dx = 0$$

Clamped-clamped $$\int_0^L \frac{d\phi_n}{dx}\frac{d^3\phi_m}{dx^3}\,dx = 0$$

Free-pinned $$\int_0^L \frac{d\phi_n}{dx}\frac{d^3\phi_m}{dx^3}\,dx = 0$$

Free-free $$\int_0^L \frac{d\phi_n}{dx}\frac{d^3\phi_m}{dx^3}\,dx = 0$$

24. $$\int_0^L \frac{d^2\phi_n}{dx^2}\frac{d^3\phi_m}{dx^3}\,dx = \frac{\beta_m^4}{\beta_n^4-\beta_m^4}\left[\frac{d\phi_m}{dx}\frac{d^3\phi_n}{dx^3}+\frac{\beta_n^4}{\beta_m^4}\frac{d\phi_n}{dx}\frac{d^3\phi_m}{dx^3}-\beta_n^4\phi_m\phi_n-\frac{d^2\phi_m}{dx^2}\frac{d^2\phi_n}{dx^2}\right]_0^L$$

$m\neq n$

Clamped-free $$\int_0^L \frac{d^2\phi_n}{dx^2}\frac{d^3\phi_m}{dx^3}\,dx = \frac{4\beta_m^4\beta_n^2}{\beta_n^4-\beta_m^4}\,[\beta_m^2-(-1)^{m+n}\beta_n^2]$$

Clamped-pinned $$\int_0^L \frac{d^2\phi_n}{dx^2}\frac{d^3\phi_m}{dx^3}\,dx = \frac{\beta_m^3\beta_n^3}{\beta_n^4-\beta_m^4}\;-(-1)^{m+n}(\beta_n^2+\beta_m^2)\sqrt{(\sigma_m^2+1)(\sigma_n^2+1)}$$
$$+(-1)^m(\beta_n^2-\beta_m^2)\sqrt{(\sigma_m^2+1)(\sigma_n^2-1)}$$
$$-(-1)^n(\beta_n^2-\beta_m^2)\sqrt{(\sigma_m^2-1)(\sigma_n^2+1)}$$
$$+(\beta_n^2+\beta_m^2)\sqrt{(\sigma_m^2-1)(\sigma_n^2-1)}-\frac{4\beta_m^3}{\beta_n}$$

Clamped-clamped $$\int_0^L \frac{d^2\phi_n}{dx^2}\frac{d^3\phi_m}{dx^3}\,dx = \frac{4\beta_m^6\beta_n^2}{\beta_n^4 - \beta_m^4}\,[1 - (-1)^{m+n}]$$

Free-pinned $$\int_0^L \frac{d^2\phi_n}{dx^2}\frac{d^3\phi_m}{dx^3}\,dx = \frac{\beta_m^3\beta_n^3}{\beta_n^4 - \beta_m^4}\,[-(-1)^{m+n}(\beta_n^2+\beta_m^2)\sqrt{(\sigma_m^2+1)}$$
$$-(-1)^m(\beta_n^2-\beta_m^2)\sqrt{(\sigma_m^2+1)(\sigma_n^2-1)}$$
$$+(-1)^n(\beta_n^2-\beta_m^2)\sqrt{(\sigma_m^2-1)(\sigma_n^2+1)}$$
$$+(\beta_n^2+\beta_m^2)\sqrt{(\sigma_m^2-1)(\sigma_n^2-1)} + 4\beta_m\beta_n]$$

Free-free $$\int_0^L \frac{d^2\phi_n}{dx^2}\frac{d^3\phi_m}{dx^3}\,dx = \frac{4\beta_m^4\beta_n^4}{\beta_n^4 - \beta_m^4}\,[1 - (-1)^{m+n}]$$

25. $$\int_0^L x\phi_n\,dx = \frac{1}{\beta_n^4}\left[x\frac{d^3\phi_n}{dx^3} - \frac{d^2\phi_n}{dx^2}\right]_0^L$$

Clamped-free $$\int_0^L x\phi_n\,dx = \frac{2}{\beta_n^2}$$

Clamped-pinned $$\int_0^L x\phi_n\,dx = \frac{1}{\beta_n^2}\,(2 - \beta_n[(-1)^n\sqrt{\sigma_n^2+1} + \sqrt{\sigma_n^2-1}]\,L)$$

Clamped-clamped $$\int_0^L x\phi_n\,dx = \frac{2}{\beta_n^2}\,[1 + (-1)^n - (-1)^n\sigma_n\beta_n L]$$

Free-pinned $$\int_0^L x\phi_n\,dx = \frac{L}{\beta_n}\,[(-1)^n\sqrt{\sigma_n^2+1} - \sqrt{\sigma_n^2-1}]$$

Free-free $$\int_0^L x\phi_n\,dx = 0$$

26. $$\int_0^L x\frac{d\phi_n}{dx}\,dx = \left[x\phi_n - \frac{1}{\beta_n^4}\frac{d^3\phi_n}{dx^3}\right]_0^L$$

Clamped-free $$\int_0^L x\frac{d\phi_n}{dx}\,dx = \frac{2}{\beta_n}\,[\sigma_n + (-1)^n\beta_n L]$$

Clamped-pinned $$\int_0^L x\frac{d\phi_n}{dx}\,dx = \frac{1}{\beta_n}\,[(-1)^n\sqrt{\sigma_n^2+1} + \sqrt{\sigma_n^2-1} - 2\sigma_n]$$

Clamped-clamped $$\int_0^L x\frac{d\phi_n}{dx}\,dx = \frac{2\sigma_n}{\beta_n}\,[(-1)^n - 1]$$

Free-pinned $$\int_0^L x\frac{d\phi_n}{dx}\,dx = \frac{1}{\beta_n}\,[\sqrt{\sigma_n^2-1} - (-1)^n\sqrt{\sigma_n^2+1}]$$

Free-free $\displaystyle\int_0^L x \frac{d\phi_n}{dx}\, dx = -2(-1)^n L$

27. $$\int_0^L x \frac{d^2\phi_n}{dx^2}\, dx = \left[x \frac{d\phi_n}{dx} - \phi_n \right]_0^L$$

Clamped-free $\displaystyle\int_0^L x \frac{d^2\phi_n}{dx^2}\, dx = 2(-1)^n (1 - \sigma_n \beta_n L)$

Clamped-pinned $\displaystyle\int_0^L x \frac{d^2\phi_n}{dx^2}\, dx = \beta_n L[(-1)^n \sqrt{\sigma_n^2 + 1} - \sqrt{\sigma_n^2 - 1}\,]$

Clamped-clamped $\displaystyle\int_0^L x \frac{d^2\phi_n}{dx^2}\, dx = 0$

Free-pinned $\displaystyle\int_0^L x \frac{d^2\phi_n}{dx^2}\, dx = 2 - \beta_n L[(-1)^n \sqrt{\sigma_n^2 + 1} + \sqrt{\sigma_n^2 - 1}\,]$

Free-free $\displaystyle\int_0^L x \frac{d^2\phi_n}{dx^2}\, dx = 2[1 + (-1)^n - (-1)^n \sigma_n \beta_n L]$

28. $$\int_0^L x \frac{d^3\phi_n}{dx^3}\, dx = \left[x \frac{d^2\phi_n}{dx^2} - \frac{d\phi_n}{dx} \right]_0^L$$

Clamped-free $\displaystyle\int_0^L x \frac{d^3\phi_n}{dx^3}\, dx = 2(-1)^n \sigma_n \beta_n$

Clamped-pinned $\displaystyle\int_0^L x \frac{d^3\phi_n}{dx^3}\, dx = \beta_n [\sqrt{\sigma_n^2 - 1} - (-1)^n \sqrt{\sigma_n^2 + 1}\,]$

Clamped-clamped $\displaystyle\int_0^L x \frac{d^3\phi_n}{dx^3}\, dx = -2(-1)^n \beta_n^2 L$

Free-pinned $\displaystyle\int_0^L x \frac{d^3\phi_n}{dx^3}\, dx = \beta_n [\sqrt{\sigma_n^2 + 1}\,(-1)^n + \sqrt{\sigma_n^2 - 1} - 2\sigma_n]$

Free-free $\displaystyle\int_0^L x \frac{d^3\phi_n}{dx^3}\, dx = 2\sigma_n \beta_n [(-1)^n - 1]$

29. $$\int_0^L x^2 \phi_n\, dx = \frac{1}{\beta_n^4} \left[x^2 \frac{d^3\phi_n}{dx^3} - 2x \frac{d^2\phi_n}{dx^2} + 2 \frac{d\phi_n}{dx} \right]_0^L$$

Clamped-free $\displaystyle\int_0^L x^2 \phi_n\, dx = \frac{-4(-1)^n \sigma_n}{\beta_n^3}$

Clamped-pinned $\int_0^L x^2\phi_n\,dx = \frac{1}{\beta_n^3}(\beta_n^2L^2[-(-1)^n\sqrt{\sigma_n^2+1}-\sqrt{\sigma_n^2-1}]$

$+2[(-1)^n\sqrt{\sigma_n^2+1}-\sqrt{\sigma_n^2-1}])$

Clamped-clamped $\int_0^L x^2\phi_n\,dx = \frac{2(-1)^nL}{\beta_n^2}[2-\sigma_n\beta_nL]$

Free-pinned $\int_0^L x^2\phi_n\,dx = \frac{1}{\beta_n^3}(\beta_n^2L^2[(-1)^n\sqrt{\sigma_n^2+1}-\sqrt{\sigma_n^2-1}]$

$-2[(-1)^n\sqrt{\sigma_n^2+1}+\sqrt{\sigma_n^2-1}]+4\sigma_n)$

Free-free $\int_0^L x^2\phi_n\,dx = \frac{4\sigma_n}{\beta_n^3}[1-(-1)^n]$

30. $$\int_0^L x^2\frac{d\phi_n}{dx}\,dx = \frac{1}{\beta_n^4}\left[\beta_n^4x^2\phi_n - 2x\frac{d^3\phi_n}{dx^3} + 2\frac{d^2\phi_n}{dx^2}\right]_0^L$$

Clamped-free $\int_0^L x^2\frac{d\phi_n}{dx}\,dx = -\frac{1}{\beta_n^2}[2+(-1)^n\beta_n^2L^2]$

Clamped-pinned $\int_0^L x^2\frac{d\phi_n}{dx}\,dx = \frac{2}{\beta_n^2}(\beta_nL[(-1)^n\sqrt{\sigma_n^2+1}+\sqrt{\sigma_n^2-1}]-2)$

Clamped-clamped $\int_0^L x^2\frac{d\phi_n}{dx}\,dx = \frac{4}{\beta_n^2}[(-1)^n\sigma_n\beta_nL-(-1)^n-1]$

Free-pinned $\int_0^L x^2\frac{d\phi_n}{dx}\,dx = -\frac{2L}{\beta_n}[(-1)^n\sqrt{\sigma_n^2+1}-\sqrt{\sigma_n^2-1}]$

Free-free $\int_0^L x^2\frac{d\phi_n}{dx}\,dx = -2(-1)^nL^2$

31. $$\int_0^L x^2\frac{d^2\phi_n}{dx^2}\,dx = \left[x^2\frac{d\phi_n}{dx} - 2x\phi_n + \frac{2}{\beta_n^4}\frac{d^3\phi_n}{dx^3}\right]_0^L$$

Clamped-free $\int_0^L x^2\frac{d^2\phi_n}{dx^2}\,dx = \frac{2}{\beta_n}[2\sigma_n+2(-1)^n\beta_nL-(-1)^n\sigma_n\beta_n^2L^2]$

Clamped-pinned $\int_0^L x^2\frac{d^2\phi_n}{dx^2}\,dx = \frac{1}{\beta_n}(\beta_n^2L^2[(-1)^n\sqrt{\sigma_n^2+1}-\sqrt{\sigma_n^2-1}]$

$-2[(-1)^n\sqrt{\sigma_n^2+1}+\sqrt{\sigma_n^2-1}]+4\sigma_n)$

Clamped-clamped $\int_0^L x^2\frac{d^2\phi_n}{dx^2}\,dx = \frac{4\sigma_n}{\beta_n}[1-(-1)^n]$

Free-pinned $\int_0^L x^2 \frac{d^2\phi_n}{dx^2}\,dx = \frac{1}{\beta_n}(2[(-1)^n\sqrt{\sigma_n^2+1} - \sqrt{\sigma_n^2-1}]$

$- \beta_n^2 L^2[(-1)^n\sqrt{\sigma_n^2+1} + \sqrt{\sigma_n^2-1}])$

Free-free $\int_0^L x^2 \frac{d^2\phi_n}{dx^2}\,dx = 2(-1)^n L(2 - \sigma_n\beta_n L)$

32. $$\int_0^L x^2 \frac{d^3\phi_n}{dx^3}\,dx = \left[x^2\frac{d^2\phi_n}{dx^2} - 2x\frac{d\phi_n}{dx} + 2\phi_n\right]_0^L$$

Clamped-free $\int_0^L x^2 \frac{d^3\phi_n}{dx^3}\,dx = 4(-1)^n(\sigma_n\beta_n L - 1)$

Clamped-pinned $\int_0^L x^2 \frac{d^3\phi_n}{dx^3}\,dx = -2L[(-1)^n\sqrt{\sigma_n^2+1} - \sqrt{\sigma_n^2-1}]$

Clamped-clamped $\int_0^L x^2 \frac{d^3\phi_n}{dx^3}\,dx = -2(-1)^n\beta_n^2 L^2$

Free-pinned $\int_0^L x^2 \frac{d^3\phi_n}{dx^3}\,dx = 4[(-1)^n\sigma_n\beta_n L - 4(-1)^n - 4]$

Free-free $\int_0^L x^2 \frac{d^3\phi_n}{dx^3}\,dx = 2(\beta_n L[(-1)^n\sqrt{\sigma_n^2+1} + \sqrt{\sigma_n^2-1}] - 2)$

33. $$\int_0^L x^3\phi_n\,dx = \frac{1}{\beta_n^4}\left[x^3\frac{d^3\phi_n}{dx^3} - 3x^2\frac{d^2\phi_n}{dx^2} + 6x\frac{d\phi_n}{dx} - 6\phi_n\right]_0^L$$

Clamped-free $\int_0^L x^3\phi_n\,dx = \frac{12(-1)^n}{\beta_n^4}(1 - \sigma_n\beta_n L)$

Clamped-pinned $\int_0^L x^3\phi_n\,dx = \frac{L}{\beta_n^3}(6[(-1)^n\sqrt{\sigma_n^2+1} - \sqrt{\sigma_n^2-1}]$

$- \beta_n^2 L^2[(-1)^n\sqrt{\sigma_n^2+1} + \sqrt{\sigma_n^2-1}])$

Clamped-clamped $\int_0^L x^3\phi_n\,dx = \frac{2(-1)^n L^2}{\beta_n^2}(3 - \sigma_n\beta_n L)$

Free-pinned $\int_0^L x^3\phi_n\,dx = \frac{1}{\beta_n^4}(\beta_n^3 L^3[(-1)^n\sqrt{\sigma_n^2+1} - \sqrt{\sigma_n^2-1}]$

$- 6\beta_n L[(-1)^n\sqrt{\sigma_n^2+1} + \sqrt{\sigma_n^2-1}] + 12)$

Free-free $\int_0^L x^3\phi_n\,dx = \frac{12}{\beta_n^4}[1 + (-1)^n - (-1)^n\sigma_n\beta_n L]$

34. $$\int_0^L x^3 \frac{d\phi_n}{dx}\,dx = \frac{1}{\beta_n^4}\left[\beta_n^4 x^3 \phi_n - 3x^2 \frac{d^3\phi_n}{dx^3} + 6x \frac{d^2\phi_n}{dx^2} - 6\frac{d\phi_n}{dx}\right]_0^L$$

Clamped-free $$\int_0^L x^3 \frac{d\phi_n}{dx}\,dx = \frac{2(-1)^n}{\beta_n^3}(6\sigma_n - \beta_n^3 L^3)$$

Clamped-pinned $$\int_0^L x^3 \frac{d\phi_n}{dx}\,dx = \frac{3}{\beta_n^3}\left(\beta_n^2 L^2 [(-1)^n \sqrt{\sigma_n^2+1} + \sqrt{\sigma_n^2-1}] - 2[(-1)^n \sqrt{\sigma_n^2+1} - \sqrt{\sigma_n^2-1}]\right)$$

Clamped-clamped $$\int_0^L x^3 \frac{d\phi_n}{dx}\,dx = \frac{6(-1)^n L}{\beta_n^2}(\sigma_n \beta_n L - 2)$$

Free-pinned $$\int_0^L x^3 \frac{d\phi_n}{dx}\,dx = \frac{3}{\beta_n^3}\left(2[(-1)^n \sqrt{\sigma_n^2+1} + \sqrt{\sigma_n^2-1}] - \beta_n^2 L^2 [(-1)^n \sqrt{\sigma_n^2+1} - \sqrt{\sigma_n^2-1}] - 4\sigma_n\right)$$

Free-free $$\int_0^L x^3 \frac{d\phi_n}{dx}\,dx = \frac{2}{\beta_n^3}\left(6\sigma_n[(-1)^n - 1] - (-1)^n \beta_n^3 L^3\right)$$

35. $$\int_0^L x^3 \frac{d^2\phi_n}{dx^2}\,dx = \frac{1}{\beta_n^4}\left[\beta_n^4 x^3 \frac{d\phi_n}{dx} - 3\beta_n^4 x^2 \phi_n + 6x \frac{d^3\phi_n}{dx^3} - 6\frac{d^2\phi_n}{dx^2}\right]_0^L$$

Clamped-free $$\int_0^L x^3 \frac{d^2\phi_n}{dx^2}\,dx = \frac{2}{\beta_n^2}[3(-1)^n \beta_n^2 L^2 - (-1)^n \sigma_n \beta_n^3 L^3 + 6]$$

Clamped-pinned $$\int_0^L x^3 \frac{d^2\phi_n}{dx^2}\,dx = \frac{1}{\beta_n^2}\left(\beta_n^3 L^3 [(-1)^n \sqrt{\sigma_n^2+1} - \sqrt{\sigma_n^2-1}] - 6\beta_n L[(-1)^n \sqrt{\sigma_n^2+1} + \sqrt{\sigma_n^2-1}] + 12\right)$$

Clamped-clamped $$\int_0^L x^3 \frac{d^2\phi_n}{dx^2}\,dx = \frac{12}{\beta_n^2}[1 + (-1)^n - (-1)^n \sigma_n \beta_n L]$$

Free-pinned $$\int_0^L x^3 \frac{d^2\phi_n}{dx^2}\,dx = \frac{L}{\beta_n}\left(6[(-1)^n \sqrt{\sigma_n^2+1} - \sqrt{\sigma_n^2-1}] - \beta_n^2 L^2 [(-1)^n \sqrt{\sigma_n^2+1} + \sqrt{\sigma_n^2-1}]\right)$$

Free-free $$\int_0^L x^3 \frac{d^2\phi_n}{dx^2}\,dx = 2(-1)^n L^2(3 - \sigma_n \beta_n L)$$

36. $$\int_0^L x^3 \frac{d^3\phi_n}{dx^3}\,dx = \frac{1}{\beta_n^4}\left[\beta_n^4 x^3 \frac{d^2\phi_n}{dx^2} - 3\beta_n^4 x^2 \frac{d\phi_n}{dx} + 6\beta_n^4 x \phi_n - 6\frac{d^3\phi_n}{dx^3}\right]_0^L$$

Clamped-free $$\int_0^L x^3 \frac{d^3\phi_n}{dx^3}\,dx = \frac{6}{\beta_n}[(-1)^n \sigma_n\beta_n^2L^2 - 2(-1)^n \beta_n L - 2\sigma_n]$$

Clamped-pinned $$\int_0^L x^3 \frac{d^3\phi_n}{dx^3}\,dx = \frac{3}{\beta_n}(2[(-1)^n \sqrt{\sigma_n^2+1} + \sqrt{\sigma_n^2-1}] - \beta_n^2L^2[(-1)^n \sqrt{\sigma_n^2+1} - \sqrt{\sigma_n^2-1}] - 4\sigma_n)$$

Clamped-clamped $$\int_0^L x^3 \frac{d^3\phi_n}{dx^3}\,dx = \frac{2}{\beta_n}(6\sigma_n[(-1)^n - 1] - (-1)^n \beta_n^3L^3)$$

Free-pinned $$\int_0^L x^3 \frac{d^3\phi_n}{dx^3}\,dx = \frac{3}{\beta_n}(\beta_n^2L^2[(-1)^n \sqrt{\sigma_n^2+1} + \sqrt{\sigma_n^2-1}] - 2[(-1)^n \sqrt{\sigma_n^2+1} - \sqrt{\sigma_n^2-1}])$$

Free-free $$\int_0^L x^3 \frac{d^3\phi_n}{dx^3}\,dx = 6(-1)^n(\sigma_n\beta_n L - 2)$$

37. $$\int_0^L x^4\phi_n\,dx = \frac{1}{\beta_n^4}\left[x^4\frac{d^3\phi_n}{dx^3} - 4x^3\frac{d^2\phi_n}{dx^2} + 12x^2\frac{d\phi_n}{dx} - 24x\phi_n + \frac{24}{\beta_n^4}\frac{d^3\phi_n}{dx^3}\right]_0^L$$

Clamped-free $$\int_0^L x^4\phi_n\,dx = \frac{24}{\beta_n^5}[2\sigma_n + 2(-1)^n\beta_n L - (-1)^n\sigma_n\beta_n^2L^2]$$

Clamped-pinned $$\int_0^L x^4\phi_n\,dx = \frac{1}{\beta_n^5}(12\beta_n^2L^2[(-1)^n \sqrt{\sigma_n^2+1} - \sqrt{\sigma_n^2-1}] - [\beta_n^4L^4 + 24][(-1)^n \sqrt{\sigma_n^2+1} + \sqrt{\sigma_n^2-1}] + 48\sigma_n)$$

Clamped-clamped $$\int_0^L x^4\phi_n\,dx = \frac{2}{\beta_n^5}(4(-1)^n\beta_n^3L^3 + 24\sigma_n[1 - (-1)^n] - (-1)^n\sigma_n\beta_n^4L^4)$$

Free-pinned $$\int_0^L x^4\phi_n\,dx = \frac{1}{\beta_n^5}([\beta_n^4L^4 + 24][(-1)^n \sqrt{\sigma_n^2+1} - \sqrt{\sigma_n^2-1}] - 12\beta_n^2L^2[(-1)^n \sqrt{\sigma_n^2+1} + \sqrt{\sigma_n^2-1}])$$

Free-free $$\int_0^L x^4\phi_n\,dx = \frac{24(-1)^n L}{\beta_n^4}[2 - \sigma_n\beta_n L]$$

38. $$\int_0^L x^4\frac{d\phi_n}{dx}\,dx = \frac{1}{\beta_n^4}\left[\beta_n^4x^4\phi_n - 4x^3\frac{d^3\phi_n}{dx^3} + 12x^2\frac{d^2\phi_n}{dx^2} - 24x\frac{d\phi_n}{dx} + 24\phi_n\right]_0^L$$

Clamped-free $$\int_0^L x^4\frac{d\phi_n}{dx}\,dx = \frac{2(-1)^n}{\beta_n^4}[24(\sigma_n\beta_n L - 1) - \beta_n^4L^4]$$

Clamped-pinned $$\int_0^L x^4 \frac{d\phi_n}{dx}\,dx = \frac{4L}{\beta_n^3}\left(\beta_n^2 L^2[(-1)^n\sqrt{\sigma_n^2+1} + \sqrt{\sigma_n^2-1}] - 6[(-1)^n\sqrt{\sigma_n^2+1} - \sqrt{\sigma_n^2-1}]\right)$$

Clamped-clamped $$\int_0^L x^4 \frac{d\phi_n}{dx}\,dx = \frac{8(-1)^n L^2}{\beta_n^2}(\sigma_n\beta_n L - 3)$$

Free-pinned $$\int_0^L x^4 \frac{d\phi_n}{dx}\,dx = \frac{4}{\beta_n^4}\left(6\beta_n[(-1)^n\sqrt{\sigma_n^2+1} + \sqrt{\sigma_n^2-1}]L - \beta_n^3 L^3[(-1)^n\sqrt{\sigma_n^2+1} - \sqrt{\sigma_n^2-1}] - 12\right)$$

Free-free $$\int_0^L x^4 \frac{d\phi_n}{dx}\,dx = \frac{2(-1)^n}{\beta_n^4}\left(24\sigma_n\beta_n L - \beta_n^4 L^4 - 24[1+(-1)^n]\right)$$

39. $$\int_0^L x^4 \frac{d^2\phi_n}{dx^2}\,dx = \frac{1}{\beta_n^4}\left[\beta_n^4 x^4 \frac{d\phi_n}{dx} - 4\beta_n^4 x^3\phi_n + 12x^2\frac{d^3\phi_n}{dx^3} - 24x\frac{d^2\phi_n}{dx^2} + 24\frac{d\phi_n}{dx}\right]_0^L$$

Clamped-free $$\int_0^L x^4 \frac{d^2\phi_n}{dx^2}\,dx = \frac{2(-1)^n}{\beta_n^3}\left(4\beta_n^3 L^3 - \sigma_n\beta_n^4 L^4 - 24\sigma_n\right)$$

Clamped-pinned $$\int_0^L x^4 \frac{d^2\phi_n}{dx^2}\,dx = \frac{1}{\beta_n^3}\left([\beta_n^4 L^4 + 24][(-1)^n\sqrt{\sigma_n^2+1} - \sqrt{\sigma_n^2-1}] - 12\beta_n^2 L^2[(-1)^n\sqrt{\sigma_n^2+1} + \sqrt{\sigma_n^2-1}]\right)$$

Clamped-clamped $$\int_0^L x^4 \frac{d^2\phi_n}{dx^2}\,dx = \frac{24(-1)^n L}{\beta_n^2}(2 - \sigma_n\beta_n L)$$

Free-pinned $$\int_0^L x^4 \frac{d^2\phi_n}{dx^2}\,dx = \frac{1}{\beta_n^3}\left(12\beta_n^2 L^2[(-1)^n\sqrt{\sigma_n^2+1} - \sqrt{\sigma_n^2-1}] - [\beta_n^4 L^4 + 24][(-1)^n\sqrt{\sigma_n^2+1} + \sqrt{\sigma_n^2-1}] + 48\sigma_n\right)$$

Free-free $$\int_0^L x^4 \frac{d^2\phi_n}{dx^2}\,dx = \frac{2}{\beta_n^3}\left(4(-1)^n\beta_n^3 L^3 + 24\sigma_n[1-(-1)^n] - (-1)\,\sigma_n\beta_n^4 L^4\right)$$

40. $$\int_0^L x^4 \frac{d^3\phi_n}{dx^3}\,dx = \frac{1}{\beta_n^4}\left[\beta_n^4 x^4 \frac{d^2\phi_n}{dx^2} - 4\beta_n^4 x^3\frac{d\phi_n}{dx} + 12\beta_n^4 x^2\phi_n - 24x\frac{d^3\phi_n}{dx^3} + 24\frac{d^2\phi_n}{dx^2}\right]_0^L$$

Clamped-free $$\int_0^L x^4 \frac{d^3\phi_n}{dx^3}\,dx = \frac{8}{\beta_n^2}\left[(-1)^n\sigma_n\beta_n^3 L^3 - 3(-1)^n\beta_n^2 L^2 - 6\right]$$

Clamped-pinned $$\int_0^L x^4 \frac{d^3\phi_n}{dx^3}\,dx = \frac{4}{\beta_n^2}\left(6\beta_n L[(-1)^n\sqrt{\sigma_n^2+1} + \sqrt{\sigma_n^2-1}] - \beta_n^3 L^3[(-1)^n\sqrt{\sigma_n^2+1} - \sqrt{\sigma_n^2-1}] - 12\right)$$

Clamped-clamped $\int_0^L x^4 \frac{d^3\phi_n}{dx^3}\,dx = \frac{2}{\beta_n^4}(24(-1)^n \sigma_n\beta_n L - 24[1+(-1)^n] - (-1)^n \beta_n^4 L^4)$

Free-pinned $\int_0^L x^4 \frac{d^3\phi_n}{dx^3}\,dx = \frac{4L}{\beta_n}(\beta_n^2 L^2[(-1)^n\sqrt{\sigma_n^2+1} + \sqrt{\sigma_n^2-1}]$

$- 6[(-1)^n\sqrt{\sigma_n^2+1} - \sqrt{\sigma_n^2-1}])$

Free-free $\int_0^L x^4 \frac{d^3\phi_n}{dx^3}\,dx = 8(-1)^n L^2(\sigma_n\beta_n L - 3)$

41. $$\int_0^L x^p\phi_n\,dx = \frac{1}{\beta_n^4}\left[x^p\frac{d^3\phi_n}{dx^3} - \frac{dx^p}{dx}\frac{d^2\phi_n}{dx^2} + \frac{d^2x^p}{dx^2}\frac{d\phi_n}{dx} - \frac{d^3x^p}{dx^3}\phi_n\right]_0^L + \frac{1}{\beta_n^4}\int_0^L \frac{d^4x^p}{dx^4}\phi_n\,dx$$

$p = 1, 2, 3, \ldots$

Formula 41 is a recurrence formula which may be used for any integer p. It may also be modified to evaluate integrals of the type $\int_0^L x^p\psi_m\,dx$ where ψ_m is a derivative of a characteristic function.

For Example: $\psi_m = \frac{d^2\phi_n}{dx^2}$

$$\int_0^L x^p\frac{d^2\phi_n}{dx^2}\,dx = \frac{1}{\beta_n^4}\left[\beta_n^4 x^p\frac{d\phi_n}{dx} - \beta_n^4\frac{dx^p}{dx}\phi_n + \frac{d^2x^p}{dx^2}\frac{d^3\phi_n}{dx^3} - \frac{d^3x^p}{dx^3}\frac{d^2\phi_n}{dx^2}\right]_0^L$$

$$+ \frac{1}{\beta_n^4}\int_0^L \frac{d^4x^p}{dx^4}\frac{d^2\phi_n}{dx^2}\,dx$$

42. $$\int_0^L \phi_n\psi_m\,dx = \frac{1}{\beta_n^4-\lambda_m^4}\left[\psi_m\frac{d^3\phi_n}{dx^3} - \phi_n\frac{d^3\psi_m}{dx^3} + \frac{d\phi_n}{dx}\frac{d^2\psi_m}{dx^2} - \frac{d\psi_m}{dx}\frac{d^2\phi_n}{dx^2}\right]_0^L$$

ϕ_n, ψ_m are characteristic functions; β_m is the characteristic value associated with ϕ_n, λ_m is the characteristic value associated with ψ_m.

Formula 42 was derived on the assumption that ψ_m is a characteristic function satisfying the differential equation $d^4\psi_m/dx^4 = \lambda_m^4\psi_m$. However if ψ_m is a function which is integrable in the interval $\sigma \leqslant x \leqslant L$ we may write:

43. $$\int_0^L \phi_n\psi_m\,dx = \frac{1}{\beta_n^4}\left[\psi_m\frac{d^3\phi_n}{dx^3} - \frac{d\psi_m}{dx}\frac{d^2\phi_n}{dx^2} + \frac{d^2\psi_m}{dx^2}\frac{d\phi_n}{dx} - \frac{d^3\psi_m}{dx^3}\phi_n\right]_0^L$$

$$+ \frac{1}{\beta_n^4}\int_0^L \frac{d^4\psi_m}{dx^4}\phi_n\,dx$$

Formula 43 is the fundamental integral from which all the previous integrals 1–42, inclusive, may be derived.

The substitutions in formula 43 are simple and straightforward in each previous case except integrals 5, 6, 7, and 8. The procedure to be followed in the case of 5 is given below; 6, 7, and 8 are evaluated in a similar manner.

Derivation of Formula 5 from Formula 43

Take formula 43 in the form of 42:

$$\int_0^L \phi_n \psi_m \, dx = \frac{1}{\beta_n^4 - \lambda_m^4}\left[\psi_m \frac{d^3\phi_n}{dx^3} - \phi_n \frac{d^3\psi_m}{dx^3} + \frac{d\phi_n}{dx}\frac{d^2\psi_m}{dx^2} - \frac{d\psi_m}{dx}\frac{d^2\phi_n}{dx^2}\right]_0^l$$

Make the following substitutions:

$$\beta_n = \beta_m + \Delta\beta_m, \quad \phi_n = \phi_m + \Delta\phi_m, \quad \psi_m = \phi_m, \quad \lambda_m = \beta_m.$$

Then we have $\beta_n^4 - \lambda_m^4 = (\beta_m + \Delta\beta_m)^4 - \beta_m^4 = 4\beta_m^3 \Delta\beta_m$, neglecting all increments raised to powers higher than 1. Therefore,

$$4\beta_m^3 \Delta\beta_m \int_0^L (\phi_m + \Delta\phi_m)\,\phi_m \, dx$$

$$= \left[\phi_m \frac{d^3\phi_m}{dx^3} + \phi_m \frac{d^3\Delta\phi_m}{dx^3} - \phi_m \frac{d^3\phi_m}{dx^3} - \Delta\phi_m \frac{d^3\phi_m}{dx^3} + \frac{d\phi_m}{dx}\frac{d^2\phi_m}{dx^2} + \frac{d\Delta\phi_m}{dx}\frac{d^2\phi_m}{dx^2}\right.$$

$$\left. - \frac{d\phi_m}{dx}\frac{d^2\phi_m}{dx^2} - \frac{d\phi_m}{dx}\frac{d^2\Delta\phi_m}{dx^2}\right]_0^L$$

Dividing by $\Delta\beta_m$ and taking the limit as $\Delta\beta_m \to 0$, we have the following relations:

$$\lim_{\Delta\beta_m \to 0} \frac{1}{\Delta\beta_m}\frac{d^3\Delta\phi_m}{dx^3} = \frac{3}{\beta_m}\frac{d^3\phi_m}{dx^3} + \beta_m^3 x \phi_m$$

$$\lim_{\Delta\beta_m \to 0} \frac{\Delta\phi_m}{\Delta\beta_m} = \frac{x}{\beta_m}\frac{d\phi_m}{dx}$$

$$\lim_{\Delta\beta_m \to 0} \frac{1}{\Delta\beta_m}\frac{d^2\Delta\phi_m}{dx^2} = \frac{2}{\beta_m}\frac{d^2\phi_m}{dx^2} + \frac{x}{\beta_m}\frac{d^3\phi_m}{dx^3}$$

$$\lim_{\Delta\beta_m \to 0} \frac{1}{\Delta\beta_m}\frac{d\Delta\phi_m}{dx} = \frac{1}{\beta_m}\frac{d\phi_m}{dx} + \frac{x}{\beta_m}\frac{d^2\phi_m}{dx^2}$$

Therefore

$$\int_0^L \phi_n^2 \, dx = \frac{1}{4}\left[\frac{3}{\beta_n^4}\phi_n \frac{d^3\phi_n}{dx^3} + x\phi_n^2 - \frac{2x}{\beta_n^4}\frac{d\phi_n}{dx}\frac{d^3\phi_n}{dx^3} - \frac{1}{\beta_n^4}\frac{d\phi_n}{dx}\frac{d^2\phi_n}{dx^2} + \frac{x}{\beta_n^4}\left(\frac{d^2\phi_n}{dx^2}\right)^2\right]_0^L$$

In the following group of integrals ϕ_n and ψ_m are characteristic functions satisfying different boundary conditions; β_n and λ_m are characteristic values, and σ_n and μ_m are constants associated with the respective characteristic functions.

44. ϕ_n, characteristic function for *clamped-free* beam. ψ_m, characteristic function for beam with any other boundary conditions.

$$\int_0^L \phi_n \psi_m \, dx = \frac{1}{\beta_n^4 - \lambda_m^4}\left[2(-1)^n \left(\frac{d^3\psi_m}{dx^3}\right)_{x=L} - 2(-1)^n \sigma_n \beta_n \left(\frac{d^2\psi_m}{dx^2}\right)_{x=L}\right.$$

$$\left. + 2\sigma_n \beta_n^3 \psi_m(0) + 2\beta_n^2 \left(\frac{d\psi_m}{dx}\right)_{x=0}\right]$$

45. ϕ_n, characteristic function for *clamped-pinned* beam. ψ_m, characteristic function for beam with any other boundary conditions.

$$\int_0^L \phi_n \psi_m \, dx = \frac{\beta_n}{\beta_n^4 - \lambda_m^4} \left[-\beta_n^2 ((-1)^n \sqrt{\sigma_n^2 + 1} + \sqrt{\sigma_n^2 - 1}) \psi_m(L) \right.$$

$$+ ((-1)^n \sqrt{\sigma_n^2 + 1} - \sqrt{\sigma_n^2 - 1}) \left(\frac{d^2 \psi_m}{dx^2} \right)_{x=L}$$

$$\left. + 2\sigma_n \beta_n^2 \psi_m(0) + 2\beta_n \left(\frac{d\psi_m}{dx} \right)_{x=0} \right]$$

46. ϕ_n, characteristic function for *clamped-clamped* beam. ψ_m, characteristic function for beam with any other boundary conditions.

$$\int_0^L \phi_n \psi_m \, dx = \frac{1}{\beta_n^4 - \lambda_m^4}$$

$$\cdot 2\beta_n^2 \left[-(-1)^n \sigma_n \beta_n \psi_m(L) + (-1)^n \left(\frac{d\psi_m}{dx} \right)_{x=L} + \sigma_n \beta_n \psi_m(0) + \left(\frac{d\psi_m}{dx} \right)_{x=0} \right]$$

47. ϕ_n, characteristic function for *free-pinned* beam. ψ_m, characteristic function for beam with any other boundary conditions.

$$\int_0^L \phi_n \psi_m \, dx = \frac{1}{\beta_n^4 - \lambda_m^4} \left(\beta_n^3 [(-1)^n \sqrt{\sigma_n^2 + 1} - \sqrt{\sigma_n^2 - 1}] \psi_m(L) \right.$$

$$\left. - \beta_n [(-1)^n \sqrt{\sigma_n^2 + 1} + \sqrt{\sigma_n^2 - 1}] \left(\frac{d^2 \psi_m}{dx^2} \right)_{x=L} + 2 \left(\frac{d^3 \psi_m}{dx^3} \right)_{x=0} - 2\sigma_n \beta_n \left(\frac{d^2 \psi_m}{dx^2} \right)_{x=0} \right)$$

48. ϕ_n, characteristic function for *free-free* beam. ψ_m, characteristic function for beam with any other boundary conditions.

$$\int_0^L \phi_n \psi_m \, dx = \frac{2}{\beta_n^4 - \lambda_m^4} \left[(-1)^n \left(\frac{d^3 \psi_m}{dx^3} \right)_{x=L} \right.$$

$$\left. - (-1)^n \sigma_n \beta_n \left(\frac{d^2 \psi_m}{dx^2} \right)_{x=L} + \left(\frac{d^3 \psi_m}{dx^3} \right)_{x=0} + \sigma_n \beta_n \left(\frac{d^2 \psi_m}{dx^2} \right)_{x=0} \right]$$

49. ϕ_n, characteristic function for *pinned-pinned* beam. ψ_m, characteristic function for beam with any other boundary conditions.

$$\beta_n = \frac{n\pi}{L}$$

$$\int_0^L \phi_n \psi_m \, dx = \frac{\sqrt{2}\, \dfrac{n\pi}{L}}{\dfrac{n^4 \pi^4}{L^4} - \lambda_m^4}$$

$$\cdot \left[-(-1)^n \frac{n^2 \pi^2}{L^2} \psi_m(L) + (-1)^n \left(\frac{d^2 \psi_m}{dx^2} \right)_{x=L} + \frac{n^2 \pi^2}{2} \psi_m(0) - \left(\frac{d^2 \psi_m}{dx^2} \right)_{x=0} \right]$$

50. ϕ_n, characteristic function for *clamped-free* beam. ψ_m, characteristic function for *clamped-pinned* beam.

$$\int_0^L \phi_n \psi_m \, dx = \frac{-2(-1)^n \lambda_m^3}{\beta_n^4 - \lambda_m^4} [(-1)^m \sqrt{\mu_m^2 + 1} - \sqrt{\mu_m^2 - 1}]$$

51. ϕ_n, characteristic function for *clamped-free* beam. ψ_m, characteristic function for *clamped-clamped* beam.

$$\int_0^L \phi_n \psi_m \, dx = \frac{4(-1)^{m+n} \lambda_m^2}{\beta_n^4 - \lambda_m^4} (\sigma_n \beta_n - \mu_m \lambda_m)$$

52. ϕ_n, characteristic function for *clamped-free* beam. ψ_m, characteristic function for *free-pinned* beam.

$$\int_0^L \phi_n \psi_m \, dx = \frac{2}{\beta_n^4 - \lambda_m^4} ((-1)^n \lambda_m^3 [(-1)^m \sqrt{\mu_m^2 + 1} - \sqrt{\mu_m^2 - 1}] + 2\sigma_n \beta_n^3 - 2\beta_n^2 \mu_m \lambda_m)$$

53. ϕ_n, characteristic function for *clamped-free* beam. ψ_m, characteristic function for *free-free* beam.

$$\int_0^L \phi_n \psi_m \, dx = \frac{4\beta_n^2}{\beta_n^4 - \lambda_m^4} (\sigma_n \beta_n - \mu_m \lambda_m)$$

54. ϕ_n, characteristic function for *clamped-free* beam. ψ_m, characteristic function for *pinned-pinned* beam.

$$\int_0^L \phi_n \psi_m \, dx = \frac{2\sqrt{2}\,\dfrac{m\pi}{L}}{\beta_n^4 - \dfrac{m^4\pi^4}{L^4}} \left[\beta_n^2 - (-1)^{m+n} \frac{m^2\pi^2}{L^2}\right]$$

55. ϕ_n, characteristic function for *clamped-pinned* beam. ψ_m, characteristic function for *clamped-clamped* beam.

$$\int_0^L \phi_n \psi_m \, dx = \frac{-2(-1)^m \beta_n \lambda_m^2}{\beta_n^4 - \lambda_m^4} [(-1)^n \sqrt{\sigma_n^2 + 1} - \sqrt{\sigma_n^2 - 1}]$$

56. ϕ_n, characteristic function for *clamped-pinned* beam. ψ_m, characteristic function for *free-pinned* beam.

$$\int_0^L \phi_n \psi_m \, dx = \frac{4\beta_n^2}{\beta_n^4 - \lambda_m^4} (\sigma_n \beta_n - \mu_m \lambda_m)$$

57. ϕ_n, characteristic function for *clamped-pinned* beam. ψ_m, characteristic function for *free-free* beam.

$$\int_0^L \phi_n \psi_m \, dx = \frac{2\beta_n^2}{\beta_n^4 - \lambda_m^4} ((-1)^m \beta_n [(-1)^n \sqrt{\sigma_n^2 + 1} + \sqrt{\sigma_n^2 - 1}] + 2\sigma_n \beta_n - 2\mu_m \lambda_m)$$

58. ϕ_n, characteristic function for *clamped-pinned* beam. ψ_m, characteristic function for *pinned-pinned* beam.

$$\int_0^L \phi_n \psi_m \, dx = \frac{2\sqrt{2}\,\beta_n^2}{\beta_n^4 - \dfrac{m^4\pi^4}{L^4}} \frac{m\pi}{L}$$

59. ϕ_n, characteristic function for *clamped-clamped* beam. ψ_m, characteristic function for *free-pinned* beam.

$$\int_0^L \phi_n \psi_m \, dx = \frac{2\beta_n^2}{\beta_n^4 - \lambda_m^4} \left(-(-1)^n \lambda_m \left[(-1)^m \sqrt{\mu_m^2 + 1} + \sqrt{\mu_m^2 - 1}\right] + 2\sigma_n \beta_n - 2\mu_m \lambda_m\right)$$

60. ϕ_n, characteristic function for *clamped-clamped* beam. ψ_m, characteristic function for *free-free* beam.

$$\int_0^L \phi_n \psi_m \, dx = \frac{4\beta_n^2}{\beta_n^4 - \lambda_m^4} \left[1 + (-1)^{m+n}\right] \left[\sigma_n \beta_n - \mu_m \lambda_m\right]$$

61. ϕ_n, characteristic function for *clamped-clamped* beam. ψ_m, characteristic function for *pinned-pinned* beam.

$$\int_0^L \phi_n \psi_m \, dx = \left[1 + (-1)^{m+n}\right] \frac{2\sqrt{2}\,\beta_n^2 \dfrac{m\pi}{L}}{\beta_n^4 - \dfrac{m^4\pi^4}{L^4}}$$

62. ϕ_n, characteristic function for *free-pinned* beam. ψ_m, characteristic function for *free-free* beam.

$$\int_0^L \phi_n \psi_m \, dx = \frac{-2(-1)^m \beta_n^3}{\beta_n^4 - \lambda_m^4} \left[(-1)^n \sqrt{\sigma_n^2 + 1} - \sqrt{\sigma_n^2 - 1}\right]$$

63. ϕ_n, characteristic function for *free-pinned* beam. ψ_m, characteristic function for *pinned-pinned* beam.

$$\int_0^L \phi_n \psi_m \, dx = \frac{-2\sqrt{2}}{\beta_n^4 - \dfrac{m^4\pi^4}{L^4}} \frac{m^3\pi^3}{L^3}$$

64. ϕ_n, characteristic function for *free-free* beam. ψ_m, characteristic function for *pinned-pinned* beam.

$$\int_0^L \phi_n \psi_m \, dx = -2\sqrt{2}\left[1 + (-1)^{m+n}\right] \frac{\dfrac{m^3\pi^3}{L^3}}{\beta_n^4 - \dfrac{m^4\pi^4}{L^4}}$$

65. $$\int_0^L x\phi_n^2\,dx = \frac{1}{\beta_n^4}\left[\frac{3x}{4}\phi_n\frac{d^3\phi_n}{dx^3} - \phi_n\frac{d^2\phi_n}{dx^2} + \frac{5}{8}\left(\frac{d\phi_n}{dx}\right)^2 - \frac{x}{4}\frac{d\phi_n}{dx}\frac{d^2\phi_n}{dx^2}\right.$$
$$\left. + \frac{x^2}{8}\left(\frac{d^2\phi_n}{dx^2}\right)^2 - \frac{x^2}{4}\frac{d\phi_n}{dx}\frac{d^3\phi_n}{dx^3} + \frac{\beta_n^4}{8}x^2\phi_n^2 + \frac{1}{8\beta_n^4}\left(\frac{d^3\phi_n}{dx^3}\right)^2\right]_0^L$$

Clamped-free $$\int_0^L x\phi_n^2\,dx = \frac{1}{\beta_n^2}\left(2\sigma_n^2 + \frac{\beta_n^2 L^2}{2}\right)$$

Clamped-pinned $$\int_0^L x\phi_n^2\,dx = \frac{1}{\beta_n^2}\left[\sigma_n^2 + \frac{\beta_n^2 L^2}{2} - (-1)^n\sqrt{\sigma_n^4 - 1}\right]$$

Clamped-clamped $$\int_0^L x\phi_n^2\,dx = \frac{L^2}{2}$$

Free-pinned $$\int_0^L x\phi_n^2\,dx = \frac{1}{\beta_n^2}\left[-\sigma_n^2 + \frac{\beta_n^2 L^2}{2} + (-1)^n\sqrt{\sigma_n^4 - 1}\right]$$

Free-free $$\int_0^L x\phi_n^2\,dx = \frac{L^2}{2}$$

66. $$\int_0^L x\left(\frac{d\phi_n}{dx}\right)^2 dx = \frac{1}{8}\left[6x\phi_n\frac{d\phi_n}{dx} + x^2\left(\frac{d\phi_n}{dx}\right)^2 - 2x^2\phi_n\frac{d^2\phi_n}{dx^2} - \frac{2x}{\beta_n^4}\frac{d^2\phi_n}{dx^2}\frac{d^3\phi_n}{dx^3}\right.$$
$$\left. + \frac{x^2}{\beta_n^4}\left(\frac{d^3\phi_n}{dx^3}\right)^2 - 3\phi_n^2 + \frac{1}{\beta_n^4}\left(\frac{d^2\phi_n}{dx^2}\right)^2\right]_0^L$$

Clamped-free $$\int_0^L x\left(\frac{d\phi_n}{dx}\right)^2 dx = \frac{1}{2}(3\sigma_n\beta_n L + \sigma_n^2\beta_n^2 L^2 - 4)$$

Clamped-pinned $$\int_0^L x\left(\frac{d\phi_n}{dx}\right)^2 dx = \frac{1}{2}(\sigma_n^2\beta_n^2 L^2 - 1)$$

Clamped-clamped $$\int_0^L x\left(\frac{d\phi_n}{dx}\right)^2 dx = \frac{\sigma_n\beta_n L}{2}(\sigma_n\beta_n L - 2)$$

Free-pinned $$\int_0^L x\left(\frac{d\phi_n}{dx}\right)^2 dx = \frac{1}{2}(\sigma_n^2\beta_n^2 L^2 + 3)$$

Free-free $$\int_0^L x\left(\frac{d\phi_n}{dx}\right)^2 dx = \frac{\sigma_n\beta_n L}{2}(\sigma_n\beta_n L + 6)$$

67. $$\int_0^L x\left(\frac{d^2\phi_n}{dx^2}\right)^2 dx = \frac{1}{8}\left[6x\frac{d\phi_n}{dx}\frac{d^2\phi_n}{dx^2} - 3\left(\frac{d\phi_n}{dx}\right)^2 + x^2\left(\frac{d^2\phi_n}{dx^2}\right)^2 - 2x^2\frac{d\phi_n}{dx}\frac{d^3\phi_n}{dx^3}\right.$$
$$\left. - 2x\phi_n\frac{d^3\phi_n}{dx^3} + \beta_n^4 x^2\phi_n^2 + \frac{1}{\beta_n^4}\left(\frac{d^3\phi_n}{dx^3}\right)^2\right]_0^L$$

Clamped-free $\displaystyle\int_0^L x\left(\frac{d^2\phi_n}{dx^2}\right)^2 dx = \frac{\beta_n^2}{2}(\beta_n^2 L^2 - 4\sigma_n^2)$

Clamped-pinned $\displaystyle\int_0^L x\left(\frac{d^2\phi_n}{dx^2}\right)^2 dx = \beta_n^2\left[-\sigma_n^2 + \frac{\beta_n^2 L^2}{2} + (-1)^n\sqrt{\sigma_n^4 - 1}\right]$

Clamped-clamped $\displaystyle\int_0^L x\left(\frac{d^2\phi_n}{dx^2}\right)^2 dx = \frac{\beta_n^4 L^2}{2}$

Free-pinned $\displaystyle\int_0^L x\left(\frac{d^2\phi_n}{dx^2}\right)^2 dx = \beta_n^2\left[\sigma_n^2 + \frac{\beta_n^2 L^2}{2} - (-1)^n\sqrt{\sigma_n^4 - 1}\right]$

Free-free $\displaystyle\int_0^L x\left(\frac{d^2\phi_n}{dx^2}\right)^2 dx = \frac{\beta_n^4 L^2}{2}$

68. $$\int_0^L x\left(\frac{d^3\phi_n}{dx^3}\right)^2 dx = \frac{1}{8}\left[-2\beta_n^4 x\phi_n \frac{d\phi_n}{dx} + \beta_n^4 x^2\left(\frac{d\phi_n}{dx}\right)^2 - 2\beta_n^4 x^2\phi_n \frac{d^2\phi_n}{dx^2} + 6x\frac{d^2\phi_n}{dx^2}\frac{d^3\phi_n}{dx^3} + x^2\left(\frac{d^3\phi_n}{dx^3}\right)^2 + \beta_n^4\phi_n^2 - 3\left(\frac{d^2\phi_n}{dx^2}\right)^2\right]_0^L$$

Clamped-free $\displaystyle\int_0^L x\left(\frac{d^3\phi_n}{dx^3}\right)^2 dx = \frac{\beta_n^4}{2}(\sigma_n^2\beta_n^2 L^2 - 2\sigma_n\beta_n L + 4)$

Clamped-pinned $\displaystyle\int_0^L x\left(\frac{d^3\phi_n}{dx^3}\right)^2 dx = \frac{\beta_n^4}{2}(\sigma_n^2\beta_n^2 L^2 + 3)$

Clamped-clamped $\displaystyle\int_0^L x\left(\frac{d^3\phi_n}{dx^3}\right)^2 dx = \frac{\sigma_n\beta_n^5 L}{2}(\sigma_n\beta_n L + 6)$

Free-pinned $\displaystyle\int_0^L x\left(\frac{d^3\phi_n}{dx^3}\right)^2 dx = \frac{\beta_n^4}{2}(\sigma_n^2\beta_n^2 L^2 - 1)$

Free-free $\displaystyle\int_0^L x\cdot\left(\frac{d^3\phi_n}{dx^3}\right)^2 dx = \frac{\sigma_n\beta_n^5 L}{2}(\sigma_n\beta_n L - 2)$

69. $$\int_0^L x\phi_n \frac{d\phi_n}{dx} dx = \frac{1}{8\beta_n^4}\left[3\beta_n^2 x\phi_n^2 - 3\phi_n\frac{d^3\phi_n}{dx^3} + 2x\frac{d\phi_n}{dx}\frac{d^3\phi_n}{dx^3} + \frac{d\phi_n}{dx}\frac{d^2\phi_n}{dx^2} - x\left(\frac{d^2\phi_n}{dx^2}\right)^2\right]_0^L$$

Clamped-free $\displaystyle\int_0^L x\phi_n\frac{d\phi_n}{dx} dx = \frac{3}{2}L$

Clamped-pinned $\displaystyle\int_0^L x\phi_n\frac{d\phi_n}{dx} dx = -\frac{L}{2}$

Clamped-clamped $\int_0^L x\phi_n \frac{d\phi_n}{dx}\,dx = -\frac{L}{2}$

Free-pinned $\int_0^L x\phi_n \frac{d\phi_n}{dx}\,dx = -\frac{L}{2}$

Free-free $\int_0^L x\phi_n \frac{d\phi_n}{dx}\,dx = \frac{3}{2}L$

70. $$\int_0^L x\phi_n \frac{d^2\phi_n}{dx^2}\,dx = \frac{1}{8}\left[2x\phi_n \frac{d\phi_n}{dx} - x^2\left(\frac{d\phi_n}{dx}\right)^2 + 2x^2\phi_n \frac{d^2\phi_n}{dx^2} + \frac{2x}{\beta_n^4}\frac{d^2\phi_n}{dx^2}\frac{d^3\phi_n}{dx^3} - \frac{x^2}{\beta_n^4}\left(\frac{d^3\phi_n}{dx^3}\right)^2 - \phi_n^2 - \frac{1}{\beta_n^4}\left(\frac{d^2\phi_n}{dx^2}\right)^2\right]_0^L$$

Clamped-free $\int_0^L x\phi_n \frac{d^2\phi_n}{dx^2}\,dx = \frac{\sigma_n\beta_n L}{2}(2 - \sigma_n\beta_n L)$

Clamped-pinned $\int_0^L x\phi_n \frac{d^2\phi_n}{dx^2}\,dx = \frac{1}{2}(1 - \sigma_n^2\beta_n^2 L^2)$

Clamped-clamped $\int_0^L x\phi_n \frac{d^2\phi_n}{dx^2}\,dx = \frac{\sigma_n\beta_n L}{2}(2 - \sigma_n\beta_n L)$

Free-pinned $\int_0^L x\phi_n \frac{d^2\phi_n}{dx^2}\,dx = \frac{1}{2}(1 - \sigma_n^2\beta_n^2 L^2)$

Free-free $\int_0^L x\phi_n \frac{d^2\phi_n}{dx^2}\,dx = \frac{\sigma_n\beta_n L}{2}(2 - \sigma_n\beta_n L)$

71. $$\int_0^L x\phi_n \frac{d^3\phi_n}{dx^3}\,dx = \frac{1}{4}\left[x\phi_n \frac{d^2\phi_n}{dx^2} + \frac{1}{2}\phi_n \frac{d\phi_n}{dx} - \frac{1}{2}x\left(\frac{d\phi_n}{dx}\right)^2 - \frac{3}{2\beta_n^4}\frac{d^2\phi_n}{dx^2}\frac{d^3\phi_n}{dx^3} + \frac{3x}{2\beta_n^4}\left(\frac{d^3\phi_n}{dx^3}\right)^2\right]_0^L$$

Clamped-free $\int_0^L x\phi_n \frac{d^3\phi_n}{dx^3}\,dx = -\frac{\sigma_n\beta_n}{2}(2 + \sigma_n\beta_n L)$

Clamped-pinned $\int_0^L x\phi_n \frac{d^3\phi_n}{dx^3}\,dx = \frac{\beta_n}{2}[\sigma_n^2\beta_n L + 2(-1)^n\beta_n L\sqrt{\sigma_n^4 - 1} - 3\sigma_n]$

Clamped-clamped $\int_0^L x\phi_n \frac{d^3\phi_n}{dx^3}\,dx = \frac{3\sigma_n\beta_n}{2}(\sigma_n\beta_n L - 2)$

Free-pinned $\int_0^L x\phi_n \frac{d^3\phi_n}{dx^3}\,dx = \frac{\beta_n}{2}[\sigma_n^2\beta_n L + \sigma_n - 2(-1)^n\beta_n L\sqrt{\sigma_n^4 - 1}]$

Free-free $\displaystyle\int_0^L x\phi_n \frac{d^3\phi_n}{dx^3}\,dx = \frac{\sigma_n\beta_n}{2}(2 - \sigma_n\beta_n L)$

72. $$\int_0^L x \frac{d\phi_n}{dx}\frac{d^2\phi_n}{dx^2}\,dx = \frac{1}{8}\left[3x\left(\frac{d\phi_n}{dx}\right)^2 - 3\phi_n\frac{d\phi_n}{dx} + 2x\phi_n\frac{d^2\phi_n}{dx^2} + \frac{1}{\beta_n^4}\frac{d^2\phi_n}{dx^2}\frac{d^3\phi_n}{dx^3} - \frac{x}{\beta_n^4}\left(\frac{d^3\phi_n}{dx^3}\right)^2\right]_0^L$$

Clamped-free $\displaystyle\int_0^L x\frac{d\phi_n}{dx}\frac{d^2\phi_n}{dx^2}\,dx = \frac{\sigma_n\beta_n}{2}(3\sigma_n\beta_n L - 2)$

Clamped-pinned $\displaystyle\int_0^L x\frac{d\phi_n}{dx}\frac{d^2\phi_n}{dx^2}\,dx = \frac{\beta_n}{2}\left[\sigma_n^2\beta_n L + \sigma_n - 2(-1)^n\beta_n L\sqrt{\sigma_n^4 - 1}\right]$

Clamped-clamped $\displaystyle\int_0^L x\frac{d\phi_n}{dx}\frac{d^2\phi_n}{dx^2}\,dx = \frac{\sigma_n\beta_n}{2}(2 - \sigma_n\beta_n L)$

Free-pinned $\displaystyle\int_0^L x\frac{d\phi_n}{dx}\frac{d^2\phi_n}{dx^2}\,dx = \frac{\beta_n}{2}\left[\sigma_n^2\beta_n L + 2(-1)^n\beta_n L\sqrt{\sigma_n^4 - 1} - 3\sigma_n\right]$

Free-free $\displaystyle\int_0^L x\frac{d\phi_n}{dx}\frac{d^2\phi_n}{dx^2}\,dx = \frac{3\sigma_n\beta_n}{2}(\sigma_n\beta_n L - 2)$

73. $$\int_0^L x\frac{d\phi_n}{dx}\frac{d^3\phi_n}{dx^3}\,dx = \frac{1}{8}\left[2x\frac{d\phi_n}{dx}\frac{d^2\phi_n}{dx^2} - x^2\left(\frac{d^2\phi_n}{dx^2}\right)^2 + 2x^2\frac{d\phi_n}{dx}\frac{d^3\phi_n}{dx^3} + 2x\phi_n\frac{d^3\phi_n}{dx^3} - \beta_n^4 x^2\phi_n^2 - \left(\frac{d\phi_n}{dx}\right)^2 - \frac{1}{\beta_n^4}\left(\frac{d^3\phi_n}{dx^3}\right)^2\right]_0^L$$

Clamped-free $\displaystyle\int_0^L x\frac{d\phi_n}{dx}\frac{d^3\phi_n}{dx^3}\,dx = -\frac{\beta_n^4 L^2}{2}$

Clamped-pinned $\displaystyle\int_0^L x\frac{d\phi_n}{dx}\frac{d^3\phi_n}{dx^3}\,dx = -\frac{\beta_n^4 L^2}{2}$

Clamped-clamped $\displaystyle\int_0^L x\frac{d\phi_n}{dx}\frac{d^3\phi_n}{dx^3}\,dx = -\frac{\beta_n^4 L^2}{2}$

Free-pinned $\displaystyle\int_0^L x\frac{d\phi_n}{dx}\frac{d^3\phi_n}{dx^3}\,dx = -\frac{\beta_n^4 L^2}{2}$

Free-free $\displaystyle\int_0^L x\frac{d\phi_n}{dx}\frac{d^3\phi_n}{dx^3}\,dx = -\frac{\beta_n^4 L^2}{2}$

74. $$\int_0^L x \frac{d^2\phi_n}{dx^2}\frac{d^3\phi_n}{dx^3}\,dx = \frac{1}{8}\left[3x\left(\frac{d^2\phi_n}{dx^2}\right)^2 - 3\frac{d^2\phi_n}{dx^2}\frac{d\phi_n}{dx} + 2x\frac{d\phi_n}{dx}\frac{d^3\phi_n}{dx^3} + \phi_n\frac{d^3\phi_n}{dx^3} - \beta_n^4 x\phi_n^2\right]_0^L$$

Clamped-free $$\int_0^L x \frac{d^2\phi_n}{dx^2}\frac{d^3\phi_n}{dx^3}\,dx = -\frac{\beta_n^4 L}{2}$$

Clamped-pinned $$\int_0^L x \frac{d^2\phi_n}{dx^2}\frac{d^3\phi_n}{dx^3}\,dx = -\frac{\beta_n^4 L}{2}$$

Clamped-clamped $$\int_0^L x \frac{d^2\phi_n}{dx^2}\frac{d^3\phi_n}{dx^3}\,dx = \frac{3\beta_n^4 L}{2}$$

Free-pinned $$\int_0^L x \frac{d^2\phi_n}{dx^2}\frac{d^3\phi_n}{dx^3}\,dx = -\frac{\beta_n^4 L}{2}$$

Free-free $$\int_0^L x \frac{d^2\phi_n}{dx^2}\frac{d^3\phi_n}{dx^3}\,dx = -\frac{\beta_n^4 L}{2}$$

75. $$\int_0^L x\phi_n\phi_m\,dx = \frac{1}{\beta_n^4 - \beta_m^4}\left[x\left(\phi_m\frac{d^3\phi_n}{dx^3} - \phi_n\frac{d^3\phi_m}{dx^3} - \frac{d\phi_m}{dx}\frac{d^2\phi_n}{dx^2} + \frac{d\phi_n}{dx}\frac{d^2\phi_m}{dx^2}\right)\right]_0^L$$
$$- \frac{1}{(\beta_n^4 - \beta_m^4)^2}\left[(\beta_n^4 + 3\beta_m^4)\phi_m\frac{d^2\phi_n}{dx^2} + (3\beta_n^4 + \beta_m^4)\phi_n\frac{d^2\phi_m}{dx^2}\right.$$
$$\left. - 2(\beta_n^4 + \beta_m^4)\frac{d\phi_m}{dx}\frac{d\phi_n}{dx} - 4\frac{d^3\phi_m}{dx^3}\frac{d^3\phi_n}{dx^3}\right]_0^L \qquad m \neq n$$

Clamped-free $$\int_0^L x\phi_n\phi_m\,dx = \frac{8(-1)^{m+n}\sigma_m\beta_m\sigma_n\beta_n}{(\beta_n^4 - \beta_m^4)^2}[(-1)^n\beta_n^2 - (-1)^m\beta_m^2]^2$$

Clamped-pinned $$\int_0^L x\phi_n\phi_m\,dx = \frac{2\beta_n\beta_m}{(\beta_n^4 - \beta_m^4)^2}[(\beta_n^2 + \beta_m^2)(-1)^{m+n}\sqrt{(\sigma_m^2 + 1)(\sigma_n^2 + 1)}$$
$$- (\beta_n^2 - \beta_m^2)^2(-1)^m\sqrt{(\sigma_m^2 + 1)(\sigma_n^2 - 1)}$$
$$- (\beta_n^2 - \beta_m^2)^2(-1)^n\sqrt{(\sigma_m^2 - 1)(\sigma_n^2 + 1)}$$
$$+ (\beta_n^2 + \beta_m^2)^2\sqrt{(\sigma_m^2 - 1)(\sigma_n^2 - 1)} - 8\sigma_n\sigma_m\beta_n^2\beta_m^2]$$

Clamped-clamped $$\int_0^L x\phi_n\phi_m\,dx = \frac{16\sigma_m\beta_m^3\sigma_n\beta_n^3}{(\beta_n^4 - \beta_m^4)^2}[(-1)^{m+n} - 1]$$

Free-pinned $$\int_0^L x\phi_n\phi_m\,dx = \frac{2\beta_n\beta_m}{(\beta_n^4 - \beta_m^4)^2}[(-1)^{m+n}(\beta_n^2 + \beta_m^2)\sqrt{(\sigma_m^2 + 1)(\sigma_n^2 + 1)}$$
$$+ (-1)^n(\beta_n^2 - \beta_m^2)^2\sqrt{(\sigma_m^2 - 1)(\sigma_n^2 + 1)}$$
$$+ (-1)^m(\beta_n^2 - \beta_m^2)^2\sqrt{(\sigma_m^2 + 1)(\sigma_n^2 - 1)}$$
$$+ (\beta_n^2 + \beta_m^2)^2\sqrt{(\sigma_m^2 - 1)(\sigma_n^2 - 1)} - 4\sigma_n\sigma_m(\beta_n^4 + \beta_m^4)]$$

Free-free $$\int_0^L x\phi_n\phi_m\,dx = \frac{8\sigma_m\beta_m\sigma_n\beta_n(\beta_n^4+\beta_m^4)}{(\beta_n^4-\beta_m^4)^2}[(-1)^{m+n}-1]$$

76. $$\int_0^L x\frac{d\phi_n}{dx}\frac{d\phi_m}{dx}\,dx = \frac{1}{\beta_n^4-\beta_m^4}\left[x\left(\beta_n^4\phi_n\frac{d\phi_m}{dx}-\beta_m^4\phi_m\frac{d\phi_n}{dx}-\frac{d^2\phi_m}{dx^2}\frac{d^3\phi_n}{dx^3}-\frac{d^2\phi_n}{dx^2}\frac{d^3\phi_m}{dx^3}\right)\right]_0^L$$

$$-\frac{1}{(\beta_n^4-\beta_m^4)^2}\left[(\beta_n^4+3\beta_m^4)\frac{d\phi_m}{dx}\frac{d^3\phi_n}{dx^3}+(3\beta_n^4+\beta_m^4)\frac{d\phi_n}{dx}\frac{d^3\phi_m}{dx^3}\right.$$

$$\left.-2(\beta_n^4+\beta_m^4)\frac{d^2\phi_m}{dx^2}\frac{d^2\phi_n}{dx^2}-4\beta_m^4\beta_n^4\phi_m\phi_n\right]_0^L$$

Clamped-free $$\int_0^L x\frac{d\phi_n}{dx}\frac{d\phi_m}{dx}\,dx = \frac{4(-1)^{m+n}\beta_m\beta_n L}{\beta_n^4-\beta_m^4}(\sigma_m\beta_n^3-\sigma_n\beta_m^3)$$

$$-\frac{8\beta_m^2\beta_n^2}{(\beta_n^4-\beta_m^4)^2}[(-1)^m\beta_m^2-(-1)^n\beta_n^2]^2$$

Clamped-pinned $$\int_0^L x\frac{d\phi_n}{dx}\frac{d\phi_m}{dx}\,dx = \frac{\beta_m\beta_n}{(\beta_n^4-\beta_m^4)^2}\Big[(-1)^{m+n}(\beta_n^2+\beta_m^2)^3\sqrt{(\sigma_m^2+1)(\sigma_n^2+1)}$$

$$+(-1)^m(\beta_n^2-\beta_m^2)^3\sqrt{(\sigma_m^2+1)(\sigma_n^2-1)}$$

$$-(-1)^n(\beta_n^2-\beta_m^2)^3\sqrt{(\sigma_m^2-1)(\sigma_n^2+1)}$$

$$-(\beta_n^2+\beta_m^2)^3\sqrt{(\sigma_m^2-1)(\sigma_n^2-1)}$$

$$-8\beta_m\beta_n(\beta_n^4+\beta_m^4)\Big]$$

Clamped-clamped $$\int_0^L x\frac{d\phi_n}{dx}\frac{d\phi_m}{dx}\,dx = \frac{4(-1)^{m+n}\beta_m^2\beta_n^2 L}{\beta_n^4-\beta_m^4}(\sigma_m\beta_m-\sigma_n\beta_n)$$

$$-\frac{8[1-(-1)^{m+n}]\beta_m^2\beta_n^2(\beta_n^2+\beta_m^4)}{(\beta_n^4-\beta_m^4)^2}$$

Free-pinned $$\int_0^L x\frac{d\phi_n}{dx}\frac{d\phi_m}{dx}\,dx = \frac{\beta_m\beta_n}{(\beta_n^4-\beta_m^4)^2}[(-1)^{m+n}(\beta_n^2+\beta_m^2)^3\sqrt{(\sigma_m^2+1)(\sigma_n^2+1)}$$

$$-(-1)^m(\beta_n^2-\beta_m^2)^3\sqrt{(\sigma_m^2+1)(\sigma_n^2-1)}$$

$$+(-1)^n(\beta_n^2-\beta_m^2)^3\sqrt{(\sigma_m^2-1)(\sigma_n^2+1)}$$

$$-(\beta_n^2+\beta_m^2)^3\sqrt{(\sigma_m^2-1)(\sigma_n^2-1)}-16\beta_m^3\beta_n^3]$$

Free-free $$\int_0^L x\frac{d\phi_n}{dx}\frac{d\phi_m}{dx}\,dx = \frac{4(-1)^{m+n}\beta_m\beta_n L}{\beta_n^4-\beta_m^4}(\sigma_m\beta_n^3-\sigma_n\beta_m^3)$$

$$+\frac{16\beta_m^4\beta_n^4}{(\beta_n^4-\beta_m^4)^2}[(-1)^{m+n}-1]$$

77. $$\int_0^L x \frac{d^2\phi_n}{dx^2}\frac{d^2\phi_m}{dx^2}dx = \frac{1}{\beta_n^4-\beta_m^4}\left[x\left(\beta_n^4 \frac{d\phi_n}{dx}\frac{d^2\phi_m}{dx^2} - \beta_m^4 \frac{d\phi_m}{dx}\frac{d^2\phi_n}{dx^2} - \beta_n^4\phi_n\frac{d^3\phi_m}{dx^3} + \beta_m^4\phi_m\frac{d^3\phi_n}{dx^3}\right)\right]_0^L - \frac{1}{(\beta_n^4-\beta_m^4)^2}\left[-4\beta_m^4\beta_n^4\frac{d\phi_m}{dx}\frac{d\phi_n}{dx} + \beta_m^4(3\beta_n^4+\beta_m^4)\phi_m\frac{d^2\phi_n}{dx^2} + \beta_n^4(\beta_n^4+3\beta_m^4)\phi_n\frac{d^2\phi_m}{dx^2} - 2(\beta_n^4+\beta_m^4)\frac{d^3\phi_m}{dx^3}\frac{d^3\phi_n}{dx^3}\right]_0^L \quad m \neq n$$

Clamped-free $$\int_0^L x\frac{d^2\phi_n}{dx^2}\frac{d^2\phi_m}{dx^2}dx = -\frac{8\sigma_m\beta_m^3\sigma_n\beta_n^3}{(\beta_n^4-\beta_m^4)^2}[(-1)^m\beta_m^2 - (-1)^n\beta_n^2]^2$$

Clamped-pinned $$\int_0^L x\frac{d^2\phi_n}{dx^2}\frac{d^2\phi_m}{dx^2}dx = \frac{2\beta_m^3\beta_n^3}{(\beta_n^4-\beta_m^4)^2} \cdot [(-1)^{m+n}(\beta_m^2+\beta_n^2)^2\sqrt{(\sigma_m^2+1)(\sigma_n^2+1)} + (-1)^m(\beta_n^2-\beta_m^2)^2\sqrt{(\sigma_m^2+1)(\sigma_n^2-1)} + (-1)^n(\beta_n^2-\beta_m^2)^2\sqrt{(\sigma_m^2-1)(\sigma_n^2+1)} + (\beta_n^2+\beta_m^2)^2\sqrt{(\sigma_m^2-1)(\sigma_n^2-1)} - 4\sigma_m\sigma_n(\beta_n^4+\beta_m^4)]$$

Clamped-clamped $$\int_0^L x\frac{d^2\phi_n}{dx^2}\frac{d^2\phi_m}{dx^2}dx = \frac{8\sigma_m\beta_m^3\sigma_n\beta_n^3(\beta_n^4+\beta_m^4)}{(\beta_n^4-\beta_m^4)^2}[(-1)^{m+n}-1]$$

Free-pinned $$\int_0^L x\frac{d^2\phi_n}{dx^2}\frac{d^2\phi_m}{dx^2}dx = \frac{2\beta_m^3\beta_n^3}{(\beta_n^4-\beta_m^4)^2} \cdot [(-1)^{m+n}(\beta_n^2+\beta_m^2)^2\sqrt{(\sigma_m^2+1)(\sigma_n^2+1)} - (-1)^m(\beta_n^2-\beta_m^2)^2\sqrt{(\sigma_m^2+1)(\sigma_n^2-1)} - (-1)^n(\beta_n^2-\beta_m^2)^2\sqrt{(\sigma_m^2-1)(\sigma_n^2+1)} + (\beta_n^2+\beta_m^2)^2\sqrt{(\sigma_m^2-1)(\sigma_n^2-1)} - 8\sigma_m\beta_m^2\sigma_n\beta_n^2]$$

Free-free $$\int_0^L x\frac{d^2\phi_n}{dx^2}\frac{d^2\phi_m}{dx^2}dx = \frac{16\sigma_m\beta_m^5\sigma_n\beta_n^5}{(\beta_n^4-\beta_m^4)^2}[(-1)^{m+n}-1]$$

78. $$\int_0^L x\frac{d^3\phi_m}{dx^3}\frac{d^3\phi_n}{dx^3}dx = \frac{1}{\beta_n^4-\beta_m^4}\left[x\left(\beta_n^4\frac{d^2\phi_n}{dx^2}\frac{d^3\phi_m}{dx^3} - \beta_m^4\frac{d^2\phi_m}{dx^2}\frac{d^3\phi_n}{dx^3} - \beta_m^4\beta_n^4\phi_m\frac{d\phi_n}{dx} + \beta_m^4\beta_n^4\phi_n\frac{d\phi_m}{dx}\right)\right]_0^L - \frac{1}{(\beta_n^4-\beta_m^4)^2}\left[-4\beta_n^4\beta_m^4\frac{d^2\phi_m}{dx^2}\frac{d^2\phi_n}{dx^2} + \beta_m^4(3\beta_n^4+\beta_m^4)\frac{d\phi_m}{dx}\frac{d^3\phi_n}{dx^3} + \beta_n^4(\beta_n^4+3\beta_m^4)\frac{d\phi_n}{dx}\frac{d^3\phi_m}{dx^3} - 2\beta_m^4\beta_n^4(\beta_m^4+\beta_n^4)\phi_m\phi_n\right]_0^L$$

Clamped-free

$$\int_0^L x \frac{d^3\phi_m}{dx^3} \frac{d^3\phi_n}{dx^3} dx = \frac{4(-1)^{m+n}\beta_m^4 \beta_n^4}{\beta_n^4 - \beta_m^4} (\beta_m \sigma_m - \beta_n \sigma_n) + \frac{8(-1)^{m+n}\beta_m^4 \beta_n^4}{(\beta_n^4 - \beta_m^4)^2} [(-1)^m \beta_m^2 - (-1)^n \beta_n^2]^2$$

Clamped-pinned

$$\int_0^L x \frac{d^3\phi_m}{dx^3} \frac{d^3\phi_n}{dx^3} dx = \frac{\beta_m^3 \beta_n^3}{(\beta_n^4 - \beta_m^4)^2} \cdot \Big[(-1)^{m+n}(\beta_n^2 + \beta_m^2)^3 \sqrt{(\sigma_m^2 + 1)(\sigma_n^2 + 1)} - (-1)^m (\beta_n^2 - \beta_m^2)^3 \sqrt{(\sigma_m^2 + 1)(\sigma_n^2 - 1)} + (-1)^n (\beta_n^2 - \beta_m^2)^3 \sqrt{(\sigma_m^2 - 1)(\sigma_n^2 + 1)} - (\beta_n^2 + \beta_m^2)^3 \sqrt{(\sigma_m^2 - 1)(\sigma_n^2 - 1)} - 16\beta_m^3 \beta_n^3\Big]$$

Clamped-clamped

$$\int_0^L x \frac{d^3\phi_m}{dx^3} \frac{d^3\phi_n}{dx^3} dx = \frac{4(-1)^{m+n}\beta_m^3 \beta_n^3 L}{\beta_n^4 - \beta_m^4} (\sigma_m \beta_n^3 - \sigma_n \beta_m^3) + \frac{16\beta_m^6 \beta_n^6}{(\beta_n^4 - \beta_m^4)^2} [(-1)^{m+n} - 1]$$

Free-pinned

$$\int_0^L x \frac{d^3\phi_m}{dx^3} \frac{d^3\phi_n}{dx^3} dx = \frac{\beta_m^3 \beta_n^3}{(\beta_n^4 - \beta_m^4)^2} \cdot \Big[(-1)^{m+n}(\beta_n^2 + \beta_m^2)^3 \sqrt{(\sigma_m^2 + 1)(\sigma_n^2 + 1)} + (-1)^m (\beta_n^2 - \beta_m^2)^3 \sqrt{(\sigma_m^2 + 1)(\sigma_n^2 - 1)} - (-1)^n (\beta_n^2 - \beta_m^2)^3 \sqrt{(\sigma_m^2 - 1)(\sigma_n^2 + 1)} - (\beta_n^2 + \beta_m^2)^3 \sqrt{(\sigma_m^2 - 1)(\sigma_n^2 - 1)} - 8\beta_m \beta_n (\beta_n^4 + \beta_m^4)\Big]$$

Free-free

$$\int_0^L x \frac{d^3\phi_m}{dx^3} \frac{d^3\phi_n}{dx^3} dx = \frac{4(-1)^{m+n}\beta_m^4 \beta_n^4 L(\sigma_m \beta_m - \sigma_n \beta_n)}{\beta_n^4 - \beta_m^4} + \frac{8\beta_m^4 \beta_n^4 (\beta_m^4 + \beta_n^4)}{(\beta_n^4 - \beta_m^4)^2} [(-1)^{m+n} - 1]$$

79. Recurrence formula. ϕ_n, ψ_m are characteristic functions; β_n, λ_m are the characteristic values associated with these characteristic functions. $\theta(x)$ is any function integrable on the interval $0 \leqslant x \leqslant L$.

$$\int_0^L \theta \psi_m \phi_n \, dx = \frac{1}{\beta_n^4 - \lambda_m^4} \left[\theta \left(\psi_m \frac{d^3\phi_n}{dx^3} - \phi_n \frac{d^3\psi_m}{dx^3} + \frac{d\phi_n}{dx}\frac{d^2\psi_m}{dx^2} - \frac{d\psi_m}{dx}\frac{d^2\phi_n}{dx^2}\right)\right]_0^L$$

$$- \frac{1}{(\beta_n^4 - \lambda_m^4)^2} \left(\frac{d\theta}{dx}\left[(\beta_n^4 + 3\lambda_m^4)\psi_m \frac{d^2\phi_n}{dx^2} - 4\frac{d^3\phi_n}{dx^3}\frac{d^3\psi_m}{dx^3} + (3\beta_n^4 + \lambda_m^4)\phi_n \frac{d^2\psi_m}{dx^2} - 2(\beta_n^4 + \lambda_m^4)\frac{d\psi_m}{dx}\frac{d\phi_n}{dx}\right]\right)_0^L$$

$$+\frac{1}{(\beta_n^4-\lambda_m^4)^3}\left(\frac{d^2\theta}{dx^2}\left[(\beta_n^8+12\beta_n^4\lambda_m^4+3\lambda_m^8)\,\psi_m\,\frac{d\phi_n}{dx}\right.\right.$$

$$-(10\beta_n^4+6\lambda_m^4)\frac{d^2\phi_n}{dx^2}\frac{d^3\psi_m}{dx^3}+(6\beta_n^4+10\lambda_m^4)\frac{d^3\phi_n}{dx^3}\frac{d^3\psi_m}{dx^3}$$

$$\left.\left.-(3\beta_n^8+12\beta_n^4\lambda_m^4+\lambda_m^8)\,\phi_n\,\frac{d\psi_m}{dx}\right]\right)_0^L$$

$$-\frac{1}{(\beta_n^4-\lambda_m^4)^4}\left(\frac{d^3\theta}{dx^3}\left[(\beta_n^{12}+31\beta_n^8\lambda_m^4+31\beta_n^4\lambda_m^8+\lambda_m^{12})\,\phi_n\psi_m\right.\right.$$

$$-2(10\beta_n^8+15\beta_n^4\lambda_m^4+2\lambda_m^8)\frac{d\phi_n}{dx}\frac{d^3\psi_m}{dx^3}$$

$$+2(5\beta_n^8+21\beta_n^4\lambda_m^4+2\lambda_m^8)\frac{d^2\phi_n}{dx^2}\frac{d^2\psi_m}{dx^2}$$

$$\left.\left.-4(\beta_n^8+10\beta_n^4\lambda_m^4+5\lambda_m^8)\frac{d^3\phi_n}{dx^3}\frac{d\psi_m}{dx}\right)\right]_0^L$$

$$+\frac{1}{(\beta_n^4-\lambda_m^4)^4}\int_0^L\frac{d^4\theta}{dx^4}\left[(\beta_n^{12}+31\beta_n^8\lambda_m^4+4\lambda_m^{12})\,\phi_n\psi_m\right.$$

$$-2(10\beta_n^8+15\beta_n^4\lambda_m^4+2\lambda_m^8)\frac{d\phi_n}{dx}\frac{d^3\psi_m}{dx^3}$$

$$+2(5\beta_n^8+21\beta_n^4\lambda_m^4+2\lambda_m^8)\frac{d^2\phi_n}{dx^2}\frac{d^2\psi_m}{dx^2}$$

$$\left.-4(\beta_n^8+10\beta_n^4\lambda_m^4+5\lambda_m^8)\frac{d^3\phi_n}{dx^3}\frac{d\psi_m}{dx}\right]dx$$

INDEX